$434.95 per copy (in United States)
Price is subject to change without prior notice.

0304

The 2014 *Facilities Maintenance & Repair Cost Data*, 21st annual edition, and all of the RSMeans annual cost data books for 2014, are dedicated to the memory of our respected colleague and friend, Paula Reale-Camelio.

RSMeans Facilities Maintenance & Repair Cost Data

21ST ANNUAL EDITION

2014

RSMeans
A division of Reed Construction Data, LLC
Construction Publishers & Consultants
700 Longwater Drive
Norwell, MA 02061
USA
1-800-334-3509
www.rsmeans.com

Copyright 2013 by RSMeans
A division of Reed Construction Data, LLC
All rights reserved.

Printed in the United States of America
ISSN 1074-0953
ISBN 978-1-940238-07-4

The authors, editors, and engineers of RSMeans apply diligence and judgment in locating and using reliable sources for the information published. However, RSMeans makes no express or implied warranty or guarantee in connection with the content of the information contained herein, including the accuracy, correctness, value, sufficiency, or completeness of the data, methods, and other information contained herein. RSMeans makes no express or implied warranty of merchantability or fitness for a particular purpose. RSMeans shall have no liability to any customer or third party for any loss, expense, or damage, including consequential, incidental, special, or punitive damage, including lost profits or lost revenue, caused directly or indirectly by any error or omission, or arising out of, or in connection with, the information contained herein.

No part of this publication may be reproduced, stored in a retrieval system, or transmitted in any form or by any means without prior written permission of RSMeans. The cost data contained in this publication is valuable and proprietary information of RSMeans and others and there is no transfer to you of any ownership rights in the cost data or any license granted to you to create derivative works based on or which utilize the cost data.

Senior Editor
Stephen C. Plotner *(3, 5)**

Contributing Editors
Christopher Babbitt
Adrian C. Charest, PE *(26, 27, 28, 48)*
Gary W. Christensen
David G. Drain, PE *(33)*
Cheryl Elsmore
David Fiske
Robert Fortier, PE *(2, 31, 32, 34, 35, 46)*
Robert J. Kuchta *(8)*
Thomas Lane
Robert C. McNichols
Genevieve Medeiros
Melville J. Mossman, PE *(21, 22, 23)*
Jeannene D. Murphy
Marilyn Phelan, AIA *(9, 10, 11, 12)*
Siobhan Soucie
Phillip R. Waier, PE *(1,13)*

Cover Design
Wendy McBay
Paul Robertson
Wayne Engebretson

Chief Operating Officer
Richard Remington

Vice President, Business Development
Steven Ritchie

Vice President, Finance & Operations
Renée Pinczes

Senior Director, Inside Sales
Derek F. Dean

Sales Manager
Reid Clanton

Director, Engineering
Bob Mewis, CCC *(4, 6, 7, 14, 41, 44)*

Senior Director, Product Management
Jennifer Johnson

Senior Product Marketing Manager
Andrea Sillah

Senior Director, Operations
Rodney L. Sharples

Production Manager
Michael Kokernak

Production
Jill Goodman
Jonathan Forgit
Debbie Panarelli
Sara Rutan
Mary Lou Geary

Technical Support
Christopher Anderson
Gary L. Hoitt
Kathryn S. Rodriguez

**Numbers in italics are the divisional responsibilities for each editor. Please contact the designated editor directly with any questions.*

This book is printed on 90% recycled paper using soy-based ink and is entirely recyclable.

Related RSMeans Products and Services

The engineers at RSMeans suggest the following products and services as companion information resources to *RSMeans Facilities Maintenance & Repair Cost Data:*

Construction Cost Data Books
Facilities Construction Cost Data 2014
Electrical Cost Data 2014
Mechanical Cost Data 2014
Plumbing Cost Data 2014

Reference Books
Facilities Planning and Relocation
Life Cycle Costing for Facilities
Cost Planning & Estimating for Facilities Maintenance
Job Order Contracting
Estimating Building Costs
RSMeans Estimating Handbook
Building Security: Strategies & Costs

Seminars and In-House Training
Facilities Maintenance & Repair Estimating
Unit Price Estimating
RSMeans Online® Training
RSMeans Data for Job Order Contracting (JOC)
Mechanical & Electrical Estimating
Scheduling with MSProject for Construction Professionals

RSMeans Online Bookstore
Visit RSMeans at **www.rsmeans.com** for a one-stop portal for the most reliable and current resources available on the market. Here you can learn about our more than 20 annual cost data books, also available on CD and online versions, as well as our library of reference books and RSMeans seminars.

RSMeans Electronic Data
Get the information found in RSMeans cost books electronically at **RSMeansOnline.com**.

RSMeans Business Solutions
Cost engineers and analysts conduct facility life cycle and benchmark research studies and predictive cost modeling, as well as offer consultation services for conceptual estimating, real property management, and job order contracting. Research studies are designed with the application of proprietary cost and project data from Reed Construction Data and RSMeans' extensive North American databases. Analysts offer building product manufacturers qualitative and quantitative market research, as well as time/motion studies for new products, and Web-based dashboards for market opportunity analysis. Clients are from both the public and private sectors, including federal, state, and municipal agencies; corporations; institutions; construction management firms; hospitals; and associations.

RSMeans for Job Order Contracting (JOC)
Best practice job order contracting (JOC) services support the owner in the development of their program. RSMeans develops all required contracting documents, works with stakeholders, and develops qualified and experienced contractor lists. RSMeans JOCWorks® software is the best of class in the functional tools that meet JOC specific requirements. RSMeans data is organized to meet JOC cost estimating requirements.

Construction Costs for Software Applications
More than 25 unit price and assemblies cost databases are available through a number of leading estimating and facilities management software providers (listed below). For more information see the "Other RSMeans Products" pages at the back of this publication.

RSMeansData™ is also available to federal, state, and local government agencies as multi-year, multi-seat licenses.

- 4Clicks-Solutions, LLC
- Assetworks
- Beck Technology
- BSD – Building Systems Design, Inc.
- CMS – Construction Management Software
- Corecon Technologies, Inc.
- CorVet Systems
- Hard Dollar Corporation
- Maxwell Systems
- MC^2
- MOCA Systems
- Parsons Corporation
- ProEst
- Refined Data
- Sage Timberline Office
- UDA Technologies, Inc.
- US Cost, Inc.
- VFA – Vanderweil Facility Advisers
- WinEstimator, Inc.
- R & K Solutions
- Tririga

Table of Contents

Foreword	iv
How the Book Is Built: An Overview	v
How to Use the Book: The Details	viii
Maintenance & Repair	1
Preventative Maintenance	377
General Maintenance	555
Facilities Audits	569
Auditing the Facility	571
Audits Checklists	575
Reference Section	597
Construction Equipment Rental Costs	599
Crew Listings	611
Travel Costs	617
Reference Tables	618
City Adjustments	627
Location Factors	642
Reference Aids	648
Life Cycle Costing	663
Abbreviations	668
Other RSMeans Products and Services	679
In-House Contractor's Overhead & Profit	Inside Back Cover

Foreword

Our Mission

Since 1942, RSMeans has been actively engaged in construction cost publishing and consulting throughout North America.

Today, more than 70 years after RSMeans began, our primary objective remains the same: to provide you, the construction and facilities professional, with the most current and comprehensive construction cost data possible.

Whether you are a contractor, owner, architect, engineer, facilities manager, or anyone else who needs a reliable construction cost estimate, you'll find this publication to be a highly useful and necessary tool.

With the constant flow of new construction methods and materials today, it's difficult to find the time to look at and evaluate all the different construction cost possibilities. In addition, because labor and material costs keep changing, last year's cost information is not a reliable basis for today's estimate or budget.

That's why so many construction professionals turn to RSMeans. We keep track of the costs for you, along with a wide range of other key information, from city cost indexes . . . to productivity rates . . . to crew composition . . . to contractor's overhead and profit rates.

RSMeans performs these functions by collecting data from all facets of the industry and organizing it in a format that is instantly accessible to you. From the preliminary budget to the detailed unit price estimate, you'll find the data in this book useful for all phases of construction cost determination.

The Staff, the Organization, and Our Services

When you purchase one of RSMeans' publications, you are, in effect, hiring the services of a full-time staff of construction and engineering professionals.

Our thoroughly experienced and highly qualified staff works daily at collecting, analyzing, and disseminating comprehensive cost information for your needs. These staff members have years of practical construction experience and engineering training prior to joining the firm. As a result, you can count on them not only for accurate cost figures, but also for additional background reference information that will help you create a realistic estimate.

The RSMeans organization is always prepared to help you solve construction problems through its variety of data solutions, including online, CD, and print book formats, as well as cost estimating expertise available via our business solutions, training, and seminars.

Besides a full array of construction cost estimating books, RSMeans also publishes a number of other reference works for the construction industry. Subjects include construction estimating and project and business management, special topics such as green building and job order contracting, and a library of facility management references.

In addition, you can access all of our construction cost data electronically in convenient CD format or on the web. Visit **RSMeansOnline.com** for more information on our 24/7 online cost data.

What's more, you can increase your knowledge and improve your construction estimating and management performance with an RSMeans construction seminar or in-house training program. These two-day seminar programs offer unparalleled opportunities for everyone in your organization to become updated on a wide variety of construction-related issues.

RSMeans is also a worldwide provider of construction cost management and analysis services for commercial and government owners.

In short, RSMeans can provide you with the tools and expertise for constructing accurate and dependable construction estimates and budgets in a variety of ways.

Robert Snow Means Established a Tradition of Quality That Continues Today

Robert Snow Means spent years building RSMeans, making certain he always delivered a quality product.

Today, at RSMeans, we do more than talk about the quality of our data and the usefulness of our books. We stand behind all of our data, from historical cost indexes to construction materials and techniques to current costs.

If you have any questions about our products or services, please call us toll-free at 1-800-334-3509. Our customer service representatives will be happy to assist you. You can also visit our website at **www.rsmeans.com**

How the Book is Built: An Overview

A Complete Reference for the Facility Manager

This book was designed as a complete reference and cost data source for facility managers, owners, engineers, and contractors, or anyone who manages real estate. A complete facilities management program starts with a building audit. The audit describes the structures, their characteristics and major equipment, and apparent deficiencies. An audit is not a prerequisite to using any data contained in this book, but it does provide an organizational framework. A list of equipment and deficiencies forms the basis for a repair budget and preventive maintenance program. An audit also provides facility size and usage criteria on which the general maintenance (cleaning) program will be based. Together, the five sections of this book provide a framework and definitive data for developing a complete program for all facets of facilities maintenance.

Even if you have a regular building audit and preventive maintenance program in place, the information in this book will be useful.

For the occasional user, the book provides:

- a reference for time and material requirements for Maintenance & Repair tasks to fill in where your organization lacks its own historical data;

- Preventive Maintenance (PM) checklists with labor-hour standards that can be used to benchmark your program;

- a quick reference for General Maintenance (cleaning) productivity rates to analyze proposals from outside contractors;

- an explanation of Life Cycle Cost Analysis as a tool to assist in making the correct budget decisions; and

- data to validate line items in budgets.

For the frequent user, the book supplies all of the above, plus:

- guidance for getting the audit under way;

- a foundation for zero-based budgeting of maintenance & repair, preventive maintenance, and general maintenance;

- cost data that can be used to estimate preventive maintenance, deferred maintenance (repairs), and general maintenance (cleaning) programs; and

- detailed task descriptions for establishing a complete PM program.

How to Use the Book: The Details

This section contains an in-depth explanation of how the book is arranged . . . and how you can use it to determine a reliable construction cost estimate. It includes information about how we develop our cost figures and how to completely prepare your estimate.

Section 1: Maintenance & Repair (M&R)

The Maintenance & Repair section is a listing of common maintenance tasks performed at facilities. The tasks include removal and replacement, repair, and refinishing. This section is organized according to the UNIFORMAT II system. (UNIFORMAT was originally developed for the General Services Administration as a systems approach to organizing building construction and has since become widely used in the building industry. It is now maintained and published by CSI, the Construction Specifications Institute.)

The purpose of Section 1 is to provide cost data and an approximate frequency of occurrence for each maintenance and repair task. This data is used by facility managers, owners, plant engineers, and contractors to prepare estimates for deferred maintenance projects. The data is also used to estimate labor-hours and material costs for in-house staff. The frequency listing facilitates preparation of a zero-based budget.

Under the System Description column, each task, including its components, is described. In some cases additional information (such as "2% of total roof area") appears in parentheses. The next column, Frequency, refers to how often one should expect, and therefore estimate, that this task will have to be performed. Labor hours have been enhanced, where appropriate, by 30% to include setup and delay time. In some cases, the list of steps comprising the task includes most of the potential repairs that could be made on a system. For those tasks, individual items, as appropriate, can be selected from the all-encompassing list.

Section 2: Preventive Maintenance (PM)

The Preventive Maintenance section provides the framework for a complete PM program. The establishment of a program is frequently hindered by the lack of a comprehensive

Maintenance & Repair

B30 ROOFING			B3013		Roof Covering					
B3013 105		Built-Up Roofing								
	System Description	Freq. (Years)	Crew	Unit	Labor Hours	2014 Bare Costs			Total In-House	Total w/O&P
						Material	Labor	Equipment	Total	
0600	Place new BUR membrane over existing	20	G-5	Sq.						
	Set up, secure and take down ladder				.020		.78		.78	1.17	1.41
	Sweep / spud ballast clean				.500		19.60		19.60	29	35.50
	Vent existing membrane				.130		5.10		5.10	7.60	9.15
	Cut out buckled flashing				.024		.94		.94	1.40	1.70
	Install 2 ply membrane flashing				.027	.39	1.04		1.43	1.98	2.36
	Install 4 ply bituminous roofing				3.733	134	137	37	308	390	455
	Reinstall ballast				.381		14.90		14.90	22.50	27
	Clean up				.390		15.30		15.30	23	27.50
	Total				5.205	134.39	194.66	37	366.05	476.65	559.62
0700	Total BUR roof replacement	28	G-1	Sq.						
	Set up, secure and take down ladder				.020		.78		.78	1.17	1.41
	Sweep / spud ballast clean				.500		19.60		19.60	29	35.50
	Remove built-up roofing				2.500		93		93	123	152
	Remove insulation board				1.026		38		38	51	62
	Remove flashing				.024		.94		.94	1.40	1.70
	Install 2" perlite insulation				.879	99	34		133	160	186
	Install 2 ply membrane flashing				.027	.39	1.04		1.43	1.98	2.36
	Install 4 ply bituminous membrane				2.800	134	102	27.50	263.50	330	385
	Clean up				1.000		39		39	58.50	70.50
	Total				8.776	233.39	328.36	27.50	589.25	756.05	896.47
B3013 120		Modified Bituminous / Thermoplastic								
	System Description	Freq. (Years)	Crew	Unit	Labor Hours	2014 Bare Costs			Total In-House	Total w/O&P
						Material	Labor	Equipment	Total	
0100	Debris removal by hand & visual inspection	1	2 ROFC	M.S.F.						
	Set up, secure and take down ladder				.052		2.03		2.03	3.03	3.67
	Pick up trash / debris & clean up				.327		12.80		12.80	19.05	23
	Visual inspection				.327		12.80		12.80	19.05	23
	Total				.705		27.63		27.63	41.13	49.67

v

list of equipment, actual PM steps, and budget documentation. This section fulfills all these needs. The facility manager, plant engineer, or owner can use these schedules to establish labor-hours and a budget. The maintenance contractor can use these PM checklists as the basis for a comprehensive maintenance proposal or as an estimating aid when bidding PM contracts. Selected pages, with preventive maintenance steps, can be photocopied and assembled into a customized PM program. Copies of individual sheets can also be distributed to maintenance personnel to identify the procedures required. The hours listed are predicated on work being performed by experienced technicians familiar with the PM tasks and equipped with the proper tools and materials.

The PM section lists the tasks and their frequency, whether weekly, monthly, quarterly, semi-annually, or annually. The frequency of those procedures is based on non-critical usage (e.g., in "normal use" situations, versus in facilities such as surgical suites or computer rooms that demand absolute adherence to a limited range of environmental conditions). The labor-hours to perform each item on the checklist are listed in the next column. Beneath the table is cost data for the PM schedule. The data is shown both annually and annualized, to provide the facility manager with an estimating range. If all tasks on the schedule are performed once a year, the *annually* line should be used in the PM estimate. The *annualized* data is used when all items on the schedule are performed at the frequency shown.

Section 3: General Maintenance (GM)

This section provides labor-hour estimates and costs to perform day-to-day cleaning. The data is used to estimate cleaning times, compare and assess estimates submitted by cleaning companies, or to budget in-house staff. The information is divided into *Interior* and *Exterior* maintenance. A common maintenance laborer (*Clam*) is used to perform these tasks.

Section 4: Facilities Audits

The Facilities Audit is generally the basis from which the Maintenance & Repair, Preventive Maintenance, and General Maintenance estimates are prepared. The audit provides the following:

1. A list of deficiencies for the Maintenance & Repair estimate
2. A list of equipment and other items from which to prepare a Preventive Maintenance estimate
3. A list of facilities' usage and size requirements to prepare a General Maintenance estimate

The Introduction to Facilities Audits in this section explains the rationale for the audit and the steps required to complete it successfully.

This section also provides forms for listing all of the organization's facilities, and a separate detailed form for the specifics of each facility. Checklists are provided for major building components. Each checklist should be accompanied by a blank audit form. This form is used to record the audit findings, prioritize the deficiencies, and estimate the cost to remedy these deficiencies.

Preventive Maintenance

General Maintenance

Reference Section

This section includes information on Equipment Rental Costs, Crew Listings, Travel Costs, Reference Tables, Reference Tables, Life Cycle Costing, and a listing of Abbreviations.

Equipment Rental Costs: This section contains the average costs to rent and operate hundreds of pieces of construction equipment.

Crew Listings: This section lists all of the crews referenced in the book. For the purposes of this book, a crew is composed of more than one trade classification and/or the addition of power equipment to any trade classification. Power equipment is included in the cost of the crew. Costs are shown both with bare labor rates and with the installing contractor's overhead and profit added. For each, the total crew cost per eight-hour day and the composite cost per labor-hour are listed.

Travel Costs: This chart provides labor-hour costs for round-trip travel between a base of operation and the project site.

Reference Tables: In this section, you'll find reference tables, explanations, estimating information that support how we develop the unit price data, technical data, and estimating procedures.

Life Cycle Costing: This section provides definitions and basic equations for performing life cycle analysis. A sample problem is solved to demonstrate the methodology.

Abbreviations: A listing of the abbreviations used throughout this book, along with the terms they represent, is included in this section.

The Design Assumptions of this Book

This book is designed to be as easy to use as possible. To that end, we have made certain assumptions and limited its scope.

1. Unlike any other RSMeans publication, *Facilities Maintenance & Repair Cost Data* was designed for estimating a wide range of maintenance tasks in diverse environments. Due to this diversity, the level of accuracy of the data is +/− 20%.
2. We have established material prices based on a national average.
3. We have computed labor costs based on a 30-city national average of union wage rates.
4. Except where major equipment or component replacement is described, the projects in this book are small; therefore, material prices and labor-hours have been enhanced to reflect the increased costs of small-scale work.
5. The PM frequencies are based on non-critical applications. Increased frequencies are required for critical environments.

How to Use the Book: The Details

What's Behind the Numbers? The Development of Cost Data

The staff at RSMeans continually monitors developments in the construction industry in order to ensure reliable, thorough, and up-to-date cost information. While overall construction costs may vary relative to general economic conditions, price fluctuations within the industry are dependent upon many factors. Individual price variations may, in fact, be opposite to overall economic trends. Therefore, costs are constantly tracked and complete updates are published yearly. Also, new items are frequently added in response to changes in materials and methods.

Costs—$ (U.S.)

All costs represent U.S. national averages and are given in U.S. dollars. The RSMeans City Cost Indexes can be used to adjust costs to a particular location. The City Cost Indexes for Canada can be used to adjust U.S. national averages to local costs in Canadian dollars. No exchange rate conversion is necessary.

[G] The processes or products identified by the green symbol in our publications have been determined to be environmentally responsible and/or resource-efficient solely by the RSMeans engineering staff. The inclusion of the green symbol does not represent compliance with any specific industry association or standard.

Material Costs

The RSMeans staff contacts manufacturers, dealers, distributors, and contractors all across the U.S. and Canada to determine national average material costs. If you have access to current material costs for your specific location, you may wish to make adjustments to reflect differences from the national average. Included within material costs are fasteners for a normal installation. RSMeans engineers use manufacturers' recommendations, written specifications, and/or standard construction practice for size and spacing of fasteners. Adjustments to material costs may be required for your specific application or location. The manufacturer's warranty is assumed. Extended warranties are not included in the material costs. Material costs do not include sales tax.

Labor Costs

Labor costs are based on the average of wage rates from 30 major U.S. cities. Rates are determined from labor union agreements or prevailing wages for construction trades for the current year. Rates, along with overhead and profit markups, are listed on the inside back cover of this book.

- If wage rates in your area vary from those used in this book, or if rate increases are expected within a given year, labor costs should be adjusted accordingly.

Labor costs reflect productivity based on actual working conditions. In addition to actual installation, these figures include time spent during a normal weekday on tasks such as, material receiving and handling, mobilization at site, site movement, breaks, and cleanup.

Productivity data is developed over an extended period so as not to be influenced by abnormal variations and reflects a typical average.

Equipment Costs

Equipment costs include not only rental, but also operating costs for equipment under normal use. The operating costs include parts and labor for routine servicing such as repair and replacement of pumps, filters, and worn lines. Normal operating expendables, such as fuel, lubricants, tires, and electricity (where applicable), are also included. Extraordinary operating expendables with highly variable wear patterns, such as diamond bits and blades, are excluded. These costs are included under materials. Equipment rental rates are obtained from industry sources throughout North America—contractors, suppliers, dealers, manufacturers, and distributors.

Rental rates can also be treated as reimbursement costs for contractor-owned equipment. Owned equipment costs include depreciation, loan payments, interest, taxes, insurance, storage and major repairs.

Equipment costs do not include operators' wages; nor do they include the cost to move equipment to a job site (mobilization) or from a job site (demobilization).

Equipment Cost/Day——The cost of power equipment required for each crew is included in the Crew Listings in the Reference Section (small tools that are considered as essential everyday tools are not listed out separately). The Crew Listings itemize specialized tools and heavy equipment along with labor trades. The daily cost of itemized equipment included in a crew is based on dividing the weekly bare rental rate by 5 (number of working days per week) and then adding the hourly operating cost times 8 (the number of hours per day). This Equipment Cost/Day is shown in the last column of the Equipment Rental Cost pages in the Reference Section.

Mobilization/Demobilization——The cost to move construction equipment from an equipment yard or rental company to the job site and back again is not included in equipment costs. Mobilization (to the site) and demobilization (from the site) costs can be found in the Unit Price Section. If a piece of equipment is already at the job site, it is not appropriate to utilize mob./demob. costs again in an estimate.

Overhead and Profit

Total Cost including O&P for the Installing Contractor is shown in the last column on the Unit Price and/or the Assemblies pages of this book. This figure is the sum of the bare material cost plus 10% for profit, the bare labor cost plus total overhead and profit, and the bare equipment cost plus 10% for profit. Details for the calculation of Overhead and Profit on labor are shown on the inside back cover and in the Reference Section of this book. (See "How RSMeans Data Works" for an example of this calculation.)

General Conditions

Cost data in this book is presented in two ways: Bare Costs and Total Cost including O&P (Overhead and Profit). General Conditions, or General Requirements, of the contract should also be added to the Total Cost including O&P when applicable. Costs for General Conditions are listed in Division 1 of the Unit Price Section and the Reference Section of this book. General Conditions for the *Installing Contractor* may range from 0% to 10% of the Total Cost including O&P. For the *General* or *Prime Contractor*, costs for General Conditions may range from 5% to 15% of the Total Cost including O&P, with a figure of 10% as the most typical allowance. If applicable, the Assemblies and Models sections of this book use costs that include the installing contractor's overhead and profit (O&P).

Factors Affecting Costs

Costs can vary depending upon a number of variables. Here's how we have handled the main factors affecting costs.

Quality——The prices for materials and the workmanship upon which productivity is based represent sound construction work. They are also in line with U.S. government specifications.

Overtime——We have made no allowance for overtime. If you anticipate premium time or work beyond normal working hours, be sure to make an appropriate adjustment to your labor costs.

Productivity——The productivity, daily output, and labor-hour figures for each line item are based on working an eight-hour day in daylight hours in moderate temperatures. For work that extends beyond normal work hours or is performed under adverse conditions, productivity may decrease. (See "How to Use the General Maintenance Pages" for more on productivity.)

Size of Project——The size, scope of work, and type of construction project will have a significant impact on cost. Economies of scale can reduce costs for large projects. Unit costs can often run higher for small projects.

Location——Material prices in this book are for metropolitan areas. However, in dense urban areas, traffic and site storage limitations may increase costs. Beyond a 20-mile radius of large cities, extra trucking or transportation charges may also increase the material costs slightly. On the other hand, lower wage rates may be in effect. Be sure to consider both of these factors when preparing an estimate, particularly if the job site is located in a central city or remote rural location. In addition, highly specialized subcontract items may require travel and per-diem expenses for mechanics.

Other Factors——
- season of year
- contractor management
- weather conditions
- local union restrictions
- building code requirements
- availability of:
  - adequate energy
  - skilled labor
  - building materials
- owner's special requirements/restrictions
- safety requirements
- environmental considerations

Unpredictable Factors——General business conditions influence "in-place" costs of all items. Substitute materials and construction methods may have to be employed. These may affect the installed cost and/or life cycle costs. Such factors may be difficult to evaluate and cannot necessarily be predicted on the basis of the job's location in a particular section of the country. Thus, where these factors apply, you may find significant but unavoidable cost variations for which you will have to apply a measure of judgment to your estimate.

Rounding of Costs

In general, all unit prices in excess of $5.00 have been rounded to make them easier to use and still maintain adequate precision of the results. The rounding rules we have chosen are in the following table.

Prices from . . .	Rounded to the nearest . . .
$.01 to $5.00	$.01
$5.01 to $20.00	$.05
$20.01 to $100.00	$.50
$100.01 to $300.00	$1.00
$300.01 to $1,000.00	$5.00
$1,000.01 to $10,000.00	$25.00
$10,000.01 to $50,000.00	$100.00
$50,000.01 and above	$500.00

How Subcontracted Items Affect Costs

A considerable portion of all large construction jobs is usually subcontracted. In fact, the percentage done by subcontractors is constantly increasing and may run over 90%. Since the workers employed by these companies do nothing else but install their particular product, they soon become expert in that line. The result is, installation by these firms is accomplished so efficiently that the total in-place cost, even adding the general contractor's overhead and profit, is no more, and often less, than if the principal contractor had handled the installation himself/herself. Companies that deal with construction specialties are anxious to have their product perform well and, consequently, the installation will be the best possible.

Contingencies

The allowance for contingencies generally provides for unforeseen construction difficulties. On alterations or repair jobs, 20% is not too much. If drawings are final and only field contingencies are being considered, 2% or 3% is probably sufficient, and often nothing need be added. Contractually, changes in plans will be covered by extras. The contractor should consider inflationary price trends and possible material shortages during the course of the job. These escalation factors are dependent upon both economic conditions and the anticipated time between the estimate and actual construction. If drawings are not complete or approved, or a budget cost is wanted, it is wise to add 5% to 10%. Contingencies, then, are a matter of judgment. Additional allowances for contingencies are shown in Division 1.

Important Estimating Considerations

The "productivity," or daily output, of each craftsman includes mobilization and cleanup time, break time and plan layout time, as well as an allowance to carry stock from the storage trailer or location on the job site up to 200' into the building and to the first or second floor. If material has to be transported over greater distances or to higher floors, an additional allowance should be considered by the estimator. An allowance has also been included in the piping and fittings installation time for leak checking and minor tightening. Equipment installation time includes the following applicable items: positioning, leveling and securing the unit in place, connecting all associated piping, ducts, vents, etc., which shall have been estimated separately, connecting to an adjacent power source, filling/bleeding, startup, adjusting the controls up and down to ensure proper response, setting the integral controls/valves/regulators/thermostats for proper operation (does not include external building type control systems, DDC systems, etc.), explaining/training owner's operator, and warranty. A reasonable breakdown of the labor costs is as follows:

1. Movement into building, installation/setting of equipment 35%
2. Connection to piping/duct/power, etc. 25%
3. Filling/flushing/cleaning/touchup, etc. 15%
4. Startup/running adjustments 5%
5. Training owner's representative 5%
6. Warranty/call back/service 15%

Note that cranes or other lifting equipment are not included on any lines in the Mechanical divisions. For example, if a crane is required to lift a heavy piece of pipe into place high above a gym floor, or to put a rooftop unit on the roof of a four-story building, etc., it must be added. Due to the potential for extreme variation—from nothing additional required to a major crane or helicopter—we feel that including a nominal amount for "lifting contingency" would be useless and detract from the accuracy of the estimate. When using equipment rental from RSMeans, remember to include the cost of the operator(s).

Estimating Labor-Hours

The labor-hours expressed in this publication are based on Average Installation time, using an efficiency level of approximately 60%–65% (see item 7), which has been found reasonable and acceptable by many contractors. The book uses this national efficiency average to establish a consistent benchmark. For bid situations, adjustments to this efficiency level should be the responsibility of the contractor bidding the project. The unit labor-hour is divided in the following manner. A typical day for a journeyman might be:

1.	Study Plans	3%	14.4 min.
2.	Material Procurement	3%	14.4 min.
3.	Receiving and Storing	3%	14.4 min.
4.	Mobilization	5%	24.0 min.
5.	Site Movement	5%	24.0 min.
6.	Layout and Marking	8%	38.4 min.
7.	Actual Installation	64%	307.2 min.
8.	Cleanup	3%	14.4 min.
9.	Breaks, Nonproductive	6%	28.8 min.
		100%	480.0 min.

If any of the percentages expressed in this breakdown do not apply to the particular work or project situation, then that percentage or a portion of it may be deducted from or added to labor-hours.

Final Checklist

Estimating can be a straightforward process provided you remember the basics. Here's a checklist of some of the steps you should remember to complete before finalizing your estimate.

Did you remember to . . .

- factor in the City Cost Index for your locale?
- take into consideration which items have been marked up and by how much?
- mark up the entire estimate sufficiently for your purposes?
- read the background information on techniques and technical matters that could impact your project time span and cost?
- include all components of your project in the final estimate?
- double check your figures for accuracy?
- call RSMeans if you have any questions about your estimate or the data you've found in our publications? Remember, RSMeans stands behind its publications. If you have any questions about your estimate . . . about the costs you've used from our books . . . or even about the technical aspects of the job that may affect your estimate, feel free to call the RSMeans editors at 1-800-334-3509.

Free Quarterly Updates

Stay up-to-date throughout 2014 with RSMeans' free cost data updates four times a year. Sign up online to make sure you have access to the newest data. Every quarter we provide the city cost adjustment factors for hundreds of cities and key materials. Register at: **www.rsmeans2014updates.com**

Maintenace & Repair

Table of Contents

A10 Foundations
A1033 Slab on Grade
A1033 110 Concrete, Unfinished 7

B10 Superstructure
B1013 Floor Construction
B1013 510 Concrete Stairs 8
B1013 520 Masonry Steps 8
B1013 530 Wood Stairs 9
B1013 540 Railing, Wood 9
B1013 550 Metal Stairs 10
B1013 560 Railing, Metal 10
B1013 570 Wrought Iron Exterior Balustrade .. 11
B1013 580 Metal Fire Escapes 11
B1013 910 Steel Decking 12
B1013 920 Metal Grating 12

B20 Exterior Closure
B2013 Exterior Walls
B2013 104 Copings 13
B2013 109 Concrete Block 14
B2013 114 Concrete Block Wall, Painted 15
B2013 118 Terra Cotta 16
B2013 119 Clay Brick 17
B2013 121 Clay Brick, Painted 18
B2013 123 Field Stone 20
B2013 140 Glass Block 21
B2013 141 Aluminum Siding 22
B2012 142 Steel Siding 24
B2013 143 Masonite Panel, Sealed 26
B2013 145 Wood Clapboard Finished, 1 Coat . 26
B2013 147 Wood Shingles, Unfinished .. 27
B2013 148 Wood Shingles, Finished 28
B2013 150 Fiberglass Panel, Rigid 30
B2013 152 Synthetic Veneer Plaster ... 31
B2013 156 Exterior Soffit Gypsum Board, Painted 31
B2013 157 Overhang, Exterior Entry ... 33
B2013 158 Recaulking 33
B2013 310 Wood Louvers and Shutters .. 34
B2013 320 Aluminum Louver 36
B2013 330 Steel Louver 37

B2023 Exterior Windows
B2023 102 Steel Frame, Operating 39
B2023 103 Steel Frame, Fixed 42
B2023 110 Aluminum Window, Operating ... 44
B2023 111 Aluminum Window, Fixed 46
B2023 112 Wood Frame, Operating 48
B2023 113 Wood Frame, Fixed 51
B2023 116 Aluminum Frame Storm Window .. 53
B2023 117 Steel Frame Storm Window ... 54
B2023 119 Wood Frame Storm Window ... 56
B2023 122 Vinyl Clad Wood Window, Operating 57
B2023 123 Vinyl Clad Wood Window, Fixed .. 58
B2023 150 Glass Block, Fixed 59
B2023 160 Aluminum Shutter 60
B2023 162 Steel Shutter 61
B2023 164 Wood Shutter 62
B2023 166 Metal Window Grating 64

B2023 168 Metal Wire Mesh Cover 65
B2023 170 Aluminum Window Screen 66
B2023 172 Steel Frame Screen Window .. 67
B2023 174 Wood Frame Screen Window ... 68
B2023 324 Plate Glass 69

B2033 Exterior Doors
B2033 111 Glazed Aluminum 70
B2033 130 Glazed Wood 71
B2033 206 Solid Core, Painted 72
B2033 210 Aluminum Doors 74
B2033 220 Steel, Painted 77
B2033 221 Steel, Unpainted 80
B2033 410 Steel Single, Roll-Up 83
B2033 412 Steel Double, Roll-Up 83
B2033 413 Steel, Single Leaf, Spring Garage Door 84
B2033 420 Aluminum Single, Roll-Up ... 84
B2033 422 Aluminum Double, Roll-Up ... 85
B2033 423 Aluminum, Single Leaf, Spring Garage Door 85
B2033 430 Wood Single, Roll-Up 86
B2033 432 Wood Double, Roll-Up 86
B2033 433 Wood, Single Leaf, Spring Garage Door 87
B2033 513 Electric Bifolding Hangar Door .. 87
B2033 810 Hinges, Brass 88
B2033 820 Lockset, Brass 88
B2033 830 Door Closer, Brass 88
B2033 840 Door Opener 89
B2033 850 Deadbolt, Brass 89
B2033 860 Weatherstripping, Brass 89
B2033 870 Exit Bolt, Metal 90
B2033 905 Hollow Core, Painted 90
B2033 925 Metal Grated 91
B2033 926 Metal Wire Mesh Door 92
B2033 927 Louvered Door 93
B2033 935 Screen/Storm Aluminum, Plain/ Anodized 95
B2033 960 Aluminized Steel Pedestrian Gate .. 95
B2033 961 Steel Pedestrian Gate 96
B2033 962 Wood Pedestrian Gate 96
B2033 963 Wrought Iron Gate 97

B30 Roofing
B3013 Roof Covering
B3013 105 Built-Up Roofing 98
B3013 120 Modified Bituminous/ Thermoplastic 99
B3013 125 Thermosetting 101
B3013 127 EPDM (Ethylene Propylene Diene Monomer) 102
B3013 130 Metal Panel Roofing 102
B3013 140 Slate Tile Roofing 104
B3013 142 Mineral Fiber Steep Roofing .. 104
B3013 144 Clay Tile Roofing 105
B3013 146 Asphalt Shingle Roofing 106
B3013 160 Roll Roofing 107
B3013 165 Corrugated Fiberglass Panel Roofing 108
B3013 170 Roof Edges 109
B3013 620 Gutters and Downspouts 110

B3023 Roof Openings
B3023 110 Single Unit Glass Skylight .. 113
B3023 120 Hatches 113
B3023 130 Smoke Hatches 114

C10 Interior Construction
C1013 Partitions
C1013 110 Concrete Block, Painted 115
C1013 115 Glazed CMU Interior Wall Finish .. 115
C1013 120 Glass Block 116
C1013 125 Stone Veneer 116
C1013 230 Demountable Partitions 117
C1013 544 Channel Frame Wire Mesh Wall ... 117
C1013 730 Plate Glass Wall - Interior ... 117

C1023 Interior Doors
C1023 108 Fully Glazed Wooden Doors ... 119
C1023 110 Steel Painted Interior Doors ... 119
C1023 111 Steel Unpainted Interior Doors ... 122
C1023 112 Bi-Fold Interior Doors 124
C1023 114 Aluminum Interior Doors 124
C1023 116 Fully Glazed Aluminum Doors ... 126
C1023 120 Hollow Core Interior Doors . 127
C1023 121 Solid Core Interior Doors .. 127
C1023 129 Interior Gates 129
C1023 324 Hinges, Brass 129
C1023 325 Lockset, Brass 129
C1023 326 Door Closer, Brass 130
C1023 327 Deadbolt, Brass 130
C1023 328 Weatherstripping, Brass 130
C1023 329 Panic Bar 130

C1033 Fittings
C1033 110 Toilet Partitions 131
C1033 210 Metal Lockers 132

C20 Stairs
C2013 Stair Construction
C2013 110 Concrete Steps 133
C2013 115 Masonry Steps, Painted 133
C2013 130 Metal Steps 133
C2013 435 Metal Interior Stair Railing ... 134
C2013 437 Iron Interior Stair Railing ... 134
C2013 440 Wood Steps 135
C2013 445 Wood Interior Stair Railing .. 135

C2023 Stair Finishes
C2023 110 Carpeted Steps 136
C2023 120 Rubber Steps 136
C2023 130 Terrazzo Steps 136

C30 Interior Finishes
C3013 Wall Finishes
C3013 106 Vinyl Wall Covering 137
C3013 202 Wallpaper 137
C3013 206 Fabric Interior Wall Finish .. 138
C3013 208 Cork Tile 138
C3013 210 Acoustical Tile 139
C3013 212 Stucco 139
C3013 213 Plaster 140
C3013 214 Drywall 141
C3013 215 Fiberglass Panels, Rigid ... 142
C3013 220 Tile 143
C3013 230 Plywood Paneling 143
C3013 234 Wainscot 144
C3013 240 Stainless Steel Interior Finish .. 144
C3013 410 Fireplace Mantles, Wood 144

Table of Contents (cont.)

C3023 Floor Finishes
C3023 112 Concrete, Finished 145
C3023 405 Epoxy Flooring 145
C3023 410 Vinyl 145
C3023 412 Vinyl Sheet 145
C3023 414 Rubber Tile 146
C3023 418 Rubber/Vinyl Trim 146
C3023 420 Ceramic Tile 146
C3023 428 Ceramic Trim 147
C3023 430 Terrazzo 147
C3023 438 Terrazzo Trim 147
C3023 440 Quarry Tile 148
C3023 450 Brick 148
C3023 460 Marble 148
C3023 470 Plywood 149
C3023 472 Wood Parquet 149
C3023 474 Maple Strip 149
C3023 478 Wood Trim 150
C3023 510 Carpet 150

C3033 Ceiling Finishes
C3033 105 Plaster 151
C3033 107 Gypsum Wall Board 151
C3033 108 Acoustic Tile 152
C3033 109 Acoustic Tile, Fire Rated 154
C3033 120 Wood 154

D20 Plumbing
D2013 Plumbing Fixtures
D2013 110 Tankless Water Closet 155
D2013 130 Flush-Tank Water Closet 156
D2013 210 Urinal 158
D2013 310 Lavatory, Iron, Enamel 159
D2013 330 Lavatory, Vitreous China 160
D2013 350 Lavatory, Enameled Steel 161
D2013 410 Sink, Iron Enamel 162
D2013 420 Sink, Enameled Steel 163
D2013 430 Sink, Stainless Steel 164
D2013 440 Sink, Plastic 165
D2013 444 Laboratory Sink 167
D2013 450 Laundry Sink, Plastic 167
D2013 455 Laundry Sink, Stone 168
D2013 460 Service/Utility Sink 169
D2013 470 Group Washfountain 170
D2013 510 Bathtub, Cast Iron Enamel 170
D2013 530 Bathtub, Enameled Steel 171
D2013 550 Bathtub, Fiberglass 171
D2013 710 Shower, Terrazzo 172
D2013 730 Shower, Enameled Steel 173
D2013 750 Shower, Fiberglass 174
D2013 770 Shower, Misc. 175
D2013 810 Drinking Fountain 176
D2013 910 Emergency Shower Station 177
D2013 920 Emergency Eye Wash 177

D2023 Domestic Water Distribution
D2023 110 Pipe & Fittings, Copper 178
D2023 120 Pipe & Fittings, Steel/Iron, Threaded 179
D2023 130 Solar Piping: Pipe & Fittings, PVC . 179
D2023 150 Valve, Non-Drain, Less Than 1-1/2" 181
D2023 152 Valve, Non-Drain, 2" 182
D2023 154 Valve, Non-Drain, 3" 182
D2023 156 Valve, Non-Drain, 4" and Larger ... 182
D2023 160 Insulation, Pipe 183
D2023 210 Water Heater, Gas/Oil, 30 Gallon .. 184
D2023 212 Water Heater, Gas/Oil, 70 Gallon .. 185
D2023 214 Water Heater, Gas/Oil, 1150 GPH . 185
D2023 220 Water Heater, Electric, 120 Gal. ... 186
D2023 222 Water Heater, Electric, 300 Gal. ... 186
D2023 224 Water Heater, Electric, 1000 Gal. .. 187
D2023 226 Water Heater, Electric, 2000 Gal. .. 187
D2023 230 Steam Converter, Domestic Hot Water 188
D2023 240 Storage Tank, Domestic Hot Water . 188
D2023 245 Solar Storage Tank, 1000 Gallon .. 188
D2023 250 Expansion Chamber 189
D2023 260 Circulation Pump, 1/12 HP 189
D2023 261 Circulation Pump, 1/8 HP 190
D2023 262 Circulation Pump, 1/6 HP 190
D2023 264 Circulation Pump, 1/2 HP 191
D2023 266 Circulation Pump, Bronze 1 HP . 191
D2023 267 Circulation Pump, C.I. 1-1/2 HP . 192
D2023 268 Circulation Pump, 3 HP 192
D2023 310 Hose Bibb 193
D2023 320 Water Meter 193
D2023 370 Water Softener 194

D2033 Sanitary Waste
D2033 110 Pipe & Fittings, Cast Iron 195
D2033 130 Pipe & Fittings, PVC 195
D2033 305 Pipe & Fittings 196
D2033 310 Floor Drain w/o Bucket 197
D2033 330 Floor Drain with Bucket 197

D2043 Rain Water Drainage
D2043 110 Distribution: Gutters, Pipe ... 198
D2043 210 Drain: Roof, Scupper, Area 198
D2043 310 Rainwater Sump Pump 199

D2093 Other Plumbing Systems
D2093 910 Pipe & Fittings, Industrial Gas ... 200
D2093 920 Pipe & Fittings, Anesthesia ... 200
D2093 930 Pipe & Fittings, Oxygen 200
D2093 940 Pipe & Fittings, Compressed Air . 201
D2093 946 Compressed Air Systems, Compressors 202

D30 HVAC
D3013 Energy Supply
D3013 110 Fuel Oil Storage Tank, 275 Gallon . 205
D3013 150 Fuel Level Meter 205
D3013 160 Oil Filter 205
D3013 170 Fuel Oil Storage: Pipe & Fittings, Copper 206
D3013 210 Natural Gas: Pipe & Fittings, Steel/Iron 207
D3013 240 Natural Gas: Pressure Reducing Valve 212
D3013 260 LPG Distribution: Pipe & Fittings, Steel/Iron 213
D3013 601 Solar Panel, 3' x 8' 216

D3023 Heat Generating Systems
D3023 180 Boiler, Gas 217
D3023 182 Boiler, Coal 219
D3023 184 Boiler, Oil 219
D3023 186 Boiler, Gas/Oil 222
D3023 198 Blowoff System 223
D3023 292 Chemical Feed System 224
D3023 294 Feed Water Supply 224
D3023 296 Deaerator 225
D3023 298 Separators for 900 Ton Chilled Water System 225
D3023 310 Metal Flue/Chimney 225
D3023 388 Pneumatic Coal Spreader 226
D3023 390 Fuel Oil Equipment 227

D3033 Cooling Generating Systems
D3033 115 Cooling Tower 228
D3033 130 Chiller, Water Cooled, Recip. . 229
D3033 135 Chiller, Air Cooled, Reciprocating . 231
D3033 137 Chiller, Water Cooled, Scroll ... 232
D3033 140 Chiller, Hermetic Centrifugal ... 234
D3033 142 Chiller, Open Centrifugal 236
D3033 145 Chiller, Absorption 237
D3033 210 Air Cooled Condenser 238
D3033 260 Evaporative Condenser 240

D3043 Distribution Systems
D3043 120 Fan Coil 242
D3043 122 Fan Coil, DX A/C, Cooling Only ... 244
D3043 124 Fan Coil, DX A/C w/Heat 248
D3043 128 Unit Ventilator 249
D3043 140 Duct Heater 250
D3043 210 Draft Fan 250
D3043 220 Exhaust Fan 251
D3043 250 Fireplaces, Clay Flue 253
D3043 252 Fireplaces, Metal Flue 253
D3043 310 Steam Converter, Commercial ... 254
D3043 320 Flash Tank, 24 Gallon 254
D3043 330 Steam Regulator Valve 254
D3043 340 Condensate Meter 255
D3043 350 Steam Traps 256
D3043 410 Radiator Valve 256
D3043 420 Cast Iron Radiator, 10' Section ... 257
D3043 430 Baseboard Radiation, 10' Section .. 257
D3043 440 Finned Radiator, Wall, 10' Section .. 258
D3043 450 Duct Coil, 1-Row, Hot Water ... 258
D3043 510 Pipe & Fntgs., Steel/Iron, Flanged . 259
D3043 520 Valves 259
D3043 530 Circulator Pump 261
D3043 540 Expansion Tank 262
D3043 550 Pipe Insulation 263

D3053 Terminal & Package Units
D3053 110 Unit Heater 269
D3053 112 Infrared Heater Suspended, Commercial 270
D3053 114 Standard Suspended Heater 271
D3053 116 Explosionproof Industrial Heater . 271
D3053 150 Wall Mounted/Recessed Heater, with Fan 272

Table of Contents (cont.)

- D3053 160 Convector Suspended, Commercial 272
- D3050 170 Terminal Reheat 273
- D3053 245 Heat Pump 274
- D3053 265 Air Conditioner, Window, 1 Ton ... 277
- D3053 266 Air Conditioner, Thru-The-Wall 278
- D3053 272 Air Conditioner, DX Package 278
- D3053 274 Computer Room A/C Units, Air Cooled 281
- D3053 276 Computer Room A/C Units, Chilled Water 285
- D3053 278 Multi-Zone Air Conditioner 287
- D3053 280 Single Zone Air Conditioner 289
- D3053 282 Multi-Zone Variable Volume 293
- D3053 284 Single Zone Variable Volume 296
- D3053 286 Central Station Air Conditioning Air Handling Unit 298
- D3053 310 Residential Furnace, Gas 300
- D3053 320 Residential Furnace, Oil 301
- D3053 330 Residential Furnace, Electric 303
- D3053 410 Baseboard Heating Units 304
- D3053 420 Cast Iron Radiator, 10 Section, 1 Side 304
- D3053 710 Fireplaces, Firebrick 304

D40 Fire Protection
D4013 Sprinklers
- D4013 110 Backflow Preventer 305
- D4013 310 Sprinkler Sys., Fire Suppression ... 306
- D4013 410 Fire Pump 307

D50 Electrical
D5013 Electrical Service/Distribution
- D5013 110 Primary Transformer, Liquid Filled . 308
- D5013 120 Primary Transformer, Dry 308
- D5013 210 Switchgear, Mainframe 309
- D5013 216 Fuses 309
- D5013 220 Switchgear, Indoor, Less Than 600 V 309
- D5013 224 Switchgear, Indoor, 600 V 312
- D5013 230 Meters 312
- D5013 240 Inverter 312
- D5013 250 Rectifier, Up To 600 V 313
- D5013 264 Motor Starter, Up To 600 V 313
- D5013 266 Motor Starter, 600 V 314
- D5013 272 Secondary Transformer, Liquid Filled 315
- D5013 274 Secondary Transformer, Dry 315
- D5013 280 Lighting Panel, Indoor 316

D5023 Lighting & Branch Wiring
- D5023 110 Wireway 317
- D5023 112 Conduit EMT 317
- D5023 120 Cable, Non-Metallic (NM) Sheathed 317
- D5023 122 Cable, Service 318
- D5023 124 Cable, Armored 318
- D5023 126 Branch Wiring, With Junction Box . 319
- D5023 128 Branch Wiring, 600 V 319
- D5023 130 Circuit Breaker 319
- D5023 132 Safety Switch, Heavy Duty 324
- D5023 140 Safety Switch, General Duty 326
- D5023 150 Receptacles & Plugs 327
- D5023 154 4-Pin Receptacle 327
- D5023 210 Contactors & Relays 328
- D5023 220 Wiring Devices, Switches 328
- D5023 222 Switch, Pull Cord 329
- D5023 230 Light Dimming Panel 329
- D5023 240 Incandescent Lighting Fixtures ... 330
- D5023 250 Quartz Fixture 331
- D5023 260 Fluorescent Lighting Fixt., 80 W . 331
- D5023 270 Metal Halide Fixture, 175 W 332
- D5023 280 H.P. Sodium Fixture, 250 W 334

D5033 Communications & Security
- D5033 310 Telephone Cable 335
- D5033 410 Master Clock Control 335
- D5033 510 TV Cable Outlet 336
- D5033 610 Door Bell 336
- D5033 620 Sound System Components 337
- D5033 630 Intercom System Components ... 338
- D5033 640 Security System Components ... 339
- D5033 710 Smoke Detector 339
- D5033 712 Heat Detector 340
- D5033 720 Manual Pull Station 340
- D5033 760 Fire Alarm Control Panel 341
- D5033 766 Annunciation Panel 342
- D5033 770 Fire Alarm Bell 342

D5093 Other Electrical Systems
- D5093 110 Electrical Service Ground 343
- D5093 120 Building Structure Ground 343
- D5093 130 Lightning Protection System 344
- D5093 140 Lightning Ground Rod 344
- D5093 150 Computer Ground System 344
- D5093 190 Special Ground System 345
- D5093 210 Generator, Gasoline, 175 kW ... 345
- D5093 220 Generator, Diesel, 750 kW 346
- D5093 230 Transfer Switch 346
- D5093 240 Emergency Lighting Fixture 347
- D5093 250 Exit Light 347
- D5093 255 Exit Light L.E.D. 348
- D5093 260 Battery, Wet 348
- D5093 265 Battery, Dry 349
- D5093 270 Battery Charger 349
- D5093 280 UPS Battery 349
- D5093 920 Communications Components ... 350

E10 Equipment
E1023 Institutional Equipment
- E1023 710 Laboratory Equipment 351

E1033 Vehicular Equipment
- E1033 110 Automotive Equipment 352

E1093 Other Equipment
- E1093 310 Loading Dock Equipment 353
- E1093 315 Dishwasher 353
- E1093 316 Waste Disposal, Residential 353
- E1093 320 Waste Handling Equipment 354
- E1093 510 Darkroom Dryer 354
- E1093 610 Dust Collector 355
- E1093 910 Pump Systems 355

G10 Site Preparation
G1023 Site Demolition & Relocations
- G1023 210 Underground Storage Tank Removal 356

G20 Site Improvements
G2023 Parking Lots
- G2023 210 Parking Lot Repairs 358
- G2023 310 General 358

G2033 Pedestrian Paving
- G2033 130 Asphalt Sidewalk & Curb 359
- G2033 140 Concrete Sidewalk & Curb 359
- G2033 150 Patios 360
- G2033 250 Handicap Ramp 360

G2043 Site Development
- G2043 105 Chain Link Fence & Gate Repairs . 361
- G2043 110 Wood Fence 363
- G2043 710 Bleachers, Exterior 363
- G2043 750 Tennis Court Resurfacing 364
- G1043 810 Flag Pole 364

G30 Site Mechanical Utilities
G3013 Water Supply
- G3013 400 Ground Level Water Storage Tank 365
- G3013 405 Ground Level Water Storage Standpipe 367
- G3013 410 Fire Hydrants 368
- G3013 470 Post Indicator Valve 369
- G3013 510 Elevated Water Storage Tank 369

G3063 410 Fuel Distribution
- G3063 410 Ground Level Fuel Storage Tank 371

G40 Site Electrical Utilities
G4013 Electrical Distribution
- G4013 210 Overhead Service Cables 372

G4023 Site Lighting
- G4023 210 Outdoor Pole Lights 373

H10 Construction Aids
H1043 Construction Aids
- H1043 200 Temporary Cranes 375

How to Use the Maintenance & Repair Assemblies Cost Tables

The following is a detailed explanation of a sample Maintenance & Repair Assemblies Cost Table. Next to each bold number that follows is the described item with the appropriate component of the sample entry following in paranthesis.

Total system costs, as well as the individual component costs, are shown. In most cases, the intent is for the user to apply the total system costs. However, changes and adjustments to the components or partial use of selected components is also appropriate. In particular, selected equipment system tables in the mechanical section include complete listings of operations that are meant to be chosen from rather than used in total.

1 System/Line Numbers (B3013 105 0700)

Each Maintenance & Repair Assembly has been assigned a unique identification number based on the UNIFORMAT II classification system.

UNIFORMAT II Major Group

B3013   105   0700

UNIFORMAT II Level 3

Means Major Classification

Means Individual Line Number

B30 ROOFING — B3013 Roof Covering

B3013 105 Built-Up Roofing

System Description	Freq. (Years)	Crew	Unit	Labor Hours	2014 Bare Costs Material	2014 Bare Costs Labor	2014 Bare Costs Equipment	2014 Bare Costs Total	Total In-House	Total w/O&P
0600 Place new BUR membrane over existing	20	G-5	Sq.							
Set up, secure and take down ladder				.020		.78		.78	1.17	1.41
Sweep / spud ballast clean				.500		19.60		19.60	29	35.50
Vent existing membrane				.130		5.10		5.10	7.60	9.15
Cut out buckled flashing				.024		.94		.94	1.40	1.70
Install 2 ply membrane flashing				.027	.39	1.04		1.43	1.98	2.36
Install 4 ply bituminous roofing				3.733	134	137	37	308	390	455
Reinstall ballast				.381		14.90		14.90	22.50	27
Clean up				.390		15.30		15.30	23	27.50
Total				5.205	134.39	194.66	37	366.05	476.65	559.62
0700 Total BUR roof replacement	28	G-1	Sq.							
Set up, secure and take down ladder				.020		.78		.78	1.17	1.41
Sweep / spud ballast clean				.500		19.60		19.60	29	35.50
Remove built-up roofing				2.500		93		93	123	152
Remove insulation board				1.026		38		38	51	62
Remove flashing				.024		.94		.94	1.40	1.70
Install 2" perlite insulation				.879	99	34		133	160	186
Install 2 ply membrane flashing				.027	.39	1.04		1.43	1.98	2.36
Install 4 ply bituminous membrane				2.800	134	102	27.50	263.50	330	385
Clean up				1.000		39		39	58.50	70.50
Total				8.776	233.39	328.36	27.50	589.25	756.05	896.47

B3013 120 Modified Bituminous / Thermoplastic

System Description	Freq. (Years)	Crew	Unit	Labor Hours	2014 Bare Costs Material	2014 Bare Costs Labor	2014 Bare Costs Equipment	2014 Bare Costs Total	Total In-House	Total w/O&P
0100 Debris removal by hand & visual inspection	1	2 ROFC	M.S.F.							
Set up, secure and take down ladder				.052		2.03		2.03	3.03	3.67
Pick up trash / debris & clean up				.327		12.80		12.80	19.05	23
Visual inspection				.327		12.80		12.80	19.05	23
Total				.705		27.63		27.63	41.13	49.67
0200 Non - destructive moisture inspection	5	2 ROFC	M.S.F.							
Set up, secure and take down ladder				.052		2.03		2.03	3.03	3.67
Infrared inspection of roof membrane				2.133		83.50		83.50	125	151
Total				2.185		85.53		85.53	128.03	154.67

2. System Description
A one-line description of each system is given, followed by a description of each component that makes up the system. In selected items, a percentage factor for planning annual costs is also given.

3. Frequency (28)
The projected frequency of occurrence in years for each system is listed. For example, **28** indicates that this operation should be budgeted for once every 28 years. If the frequency is less than 1, then the operation should be planned for more than once a year. For example, a value of **.5** indicates a planned occurrence of 2 times a year.

4. Crew (G-1)
The "Crew" column designates the typical trade or crew used for the task, although other trades may be used for minor portions of the work. If the task can be accomplished by one trade and requires no power equipment, that trade and the number of workers are listed (for example, "4 Rofc"). If the task requires a composite crew, a crew code designation is listed (for example, "G-1"). You'll find full details on all composite crews in the Crew Listings.
- For a complete list of all trades utilized in this book and their abbreviations, see the inside back cover of the book.

Crews

Crew No.	Bare Costs		In-House Costs		Incl. Subs O&P		Cost Per Labor-Hour		
Crew B-80	Hr.	Daily	Hr.	Daily	Hr.	Daily	Bare Costs	In House	Incl. O&P
1 Labor Foreman (outside)	$38.65	$309.20	$51.50	$412.00	$63.50	$508.00	$39.71	$52.11	$64.41
1 Laborer	36.65	293.20	48.85	390.80	60.20	481.60			
1 Truck Driver (light)	36.50	292.00	48.05	384.40	59.35	474.80			
1 Equip. Oper. (light)	47.05	376.40	60.05	480.40	74.60	596.80			
1 Flatbed Truck, Gas, 3 Ton		333.00		333.00		366.30			
1 Earth Auger, Truck-Mtd.		425.00		425.00		467.50	23.69	23.69	26.06
32 L.H. Daily Totals		$2028.80		$2425.60		$2895.00	$63.40	$75.80	$90.47

5. Unit of Measure for Each System (Sq.)
The abbreviated designation indicates the unit of measure, as defined by industry standards, upon which the price of the component is base. In this example, "Cost per Square" is the unit of measure for this system or "assembly."

6. Labor-Hours (8.776)
The "Labor-Hours" figure represents the number of labor-hours required to perform one unit of work. To calculate the number of labor-hours required for a particular task, multiply the quantity of the item times the number of labor-hours shown. For example:

Quantity	x	Productivity Rate	=	Duration
30 Sq.	x	8.776 Labor-Hours/Sq.	=	263 Labor-Hours

7. Bare Costs:

Material (233.39)
This figure for the unit material cost for the line item is the "bare" material cost with no overhead and profit allowances included. *Costs shown reflect national average material prices for the current year and generally include delivery to the facility. Small purchases may require an additional deliver charge. No sales taxes are included.*

Labor (328.36)
The unit labor costs is derived by multiplying bare labor-hour costs for Crew G-1 and other trades by labor-hour units. In this case, the bare labor-hour cost for the G-1 crew is found in the Crew Section. (If a trade is listed or used for minor portions of the work, the hourly labor cost—the wage rage—is found on the inside back cover of this book).

Equipment (27.50)
Equipment costs for each crew are listed in the description of each crew. The unit equipment cost is derived by multiplying the bare equipment hourly cost by the labor-hour units.

Total (589.25)
The total of the bare costs is the arithmetic total of the three previous columns: mat., labor, and equip.

Material	+	Labor	+	Equip.	=	Total
$233.39	+	$328.36	+	$27.50	=	$589.25

8 Total In-House Costs (756.05)

"Total In-House Costs" include suggested markups to bare costs for the direct overhead requirements of in-house maintenance staff. The figure in this column is the sum of three components: the bare material cost plus 10% (for purchasing and handling small quantities); the bare labor cost plus workers' compensation and average fixed overhead (per the labor rate table on the inside back cover of this books or, if a crew is listed, from the Crew Listings); and the bare equipment cost.

9 Total Costs Including O&P (896.47)

"Total Costs Including O&P" include suggested markups to bare costs for an outside subcontractor's overhead and profit requirements. The figure in this column is the sum of three components: the bare material cost plus 25% for profit; the bare labor cost plus total overhead and profit (per the labor rate table on the inside back cover of the book or, if a crew is listed, from the Crew Listings); and the bare equipment cost plus 10% for profit.

A10 FOUNDATIONS

A1033 Slab on Grade

A1033 110 Concrete, Unfinished

System Description	Freq. (Years)	Crew	Unit	Labor Hours	2014 Bare Costs				Total In-House	Total w/O&P
					Material	Labor	Equipment	Total		
0010 Minor repairs to concrete floor	15	1 CEFI	S.F.							
Repair spall				.612		22.50	3.57	26.07	34	41
Repair concrete flooring				.061	2.84	2.70		5.54	6.55	7.80
Total				.673	2.84	25.20	3.57	31.61	**40.55**	**48.80**
0020 Replace unfinished concrete floor	75	2 CEFI	C.S.F.							
Break up concrete slab				1.664		66.80	25.36	92.16	113	136
Load into truck				1.950		72.75	42	114.75	139	166
Rebar #3 to #7				.139	10.50	7		17.50	21	24.50
Edge form				.693	3.10	30.20		33.30	43.50	53.50
Slab on grade 4" in place with finish				1.577	96.75	66	.75	163.50	192	227
Expansion joint				.003	.04	.12		.16	.21	.26
Cure concrete				.109	5.48	4		9.48	11.50	13.45
Total				6.135	115.87	246.87	68.11	430.85	**520.21**	**620.71**

B10 SUPERSTRUCTURE   B1013   Floor Construction

B1013 510   Concrete Stairs

	System Description	Freq. (Years)	Crew	Unit	Labor Hours	2014 Bare Costs				Total In-House	Total w/O&P
						Material	Labor	Equipment	Total		
1010	**Repair concrete steps**	30	1 CEFI	S.F.							
	Repair spalls in steps				.104	11.70	4.58		16.28	18.70	22
	Total				.104	11.70	4.58		16.28	**18.70**	**22**
1020	**Replace concrete steps**	75	1 CEFI	S.F.							
	Remove concrete				.520		19.25	3.04	22.29	28.50	35
	Install forms				.328	1.31	14.65		15.96	21	25.50
	Pour concrete				.567	2.52	21.50	.78	24.80	32	39.50
	Total				1.416	3.83	55.40	3.82	63.05	**81.50**	**100**
1030	**Refinish concrete steps**	3	1 PORD	S.F.							
	Prepare surface				.016		.63		.63	.82	1.02
	Refinish surface, brushwork, 1 coat				.013	.08	.49		.57	.73	.89
	Total				.029	.08	1.12		1.20	**1.55**	**1.91**
2030	**Refinish concrete landing**	3	1 PORD	S.F.							
	Prepare surface				.016		.63		.63	.82	1.02
	Refinish surface, brushwork, 1 coat				.013	.08	.49		.57	.73	.89
	Total				.029	.08	1.12		1.20	**1.55**	**1.91**

B1013 520   Masonry Steps

	System Description	Freq. (Years)	Crew	Unit	Labor Hours	2014 Bare Costs				Total In-House	Total w/O&P
						Material	Labor	Equipment	Total		
1010	**Repair masonry steps**	4	1 BRIC	S.F.							
	Remove damaged masonry				.074		2.72		2.72	3.63	4.47
	Clean area				.019	.06	.85		.91	1.18	1.46
	Replace with new masonry				.236	3.40	9.95		13.35	16.85	20.50
	Total				.329	3.46	13.52		16.98	**21.66**	**26.43**

B10 SUPERSTRUCTURE B1013 Floor Construction

B1013 530 Wood Stairs

							2014 Bare Costs				Total In-House	Total w/O&P
		Freq. (Years)	Crew	Unit	Labor Hours		Material	Labor	Equipment	Total		
	System Description											
1030	**Refinish wood steps**	3	1 PORD	S.F.								
	Prepare surface				.016			.63		.63	.82	1.02
	Wash / dry surface				.001			.05		.05	.06	.07
	Refinish surface, brushwork, 1 coat				.013		.08	.49		.57	.73	.89
	Total				.030		.08	1.17		1.25	1.61	1.98
1040	**Replace wood steps**	30	2 CARP	S.F.								
	Remove stairs (to 6' wide)				.100			3.66		3.66	4.89	6
	New stringers (to 5 risers)				.074		1.21	3.39		4.60	5.85	7.05
	Risers				.083		.76	3.82		4.58	5.95	7.25
	Treads				.123		2.01	5.65		7.66	9.75	11.75
	Total				.380		3.98	16.52		20.50	26.44	32.05
2030	**Refinish wood landing**	3	1 PORD	S.F.								
	Prepare surface				.016			.63		.63	.82	1.02
	Wash / dry surface				.001			.05		.05	.06	.07
	Refinish surface, brushwork, 1 coat				.013		.08	.49		.57	.73	.89
	Total				.030		.08	1.17		1.25	1.61	1.98

B1013 540 Railing, Wood

							2014 Bare Costs				Total In-House	Total w/O&P
	System Description	Freq. (Years)	Crew	Unit	Labor Hours		Material	Labor	Equipment	Total		
1010	**Replace wood balustrade**	20	1 CARP	L.F.								
	Remove damaged balusters and handrail				.087			3.18		3.18	4.23	5.20
	Install new balusters and handrail				.473		26	21.50		47.50	57.50	68
	Total				.559		26	24.68		50.68	61.73	73.20
1030	**Refinish wood balustrade**	7	1 PORD	L.F.								
	Prepare balusters and handrails for painting				.052			2.06		2.06	2.66	3.30
	Wash / dry surface				.001			.05		.05	.06	.07
	Paint balusters and handrails, brushwork, primer + 1 coat				.089		.75	3.52		4.27	5.40	6.60
	Total				.142		.75	5.63		6.38	8.12	9.97

B10 SUPERSTRUCTURE B1013 Floor Construction

B1013 550 Metal Stairs

	System Description	Freq. (Years)	Crew	Unit	Labor Hours	2014 Bare Costs				Total In-House	Total w/O&P
						Material	Labor	Equipment	Total		
1010	**Repair metal steps**	15	1 SSWK	S.F.							
	Remove damaged stair tread				.099		5.06		5.06	7.70	9.30
	New stair tread				.104	42.61	5.32		47.93	55	63
	Total				.203	42.61	10.38		52.99	62.70	72.30
1040	**Replace metal steps**	40	2 SSWK	S.F.							
	Remove stairs				.168		8.60		8.60	13.15	15.80
	New stairs				.239	157.30	12.01	6.58	175.89	197	225
	Total				.408	157.30	20.61	6.58	184.49	210.15	240.80
1050	**Refinish metal steps**	3	1 PORD	S.F.							
	Prepare metal railing for painting				.019		.76		.76	.98	1.21
	Wash / dry surface				.001		.05		.05	.06	.07
	Paint metal steps, brushwork, primer + 1 coat				.015	.11	.61		.72	.91	1.12
	Total				.036	.11	1.42		1.53	1.95	2.40
2030	**Refinish metal landing**	3	1 PORD	S.F.							
	Prepare metal railing for painting				.019		.76		.76	.98	1.21
	Wash / dry surface				.001		.05		.05	.06	.07
	Paint metal landing, brushwork, primer + 1 coat				.015	.11	.61		.72	.91	1.12
	Total				.036	.11	1.42		1.53	1.95	2.40

B1013 560 Railing, Metal

	System Description	Freq. (Years)	Crew	Unit	Labor Hours	2014 Bare Costs				Total In-House	Total w/O&P
						Material	Labor	Equipment	Total		
1010	**Replace metal hand rail**	30	1 SSWK	L.F.							
	Remove damaged railings				.087		3.18		3.18	4.23	5.20
	Install new metal railings				.163	25	8.40	.72	34.12	41	47.50
	Total				.250	25	11.58	.72	37.30	45.23	52.70
1030	**Refinish metal hand rail**	7	1 PORD	L.F.							
	Prepare metal railing for painting				.019		.76		.76	.98	1.21
	Wash / dry surface				.001		.05		.05	.06	.07
	Paint metal railing, brushwork, primer + 1 coat				.015	.11	.61		.72	.91	1.12
	Total				.036	.11	1.42		1.53	1.95	2.40

B10 SUPERSTRUCTURE

B1013 Floor Construction

B1013 570 Wrought Iron Exterior Balustrade

	System Description	Freq. (Years)	Crew	Unit	Labor Hours	2014 Bare Costs				Total In-House	Total w/O&P
						Material	Labor	Equipment	Total		
1010	**Repair wrought iron balustrade**	20	1 SSWK	L.F.							
	Cut / remove damaged section				.325		11.90		11.90	15.90	19.55
	Reset new section				.867	79	44.50		123.50	154	180
	Total				1.192	79	56.40		135.40	**169.90**	**199.55**
1030	**Refinish wrought iron balustrade**	7	1 PORD	L.F.							
	Prepare metal balustrade for painting				.019		.76		.76	.98	1.21
	Wash / dry surface				.001		.05		.05	.06	.07
	Paint metal balustrade, brushwork, primer + 1 coat				.015	.11	.61		.72	.91	1.12
	Total				.036	.11	1.42		1.53	**1.95**	**2.40**

B1013 580 Metal Fire Escapes

	System Description	Freq. (Years)	Crew	Unit	Labor Hours	2014 Bare Costs				Total In-House	Total w/O&P
						Material	Labor	Equipment	Total		
1010	**Refinish fire escape balcony, 2' wide**	7	1 PORD	L.F.							
	Prepare metal fire escape balcony for painting				.100		3.96		3.96	5.10	6.35
	Brush on primer coat				.100	2.60	3.96		6.56	7.95	9.60
	Brush on finish coat				.100	2.27	3.96		6.23	7.55	9.20
	Total				.300	4.87	11.88		16.75	**20.60**	**25.15**
1020	**Replace fire escape balcony, 2' wide**	25	2 SSWK	L.F.							
	Remove metal fire escape balcony				.500		25.50		25.50	39	47
	Install new metal fire escape balcony				2.000	58.50	102		160.50	220	261
	Total				2.500	58.50	127.50		186	**259**	**308**
2010	**Refinish fire escape stair and platform**	7	1 PORD	Flight							
	Prepare fire escape stair and platform for painting				2.400		95		95	123	152
	Brush on primer coat				2.400	62.28	95.04		157.32	191	231
	Brush on finish coat				2.400	54.36	95.04		149.40	182	221
	Total				7.200	116.64	285.08		401.72	**496**	**604**
2020	**Replace fire escape stair and platform**	25	2 SSWK	Flight							
	Remove metal fire escape stair and platform				14.995		765		765	1,175	1,400
	Install new metal fire escape stair and platform				59.925	1,075	3,050		4,125	5,850	6,975
	Total				74.920	1,075	3,815		4,890	**7,025**	**8,375**

B10 SUPERSTRUCTURE | B1013 Floor Construction

B1013 580 Metal Fire Escapes

	System Description	Freq. (Years)	Crew	Unit	Labor Hours	2014 Bare Costs				Total In-House	Total w/O&P
						Material	Labor	Equipment	Total		
3010	**Refinish fire escape ladder**	7	1 PORD	V.L.F.							
	Prepare metal fire escape ladder for painting				.020		.79		.79	1.02	1.27
	Brush on primer coat				.020	.35	.79		1.14	1.40	1.70
	Brush on finish coat				.020	.30	.79		1.09	1.36	1.65
	Total				.060	.65	2.37		3.02	**3.78**	**4.62**
3020	**Replace fire escape ladder**	25	2 SSWK	V.L.F.							
	Remove metal fire escape ladder				.250		12.80		12.80	19.50	23.50
	Install new metal fire escape ladder				.500	37	25.50		62.50	79.50	93.50
	Total				.750	37	38.30		75.30	**99**	**117**

B1013 910 Steel Decking

	System Description	Freq. (Years)	Crew	Unit	Labor Hours	2014 Bare Costs				Total In-House	Total w/O&P
						Material	Labor	Equipment	Total		
0020	**Replace steel decking**	30	2 SSWK	S.F.							
	Remove decking				.011		.61	.20	3.31	3.88	4.46
	Assemble and weld decking				.011	2.50	.61	.20	3.31	3.88	4.46
	Total				.023	5	1.22	.40	6.62	**7.76**	**8.92**

B1013 920 Metal Grating

	System Description	Freq. (Years)	Crew	Unit	Labor Hours	2014 Bare Costs				Total In-House	Total w/O&P
						Material	Labor	Equipment	Total		
0010	**Metal floor grating repairs - (2% of grating)**	10	1 SSWK	S.F.							
	Misc. welding of floor grating				.348	1.08	18.45	6.15	25.68	35.50	42
	Total				.348	1.08	18.45	6.15	25.68	**35.50**	**42**
0020	**Replace metal floor grating**	30	2 SSWK	S.F.							
	Remove grating				.011		.61	.20	3.31	3.88	4.46
	Assemble and weld grating				.059	19.65	3.07	.26	22.98	26.50	30.50
	Total				.071	22.15	3.68	.46	26.29	**30.38**	**34.96**

B20 EXTERIOR CLOSURE B2013 Exterior Walls

B2013 104 Copings

	System Description	Freq. (Years)	Crew	Unit	Labor Hours	2014 Bare Costs				Total In-House	Total w/O&P
						Material	Labor	Equipment	Total		
2010	**Replace precast concrete coping, 12" wide**	50	2 BRIC	L.F.							
	Set up and secure scaffold, 3 floors to roof				.429		19.73		19.73	26	32.50
	Remove coping				.050		1.83		1.83	2.44	3.01
	Install new precast concrete coping				.229	18.45	9.45		27.90	32.50	38.50
	Remove scaffold				.429		19.73		19.73	26	32.50
	Total				1.137	18.45	50.74		69.19	86.94	106.51
2020	**Replace limestone coping, 12" wide**	50	2 BRIC	L.F.							
	Set up and secure scaffold, 3 floors to roof				.429		19.73		19.73	26	32.50
	Remove coping				.050		1.83		1.83	2.44	3.01
	Install new limestone coping				.178	14.70	7.35		22.05	26	30.50
	Remove scaffold				.429		19.73		19.73	26	32.50
	Total				1.086	14.70	48.64		63.34	80.44	98.51
2030	**Replace marble coping, 12" wide**	50	2 BRIC	L.F.							
	Set up and secure scaffold, 3 floors to roof				.429		19.73		19.73	26	32.50
	Remove coping				.050		1.83		1.83	2.44	3.01
	Install new marble coping				.200	15.70	8.25		23.95	28	33
	Remove scaffold				.429		19.73		19.73	26	32.50
	Total				1.108	15.70	49.54		65.24	82.44	101.01
2040	**Replace terra cotta coping, 12" wide**	50	2 BRIC	L.F.							
	Set up and secure scaffold, 3 floors to roof				.429		19.73		19.73	26	32.50
	Remove coping				.050		1.83		1.83	2.44	3.01
	Install new terra cotta coping				.200	8.40	8.25		16.65	20	24
	Remove scaffold				.429		19.73		19.73	26	32.50
	Total				1.108	8.40	49.54		57.94	74.44	92.01

B20 EXTERIOR CLOSURE

B2013 Exterior Walls

B2013 109 Concrete Block

	System Description	Freq. (Years)	Crew	Unit	Labor Hours	2014 Bare Costs				Total In-House	Total w/O&P
						Material	Labor	Equipment	Total		
1010	**Repair 8" concrete block wall - 1st floor**	25	1 BRIC	S.F.							
	Set up, secure and take down ladder				.130		5.95		5.95	7.80	9.65
	Remove damaged block				.183		6.72		6.72	8.90	11
	Wire truss reinforcing				.005	.25	.24		.49	.58	.70
	Install block				.111	2.84	4.68		7.52	9.25	11.15
	Total				.429	3.09	17.59		20.68	26.53	32.50
1020	**Replace 8" concrete block wall - 1st floor**	60	2 BRIC	C.S.F.							
	Set up and secure scaffold				.750		34.50		34.50	46	56.50
	Remove damaged block				4.571		168		168	223	275
	Wire truss reinforcing				.260	12.25	11.75		24	29	34.50
	Install block				11.111	284	468		752	925	1,125
	Remove scaffold				.750		34.50		34.50	46	56.50
	Total				17.442	296.25	716.75		1,013	1,269	1,547.50
1030	**Waterproof concrete block wall - 1st floor**	10	1 ROFC	C.S.F.							
	Steam clean				2.971		110	17	127	164	200
	Spray surface, silicone				.260	35	10		45	54	62
	Total				3.231	35	120	17	172	218	262
2020	**Replace 8" concrete block wall - 2nd floor**	60	2 BRIC	C.S.F.							
	Set up and secure scaffold				1.500		69		69	91.50	113
	Remove block				4.571		168		168	223	275
	Wire truss reinforcing				.260	12.25	11.75		24	29	34.50
	Install block				11.111	284	468		752	925	1,125
	Remove scaffold				1.500		69		69	91.50	113
	Total				18.942	296.25	785.75		1,082	1,360	1,660.50
2030	**Waterproof concrete block wall - 2nd floor**	10	G-8	C.S.F.							
	Position aerial lift truck, raise & lower platform				.250		9.80	20	29.80	34.50	39.50
	Steam clean				11.765		431	111	542	685	830
	Spray surface, silicone				.260	35	10	21	66	75	85
	Total				12.275	35	450.80	152	637.80	794.50	954.50

B20 EXTERIOR CLOSURE   B2013   Exterior Walls

B2013 109   Concrete Block

	System Description	Freq. (Years)	Crew	Unit	Labor Hours	2014 Bare Costs				Total In-House	Total w/O&P
						Material	Labor	Equipment	Total		
3020	**Replace 8" concrete block wall - 3rd floor**	60	2 BRIC	C.S.F.							
	Set up and secure scaffold				1.500		69		69	91.50	113
	Remove damaged block				4.571		168		168	223	275
	Wire truss reinforcing				.260	12.25	11.75		24	29	34.50
	Install block				11.111	284	468		752	925	1,125
	Remove scaffold				1.500		69		69	91.50	113
	Total				18.942	296.25	785.75		1,082	1,360	1,660.50
3030	**Waterproof concrete block wall - 3rd floor**	10	G-8	C.S.F.							
	Position aerial lift truck, raise & lower platform				.300		11.75	24	35.75	41.50	47.50
	Steam clean				11.765		431	111	542	685	830
	Spray surface, silicone				.260	35	10	21	66	75	85
	Total				12.325	35	452.75	156	643.75	801.50	962.50

B2013 114   Concrete Block Wall, Painted

	System Description	Freq. (Years)	Crew	Unit	Labor Hours	2014 Bare Costs				Total In-House	Total w/O&P
						Material	Labor	Equipment	Total		
1030	**Point & refinish block wall - 1st floor**	25	1 BRIC	C.S.F.							
	Set up and secure scaffold				.750		34.50		34.50	46	56.50
	Steam clean				2.600		95		95	127	157
	Clean & point block				3.924		179		179	235	291
	Paint surface, brushwork, 1 coat				1.040	13	41.60		54.60	67.50	82.50
	Remove scaffold				.750		34.50		34.50	46	56.50
	Total				9.064	13	384.60		397.60	521.50	643.50
2030	**Point & refinish block wall - 2nd floor**	25	1 BRIC	C.S.F.							
	Set up and secure scaffold				1.500		69		69	91.50	113
	Steam clean				2.600		95		95	127	157
	Clean & point block				3.924		179		179	235	291
	Paint surface, brushwork, 1 coat				1.040	13	41.60		54.60	67.50	82.50
	Remove scaffold				1.500		69		69	91.50	113
	Total				10.564	13	453.60		466.60	612.50	756.50

B20 EXTERIOR CLOSURE

B2013 Exterior Walls

B2013 114 Concrete Block Wall, Painted

	System Description	Freq. (Years)	Crew	Unit	Labor Hours	2014 Bare Costs				Total In-House	Total w/O&P
						Material	Labor	Equipment	Total		
3020	**Point & refinish block wall - 3rd floor**	25	1 BRIC	C.S.F.							
	Set up and secure scaffold				2.250		103.50		103.50	137	170
	Steam clean				2.600		95		95	127	157
	Clean & point block				3.924		179		179	235	291
	Paint surface, brushwork, 1 coat				1.040	13	41.60		54.60	67.50	82.50
	Remove scaffold				2.250		103.50		103.50	137	170
	Total				12.064	13	522.60		535.60	703.50	870.50

B2013 118 Terra Cotta

	System Description	Freq. (Years)	Crew	Unit	Labor Hours	2014 Bare Costs				Total In-House	Total w/O&P
						Material	Labor	Equipment	Total		
1020	**Replace terra cotta wall - 1st floor**	50	2 BRIC	C.S.F.							
	Set up and secure scaffold				.750		34.50		34.50	46	56.50
	Remove terra cotta				1.040		38.50	6.05	44.55	57.50	70
	Replace terra cotta				10.400	545	438		983	1,175	1,400
	Remove scaffold				.750		34.50		34.50	46	56.50
	Total				12.940	545	545.50	6.05	1,096.55	1,324.50	1,583
2010	**Replace terra cotta wall - 2nd floor**	50	2 BRIC	C.S.F.							
	Set up and secure scaffold				1.500		69		69	91.50	113
	Remove terra cotta				1.040		38.50	6.05	44.55	57.50	70
	Replace terra cotta				10.400	545	438		983	1,175	1,400
	Remove scaffold				1.500		69		69	91.50	113
	Total				14.440	545	614.50	6.05	1,165.55	1,415.50	1,696
2020	**Replace terra cotta wall - 3rd floor**	50	2 BRIC	C.S.F.							
	Set up and secure scaffold				2.250		103.50		103.50	137	170
	Remove terra cotta				1.040		38.50	6.05	44.55	57.50	70
	Replace terra cotta				10.400	545	438		983	1,175	1,400
	Remove scaffold				2.250		103.50		103.50	137	170
	Total				15.940	545	683.50	6.05	1,234.55	1,506.50	1,810

B20 EXTERIOR CLOSURE — B2013 Exterior Walls

B2013 119 Clay Brick

	System Description	Freq. (Years)	Crew	Unit	Labor Hours	2014 Bare Costs				Total In-House	Total w/O&P
						Material	Labor	Equipment	Total		
1010	**Repair brick wall - 1st floor**	25	1 BRIC	S.F.							
	Set up, secure and take down ladder				.130		5.95		5.95	7.80	9.65
	Remove damage				.451		16.51		16.51	22	27
	Install brick				.236	3.40	9.95		13.35	16.85	20.50
	Total				.817	3.40	32.41		35.81	**46.65**	**57.15**
1020	**Point brick wall - 1st floor**	25	2 BRIC	C.S.F.							
	Set up and secure scaffold				1.333		52.50		52.50	68.50	84.50
	Steam clean				2.971		110	17	127	164	200
	Clean & point brick				10.000	5	456		461	605	745
	Remove scaffold				1.333		52.50		52.50	68.50	84.50
	Total				15.638	5	671	17	693	**906**	**1,114**
1030	**Waterproof brick wall - 1st floor**	10	1 ROFC	C.S.F.							
	Steam clean				2.971		110	17	127	164	200
	Spray surface, silicone				.260	35	10		45	54	62
	Total				3.231	35	120	17	172	**218**	**262**
2010	**Replace brick wall - 2nd floor**	75	2 BRIC	C.S.F.							
	Set up and secure scaffold				1.500		69		69	91.50	113
	Remove damage				10.401		381		381	505	625
	Install brick				23.636	340	995		1,335	1,675	2,050
	Remove scaffold				1.500		69		69	91.50	113
	Total				37.037	340	1,514		1,854	**2,363**	**2,901**
2020	**Point brick wall - 2nd floor**	25	2 BRIC	C.S.F.							
	Set up and secure scaffold				1.500		69		69	91.50	113
	Steam clean				2.971		110	17	127	164	200
	Clean & point brick				10.000	5	456		461	605	745
	Remove scaffold				1.500		69		69	91.50	113
	Total				15.971	5	704	17	726	**952**	**1,171**
2030	**Waterproof brick wall - 2nd floor**	10	G-8	C.S.F.							
	Position aerial lift truck, raise & lower platform				.250		9.80	20	29.80	34.50	39.50
	Steam clean				11.765		431	111	542	685	830
	Spray surface, silicone				.260	35	10	21	66	75	85
	Total				12.275	35	450.80	152	637.80	**794.50**	**954.50**

B20 EXTERIOR CLOSURE — B2013 Exterior Walls

B2013 119 Clay Brick

	System Description	Freq. (Years)	Crew	Unit	Labor Hours	2014 Bare Costs Material	2014 Bare Costs Labor	2014 Bare Costs Equipment	2014 Bare Costs Total	Total In-House	Total w/O&P
3010	**Replace brick wall - 3rd floor**	75	2 BRIC	C.S.F.							
	Set up and secure scaffold				1.500		69		69	91.50	113
	Remove damage				10.401		381		381	505	625
	Install brick				23.636	340	995		1,335	1,675	2,050
	Remove scaffold				1.500		69		69	91.50	113
	Total				37.037	340	1,514		1,854	2,363	**2,901**
3020	**Point brick wall - 3rd floor**	25	2 BRIC	C.S.F.							
	Set up and secure scaffold				1.500		69		69	91.50	113
	Steam clean				2.971		110	17	127	164	200
	Clean & point brick				10.000	5	456		461	605	745
	Remove scaffold				1.500		69		69	91.50	113
	Total				15.971	5	704	17	726	952	**1,171**
3030	**Waterproof brick wall - 3rd floor**	10	G-8	C.S.F.							
	Position aerial lift truck, raise & lower platform				.300		11.75	24	35.75	41.50	47.50
	Steam clean				11.765		431	111	542	685	830
	Spray surface, silicone				.260	35	10	21	66	75	85
	Total				12.325	35	452.75	156	643.75	801.50	**962.50**

B2013 121 Clay Brick, Painted

	System Description	Freq. (Years)	Crew	Unit	Labor Hours	2014 Bare Costs Material	2014 Bare Costs Labor	2014 Bare Costs Equipment	2014 Bare Costs Total	Total In-House	Total w/O&P
1010	**Repair painted brick wall - 1st floor**	25	1 BRIC	S.F.							
	Set up, secure and take down ladder				.130		5.95		5.95	7.80	9.65
	Remove damage				.451		16.51		16.51	22	27
	Install brick				.236	3.40	9.95		13.35	16.85	20.50
	Paint surface, brushwork, 1 coat				.010	.13	.42		.55	.67	.82
	Total				.827	3.53	32.83		36.36	47.32	**57.97**

B20 EXTERIOR CLOSURE B2013 Exterior Walls

B2013 121 Clay Brick, Painted

	System Description	Freq. (Years)	Crew	Unit	Labor Hours	2014 Bare Costs				Total In-House	Total w/O&P
						Material	Labor	Equipment	Total		
1030	**Point brick wall - 1st floor**	25	1 BRIC	C.S.F.							
	Set up and secure scaffold				.500		23		23	30.50	38
	Steam clean				2.971		110	17	127	164	200
	Clean & point brick				10.000	5	456		461	605	745
	Paint surface, brushwork, 1 coat				1.040	13	41.60		54.60	67.50	82.50
	Remove scaffold				.500		23		23	30.50	38
	Total				15.011	18	653.60	17	688.60	897.50	1,103.50
2010	**Replace brick wall - 2nd floor**	75	1 BRIC	C.S.F.							
	Set up and secure scaffold				1.500		69		69	91.50	113
	Remove damage				10.401		381		381	505	625
	Install brick				23.636	340	995		1,335	1,675	2,050
	Paint surface, brushwork, 1 coat				1.040	13	41.60		54.60	67.50	82.50
	Remove scaffold				1.500		69		69	91.50	113
	Total				38.077	353	1,555.60		1,908.60	2,430.50	2,983.50
2020	**Point and refinish brick wall - 2nd floor**	25	1 BRIC	C.S.F.							
	Set up and secure scaffold				1.500		69		69	91.50	113
	Steam clean				2.971		110	17	127	164	200
	Clean & point brick				10.000	5	456		461	605	745
	Paint surface, brushwork, 1 coat				1.040	13	41.60		54.60	67.50	82.50
	Remove scaffold				1.500		69		69	91.50	113
	Total				17.011	18	745.60	17	780.60	1,019.50	1,253.50
3010	**Replace brick wall - 3rd floor**	75	1 BRIC	C.S.F.							
	Set up and secure scaffold				1.500		69		69	91.50	113
	Remove damage				10.401		381		381	505	625
	Install brick				23.636	340	995		1,335	1,675	2,050
	Paint surface, brushwork, 1 coat				.010	.13	.42		.55	.67	.82
	Remove scaffold				1.500		69		69	91.50	113
	Total				37.047	340.13	1,514.42		1,854.55	2,363.67	2,901.82

B20 EXTERIOR CLOSURE   B2013   Exterior Walls

B2013 121   Clay Brick, Painted

	System Description	Freq. (Years)	Crew	Unit	Labor Hours	2014 Bare Costs				Total In-House	Total w/O&P
						Material	Labor	Equipment	Total		
3020	**Point and refinish brick wall - 3rd floor**	25	1 BRIC	C.S.F.							
	Set up and secure scaffold				1.500		69		69	91.50	113
	Steam clean				2.971		110	17	127	164	200
	Clean & point brick				10.000	5	456		461	605	745
	Paint surface, brushwork, 1 coat				1.040	13	41.60		54.60	67.50	82.50
	Remove scaffold				1.500		69		69	91.50	113
	Total				17.011	18	745.60	17	780.60	1,019.50	1,253.50

B2013 123   Field Stone

	System Description	Freq. (Years)	Crew	Unit	Labor Hours	2014 Bare Costs				Total In-House	Total w/O&P
						Material	Labor	Equipment	Total		
1010	**Repair field stone wall - 1st floor**	25	1 BRIC	S.F.							
	Set up, secure and take down ladder				.390		14.55		14.55	19.40	24
	Remove damaged stones				.173		6.40	1.02	7.42	9.60	11.70
	Reset stones				.416	19	18.75	6.25	44	51.50	61
	Total				.979	19	39.70	7.27	65.97	80.50	96.70
1020	**Point stone wall, 8" x 8" stones - 1st floor**	25	2 BRIC	C.S.F.							
	Set up and secure scaffold				1.333		52.50		52.50	68.50	84.50
	Steam clean				2.971		110	17	127	164	200
	Clean & point stonework joints				5.714	68	261		329	415	510
	Remove scaffold				1.333		52.50		52.50	68.50	84.50
	Total				11.352	68	476	17	561	716	879
2020	**Point stone wall, 8" x 8" stones - 2nd floor**	25	2 BRIC	C.S.F.							
	Set up and secure scaffold				1.500		69		69	91.50	113
	Steam clean				2.971		110	17	127	164	200
	Clean & point stonework joints				5.714	68	261		329	415	510
	Remove scaffold				1.500		69		69	91.50	113
	Total				11.685	68	509	17	594	762	936

B20 EXTERIOR CLOSURE — B2013 Exterior Walls

B2013 123 Field Stone

	System Description	Freq. (Years)	Crew	Unit	Labor Hours	2014 Bare Costs				Total In-House	Total w/O&P
						Material	Labor	Equipment	Total		
3020	**Point stone wall, 8" x 8" stones - 3rd floor**	25	2 BRIC	C.S.F.							
	Set up and secure scaffold				1.500		69		69	91.50	113
	Steam clean				2.971		110	17	127	164	200
	Clean & point stonework joints				5.714	68	261		329	415	510
	Remove scaffold				1.500		69		69	91.50	113
	Total				11.685	68	509	17	594	762	936

B2013 140 Glass Block

	System Description	Freq. (Years)	Crew	Unit	Labor Hours	2014 Bare Costs				Total In-House	Total w/O&P
						Material	Labor	Equipment	Total		
1020	**Replace glass block wall - 1st floor**	75	2 BRIC	C.S.F.							
	Set up and secure scaffold				.750		34.50		34.50	46	56.50
	Remove blocks				5.200		193	30	223	287	350
	Replace blocks				45.217	2,900	1,905		4,805	5,700	6,725
	Remove scaffold				.750		34.50		34.50	46	56.50
	Total				51.917	2,900	2,167	30	5,097	6,079	7,188
2020	**Replace glass block wall - 2nd floor**	75	2 BRIC	C.S.F.							
	Set up and secure scaffold				1.500		69		69	91.50	113
	Remove blocks				5.200		193	30	223	287	350
	Replace blocks				45.217	2,900	1,905		4,805	5,700	6,725
	Remove scaffold				1.500		69		69	91.50	113
	Total				53.417	2,900	2,236	30	5,166	6,170	7,301
3020	**Replace glass block wall - 3rd floor**	75	2 BRIC	C.S.F.							
	Set up and secure scaffold				2.250		103.50		103.50	137	170
	Remove blocks				5.200		193	30	223	287	350
	Replace blocks				45.217	2,900	1,905		4,805	5,700	6,725
	Remove scaffold				2.250		103.50		103.50	137	170
	Total				54.917	2,900	2,305	30	5,235	6,261	7,415

B20 EXTERIOR CLOSURE

B2013 Exterior Walls

B2013 141 Aluminum Siding

	System Description	Freq. (Years)	Crew	Unit	Labor Hours	2014 Bare Costs				Total In-House	Total w/O&P
						Material	Labor	Equipment	Total		
1010	**Replace aluminum siding - 1st floor**	35	1 CARP	C.S.F.							
	Set up, secure and take down ladder				.520		23.80		23.80	32	39
	Remove damaged siding				1.434		53		53	70	86
	Install siding				5.368	163	245		408	495	595
	Total				7.322	163	321.80		484.80	**597**	**720**
1020	**Refinish aluminum siding - 1st floor**	20	1 PORD	C.S.F.							
	Set up, secure and take down ladder				1.300		51.50		51.50	66.50	82.50
	Prepare surface				1.159		46		46	59	74
	Refinish surface, brushwork, primer + 1 coat				1.231	13	49		62	77	94.50
	Total				3.690	13	146.50		159.50	**202.50**	**251**
2010	**Replace aluminum siding - 2nd floor**	35	1 CARP	C.S.F.							
	Set up and secure scaffold				1.500		69		69	91.50	113
	Remove damaged siding				1.434		53		53	70	86
	Install siding				5.368	163	245		408	495	595
	Remove scaffold				1.500		69		69	91.50	113
	Total				9.802	163	436		599	**748**	**907**
2020	**Refinish aluminum siding - 2nd floor**	20	1 PORD	C.S.F.							
	Set up and secure scaffold				1.500		69		69	91.50	113
	Prepare surface				1.159		46		46	59	74
	Refinish surface, brushwork, primer + 1 coat				1.231	13	49		62	77	94.50
	Remove scaffold				1.500		69		69	91.50	113
	Total				5.390	13	233		246	**319**	**394.50**
3010	**Replace aluminum siding - 3rd floor**	35	1 CARP	C.S.F.							
	Set up and secure scaffold				2.250		103.50		103.50	137	170
	Remove damaged siding				1.434		53		53	70	86
	Install siding				5.368	163	245		408	495	595
	Remove scaffold				2.250		103.50		103.50	137	170
	Total				11.302	163	505		668	**839**	**1,021**

B20 EXTERIOR CLOSURE | B2013 | Exterior Walls

B2013 141 | Aluminum Siding

							2014 Bare Costs				
	System Description	Freq. (Years)	Crew	Unit	Labor Hours	Material	Labor	Equipment	Total	Total In-House	Total w/O&P
3020	**Refinish aluminum siding - 3rd floor**	20	1 PORD	C.S.F.							
	Set up and secure scaffold				2.250		103.50		103.50	137	170
	Prepare surface				1.159		46		46	59	74
	Refinish surface, brushwork, primer + 1 coat				1.231	13	49		62	77	94.50
	Remove scaffold				2.250		103.50		103.50	137	170
	Total				6.890	13	302		315	410	508.50
4120	**Spray refinish aluminum siding - 1st floor**	20	1 PORD	C.S.F.							
	Set up, secure and take down ladder				1.300		51.50		51.50	66.50	82.50
	Prepare surface				1.159		46		46	59	74
	Protect/mask unpainted surfaces				.031	.30	1.20		1.50	1.90	2.38
	Spray full primer coat				.444	8	18		26	38	44
	Spray finish coat				.500	8	20		28	41	49
	Total				3.434	16.30	136.70		153	206.40	251.88
4220	**Spray refinish aluminum siding - 2nd floor**	20	2 PORD	C.S.F.							
	Set up and secure scaffold				1.500		69		69	91.50	113
	Prepare surface				1.159		46		46	59	74
	Ptotect/mask unpainted surfaces				.031	.30	1.20		1.50	1.90	2.38
	Spray full primer coat				.444	8	18		26	38	44
	Spray finish coat				.500	8	20		28	41	49
	Remove scaffold				1.500		69		69	91.50	113
	Total				5.134	16.30	223.20		239.50	322.90	395.38
4320	**Spray refinish aluminum siding - 3rd floor**	20	2 PORD	C.S.F.							
	Set up and secure scaffold				2.250		103.50		103.50	137	170
	Prepare surface				1.159		46		46	59	74
	Protect/mask unpainted surfaces				.031	.30	1.20		1.50	1.90	2.38
	Spray full primer coat				.444	8	18		26	38	44
	Spray finish coat				.500	8	20		28	41	49
	Remove scaffold				2.250		103.50		103.50	137	170
	Total				6.634	16.30	292.20		308.50	413.90	509.38

B20 EXTERIOR CLOSURE | B2013 Exterior Walls

B2013 142 Steel Siding

	System Description	Freq. (Years)	Crew	Unit	Labor Hours	2014 Bare Costs				Total In-House	Total w/O&P
						Material	Labor	Equipment	Total		
1010	**Replace steel siding - 1st floor**	35	1 CARP	C.S.F.							
	Set up, secure and take down ladder				.520		23.80		23.80	32	39
	Remove damaged siding				1.434		53		53	70	86
	Install siding				5.266	191	241		432	520	625
	Total				7.220	191	317.80		508.80	**622**	**750**
1020	**Refinish steel siding - 1st floor**	20	1 PORD	C.S.F.							
	Set up, secure and take down ladder				1.300		51.50		51.50	66.50	82.50
	Prepare surface				1.159		46		46	59	74
	Refinish surface, brushwork, primer + 1 coat				1.231	13	49		62	77	94.50
	Total				3.690	13	146.50		159.50	**202.50**	**251**
2010	**Replace steel siding - 2nd floor**	35	1 CARP	C.S.F.							
	Set up and secure scaffold				1.500		69		69	91.50	113
	Remove damaged siding				1.434		53		53	70	86
	Install siding				5.266	191	241		432	520	625
	Remove scaffold				1.500		69		69	91.50	113
	Total				9.700	191	432		623	**773**	**937**
2020	**Refinish steel siding - 2nd floor**	20	1 PORD	C.S.F.							
	Set up and secure scaffold				1.500		69		69	91.50	113
	Prepare surface				1.159		46		46	59	74
	Refinish surface, brushwork, primer + 1 coat				1.231	13	49		62	77	94.50
	Remove scaffold				1.500		69		69	91.50	113
	Total				5.390	13	233		246	**319**	**394.50**
3010	**Replace steel siding - 3rd floor**	35	1 CARP	C.S.F.							
	Set up and secure scaffold				11.302		103.50		103.50	137	170
	Remove damaged siding				2.250		53		53	70	86
	Install siding				1.434	191	241		432	520	625
	Remove scaffold				5.266		103.50		103.50	137	170
					2.250						
	Total				11.200	191	501		692	**864**	**1,051**

B20 EXTERIOR CLOSURE   B2013   Exterior Walls

B2013 142   Steel Siding

		Freq. (Years)	Crew	Unit	Labor Hours	2014 Bare Costs				Total In-House	Total w/O&P
	System Description					Material	Labor	Equipment	Total		
3020	**Refinish steel siding - 3rd floor**	20	1 PORD	C.S.F.							
	Set up and secure scaffold				2.250		103.50		103.50	137	170
	Prepare surface				1.159		46		46	59	74
	Refinish surface, brushwork, primer + 1 coat				1.231	13	49		62	77	94.50
	Remove scaffold				2.250		103.50		103.50	137	170
	Total				6.890	13	302		315	410	508.50
4120	**Spray refinish steel siding - 1st floor**	20	1 PORD	C.S.F.							
	Set up, secure and take down ladder				1.300		51.50		51.50	66.50	82.50
	Prepare surface				1.159		46		46	59	74
	Protect/mask unpainted surfaces				.031	.30	1.20		1.50	1.90	2.38
	Spray full primer coat				.444	8	18		26	38	44
	Spray finish coat				.500	8	20		28	41	49
	Total				3.434	16.30	136.70		153	206.40	251.88
4220	**Spray refinish steel siding - 2nd floor**	20	2 PORD	C.S.F.							
	Set up and secure scaffold				1.500		69		69	91.50	113
	Prepare surface				1.159		46		46	59	74
	Protect/mask unpainted surfaces				.031	.30	1.20		1.50	1.90	2.38
	Spray full primer coat				.444	8	18		26	38	44
	Spray finish coat				.500	8	20		28	41	49
	Remove scaffold				1.500		69		69	91.50	113
	Total				5.134	16.30	223.20		239.50	322.90	395.38
4320	**Spray refinish steel siding - 3rd floor**	20	2 PORD	C.S.F.							
	Set up and secure scaffold				2.250		103.50		103.50	137	170
	Prepare surface				1.159		46		46	59	74
	Protect/mask unpainted surfaces				.031	.30	1.20		1.50	1.90	2.38
	Spray full primer coat				.444	8	18		26	38	44
	Spray finish coat				.500	8	20		28	41	49
	Remove scaffold				2.250		103.50		103.50	137	170
	Total				6.634	16.30	292.20		308.50	413.90	509.38

B20 EXTERIOR CLOSURE   B2013   Exterior Walls

B2013 143   Masonite Panel, Sealed

	System Description	Freq. (Years)	Crew	Unit	Labor Hours	2014 Bare Costs				Total In-House	Total w/O&P
						Material	Labor	Equipment	Total		
1010	**Replace hardboard panels - 1st floor**	12	1 CARP	C.S.F.							
	Set up, secure and take down ladder				.520		23.80		23.80	32	39
	Remove damaged panel				1.434		53		53	70	86
	Cut new hardboard panel to fit				2.773	242	127		369	435	510
	Total				4.727	242	203.80		445.80	**537**	**635**
2010	**Replace hardboard panels - 2nd floor**	12	1 CARP	C.S.F.							
	Set up and secure scaffold				1.500		69		69	91.50	113
	Remove damaged panel				1.434		53		53	70	86
	Cut new hardboard panel to fit				2.773	242	127		369	435	510
	Remove scaffold				1.500		69		69	91.50	113
	Total				7.207	242	318		560	**688**	**822**
3010	**Replace hardboard panels - 3rd floor**	12	1 CARP	C.S.F.							
	Set up and secure scaffold				2.250		103.50		103.50	137	170
	Remove damaged panel				1.434		53		53	70	86
	Cut new hardboard panel to fit				2.773	242	127		369	435	510
	Remove scaffold				2.250		103.50		103.50	137	170
	Total				8.707	242	387		629	**779**	**936**

B2013 145   Wood, Clapboard Finished, 1 Coat

	System Description	Freq. (Years)	Crew	Unit	Labor Hours	2014 Bare Costs				Total In-House	Total w/O&P
						Material	Labor	Equipment	Total		
1010	**Replace & finish wood clapboards - 1st floor**	25	1 CARP	C.S.F.							
	Set up, secure and take down ladder				.520		23.80		23.80	32	39
	Remove damage				2.600		119		119	159	196
	Replace wood clapboards				2.711	318.46	124.62		443.08	515	605
	Refinish surface, brushwork, primer + 1 coat				2.000	20	79		99	124	152
	Total				7.831	338.46	346.42		684.88	**830**	**992**

B20 EXTERIOR CLOSURE B2013 Exterior Walls

B2013 145 Wood, Clapboard Finished, 1 Coat

	System Description	Freq. (Years)	Crew	Unit	Labor Hours	2014 Bare Costs				Total In-House	Total w/O&P
						Material	Labor	Equipment	Total		
2010	**Replace & finish wood clapboards - 2nd floor**	25	1 CARP	C.S.F.							
	Set up and secure scaffold				1.500		69		69	91.50	113
	Remove damage				2.600		119		119	159	196
	Replace wood clapboards				3.524	414	162		576	670	785
	Refinish surface, brushwork, primer + 1 coat				2.000	20	79		99	124	152
	Remove scaffold				1.500		69		69	91.50	113
	Total				11.124	434	498		932	**1,136**	**1,359**
3010	**Replace & finish wood clapboards - 3rd floor**	25	1 CARP	C.S.F.							
	Set up and secure scaffold				2.250		103.50		103.50	137	170
	Remove damage				2.600		119		119	159	196
	Replace wood clapboards				3.489	409.90	160.40		570.30	665	775
	Refinish surface, brushwork, primer + 1 coat				2.000	20	79		99	124	152
	Remove scaffold				2.250		103.50		103.50	137	170
	Total				12.589	429.90	565.40		995.30	**1,222**	**1,463**

B2013 147 Wood Shingles, Unfinished

	System Description	Freq. (Years)	Crew	Unit	Labor Hours	2014 Bare Costs				Total In-House	Total w/O&P
						Material	Labor	Equipment	Total		
1010	**Repair wood shingles - 1st floor**	12	1 CARP	S.F.							
	Set up, secure and take down ladder				.052		2.38		2.38	3.18	3.92
	Remove damaged shingles				.039		1.43		1.43	1.91	2.36
	Replace shingles				.056	2.25	2.57		4.82	5.90	7.05
	Total				.147	2.25	6.38		8.63	**10.99**	**13.33**
1020	**Replace wood shingles - 1st floor**	40	2 CARP	C.S.F.							
	Set up and secure scaffold				.750		34.50		34.50	46	56.50
	Remove damaged shingles				2.600		95		95	127	157
	Replace shingles				4.324	173	198		371	455	540
	Remove scaffold				.750		34.50		34.50	46	56.50
	Total				8.424	173	362		535	**674**	**810**

B20 EXTERIOR CLOSURE

B2013 Exterior Walls

B2013 147 Wood Shingles, Unfinished

	System Description	Freq. (Years)	Crew	Unit	Labor Hours	2014 Bare Costs				Total In-House	Total w/O&P
						Material	Labor	Equipment	Total		
2010	**Replace wood shingles - 2nd floor**	40	1 CARP	C.S.F.							
	Set up and secure scaffold				1.000		46		46	61	75.50
	Remove damaged shingles				2.600		95		95	127	157
	Replace shingles				4.324	173	198		371	455	540
	Remove scaffold				1.000		46		46	61	75.50
	Total				8.924	173	385		558	**704**	**848**
3010	**Replace wood shingles - 3rd floor**	40	1 CARP	C.S.F.							
	Set up and secure scaffold				1.500		69		69	91.50	113
	Remove damaged shingles				2.600		95		95	127	157
	Replace shingles				4.324	173	198		371	455	540
	Remove scaffold				1.500		69		69	91.50	113
	Total				9.924	173	431		604	**765**	**923**

B2013 148 Wood Shingles, Finished

	System Description	Freq. (Years)	Crew	Unit	Labor Hours	2014 Bare Costs				Total In-House	Total w/O&P
						Material	Labor	Equipment	Total		
1020	**Replace & finish wood shingles - 1st floor**	45	1 CARP	C.S.F.							
	Set up, secure and take down ladder				1.500		69		69	91.50	113
	Remove damage				2.600		95		95	127	157
	Replace shingles				4.324	173	198		371	455	540
	Refinish surface, brushwork, primer + 1 coat				2.000	20	79		99	124	152
	Total				10.424	193	441		634	**797.50**	**962**
1030	**Refinish wood shingles - 1st floor**	5	1 PORD	C.S.F.							
	Set up and secure scaffold				.750		34.50		34.50	46	56.50
	Prepare surface				1.667		66		66	85	106
	Refinish surface, brushwork, primer + 1 coat				2.000	20	79		99	124	152
	Remove scaffold				.750		34.50		34.50	46	56.50
	Total				5.167	20	214		234	**301**	**371**

B20 EXTERIOR CLOSURE — B2013 Exterior Walls

B2013 148 Wood Shingles, Finished

	System Description	Freq. (Years)	Crew	Unit	Labor Hours	2014 Bare Costs				Total In-House	Total w/O&P
						Material	Labor	Equipment	Total		
2020	**Refinish wood shingles - 2nd floor**	5	1 PORD	C.S.F.							
	Set up and secure scaffold				1.500		69		69	91.50	113
	Prepare surface				1.667		66		66	85	106
	Refinish surface, brushwork, primer + 1 coat				2.000	20	79		99	124	152
	Remove scaffold				1.500		69		69	91.50	113
	Total				6.667	20	283		303	**392**	**484**
3020	**Refinish wood shingles - 3rd floor**	5	1 PORD	C.S.F.							
	Set up and secure scaffold				2.250		103.50		103.50	137	170
	Prepare surface				1.667		66		66	85	106
	Refinish surface, brushwork, primer + 1 coat				2.000	20	79		99	124	152
	Remove scaffold				2.250		103.50		103.50	137	170
	Total				8.167	20	352		372	**483**	**598**
4120	**Spray refinish wood siding - 1st floor**	5	1 PORD	C.S.F.							
	Set up, secure and take down ladder				1.300		51.50		51.50	66.50	82.50
	Prepare surface				1.159		46		46	59	74
	Protect/mask unpainted surfaces				.031	.30	1.20		1.50	1.90	2.38
	Spray full primer coat				.410	12	16		28	34	41
	Spray finish coat				.410	19	16		35	42	50
	Total				3.310	31.30	130.70		162	**203.40**	**249.88**
4220	**Spray refinish wood siding - 2nd floor**	5	2 PORD	C.S.F.							
	Set up and secure scaffold				1.500		69		69	91.50	113
	Prepare surface				1.159		46		46	59	74
	Protect/mask unpainted surfaces				.031	.30	1.20		1.50	1.90	2.38
	Spray full primer coat				.410	12	16		28	34	41
	Spray finish coat				.410	19	16		35	42	50
	Remove scaffold				1.500		69		69	91.50	113
	Total				5.010	31.30	217.20		248.50	**319.90**	**393.38**

B20 EXTERIOR CLOSURE

B2013 Exterior Walls

B2013 148 Wood Shingles, Finished

	System Description	Freq. (Years)	Crew	Unit	Labor Hours	2014 Bare Costs				Total In-House	Total w/O&P
						Material	Labor	Equipment	Total		
4320	**Spray refinish wood siding - 3rd floor**	5	2 PORD	C.S.F.							
	Set up and secure scaffold				2.250		103.50		103.50	137	170
	Prepare surface				1.159		46		46	59	74
	Protect/mask unpainted surfaces				.031	.30	1.20		1.50	1.90	2.38
	Spray full primer coat				.410	12	16		28	34	41
	Spray finish coat				.410	19	16		35	42	50
	Remove scaffold				2.250		103.50		103.50	137	170
	Total				6.510	31.30	286.20		317.50	410.90	507.38

B2013 150 Fiberglass Panel, Rigid

	System Description	Freq. (Years)	Crew	Unit	Labor Hours	2014 Bare Costs				Total In-House	Total w/O&P
						Material	Labor	Equipment	Total		
1010	**Replace fiberglass panels - 1st floor**	9	1 CARP	C.S.F.							
	Set up, secure and take down ladder				.520		23.80		23.80	32	39
	Remove damaged fiberglass panel				1.434		53		53	70	86
	Install new fiberglass panel				4.727	360	216		576	675	795
	Total				6.681	360	292.80		652.80	777	920
2010	**Replace fiberglass panels - 2nd floor**	9	1 CARP	C.S.F.							
	Set up and secure scaffold				1.500		69		69	91.50	113
	Remove damaged fiberglass panel				1.434		53		53	70	86
	Install new fiberglass panel				4.727	360	216		576	675	795
	Remove scaffold				1.500		69		69	91.50	113
	Total				9.161	360	407		767	928	1,107
3010	**Replace fiberglass panels - 3rd floor**	9	1 CARP	C.S.F.							
	Set up and secure scaffold				2.250		103.50		103.50	137	170
	Remove damaged fiberglass panel				1.434		53		53	70	86
	Install new fiberglass panel				4.727	360	216		576	675	795
	Remove scaffold				2.250		103.50		103.50	137	170
	Total				10.661	360	476		836	1,019	1,221

B20 EXTERIOR CLOSURE — B2013 Exterior Walls

B2013 152 Synthetic Veneer Plaster

	System Description	Freq. (Years)	Crew	Unit	Labor Hours	2014 Bare Costs				Total In-House	Total w/O&P
						Material	Labor	Equipment	Total		
1030	**Refinish veneer plaster - 1st floor**	10	1 BRIC	C.S.F.							
	Set up and secure scaffold				.750		34.50		34.50	46	56.50
	Pressure wash surface				.260		10	2	12	15	19
	Refinish surface				1.270	150	50		200	230	269
	Remove scaffold				.750		34.50		34.50	46	56.50
	Total				3.030	150	129	2	281	337	401
2020	**Refinish veneer plaster - 2nd floor**	10	1 BRIC	C.S.F.							
	Set up and secure scaffold				1.500		69		69	91.50	113
	Pressure wash surface				.260		10	2	12	15	19
	Refinish surface				1.270	150	50		200	230	269
	Remove scaffold				1.500		69		69	91.50	113
	Total				4.530	150	198	2	350	428	514
3020	**Refinish veneer plaster - 3rd floor**	10	1 BRIC	C.S.F.							
	Set up and secure scaffold				2.250		103.50		103.50	137	170
	Pressure wash surface				.260		10	2	12	15	19
	Refinish surface				1.270	150	50		200	230	269
	Remove scaffold				2.250		103.50		103.50	137	170
	Total				6.030	150	267	2	419	519	628

B2013 156 Exterior Soffit Gypsum Board, Painted

	System Description	Freq. (Years)	Crew	Unit	Labor Hours	2014 Bare Costs				Total In-House	Total w/O&P
						Material	Labor	Equipment	Total		
1030	**Refinish exterior painted gypsum board - 1st floor soffit**	5	1 PORD	C.S.F.							
	Set up and secure scaffold				.750		34.50		34.50	46	56.50
	Prepare surface				1.159		46		46	59	74
	Paint surface, roller, primer + 1 coat				1.231	13	49		62	77	94.50
	Remove scaffold				.750		34.50		34.50	46	56.50
	Total				3.890	13	164		177	228	281.50

B20 EXTERIOR CLOSURE — B2013 Exterior Walls

B2013 156 Exterior Soffit Gypsum Board, Painted

	System Description	Freq. (Years)	Crew	Unit	Labor Hours	2014 Bare Costs				Total In-House	Total w/O&P
						Material	Labor	Equipment	Total		
1040	**Replace exterior painted gypsum board - 1st floor soffit**	25	2 CARP	C.S.F.							
	Set up and secure scaffold				.750		34.50		34.50	46	56.50
	Remove old exterior gypsum board				2.737		100		100	134	165
	Install new exterior gypsum board				3.467	77	159		236	297	355
	Paint surface, roller, primer + 1 coat				1.231	13	49		62	77	94.50
	Remove scaffold				.750		34.50		34.50	46	56.50
	Total				8.935	90	377		467	600	**727.50**
2010	**Replace exterior painted gypsum board - 2nd floor soffit**	25	1 CARP	C.S.F.							
	Set up and secure scaffold				1.500		69		69	91.50	113
	Remove old exterior gypsum board				2.737		100		100	134	165
	Install new exterior gypsum board				3.467	77	159		236	297	355
	Paint surface, roller, primer + 1 coat				2.462	26	98		124	154	189
	Remove scaffold				1.500		69		69	91.50	113
	Total				11.666	103	495		598	768	**935**
2020	**Refinish exterior painted gypsum board - 2nd floor soffit**	5	1 PORD	C.S.F.							
	Set up and secure scaffold				1.500		69		69	91.50	113
	Prepare surface				1.159		46		46	59	74
	Paint surface, roller, primer + 1 coat				1.231	13	49		62	77	94.50
	Remove scaffold				1.500		69		69	91.50	113
	Total				5.390	13	233		246	319	**394.50**
3010	**Replace exterior painted gypsum board - 3rd floor soffit**	25	1 CARP	C.S.F.							
	Set up and secure scaffold				2.250		103.50		103.50	137	170
	Remove old exterior gypsum board				2.737		100		100	134	165
	Install new exterior gypsum board				3.467	77	159		236	297	355
	Paint surface, roller, primer + 1 coat				3.693	39	147		186	231	283
	Remove scaffold				2.250		103.50		103.50	137	170
	Total				14.397	116	613		729	936	**1,143**
3020	**Refinish exterior painted gypsum board - 3rd floor soffit**	5	1 PORD	C.S.F.							
	Set up and secure scaffold				2.250		103.50		103.50	137	170
	Prepare surface				1.159		46		46	59	74
	Paint surface, roller, primer + 1 coat				1.231	13	49		62	77	94.50
	Remove scaffold				2.250		103.50		103.50	137	170
	Total				6.890	13	302		315	410	**508.50**

B20 EXTERIOR CLOSURE		B2013	Exterior Walls									

B2013 157 Overhang, Exterior Entry

	System Description	Freq. (Years)	Crew	Unit	Labor Hours	2014 Bare Costs				Total In-House	Total w/O&P
						Material	Labor	Equipment	Total		
1030	**Refinish wood overhang**	5	1 PORD	S.F.							
	Prepare surface				.016		.63		.63	.82	1.02
	Wash / dry surface				.001		.05		.05	.06	.07
	Refinish surface, brushwork, primer + 1 coat				.020	.20	.79		.99	1.24	1.52
	Total				.037	.20	1.47		1.67	2.12	2.61

B2013 158 Recaulking

	System Description	Freq. (Years)	Crew	Unit	Labor Hours	2014 Bare Costs				Total In-House	Total w/O&P
						Material	Labor	Equipment	Total		
1010	**Recaulk expansion & control joints**	20	1 BRIC	L.F.							
	Set up, secure and take down ladder				.130		5.95		5.95	7.80	9.65
	Cut out and recaulk, silicone				.067	3.75	3.04		6.79	8.10	9.65
	Total				.197	3.75	8.99		12.74	15.90	19.30
2010	**Recaulk door**	20	1 BRIC	L.F.							
	Set up, secure and take down ladder				.130		5.95		5.95	7.80	9.65
	Cut out and recaulk, silicone				.067	3.75	3.04		6.79	8.10	9.65
	Total				.197	3.75	8.99		12.74	15.90	19.30
3010	**Recaulk window, 4' x 6' - 1st floor**	20	2 BRIC	Ea.							
	Set up and secure scaffold				1.333		61		61	80	98.50
	Cut out and recaulk, silicone				1.600	90	72.96		162.96	194	231
	Remove scaffold				1.333		61		61	80	98.50
	Total				4.267	90	194.96		284.96	354	428
3020	**Recaulk window, 4' x 6' - 2nd floor**	20	2 BRIC	Ea.							
	Set up and secure scaffold				1.500		69		69	91.50	113
	Cut out and recaulk, silicone				1.600	90	72.96		162.96	194	231
	Remove scaffold				1.500		69		69	91.50	113
	Total				4.600	90	210.96		300.96	377	457
3030	**Recaulk window, 4' x 6' - 3rd floor**	20	2 BRIC	Ea.							
	Set up and secure scaffold				1.500		69		69	91.50	113
	Cut out and recaulk, silicone				1.600	90	72.96		162.96	194	231
	Remove scaffold				1.500		69		69	91.50	113
	Total				4.600	90	210.96		300.96	377	457

B20 EXTERIOR CLOSURE   B2013   Exterior Walls

B2013 310   Wood Louvers and Shutters

	System Description	Freq. (Years)	Crew	Unit	Labor Hours	2014 Bare Costs Material	Labor	Equipment	Total	Total In-House	Total w/O&P
1010	**Repair wood louver in frame 1st floor**	6	1 CARP	Ea.							
	Set up, secure and take down ladder				.001		.05		.05	.07	.09
	Remove louver				.473		17.35		17.35	23	28.50
	Remove damaged stile				.078		3.60		3.60	4.80	5.90
	Remove damaged slat				.039		1.78		1.78	2.36	2.92
	Install new slat				.077	1.02	3.54		4.56	5.85	7.10
	Install new stile				.157	2.37	7.20		9.57	12.20	14.80
	Reinstall louver				.075		3.43		3.43	4.57	5.65
	Total				.900	3.39	36.95		40.34	52.85	64.96
1030	**Refinish wood louver - 1st floor**	5	1 PORD	Ea.							
	Set up and secure scaffold				.444		17.60		17.60	23	28
	Prepare louver for painting				.434		17.12		17.12	22	27.50
	Paint louvered surface (to 16 S.F.), brushwork, primer + 1 coat				.800	2.30	31.50		33.80	43.50	54
	Remove scaffold				.444		17.60		17.60	23	28
	Total				2.123	2.30	83.82		86.12	111.50	137.50
1040	**Replace wood louver - 1st floor**	45	1 CARP	Ea.							
	Set up and secure scaffold				.444		20.50		20.50	27	33.50
	Remove louver				.473		17.35		17.35	23	28.50
	Install new wood louver				.520	141	23.75		164.75	188	216
	Remove scaffold				.444		20.50		20.50	27	33.50
	Total				1.882	141	82.10		223.10	265	311.50
2010	**Repair wood louver - 2nd floor**	6	1 CARP	Ea.							
	Set up and secure scaffold				.889		41		41	54.50	67
	Remove louver				.473		17.35		17.35	23	28.50
	Remove damaged stile				.078		3.60		3.60	4.80	5.90
	Remove damaged slat				.039		1.78		1.78	2.36	2.92
	Install new slat				.077	1.02	3.54		4.56	5.85	7.10
	Install new stile				.157	2.37	7.20		9.57	12.20	14.80
	Reinstall louver				.075		3.43		3.43	4.57	5.65
	Remove scaffold				.889		41		41	54.50	67
	Total				2.677	3.39	118.90		122.29	161.78	198.87

B20 EXTERIOR CLOSURE   B2013   Exterior Walls

B2013 310   Wood Louvers and Shutters

	System Description	Freq. (Years)	Crew	Unit	Labor Hours	Material	2014 Bare Costs Labor	Equipment	Total	Total In-House	Total w/O&P
2030	**Refinish wood louver - 2nd floor**	4	1 PORD	Ea.							
	Set up and secure scaffold				.889		35		35	45.50	56.50
	Prepare louver for painting				.434		17.12		17.12	22	27.50
	Paint louvered surface (to 16 S.F.), brushwork, primer + 1 coat				.800	2.30	31.50		33.80	43.50	54
	Remove scaffold				.889		35		35	45.50	56.50
	Total				3.012	2.30	118.62		120.92	156.50	194.50
2040	**Replace wood louver - 2nd floor**	45	1 CARP	Ea.							
	Set up and secure scaffold				.889		41		41	54.50	67
	Remove louver				.473		17.35		17.35	23	28.50
	Install new wood louver				.520	141	23.75		164.75	188	216
	Remove scaffold				.889		41		41	54.50	67
	Total				2.771	141	123.10		264.10	320	378.50
3010	**Repair wood louver - 3rd floor**	6	1 CARP	Ea.							
	Set up and secure scaffold				1.333		61		61	81.50	100
	Remove louver				.473		17.35		17.35	23	28.50
	Remove damaged stile				.078		3.60		3.60	4.80	5.90
	Remove damaged slat				.039		1.78		1.78	2.36	2.92
	Install new slat				.077	1.02	3.54		4.56	5.85	7.10
	Install new stile				.157	2.37	7.20		9.57	12.20	14.80
	Reinstall louver				.075		3.43		3.43	4.57	5.65
	Remove scaffold				1.333		61		61	81.50	100
	Total				3.565	3.39	158.90		162.29	215.78	264.87
3030	**Refinish wood louver - 3rd floor**	4	1 PORD	Ea.							
	Set up and secure scaffold				1.333		52.50		52.50	68.50	84.50
	Prepare louver for painting				.434		17.12		17.12	22	27.50
	Paint louvered surface (to 16 S.F.), brushwork, primer + 1 coat				.800	2.30	31.50		33.80	43.50	54
	Remove scaffold				1.333		52.50		52.50	68.50	84.50
	Total				3.901	2.30	153.62		155.92	202.50	250.50
3040	**Replace wood louver - 3rd floor**	45	1 CARP	Ea.							
	Set up and secure scaffold				1.333		61		61	81.50	100
	Remove louver				.473		17.35		17.35	23	28.50
	Install new wood louver				.520	141	23.75		164.75	188	216
	Remove scaffold				1.333		61		61	81.50	100
	Total				3.660	141	163.10		304.10	374	444.50

B20 EXTERIOR CLOSURE   B2013   Exterior Walls

B2013 320   Aluminum Louver

							2014 Bare Costs				Total In-House	Total w/O&P
	System Description	Freq. (Years)	Crew	Unit	Labor Hours	Material	Labor	Equipment	Total			
1030	**Refinish aluminum louver - 1st floor**	5	1 PORD	Ea.								
	Set up and secure scaffold				.444		17.60		17.60		23	28
	Prepare louver for painting				.434		17.12		17.12		22	27.50
	Paint louvered surface (to 16 S.F.), brushwork, primer + 1 coat				.800	2.30	31.50		33.80		43.50	54
	Remove scaffold				.444		17.60		17.60		23	28
	Total				2.123	2.30	83.82		86.12		**111.50**	**137.50**
1040	**Replace aluminum louver - 1st floor**	60	1 CARP	Ea.								
	Set up and secure scaffold				.444		20.50		20.50		27	33.50
	Remove louver				.473		17.35		17.35		23	28.50
	Install new louver				1.783	246	97.50		343.50		395	460
	Remove scaffold				.444		20.50		20.50		27	33.50
	Total				3.145	246	155.85		401.85		**472**	**555.50**
2020	**Refinish aluminum louver - 2nd floor**	12	1 PORD	Ea.								
	Set up and secure scaffold				.889		41		41		54.50	67
	Prepare louver for painting				.434		17.12		17.12		22	27.50
	Paint louvered surface (to 16 S.F.), brushwork, primer + 1 coat				.800	2.30	31.50		33.80		43.50	54
	Remove scaffold				.889		41		41		54.50	67
	Total				3.012	2.30	130.62		132.92		**174.50**	**215.50**
2040	**Replace aluminum louver - 2nd floor**	60	1 CARP	Ea.								
	Set up and secure scaffold				.889		41		41		54.50	67
	Remove louver				.473		17.35		17.35		23	28.50
	Install new louver				1.783	246	97.50		343.50		395	460
	Remove scaffold				.889		41		41		54.50	67
	Total				4.034	246	196.85		442.85		**527**	**622.50**
3030	**Refinish aluminum louver - 3rd floor**	5	1 PORD	Ea.								
	Set up and secure scaffold				1.333		52.50		52.50		68.50	84.50
	Prepare louver for painting				.434		17.12		17.12		22	27.50
	Paint louvered surface (to 16 S.F.), brushwork, primer + 1 coat				.800	2.30	31.50		33.80		43.50	54
	Remove scaffold				1.333		52.50		52.50		68.50	84.50
	Total				3.901	2.30	153.62		155.92		**202.50**	**250.50**

B20 EXTERIOR CLOSURE | B2013 Exterior Walls

B2013 320 Aluminum Louver

	System Description	Freq. (Years)	Crew	Unit	Labor Hours	2014 Bare Costs				Total In-House	Total w/O&P
						Material	Labor	Equipment	Total		
3040	**Replace aluminum louver - 3rd floor**	60	1 CARP	Ea.							
	Set up and secure scaffold				1.333		61		61	81.50	100
	Remove louver				.473		17.35		17.35	23	28.50
	Install new louver				1.783	246	97.50		343.50	395	460
	Remove scaffold				1.333		61		61	81.50	100
	Total				4.922	246	236.85		482.85	581	688.50

B2013 330 Steel Louver

	System Description	Freq. (Years)	Crew	Unit	Labor Hours	2014 Bare Costs				Total In-House	Total w/O&P
						Material	Labor	Equipment	Total		
1030	**Refinish steel louver - 1st floor**	5	1 PORD	Ea.							
	Set up and secure scaffold				.444		17.60		17.60	23	28
	Prepare louver for painting				.434		17.12		17.12	22	27.50
	Paint louvered surface (to 16 S.F.), brushwork, primer + 1 coat				.800	2.30	31.50		33.80	43.50	54
	Remove scaffold				.444		17.60		17.60	23	28
	Total				2.123	2.30	83.82		86.12	111.50	137.50
1040	**Replace steel louver - 1st floor**	40	1 CARP	Ea.							
	Set up and secure scaffold				.444		20.50		20.50	27	33.50
	Remove louver				.473		17.35		17.35	23	28.50
	Install new louver				1.783	246	97.50		343.50	395	460
	Remove scaffold				.444		20.50		20.50	27	33.50
	Total				3.145	246	155.85		401.85	472	555.50
2030	**Refinish steel louver - 2nd floor**	5	1 PORD	Ea.							
	Set up and secure scaffold				.889		35		35	45.50	56.50
	Prepare louver for painting				.434		17.12		17.12	22	27.50
	Paint louvered surface (to 16 S.F.), brushwork, primer + 1 coat				.800	2.30	31.50		33.80	43.50	54
	Remove scaffold				.889		35		35	45.50	56.50
	Total				3.012	2.30	118.62		120.92	156.50	194.50

B20 EXTERIOR CLOSURE | B2013 Exterior Walls

B2013 330 Steel Louver

		System Description	Freq. (Years)	Crew	Unit	Labor Hours	2014 Bare Costs					Total In-House	Total w/O&P
							Material	Labor	Equipment	Total		
2040		**Replace steel louver - 2nd floor**	40	1 CARP	Ea.							
		Set up and secure scaffold				.889		41		41	54.50	67
		Remove louver				.473		17.35		17.35	23	28.50
		Install new louver				1.783	246	97.50		343.50	395	460
		Remove scaffold				.889		41		41	54.50	67
		Total				4.034	246	196.85		442.85	527	622.50
3030		**Refinish steel louver - 3rd floor**	5	1 PORD	Ea.							
		Set up and secure scaffold				1.333		52.50		52.50	68.50	84.50
		Prepare louver for painting				.434		17.12		17.12	22	27.50
		Paint louvered surface (to 16 S.F.), brushwork, primer + 1 coat				.800	2.30	31.50		33.80	43.50	54
		Remove scaffold				1.333		52.50		52.50	68.50	84.50
		Total				3.901	2.30	153.62		155.92	202.50	250.50
3040		**Replace steel louver - 3rd floor**	40	1 CARP	Ea.							
		Set up and secure scaffold				1.333		61		61	81.50	100
		Remove louver				.473		17.35		17.35	23	28.50
		Install new louver				1.783	246	97.50		343.50	395	460
		Remove scaffold				1.333		61		61	81.50	100
		Total				4.922	246	236.85		482.85	581	688.50

B20 EXTERIOR CLOSURE		B2023	Exterior Windows								
B2023 102		Steel Frame, Operating									

	System Description	Freq. (Years)	Crew	Unit	Labor Hours	2014 Bare Costs				Total In-House	Total w/O&P
						Material	Labor	Equipment	Total		
1010	**Replace glass - 1st floor (1% of glass)**	1	1 CARP	S.F.							
	Remove damaged glass				.052		1.90		1.90	2.54	3.13
	Install new glass				.043	5.80	1.91		7.71	8.90	10.35
	Total				.095	5.80	3.81		9.61	11.44	13.48
1020	**Repair 3'-9" x 5'-5" steel frame window - 1st flr.**	20	1 CARP	Ea.							
	Set up and secure scaffold				.444		20.50		20.50	27	33.50
	Remove window gaskets				.250		11.45		11.45	15.30	18.85
	Install new gaskets				.564	105.46	25.86		131.32	150	174
	Misc. hardware replacement				.333	86	15.30		101.30	115	133
	Remove scaffold				.444		20.50		20.50	27	33.50
	Total				2.037	191.46	93.61		285.07	334.30	392.85
1030	**Refinish 3'-9" x 5'-5" st. frame window - 1st flr.**	5	1 PORD	Ea.							
	Set up, secure and take down ladder				.267		10.56		10.56	13.80	16.80
	Prepare window for painting				.473		18.70		18.70	24	30
	Paint window, brushwork, primer + 1 coat				1.000	2.09	39.50		41.59	53.50	66
	Total				1.739	2.09	68.76		70.85	91.30	112.80
1040	**Replace 3'-9" x 5'-5" st. frame window - 1st flr.**	45	1 CARP	Ea.							
	Set up and secure scaffold				.444		20.50		20.50	27	33.50
	Remove window				.650		24		24	32	39
	Install new window with frame & glazing (to 21 S.F.)				2.078	1,225	106		1,331	1,500	1,725
	Install sealant				.764	6.79	34.85		41.64	53.50	65
	Remove scaffold				.444		20.50		20.50	27	33.50
	Total				4.381	1,231.79	205.85		1,437.64	1,639.50	1,896
1050	**Finish new 3'-9" x 5'-5" st. fr. window - 1st flr.**	45	1 PORD	Ea.							
	Set up, secure and take down ladder				.267		10.56		10.56	13.80	16.80
	Prepare window for painting				.236		9.35		9.35	12	15
	Paint window, brushwork, primer + 1 coat				1.000	2.09	39.50		41.59	53.50	66
	Total				1.503	2.09	59.41		61.50	79.30	97.80

B20 EXTERIOR CLOSURE B2023 Exterior Windows

B2023 102 Steel Frame, Operating

	System Description	Freq. (Years)	Crew	Unit	Labor Hours	Material	2014 Bare Costs Labor	2014 Bare Costs Equipment	2014 Bare Costs Total	Total In-House	Total w/O&P
2010	**Replace glass - 2nd floor (1% of glass)**	1	1 CARP	S.F.							
	Set up and secure scaffold				.889		41		41	54.50	67
	Remove damaged glass				.052		1.90		1.90	2.54	3.13
	Install new glass				.043	5.80	1.91		7.71	8.90	10.35
	Remove scaffold				.889		41		41	54.50	67
	Total				1.873	5.80	85.81		91.61	120.44	147.48
2020	**Repair 3'-9" x 5'-5" steel frame window - 2nd flr.**	20	1 CARP	Ea.							
	Set up and secure scaffold				.889		41		41	54.50	67
	Remove window gaskets				.250		11.45		11.45	15.30	18.85
	Install new gaskets				.564	105.46	25.86		131.32	150	174
	Misc. hardware replacement				.333	86	15.30		101.30	115	133
	Remove scaffold				.889		41		41	54.50	67
	Total				2.925	191.46	134.61		326.07	389.30	459.85
2030	**Refinish 3'-9" x 5'-5" steel frame window - 2nd flr.**	5	1 PORD	Ea.							
	Set up and secure scaffold				.889		35		35	45.50	56.50
	Prepare window for painting				.473		18.70		18.70	24	30
	Paint window, brushwork, primer + 1 coat				1.000	2.09	39.50		41.59	53.50	66
	Remove scaffold				.889		35		35	45.50	56.50
	Total				3.251	2.09	128.20		130.29	168.50	209
2040	**Replace 3'-9" x 5'-5" steel frame window - 2nd flr.**	45	1 CARP	Ea.							
	Set up and secure scaffold				.889		41		41	54.50	67
	Remove window				.650		24		24	32	39
	Install new window with frame & glazing (to 21 S.F.)				2.078	1,225	106		1,331	1,500	1,725
	Install sealant				.764	6.79	34.85		41.64	53.50	65
	Remove scaffold				.889		41		41	54.50	67
	Total				5.270	1,231.79	246.85		1,478.64	1,694.50	1,963
2050	**Finish new 3'-9" x 5'-5" st. fr. window - 2nd flr.**	45	1 PORD	Ea.							
	Set up and secure scaffold				.889		35		35	45.50	56.50
	Prepare window for painting				.236		9.35		9.35	12	15
	Paint window, brushwork, primer + 1 coat				1.000	2.09	39.50		41.59	53.50	66
	Remove scaffold				.889		35		35	45.50	56.50
	Total				3.014	2.09	118.85		120.94	156.50	194

B20 EXTERIOR CLOSURE — B2023 Exterior Windows

B2023 102 Steel Frame, Operating

	System Description	Freq. (Years)	Crew	Unit	Labor Hours	2014 Bare Costs				Total In-House	Total w/O&P
						Material	Labor	Equipment	Total		
3010	**Replace glass - 3rd floor (1% of glass)**	1	1 CARP	S.F.							
	Set up and secure scaffold				.123		5.65		5.65	7.50	9.25
	Remove damaged glass				.052		1.90		1.90	2.54	3.13
	Install new glass				.043	5.80	1.91		7.71	8.90	10.35
	Remove scaffold				.123		5.65		5.65	7.50	9.25
	Total				.341	5.80	15.11		20.91	26.44	31.98
3020	**Repair 3'-9" x 5'-5" steel frame window - 3rd flr.**	20	1 CARP	Ea.							
	Set up and secure scaffold				1.333		61		61	81.50	100
	Remove window gaskets				.250		11.45		11.45	15.30	18.85
	Install new gaskets				.564	105.46	25.86		131.32	150	174
	Misc. hardware replacement				.333	86	15.30		101.30	115	133
	Remove scaffold				1.333		61		61	81.50	100
	Total				3.814	191.46	174.61		366.07	443.30	525.85
3030	**Refinish 3'-9" x 5'-5" steel frame window - 3rd flr.**	5	1 PORD	Ea.							
	Set up and secure scaffold				1.333		52.50		52.50	68.50	84.50
	Prepare window for painting				.473		18.70		18.70	24	30
	Paint window, brushwork, primer + 1 coat				1.000	2.09	39.50		41.59	53.50	66
	Remove scaffold				1.333		52.50		52.50	68.50	84.50
	Total				4.139	2.09	163.20		165.29	214.50	265
3040	**Replace 3'-9" x 5'-5" steel frame window - 3rd flr.**	45	1 CARP	Ea.							
	Set up and secure scaffold				1.333		61		61	81.50	100
	Remove window				.650		24		24	32	39
	Install new window with frame & glazing (to 21 S.F.)				2.078	1,225	106		1,331	1,500	1,725
	Install sealant				.764	6.79	34.85		41.64	53.50	65
	Remove scaffold				1.333		61		61	81.50	100
	Total				6.159	1,231.79	286.85		1,518.64	1,748.50	2,029
3050	**Finish new 3'-9" x 5'-5" st. fr. window - 3rd flr.**	45	1 PORD	Ea.							
	Set up and secure scaffold				1.333		52.50		52.50	68.50	84.50
	Prepare window for painting				.236		9.35		9.35	12	15
	Paint window, brushwork, primer + 1 coat				1.000	2.09	39.50		41.59	53.50	66
	Remove scaffold				1.333		52.50		52.50	68.50	84.50
	Total				3.903	2.09	153.85		155.94	202.50	250

B20 EXTERIOR CLOSURE B2023 Exterior Windows

B2023 103 Steel Frame, Fixed

	System Description	Freq. (Years)	Crew	Unit	Labor Hours	2014 Bare Costs Material	2014 Bare Costs Labor	2014 Bare Costs Equipment	2014 Bare Costs Total	Total In-House	Total w/O&P
1010	**Replace glass - 1st floor (1% of glass)**	1	1 CARP	S.F.							
	Remove damaged glass				.052		1.90		1.90	2.54	3.13
	Install new glass				.043	5.80	1.91		7.71	8.90	10.35
	Total				.095	5.80	3.81		9.61	11.44	13.48
1020	**Repair 2'-0" x 3'-0" steel frame window - 1st flr.**	20	1 CARP	Ea.							
	Set up and secure scaffold				.444		20.50		20.50	27	33.50
	Remove window gaskets				.250		11.45		11.45	15.30	18.85
	Install new gaskets				.308	57.50	14.10		71.60	82	95
	Remove scaffold				.444		20.50		20.50	27	33.50
	Total				1.447	57.50	66.55		124.05	151.30	180.85
1030	**Refinish 2'-0" x 3'-0" st. frame window - 1st flr.**	5	1 PORD	Ea.							
	Set up, secure and take down ladder				.267		10.56		10.56	13.80	16.80
	Prepare window for painting				.473		18.70		18.70	24	30
	Paint window, brushwork, primer + 1 coat				1.000	2.09	39.50		41.59	53.50	66
	Total				1.739	2.09	68.76		70.85	91.30	112.80
1040	**Replace 2'-0" x 3'-0" steel frame window - 1st flr.**	45	1 CARP	Ea.							
	Set up and secure scaffold				.444		20.50		20.50	27	33.50
	Remove window				.650		24		24	32	39
	Install new window with frame & glazing (2'-0" x 3'-0")				2.078	205	106		311	385	450
	Install sealant				.764	6.79	34.85		41.64	53.50	65
	Remove scaffold				.444		20.50		20.50	27	33.50
	Total				4.381	211.79	205.85		417.64	524.50	621
2010	**Replace glass - 2nd floor (1% of glass)**	1	1 CARP	S.F.							
	Set up and secure scaffold				.044		2.05		2.05	2.73	3.35
	Remove damaged glass				.052		1.90		1.90	2.54	3.13
	Install new glass				.043	5.80	1.91		7.71	8.90	10.35
	Remove scaffold				.044		2.05		2.05	2.73	3.35
	Total				.184	5.80	7.91		13.71	16.90	20.18

B20 EXTERIOR CLOSURE | B2023 | Exterior Windows

B2023 103 | Steel Frame, Fixed

	System Description	Freq. (Years)	Crew	Unit	Labor Hours	2014 Bare Costs				Total In-House	Total w/O&P
						Material	Labor	Equipment	Total		
2020	**Repair 2'-0" x 3'-0" steel frame window - 2nd flr.**	20	1 CARP	Ea.							
	Set up and secure scaffold				.889		41		41	54.50	67
	Remove window gaskets				.250		11.45		11.45	15.30	18.85
	Install new gaskets				.308	57.50	14.10		71.60	82	95
	Remove scaffold				.889		41		41	54.50	67
	Total				2.335	57.50	107.55		165.05	206.30	247.85
2030	**Refinish 2'-0" x 3'-0" steel frame window - 2nd flr.**	5	1 PORD	Ea.							
	Set up and secure scaffold				.889		35		35	45.50	56.50
	Prepare window for painting				.473		18.70		18.70	24	30
	Paint window, brushwork, primer + 1 coat				1.000	2.09	39.50		41.59	53.50	66
	Remove scaffold				.889		35		35	45.50	56.50
	Total				3.251	2.09	128.20		130.29	168.50	209
2040	**Replace 2'-0" x 3'-0" steel frame window - 2nd flr.**	45	1 CARP	Ea.							
	Set up and secure scaffold				.889		41		41	54.50	67
	Remove window				.650		24		24	32	39
	Install new window with frame & glazing (2'-0" x 3'-0")				2.078	205	106		311	385	450
	Install sealant				.764	6.79	34.85		41.64	53.50	65
	Remove scaffold				.889		41		41	54.50	67
	Total				5.270	211.79	246.85		458.64	579.50	688
3010	**Replace glass - 3rd floor (1% of glass)**	1	1 CARP	S.F.							
	Set up and secure scaffold				.123		5.65		5.65	7.50	9.25
	Remove damaged glass				.052		1.90		1.90	2.54	3.13
	Install new glass				.043	5.80	1.91		7.71	8.90	10.35
	Remove scaffold				.123		5.65		5.65	7.50	9.25
	Total				.341	5.80	15.11		20.91	26.44	31.98
3020	**Repair 2'-0" x 3'-0" steel frame window - 3rd flr.**	20	1 CARP	Ea.							
	Set up and secure scaffold				1.333		61		61	81.50	100
	Remove window gaskets				.250		11.45		11.45	15.30	18.85
	Install new gaskets				.308	57.50	14.10		71.60	82	95
	Remove scaffold				1.333		61		61	81.50	100
	Total				3.224	57.50	147.55		205.05	260.30	313.85

B20 EXTERIOR CLOSURE

B2023 Exterior Windows

B2023 103 Steel Frame, Fixed

	System Description	Freq. (Years)	Crew	Unit	Labor Hours	2014 Bare Costs				Total In-House	Total w/O&P
						Material	Labor	Equipment	Total		
3030	**Refinish 2'-0" x 3'-0" steel frame window - 3rd floor**	5	1 PORD	Ea.							
	Set up and secure scaffold				1.333		52.50		52.50	68.50	84.50
	Prepare window for painting				.473		18.70		18.70	24	30
	Paint window, brushwork, primer + 1 coat				1.000	2.09	39.50		41.59	53.50	66
	Remove scaffold				1.333		52.50		52.50	68.50	84.50
	Total				4.139	2.09	163.20		165.29	214.50	265
3040	**Replace 2'-0" x 3'-0" steel frame window - 3rd floor**	45	1 CARP	Ea.							
	Set up and secure scaffold				1.333		61		61	81.50	100
	Remove window				.650		24		24	32	39
	Install new window with frame & glazing (2'-0" x 3'-0")				2.078	205	106		311	385	450
	Install sealant				.764	6.79	34.85		41.64	53.50	65
	Remove scaffold				1.333		61		61	81.50	100
	Total				6.159	211.79	286.85		498.64	633.50	754

B2023 110 Aluminum Window, Operating

	System Description	Freq. (Years)	Crew	Unit	Labor Hours	2014 Bare Costs				Total In-House	Total w/O&P
						Material	Labor	Equipment	Total		
1010	**Replace glass - 1st flr. (1% of glass)**	1	1 CARP	S.F.							
	Remove damaged glass				.052		1.90		1.90	2.54	3.13
	Install new glass				.043	5.80	1.91		7.71	8.90	10.35
	Total				.095	5.80	3.81		9.61	11.44	13.48
1020	**Repair 3' x 4' aluminum window - 1st floor**	20	1 CARP	Ea.							
	Set up and secure scaffold				.444		20.50		20.50	27	33.50
	Remove window gaskets				.250		11.45		11.45	15.30	18.85
	Install new gaskets				.431	80.50	19.74		100.24	115	133
	Misc. hardware replacement				.333	86	15.30		101.30	115	133
	Remove scaffold				.444		20.50		20.50	27	33.50
	Total				1.903	166.50	87.49		253.99	299.30	351.85

B20 EXTERIOR CLOSURE		B2023		Exterior Windows							
B2023 110		Aluminum Window, Operating									

	System Description	Freq. (Years)	Crew	Unit	Labor Hours	2014 Bare Costs				Total In-House	Total w/O&P
						Material	Labor	Equipment	Total		
1030	**Replace 3' x 4' aluminum window - 1st floor**	50	1 CARP	Ea.							
	Set up and secure scaffold				.444		20.50		20.50	27	33.50
	Remove window				.650		24		24	32	39
	Install new window with frame & glazing (to 12 S.F.)				2.078	490	106		596	700	810
	Install sealant				.555	4.93	25.33		30.26	39	47.50
	Remove scaffold				.444		20.50		20.50	27	33.50
	Total				4.173	494.93	196.33		691.26	825	963.50
2010	**Replace glass - 2nd flr. (1% of glass)**	1	1 CARP	S.F.							
	Set up and secure scaffold				.044		2.05		2.05	2.73	3.35
	Remove damaged glass				.052		1.90		1.90	2.54	3.13
	Install new glass				.043	5.80	1.91		7.71	8.90	10.35
	Remove scaffold				.044		2.05		2.05	2.73	3.35
	Total				.184	5.80	7.91		13.71	16.90	20.18
2020	**Repair 3' x 4' aluminum window - 2nd floor**	20	1 CARP	Ea.							
	Set up and secure scaffold				.889		41		41	54.50	67
	Remove window gaskets				.250		11.45		11.45	15.30	18.85
	Install new gaskets				.431	80.50	19.74		100.24	115	133
	Misc. hardware replacement				.333	86	15.30		101.30	115	133
	Remove scaffold				.889		41		41	54.50	67
	Total				2.792	166.50	128.49		294.99	354.30	418.85
2030	**Replace 3' x 4' aluminum window - 2nd floor**	50	1 CARP	Ea.							
	Set up and secure scaffold				.889		41		41	54.50	67
	Remove window				.650		24		24	32	39
	Install new window with frame & glazing (to 12 S.F.)				2.078	490	106		596	700	810
	Install sealant				.555	4.93	25.33		30.26	39	47.50
	Remove scaffold				.889		41		41	54.50	67
	Total				5.062	494.93	237.33		732.26	880	1,030.50
3010	**Replace glass - 3rd floor (1% of glass)**	1	1 CARP	S.F.							
	Set up and secure scaffold				.123		5.65		5.65	7.50	9.25
	Remove damaged glass				.052		1.90		1.90	2.54	3.13
	Install new glass				.043	5.80	1.91		7.71	8.90	10.35
	Remove scaffold				.123		5.65		5.65	7.50	9.25
	Total				.341	5.80	15.11		20.91	26.44	31.98

B20 EXTERIOR CLOSURE   B2023   Exterior Windows

B2023 110   Aluminum Window, Operating

	System Description	Freq. (Years)	Crew	Unit	Labor Hours	2014 Bare Costs				Total In-House	Total w/O&P
						Material	Labor	Equipment	Total		
3020	**Repair 3' x 4' aluminum window - 3rd floor**	20	1 CARP	Ea.							
	Set up and secure scaffold				1.333		61		61	81.50	100
	Remove window gaskets				.250		11.45		11.45	15.30	18.85
	Install new gaskets				.431	80.50	19.74		100.24	115	133
	Misc. hardware replacement				.333	86	15.30		101.30	115	133
	Remove scaffold				1.333		61		61	81.50	100
	Total				3.681	166.50	168.49		334.99	408.30	484.85
3030	**Replace 3' x 4' aluminum window - 3rd floor**	50	1 CARP	Ea.							
	Set up and secure scaffold				1.333		61		61	81.50	100
	Remove window				.650		24		24	32	39
	Install new window with frame & glazing (to 12 S.F.)				2.078	490	106		596	700	810
	Install sealant				.555	4.93	25.33		30.26	39	47.50
	Remove scaffold				1.333		61		61	81.50	100
	Total				5.950	494.93	277.33		772.26	934	1,096.50

B2023 111   Aluminum Window, Fixed

	System Description	Freq. (Years)	Crew	Unit	Labor Hours	2014 Bare Costs				Total In-House	Total w/O&P
						Material	Labor	Equipment	Total		
1010	**Replace glass - 1st floor (1% of glass)**	1	1 CARP	S.F.							
	Remove damaged glass				.052		1.90		1.90	2.54	3.13
	Install new glass				.043	5.80	1.91		7.71	8.90	10.35
	Total				.095	5.80	3.81		9.61	11.44	13.48
1015	**Replace glass - 1st floor, 1" insulating panel with heat reflective glass**	30	2 GLAZ	S.F.							
	Remove damaged glass				.052		1.90		1.90	2.54	3.13
	Tinted insulated glass				.213	23.50	9.40		32.90	38	44.50
	Total				.265	23.50	11.30		34.80	40.54	47.63
1020	**Repair 2'-0" x 3'-0" aluminum window - 1st floor**	20	1 CARP	Ea.							
	Set up and secure scaffold				.444		20.50		20.50	27	33.50
	Remove window gaskets				.250		11.45		11.45	15.30	18.85
	Install new gaskets				.308	57.50	14.10		71.60	82	95
	Remove scaffold				.444		20.50		20.50	27	33.50
	Total				1.447	57.50	66.55		124.05	151.30	180.85

B20 EXTERIOR CLOSURE   B2023   Exterior Windows

B2023 111   Aluminum Window, Fixed

	System Description	Freq. (Years)	Crew	Unit	Labor Hours	2014 Bare Costs				Total In-House	Total w/O&P
						Material	Labor	Equipment	Total		
1030	**Replace 2'-0" x 3'-0" aluminum window - 1st floor**	50	1 CARP	Ea.							
	Set up and secure scaffold				.444		20.50		20.50	27	33.50
	Remove window				.650		24		24	32	39
	Install new window with frame & glazing (2'-0" x 3'-0")				2.078	205	106		311	385	450
	Install sealant				.555	4.93	25.33		30.26	39	47.50
	Remove scaffold				.444		20.50		20.50	27	33.50
	Total				4.173	209.93	196.33		406.26	510	603.50
2010	**Replace glass - 2nd floor (1% of glass)**	1	1 CARP	S.F.							
	Set up and secure scaffold				.889		41		41	54.50	67
	Remove damaged glass				.052		1.90		1.90	2.54	3.13
	Install new glass				.043	5.80	1.91		7.71	8.90	10.35
	Remove scaffold				.889		41		41	54.50	67
	Total				1.873	5.80	85.81		91.61	120.44	147.48
2020	**Repair 2'-0" x 3'-0" aluminum window - 2nd floor**	20	1 CARP	Ea.							
	Set up and secure scaffold				.889		41		41	54.50	67
	Remove window gaskets				.250		11.45		11.45	15.30	18.85
	Install new gaskets				.308	57.50	14.10		71.60	82	95
	Remove scaffold				.889		41		41	54.50	67
	Total				2.335	57.50	107.55		165.05	206.30	247.85
2030	**Replace 2'-0" x 3'-0" aluminum window - 2nd floor**	50	1 CARP	Ea.							
	Set up and secure scaffold				.889		41		41	54.50	67
	Remove window				.650		24		24	32	39
	Install new window with frame & glazing (2'-0" x 3'-0")				2.078	205	106		311	385	450
	Install sealant				.555	4.93	25.33		30.26	39	47.50
	Remove scaffold				.889		41		41	54.50	67
	Total				5.062	209.93	237.33		447.26	565	670.50
3010	**Replace glass - 3rd floor (1% of glass)**	1	1 CARP	S.F.							
	Set up and secure scaffold				.123		5.65		5.65	7.50	9.25
	Remove damaged glass				.052		1.90		1.90	2.54	3.13
	Install new glass				.043	5.80	1.91		7.71	8.90	10.35
	Remove scaffold				.123		5.65		5.65	7.50	9.25
	Total				.341	5.80	15.11		20.91	26.44	31.98

B20 EXTERIOR CLOSURE   B2023   Exterior Windows

B2023 111   Aluminum Window, Fixed

	System Description	Freq. (Years)	Crew	Unit	Labor Hours	2014 Bare Costs				Total In-House	Total w/O&P
						Material	Labor	Equipment	Total		
3020	**Repair 2'-0" x 3'-0" aluminum window - 3rd floor**	20	1 CARP	Ea.							
	Set up and secure scaffold				1.333		61		61	81.50	100
	Remove window gaskets				.250		11.45		11.45	15.30	18.85
	Install new gaskets				.308	57.50	14.10		71.60	82	95
	Remove scaffold				1.333		61		61	81.50	100
	Total				3.224	57.50	147.55		205.05	260.30	313.85
3030	**Replace 2'-0" x 3'-0" aluminum window - 3rd floor**	50	1 CARP	Ea.							
	Set up and secure scaffold				1.333		61		61	81.50	100
	Remove window				.650		24		24	32	39
	Install new window with frame & glazing (2'-0" x 3'-0")				2.078	205	106		311	385	450
	Install sealant				.555	4.93	25.33		30.26	39	47.50
	Remove scaffold				1.333		61		61	81.50	100
	Total				5.950	209.93	277.33		487.26	619	736.50

B2023 112   Wood Frame, Operating

	System Description	Freq. (Years)	Crew	Unit	Labor Hours	2014 Bare Costs				Total In-House	Total w/O&P
						Material	Labor	Equipment	Total		
1010	**Replace glass - 1st floor (1% of glass)**	1	1 CARP	S.F.							
	Remove damaged glass				.052		1.90		1.90	2.54	3.13
	Install new glass				.043	5.80	1.91		7.71	8.90	10.35
	Total				.095	5.80	3.81		9.61	11.44	13.48
1020	**Repair 2'-3" x 6'-0" wood frame window - 1st flr.**	15	1 CARP	Ea.							
	Set up and secure scaffold				.444		20.50		20.50	27	33.50
	Remove putty				.667		30.50		30.50	40.50	50
	Install new putty				.333	5.20	15.30		20.50	26	31.50
	Misc. hardware replacement				.333	86	15.30		101.30	115	133
	Remove scaffold				.444		20.50		20.50	27	33.50
	Total				2.222	91.20	102.10		193.30	235.50	281.50
1030	**Refinish 2'-3" x 6'-0" wood frame window - 1st flr.**	4	1 PORD	Ea.							
	Set up, secure and take down ladder				.267		10.55		10.55	13.65	16.90
	Prepare window for painting				.473		18.70		18.70	24	30
	Paint window, brushwork, primer + 1 coat				1.000	2.09	39.50		41.59	53.50	66
	Total				1.739	2.09	68.75		70.84	91.15	112.90

B20 EXTERIOR CLOSURE — B2023 Exterior Windows

B2023 112 Wood Frame, Operating

	System Description	Freq. (Years)	Crew	Unit	Labor Hours	2014 Bare Costs				Total In-House	Total w/O&P
						Material	Labor	Equipment	Total		
1040	Replace 2'-3" x 6'-0" wood frame window - 1st flr.	40	1 CARP	Ea.							
	Set up and secure scaffold				.444		20.50		20.50	27	33.50
	Remove window				.473		17.35		17.35	23	28.50
	Install new window with frame & glazing (to 14 S.F.)				1.143	320	52.50		372.50	420	485
	Install sealant				.625	5.55	28.50		34.05	43.50	53.50
	Remove scaffold				.444		20.50		20.50	27	33.50
	Total				3.130	325.55	139.35		464.90	540.50	634
2010	Replace glass - 2nd floor (1% of glass)	1	1 CARP	S.F.							
	Set up and secure scaffold				.044		2.05		2.05	2.73	3.35
	Remove damaged glass				.052		1.90		1.90	2.54	3.13
	Install new glass				.043	5.80	1.91		7.71	8.90	10.35
	Remove scaffold				.044		2.05		2.05	2.73	3.35
	Total				.184	5.80	7.91		13.71	16.90	20.18
2020	Repair 2'-3" x 6'-0" wood frame window - 2nd flr.	15	1 CARP	Ea.							
	Set up and secure scaffold				.889		41		41	54.50	67
	Remove putty				.667		30.50		30.50	40.50	50
	New putty				.333	5.20	15.30		20.50	26	31.50
	Misc. hardware replacement				.333	86	15.30		101.30	115	133
	Remove scaffold				.889		41		41	54.50	67
	Total				3.111	91.20	143.10		234.30	290.50	348.50
2030	Refinish 2'-3" x 6'-0" wood frame window - 2nd flr.	5	1 PORD	Ea.							
	Set up and secure scaffold				.889		35		35	45.50	56.50
	Prepare window for painting				.473		18.70		18.70	24	30
	Paint window, brushwork, primer + 1 coat				1.000	2.09	39.50		41.59	53.50	66
	Remove scaffold				.889		35		35	45.50	56.50
	Total				3.251	2.09	128.20		130.29	168.50	209
2040	Replace 2'-3" x 6'-0" wood frame window - 2nd flr.	40	1 CARP	Ea.							
	Set up and secure scaffold				.889		41		41	54.50	67
	Remove window				.473		17.35		17.35	23	28.50
	Install new window with frame & glazing (to 14 S.F.)				1.143	320	52.50		372.50	420	485
	Install sealant				.625	5.55	28.50		34.05	43.50	53.50
	Remove scaffold				.889		41		41	54.50	67
	Total				4.018	325.55	180.35		505.90	595.50	701

B20 EXTERIOR CLOSURE		B2023	Exterior Windows								
B2023 112		Wood Frame, Operating									

	System Description	Freq. (Years)	Crew	Unit	Labor Hours	2014 Bare Costs				Total In-House	Total w/O&P
						Material	Labor	Equipment	Total		
3010	**Replace glass - 3rd floor (1% of glass)**	1	1 CARP	Ea.							
	Set up and secure scaffold				.123		5.65		5.65	7.50	9.25
	Remove damaged glass				.052		1.90		1.90	2.54	3.13
	Install new glass				.043	5.80	1.91		7.71	8.90	10.35
	Remove scaffold				.123		5.65		5.65	7.50	9.25
	Total				.341	5.80	15.11		20.91	26.44	31.98
3020	**Repair 2'-3" x 6'-0" wood frame window - 3rd flr.**	15	1 CARP	Ea.							
	Set up and secure scaffold				1.333		61		61	81.50	100
	Remove putty				.667		30.50		30.50	40.50	50
	Install new putty				.333	5.20	15.30		20.50	26	31.50
	Misc. hardware replacement				.333	86	15.30		101.30	115	133
	Remove scaffold				1.333		61		61	81.50	100
	Total				4.000	91.20	183.10		274.30	344.50	414.50
3030	**Refinish 2'-3" x 6'-0" wood frame window - 3rd flr.**	5	1 PORD	Ea.							
	Set up and secure scaffold				1.333		52.50		52.50	68.50	84.50
	Prepare window for painting				.473		18.70		18.70	24	30
	Paint window, brushwork, primer + 1 coat				1.000	2.09	39.50		41.59	53.50	66
	Remove scaffold				1.333		52.50		52.50	68.50	84.50
	Total				4.139	2.09	163.20		165.29	214.50	265
3040	**Replace 2'-3" x 6'-0" wood frame window - 3rd flr.**	40	1 CARP	Ea.							
	Set up and secure scaffold				1.333		61		61	81.50	100
	Remove window				.473		17.35		17.35	23	28.50
	Install new window with frame & glazing (to 14 S.F.)				1.143	320	52.50		372.50	420	485
	Install sealant				.625	5.55	28.50		34.05	43.50	53.50
	Remove scaffold				1.333		61		61	81.50	100
	Total				4.907	325.55	220.35		545.90	649.50	767

B20 EXTERIOR CLOSURE | B2023 | Exterior Windows

B2023 113 | Wood Frame, Fixed

	System Description	Freq. (Years)	Crew	Unit	Labor Hours	2014 Bare Costs				Total In-House	Total w/O&P
						Material	Labor	Equipment	Total		
1010	**Replace glass - 1st floor**	1	1 CARP	S.F.							
	Remove damaged glass				.052		1.90		1.90	2.54	3.13
	Install new glass				.043	5.80	1.91		7.71	8.90	10.35
	Total				.095	5.80	3.81		9.61	11.44	13.48
1020	**Repair 3'-6" x 4'-0" wood frame window - 1st flr.**	20	1 CARP	Ea.							
	Set up and secure scaffold				.444		20.50		20.50	27	33.50
	Remove putty				.667		30.50		30.50	40.50	50
	New putty				.333	5.20	15.30		20.50	26	31.50
	Remove scaffold				.444		20.50		20.50	27	33.50
	Total				1.889	5.20	86.80		92	120.50	148.50
1030	**Refinish 3'-6" x 4'-0" wood frame window - 1st flr.**	5	1 PORD	Ea.							
	Set up, secure and take down ladder				.444		17.60		17.60	23	28
	Prepare window for painting				.473		18.70		18.70	24	30
	Paint window, brushwork, primer + 1 coat				1.000	2.09	39.50		41.59	53.50	66
	Total				1.917	2.09	75.80		77.89	100.50	124
1040	**Replace 3'-6" x 4'-0" wood frame window - 1st flr.**	40	1 CARP	Ea.							
	Set up and secure scaffold				.444		20.50		20.50	27	33.50
	Remove window				.473		17.35		17.35	23	28.50
	Install new window with frame & glazing (to 14 S.F.)				1.733	420	79.50		499.50	570	655
	Install sealant				.625	5.55	28.50		34.05	43.50	53.50
	Remove scaffold				.444		20.50		20.50	27	33.50
	Total				3.720	425.55	166.35		591.90	690.50	804
2010	**Replace glass - 2nd floor (1% of glass)**	1	1 CARP	S.F.							
	Set up and secure scaffold				.044		2.05		2.05	2.73	3.35
	Remove damaged glass				.052		1.90		1.90	2.54	3.13
	Install new glass				.043	5.80	1.91		7.71	8.90	10.35
	Remove scaffold				.044		2.05		2.05	2.73	3.35
	Total				.184	5.80	7.91		13.71	16.90	20.18

B20 EXTERIOR CLOSURE   B2023   Exterior Windows

B2023 113   Wood Frame, Fixed

	System Description	Freq. (Years)	Crew	Unit	Labor Hours	Material	2014 Bare Costs Labor	Equipment	Total	Total In-House	Total w/O&P
2020	**Repair 3'-6" x 4'-0" wood frame window - 2nd flr.**	20	1 CARP	Ea.							
	Set up and secure scaffold				.889		41		41	54.50	67
	Remove putty				.667		30.50		30.50	40.50	50
	Install new putty				.333	5.20	15.30		20.50	26	31.50
	Remove scaffold				.889		41		41	54.50	67
	Total				2.778	5.20	127.80		133	175.50	215.50
2030	**Refinish 3'-6" x 4'-0" wood frame window - 2nd flr.**	5	1 PORD	Ea.							
	Set up and secure scaffold				.889		35		35	45.50	56.50
	Prepare window for painting				.473		18.70		18.70	24	30
	Paint window, brushwork, primer + 1 coat				1.000	2.09	39.50		41.59	53.50	66
	Remove scaffold				.889		35		35	45.50	56.50
	Total				3.251	2.09	128.20		130.29	168.50	209
2040	**Replace 3'-6" x 4'-0" wood frame window - 2nd flr.**	40	1 CARP	Ea.							
	Set up and secure scaffold				.889		41		41	54.50	67
	Remove window				.473		17.35		17.35	23	28.50
	Install new window with frame & glazing (to 14 S.F.)				1.733	420	79.50		499.50	570	655
	Install sealant				.625	5.55	28.50		34.05	43.50	53.50
	Remove scaffold				.889		41		41	54.50	67
	Total				4.609	425.55	207.35		632.90	745.50	871
3010	**Replace glass - 3rd floor (1% of glass)**	1	1 CARP	S.F.							
	Set up and secure scaffold				.123		5.65		5.65	7.50	9.25
	Remove damaged glass				.052		1.90		1.90	2.54	3.13
	Install new glass				.043	5.80	1.91		7.71	8.90	10.35
	Remove scaffold				.123		5.65		5.65	7.50	9.25
	Total				.341	5.80	15.11		20.91	26.44	31.98
3020	**Repair 3'-6" x 4'-0" wood frame window - 3rd flr.**	20	1 CARP	Ea.							
	Set up and secure scaffold				1.333		61		61	81.50	100
	Remove putty				.667		30.50		30.50	40.50	50
	Install new putty				.333	5.20	15.30		20.50	26	31.50
	Remove scaffold				1.333		61		61	81.50	100
	Total				3.667	5.20	167.80		173	229.50	281.50

B20 EXTERIOR CLOSURE | B2023 | Exterior Windows

B2023 113 | Wood Frame, Fixed

	System Description	Freq. (Years)	Crew	Unit	Labor Hours	2014 Bare Costs				Total In-House	Total w/O&P
						Material	Labor	Equipment	Total		
3030	**Refinish 3'-6" x 4'-0" wood frame window - 3rd flr.**	5	1 PORD	Ea.							
	Set up and secure scaffold				1.333		52.50		52.50	68.50	84.50
	Prepare window for painting				.473		18.70		18.70	24	30
	Paint window, brushwork, primer + 1 coat				1.000	2.09	39.50		41.59	53.50	66
	Remove scaffold				1.333		52.50		52.50	68.50	84.50
	Total				4.139	2.09	163.20		165.29	214.50	265
3040	**Replace 3'-6" x 4'-0" wood frame window - 3rd flr.**	40	1 CARP	Ea.							
	Set up and secure scaffold				1.333		61		61	81.50	100
	Remove window				.473		17.35		17.35	23	28.50
	Install new window with frame & glazing (to 14 S.F.)				1.733	420	79.50		499.50	570	655
	Install sealant				.625	5.55	28.50		34.05	43.50	53.50
	Remove scaffold				1.333		61		61	81.50	100
	Total				5.498	425.55	247.35		672.90	799.50	937

B2023 116 | Aluminum Frame Storm Window

	System Description	Freq. (Years)	Crew	Unit	Labor Hours	2014 Bare Costs				Total In-House	Total w/O&P
						Material	Labor	Equipment	Total		
1010	**Replace glass - (1% of glass)**	1	1 CARP	S.F.							
	Remove glass				.052		1.90		1.90	2.54	3.13
	Install glass / storm window				.173	5.75	7.65		13.40	16.35	19.60
	Total				.225	5.75	9.55		15.30	18.89	22.73
1030	**Refinish 3'-0" x 4'-0" alum. fr. storm wdo. - 1st flr.**	10	1 PORD	Ea.							
	Set up and secure scaffold				.444		17.60		17.60	23	28
	Prepare window frame surface				.473		18.70		18.70	24	30
	Refinish window frame surface, brushwork, primer + 1 coat				.615	1.15	24.50		25.65	33	40.50
	Remove scaffold				.444		17.60		17.60	23	28
	Total				1.977	1.15	78.40		79.55	103	126.50
1040	**Replace 3'-0" x 4'-0" alum. frame storm wdo. - 1st flr.**	50	1 CARP	Ea.							
	Set up and secure scaffold				.444		20.50		20.50	27	33.50
	Remove old window				.385		14.10		14.10	18.80	23
	Install new window (to 12 S.F.)				.743	114	34		148	171	199
	Remove scaffold				.444		20.50		20.50	27	33.50
	Total				2.017	114	89.10		203.10	243.80	289

B20 EXTERIOR CLOSURE | B2023 Exterior Windows

B2023 116 Aluminum Frame Storm Window

	System Description	Freq. (Years)	Crew	Unit	Labor Hours	2014 Bare Costs				Total In-House	Total w/O&P
						Material	Labor	Equipment	Total		
2030	**Refinish 3'-0" x 4'-0" alum. fr. storm wdo. - 2nd flr.**	10	1 PORD	Ea.							
	Set up and secure scaffold				.889		35		35	45.50	56.50
	Prepare window frame surface				.473		18.70		18.70	24	30
	Refinish window frame surface, brushwork, primer + 1 coat				.615	1.15	24.50		25.65	33	40.50
	Remove scaffold				.889		35		35	45.50	56.50
	Total				2.866	1.15	113.20		114.35	148	183.50
2040	**Replace 3'-0" x 4'-0" alum. fr. storm wdo. - 2nd flr.**	50	1 CARP	Ea.							
	Set up and secure scaffold				.889		41		41	54.50	67
	Remove old window				.385		14.10		14.10	18.80	23
	Install new window (to 12 S.F.)				.743	114	34		148	171	199
	Remove scaffold				.889		41		41	54.50	67
	Total				2.906	114	130.10		244.10	298.80	356
3030	**Refinish 3'-0" x 4'-0" alum. fr. storm wdo. - 3rd flr.**	10	1 PORD	Ea.							
	Set up and secure scaffold				1.333		52.50		52.50	68.50	84.50
	Prepare window frame surface				.473		18.70		18.70	24	30
	Refinish window frame surface, brushwork, primer + 1 coat				.615	1.15	24.50		25.65	33	40.50
	Remove scaffold				1.333		52.50		52.50	68.50	84.50
	Total				3.755	1.15	148.20		149.35	194	239.50
3040	**Replace 3'-0" x 4'-0" alum. fr. storm wdo. - 3rd flr.**	50	1 CARP	Ea.							
	Set up and secure scaffold				1.333		61		61	81.50	100
	Remove old window				.385		14.10		14.10	18.80	23
	Install new window (to 12 S.F.)				.743	114	34		148	171	199
	Remove scaffold				1.333		61		61	81.50	100
	Total				3.795	114	170.10		284.10	352.80	422

B2023 117 Steel Frame Storm Window

	System Description	Freq. (Years)	Crew	Unit	Labor Hours	2014 Bare Costs				Total In-House	Total w/O&P
						Material	Labor	Equipment	Total		
1010	**Replace glass - (1% of glass)**	1	1 CARP	S.F.							
	Remove glass				.052		1.90		1.90	2.54	3.13
	Install glass / storm window				.173	5.75	7.65		13.40	16.35	19.60
	Total				.225	5.75	9.55		15.30	18.89	22.73

B20 EXTERIOR CLOSURE		B2023	Exterior Windows									
B2023 117		Steel Frame Storm Window										
								2014 Bare Costs				
	System Description	Freq. (Years)	Crew	Unit	Labor Hours	Material	Labor	Equipment	Total	Total In-House	Total w/O&P
---	---	---	---	---	---	---	---	---	---	---	---
1030	**Refinish 3'-0" x 4'-0" st. fr. storm wdo. - 1st floor**	5	1 PORD	Ea.							
	Set up and secure scaffold				.444		17.60		17.60	23	28
	Prepare window frame surface				.473		18.70		18.70	24	30
	Refinish window frame surface, brushwork, primer + 1 coat				.615	1.15	24.50		25.65	33	40.50
	Remove scaffold				.444		17.60		17.60	23	28
	Total				1.977	1.15	78.40		79.55	103	126.50
1040	**Replace 3'-0" x 4'-0" st. fr. storm wdo. - 1st floor**	25	1 CARP	Ea.							
	Set up and secure scaffold				.444		20.50		20.50	27	33.50
	Remove old window				.385		14.10		14.10	18.80	23
	Install new window (to 12 S.F.)				.743	114	34		148	171	199
	Remove scaffold				.444		20.50		20.50	27	33.50
	Total				2.017	114	89.10		203.10	243.80	289
2030	**Refinish 3'-0" x 4'-0" st. fr. storm wdo. - 2nd floor**	5	1 PORD	Ea.							
	Set up and secure scaffold				.889		35		35	45.50	56.50
	Prepare window frame surface				.473		18.70		18.70	24	30
	Refinish window frame surface, brushwork, primer + 1 coat				.615	1.15	24.50		25.65	33	40.50
	Remove scaffold				.889		35		35	45.50	56.50
	Total				2.866	1.15	113.20		114.35	148	183.50
2040	**Replace 3'-0" x 4'-0" st. fr. storm wdo. - 2nd floor**	25	1 CARP	Ea.							
	Set up and secure scaffold				.889		41		41	54.50	67
	Remove old window				.385		14.10		14.10	18.80	23
	Install new window (to 12 S.F.)				.743	114	34		148	171	199
	Remove scaffold				.889		41		41	54.50	67
	Total				2.906	114	130.10		244.10	298.80	356
3030	**Refinish 3'-0" x 4'-0" st. fr. storm wdo. - 3rd floor**	5	1 PORD	Ea.							
	Set up and secure scaffold				1.333		52.50		52.50	68.50	84.50
	Prepare window frame surface				.473		18.70		18.70	24	30
	Refinish window frame surface, brushwork, primer + 1 coat				.615	1.15	24.50		25.65	33	40.50
	Remove scaffold				1.333		52.50		52.50	68.50	84.50
	Total				3.755	1.15	148.20		149.35	194	239.50

B20 EXTERIOR CLOSURE

B2023 Exterior Windows

B2023 117 Steel Frame Storm Window

	System Description	Freq. (Years)	Crew	Unit	Labor Hours	Material	2014 Bare Costs Labor	Equipment	Total	Total In-House	Total w/O&P
3040	**Replace 3'-0" x 4'-0" st. fr. storm wdo. - 3rd floor**	25	1 CARP	Ea.							
	Set up and secure scaffold				1.333		61		61	81.50	100
	Remove old window				.385		14.10		14.10	18.80	23
	Install new window (to 12 S.F.)				.743	114	34		148	171	199
	Remove scaffold				1.333		61		61	81.50	100
	Total				3.795	114	170.10		284.10	352.80	422

B2023 119 Wood Frame Storm Window

	System Description	Freq. (Years)	Crew	Unit	Labor Hours	Material	2014 Bare Costs Labor	Equipment	Total	Total In-House	Total w/O&P
1010	**Replace glass - (1% of glass)**	1	1 CARP	S.F.							
	Remove glass				.052		1.90		1.90	2.54	3.13
	Install new float glass				.043	5.80	1.91		7.71	8.90	10.35
	Total				.095	5.80	3.81		9.61	11.44	13.48
1020	**Repair 3'-0" x 5'-0" wood frame storm window**	10	1 CARP	Ea.							
	Remove putty				.667		30.50		30.50	40.50	50
	Install new putty				.333	5.20	15.30		20.50	26	31.50
	Total				1.000	5.20	45.80		51	66.50	81.50
1030	**Refinish 3'-0" x 5'-0" wood frame storm window**	5	1 PORD	Ea.							
	Prepare window frame surface				.473		18.70		18.70	24	30
	Refinish window frame surface, brushwork, primer + 1 coat				.615	1.15	24.50		25.65	33	40.50
	Total				1.088	1.15	43.20		44.35	57	70.50
1040	**Replace 3'-0" x 5'-0" wd. fr. storm wdo. - 1st floor**	40	1 CARP	Ea.							
	Set up and secure scaffold				.444		20.50		20.50	27	33.50
	Remove old window				.473		17.35		17.35	23	28.50
	Install new window (to 15 S.F.)				.693	168	31.80		199.80	228	262
	Remove scaffold				.444		20.50		20.50	27	33.50
	Total				2.055	168	90.15		258.15	305	357.50

B20 EXTERIOR CLOSURE   B2023   Exterior Windows

B2023 119   Wood Frame Storm Window

	System Description	Freq. (Years)	Crew	Unit	Labor Hours	2014 Bare Costs Material	2014 Bare Costs Labor	2014 Bare Costs Equipment	2014 Bare Costs Total	Total In-House	Total w/O&P
2040	**Replace 3'-0" x 5'-0" wd. fr. storm wdo. - 2nd floor**	40	1 CARP	Ea.							
	Set up and secure scaffold				.889		41		41	54.50	67
	Remove old window				.473		17.35		17.35	23	28.50
	Install new window (to 15 S.F.)				.693	168	31.80		199.80	228	262
	Remove scaffold				.889		41		41	54.50	67
	Total				2.944	168	131.15		299.15	360	424.50
3040	**Replace 3'-0" x 5'-0" wd. fr. storm wdo. - 3rd floor**	40	1 CARP	Ea.							
	Set up and secure scaffold				1.333		61		61	81.50	100
	Remove old window				.473		17.35		17.35	23	28.50
	Install new window (to 15 S.F.)				.693	168	31.80		199.80	228	262
	Remove scaffold				1.333		61		61	81.50	100
	Total				3.833	168	171.15		339.15	414	490.50

B2023 122   Vinyl Clad Wood Window, Operating

	System Description	Freq. (Years)	Crew	Unit	Labor Hours	2014 Bare Costs Material	2014 Bare Costs Labor	2014 Bare Costs Equipment	2014 Bare Costs Total	Total In-House	Total w/O&P
1030	**Repl. 3'-0" x 4'-0" vinyl clad window - 1st flr.**	60	1 CARP	Ea.							
	Set up and secure scaffold				.444		20.50		20.50	27	33.50
	Remove window				.473		17.35		17.35	23	28.50
	Install new window (3'-0" x 4'-0")				1.156	395	53		448	505	580
	Remove scaffold				.444		20.50		20.50	27	33.50
	Total				2.518	395	111.35		506.35	582	675.50
2030	**Repl. 3'-0" x 4'-0" vinyl clad window - 2nd flr.**	60	1 CARP	Ea.							
	Set up and secure scaffold				.889		41		41	54.50	67
	Remove window				.473		17.35		17.35	23	28.50
	Install new window (3'-0" x 4'-0")				1.156	395	53		448	505	580
	Remove scaffold				.889		41		41	54.50	67
	Total				3.407	395	152.35		547.35	637	742.50

B20 EXTERIOR CLOSURE

B2023 Exterior Windows

B2023 122 Vinyl Clad Wood Window, Operating

	System Description	Freq. (Years)	Crew	Unit	Labor Hours	2014 Bare Costs				Total In-House	Total w/O&P
						Material	Labor	Equipment	Total		
3030	**Repl. 3'-0" x 4'-0" vinyl clad window - 3rd flr.**	60	1 CARP	Ea.							
	Set up and secure scaffold				1.333		61		61	81.50	100
	Remove window				.473		17.35		17.35	23	28.50
	Install new window (3'0" x 4'0")				1.156	395	53		448	505	580
	Remove scaffold				1.333		61		61	81.50	100
	Total				4.295	395	192.35		587.35	691	808.50

B2023 123 Vinyl Clad Wood Window, Fixed

	System Description	Freq. (Years)	Crew	Unit	Labor Hours	2014 Bare Costs				Total In-House	Total w/O&P
						Material	Labor	Equipment	Total		
1030	**Repl. 4'-0" x 4'-0" vinyl clad window - 1st flr.**	60	1 CARP	Ea.							
	Set up and secure scaffold				.444		20.50		20.50	27	33.50
	Remove window				.473		17.35		17.35	23	28.50
	Install new window (4'-0" x 4'-0")				1.733	495	79.50		574.50	650	750
	Remove scaffold				.444		20.50		20.50	27	33.50
	Total				3.095	495	137.85		632.85	727	845.50
2030	**Repl. 4'-0" x 4'-0" vinyl clad window - 2nd flr.**	60	1 CARP	Ea.							
	Set up and secure scaffold				.889		41		41	54.50	67
	Remove window				.473		17.35		17.35	23	28.50
	Install new window (4'-0" x 4'-0")				1.733	495	79.50		574.50	650	750
	Remove scaffold				.889		41		41	54.50	67
	Total				3.984	495	178.85		673.85	782	912.50
3030	**Repl. 4'-0" x 4'-0" vinyl clad window - 3rd flr.**	60	1 CARP	Ea.							
	Set up and secure scaffold				1.333		61		61	81.50	100
	Remove window				.473		17.35		17.35	23	28.50
	Install new window (4'-0" x 4'-0")				1.733	495	79.50		574.50	650	750
	Remove scaffold				1.333		61		61	81.50	100
	Total				4.873	495	218.85		713.85	836	978.50

B20 EXTERIOR CLOSURE B2023 Exterior Windows

B2023 150 Glass Block, Fixed

	System Description	Freq. (Years)	Crew	Unit	Labor Hours	2014 Bare Costs				Total In-House	Total w/O&P
						Material	Labor	Equipment	Total		
1020	**Repair window - 1st floor (2% of glass)**	8	1 BRIC	S.F.							
	Cut out damaged block				.433		16	2.54	18.54	24	29.50
	Replace glass block				.452	29	19.05		48.05	57	67.50
	Total				.886	29	35.05	2.54	66.59	81	97
1030	**Replace glass block window - 1st floor**	75	1 BRIC	S.F.							
	Set up, secure and take down ladder				.267		12.15		12.15	15.95	19.75
	Demolish glass block				.074		2.72		2.72	3.63	4.47
	Replace glass block				.452	29	19.05		48.05	57	67.50
	Total				.793	29	33.92		62.92	76.58	91.72
2020	**Repair window - 2nd floor (2% of glass)**	8	1 BRIC	S.F.							
	Set up and secure scaffold				.889		40.50		40.50	53	66
	Cut out damaged block				.433		16	2.54	18.54	24	29.50
	Replace glass block				.452	29	19.05		48.05	57	67.50
	Remove scaffold				.889		40.50		40.50	53	66
	Total				2.663	29	116.05	2.54	147.59	187	229
2030	**Replace glass block window - 2nd floor**	75	1 BRIC	S.F.							
	Set up and secure scaffold				.889		40.50		40.50	53	66
	Demolish glass block				.074		2.72		2.72	3.63	4.47
	Replace glass block				.452	29	19.05		48.05	57	67.50
	Remove scaffold				.889		40.50		40.50	53	66
	Total				2.304	29	102.77	2.54	131.77	166.63	203.97
3020	**Repair window - 3rd floor (2% of glass)**	8	1 BRIC	S.F.							
	Set up and secure scaffold				1.333		61		61	80	98.50
	Cut out damaged block				.433		16	2.54	18.54	24	29.50
	Replace glass block				.452	29	19.05		48.05	57	67.50
	Remove scaffold				1.333		61		61	80	98.50
	Total				3.552	29	157.05	2.54	188.59	241	294
3030	**Replace glass block window - 3rd floor**	75	1 BRIC	S.F.							
	Set up and secure scaffold				1.333		61		61	80	98.50
	Demolish glass block				.074		2.72		2.72	3.63	4.47
	Replace glass block				.452	29	19.05		48.05	57	67.50
	Remove scaffold				1.333		61		61	80	98.50
	Total				3.193	29	143.77	2.54	172.77	220.63	268.97

B20 EXTERIOR CLOSURE — B2023 Exterior Windows

B2023 160 Aluminum Shutter

	System Description	Freq. (Years)	Crew	Unit	Labor Hours	2014 Bare Costs				Total In-House	Total w/O&P
						Material	Labor	Equipment	Total		
1030	**Refinish aluminum shutter - 1st floor**	5	1 PORD	Ea.							
	Set up and secure scaffold				.444		17.60		17.60	23	28
	Prepare shutter for painting				.434		17.12		17.12	22	27.50
	Paint shutter surface (to 16 S.F.), brushwork, primer + 1 coat				.800	2.30	31.50		33.80	43.50	54
	Remove scaffold				.444		17.60		17.60	23	28
	Total				2.123	2.30	83.82		86.12	**111.50**	**137.50**
1040	**Replace aluminum shutter - 1st floor**	60	1 CARP	Ea.							
	Set up and secure scaffold				.444		20.50		20.50	27	33.50
	Remove shutter				.260		12		12	16	19.50
	Install new shutter				.520	37.50	23.75		61.25	73	86
	Remove scaffold				.444		20.50		20.50	27	33.50
	Total				1.669	37.50	76.75		114.25	**143**	**172.50**
2030	**Refinish aluminum shutter - 2nd floor**	5	1 PORD	Ea.							
	Set up and secure scaffold				.889		35		35	45.50	56.50
	Prepare shutter for painting				.434		17.12		17.12	22	27.50
	Paint shutter surface (to 16 S.F.), brushwork, primer + 1 coat				.800	2.30	31.50		33.80	43.50	54
	Remove scaffold				.889		35		35	45.50	56.50
	Total				3.012	2.30	118.62		120.92	**156.50**	**194.50**
2040	**Replace aluminum shutter - 2nd floor**	60	1 CARP	Ea.							
	Set up and secure scaffold				.889		41		41	54.50	67
	Remove shutter				.260		12		12	16	19.50
	Install new shutter				.520	37.50	23.75		61.25	73	86
	Remove scaffold				.889		41		41	54.50	67
	Total				2.558	37.50	117.75		155.25	**198**	**239.50**
3030	**Refinish aluminum shutter - 3rd floor**	5	1 PORD	Ea.							
	Set up and secure scaffold				1.333		52.50		52.50	68.50	84.50
	Prepare shutter for painting				.434		17.12		17.12	22	27.50
	Paint shutter surface (to 16 S.F.), brushwork, primer + 1 coat				.800	2.30	31.50		33.80	43.50	54
	Remove scaffold				1.333		52.50		52.50	68.50	84.50
	Total				3.901	2.30	153.62		155.92	**202.50**	**250.50**

B20 EXTERIOR CLOSURE	B2023	Exterior Windows									

B2023 160 Aluminum Shutter

	System Description	Freq. (Years)	Crew	Unit	Labor Hours	2014 Bare Costs				Total In-House	Total w/O&P
						Material	Labor	Equipment	Total		
3040	Replace aluminum shutter - 3rd floor	60	1 CARP	Ea.							
	Set up and secure scaffold				1.333		61		61	81.50	100
	Remove shutter				.260		12		12	16	19.50
	Install new shutter				.520	37.50	23.75		61.25	73	86
	Remove scaffold				1.333		61		61	81.50	100
	Total				3.447	37.50	157.75		195.25	252	305.50

B2023 162 Steel Shutter

	System Description	Freq. (Years)	Crew	Unit	Labor Hours	2014 Bare Costs				Total In-House	Total w/O&P
						Material	Labor	Equipment	Total		
1030	Refinish steel shutter - 1st floor	5	1 PORD	Ea.							
	Set up and secure scaffold				.444		17.60		17.60	23	28
	Prepare shutter for painting				.434		17.12		17.12	22	27.50
	Paint shutter surface (to 16 S.F.), brushwork, primer + 1 coat				.800	2.30	31.50		33.80	43.50	54
	Remove scaffold				.444		17.60		17.60	23	28
	Total				2.123	2.30	83.82		86.12	111.50	137.50
1040	Replace steel shutter - 1st floor	40	1 CARP	Ea.							
	Set up and secure scaffold				.444		20.50		20.50	27	33.50
	Remove old shutter				.308		15.90	1.38	17.28	25.50	30.50
	Install new steel shutter				.617	37.50	31.80	2.76	72.06	92.50	108
	Remove scaffold				.444		20.50		20.50	27	33.50
	Total				1.814	37.50	88.70	4.14	130.34	172	205.50
2030	Refinish steel shutter - 2nd floor	5	1 PORD	Ea.							
	Set up and secure scaffold				.889		35		35	45.50	56.50
	Prepare shutter for painting				.434		17.12		17.12	22	27.50
	Paint shutter surface (to 16 S.F.), brushwork, primer + 1 coat				.800	2.30	31.50		33.80	43.50	54
	Remove scaffold				.889		35		35	45.50	56.50
	Total				3.012	2.30	118.62		120.92	156.50	194.50

B20 EXTERIOR CLOSURE

B2023 Exterior Windows

B2023 162 Steel Shutter

	System Description	Freq. (Years)	Crew	Unit	Labor Hours	2014 Bare Costs				Total In-House	Total w/O&P
						Material	Labor	Equipment	Total		
2040	**Replace steel shutter - 2nd floor**	40	1 CARP	Ea.							
	Set up and secure scaffold				.889		41		41	54.50	67
	Remove old shutter				.308		15.90	1.38	17.28	25.50	30.50
	Install new steel shutter				.617	37.50	31.80	2.76	72.06	92.50	108
	Remove scaffold				.889		41		41	54.50	67
	Total				2.703	37.50	129.70	4.14	171.34	227	272.50
3030	**Refinish steel shutter - 3rd floor**	5	1 PORD	Ea.							
	Set up and secure scaffold				1.333		52.50		52.50	68.50	84.50
	Prepare shutter for painting				.434		17.12		17.12	22	27.50
	Paint shutter surface (to 16 S.F.), brushwork, primer + 1 coat				.800	2.30	31.50		33.80	43.50	54
	Remove scaffold				1.333		52.50		52.50	68.50	84.50
	Total				3.901	2.30	153.62		155.92	202.50	250.50
3040	**Replace steel shutter - 3rd floor**	40	1 CARP	Ea.							
	Set up and secure scaffold				1.333		61		61	81.50	100
	Remove old shutter				.308		15.90	1.38	17.28	25.50	30.50
	Install new steel shutter				.617	37.50	31.80	2.76	72.06	92.50	108
	Remove scaffold				1.333		61		61	81.50	100
	Total				3.592	37.50	169.70	4.14	211.34	281	338.50

B2023 164 Wood Shutter

	System Description	Freq. (Years)	Crew	Unit	Labor Hours	2014 Bare Costs				Total In-House	Total w/O&P
						Material	Labor	Equipment	Total		
1010	**Repair wood shutter - 1st floor**	6	1 CARP	Ea.							
	Remove shutter				.260		12		12	16	19.50
	Remove damaged panel				.157		7.20		7.20	9.60	11.80
	Install new panel				.314	4.74	14.40		19.14	24.50	29.50
	Install shutter				.520		23.75		23.75	32	39.50
	Total				1.251	4.74	57.35		62.09	82.10	100.30

B20 EXTERIOR CLOSURE — B2023 Exterior Windows

B2023 164 Wood Shutter

	System Description	Freq. (Years)	Crew	Unit	Labor Hours	2014 Bare Costs				Total In-House	Total w/O&P
						Material	Labor	Equipment	Total		
1030	**Refinish wood shutter - 1st floor**	5	1 PORD	Ea.							
	Set up and secure scaffold				.444		17.60		17.60	23	28
	Prepare shutter for painting				.434		17.12		17.12	22	27.50
	Paint shutter surface (to 16 S.F.), brushwork, primer + 1 coat				.800	2.30	31.50		33.80	43.50	54
	Remove scaffold				.444		17.60		17.60	23	28
	Total				2.123	2.30	83.82		86.12	111.50	137.50
1040	**Replace wood shutter - 1st floor**	40	1 CARP	Ea.							
	Set up and secure scaffold				.444		20.50		20.50	27	33.50
	Remove shutter				.260		12		12	16	19.50
	Install new wood shutter				.520	141	23.75		164.75	188	216
	Remove scaffold				.444		20.50		20.50	27	33.50
	Total				1.669	141	76.75		217.75	258	302.50
2010	**Repair wood shutter - 2nd floor**	6	1 CARP	Ea.							
	Set up and secure scaffold				.889		41		41	54.50	67
	Remove shutter				.260		12		12	16	19.50
	Remove damaged panel				.157		7.20		7.20	9.60	11.80
	Install new panel				.314	4.74	14.40		19.14	24.50	29.50
	Install shutter				.520		23.75		23.75	32	39.50
	Remove scaffold				.889		41		41	54.50	67
	Total				3.029	4.74	139.35		144.09	191.10	234.30
2030	**Refinish wood shutter - 2nd floor**	5	1 PORD	Ea.							
	Set up and secure scaffold				.889		35		35	45.50	56.50
	Prepare shutter for painting				.434		17.12		17.12	22	27.50
	Paint shutter surface (to 16 S.F.), brushwork, primer + 1 coat				.800	2.30	31.50		33.80	43.50	54
	Remove scaffold				.889		35		35	45.50	56.50
	Total				3.012	2.30	118.62		120.92	156.50	194.50
2040	**Replace wood shutter - 2nd floor**	40	1 CARP	Ea.							
	Set up and secure scaffold				.889		41		41	54.50	67
	Remove shutter				.260		12		12	16	19.50
	Install new wood shutter				.520	141	23.75		164.75	188	216
	Remove scaffold				.889		41		41	54.50	67
	Total				2.558	141	117.75		258.75	313	369.50

B20 EXTERIOR CLOSURE

B2023 Exterior Windows

B2023 164 Wood Shutter

	System Description	Freq. (Years)	Crew	Unit	Labor Hours	2014 Bare Costs				Total In-House	Total w/O&P
						Material	Labor	Equipment	Total		
3010	**Repair wood shutter - 3rd floor**	6	1 CARP	Ea.							
	Set up and secure scaffold				1.333		61		61	81.50	100
	Remove shutter				.260		12		12	16	19.50
	Remove damaged panel				.157		7.20		7.20	9.60	11.80
	Install new panel				.314	4.74	14.40		19.14	24.50	29.50
	Install shutter				.520		23.75		23.75	32	39.50
	Remove scaffold				1.333		61		61	81.50	100
	Total				3.917	4.74	179.35		184.09	**245.10**	**300.30**
3030	**Refinish wood shutter - 3rd floor**	5	1 PORD	Ea.							
	Set up and secure scaffold				1.333		52.50		52.50	68.50	84.50
	Prepare shutter for painting				.434		17.12		17.12	22	27.50
	Paint shutter surface (to 16 S.F.), brushwork, primer + 1 coat				.800	2.30	31.50		33.80	43.50	54
	Remove scaffold				1.333		52.50		52.50	68.50	84.50
	Total				3.901	2.30	153.62		155.92	**202.50**	**250.50**
3040	**Replace wood shutter - 3rd floor**	40	1 CARP	Ea.							
	Set up and secure scaffold				1.333		61		61	81.50	100
	Remove shutter				.260		12		12	16	19.50
	Install new wood shutter				.520	141	23.75		164.75	188	216
	Remove scaffold				1.333		61		61	81.50	100
	Total				3.447	141	157.75		298.75	**367**	**435.50**

B2023 166 Metal Window Grating

	System Description	Freq. (Years)	Crew	Unit	Labor Hours	2014 Bare Costs				Total In-House	Total w/O&P
						Material	Labor	Equipment	Total		
1020	**Refinish 3'-0" x 4'-0" metal window grating**	5	1 PORD	Ea.							
	Prepare window grating surface				.308		12.15		12.15	15.75	19.50
	Paint grated surface, brushwork, primer + 1 coat				.167	1.68	6.48		8.16	10.55	12.65
	Total				.474	1.68	18.63		20.31	**26.30**	**32.15**
1030	**Replace 3'-0" x 4'-0" metal wdo. grating - 1st floor**	30	1 SSWK	Ea.							
	Remove old grating				.800		29.50		29.50	39	48
	Fabricate and install new window grating				6.275	314	320		634	830	985
	Total				7.075	314	349.50		663.50	**869**	**1,033**

B20 EXTERIOR CLOSURE   B2023   Exterior Windows

B2023 166   Metal Window Grating

	System Description	Freq. (Years)	Crew	Unit	Labor Hours	2014 Bare Costs				Total In-House	Total w/O&P
						Material	Labor	Equipment	Total		
2030	**Replace 3'-0" x 4'-0" metal wdo. grating - 2nd floor**	30	1 SSWK	Ea.							
	Set up and secure scaffold				.889		45.50		45.50	69.50	83.50
	Remove old grating				.800		29.50		29.50	39	48
	Fabricate and install new window grating				6.275	314	320		634	830	985
	Remove scaffold				.889		45.50		45.50	69.50	83.50
	Total				8.852	314	440.50		754.50	1,008	1,200
3030	**Replace 3'-0" x 4'-0" metal wdo. grating - 3rd floor**	30	1 SSWK	Ea.							
	Set up and secure scaffold				1.333		68		68	104	125
	Remove old window grating				.800		29.50		29.50	39	48
	Fabricate and install new window grating				6.275	314	320		634	830	985
	Remove scaffold				1.333		68		68	104	125
	Total				9.741	314	485.50		799.50	1,077	1,283

B2023 168   Metal Wire Mesh Cover

	System Description	Freq. (Years)	Crew	Unit	Labor Hours	2014 Bare Costs				Total In-House	Total w/O&P
						Material	Labor	Equipment	Total		
1020	**Refinish 3'-0" x 4'-0" mtl. wire mesh wdo. cover**	5	1 PORD	Ea.							
	Prepare metal mesh surface				.103		4.06		4.06	5.25	6.50
	Paint metal mesh surface, brushwork, primer + 1 coat				.167	1.68	6.48		8.16	10.55	12.65
	Total				.269	1.68	10.54		12.22	15.80	19.15
1030	**Repl. 3'-0" x 4'-0" mtl. wire mesh wdo. cover - 1st flr.**	30	1 SSWK	Ea.							
	Set up and secure scaffold				.444		22.50		22.50	34.50	41.50
	Remove old mesh				.800		29.50		29.50	39	48
	Install new wire mesh & frame (to 12 S.F.)				.452	60	20.75		80.75	93.50	109
	Remove scaffold				.444		22.50		22.50	34.50	41.50
	Total				2.141	60	95.25		155.25	201.50	240
2030	**Repl. 3'-0" x 4'-0" mtl. wire mesh wdo. cover - 2nd flr.**	30	1 SSWK	Ea.							
	Set up and secure scaffold				.889		45.50		45.50	69.50	83.50
	Remove old mesh				.800		29.50		29.50	39	48
	Install new wire mesh & frame (to 12 S.F.)				.452	60	20.75		80.75	93.50	109
	Remove scaffold				.889		45.50		45.50	69.50	83.50
	Total				3.030	60	141.25		201.25	271.50	324

B20 EXTERIOR CLOSURE | B2023 | Exterior Windows

B2023 168 | Metal Wire Mesh Cover

	System Description	Freq. (Years)	Crew	Unit	Labor Hours	2014 Bare Costs				Total In-House	Total w/O&P
						Material	Labor	Equipment	Total		
3030	Repl. 3'-0" x 4'-0" mtl. wire mesh wdo. cover - 3rd flr.	30	1 SSWK	Ea.							
	Set up and secure scaffold				1.333		68		68	104	125
	Remove old mesh				.800		29.50		29.50	39	48
	Install new wire mesh & frame (to 12 S.F.)				.452	60	20.75		80.75	93.50	109
	Remove scaffold				1.333		68		68	104	125
	Total				3.919	60	186.25		246.25	340.50	407

B2023 170 | Aluminum Window Screen

	System Description	Freq. (Years)	Crew	Unit	Labor Hours	2014 Bare Costs				Total In-House	Total w/O&P
						Material	Labor	Equipment	Total		
1010	Repair alum. window screen	20	1 CARP	S.F.							
	Remove screen from mountings				.052		2.66		2.66	4.05	4.88
	Repair damaged screen				.017	3.08	.89		3.97	4.74	5.45
	Remount screen				.104		5.30		5.30	8.10	9.75
	Total				.173	3.08	8.85		11.93	16.89	20.08
1020	Refinish 3'-0" x 4'-0" alum. window screen	5	1 PORD	Ea.							
	Prepare screen frame surface				.473		18.70		18.70	24	30
	Refinish screen frame surface, brushwork, primer + 1 coat				.615	1.15	24.50		25.65	33	40.50
	Total				1.088	1.15	43.20		44.35	57	70.50
1030	Replace 3'-0" x 4'-0" alum. wdo. screen - 1st flr.	50	1 CARP	Ea.							
	Set up and secure scaffold				.444		20.50		20.50	27	33.50
	Remove screen from mountings				.624		31.92		31.92	48.50	58.50
	Install new screen				1.248	258	63.60		321.60	385	440
	Remove scaffold				.444		20.50		20.50	27	33.50
	Total				2.761	258	136.52		394.52	487.50	565.50
2030	Replace 3'-0" x 4'-0" alum. wdo. screen - 2nd flr.	50	1 CARP	Ea.							
	Set up and secure scaffold				.889		41		41	54.50	67
	Remove screen from mountings				.624		31.92		31.92	48.50	58.50
	Install new screen				1.248	258	63.60		321.60	385	440
	Remove scaffold				.889		41		41	54.50	67
	Total				3.650	258	177.52		435.52	542.50	632.50

B20 EXTERIOR CLOSURE | B2023 Exterior Windows

B2023 170 Aluminum Window Screen

	System Description	Freq. (Years)	Crew	Unit	Labor Hours	2014 Bare Costs Material	2014 Bare Costs Labor	2014 Bare Costs Equipment	2014 Bare Costs Total	Total In-House	Total w/O&P
3030	**Replace 3'-0" x 4'-0" alum. wdo. screen - 3rd flr.**	50	1 CARP	Ea.							
	Set up and secure scaffold				1.333		61		61	81.50	100
	Remove screen from mountings				.624		31.92		31.92	48.50	58.50
	Install new screen				1.248	258	63.60		321.60	385	440
	Remove scaffold				1.333		61		61	81.50	100
	Total				4.539	258	217.52		475.52	596.50	698.50

B2023 172 Steel Frame Screen Window

	System Description	Freq. (Years)	Crew	Unit	Labor Hours	2014 Bare Costs Material	2014 Bare Costs Labor	2014 Bare Costs Equipment	2014 Bare Costs Total	Total In-House	Total w/O&P
1010	**Repair steel frame screen window**	10	1 CARP	S.F.							
	Remove screen from mountings				.052		2.66		2.66	4.05	4.88
	Repair damaged screen				.017	3.08	.89		3.97	4.74	5.45
	Remount screen				.104		5.30		5.30	8.10	9.75
	Total				.173	3.08	8.85		11.93	16.89	20.08
1020	**Refinish 3'-0" x 4'-0" steel frame screen window**	5	1 PORD	Ea.							
	Prepare screen frame surface				.473		18.70		18.70	24	30
	Refinish screen frame surface, brushwork, primer + 1 coat				.615	1.15	24.50		25.65	33	40.50
	Total				1.088	1.15	43.20		44.35	57	70.50
1030	**Replace 3'-0" x 4'-0" st. fr. screen wdo. - 1st flr.**	40	1 CARP	Ea.							
	Set up and secure scaffold				.444		20.50		20.50	27	33.50
	Remove screen from mountings				.624		31.92		31.92	48.50	58.50
	Install new screen				1.248	50.88	63.60		114.48	153	181
	Remove scaffold				.444		20.50		20.50	27	33.50
	Total				2.761	50.88	136.52		187.40	255.50	306.50
2030	**Replace 3'-0" x 4'-0" st. fr. screen wdo. - 2nd flr.**	40	1 CARP	Ea.							
	Set up and secure scaffold				.889		41		41	54.50	67
	Remove screen from mountings				.624		31.92		31.92	48.50	58.50
	Install new screen				1.248	50.88	63.60		114.48	153	181
	Remove scaffold				.889		41		41	54.50	67
	Total				3.650	50.88	177.52		228.40	310.50	373.50

B20 EXTERIOR CLOSURE B2023 Exterior Windows

B2023 172 Steel Frame Screen Window

	System Description	Freq. (Years)	Crew	Unit	Labor Hours	2014 Bare Costs				Total In-House	Total w/O&P
						Material	Labor	Equipment	Total		
3030	**Replace 3'-0" x 4'-0" st. fr. screen wdo. - 3rd flr.**	40	1 CARP	Ea.							
	Set up and secure scaffold				1.333		61		61	81.50	100
	Remove screen from mountings				.624		31.92		31.92	48.50	58.50
	Install new screen				1.248	50.88	63.60		114.48	153	181
	Remove scaffold				1.333		61		61	81.50	100
	Total				4.539	50.88	217.52		268.40	364.50	439.50

B2023 174 Wood Frame Screen Window

	System Description	Freq. (Years)	Crew	Unit	Labor Hours	2014 Bare Costs				Total In-House	Total w/O&P
						Material	Labor	Equipment	Total		
1010	**Repair wood frame screen window**	10	1 CARP	S.F.							
	Remove screen from mountings				.052		2.66		2.66	4.05	4.88
	Repair damaged screen				.017	3.08	.89		3.97	4.74	5.45
	Remount screen				.104		5.30		5.30	8.10	9.75
	Total				.173	3.08	8.85		11.93	16.89	20.08
1020	**Refinish 3'-0" x 4'-0" wood frame screen window**	5	1 PORD	Ea.							
	Prepare screen frame surface				.473		18.70		18.70	24	30
	Refinish screen frame surface, brushwork, primer + 1 coat				.615	1.15	24.50		25.65	33	40.50
	Total				1.088	1.15	43.20		44.35	57	70.50
1030	**Replace 3'-0" x 4'-0" wd. fr. screen wdo. - 1st flr.**	40	1 CARP	Ea.							
	Set up and secure scaffold				.444		20.50		20.50	27	33.50
	Remove screen from mountings				.624		31.92		31.92	48.50	58.50
	Install new screen				.666	97.80	30.48		128.28	148	172
	Remove scaffold				.444		20.50		20.50	27	33.50
	Total				2.179	97.80	103.40		201.20	250.50	297.50
2030	**Replace 3'-0" x 4'-0" wd. fr. screen wdo. - 2nd flr.**	40	1 CARP	Ea.							
	Set up and secure scaffold				.889		41		41	54.50	67
	Remove screen from mountings				.624		31.92		31.92	48.50	58.50
	Install new screen				.666	97.80	30.48		128.28	148	172
	Remove scaffold				.889		41		41	54.50	67
	Total				3.067	97.80	144.40		242.20	305.50	364.50

B20 EXTERIOR CLOSURE | B2023 Exterior Windows

B2023 174 Wood Frame Screen Window

	System Description	Freq. (Years)	Crew	Unit	Labor Hours	2014 Bare Costs Material	2014 Bare Costs Labor	2014 Bare Costs Equipment	2014 Bare Costs Total	Total In-House	Total w/O&P
3030	**Replace 3'-0" x 4'-0" wd. fr. screen wdo. - 3rd flr.**	40	1 CARP	Ea.							
	Set up and secure scaffold				1.333		61		61	81.50	100
	Remove screen from mountings				.624		31.92		31.92	48.50	58.50
	Install new screen				.666	97.80	30.48		128.28	148	172
	Remove scaffold				1.333		61		61	81.50	100
	Total				3.956	97.80	184.40		282.20	359.50	430.50

B2023 324 Plate Glass

	System Description	Freq. (Years)	Crew	Unit	Labor Hours	2014 Bare Costs Material	2014 Bare Costs Labor	2014 Bare Costs Equipment	2014 Bare Costs Total	Total In-House	Total w/O&P
1020	**Replace plate glass storefront - 1st floor**	50	2 CARP	C.S.F.							
	Remove plate glass and frame				3.467		159		159	212	261
	Replace plate glass and frame				16.000	2,600	705		3,305	3,800	4,400
	Total				19.467	2,600	864		3,464	4,012	4,661
2020	**Replace plate glass storefront - 2nd floor**	50	2 CARP	C.S.F.							
	Set up and secure scaffold				1.500		69		69	91.50	113
	Remove plate glass and frame				3.467		159		159	212	261
	Replace plate glass and frame				16.000	2,600	705		3,305	3,800	4,400
	Remove scaffold				1.500		69		69	91.50	113
	Total				22.467	2,600	1,002		3,602	4,195	4,887
3020	**Replace plate glass storefront - 3rd floor**	50	2 CARP	C.S.F.							
	Set up and secure scaffold				2.250		103.50		103.50	137	170
	Remove plate glass and frame				3.467		159		159	212	261
	Replace plate glass and frame				16.000	2,600	705		3,305	3,800	4,400
	Remove scaffold				2.250		103.50		103.50	137	170
	Total				23.967	2,600	1,071		3,671	4,286	5,001

B20 EXTERIOR CLOSURE   B2033   Exterior Doors

B2033 111   Glazed Aluminum

	System Description	Freq. (Years)	Crew	Unit	Labor Hours	2014 Bare Costs Material	2014 Bare Costs Labor	2014 Bare Costs Equipment	2014 Bare Costs Total	Total In-House	Total w/O&P
1020	**Repair aluminum storefront door**	12	1 CARP	Ea.							
	Remove weatherstripping				.015		.67		.67	.90	1.11
	Remove door closer				.167		7.65		7.65	10.20	12.55
	Remove door hinge				.167		7.65		7.65	10.20	12.55
	Install new weatherstrip				3.478	46.50	159		205.50	264	320
	Install new door closer				.868	94	39.75		133.75	158	183
	Install new hinge (50% of total hinges / 12 years)				.167		7.65		7.65	10.20	12.55
	Hinge, 5 x 5, brass base					61			61	67	76.50
	Oil / lubricate door closer				.065		2.96		2.96	3.94	4.86
	Oil / lubricate hinges				.065		2.96		2.96	3.94	4.86
	Total				4.990	201.50	228.29		429.79	528.38	627.98
1030	**Replace 3'-0" x 7'-0" aluminum storefront doors**	50	1 CARP	Ea.							
	Remove door				.650		24		24	32	39
	Remove door frame				1.300		59.50		59.50	79.50	98
	Install new door, frame, and hardware				10.458	820	535		1,355	1,725	2,000
	Install new glass				3.640	182.70	160.65		343.35	410	490
	Total				16.048	1,002.70	779.15		1,781.85	2,246.50	2,627
2010	**Replace insulating glass - (3% of glass)**	1	1 CARP	S.F.							
	Remove damaged glass				.052		1.90		1.90	2.54	3.13
	Install new insulating glass				.277	30.50	12.20		42.70	49.50	58
	Total				.329	30.50	14.10		44.60	52.04	61.13
2020	**Repair aluminum storefront sliding door**	12	1 CARP	Ea.							
	Remove weatherstripping				.015		.67		.67	.90	1.11
	Oil / lubricate door closer				.065		2.96		2.96	3.94	4.86
	Oil / inspect wheel & track				.065		2.96		2.96	3.94	4.86
	Install new door track / glide wheels (25% of total / 12 years)				.440	13.13	20.13		33.26	41.50	49.50
	Install new weatherstrip				4.167	52	191		243	310	380
	Total				4.750	65.13	217.72		282.85	360.28	440.33

B20 EXTERIOR CLOSURE — B2033 Exterior Doors

B2033 130 Glazed Wood

	System Description	Freq. (Years)	Crew	Unit	Labor Hours	2014 Bare Costs				Total In-House	Total w/O&P
						Material	Labor	Equipment	Total		
1020	**Repair glazed wood doors**	12	1 CARP	Ea.							
	Remove weatherstripping				.015		.67		.67	.90	1.11
	Remove door closer				.167		7.65		7.65	10.20	12.55
	Remove door hinge				.167		7.65		7.65	10.20	12.55
	Install new weatherstrip				3.478	46.50	159		205.50	264	320
	Install new door closer				.868	94	39.75		133.75	158	183
	Install new hinge (50% of total hinges / 12 years)				.167		7.65		7.65	10.20	12.55
	Hinge, 5 x 5, brass base					61			61	67	76.50
	Oil / lubricate door closer				.065		2.96		2.96	3.94	4.86
	Oil / lubricate hinges				.065		2.96		2.96	3.94	4.86
	Total				4.990	201.50	228.29		429.79	528.38	627.98
1040	**Replace 3'-0" x 7'-0" glazed wood door**	40	1 CARP	Ea.							
	Remove door				.650		24		24	32	39
	Remove door frame				.743		34		34	45.50	56
	Install new door frame				.943	132.60	43.18		175.78	203	237
	Install new oak door sill				.832	68.40	38.20		106.60	126	148
	Install exterior door				2.105	825	96.50		921.50	1,025	1,200
	Total				5.274	1,026	235.88		1,261.88	1,431.50	1,680
1050	**Refinish glazed wood doors**	4	1 PORD	Ea.							
	Paint door, door frame & trim, brushwork, primer + 2 coats				1.760	16.40	69.50		85.90	108	133
	Total				1.760	16.40	69.50		85.90	108	133
2010	**Replace insulating glass - (8% of glass)**	1	1 CARP	S.F.							
	Remove damaged glass				.052		1.90		1.90	2.54	3.13
	Install new insulated glass				.277	30.50	12.20		42.70	49.50	58
	Total				.329	30.50	14.10		44.60	52.04	61.13
2020	**Repair glazed wood sliding door**	14	1 CARP	Ea.							
	Remove weatherstripping				.015		.67		.67	.90	1.11
	Oil / lubricate door closer				.065		2.96		2.96	3.94	4.86
	Oil / inspect wheel & track				.065		2.96		2.96	3.94	4.86
	Install new door track / glide wheels (25% of total / 14 years)				.440	13.13	20.13		33.26	41.50	49.50
	Install new weatherstrip				4.167	52	191		243	310	380
	Total				4.750	65.13	217.72		282.85	360.28	440.33

B20 EXTERIOR CLOSURE | B2033 | Exterior Doors

B2033 130 | Glazed Wood

	System Description	Freq. (Years)	Crew	Unit	Labor Hours	2014 Bare Costs				Total In-House	Total w/O&P
						Material	Labor	Equipment	Total		
2030	**Prepare and refinish glazed wood sliding door**	4	1 PORD	Ea.							
	Prepare surface				.310		12.25		12.25	15.90	19.65
	Paint door surface, brushwork, primer + 2 coats				.310	7.02	12.22		19.24	23.50	28.50
	Total				.621	7.02	24.47		31.49	39.40	48.15
2040	**Replace 6'-0" x 7'-0" glazed wood sliding door**	40	2 CARP	Ea.							
	Remove sliding door				1.733		79.50		79.50	106	131
	Install new sliding door				5.200	1,425	238		1,663	1,875	2,175
	Total				6.933	1,425	317.50		1,742.50	**1,981**	**2,306**
2050	**Refinish glazed wood sliding door & frame**	4	1 PORD	Ea.							
	Prepare surface				.465		18.38		18.38	24	29.50
	Paint door, door frame & trim				1.760	16.40	69.50		85.90	108	133
	Total				2.225	16.40	87.88		104.28	**132**	**162.50**

B2033 206 | Solid Core, Painted

	System Description	Freq. (Years)	Crew	Unit	Labor Hours	2014 Bare Costs				Total In-House	Total w/O&P
						Material	Labor	Equipment	Total		
1010	**Repair solid core, painted, door**	12	1 CARP	Ea.							
	Remove weatherstripping				.015		.67		.67	.90	1.11
	Remove door hinge				.250		11.48		11.48	15.30	18.85
	Install new weatherstrip				3.478	46.50	159		205.50	264	320
	Install new hinge (50% of total hinges / 12 years)				.250		11.48		11.48	15.30	18.85
	Hinge, 5 x 5, brass base					91.50			91.50	101	114
	Oil / lubricate hinges				.065		2.96		2.96	3.94	4.86
	Total				4.058	138	185.59		323.59	**400.44**	**477.67**
1020	**Prepare and refinish solid core, painted, door**	4	1 PORD	Ea.							
	Prepare door surface				.333		13.20		13.20	17.05	21
	Paint door surface, brushwork, primer + 2 coats				.549	12.42	21.62		34.04	42	50.50
	Total				.883	12.42	34.82		47.24	**59.05**	**71.50**

B20 EXTERIOR CLOSURE   B2033   Exterior Doors

B2033 206   Solid Core, Painted

	System Description	Freq. (Years)	Crew	Unit	Labor Hours	2014 Bare Costs				Total In-House	Total w/O&P
						Material	Labor	Equipment	Total		
1030	**Replace 3'-0" x 7'-0" solid core, painted, door**	40	1 CARP	Ea.							
	Remove door				.650		24		24	32	39
	Remove door frame				.743		34		34	45.50	56
	Install new door frame				.943	132.60	43.18		175.78	203	237
	Install new oak door sill				.832	68.40	38.20		106.60	126	148
	Install new solid core door				1.143	209	52.50		261.50	300	345
	Install new weatherstrip				4.167	52	191		243	310	380
	Install new lockset				.867	63.50	39.50		103	123	145
	Total				9.345	525.50	422.38		947.88	1,139.50	1,350
2010	**Repair solid core sliding wood door**	14	1 CARP	Ea.							
	Remove weatherstripping				.015		.67		.67	.90	1.11
	Oil / lubricate door closer				.065		2.96		2.96	3.94	4.86
	Oil / inspect wheel & track				.065		2.96		2.96	3.94	4.86
	Install new door track / glide wheels (25% of total / 14 years)				.440	13.13	20.13		33.26	41.50	49.50
	Install new weatherstrip				4.167	52	191		243	310	380
	Total				4.750	65.13	217.72		282.85	360.28	440.33
2030	**Replace 3'-0" x 7'-0" solid core sliding wood door**	40	1 CARP	Ea.							
	Remove sliding door				1.733		79.50		79.50	106	131
	Install new sliding door frame				.998	118.80	45.72		164.52	192	224
	Install new door track / glide wheels				1.760	52.50	80.50		133	165	199
	Install new solid core door				1.143	209	52.50		261.50	300	345
	Install new weatherstrip				3.478	46.50	159		205.50	264	320
	Install new lockset				.867	63.50	39.50		103	123	145
	Total				9.980	490.30	456.72		947.02	1,150	1,364
3010	**Replace sliding door glass - (3% of glass)**	1	1 CARP	S.F.							
	Remove damaged glass				.052		1.90		1.90	2.54	3.13
	Install new glass				.173	8.70	7.65		16.35	19.60	23.50
	Total				.225	8.70	9.55		18.25	22.14	26.63

B20 EXTERIOR CLOSURE — B2033 Exterior Doors

B2033 206 Solid Core, Painted

	System Description	Freq. (Years)	Crew	Unit	Labor Hours	2014 Bare Costs Material	2014 Bare Costs Labor	2014 Bare Costs Equipment	2014 Bare Costs Total	Total In-House	Total w/O&P
3040	Repl. 3'-0" x 7'-0" solid core, w/ safety glass, ptd. dr.	40	1 CARP	Ea.							
	Remove door				.650		24		24	32	39
	Remove door frame				.743		34		34	45.50	56
	Install new door frame				.943	132.60	43.18		175.78	203	237
	Install new oak door sill				.832	68.40	38.20		106.60	126	148
	Install new solid core door				1.143	209	52.50		261.50	300	345
	Install new glass (3% of total glass/year)				.201	10.09	8.87		18.96	22.50	27
	Install new weatherstrip				4.167	52	191		243	310	380
	Install new lockset				.867	63.50	39.50		103	123	145
	Total				9.546	535.59	431.25		966.84	1,162	1,377

B2033 210 Aluminum Doors

	System Description	Freq. (Years)	Crew	Unit	Labor Hours	2014 Bare Costs Material	2014 Bare Costs Labor	2014 Bare Costs Equipment	2014 Bare Costs Total	Total In-House	Total w/O&P
1010	Repair aluminum door	12	1 CARP	Ea.							
	Remove door closer				.167		7.65		7.65	10.20	12.55
	Remove door hinge				.250		11.48		11.48	15.30	18.85
	Install new door closer				1.735	188	79.50		267.50	315	365
	Install new hinge (50% of total hinges / 12 years)				.250		11.48		11.48	15.30	18.85
	Hinge, 5 x 5, brass base					91.50			91.50	101	114
	Oil / lubricate door closer				.065		2.96		2.96	3.94	4.86
	Oil / lubricate hinges				.065		2.96		2.96	3.94	4.86
	Total				2.531	279.50	116.03		395.53	464.68	538.97
1020	Replace 3'-0" x 7'-0" flush aluminum door	50	1 CARP	Ea.							
	Remove door				.650		24		24	32	39
	Remove cylinder lock				.400		18.35		18.35	24.50	30
	Reinstall cylinder lock				1.156		53		53	70.50	87
	Install new door				1.224	1,400	56		1,456	1,625	1,850
	Total				3.431	1,400	151.35		1,551.35	1,752	2,006
2010	Replace wire glass - (3% of glass)	1	1 CARP	S.F.							
	Remove damaged glass				.052		1.90		1.90	2.54	3.13
	Install new wire glass				.179	27.26	7.89		35.15	40.50	47
	Total				.231	27.26	9.79		37.05	43.04	50.13

B20 EXTERIOR CLOSURE B2033 Exterior Doors

B2033 210 Aluminum Doors

		System Description	Freq. (Years)	Crew	Unit	Labor Hours	2014 Bare Costs				Total In-House	Total w/O&P
							Material	Labor	Equipment	Total		
2020		**Repair aluminum frame and door**	12	1 CARP	Ea.							
		Remove weatherstripping				.015		.67		.67	.90	1.11
		Remove door closer				.167		7.65		7.65	10.20	12.55
		Remove door hinge				.250		11.48		11.48	15.30	18.85
		Install new weatherstrip				3.478	46.50	159		205.50	264	320
		Install new door closer				1.735	188	79.50		267.50	315	365
		Install new hinge (50% of total hinges / 12 years)				.250		11.48		11.48	15.30	18.85
		Hinge, 5 x 5, brass base					91.50			91.50	101	114
		Oil / lubricate door closer				.065		2.96		2.96	3.94	4.86
		Oil / lubricate hinges				.065		2.96		2.96	3.94	4.86
		Total				6.024	326	275.70		601.70	729.58	860.08
2030		**Replace 3'-0" x 7'-0" aluminum door, incl. vision lite**	50	1 CARP	Ea.							
		Remove door closer				.167		7.65		7.65	10.20	12.55
		Remove door				.650		24		24	32	39
		Install new door				1.224	1,400	56		1,456	1,625	1,850
		Reinstall door closer				1.735		79.50		79.50	106	131
		Install new glass				.217	10.88	9.56		20.44	24.50	29
		Total				3.993	1,410.88	176.71		1,587.59	1,797.70	2,061.55
3010		**Replace float glass - (3% of glass)**	1	1 CARP	S.F.							
		Remove damaged glass				.052		1.90		1.90	2.54	3.13
		Install new glass				.201	10.09	8.87		18.96	22.50	27
		Total				.253	10.09	10.77		20.86	25.04	30.13
3020		**Repair aluminum sliding door**	12	1 CARP	Ea.							
		Remove weatherstripping				.015		.67		.67	.90	1.11
		Oil / lubricate door closer				.065		2.96		2.96	3.94	4.86
		Install new door closer (25% of total / 12 years)				.434	47	19.88		66.88	79	91.50
		Oil / inspect wheel & track				.065		2.96		2.96	3.94	4.86
		Install new door track / glide wheels (25% of total / 12 years)				.440	13.13	20.13		33.26	41.50	49.50
		Install new weatherstrip				4.167	52	191		243	310	380
		Total				5.184	112.13	237.60		349.73	439.28	531.83

B20 EXTERIOR CLOSURE   B2033   Exterior Doors

B2033 210   Aluminum Doors

	System Description	Freq. (Years)	Crew	Unit	Labor Hours	Material	Labor	Equipment	Total	Total In-House	Total w/O&P
							2014 Bare Costs				
3030	**Replace 3'-0" x 7'-0" aluminum sliding door**	50	1 CARP	Ea.							
	Remove door				.650		24		24	32	39
	Install new door track / glide wheels				1.760	52.50	80.50		133	165	199
	Install new aluminum door				1.224	1,400	56		1,456	1,625	1,850
	Total				3.634	1,452.50	160.50		1,613	**1,822**	**2,088**
4020	**Repair flush aluminum, wood core, door**	12	1 CARP	Ea.							
	Remove weatherstripping				.015		.67		.67	.90	1.11
	Remove door closer				.167		7.65		7.65	10.20	12.55
	Remove door hinge				.250		11.48		11.48	15.30	18.85
	Install new weatherstrip				3.478	46.50	159		205.50	264	320
	Install new door closer				1.735	188	79.50		267.50	315	365
	Install new hinge (50% of total hinges / 12 years)				.250		11.48		11.48	15.30	18.85
	Hinge, 5 x 5, brass base					91.50			91.50	101	114
	Oil / lubricate door closer				.065		2.96		2.96	3.94	4.86
	Oil / lubricate hinges				.065		2.96		2.96	3.94	4.86
	Total				6.024	326	275.70		601.70	**729.58**	**860.08**
4030	**Replace 3'-0" x 7'-0" aluminum, wood core, door**	50	1 CARP	Ea.							
	Remove door				.650		24		24	32	39
	Remove door closer				.167		7.65		7.65	10.20	12.55
	Install new aluminum door				1.224	1,400	56		1,456	1,625	1,850
	Install door closer				1.735		79.50		79.50	106	131
	Total				3.777	1,400	167.15		1,567.15	**1,773.20**	**2,032.55**
5010	**Replace tempered glass - (3% of glass)**	1	1 CARP	S.F.							
	Remove damaged glass				.052		1.90		1.90	2.54	3.13
	Install new glass				.201	10.09	8.87		18.96	22.50	27
	Total				.253	10.09	10.77		20.86	**25.04**	**30.13**

B20 EXTERIOR CLOSURE

B2033 Exterior Doors

B2033 210 Aluminum Doors

	System Description	Freq. (Years)	Crew	Unit	Labor Hours	2014 Bare Costs				Total In-House	Total w/O&P
						Material	Labor	Equipment	Total		
5020	**Repair aluminum, insulated, door**	12	1 CARP	Ea.							
	Remove weatherstripping				.015		.67		.67	.90	1.11
	Remove door closer				.167		7.65		7.65	10.20	12.55
	Remove door hinge				.250		11.48		11.48	15.30	18.85
	Install new weatherstrip				3.478	46.50	159		205.50	264	320
	Install new door closer				1.735	188	79.50		267.50	315	365
	Install new hinge (50% of total hinges / 12 years)				.250		11.48		11.48	15.30	18.85
	Hinge, 5 x 5, brass base					91.50			91.50	101	114
	Oil / lubricate door closer				.065		2.96		2.96	3.94	4.86
	Oil / lubricate hinges				.065		2.96		2.96	3.94	4.86
	Total				6.024	326	275.70		601.70	729.58	860.08
5030	**Replace 3'-0" x 7'-0" aluminum, insulated, door**	50	1 CARP	Ea.							
	Remove door				.650		24		24	32	39
	Remove door closer				.167		7.65		7.65	10.20	12.55
	Install new aluminum door				1.224	1,400	56		1,456	1,625	1,850
	Install door closer				1.735		79.50		79.50	106	131
	Total				3.777	1,400	167.15		1,567.15	1,773.20	2,032.55

B2033 220 Steel, Painted

	System Description	Freq. (Years)	Crew	Unit	Labor Hours	2014 Bare Costs				Total In-House	Total w/O&P
						Material	Labor	Equipment	Total		
1010	**Repair steel, painted, door**	14	1 CARP	Ea.							
	Remove weatherstripping				.015		.67		.67	.90	1.11
	Remove door closer				.167		7.65		7.65	10.20	12.55
	Remove door hinge				.250		11.48		11.48	15.30	18.85
	Install new weatherstrip				3.478	46.50	159		205.50	264	320
	Install new door closer				1.735	188	79.50		267.50	315	365
	Install new hinge (50% of total hinges / 14 years)				.250		11.48		11.48	15.30	18.85
	Hinge, 5 x 5, brass base					91.50			91.50	101	114
	Oil / lubricate door closer				.065		2.96		2.96	3.94	4.86
	Oil / lubricate hinges				.065		2.96		2.96	3.94	4.86
	Total				6.024	326	275.70		601.70	729.58	860.08

B20 EXTERIOR CLOSURE | B2033 | Exterior Doors

B2033 220 | Steel, Painted

	System Description	Freq. (Years)	Crew	Unit	Labor Hours	2014 Bare Costs				Total In-House	Total w/O&P
						Material	Labor	Equipment	Total		
1020	**Refinish 3'-0" x 7'-0" steel, painted, door**	4	1 PORD	Ea.							
	Paint on finished metal surface, brushwork, primer + 1 coat				.800	4.37	31.50		35.87	46	56.50
	Total				.800	4.37	31.50		35.87	**46**	**56.50**
1030	**Replace 3'-0" x 7'-0" steel, painted, door**	45	1 CARP	Ea.							
	Remove door				.650		24		24	32	39
	Remove cylinder lock				.400		18.35		18.35	24.50	30
	Install new door				1.224	435	56		491	555	635
	Reinstall cylinder lock				1.156		53		53	70.50	87
	Total				3.431	435	151.35		586.35	**682**	**791**
2010	**Replace tempered glass - (3% of glass)**	1	1 CARP	S.F.							
	Remove damaged glass				.052		1.90		1.90	2.54	3.13
	Install new glass				.201	10.09	8.87		18.96	22.50	27
	Total				.253	10.09	10.77		20.86	**25.04**	**30.13**
2030	**Prepare and refinish 3'-0" x 7'-0" steel ptd. door**	4	1 PORD	Ea.							
	Prepare surface				.333		13.20		13.20	17.05	21
	Refinish door surface, brushwork, primer + 1 coat				1.333	4.11	52.50		56.61	73	89.50
	Total				1.667	4.11	65.70		69.81	**90.05**	**110.50**
2040	**Replace 3'-0" x 7'-0" steel, w/ wire glass, ptd. door**	45	1 CARP	Ea.							
	Remove door				.650		24		24	32	39
	Remove door closer				.167		7.65		7.65	10.20	12.55
	Remove door frame & trim				1.300		59.50		59.50	79.50	98
	Install new steel door				1.224	625	56		681	760	875
	Install new steel frame				1.301	150	59.50		209.50	244	286
	Install new wire glass				.179	31.90	7.89		39.79	45	52.50
	Reinstall door closer				1.735		79.50		79.50	106	131
	Total				6.556	806.90	294.04		1,100.94	**1,276.70**	**1,494.05**

B20 EXTERIOR CLOSURE — B2033 Exterior Doors

B2033 220 Steel, Painted

	System Description	Freq. (Years)	Crew	Unit	Labor Hours	2014 Bare Costs Material	2014 Bare Costs Labor	2014 Bare Costs Equipment	2014 Bare Costs Total	Total In-House	Total w/O&P
3010	**Repair steel sliding painted door**	14	1 CARP	Ea.							
	Remove weatherstripping				.015		.67		.67	.90	1.11
	Oil / lubricate door closer				.065		2.96		2.96	3.94	4.86
	Install new door closer (25% of total / 14 years)				.434	47	19.88		66.88	79	91.50
	Oil / inspect wheel & track				.065		2.96		2.96	3.94	4.86
	Install new door track / glide wheels (25% of total / 14 years)				.440	13.13	20.13		33.26	41.50	49.50
	Install new weatherstrip				4.167	52	191		243	310	380
	Total				5.184	112.13	237.60		349.73	439.28	531.83
3020	**Refinish 3'-0" x 7'-0" steel sliding painted door**	4	1 PORD	Ea.							
	Prepare surface				.333		13.20		13.20	17.05	21
	Refinish door surface, brushwork, primer + 1 coat				1.333	4.11	52.50		56.61	73	89.50
	Total				1.667	4.11	65.70		69.81	90.05	110.50
3030	**Replace 3'-0" x 7'-0" steel sliding painted door**	45	1 CARP	Ea.							
	Remove door				.650		24		24	32	39
	Remove door closer				.167		7.65		7.65	10.20	12.55
	Install new steel door				1.224	625	56		681	760	875
	Install door closer				1.735		79.50		79.50	106	131
	Total				3.777	625	167.15		792.15	908.20	1,057.55
4010	**Replace steel sliding door glass - (3% of glass)**	1	1 CARP	S.F.							
	Remove damaged glass				.052		1.90		1.90	2.54	3.13
	Install new glass				.201	10.09	8.87		18.96	22.50	27
	Total				.253	10.09	10.77		20.86	25.04	30.13
4020	**Repair steel, insulated core, painted door**	14	1 CARP	Ea.							
	Remove weatherstripping				.015		.67		.67	.90	1.11
	Remove door closer				.167		7.65		7.65	10.20	12.55
	Remove door hinge				.167		7.65		7.65	10.20	12.55
	Install new weatherstrip				3.478	46.50	159		205.50	264	320
	Install new door closer				1.735	188	79.50		267.50	315	365
	Install new hinge (50% of total hinges / 14 years)				.167		7.65		7.65	10.20	12.55
	Hinge, 5 x 5, brass base					61			61	67	76.50
	Oil / lubricate door closer				.065		2.96		2.96	3.94	4.86
	Oil / lubricate hinges				.065		2.96		2.96	3.94	4.86
	Total				5.857	295.50	268.04		563.54	685.38	809.98

B20 EXTERIOR CLOSURE | B2033 | Exterior Doors

B2033 220 Steel, Painted

	System Description	Freq. (Years)	Crew	Unit	Labor Hours	2014 Bare Costs Material	2014 Bare Costs Labor	2014 Bare Costs Equipment	2014 Bare Costs Total	Total In-House	Total w/O&P
4030	**Refinish 3'-0" x 7'-0" steel, insul. core, ptd. door**	4	1 PORD	Ea.							
	Prepare surface				.333		13.20		13.20	17.05	21
	Refinish door surface, brushwork, primer + 1 coat				1.333	4.11	52.50		56.61	73	89.50
	Total				1.667	4.11	65.70		69.81	90.05	110.50
4040	**Replace 3'-0" x 7'-0" steel, insul. core, ptd. door**	45	1 CARP	Ea.							
	Remove door				.650		24		24	32	39
	Remove door closer				.167		7.65		7.65	10.20	12.55
	Remove door frame & trim				1.300		59.50		59.50	79.50	98
	Install new steel door				1.224	625	56		681	760	875
	Install new steel frame				1.301	150	59.50		209.50	244	286
	Install new insulated glass				.322	35.38	14.15		49.53	57.50	67
	Install door closer				1.735		79.50		79.50	106	131
	Total				6.699	810.38	300.30		1,110.68	1,289.20	1,508.55

B2033 221 Steel, Unpainted

	System Description	Freq. (Years)	Crew	Unit	Labor Hours	2014 Bare Costs Material	2014 Bare Costs Labor	2014 Bare Costs Equipment	2014 Bare Costs Total	Total In-House	Total w/O&P
1010	**Repair steel, unpainted, doors**	14	1 CARP	Ea.							
	Remove weatherstripping				.015		.67		.67	.90	1.11
	Remove door closer				.167		7.65		7.65	10.20	12.55
	Remove door hinge				.167		7.65		7.65	10.20	12.55
	Install new weatherstrip				3.478	46.50	159		205.50	264	320
	Install new door closer				1.735	188	79.50		267.50	315	365
	Install new hinge (50% of total hinges / 14 years)				.167		7.65		7.65	10.20	12.55
	Hinge, 5 x 5, brass base					61			61	67	76.50
	Oil / lubricate door closer				.065		2.96		2.96	3.94	4.86
	Oil / lubricate hinges				.065		2.96		2.96	3.94	4.86
	Total				5.857	295.50	268.04		563.54	685.38	809.98
1020	**Replace 3'-0" x 7'-0" steel, unpainted, doors**	45	1 CARP	Ea.							
	Remove door				.650		24		24	32	39
	Remove door frame & trim				1.300		59.50		59.50	79.50	98
	Install new steel door				1.224	640	56		696	780	890
	Install new steel frame				1.301	150	59.50		209.50	244	286
	Total				4.475	790	199		989	1,135.50	1,313

B20 EXTERIOR CLOSURE — B2033 — Exterior Doors

B2033 221 — Steel, Unpainted

	System Description	Freq. (Years)	Crew	Unit	Labor Hours	2014 Bare Costs Material	2014 Bare Costs Labor	2014 Bare Costs Equipment	2014 Bare Costs Total	Total In-House	Total w/O&P
2010	**Replace wire glass - (3% of glass)**	1	1 CARP	S.F.							
	Remove damaged glass				.052		1.90		1.90	2.54	3.13
	Install new wire glass				.179	27.26	7.89		35.15	40.50	47
	Total				.231	27.26	9.79		37.05	43.04	50.13
2020	**Repair steel, with safety glass, unptd. door**	14	1 CARP	Ea.							
	Remove weatherstripping				.015		.67		.67	.90	1.11
	Remove door closer				.167		7.65		7.65	10.20	12.55
	Remove door hinge				.167		7.65		7.65	10.20	12.55
	Install new weatherstrip				3.478	46.50	159		205.50	264	320
	Install new door closer				1.735	188	79.50		267.50	315	365
	Install new hinge (50% of total hinges / 14 years)				.167		7.65		7.65	10.20	12.55
	Hinge, 5 x 5, brass base					61			61	67	76.50
	Oil / lubricate door closer				.065		2.96		2.96	3.94	4.86
	Oil / lubricate hinges				.065		2.96		2.96	3.94	4.86
	Total				5.857	295.50	268.04		563.54	685.38	809.98
2030	**Repl. 3'-0" x 7'-0" steel, w/ safety glass, unptd. dr.**	45	1 CARP	Ea.							
	Remove door				.650		24		24	32	39
	Remove door closer				.167		7.65		7.65	10.20	12.55
	Remove door frame & trim				1.300		59.50		59.50	79.50	98
	Install new steel door				1.301	510	59.50		569.50	640	735
	Install new steel frame				1.301	150	59.50		209.50	244	286
	Install new wire glass				.308	55	13.60		68.60	78	91
	Install door closer				1.735		79.50		79.50	106	131
	Total				6.762	715	303.25		1,018.25	1,189.70	1,392.55
3010	**Repair steel sliding unpainted door**	14	1 CARP	Ea.							
	Remove weatherstripping				.015		.67		.67	.90	1.11
	Oil / lubricate door closer				.065		2.96		2.96	3.94	4.86
	Install new door closer (25% of total / 14 years)				.434	47	19.88		66.88	79	91.50
	Oil / inspect wheel & track				.065		2.96		2.96	3.94	4.86
	New door track / glide wheels (25% of total / 14 years)				.440	13.13	20.13		33.26	41.50	49.50
	Total				1.018	60.13	46.60		106.73	129.28	151.83

B20 EXTERIOR CLOSURE   B2033   Exterior Doors

B2033 221   Steel, Unpainted

	System Description	Freq. (Years)	Crew	Unit	Labor Hours	Material	Labor	Equipment	Total	Total In-House	Total w/O&P
							2014 Bare Costs				
3020	**Replace 3'-0" x 7'-0" steel sliding unpainted door**	45	1 CARP	Ea.							
	Remove door				.650		24		24	32	39
	Remove door closer				.167		7.65		7.65	10.20	12.55
	Install new steel door				1.224	640	56		696	780	890
	Reinstall door closer				1.735		79.50		79.50	106	131
	Total				3.776	640	167.15		807.15	928.20	1,072.55
4010	**Replace tempered glass - (3% of glass)**	1	1 CARP	S.F.							
	Remove damaged glass				.052		1.90		1.90	2.54	3.13
	Install new glass				.201	10.09	8.87		18.96	22.50	27
	Total				.253	10.09	10.77		20.86	25.04	30.13
4020	**Repair steel, insulated core, unpainted door**	14	1 CARP	Ea.							
	Remove weatherstripping				.015		.67		.67	.90	1.11
	Remove door closer				.167		7.65		7.65	10.20	12.55
	Remove door hinge				.167		7.65		7.65	10.20	12.55
	Install new weatherstrip				3.478	46.50	159		205.50	264	320
	Install new door closer (50% of total / 14 years)				.868	94	39.75		133.75	158	183
	Install new hinge (50% of total hinges / 14 years)				.167		7.65		7.65	10.20	12.55
	Hinge, 5 x 5, brass base					61			61	67	76.50
	Oil / lubricate door closer				.065		2.96		2.96	3.94	4.86
	Oil / lubricate hinges				.065		2.96		2.96	3.94	4.86
	Total				4.990	201.50	228.29		429.79	528.38	627.98
4030	**Repl. 3'-0" x 7'-0" steel, insul. core, unptd. dr.**	45	1 CARP	Ea.							
	Remove door				.650		24		24	32	39
	Remove door closer				.167		7.65		7.65	10.20	12.55
	Remove door frame & trim				1.300		59.50		59.50	79.50	98
	Install new insulated steel door				1.301	685	59.50		744.50	835	955
	Install new steel frame				1.301	150	59.50		209.50	244	286
	Install new insulated glass				.322	35.38	14.15		49.53	57.50	67
	Reinstall door closer				1.735		79.50		79.50	106	131
	Total				6.776	870.38	303.80		1,174.18	1,364.20	1,588.55

B20 EXTERIOR CLOSURE   B2033   Exterior Doors

B2033 410   Steel Single, Roll-Up

	System Description	Freq. (Years)	Crew	Unit	Labor Hours	2014 Bare Costs				Total In-House	Total w/O&P
						Material	Labor	Equipment	Total		
1010	Repair 12' x 12' steel roll-up door	10	1 CARP	Ea.							
	Remove overhead type door section				1.303		59.75		59.75	80	99
	Install new roll-up door section				4.348	468.75	222.50		691.25	855	990
	Total				5.651	468.75	282.25		751	935	1,089
1020	Refinish 12' x 12' steel roll-up door	5	1 PORD	Ea.							
	Set up, secure and take down ladder				.258		10.20		10.20	13.20	16.35
	Prepare door surface				2.000		79.20		79.20	102	126
	Paint overhead door, brushwork, primer + 1 coat, 1 side				1.493	31.50	59.22		90.72	112	134
	Total				3.751	31.50	148.62		180.12	227.20	276.35
1030	Replace 12' x 12' steel roll-up door	35	2 CARP	Ea.							
	Remove overhead type door				5.212		239		239	320	395
	Install new roll-up door				17.391	1,875	890		2,765	3,425	3,975
	Total				22.603	1,875	1,129		3,004	3,745	4,370

B2033 412   Steel Double, Roll-Up

	System Description	Freq. (Years)	Crew	Unit	Labor Hours	2014 Bare Costs				Total In-House	Total w/O&P
						Material	Labor	Equipment	Total		
1020	Refinish 12' x 24' steel double roll-up door	5	1 PORD	Ea.							
	Set up, secure and take down ladder				.533		21		21	27.50	34
	Prepare door surface				4.000		158.40		158.40	205	252
	Paint overhead door, brushwork, primer + 1 coat, 1 side				2.986	63	118.44		181.44	224	268
	Total				7.519	63	297.84		360.84	456.50	554
1040	Replace 12' x 24' steel double roll-up door	35	2 CARP	Ea.							
	Remove overhead type door				10.423		478		478	640	790
	Install new door				34.783	3,750	1,780		5,530	6,850	7,950
	Total				45.206	3,750	2,258		6,008	7,490	8,740

B20 EXTERIOR CLOSURE — B2033 Exterior Doors

B2033 413 Steel, Single Leaf, Spring Garage Door

	System Description	Freq. (Years)	Crew	Unit	Labor Hours	2014 Bare Costs Material	2014 Bare Costs Labor	2014 Bare Costs Equipment	Total	Total In-House	Total w/O&P
1010	Repair 9' x 7' steel, single leaf, garage door	10	1 CARP	Ea.							
	Repair damaged track				.068		3.14		3.14	4.18	5.15
	Remove cremone lockset				.400		18.35		18.35	24.50	30
	Install new cremone lockset				1.040	108	47.50		155.50	182	214
	Retension door coil spring				1.143		52.50		52.50	70	86
	Total				2.651	108	121.49		229.49	280.68	335.15
1020	Refinish 9' x 7' steel, single leaf, garage door	5	1 PORD	Ea.							
	Set up, secure and take down ladder				.258		10.20		10.20	13.20	16.35
	Prepare surface				2.000		79.20		79.20	102	126
	Paint garage door, brushwork, primer + 1 coat, 1 side				1.493	31.50	59.22		90.72	112	134
	Total				3.751	31.50	148.62		180.12	227.20	276.35
1030	Replace 9' x 7' steel, single leaf, garage door	35	2 CARP	Ea.							
	Remove door				2.602		119		119	159	196
	Install new door				2.602	965	119		1,084	1,225	1,400
	Total				5.203	965	238		1,203	1,384	1,596

B2033 420 Aluminum Single, Roll-Up

	System Description	Freq. (Years)	Crew	Unit	Labor Hours	2014 Bare Costs Material	2014 Bare Costs Labor	2014 Bare Costs Equipment	Total	Total In-House	Total w/O&P
1010	Repair 12' x 12' aluminum single roll-up door	10	1 CARP	Ea.							
	Remove door section				1.303		59.75		59.75	80	99
	Install new door section				3.478	675	160		835	955	1,100
	Total				4.781	675	219.75		894.75	1,035	1,199
1020	Refinish 12' x 12' aluminum single roll-up door	5	1 PORD	Ea.							
	Set up, secure and take down ladder				.258		10.20		10.20	13.20	16.35
	Prepare surface				2.000		79.20		79.20	102	126
	Paint overhead type door				1.493	31.50	59.22		90.72	112	134
	Total				3.751	31.50	148.62		180.12	227.20	276.35
1030	Replace 12' x 12' aluminum single roll-up door	35	2 CARP	Ea.							
	Paint overhead door, brushwork, primer + 1 coat, 1 side				5.212		239		239	320	395
	Install new door				13.913	2,700	640		3,340	3,825	4,425
	Total				19.125	2,700	879		3,579	4,145	4,820

B20 EXTERIOR CLOSURE   B2033   Exterior Doors

B2033 422   Aluminum Double, Roll-Up

	System Description	Freq. (Years)	Crew	Unit	Labor Hours	Material	Labor	2014 Bare Costs Equipment	Total	Total In-House	Total w/O&P
1010	**Repair 12' x 24' aluminum double roll-up door**	10	1 CARP	Ea.							
	Remove door section				2.606		119.50		119.50	160	198
	Install new door section				6.957	1,350	320		1,670	1,925	2,225
	Total				9.562	1,350	439.50		1,789.50	2,085	2,423
1020	**Refinish 12' x 24' aluminum double roll-up door**	5	1 PORD	Ea.							
	Set up, secure and take down ladder				.533		21		21	27.50	34
	Prepare surface				4.000		158.40		158.40	205	252
	Paint overhead door, brushwork, primer + 1 coat, 1 side				2.986	63	118.44		181.44	224	268
	Total				7.519	63	297.84		360.84	456.50	554
1030	**Replace 12' x 24' aluminum double roll-up door**	35	2 CARP	Ea.							
	Remove overhead type door				10.423		478		478	640	790
	Install new door				27.826	5,400	1,280		6,680	7,650	8,850
	Total				38.250	5,400	1,758		7,158	8,290	9,640

B2033 423   Aluminum, Single Leaf, Spring Garage Door

	System Description	Freq. (Years)	Crew	Unit	Labor Hours	Material	Labor	2014 Bare Costs Equipment	Total	Total In-House	Total w/O&P
1010	**Repair 9' x 7' alum., single leaf, garage door**	10	1 CARP	Ea.							
	Repair damaged track				.684		31.40		31.40	42	51.50
	Remove cremone lockset				.400		18.35		18.35	24.50	30
	Install new cremone lockset				1.040	108	47.50		155.50	182	214
	Retension door coil spring				1.143		52.50		52.50	70	86
	Total				3.267	108	149.75		257.75	318.50	381.50
1020	**Refinish 9' x 7' alum., single leaf, garage dr.**	5	1 PORD	Ea.							
	Set up, secure and take down ladder				.258		10.20		10.20	13.20	16.35
	Prepare surface				2.000		79.20		79.20	102	126
	Paint garage door, brushwork, primer + 1 coat, 1 side				.747	15.75	29.61		45.36	56	67
	Total				3.005	15.75	119.01		134.76	171.20	209.35
1030	**Replace 9' x 7' alum., single leaf, garage dr.**	35	2 CARP	Ea.							
	Remove door				2.602		119		119	159	196
	Install new door				2.602	965	119		1,084	1,225	1,400
	Total				5.203	965	238		1,203	1,384	1,596

B20 EXTERIOR CLOSURE | B2033 Exterior Doors

B2033 430 Wood Single, Roll-Up

	System Description	Freq. (Years)	Crew	Unit	Labor Hours	2014 Bare Costs				Total In-House	Total w/O&P
						Material	Labor	Equipment	Total		
1010	Repair 12' x 12' wood single roll-up door	8	1 CARP	Ea.							
	Remove door section				1.303		59.75		59.75	80	99
	Install new door section				3.478	506.25	160		666.25	770	895
	Total				4.781	506.25	219.75		726	850	994
1020	Refinish 12' x 12' wood single roll-up door	5	1 PORD	Ea.							
	Set up, secure and take down ladder				.258		10.20		10.20	13.20	16.35
	Prepare door surface				2.000		79.20		79.20	102	126
	Paint overhead door, brushwork, primer + 1 coat, 1 side				3.150	15.12	124.74		139.86	178	219
	Total				5.408	15.12	214.14		229.26	293.20	361.35
1030	Replace 12' x 12' wood single roll-up door	16	2 CARP	Ea.							
	Remove overhead type door				5.212		239		239	320	395
	Install new door				13.913	2,025	640		2,665	3,075	3,575
	Total				19.125	2,025	879		2,904	3,395	3,970

B2033 432 Wood Double, Roll-Up

	System Description	Freq. (Years)	Crew	Unit	Labor Hours	2014 Bare Costs				Total In-House	Total w/O&P
						Material	Labor	Equipment	Total		
1010	Repair 12' x 24' wood double roll-up door	8	1 CARP	Ea.							
	Remove door section				2.606		119.50		119.50	160	198
	Install new door section				6.957	1,012.50	320		1,332.50	1,550	1,800
	Total				9.562	1,012.50	439.50		1,452	1,710	1,998
1020	Refinish 12' x 24' wood double roll-up door	5	1 PORD	Ea.							
	Set up, secure and take down ladder				.533		21		21	27.50	34
	Prepare surface				4.000		158.40		158.40	205	252
	Paint overhead door, brushwork, primer + 1 coat, 1 side				6.300	30.24	249.48		279.72	355	440
	Total				10.833	30.24	428.88		459.12	587.50	726
1030	Replace 12' x 24' wood double roll-up door	16	2 CARP	Ea.							
	Remove overhead type door				10.423		478		478	640	790
	Install new door				27.826	4,050	1,280		5,330	6,150	7,175
	Total				38.250	4,050	1,758		5,808	6,790	7,965

B20 EXTERIOR CLOSURE — B2033 Exterior Doors

B2033 433 Wood, Single Leaf, Spring Garage Door

	System Description	Freq. (Years)	Crew	Unit	Labor Hours	Material	Labor	Equipment	Total	Total In-House	Total w/O&P
1010	**Repair 9' x 7' wood, single leaf, garage door**	8	1 CARP	Ea.							
	Repair damaged track				.068		3.14		3.14	4.18	5.15
	Remove cremone lockset				.400		18.35		18.35	24.50	30
	Install new cremone lockset				1.040	108	47.50		155.50	182	214
	Retension door coil spring				1.143		52.50		52.50	70	86
	Total				2.651	108	121.49		229.49	280.68	335.15
1020	**Refinish 9' x 7' wood, single leaf, garage dr.**	5	1 PORD	Ea.							
	Set up, secure and take down ladder				.258		10.20		10.20	13.20	16.35
	Prepare surface				2.000		79.20		79.20	102	126
	Paint garage door, brushwork, primer + 1 coat, 1 side				3.150	15.12	124.74		139.86	178	219
	Total				5.408	15.12	214.14		229.26	293.20	361.35
1030	**Replace 9' x 7' wood, single leaf, garage door**	16	2 CARP	Ea.							
	Remove door				2.602		119		119	159	196
	Install new door				2.602	700	119		819	930	1,075
	Total				5.203	700	238		938	1,089	1,271

B2033 513 Electric Bifolding Hangar Door

	System Description	Freq. (Years)	Crew	Unit	Labor Hours	Material	Labor	Equipment	Total	Total In-House	Total w/O&P
0120	**Remove and replace electric bi-folding hangar door motor**	15	2 SKWK	Ea.							
	Hangar door, electric bi-folding door, replace motor				3.200	505	151		656	755	880
	Total				3.200	505	151		656	755	880
0140	**Remove and replace electric bi-folding hangar door cables**	15	2 SKWK	Ea.							
	Hangar door, elec bi-folding door, replace cables				8.000	89	380		469	595	725
	Total				8.000	89	380		469	595	725
0160	**Remove and replace electric bi-folding hangar door**	20	2 SKWK	Ea.							
	Hangar door demo				77.568		3,968		3,968	6,050	7,275
	Hangar door, bi-fldg dr, remove and replace 20' x 80'				116.368	34,400	5,952		40,352	47,200	54,000
	Total				193.936	34,400	9,920		44,320	53,250	61,275

B20 EXTERIOR CLOSURE B2033 Exterior Doors

B2033 810 Hinges, Brass

						2014 Bare Costs				
	System Description	Freq. (Years)	Crew	Unit	Labor Hours	Material	Labor	Equipment	Total	Total In-House	Total w/O&P
1010	Replace brass door hinge	60	1 CARP	Ea.							
	Remove door				.650		24		24	32	39
	Remove door hinge				.167		7.65		7.65	10.20	12.55
	Install new hinge				.167		7.65		7.65	10.20	12.55
	Full mortise hinge, 5 x 5, brass base				1.388	61			61	67	76.50
	Reinstall exterior door						63.50		63.50	85	105
	Total				2.371	61	102.80		163.80	204.40	245.60

B2033 820 Lockset, Brass

						2014 Bare Costs				
	System Description	Freq. (Years)	Crew	Unit	Labor Hours	Material	Labor	Equipment	Total	Total In-House	Total w/O&P
1010	Replace brass door lockset	30	1 CARP	Ea.							
	Remove mortise lock set				.421		19.30		19.30	25.50	31.50
	Mortise lock				1.301	330	59.50		389.50	440	510
	Total				1.722	330	78.80		408.80	465.50	541.50

B2033 830 Door Closer, Brass

						2014 Bare Costs				
	System Description	Freq. (Years)	Crew	Unit	Labor Hours	Material	Labor	Equipment	Total	Total In-House	Total w/O&P
1010	Replace brass door closer	15	1 CARP	Ea.							
	Remove door closer				.167		7.65		7.65	10.20	12.55
	Install new door closer				1.735	188	79.50		267.50	315	365
	Total				1.902	188	87.15		275.15	325.20	377.55

B20 EXTERIOR CLOSURE B2033 Exterior Doors

B2033 840 Door Opener

	System Description	Freq. (Years)	Crew	Unit	Labor Hours	2014 Bare Costs				Total In-House	Total w/O&P
						Material	Labor	Equipment	Total		
1010	**Automatic door opener on existing door**	7	1 ELEC	Ea.							
	Cutout demolition of exterior wall for actuator button				1.739		64.50	10.15	74.65	96	117
	Install 1/2" E.M.T. conduit				2.462	22.40	131.20		153.60	187	232
	Install conductor, type THWN copper, #14				1.040	13.33	55.25		68.58	83	102
	Install junction box				.533	14.20	28.50		42.70	51	62
	Install outlet				.381	8.35	20.50		28.85	34.50	42
	Install plate				.100	2.68	5.35		8.03	9.55	11.60
	Automatic opener, button operation				20.000	3,550	945		4,495	5,150	6,000
	Total				26.255	3,610.96	1,250.30	10.15	4,871.41	5,611.05	6,566.60

B2033 850 Deadbolt, Brass

	System Description	Freq. (Years)	Crew	Unit	Labor Hours	2014 Bare Costs				Total In-House	Total w/O&P
						Material	Labor	Equipment	Total		
1010	**Replace brass deadbolt**	20	1 CARP	Ea.							
	Remove deadbolt				.154		7.05		7.05	9.40	11.60
	Install new deadbolt				1.156	200	53		253	291	335
	Total				1.309	200	60.05		260.05	300.40	346.60

B2033 860 Weatherstripping, Brass

	System Description	Freq. (Years)	Crew	Unit	Labor Hours	2014 Bare Costs				Total In-House	Total w/O&P
						Material	Labor	Equipment	Total		
1010	**Replace brass door weatherstripping**	20	1 CARP	Ea.							
	Remove weatherstripping				.015		.67		.67	.90	1.11
	Install new weatherstrip				3.478	46.50	159		205.50	264	320
	Total				3.493	46.50	159.67		206.17	264.90	321.11

B20 EXTERIOR CLOSURE

B2033 Exterior Doors

B2033 870 Panic Device

	System Description	Freq. (Years)	Crew	Unit	Labor Hours	2014 Bare Costs				Total In-House	Total w/O&P
						Material	Labor	Equipment	Total		
1010	Replace door panic device	25	1 CARP	Ea.							
	Remove panic bar				.421		19.30		19.30	25.50	31.50
	Install new panic bar				2.080	865	95.50		960.50	1,075	1,250
	Total				2.501	865	114.80		979.80	1,100.50	1,281.50

B2033 905 Hollow Core, Painted

	System Description	Freq. (Years)	Crew	Unit	Labor Hours	2014 Bare Costs				Total In-House	Total w/O&P
						Material	Labor	Equipment	Total		
1010	Repair hollow core door	12	1 CARP	Ea.							
	Remove weatherstripping				.015		.67		.67	.90	1.11
	Remove door hinge				.167		7.65		7.65	10.20	12.55
	Install new weatherstrip				3.478	46.50	159		205.50	264	320
	Install new hinge (50% of total hinges / 12 years)				.167		7.65		7.65	10.20	12.55
	Hinge, 5 x 5, brass base					61			61	67	76.50
	Oil / lubricate hinges				.065		2.96		2.96	3.94	4.86
	Total				3.891	107.50	177.93		285.43	356.24	427.57
1020	Refinish 3'-0" x 7'-0" hollow core door	4	1 PORD	Ea.							
	Prepare door surface				.333		13.20		13.20	17.05	21
	Paint door surface, brushwork, primer + 1 coat				.549	12.42	21.62		34.04	42	50.50
	Total				.883	12.42	34.82		47.24	59.05	71.50
1030	Replace 3'-0" x 7'-0" hollow core door	30	1 CARP	Ea.							
	Remove door				.650		24		24	32	39
	Remove door frame				.743		34		34	45.50	56
	Install new door frame				.943	132.60	43.18		175.78	203	237
	Install new oak door sill				.832	68.40	38.20		106.60	126	148
	Install new hollow core door				1.000	109	46		155	181	212
	Install new weatherstrip				3.478	46.50	159		205.50	264	320
	Install new lockset				.867	63.50	39.50		103	123	145
	Total				8.514	420	383.88		803.88	974.50	1,157

B20 EXTERIOR CLOSURE	B2033	Exterior Doors

B2033 905 Hollow Core, Painted

	System Description	Freq. (Years)	Crew	Unit	Labor Hours	2014 Bare Costs				Total In-House	Total w/O&P
						Material	Labor	Equipment	Total		
2010	**Repair hollow core sliding wood door**	14	1 CARP	Ea.							
	Remove weatherstripping				.015		.67		.67	.90	1.11
	Oil / lubricate door closer				.065		2.96		2.96	3.94	4.86
	Oil / inspect wheel & track				.065		2.96		2.96	3.94	4.86
	Install new door track / glide wheels (25% of total / 14 years)				.440	13.13	20.13		33.26	41.50	49.50
	Install new weatherstrip				4.167	52	191		243	310	380
	Total				4.750	65.13	217.72		282.85	360.28	440.33
2030	**Replace 3'-0" x 7'-0" hollow core sliding wood door**	30	1 CARP	Ea.							
	Remove sliding door				1.733		79.50		79.50	106	131
	Install new sliding door frame				4.622	435	239	20.50	694.50	865	1,000
	Install new hollow core door				1.000	109	46		155	181	212
	Install new weatherstrip				3.478	46.50	159		205.50	264	320
	Lockset				.867	63.50	39.50		103	123	145
	Total				11.701	654	563	20.50	1,237.50	1,539	1,808

B2033 925 Metal Grated

	System Description	Freq. (Years)	Crew	Unit	Labor Hours	2014 Bare Costs				Total In-House	Total w/O&P
						Material	Labor	Equipment	Total		
1010	**Repair metal grated door**	15	1 SSWK	Ea.							
	Remove damaged metal bar				.836		42.78		42.78	65	78.50
	Oil / lubricate door closer				.065		3.30		3.30	5.05	6.05
	Oil / lubricate hinges				.065		2.96		2.96	3.94	4.86
	Metal grating repair (3% of door area / 15 years)				.320	1.30	17	5.68	23.98	33	39
	Replace damaged metal bar				1.600	13.30	83.50	14.20	111	156	185
	Total				2.885	14.60	149.54	19.88	184.02	262.99	313.41
1020	**Prepare and refinish metal grated painted door**	4	1 PORD	Ea.							
	Prepare grated door surface				.308		12.15		12.15	15.75	19.50
	Paint grated door surface, brushwork, 1 coat				.319	3.22	12.42		15.64	20	24.50
	Total				.627	3.22	24.57		27.79	35.75	44

B20 EXTERIOR CLOSURE   B2033   Exterior Doors

B2033 925   Metal Grated

	System Description	Freq. (Years)	Crew	Unit	Labor Hours	2014 Bare Costs Material	2014 Bare Costs Labor	2014 Bare Costs Equipment	2014 Bare Costs Total	Total In-House	Total w/O&P
1030	**Replace 3'-0" x 7'-0" metal grated painted door**	30	2 SSWK	Ea.							
	Remove door				.650		24		24	32	39
	Remove door closer				.167		7.65		7.65	10.20	12.55
	Remove door hinge				.500		22.95		22.95	30.50	37.50
	Install new grated door				15.686	785	800		1,585	2,075	2,450
	Install new hinge (25% of total hinges / 30 years)				.500		22.95		22.95	30.50	37.50
	Hinge, 5 x 5					91.50			91.50	101	114
	Total				17.503	876.50	877.55		1,754.05	2,279.20	2,690.55

B2033 926   Metal Wire Mesh Door

	System Description	Freq. (Years)	Crew	Unit	Labor Hours	2014 Bare Costs Material	2014 Bare Costs Labor	2014 Bare Costs Equipment	2014 Bare Costs Total	Total In-House	Total w/O&P
1010	**Repair metal wire mesh door**	15	1 SSWK	Ea.							
	Remove wire mesh				1.750		89.25		89.25	137	164
	Wire mesh					4.94			4.94	5.45	6.15
	Install wire mesh				7.000		358.05		358.05	545	660
	Oil / lubricate door closer				.065		3.30		3.30	5.05	6.05
	Oil / lubricate hinges				.065		2.96		2.96	3.94	4.86
	Total				8.879	4.94	453.56		458.50	696.44	841.06
1020	**Prepare and refinish metal wire mesh door**	4	1 PORD	Ea.							
	Prepare metal mesh surface				.312		12.35		12.35	15.95	19.80
	Paint metal mesh door surface, brushwork, 1 coat				.319	3.22	12.42		15.64	20	24.50
	Total				.631	3.22	24.77		27.99	35.95	44.30
1030	**Replace metal wire mesh door**	30	2 SSWK	Ea.							
	Remove wire mesh door				.640		32.50		32.50	50	60
	Remove wire mesh door frame				2.602		133		133	203	244
	Install new woven wire swinging door				3.471	292	159		451	535	625
	Install new metal door frame				3.200	239	165	14.20	418.20	530	620
	Install new hinge (25% of total hinges / 30 years)				.500		22.95		22.95	30.50	37.50
	Hinge, 5 x 5					91.50			91.50	101	114
	Install new deadlock night latch				1.039	45	47.50		92.50	113	135
	Total				11.451	667.50	559.95	14.20	1,241.65	1,562.50	1,835.50

B20 EXTERIOR CLOSURE

B2033 927 Louvered Door

B2033 Exterior Doors

	System Description	Freq. (Years)	Crew	Unit	Labor Hours	2014 Bare Costs Material	2014 Bare Costs Labor	2014 Bare Costs Equipment	2014 Bare Costs Total	Total In-House	Total w/O&P
1010	**Repair aluminum louvered door**	12	1 CARP	Ea.							
	Remove door				.650		24		24	32	39
	Remove weatherstripping				.015		.67		.67	.90	1.11
	Remove damaged louver				.137		6.25		6.25	8.35	10.30
	Install new louver				.274	15	12.55		27.55	33	39.50
	Oil / lubricate door closer				.065		2.96		2.96	3.94	4.86
	Install new weatherstrip				3.478	46.50	159		205.50	264	320
	Reinstall exterior door				1.388		63.50		63.50	85	105
	Total				6.006	61.50	268.93		330.43	427.19	519.77
1020	**Prepare and refinish aluminum louvered door**	4	1 PORD	Ea.							
	Prepare door surface				.333		13.20		13.20	17.05	21
	Paint door surface, brushwork, primer + 1 coat				.549	12.42	21.62		34.04	42	50.50
	Total				.883	12.42	34.82		47.24	59.05	71.50
1040	**Replace 3'-0" x 7'-0" aluminum louvered door**	50	1 CARP	Ea.							
	Remove door				.650		24		24	32	39
	Remove door closer				.167		7.65		7.65	10.20	12.55
	Install new aluminum door				1.224	1,400	56		1,456	1,625	1,850
	Install new door closer				1.735	188	79.50		267.50	315	365
	Total				3.777	1,588	167.15		1,755.15	1,982.20	2,266.55
2010	**Repair steel louvered door**	14	1 CARP	Ea.							
	Remove door				.650		24		24	32	39
	Remove weatherstripping				.015		.67		.67	.90	1.11
	Remove damaged louver				.137		6.25		6.25	8.35	10.30
	Install new louver				.274	15	12.55		27.55	33	39.50
	Oil / lubricate door closer				.065		2.96		2.96	3.94	4.86
	Install new weatherstrip				3.478	46.50	159		205.50	264	320
	Reinstall exterior door				1.388		63.50		63.50	85	105
	Total				6.006	61.50	268.93		330.43	427.19	519.77
2020	**Prepare and refinish steel louvered door**	4	1 PORD	Ea.							
	Prepare door surfaces				.333		13.20		13.20	17.05	21
	Paint door surface, brushwork, primer + 1 coat				.549	12.42	21.62		34.04	42	50.50
	Total				.883	12.42	34.82		47.24	59.05	71.50

B20 EXTERIOR CLOSURE | B2033 | Exterior Doors

B2033 927 Louvered Door

	System Description	Freq. (Years)	Crew	Unit	Labor Hours	2014 Bare Costs				Total In-House	Total w/O&P
						Material	Labor	Equipment	Total		
2040	**Replace 3'-0" x 7'-0" steel louvered door**	45	1 CARP	Ea.							
	Remove door				.650		24		24	32	39
	Remove door closer				.167		7.65		7.65	10.20	12.55
	Install new steel door, incl. louver				1.224	640	56		696	780	890
	Install new door closer				1.735	188	79.50		267.50	315	365
	Total				3.776	828	167.15		995.15	1,137.20	1,306.55
3010	**Repair wood louvered door**	7	1 CARP	Ea.							
	Remove door				.650		24		24	32	39
	Remove weatherstripping				.015		.67		.67	.90	1.11
	Remove damaged louver				.137		6.25		6.25	8.35	10.30
	Install new louver				.274	15	12.55		27.55	33	39.50
	Oil / lubricate door closer				.065		2.96		2.96	3.94	4.86
	Install new weatherstrip				3.478	46.50	159		205.50	264	320
	Reinstall exterior door				1.388		63.50		63.50	85	105
	Total				6.006	61.50	268.93		330.43	427.19	519.77
3020	**Prepare and refinish wood louvered door**	4	1 PORD	Ea.							
	Prepare door surfaces				.333		13.20		13.20	17.05	21
	Paint door surface, brushwork, primer + 1 coat				.549	12.42	21.62		34.04	42	50.50
	Total				.883	12.42	34.82		47.24	59.05	71.50
3040	**Replace 3'-0" x 7'-0" wood louvered door**	40	1 CARP	Ea.							
	Remove door				.650		24		24	32	39
	Remove door closer				.167		7.65		7.65	10.20	12.55
	Install new solid core door				1.143	209	52.50		261.50	300	345
	Add for installed wood louver					281			281	310	350
	Install new door closer				1.735	188	79.50		267.50	315	365
	Total				3.695	678	163.65		841.65	967.20	1,111.55

B20 EXTERIOR CLOSURE   B2033   Exterior Doors

B2033 935   Aluminum Storm/Screen Door

	System Description	Freq. (Years)	Crew	Unit	Labor Hours	2014 Bare Costs Material	2014 Bare Costs Labor	2014 Bare Costs Equipment	2014 Bare Costs Total	Total In-House	Total w/O&P
1010	**Repair aluminum storm/screen door**	5	1 CARP	Ea.							
	Remove weatherstripping				.015		.67		.67	.90	1.11
	Remove screen / storm door threshold				.519		23.80		23.80	32	39
	Remove door closer				.167		7.65		7.65	10.20	12.55
	Install new screen / storm door closer				1.301	26.50	59.50		86	109	131
	Install new weatherstrip				.870	11.63	39.75		51.38	66	80
	Install new plain aluminum threshold				.520	8.15	24		32.15	41	49
	Total				3.391	46.28	155.37		201.65	259.10	312.66
1020	**Replace aluminum storm/screen door**	20	1 CARP	Ea.							
	Remove door				.650		24		24	32	39
	Install new anodized aluminum screen / storm door				1.487	163	68		231	270	315
	Total				2.137	163	92		255	302	354

B2033 960   Aluminized Steel Pedestrian Gate

	System Description	Freq. (Years)	Crew	Unit	Labor Hours	2014 Bare Costs Material	2014 Bare Costs Labor	2014 Bare Costs Equipment	2014 Bare Costs Total	Total In-House	Total w/O&P
1010	**Repair aluminized steel pedestrian gate**	10	1 CARP	Ea.							
	Remove gate from hinges				.286		13.10		13.10	17.45	21.50
	Oil / lubricate hinges				.065		2.96		2.96	3.94	4.86
	Install new 6' aluminized steel fencing				.360	70.50	13.17	4.17	87.84	99	114
	Reinstall gate				.615		28		28	37.50	46.50
	Total				1.326	70.50	57.23	4.17	131.90	157.89	186.86
1030	**Prepare and refinish pedestrian gate**	5	1 PORD	Ea.							
	Lay drop cloth				.090		3.55		3.55	4.60	5.70
	Prepare / brush surface				.042		1.62		1.62	2.10	2.64
	Paint gate, brushwork, 1 coat				.150	1.44	5.94		7.38	9.35	11.35
	Remove drop cloth				.090		3.55		3.55	4.60	5.70
	Total				.371	1.44	14.66		16.10	20.65	25.39
1040	**Replace pedestrian gate**	40	1 CARP	Ea.							
	Remove old gate from hinges				.286		13.10		13.10	17.45	21.50
	Install new 6' x 3' aluminized steel gate				3.117	190	114	36	340	395	465
	Total				3.403	190	127.10	36	353.10	412.45	486.50

B20 EXTERIOR CLOSURE

B2033 Exterior Doors

B2033 961 Steel Pedestrian Gate

	System Description	Freq. (Years)	Crew	Unit	Labor Hours	2014 Bare Costs				Total In-House	Total w/O&P
						Material	Labor	Equipment	Total		
1010	**Repair steel gate**	10	1 CARP	Ea.							
	Remove gate from hinges				.286		13.10		13.10	17.45	21.50
	Oil / lubricate hinges				.065		2.96		2.96	3.94	4.86
	Install new 6' galvanized steel fencing				.360	58.20	13.17	4.17	75.54	85.50	99
	Reinstall gate				.615		28		28	37.50	46.50
	Total				1.326	58.20	57.23	4.17	119.60	144.39	171.86
1030	**Prepare and refinish steel gate**	5	1 PORD	Ea.							
	Lay drop cloth				.090		3.55		3.55	4.60	5.70
	Prepare / brush surface				.042		1.62		1.62	2.10	2.64
	Paint gate, brushwork, 1 coat				.150	1.44	5.94		7.38	9.35	11.35
	Remove drop cloth				.090		3.55		3.55	4.60	5.70
	Total				.371	1.44	14.66		16.10	20.65	25.39
1040	**Replace steel gate**	40	1 CARP	Ea.							
	Remove old gate from hinges				.286		13.10		13.10	17.45	21.50
	Install new 6' x 3' galvanized steel gate				3.117	190	114	36	340	395	465
	Total				3.403	190	127.10	36	353.10	412.45	486.50

B2033 962 Wood Pedestrian Gate

	System Description	Freq. (Years)	Crew	Unit	Labor Hours	2014 Bare Costs				Total In-House	Total w/O&P
						Material	Labor	Equipment	Total		
1010	**Repair wood gate**	7	1 CARP	Ea.							
	Remove gate from hinges				.286		13.10		13.10	17.45	21.50
	Oil / lubricate hinges				.065		2.96		2.96	3.94	4.86
	Replace damaged fencing				1.171	119.10	42.90	13.56	175.56	201	234
	Reinstall gate				.615		28		28	37.50	46.50
	Total				2.136	119.10	86.96	13.56	219.62	259.89	306.86
1030	**Prepare and refinish wood gate**	4	1 PORD	Ea.							
	Lay drop cloth				.090		3.55		3.55	4.60	5.70
	Prepare / brush surface				.042		1.62		1.62	2.10	2.64
	Paint gate, brushwork, primer + 1 coat				.138	1.44	5.40		6.84	8.65	10.60
	Remove drop cloth				.090		3.55		3.55	4.60	5.70
	Total				.359	1.44	14.12		15.56	19.95	24.64

B20 EXTERIOR CLOSURE | B2033 | Exterior Doors

B2033 962 | Wood Pedestrian Gate

	System Description	Freq. (Years)	Crew	Unit	Labor Hours	2014 Bare Costs				Total In-House	Total w/O&P
						Material	Labor	Equipment	Total		
1040	**Replace wood gate**	25	1 CARP	Ea.							
	Remove old gate from hinges				.286		13.10		13.10	17.45	21.50
	New 6'-0" x 3'-0" wood gate				3.468	215	127	40	382	445	520
	Total				3.754	215	140.10	40	395.10	462.45	541.50

B2033 963 | Wrought Iron Gate

	System Description	Freq. (Years)	Crew	Unit	Labor Hours	2014 Bare Costs				Total In-House	Total w/O&P
						Material	Labor	Equipment	Total		
1010	**Repair wrought iron gate**	11	1 CARP	Ea.							
	Remove gate from hinges				.286		13.10		13.10	17.45	21.50
	Oil / lubricate hinges				.065		2.96		2.96	3.94	4.86
	Repair hand forged wrought iron				2.600	237	133.50		370.50	460	540
	Reinstall gate				.615		28		28	37.50	46.50
	Total				3.566	237	177.56		414.56	518.89	612.86
1030	**Prepare and refinish wrought iron gate**	5	1 PORD	Ea.							
	Lay drop cloth				.090		3.55		3.55	4.60	5.70
	Prepare / brush surface				.125		4.86		4.86	6.30	7.90
	Paint gate, brushwork, primer + 1 coat				.167	1.44	6.66		8.10	10.10	12.40
	Remove drop cloth				.090		3.55		3.55	4.60	5.70
	Total				.471	1.44	18.62		20.06	25.60	31.70
1040	**Replace wrought iron gate**	45	1 CARP	Ea.							
	Remove old gate from hinges				.286		13.10		13.10	17.45	21.50
	Install new hand forged wrought iron gate				3.467	316	178		494	615	720
	Total				3.753	316	191.10		507.10	632.45	741.50

B30 ROOFING

B3013 Roof Covering

B3013 105 Built-Up Roofing

						2014 Bare Costs					
	System Description	Freq. (Years)	Crew	Unit	Labor Hours	Material	Labor	Equipment	Total	Total In-House	Total w/O&P
0100	**Debris removal and visual inspection**	0.50	2 ROFC	M.S.F.							
	Set up, secure and take down ladder				.052		2.03		2.03	3.03	3.67
	Pick up trash / debris & clean-up				.327		12.80		12.80	19.05	23
	Visual inspection				.327		12.80		12.80	19.05	23
	Total				.705		27.63		27.63	41.13	49.67
0200	**Non-destructive moisture inspection**	5	2 ROFC	M.S.F.							
	Set up, secure and take down ladder				.052		2.03		2.03	3.03	3.67
	Infrared inspection of roof membrane				2.133		83.50		83.50	125	151
	Total				2.185		85.53		85.53	128.03	154.67
0300	**Minor BUR membrane repairs - (2% of roof area)**	1	G-5	Sq.							
	Set up, secure and take down ladder				1.000		39		39	58	71
	Sweep / spud ballast clean				.640		25		25	37.50	45
	Cross cut incision through bitumen				.258		10.10		10.10	15.05	18.20
	Install base sheet and 2 plies of glass mopped				3.294	106	121	32.50	259.50	330	385
	Reinstall ballast in bitumen				.390		15.30		15.30	23	27.50
	Clean up				.390		15.30		15.30	23	27.50
	Total				5.973	106	225.70	32.50	364.20	486.55	574.20
0400	**BUR flashing repairs - (2 S.F. per sq. repaired)**	1	2 ROFC	S.F.							
	Set up, secure and take down ladder				.010		.39		.39	.58	.71
	Sweep / spud ballast clean				.009		.35		.35	.53	.64
	Cut out buckled flashing				.020		.78		.78	1.17	1.41
	Install 2 ply membrane flashing				.013	.19	.52		.71	.99	1.18
	Reinstall ballast in bitumen				.005		.20		.20	.30	.37
	Clean up				.004		.16		.16	.23	.28
	Total				.062	.19	2.40		2.59	3.80	4.59
0500	**Minor BUR membrane replacement - (25% of roof area)**	15	G-5	Sq.							
	Set up, secure and take down ladder				.078		3.06		3.06	4.56	5.50
	Remove 5 ply built-up roof				3.252		120		120	161	198
	Remove roof insulation board				1.333		49		49	66	81
	Install 2" polystyrene insulation				.641	69	25		94	113	131
	Install 4 ply bituminous roofing				3.733	134	137	37	308	390	455
	Clean up				1.000		39		39	58.50	70.50
	Total				10.037	203	373.06	37	613.06	793.06	941

B30 ROOFING		B3013		Roof Covering							
B3013 105		**Built-Up Roofing**									

	System Description	Freq. (Years)	Crew	Unit	Labor Hours	2014 Bare Costs				Total In-House	Total w/O&P
						Material	Labor	Equipment	Total		
0600	**Place new BUR membrane over existing**	20	G-5	Sq.							
	Set up, secure and take down ladder				.020		.78		.78	1.17	1.41
	Sweep / spud ballast clean				.500		19.60		19.60	29	35.50
	Vent existing membrane				.130		5.10		5.10	7.60	9.15
	Cut out buckled flashing				.024		.94		.94	1.40	1.70
	Install 2 ply membrane flashing				.027	.39	1.04		1.43	1.98	2.36
	Install 4 ply bituminous roofing				3.733	134	137	37	308	390	455
	Reinstall ballast				.381		14.90		14.90	22.50	27
	Clean up				.390		15.30		15.30	23	27.50
	Total				5.205	134.39	194.66	37	366.05	476.65	559.62
0700	**Total BUR roof replacement**	28	G-1	Sq.							
	Set up, secure and take down ladder				.020		.78		.78	1.17	1.41
	Sweep / spud ballast clean				.500		19.60		19.60	29	35.50
	Remove built-up roofing				2.500		93		93	123	152
	Remove insulation board				1.026		38		38	51	62
	Remove flashing				.024		.94		.94	1.40	1.70
	Install 2" perlite insulation				.879	99	34		133	160	186
	Install 2 ply membrane flashing				.027	.39	1.04		1.43	1.98	2.36
	Install 4 ply bituminous membrane				2.800	134	102	27.50	263.50	330	385
	Clean up				1.000		39		39	58.50	70.50
	Total				8.776	233.39	328.36	27.50	589.25	756.05	896.47

B3013 120		**Modified Bituminous / Thermoplastic**									

	System Description	Freq. (Years)	Crew	Unit	Labor Hours	2014 Bare Costs				Total In-House	Total w/O&P
						Material	Labor	Equipment	Total		
0100	**Debris removal by hand & visual inspection**	1	2 ROFC	M.S.F.							
	Set up, secure and take down ladder				.052		2.03		2.03	3.03	3.67
	Pick up trash / debris & clean up				.327		12.80		12.80	19.05	23
	Visual inspection				.327		12.80		12.80	19.05	23
	Total				.705		27.63		27.63	41.13	49.67
0200	**Non - destructive moisture inspection**	5	2 ROFC	M.S.F.							
	Set up, secure and take down ladder				.052		2.03		2.03	3.03	3.67
	Infrared inspection of roof membrane				2.133		83.50		83.50	125	151
	Total				2.185		85.53		85.53	128.03	154.67

B30 ROOFING

B3013 Roof Covering

B3013 120 Modified Bituminous / Thermoplastic

	System Description	Freq. (Years)	Crew	Unit	Labor Hours	2014 Bare Costs				Total In-House	Total w/O&P
						Material	Labor	Equipment	Total		
0300	**Minor thermoplastic membrane repairs - (2% of roof area)**	1	G-5	Sq.							
	Set up, secure and take down ladder				1.000		39		39	58	71
	Clean away loose surfacing				.258		10.10		10.10	15.05	18.20
	Install 150 mil mod. bit., fully adhered				2.597	80	92	13	185	239	280
	Clean up				.390		15.30		15.30	23	27.50
	Total				4.245	80	156.40	13	249.40	335.05	396.70
0500	**Thermoplastic flashing repairs - (2 S.F. per sq. repaired)**	1	2 ROFC	S.F.							
	Set up, secure and take down ladder				.010		.39		.39	.58	.71
	Clean away loose surfacing				.003		.10		.10	.15	.18
	Remove flashing				.020		.78		.78	1.17	1.41
	Install 120 mil mod. bit. flashing				.019	.76	.66	.09	1.51	1.92	2.24
	Clean up				.004		.16		.16	.23	.28
	Total				.055	.76	2.09	.09	2.94	4.05	4.82
0600	**Minor thermoplastic membrane replacement - (25% of roof area)**	20	G-5	Sq.							
	Set up, secure and take down ladder				.078		3.06		3.06	4.56	5.50
	Remove existing membrane / insulation				4.571		179		179	267	325
	Install 2" perlite insulation				1.140	99	45		144	176	204
	Clean adjacent membrane / patch				1.143		44.50		44.50	66.50	80.50
	Install fully adhered 150 mil membrane				2.597	80	92	13	185	239	280
	Clean up				1.000		39		39	58.50	70.50
	Total				10.529	179	402.56	13	594.56	811.56	965.50
0700	**Total thermoplastic roof replacement**	25	G-1	Sq.							
	Set up, secure and take down ladder				.020		.78		.78	1.17	1.41
	Remove existing membrane / insulation				3.501		137		137	204	247
	Remove flashing				.026		1.02		1.02	1.52	1.84
	Install 2" perlite insulation				.879	99	34		133	160	186
	Install flashing				.037	1.52	1.32	.18	3.02	3.84	4.48
	Install fully adhered 180 mil membrane				2.000	89	71	10	170	214	250
	Clean up				1.000		39		39	58.50	70.50
	Total				7.463	189.52	284.12	10.18	483.82	643.03	761.23

B30 ROOFING B3013 Roof Covering

B3013 125 Thermosetting

	System Description	Freq. (Years)	Crew	Unit	Labor Hours	2014 Bare Costs				Total In-House	Total w/O&P
						Material	Labor	Equipment	Total		
0100	**Debris removal by hand & visual inspection**	1	2 ROFC	M.S.F.							
	Set up, secure and take down ladder				.052		2.03		2.03	3.03	3.67
	Pick up trash / debris & clean up				.327		12.80		12.80	19.05	23
	Visual inspection				.327		12.80		12.80	19.05	23
	Total				.705		27.63		27.63	41.13	49.67
0200	**Non - destructive moisture inspection**	5	2 ROFC	M.S.F.							
	Set up, secure and take down ladder				.052		2.03		2.03	3.03	3.67
	Infrared inspection of roof membrane				2.133		83.50		83.50	125	151
	Total				2.185		85.53		85.53	128.03	154.67
0300	**Minor thermoset membrane repairs - (2% of roof area)**	1	G-5	Sq.							
	Set up, secure and take down ladder				1.000		39		39	58	71
	Broom roof surface clean				.258		10.10		10.10	15.05	18.20
	Remove / vent damaged membrane				.258		10.10		10.10	15.05	18.20
	Install thermosetting patch				2.597	68	92	13	173	226	265
	Clean up				.390		15.30		15.30	23	27.50
	Total				4.503	68	166.50	13	247.50	337.10	399.90
0400	**Thermoset flashing repairs - (2 S.F. per sq. repaired)**	1	2 ROFC	S.F.							
	Set up, secure and take down ladder				.010		.39		.39	.58	.71
	Remove flashing				.020		.78		.78	1.17	1.41
	Prime surfaces				.003		.12		.12	.18	.21
	Install 2 ply membrane flashing				.007	.10	.26		.36	.50	.59
	Clean up				.004		.16		.16	.23	.28
	Total				.044	.10	1.71		1.81	2.66	3.20
0500	**Minor thermoset membrane replacement - (25% of roof area)**	10	G-5	Sq.							
	Set up, secure and take down ladder				.078		3.06		3.06	4.56	5.50
	Broom roof surface clean				.258		10.10		10.10	15.05	18.20
	Remove, strip-in & seal joint				.593		23		23	34.50	42
	Install base sheet and 2 plies of glass mopped				3.294	106	121	32.50	259.50	330	385
	Clean up				.390		15.30		15.30	23	27.50
	Total				4.613	106	172.46	32.50	310.96	407.11	478.20

B30 ROOFING   B3013   Roof Covering

B3013 125   Thermosetting

	System Description	Freq. (Years)	Crew	Unit	Labor Hours	2014 Bare Costs				Total In-House	Total w/O&P
						Material	Labor	Equipment	Total		
0600	Total thermoset roof replacement	20	G-1	Sq.							
	Set up, secure and take down ladder				.020		.78		.78	1.17	1.41
	Remove existing membrane / insulation				3.501		137		137	204	247
	Remove flashing				.026		1.02		1.02	1.52	1.84
	Install 2" perlite insulation				.879	99	34		133	160	186
	Install new flashing				.020	.39	.78		1.17	1.60	1.90
	Install thermosetting membrane				2.800	71	102	28	201	259	305
	Clean up				1.000		39		39	58.50	70.50
	Total				8.246	170.39	314.58	28	512.97	685.79	813.65

B3013 127   EPDM (Ethylene Propylene Diene Monomer)

	System Description	Freq. (Years)	Crew	Unit	Labor Hours	2014 Bare Costs				Total In-House	Total w/O&P
						Material	Labor	Equipment	Total		
0600	Total EPDM roof replacement	25	G-5	Sq.							
	Set up, secure and take down ladder				.020		.78		.78	1.17	1.41
	Remove existing membrane / insulation				3.501		137		137	204	247
	Remove flashing				.026		1.02		1.02	1.52	1.84
	Install 2" perlite insulation				.879	99	34		133	160	186
	Install new flashing				.020	.39	.78		1.17	1.60	1.90
	Fully adhered with adhesive				1.538	111	54.50	7.65	173.15	211	246
	Clean up				1.000		39		39	58.50	70.50
	Total				6.985	210.39	267.08	7.65	485.12	637.79	754.65

B3013 130   Metal Panel Roofing

	System Description	Freq. (Years)	Crew	Unit	Labor Hours	2014 Bare Costs				Total In-House	Total w/O&P
						Material	Labor	Equipment	Total		
0100	Debris removal by hand & visual inspection	1	2 ROFC	M.S.F.							
	Set up, secure and take down ladder				.100		3.92		3.92	5.85	7.05
	Visual inspection				.327		12.80		12.80	19.05	23
	Total				.427		16.72		16.72	24.90	30.05

B30 ROOFING B3013 Roof Covering

B3013 130 Metal Panel Roofing

	System Description	Freq. (Years)	Crew	Unit	Labor Hours	2014 Bare Costs				Total In-House	Total w/O&P
						Material	Labor	Equipment	Total		
0300	**Minor metal roof finish repairs - (2% of roof area)**	5	2 ROFC	S.F.							
	Set up, secure and take down ladder				.010		.39		.39	.58	.71
	Wire brush surface to remove oxide				.020		.78		.78	1.17	1.41
	Prime and paint prepared surface				.016	.68	.63		1.31	1.68	1.98
	Clean up				.004		.16		.16	.23	.28
	Total				.050	.68	1.96		2.64	3.66	4.38
0400	**Metal roof flashing replacement - (2 S.F. per sq. repaired)**	1	2 ROFC	S.F.							
	Set up, secure and take down ladder				.010		.39		.39	.58	.71
	Dismantle seam				.016		.63		.63	.93	1.13
	Remove metal roof panels				.060		2.34		2.34	3.51	4.23
	Remove damaged metal flashing				.019		.74		.74	1.11	1.34
	Install new aluminum flashing				.072	1.35	2.82		4.17	5.70	6.80
	Replace metal panels				.129		5.04		5.04	7.55	9.10
	Clean up				.004		.16		.16	.23	.28
	Total				.310	1.35	12.12		13.47	19.61	23.59
0500	**Minor metal roof panel replacement - (2.5% of roof area)**	20	2 ROFC	S.F.							
	Set up, secure and take down ladder				.010		.39		.39	.58	.71
	Dismantle seam				.016		.63		.63	.93	1.13
	Remove metal roof panels				.020		.78		.78	1.17	1.41
	Install new metal panel				.044	2.90	2		4.90	5.80	6.85
	Re-assemble seam				.016		.63		.63	.93	1.13
	Prime and paint surface				.016	.68	.63		1.31	1.68	1.98
	Clean up				.004		.16		.16	.23	.28
	Total				.126	3.58	5.22		8.80	11.32	13.49
0600	**Total metal roof panel replacement**	30	2 ROFC	Sq.							
	Set up, secure and take down ladder				.020		.78		.78	1.17	1.41
	Remove existing roofing panel				3.600		141		141	210	254
	Remove flashing metal				.038		1.48		1.48	2.22	2.68
	Install new aluminum flashing				.072	1.35	2.82		4.17	5.70	6.80
	Install new metal panel				4.384	290	200		490	580	685
	Clean up				.390		15.30		15.30	23	27.50
	Total				8.504	291.35	361.38		652.73	822.09	977.39

B30 ROOFING

B3013 Roof Covering

B3013 140 Slate Tile Roofing

	System Description	Freq. (Years)	Crew	Unit	Labor Hours	2014 Bare Costs				Total In-House	Total w/O&P
						Material	Labor	Equipment	Total		
0100	**Visual inspection**	3	2 ROFC	M.S.F.							
	Set up, secure and take down ladder				.100		3.92		3.92	5.85	7.05
	Visual inspection				.327		12.80		12.80	19.05	23
	Total				.427		16.72		16.72	**24.90**	**30.05**
0200	**Minor slate tile replacement - (2.5% of roof area)**	20	2 ROTS	S.F.							
	Set up, secure and take down ladder				.010		.39		.39	.58	.71
	Remove and replace damaged slates				.800	5.90	31.50		37.40	53.50	64
	Clean up				.013		.51		.51	.76	.92
	Total				.823	5.90	32.40		38.30	**54.84**	**65.63**
0300	**Slate tile flashing repairs - (2 S.F. per sq. repaired)**	20	2 ROTS	S.F.							
	Set up, secure and take down ladder				.010		.39		.39	.58	.71
	Removal of adjacent slate tiles				.100		3.92		3.92	5.85	7.05
	Remove damaged metal flashing				.052		2.03		2.03	3.03	3.67
	Install 16 oz. copper flashing				.089	6.95	3.48		10.43	12.85	14.95
	Reinstall tiles in position				.134		5.24		5.24	7.80	9.45
	Clean up				.013		.51		.51	.76	.92
	Total				.398	6.95	15.57		22.52	**30.87**	**36.75**
0400	**Total slate tile roof replacement**	70	2 ROTS	Sq.							
	Set up, secure and take down ladder				.020		.78		.78	1.17	1.41
	Remove slate shingles				1.600		59		59	79	97
	Install 16 oz. copper flashing				.178	13.90	6.96		20.86	25.50	30
	Install 15# felt				.138	9.65	5.40		15.05	18.70	22
	Install slate roofing				4.571	480	180		660	795	925
	Clean up				1.333		52		52	78	94
	Total				7.840	503.55	304.14		807.69	**997.37**	**1,169.41**

B3013 142 Mineral Fiber Steep Roofing

	System Description	Freq. (Years)	Crew	Unit	Labor Hours	2014 Bare Costs				Total In-House	Total w/O&P
						Material	Labor	Equipment	Total		
0300	**Visual inspection**	3	2 ROFC	M.S.F.							
	Set up, secure and take down ladder				.100		3.92		3.92	5.85	7.05
	Visual inspection				.327		12.80		12.80	19.05	23
	Total				.427		16.72		16.72	**24.90**	**30.05**

B30 ROOFING		B3013	Roof Covering								
B3013 144	Clay Tile Roofing										
	System Description	Freq. (Years)	Crew	Unit	Labor Hours	2014 Bare Costs				Total In-House	Total w/O&P
						Material	Labor	Equipment	Total		
0300	**Visual inspection**	3	2 ROFC	M.S.F.							
	Set up, secure and take down ladder				.100		3.92		3.92	5.85	7.05
	Visual inspection				.327		12.80		12.80	19.05	23
	Total				.427		16.72		16.72	24.90	30.05
0400	**Minor clay tile repairs - (2% of roof area)**	20	2 ROFC	S.F.							
	Set up, secure and take down ladder				.010		.39		.39	.58	.71
	Remove damaged tiles				.020		.79		.79	1.17	1.41
	Install new clay tiles				.073	5.10	2.86		7.96	9.90	11.55
	Clean up				.013		.51		.51	.76	.92
	Total				.116	5.10	4.55		9.65	12.41	14.59
0500	**Clay tile flashing repairs - (2 S.F. per sq. repaired)**	1	2 ROFC	S.F.							
	Set up, secure and take down ladder				.010		.39		.39	.58	.71
	Remove adjacent tiles				.208		8.12		8.12	12.10	14.70
	Remove damaged metal flashing				.052		2.03		2.03	3.03	3.67
	Install 16 oz copper flashing				.178	13.90	6.96		20.86	25.50	30
	Reinstall clay tiles				.073		2.86		2.86	4.25	5.15
	Clean up				.013		.51		.51	.76	.92
	Total				.533	13.90	20.87		34.77	46.22	55.15
0600	**Total clay tile roof replacement**	70	2 ROFC	Sq.							
	Set up, secure and take down ladder				.020		.78		.78	1.17	1.41
	Remove clay tile roofing				3.200		125		125	187	226
	Remove damaged metal flashing				.040		1.57		1.57	2.34	2.82
	Install 16 oz copper flashing				.178	13.90	6.96		20.86	25.50	30
	Install felt				.138	9.65	5.40		15.05	18.70	22
	Install new clay tiles				7.273	510	286		796	990	1,150
	Clean up				1.333		52		52	78	94
	Total				12.182	533.55	477.71		1,011.26	1,302.71	1,526.23

B30 ROOFING

B3013 Roof Covering

B3013 146 Asphalt Shingle Roofing

	System Description	Freq. (Years)	Crew	Unit	Labor Hours	Material	2014 Bare Costs Labor	Equipment	Total	Total In-House	Total w/O&P
0100	**Debris removal by hand & visual inspection**	1	2 ROFC	M.S.F.							
	Set up, secure and take down ladder				.100		3.92		3.92	5.85	7.05
	Visual inspection				.327		12.80		12.80	19.05	23
	Total				.427		16.72		16.72	24.90	30.05
0300	**Minor asphalt shingle repair - (2% of roof area)**	1	2 ROFC	S.F.							
	Set up, secure and take down ladder				.010		.39		.39	.58	.71
	Remove damaged shingles				.024		.87		.87	1.16	1.43
	Install asphalt shingles				.036	1.06	1.39		2.45	3.25	3.84
	Clean up				.003		.10		.10	.15	.18
	Total				.072	1.06	2.75		3.81	5.14	6.16
0500	**Asphalt shingle flashing repairs - (2 S.F. per sq. repaired)**	1	2 ROFC	S.F.							
	Set up, secure and take down ladder				.010		.39		.39	.58	.71
	Remove adjoining shingles				.046		1.70		1.70	2.26	2.78
	Remove damaged metal flashing				.019		.74		.74	1.11	1.34
	Install new aluminum flashing				.072	1.35	2.82		4.17	5.70	6.80
	Install asphalt shingles				.071	2.12	2.78		4.90	6.50	7.65
	Clean up				.003		.10		.10	.15	.18
	Total				.221	3.47	8.53		12	16.30	19.46
0600	**Install new asphalt shingles over existing**	20	2 ROFC	Sq.							
	Set up, secure and take down ladder				.020		.78		.78	1.17	1.41
	Remove shingles at eaves and ridge				.460		18		18	27	32.50
	Install new aluminum flashing				.144	2.70	5.64		8.34	11.40	13.60
	Install asphalt shingles				1.778	106	69.50		175.50	220	258
	Clean up				.258		10.10		10.10	15.05	18.20
	Total				2.660	108.70	104.02		212.72	274.62	323.71
0700	**Total asphalt shingle roof replacement**	40	2 ROFC	Sq.							
	Set up, secure and take down ladder				.020		.78		.78	1.17	1.41
	Remove existing shingles				2.353		87		87	116	143
	Remove damaged metal flashing				.038		1.48		1.48	2.22	2.68
	Install 15# felt				.138	9.65	5.40		15.05	18.70	22
	Install new aluminum flashing				.144	2.70	5.64		8.34	11.40	13.60
	Install asphalt shingles				1.778	106	69.50		175.50	220	258
	Clean up				.258		10.10		10.10	15.05	18.20
	Total				4.729	118.35	179.90		298.25	384.54	458.89

B30 ROOFING   B3013 Roof Covering

B3013 160   Roll Roofing

	System Description	Freq. (Years)	Crew	Unit	Labor Hours	2014 Bare Costs				Total In-House	Total w/O&P
						Material	Labor	Equipment	Total		
0100	**Debris removal by hand & visual inspection**	1	2 ROFC	M.S.F.							
	Set up, secure and take down ladder				.100		3.92		3.92	5.85	7.05
	Pick up trash / debris & clean up				.327		12.80		12.80	19.05	23
	Visual inspection				.327		12.80		12.80	19.05	23
	Total				.753		29.52		29.52	43.95	53.05
0300	**Minor roofing repairs - (2% of roof area)**	1	2 ROFC	S.F.							
	Set up, secure and take down ladder				.010		.39		.39	.58	.71
	Cut out damaged area				.015		.59		.59	.88	1.06
	Install roll roofing with mastic				.027	.80	.99		2.06	2.62	3.07
	Clean up				.003		.10		.10	.15	.18
	Total				.055	.80	2.07		3.14	4.23	5.02
0400	**Flashing repairs - (2 S.F. per sq. repaired)**	1	2 ROFC	S.F.							
	Set up, secure and take down ladder				.010		.39		.39	.58	.71
	Remove adjacent roofing				.015		.59		.59	.88	1.06
	Remove damaged metal flashing				.019		.74		.74	1.11	1.34
	Install aluminum flashing				.072	1.35	2.82		4.17	5.70	6.80
	Install roll roofing with mastic				.054	1.59	1.98	.54	4.11	5.25	6.15
	Clean up				.003		.10		.10	.15	.18
	Total				.173	2.94	6.62	.54	10.10	13.67	16.24
0500	**Minor replacement - (25% of roof area)**	15	2 ROFC	Sq.							
	Set up, secure and take down ladder				.078		3.06		3.06	4.56	5.50
	Remove strip-in				1.067		42		42	62.50	75.50
	Renew striping with 2 ply felts				2.917	125	107	29	261	325	380
	Clean up				.258		10.10		10.10	15.05	18.20
	Total				4.319	125	162.16	29	316.16	407.11	479.20
0600	**Total roof replacement**	20	2 ROFC	Sq.							
	Set up, secure and take down ladder				.020		.78		.78	1.17	1.41
	Remove existing shingles				1.333		52		52	78	94
	Remove flashing				.019		.74		.74	1.11	1.34
	Install aluminum flashing				.144	2.70	5.64		8.34	11.40	13.60
	Install 4 ply roll roofing				2.917	125	107	29	261	325	380
	Clean up				.258		10.10		10.10	15.05	18.20
	Total				4.691	127.70	176.26	29	332.96	431.73	508.55

B30 ROOFING

B3013 Roof Covering

B3013 165 Corrugated Fiberglass Panel Roofing

	System Description	Freq. (Years)	Crew	Unit	Labor Hours	2014 Bare Costs				Total In-House	Total w/O&P
						Material	Labor	Equipment	Total		
0100	**Debris removal & visual inspection**	5	2 ROFC	M.S.F.							
	Set up, secure and take down ladder				.100		3.92		3.92	5.85	7.05
	Visual inspection				.327		12.80		12.80	19.05	23
	Total				.427		16.72		16.72	24.90	30.05
0200	**Fiberglass roof flashing repairs - (2 S.F. per sq. repaired)**	15	2 ROFC	S.F.							
	Set up, secure and take down ladder				.010		.39		.39	.58	.71
	Cut flashing				.010		.39		.39	.58	.71
	Install new flashing				.036	.35	1.42		1.77	2.51	3.01
	Clean up				.004		.16		.16	.23	.28
	Total				.060	.35	2.36		2.71	3.90	4.71
0400	**Minor fiberglass roof panel replacement - (25% of roof area)**	15	2 ROFC	S.F.							
	Set up, secure and take down ladder				.010		.39		.39	.58	.71
	Remove damaged panel				.030		1.17		1.17	1.75	2.12
	Install new fiberglass panel				.064	2.08	2.92		5	6.05	7.30
	Install strip-in with joint seal				.029	.01	1.32		1.33	1.74	2.15
	Clean up				.004		.16		.16	.23	.28
	Total				.137	2.09	5.96		8.05	10.35	12.56
0600	**Total fiberglass roof replacement**	20	2 ROFC	Sq.							
	Set up, secure and take down ladder				.020		.78		.78	1.17	1.41
	Remove existing fiberglass panels				2.500		98		98	146	176
	Install new flashing				.073	.70	2.84		3.54	5	6
	Install new fiberglass panel				6.400	208	292		500	605	730
	Apply joint sealer to seams				.029	.01	1.32		1.33	1.74	2.15
	Clean up				.390		15.30		15.30	23	27.50
	Total				9.412	208.71	410.24		618.95	781.91	943.06

B30 ROOFING B3013 Roof Covering

B3013 170 Roof Edges

	System Description	Freq. (Years)	Crew	Unit	Labor Hours	2014 Bare Costs				Total In-House	Total w/O&P
						Material	Labor	Equipment	Total		
1010	**Replace Roof edges, aluminum, duranodic, .050" thick, 6" face**	25	2 ROFC	L.F.							
	Set up, secure, and take down ladder				.010		.39		.39	.58	.71
	Remove adjacent roofing				.015		.59		.59	.88	1.06
	Remove roof edges				.080		2.93		2.93	3.91	4.82
	2" x 8" Edge blocking				.025	1.01	1.13		2.14	2.60	3.10
	Cants, 4" x 4", treated timber, cut diagonally				.025	1.75	.96		2.71	3.37	3.93
	Aluminum gravel stop with Duranodic finish				.089	10.43	4.86		15.29	17.65	20.50
	Roof covering, splice sheet				.011	3.10	.41	.06	3.57	4.05	4.67
	Total				.255	16.29	11.27	.06	27.62	33.04	38.79
1020	**Replace Roof edges, copper, 20 oz., 8" face**	35	2 ROFC	L.F.							
	Set up, secure, and take down ladder				.010		.39		.39	.58	.71
	Remove adjacent roofing				.015		.59		.59	.88	1.06
	Remove roof edge				.080		2.93		2.93	3.91	4.82
	2" x 8" Edge blocking				.025	1.01	1.13		2.14	2.60	3.10
	Cants, 4" x 4", treated timber, cut diagonally				.025	1.75	.96		2.71	3.37	3.93
	Copper gravel stop				.139	13.90	5.44		19.34	23.50	27
	Roof covering, splice sheet				.011	3.10	.41	.06	3.57	4.05	4.67
	Total				.305	19.76	11.85	.06	31.67	38.89	45.29
1030	**Replace Roof edges, galvanized, 20 ga., 6" face**	25	2 ROFC	L.F.							
	Set up, secure, and take down ladder				.010		.39		.39	.58	.71
	Remove adjacent roofing				.015		.59		.59	.88	1.06
	Remove roof edge				.080		2.93		2.93	3.91	4.82
	2" x 8" edge blocking				.025	1.01	1.13		2.14	2.60	3.10
	Cants, 4" x 4", treated timber, cut diagonally				.025	1.75	.96		2.71	3.37	3.93
	Galvanized steel gravel stop				.092	1.88	3.62		5.50	7.45	8.85
	Roof covering, splice sheet				.011	3.10	.41	.06	3.57	4.05	4.67
	Total				.258	7.74	10.03	.06	17.83	22.84	27.14

B30 ROOFING | B3013 Roof Covering

B3013 620 Gutters and Downspouts

	System Description	Freq. (Years)	Crew	Unit	Labor Hours	2014 Bare Costs				Total In-House	Total w/O&P
						Material	Labor	Equipment	Total		
1010	**Replace aluminum gutter, enameled, 5" K type, .027" thick**	40	2 SHEE	L.F.							
	Set up, secure and take down ladder				.010		.39		.39	.58	.71
	Remove gutter				.033		1.22		1.22	1.63	2.01
	Install new gutter				.064	2.60	3.50		6.10	7.30	8.75
	Clean up				.003		.10		.10	.15	.18
	Total				.110	2.60	5.21		7.81	9.66	11.65
1020	**Replace aluminum gutter, enameled, 5" K type, .032" thick**	40	2 SHEE	L.F.							
	Set up, secure and take down ladder				.010		.39		.39	.58	.71
	Remove gutter				.033		1.22		1.22	1.63	2.01
	Install new gutter				.064	3.20	3.50		6.70	7.95	9.50
	Clean up				.003		.10		.10	.15	.18
	Total				.110	3.20	5.21		8.41	10.31	12.40
1060	**Replace aluminum downspout, 2" x 3", .024" thick**	25	2 SHEE	L.F.							
	Set up, secure and take down ladder				.010		.39		.39	.58	.71
	Remove downspout				.023		.84		.84	1.12	1.38
	Install new downspout				.044	2.05	2.43		4.48	5.35	6.40
	Clean up				.003		.10		.10	.15	.18
	Total				.080	2.05	3.76		5.81	7.20	8.67
1070	**Replace aluminum downspout, 3" x 4", .024" thick**	25	2 SHEE	L.F.							
	Set up, secure and take down ladder				.010		.39		.39	.58	.71
	Remove downspout				.023		.84		.84	1.12	1.38
	Install new downspout				.057	2.88	3.13		6.01	7.15	8.55
	Clean up				.003		.10		.10	.15	.18
	Total				.093	2.88	4.46		7.34	9	10.82
2010	**Replace galvanized steel gutter, 5" box type, 28 ga.**	40	2 SHEE	L.F.							
	Set up, secure and take down ladder				.010		.39		.39	.58	.71
	Remove gutter				.033		1.22		1.22	1.63	2.01
	Install new gutter				.064	2.18	3.50		5.68	6.85	8.25
	Clean up				.003		.10		.10	.15	.18
	Total				.110	2.18	5.21		7.39	9.21	11.15

B30 ROOFING — B3013 Roof Covering

B3013 620 Gutters and Downspouts

	System Description	Freq. (Years)	Crew	Unit	Labor Hours	2014 Bare Costs				Total In-House	Total w/O&P
						Material	Labor	Equipment	Total		
2020	**Replace galvanized steel gutter, 6" box type, 26 ga.**	40	2 SHEE	L.F.							
	Set up, secure and take down ladder				.010		.39		.39	.58	.71
	Remove gutter				.033		1.22		1.22	1.63	2.01
	Install new gutter				.064	2.67	3.50		6.17	7.40	8.85
	Clean up				.003		.10		.10	.15	.18
	Total				.110	2.67	5.21		7.88	9.76	11.75
2060	**Replace round corrugated galvanized downspout, 3" diameter**	25	2 SHEE	L.F.							
	Set up, secure and take down ladder				.010		.39		.39	.58	.71
	Remove downspout				.023		.84		.84	1.12	1.38
	Install new downspout				.042	2.08	2.30		4.38	5.20	6.25
	Clean up				.003		.10		.10	.15	.18
	Total				.078	2.08	3.63		5.71	7.05	8.52
2070	**Replace round corrugated galvanized downspout, 4" diameter**	25	2 SHEE	L.F.							
	Set up, secure and take down ladder				.010		.39		.39	.58	.71
	Remove downspout				.023		.84		.84	1.12	1.38
	Install new downspout				.055	2.61	3.02		5.63	6.70	8
	Clean up				.003		.10		.10	.15	.18
	Total				.091	2.61	4.35		6.96	8.55	10.27
3020	**Replace vinyl box gutter, 5" wide**	20	1 CARP	L.F.							
	Set up, secure and take down ladder				.119		.39		.39	.58	.71
	Remove gutter				.010		1.22		1.22	1.63	2.01
	Install new gutter				.033	1.55	3.19		4.74	5.95	7.20
	Clean up				.070		.10		.10	.15	.18
					.003						
	Total				.116	1.55	4.90		6.45	8.31	10.10
3060	**Replace vinyl downspout, 2" x 3"**	15	1 CARP	L.F.							
	Set up, secure and take down ladder				.074		.39		.39	.58	.71
	Remove downspout				.010		.84		.84	1.12	1.38
	Install new downspout				.023	2.08	2.08		4.16	4.93	5.90
	Clean up				.038		.10		.10	.15	.18
					.003						
	Total				.074	2.08	3.41		5.49	6.78	8.17

B30 ROOFING

B3013 Roof Covering

B3013 620 Gutters and Downspouts

	System Description	Freq. (Years)	Crew	Unit	Labor Hours	2014 Bare Costs				Total In-House	Total w/O&P
						Material	Labor	Equipment	Total		
4020	**Replace 6" copper box gutters, 16 oz.**	50	1 SHEE	L.F.							
	Set up, secure and take down ladder				.010		.39		.39	.58	.71
	Remove gutter				.033		1.22		1.22	1.63	2.01
	Install new gutter				.064	8.80	3.50		12.30	14.10	16.50
	Clean up				.003		.10		.10	.15	.18
	Total				.110	8.80	5.21		14.01	16.46	**19.40**
4060	**Replace copper downspouts, 2" x 3", 16 oz.**	40	1 SHEE	L.F.							
	Set up, secure and take down ladder				.010		.39		.39	.58	.71
	Remove gutter				.023		.84		.84	1.12	1.38
	Install new gutter				.042	8.80	2.30		11.10	12.60	14.65
	Clean up				.003		.10		.10	.15	.18
	Total				.078	8.80	3.63		12.43	14.45	**16.92**

B30 ROOFING B3023 Roof Openings

B3023 110 Single Unit Glass Skylight

						2014 Bare Costs				Total In-House	Total w/O&P
	System Description	Freq. (Years)	Crew	Unit	Labor Hours	Material	Labor	Equipment	Total		
1010	**Repair glass skylight glazing**	6	1 CARP	S.F.							
	Set up, secure and take down ladder				.130		5.95		5.95	7.95	9.80
	Remove glazing				.069		2.54		2.54	3.39	4.17
	Install new glazing				.252	19.35	11.50		30.85	36	42.50
	Total				.451	19.35	19.99		39.34	47.34	56.47
1020	**Replace continuous skylight and structure**	40	2 CARP	C.S.F.							
	Set up and secure scaffold				.750		34.50		34.50	46	56.50
	Remove unit skylight				5.202		238		238	320	390
	Inspect / remove flashing				3.152	193	123		316	395	465
	Replace flashing				3.152	193	123		316	395	465
	Install new unit				12.960	2,100	590		2,690	3,100	3,575
	Remove scaffold				.750		34.50		34.50	46	56.50
	Total				25.966	2,486	1,143		3,629	4,302	5,008
1040	**Replace skylight and structure, double glazed, 10 to 20 S.F.**	40	G-3	S.F.							
	Set up, secure and take down ladder				.010		.39		.39	.58	.71
	Remove skylight, plstc domes,flush/curb mtd				.081		3.70		3.70	4.79	5.95
	10 S.F. to 20 S.F., double				.102	26.50	4.64		31.14	35	40.50
	Total				.193	26.50	8.73		35.23	40.37	47.16
1080	**Replace skylight and structure, double glazed, 30 to 65 S.F.**	40	G-3	S.F.							
	Set up, secure and take down ladder				.010		.39		.39	.58	.71
	Remove skylight, plstc domes,flush/curb mtd				.081		3.70		3.70	4.79	5.95
	30 S.F. to 65 S.F., double				.069	26	3.14		29.14	32.50	37.50
	Total				.160	26	7.23		33.23	37.87	44.16

B3023 120 Hatches

						2014 Bare Costs				Total In-House	Total w/O&P
	System Description	Freq. (Years)	Crew	Unit	Labor Hours	Material	Labor	Equipment	Total		
1020	**Replace aluminum roof hatch & structure**	40	2 CARP	Ea.							
	Set up, secure and take down ladder				.130		5.95		5.95	7.95	9.80
	Remove roof hatch				1.000		36.50		36.50	49	60
	Install 2'-6" x 3'-0" aluminum hatch				3.200	1,075	146		1,221	1,375	1,575
	Total				4.330	1,075	188.45		1,263.45	1,431.95	1,644.80

B30 ROOFING			B3023		Roof Openings							

B3023 120 Hatches

	System Description	Freq. (Years)	Crew	Unit	Labor Hours	2014 Bare Costs				Total In-House	Total w/O&P
						Material	Labor	Equipment	Total		
1120	Replace galvanized roof hatch & structure	40	2 CARP	Ea.							
	Set up, secure and take down ladder				.130		5.95		5.95	7.95	9.80
	Selective demo, therm & moist protect, roof acc, roof hatch, to 10 S.F.				1.000		36.50		36.50	49	60
	Install 2'-6" x 3'-0" galvanized roof hatch				3.200	640	146		786	895	1,025
	Total				4.330	640	188.45		828.45	951.95	1,094.80

B3023 130 Smoke Hatches

	System Description	Freq. (Years)	Crew	Unit	Labor Hours	2014 Bare Costs				Total In-House	Total w/O&P
						Material	Labor	Equipment	Total		
1010	Replace galvanized smoke hatch single unit 4' x 4'	40	2 CARP	Ea.							
	Set up and secure scaffold				.750		34.50		34.50	46	56.50
	Inspect / remove flashing				.032	1.93	1.23		3.16	3.96	4.63
	Replace flashing				.032	1.93	1.23		3.16	3.96	4.63
	Galvanized steel cover and frame				2.462	1,800	112		1,912	2,125	2,425
	Remove scaffold				.750		34.50		34.50	46	56.50
	Total				4.025	1,803.86	183.46		1,987.32	2,224.92	2,547.26
1020	Replace aluminum smoke hatch single unit, 4' x 8'	40	2 CARP	Ea.							
	Set up and secure scaffold				.750		34.50		34.50	46	56.50
	Smoke vent 4' x 8'				4.000		147		147	195	241
	Inspect / remove flashing				.756	46.32	29.52		75.84	95	111
	Replace flashing				.756	46.32	29.52		75.84	95	111
	4' x 8' aluminum cover and frame				5.246	2,800	240		3,040	3,400	3,875
	Remove scaffold				.750		34.50		34.50	46	56.50
	Total				12.259	2,892.64	515.04		3,407.68	3,877	4,451

C10 INTERIOR CONSTRUCTION C1013 Partitions

C1013 110 Concrete Block, Painted

	System Description	Freq. (Years)	Crew	Unit	Labor Hours	2014 Bare Costs				Total In-House	Total w/O&P
						Material	Labor	Equipment	Total		
0010	**Repair 8" concrete block wall - (2% of walls)**	25	1 BRIC	C.S.F.							
	Remove block				6.420		238	37	275	355	430
	Replace block				11.111	284	468		752	925	1,125
	Total				17.531	284	706	37	1,027	**1,280**	**1,555**
0020	**Refinish concrete block wall**	4	1 PORD	C.S.F.							
	Prepare surface				.250		10		10	13	16
	Fill & paint block, brushwork				1.376	20	54		74	92	112
	Place and remove mask and drops				.370		15		15	19	23
	Total				1.996	20	79		99	**124**	**151**
0040	**Replace 8" concrete block wall**	75	2 BRIC	C.S.F.							
	Set up and secure scaffold				.750		34.50		34.50	46	56.50
	Remove block				6.420		238	37	275	355	430
	Replace block				11.111	284	468		752	925	1,125
	Remove scaffold				.750		34.50		34.50	46	56.50
	Total				19.031	284	775	37	1,096	**1,372**	**1,668**

C1013 115 Glazed CMU Interior Wall Finish

	System Description	Freq. (Years)	Crew	Unit	Labor Hours	2014 Bare Costs				Total In-House	Total w/O&P
						Material	Labor	Equipment	Total		
0010	**Repair 4" glazed C.M.U. wall - (2% of walls)**	25	1 BRIC	C.S.F.							
	Remove damaged blocks				5.200		193	30	223	287	350
	Install new blocks				15.072	845	635		1,480	1,775	2,075
	Total				20.272	845	828	30	1,703	**2,062**	**2,425**
0020	**Replace 4" glazed C.M.U. wall**	75	2 BRIC	C.S.F.							
	Set up and secure scaffold				.750		34.50		34.50	46	56.50
	Remove blocks				5.200		193	30	223	287	350
	Install new blocks				15.072	845	635		1,480	1,775	2,075
	Remove scaffold				.750		34.50		34.50	46	56.50
	Total				21.772	845	897	30	1,772	**2,154**	**2,538**

C10 INTERIOR CONSTRUCTION C1013 Partitions

C1013 120 Glass Block

		System Description	Freq. (Years)	Crew	Unit	Labor Hours	2014 Bare Costs				Total In-House	Total w/O&P
							Material	Labor	Equipment	Total		
0010		**Repair glass block wall - (2% of walls)**	25	1 BRIC	C.S.F.							
		Remove damaged glass block				17.829		665		665	885	1,100
		Repair / replace blocks				45.217	2,900	1,905		4,805	5,700	6,725
		Total				63.046	2,900	2,570		5,470	6,585	7,825
0020		**Replace glass block wall**	75	2 BRIC	C.S.F.							
		Set up and secure scaffold				.750		34.50		34.50	46	56.50
		Remove damaged glass block				17.829		665		665	885	1,100
		Repair / replace blocks				45.217	2,900	1,905		4,805	5,700	6,725
		Remove scaffold				.750		34.50		34.50	46	56.50
		Total				64.546	2,900	2,639		5,539	6,677	7,938

C1013 125 Stone Veneer

		System Description	Freq. (Years)	Crew	Unit	Labor Hours	2014 Bare Costs				Total In-House	Total w/O&P
							Material	Labor	Equipment	Total		
0010		**Repair 4" stone veneer wall - (2% of walls)**	25	1 BRIC	C.S.F.							
		Remove damaged stone				7.429		339		339	500	615
		Replace stone				41.600	2,000	1,755		3,755	4,500	5,350
		Total				49.029	2,051	2,094		4,145	5,000	5,965
0020		**Replace 4" stone veneer wall**	75	2 BRIC	C.S.F.							
		Set up and secure scaffold				.750		34.50		34.50	46	56.50
		Remove damaged stone				7.429		339		339	500	615
		Replace stone				41.600	2,000	1,755		3,755	4,500	5,350
		Remove scaffold				.750		34.50		34.50	46	56.50
		Total				50.529	2,051	2,163		4,214	5,092	6,078

C10 INTERIOR CONSTRUCTION | C1013 Partitions

C1013 230 Demountable Partitions

	System Description	Freq. (Years)	Crew	Unit	Labor Hours	2014 Bare Costs				Total In-House	Total w/O&P
						Material	Labor	Equipment	Total		
1010	**Remove and reinstall demountable partitions**	5	2 CARP	C.L.F.							
	Remove demountable partitions for reuse				16.000		735		735	980	1,200
	Vinyl clad gypsum wall system, material only to replace damaged					1,180			1,180	1,300	1,475
	Install new and reused vinyl clad gypsum in new configuration				34.286		1,570		1,570	2,100	2,600
	18 ga. prefinished 3'-0" x 6'-8" to 4-7/8" deep frame				3.000	492	138		630	725	840
	Reinstall doors				3.200		147		147	195	242
	Cleanup				1.000		46		46	61	75
	Total				57.486	1,672	2,636		4,308	5,361	6,432

C1013 544 Channel Frame Wire Mesh Wall

	System Description	Freq. (Years)	Crew	Unit	Labor Hours	2014 Bare Costs				Total In-House	Total w/O&P
						Material	Labor	Equipment	Total		
0010	**Repair channel frame wire mesh wall - (2% of walls)**	20	1 SSWK	C.S.F.							
	Remove damaged 1-1/2" diamond mesh panel				12.000		448		448	595	735
	Replace with new 1-1/2" diamond mesh panel				2.857	477.50	131.25		608.75	700	810
	Total				14.857	477.50	579.25		1,056.75	1,295	1,545
0020	**Refinish channel frame wire mesh wall**	5	1 PORD	C.S.F.							
	Prepare surface				.370		15		15	19	23
	Scrape surface				1.250		49		49	64	79
	Refinish surface, brushwork, primer + 1 coat				1.538	43	61		104	126	152
	Total				3.158	43	125		168	209	254

C1013 730 Plate Glass Wall - Interior

	System Description	Freq. (Years)	Crew	Unit	Labor Hours	2014 Bare Costs				Total In-House	Total w/O&P
						Material	Labor	Equipment	Total		
0010	**Repair plate glass interior wall - (2% of total)**	25	1 CARP	C.S.F.							
	Remove plate glass & frame				6.933		254		254	340	415
	Repair damaged glass				23.111	6,550	1,100		7,650	8,750	10,100
	Total				30.044	6,550	1,354		7,904	9,090	10,515

C10 INTERIOR CONSTRUCTION C1013 Partitions

C1013 730 Plate Glass Wall - Interior

	System Description	Freq. (Years)	Crew	Unit	Labor Hours	2014 Bare Costs				Total In-House	Total w/O&P
						Material	Labor	Equipment	Total		
0020	Replace plate glass interior wall	50	2 CARP	C.S.F.							
	Set up and secure scaffold				.750		34.50		34.50	46	56.50
	Remove plate glass & frame				6.933		254		254	340	415
	Repair damaged glass				23.111	6,550	1,100		7,650	8,750	10,100
	Remove scaffold				.750		34.50		34.50	46	56.50
	Total				31.544	6,550	1,423		7,973	9,182	10,628

C10 INTERIOR CONSTRUCTION | C1023 | Interior Doors

C1023 108 | Fully Glazed Wooden Doors

	System Description	Freq. (Years)	Crew	Unit	Labor Hours	Material	Labor	Equipment	Total	Total In-House	Total w/O&P
							2014 Bare Costs				
1010	**Replace insulating glass (3% of glass)**	1	1 CARP	S.F.							
	Remove damaged glass				.052		1.90		1.90	2.54	3.13
	Install new insulating glass				.277	29	12.20		41.20	48	56
	Total				.329	29	14.10		43.10	50.54	**59.13**
1020	**Repair fully glazed wood door**	10	1 CARP	Ea.							
	Remove lockset				.520		24		24	32	39
	Replace lockset				1.040	108	47.50		155.50	182	214
	Oil hinges				.065		2.96		2.96	3.94	4.86
	Oil door closer				.065		2.96		2.96	3.94	4.86
	Total				1.689	108	77.42		185.42	221.88	**262.72**
1030	**Refinish 3'-0" x 7'-0" fully glazed wood door**	4	1 PORD	Ea.							
	Prepare interior door for painting				.433		17.15		17.15	22	27.50
	Paint door, roller + brush, 1 coat				1.333	2.40	52.50		54.90	71	87.50
	Total				1.767	2.40	69.65		72.05	93	**115**
1040	**Replace 3'-0" x 7'-0" fully glazed wood door**	40	1 CARP	Ea.							
	Remove interior door				.520		19.05		19.05	25.50	31.50
	Remove interior door frame				1.300		59.50		59.50	79.50	98
	Remove door closer				.167		7.65		7.65	10.20	12.55
	Install new wood door and frame				2.600	712.50	119		831.50	940	1,075
	Reinstall door closer				1.735		79.50		79.50	106	131
	Total				6.322	712.50	284.70		997.20	1,161.20	**1,348.05**

C1023 110 | Steel Painted Interior Doors

	System Description	Freq. (Years)	Crew	Unit	Labor Hours	Material	Labor	Equipment	Total	Total In-House	Total w/O&P
							2014 Bare Costs				
1010	**Repair steel painted door**	14	1 CARP	Ea.							
	Remove lockset				.520		24		24	32	39
	Replace lockset				1.040	108	47.50		155.50	182	214
	Oil hinges				.065		2.96		2.96	3.94	4.86
	Oil door closer				.065		2.96		2.96	3.94	4.86
	Total				1.689	108	77.42		185.42	221.88	**262.72**

C10 INTERIOR CONSTRUCTION   C1023   Interior Doors

C1023 110   Steel Painted Interior Doors

	System Description	Freq. (Years)	Crew	Unit	Labor Hours	2014 Bare Costs				Total In-House	Total w/O&P
						Material	Labor	Equipment	Total		
1020	**Refinish 3'-0" x 7'-0" steel painted door**	4	1 PORD	Ea.							
	Prepare interior door for painting				.433		17.15		17.15	22	27.50
	Paint door, roller + brush, 1 coat				.800	4.37	31.50		35.87	46	56.50
	Total				1.233	4.37	48.65		53.02	68	84
1030	**Replace 3'-0" x 7'-0" steel painted door**	60	1 CARP	Ea.							
	Remove interior door				.520		19.05		19.05	25.50	31.50
	Remove interior door frame				1.300		59.50		59.50	79.50	98
	Remove door closer				.167		7.65		7.65	10.20	12.55
	Install new steel door				1.224	640	56		696	780	890
	Install new steel door frame				1.000	150	46		196	226	263
	Reinstall door closer				1.735		79.50		79.50	106	131
	Total				5.946	790	267.70		1,057.70	1,227.20	1,426.05
2010	**Safety glass repl., (3% of glass)**	1	1 CARP	S.F.							
	Remove damaged glass				.052		1.90		1.90	2.54	3.13
	Install new safety glass				.173	8.70	7.65		16.35	19.60	23.50
	Total				.225	8.70	9.55		18.25	22.14	26.63
2030	**Refinish, 3'-0" x 7'-0" steel w/ safety glass door**	4	1 PORD	Ea.							
	Prepare interior door for painting				.433		17.15		17.15	22	27.50
	Paint door, roller + brush, 1 coat				.800	4.37	31.50		35.87	46	56.50
	Total				1.233	4.37	48.65		53.02	68	84
2040	**Replace 3'-0" x 7'-0" steel door w/ vision lite, & frame**	60	1 CARP	Ea.							
	Remove interior door				.520		19.05		19.05	25.50	31.50
	Remove interior door frame				1.300		59.50		59.50	79.50	98
	Remove door closer				.167		7.65		7.65	10.20	12.55
	Install new steel door				1.224	640	56		696	780	890
	Install new steel door frame				1.301	150	59.50		209.50	244	286
	Reinstall door closer				1.735		79.50		79.50	106	131
	Vision lite					94.50			94.50	104	118
	Total				6.246	884.50	281.20		1,165.70	1,349.20	1,567.05

C10 INTERIOR CONSTRUCTION | C1023 | Interior Doors

C1023 110 | Steel Painted Interior Doors

	System Description	Freq. (Years)	Crew	Unit	Labor Hours	2014 Bare Costs				Total In-House	Total w/O&P
						Material	Labor	Equipment	Total		
3010	**Repair 3'-0" x 7'-0" steel sliding door**	14	1 CARP	Ea.							
	Oil door closer				.065		2.96		2.96	3.94	4.86
	Oil / inspect track / glide wheels				.065		2.96		2.96	3.94	4.86
	Remove door track / glide wheels (25% of total / 14 years)				.220		10.13		10.13	13.40	16.50
	Install new door track / glide wheels (25% of total / 14 years)				.440	13.13	20.13		33.26	41.50	49.50
	Total				.789	13.13	36.18		49.31	62.78	75.72
3020	**Refinish 3'-0" x 7'-0" steel sliding door**	4	1 PORD	Ea.							
	Prepare interior door for painting				.433		17.15		17.15	22	27.50
	Paint door, roller + brush, 1 coat				.800	4.37	31.50		35.87	46	56.50
	Total				1.233	4.37	48.65		53.02	68	84
3030	**Replace 3'-0" x 7'-0" steel sliding door & frame**	60	1 CARP	Ea.							
	Remove interior door				.520		19.05		19.05	25.50	31.50
	Remove interior door frame				1.300		59.50		59.50	79.50	98
	Remove door closer				.167		7.65		7.65	10.20	12.55
	Install new steel door				1.224	640	56		696	780	890
	Install new steel sliding door frame				3.200	239	165		418.20	530	620
	Install new door track / glide wheels				1.760	52.50	80.50		133	165	199
	Install door closer				1.735		79.50	14.20	79.50	106	131
	Total				9.905	931.50	467.20	14.20	1,412.90	1,696.20	1,982.05
4010	**Replace wire glass (3% of glass)**	1	1 CARP	S.F.							
	Remove damaged glass				.052		1.90		1.90	2.54	3.13
	Install new wire glass				.179	27.26	7.89		35.15	40.50	47
	Total				.231	27.26	9.79		37.05	43.04	50.13
4030	**Refinish 3'-0" x 7'-0" steel half glass door**	4	1 PORD	Ea.							
	Prepare interior door for painting				.433		17.15		17.15	22	27.50
	Paint door, roller + brush, 1 coat				.800	4.37	31.50		35.87	46	56.50
	Total				1.233	4.37	48.65		53.02	68	84

C10 INTERIOR CONSTRUCTION C1023 Interior Doors

C1023 110 Steel Painted Interior Doors

	System Description	Freq. (Years)	Crew	Unit	Labor Hours	2014 Bare Costs Material	2014 Bare Costs Labor	2014 Bare Costs Equipment	2014 Bare Costs Total	Total In-House	Total w/O&P
4040	**Replace 3'-0" x 7'-0" steel half glass door & frame**	60	1 CARP	Ea.							
	Remove interior door				.520		19.05		19.05	25.50	31.50
	Remove door closer				.167		7.65		7.65	10.20	12.55
	Remove interior door frame				1.300		59.50		59.50	79.50	98
	Install new steel door frame				1.301	150	59.50		209.50	244	286
	Install new half glass interior door				1.224	625	56		681	760	875
	Install new safety glass				1.040	52.20	45.90		98.10	118	140
	Reinstall door closer				1.735		79.50		79.50	106	131
	Total				7.287	827.20	327.10		1,154.30	1,343.20	1,574.05

C1023 111 Steel Unpainted Interior Doors

	System Description	Freq. (Years)	Crew	Unit	Labor Hours	2014 Bare Costs Material	2014 Bare Costs Labor	2014 Bare Costs Equipment	2014 Bare Costs Total	Total In-House	Total w/O&P
1010	**Repair steel unpainted door**	14	1 CARP	Ea.							
	Remove lockset				.520		24		24	32	39
	Replace lockset				1.040	108	47.50		155.50	182	214
	Oil hinges				.065		2.96		2.96	3.94	4.86
	Oil door closer				.065		2.96		2.96	3.94	4.86
	Total				1.689	108	77.42		185.42	221.88	262.72
1020	**Replace 3'-0" x 7'-0" steel unpainted door & frame**	60	1 CARP	Ea.							
	Remove interior door				.520		19.05		19.05	25.50	31.50
	Remove interior door frame				1.300		59.50		59.50	79.50	98
	Remove door closer				.167		7.65		7.65	10.20	12.55
	Install new steel door				1.224	640	56		696	780	890
	Install new steel door frame				1.000	150	46		196	226	263
	Reinstall door closer				1.735		79.50		79.50	106	131
	Total				5.946	790	267.70		1,057.70	1,227.20	1,426.05
2010	**Replace safety glass (3% of glass)**	1	1 CARP	S.F.							
	Remove damaged glass				.052		1.90		1.90	2.54	3.13
	Install new safety glass				.173	8.70	7.65		16.35	19.60	23.50
	Total				.225	8.70	9.55		18.25	22.14	26.63

C10 INTERIOR CONSTRUCTION — C1023 Interior Doors

C1023 111 Steel Unpainted Interior Doors

	System Description	Freq. (Years)	Crew	Unit	Labor Hours	2014 Bare Costs					Total In-House	Total w/O&P
						Material	Labor	Equipment	Total		
2030	**Repl. 3'-0" x 7'-0" st. half glass unpainted dr. & fr.**	60	1 CARP	Ea.							
	Remove interior door				.520		19.05		19.05	25.50	31.50
	Remove interior door frame				1.300		59.50		59.50	79.50	98
	Remove door closer				.167		7.65		7.65	10.20	12.55
	Install new half glass interior door				1.224	625	56		681	760	875
	Install new safety glass				1.560	78.30	68.85		147.15	176	209
	Install new steel door frame				1.301	150	59.50		209.50	244	286
	Reinstall door closer				1.735		79.50		79.50	106	131
	Total				7.807	853.30	350.05		1,203.35	1,401.20	1,643.05
3010	**Repair 3'-0" x 7'-0" steel sliding unpainted door**	14	1 CARP	Ea.							
	Oil door closer				.065		2.96		2.96	3.94	4.86
	Oil / inspect track / glide wheels				.065		2.96		2.96	3.94	4.86
	Remove door track / glide wheels (25% of total / 14 years)				.220		10.13		10.13	13.40	16.50
	Install new door track / glide wheels (25% of total / 14 years)				.440	13.13	20.13		33.26	41.50	49.50
	Total				.789	13.13	36.18		49.31	62.78	75.72
3020	**Replace 3'-0" x 7'-0" steel sliding unpainted interior door**	60	1 CARP	Ea.							
	Remove interior door				.520		19.05		19.05	25.50	31.50
	Remove interior door frame				1.300		59.50		59.50	79.50	98
	Remove door closer				.167		7.65		7.65	10.20	12.55
	Install new steel door				1.224	640	56		696	780	890
	Install new steel sliding door frame				3.200	239	165	14.20	418.20	530	620
	Install new door track / glide wheels				1.760	52.50	80.50		133	165	199
	Reinstall door closer				1.735		79.50		79.50	106	131
	Total				9.905	931.50	467.20	14.20	1,412.90	1,696.20	1,982.05
4010	**Replace wire glass (3% of glass)**	1	1 CARP	S.F.							
	Remove damaged glass				.052		1.90		1.90	2.54	3.13
	Install new wire glass				.179	27.26	7.89		35.15	40.50	47
	Total				.231	27.26	9.79		37.05	43.04	50.13

C10 INTERIOR CONSTRUCTION   C1023   Interior Doors

C1023 112   Bi-Fold Interior Doors

	System Description	Freq. (Years)	Crew	Unit	Labor Hours	2014 Bare Costs				Total	Total In-House	Total w/O&P
						Material	Labor	Equipment	Total			
2010	**Repair 2'-6" x 6'-8" bi-fold louvered door**	15	1 CARP	Ea.								
	Remove interior door				.520		19.05		19.05	25.50	31.50
	Remove damaged stile				.078		3.60		3.60	4.80	5.90
	Remove damaged slat				.039		1.78		1.78	2.36	2.92
	Install new slat				.077	1.02	3.54		4.56	5.85	7.10
	Install new stile				.157	2.37	7.20		9.57	12.20	14.80
	Reinstall door				.667		30.50		30.50	40.50	50
	Oil / lubricate door				.065		2.96		2.96	3.94	4.86
	Total				1.602	3.39	68.63		72.02	95.15	**117.08**
2020	**Refinish 2'-6" x 6'-8" louvered door**	8	1 PORD	Ea.								
	Prepare door for finish				.967		38.50		38.50	49.50	61.50
	Paint door, roller + brush, 1 coat				1.143	5.35	45		50.35	64.50	79
	Total				2.110	5.35	83.50		88.85	114	**140.50**
2040	**Replace 2'-6" x 6'-8" wd. louver bi-fold dr. and fr.**	24	1 CARP	Ea.								
	Remove doors				.520		19.05		19.05	25.50	31.50
	Install new door				1.600	278	73.50		351.50	405	470
	Install new interior door frame				.998	112.50	45.72		158.22	185	216
	Total				3.118	390.50	138.27		528.77	615.50	**717.50**

C1023 114   Aluminum Interior Doors

	System Description	Freq. (Years)	Crew	Unit	Labor Hours	2014 Bare Costs				Total	Total In-House	Total w/O&P
						Material	Labor	Equipment	Total			
1010	**Repair aluminum door**	12	1 CARP	Ea.								
	Remove lockset				.520		24		24	32	39
	Replace lockset				1.040	108	47.50		155.50	182	214
	Oil hinges				.065		2.96		2.96	3.94	4.86
	Oil door closer				.065		2.96		2.96	3.94	4.86
	Total				1.689	108	77.42		185.42	221.88	**262.72**

C10 INTERIOR CONSTRUCTION | C1023 | Interior Doors

C1023 114 | Aluminum Interior Doors

	System Description	Freq. (Years)	Crew	Unit	Labor Hours	2014 Bare Costs				Total In-House	Total w/O&P
						Material	Labor	Equipment	Total		
1020	**Replace 3'-0" x 7'-0" aluminum door & frame**	50	1 CARP	Ea.							
	Remove interior door				.520		19.05		19.05	25.50	31.50
	Remove interior door frame				1.300		59.50		59.50	79.50	98
	Remove door closer				.167		7.65		7.65	10.20	12.55
	Install new aluminum door				1.224	1,400	56		1,456	1,625	1,850
	Install new aluminum door frame				2.971	560	152		712	850	980
	Reinstall door closer				1.735		79.50		79.50	106	131
	Total				7.917	1,960	373.70		2,333.70	**2,696.20**	**3,103.05**
2010	**Replace safety glass (3% of glass)**	1	1 CARP	S.F.							
	Remove damaged glass				.052		1.90		1.90	2.54	3.13
	Install new safety glass				.173	8.70	7.65		16.35	19.60	23.50
	Total				.225	8.70	9.55		18.25	**22.14**	**26.63**
2030	**Replace 3'-0" x 7'-0" alum. dr., fr. & vision lite**	50	1 CARP	Ea.							
	Remove interior door				.520		19.05		19.05	25.50	31.50
	Remove interior door frame				1.300		59.50		59.50	79.50	98
	Remove door closer				.167		7.65		7.65	10.20	12.55
	Install new aluminum door				1.224	1,400	56		1,456	1,625	1,850
	Install new aluminum door frame				2.971	560	152		712	850	980
	Reinstall door closer				1.735		79.50		79.50	106	131
	Vision lite					94.50			94.50	104	118
	Total				7.917	2,054.50	373.70		2,428.20	**2,800.20**	**3,221.05**
3010	**Replace insulating glass (3% of glass)**	1	1 CARP	S.F.							
	Remove damaged glass				.052		1.90		1.90	2.54	3.13
	Install new insulating glass				.277	29	12.20		41.20	48	56
	Total				.329	29	14.10		43.10	**50.54**	**59.13**
3020	**Repair 3'-0" x 7'-0" aluminum sliding door**	14	1 CARP	Ea.							
	Oil door closer				.065		2.96		2.96	3.94	4.86
	Oil / inspect track / glide wheels				.065		2.96		2.96	3.94	4.86
	Remove door track / glide wheels (25% of total / 14 years)				.220		10.13		10.13	13.40	16.50
	Install new door track / glide wheels (25% of total / 14 years)				.440	13.13	20.13		33.26	41.50	49.50
	Total				.789	13.13	36.18		49.31	**62.78**	**75.72**

C10 INTERIOR CONSTRUCTION C1023 Interior Doors

C1023 114 Aluminum Interior Doors

	System Description	Freq. (Years)	Crew	Unit	Labor Hours	2014 Bare Costs				Total In-House	Total w/O&P
						Material	Labor	Equipment	Total		
3030	**Replace 3'-0" x 7'-0" aluminum sliding door**	50	1 CARP	Ea.							
	Remove interior door				.520		19.05		19.05	25.50	31.50
	Remove interior door frame				1.300		59.50		59.50	79.50	98
	Remove door closer				.167		7.65		7.65	10.20	12.55
	Install new aluminum door				1.224	1,400	56		1,456	1,625	1,850
	Install new aluminum door frame				2.971	560	152		712	850	980
	Install new door track / glide wheels				1.760	52.50	80.50		133	165	199
	Install door closer				1.735		79.50		79.50	106	131
	Total				9.677	2,012.50	454.20		2,466.70	2,861.20	3,302.05

C1023 116 Fully Glazed Aluminum Doors

	System Description	Freq. (Years)	Crew	Unit	Labor Hours	2014 Bare Costs				Total In-House	Total w/O&P
						Material	Labor	Equipment	Total		
1010	**Replace insulating glass (3% of glass)**	1	1 CARP	S.F.							
	Remove damaged glass				.052		1.90		1.90	2.54	3.13
	Install new insulating glass				.277	29	12.20		41.20	48	56
	Total				.329	29	14.10		43.10	50.54	59.13
1020	**Repair 3'-0" x 7'-0" fully glazed aluminum door**	12	1 CARP	Ea.							
	Remove lockset				.520		24		24	32	39
	Replace lockset				1.040	108	47.50		155.50	182	214
	Oil hinges				.065		2.96		2.96	3.94	4.86
	Oil door closer				.065		2.96		2.96	3.94	4.86
	Total				1.689	108	77.42		185.42	221.88	262.72
1030	**Replace 3'-0" x 7'-0" fully glazed aluminum door**	50	1 CARP	Ea.							
	Remove interior door				.520		19.05		19.05	25.50	31.50
	Remove interior door frame				1.300		59.50		59.50	79.50	98
	Remove door closer				.167		7.65		7.65	10.20	12.55
	Install new aluminum door and frame				10.458	820	535		1,355	1,725	2,000
	Install new safety glass				3.640	182.70	160.65		343.35	410	490
	Reinstall door closer				1.735		79.50		79.50	106	131
	Total				17.819	1,002.70	861.35		1,864.05	2,356.20	2,763.05

C10 INTERIOR CONSTRUCTION | C1023 | Interior Doors

C1023 120 | Hollow Core Interior Doors

	System Description	Freq. (Years)	Crew	Unit	Labor Hours	2014 Bare Costs				Total In-House	Total w/O&P
						Material	Labor	Equipment	Total		
1010	**Repair hollow core wood door**	7	1 CARP	Ea.							
	Remove lockset				.520		24		24	32	39
	Replace lockset				1.040	108	47.50		155.50	182	214
	Oil hinges				.065		2.96		2.96	3.94	4.86
	Oil door closer				.065		2.96		2.96	3.94	4.86
	Total				1.689	108	77.42		185.42	221.88	262.72
1020	**Refinish 3'-0" x 7'-0" hollow core wood door**	4	1 PORD	Ea.							
	Prepare interior door for painting				.433		17.15		17.15	22	27.50
	Paint door, roller + brush, 1 coat				.533	5.35	21		26.35	33	40.50
	Total				.967	5.35	38.15		43.50	55	68
1030	**Replace 3'-0" x 7'-0" hollow core wood door**	30	1 CARP	Ea.							
	Remove interior door				.520		19.05		19.05	25.50	31.50
	Remove interior door frame				1.300		59.50		59.50	79.50	98
	Install new interior wood door and frame				.941	155	43		198	228	265
	Total				2.761	155	121.55		276.55	333	394.50

C1023 121 | Solid Core Interior Doors

	System Description	Freq. (Years)	Crew	Unit	Labor Hours	2014 Bare Costs				Total In-House	Total w/O&P
						Material	Labor	Equipment	Total		
1010	**Repair solid core wood door**	11	1 CARP	Ea.							
	Remove lockset				.520		24		24	32	39
	Replace lockset				1.040	108	47.50		155.50	182	214
	Oil hinges				.065		2.96		2.96	3.94	4.86
	Oil door closer				.065		2.96		2.96	3.94	4.86
	Total				1.689	108	77.42		185.42	221.88	262.72
1020	**Refinish 3'-0" x 7'-0" solid core wood door**	4	1 PORD	Ea.							
	Prepare interior door for painting				.433		17.15		17.15	22	27.50
	Paint door, roller + brush, 1 coat				.533	5.35	21		26.35	33	40.50
	Total				.967	5.35	38.15		43.50	55	68

C10 INTERIOR CONSTRUCTION | C1023 | Interior Doors

C1023 121 | Solid Core Interior Doors

	System Description	Freq. (Years)	Crew	Unit	Labor Hours	2014 Bare Costs				Total In-House	Total w/O&P
						Material	Labor	Equipment	Total		
1030	**Replace 3'-0" x 7'-0" solid core wood door**	40	1 CARP	Ea.							
	Remove interior door				.520		19.05		19.05	25.50	31.50
	Remove interior door frame				.650		29.90		29.90	39.50	49
	Install new interior wood door & frame				.941	320	43		363	410	470
	Total				2.111	320	91.95		411.95	475	550.50
2010	**Repair solid core sliding wood door**	14	1 CARP	Ea.							
	Oil door closer				.065		2.96		2.96	3.94	4.86
	Oil / inspect track / glide wheels				.065		2.96		2.96	3.94	4.86
	Remove door track / glide wheels (25% of total / 14 years)				.220		10.13		10.13	13.40	16.50
	Install new door track / glide wheels (25% of total / 14 years)				.440	13.13	20.13		33.26	41.50	49.50
	Total				.789	13.13	36.18		49.31	62.78	75.72
2030	**Replace 3'-0" x 7'-0" solid core sliding wood door**	40	1 CARP	Ea.							
	Remove interior door				.520		19.05		19.05	25.50	31.50
	Remove interior door frame				.650		29.90		29.90	39.50	49
	Remove door closer				.167		7.65		7.65	10.20	12.55
	Install new interior wood door and frame				.941	320	43		363	410	470
	Install new door track / glide wheels				1.760	52.50	80.50		133	165	199
	Reinstall door closer				1.735		79.50		79.50	106	131
	Total				5.773	372.50	259.60		632.10	756.20	893.05
3040	**Replace 3'-0" x 7'-0" solid core, w/ safety glass dr.**	40	1 CARP	Ea.							
	Remove interior door				.520		19.05		19.05	25.50	31.50
	Remove interior door frame				.650		29.90		29.90	39.50	49
	Install new interior wood door and frame				.941	320	43		363	410	470
	Install new safety glass lite					76.50			76.50	84	95.50
	Total				2.111	396.50	91.95		488.45	559	646

C10 INTERIOR CONSTRUCTION | C1023 | Interior Doors

C1023 129 | Interior Gates

	System Description	Freq. (Years)	Crew	Unit	Labor Hours	2014 Bare Costs				Total In-House	Total w/O&P
						Material	Labor	Equipment	Total		
1030	**Prepare and refinish interior metal gate**	5	1 PORD	Ea.							
	Lay drop cloth				.090		3.55		3.55	4.60	5.70
	Prepare / brush surface				.125		4.86		4.86	6.30	7.90
	Paint gate, brushwork, primer + 1 coat				.167	1.44	6.66		8.10	10.10	12.40
	Remove drop cloth				.090		3.55		3.55	4.60	5.70
	Total				.471	1.44	18.62		20.06	**25.60**	**31.70**

C1023 324 | Hinges, Brass

	System Description	Freq. (Years)	Crew	Unit	Labor Hours	2014 Bare Costs				Total In-House	Total w/O&P
						Material	Labor	Equipment	Total		
0010	**Replace 1 1/2 pair brass hinges**	60	1 CARP	Ea.							
	Remove door				.650		24		24	32	39
	Remove hinges				.500		22.95		22.95	30.50	37.50
	Replace hinges				.500		22.95		22.95	30.50	37.50
	Hinges, 1 1/2 pair					183			183	201	229
	Install door				1.224		56		56	75	92
	Total				2.875	183	125.90		308.90	**369**	**435**

C1023 325 | Lockset, Brass

	System Description	Freq. (Years)	Crew	Unit	Labor Hours	2014 Bare Costs				Total In-House	Total w/O&P
						Material	Labor	Equipment	Total		
0010	**Replace brass lockset**	30	1 CARP	Ea.							
	Remove lockset				.400		18.35		18.35	24.50	30
	Install lockset				1.156	220	53		273	315	360
	Total				1.556	220	71.35		291.35	**339.50**	**390**

C10 INTERIOR CONSTRUCTION   C1023   Interior Doors

C1023 326   Door Closer, Brass

	System Description	Freq. (Years)	Crew	Unit	Labor Hours	2014 Bare Costs				Total In-House	Total w/O&P
						Material	Labor	Equipment	Total		
0010	Replace brass closer	15	1 CARP	Ea.							
	Remove door closer				.167		7.65		7.65	10.20	12.55
	Install new door closer				1.735	188	79.50		267.50	315	365
	Total				1.902	188	87.15		275.15	325.20	377.55

C1023 327   Deadbolt, Brass

	System Description	Freq. (Years)	Crew	Unit	Labor Hours	2014 Bare Costs				Total In-House	Total w/O&P
						Material	Labor	Equipment	Total		
0010	Replace brass deadbolt	20	1 CARP	Ea.							
	Remove deadbolt				.578		26.50		26.50	35.50	43.50
	Replace deadbolt				1.156	200	53		253	291	335
	Total				1.733	200	79.50		279.50	326.50	378.50

C1023 328   Weatherstripping, Brass

	System Description	Freq. (Years)	Crew	Unit	Labor Hours	2014 Bare Costs				Total In-House	Total w/O&P
						Material	Labor	Equipment	Total		
0010	Replace brass weatherstripping	20	1 CARP	Ea.							
	Remove weatherstripping				.015		.67		.67	.90	1.11
	Install new weatherstripping				3.478	46.50	159		205.50	264	320
	Total				3.493	46.50	159.67		206.17	264.90	321.11

C1023 329   Panic Bar

	System Description	Freq. (Years)	Crew	Unit	Labor Hours	2014 Bare Costs				Total In-House	Total w/O&P
						Material	Labor	Equipment	Total		
0010	Replace panic bar	25	1 CARP	Ea.							
	Remove panic bar				1.040		47.50		47.50	63.50	78.50
	Install panic bar				2.080	865	95.50		960.50	1,075	1,250
	Total				3.120	865	143		1,008	1,138.50	1,328.50

C10 INTERIOR CONSTRUCTION | C1033 Fittings

C1033 110 Toilet Partitions

	System Description	Freq. (Years)	Crew	Unit	Labor Hours	2014 Bare Costs				Total In-House	Total w/O&P
						Material	Labor	Equipment	Total		
1010	**Replace toilet partitions, laminate clad-overhead braced, per stall**	20	2 CARP	Ea.							
	Remove overhead braced laminate clad partitions				1.333		61		61	81.50	100
	Install overhead braced laminate clad partitions				3.478	750	159		909	1,050	1,200
	Cleanup				.500		23		23	30.50	37.50
	Total				5.312	750	243		993	1,162	1,337.50
1020	**Replace toilet partitions, painted metal-overhead braced, per stall**	20	2 CARP	Ea.							
	Remove overhead braced painted metal partitions				1.333		61		61	81.50	100
	Install overhead braced painted metal partitions				3.478	380	159		539	630	735
	Cleanup				.500		23		23	30.50	37.50
	Total				5.312	380	243		623	742	872.50
1030	**Replace toilet partitions, porcelain enamel-overhead braced, per stall**	20	2 CARP	Ea.							
	Remove overhead braced partitions				1.333		61		61	81.50	100
	Install overhead braced porcelain enamel partitions				3.478	500	159		659	765	885
	Cleanup				.500		23		23	30.50	37.50
	Total				5.312	500	243		743	877	1,022.50
1040	**Replace toilet partitions, phenolic-overhead braced, per stall**	20	2 CARP	Ea.							
	Remove overhead braced partitions				1.333		61		61	81.50	100
	Install phenolic overhead braced partition				3.478	795	159		954	1,075	1,250
	Cleanup				.500		23		23	30.50	37.50
	Total				5.312	795	243		1,038	1,187	1,387.50
1050	**Replace toilet partitions, stainless steel-overhead braced, per stall**	30	2 CARP	Ea.							
	Remove overhead braced partitions				1.333		61		61	81.50	100
	Install stainless steel overhead braced partition				3.478	1,050	159		1,209	1,375	1,575
	Cleanup				.500		23		23	30.50	37.50
	Total				5.312	1,050	243		1,293	1,487	1,712.50
1060	**Replace toilet partitions, stainless steel, ceiling hung, per stall**	30	2 CARP	Ea.							
	Remove ceiling hung partitions				1.333		61		61	81.50	100
	Install stainless steel ceiling hung partition				4.000	1,075	183		1,258	1,425	1,650
	Cleanup				.500		23		23	30.50	37.50
	Total				5.833	1,075	267		1,342	1,537	1,787.50

C10 INTERIOR CONSTRUCTION | C1033 Fittings

C1033 110 Toilet Partitions

	System Description	Freq. (Years)	Crew	Unit	Labor Hours	2014 Bare Costs				Total In-House	Total w/O&P
						Material	Labor	Equipment	Total		
2020	**Replace urinal screen, stainless steel**	30	2 CARP	Ea.							
	Remove stainless steel urinal screen				1.000		36.50		36.50	49	60
	Install stainless steel urinal screen				2.000	580	91.50		671.50	760	875
	Cleanup				.500		23		23	30.50	37.50
	Total				3.500	580	151		731	839.50	972.50
3010	**Replace shower stall, fiberglass, per stall**	30	2 SHEE	Ea.							
	Remove shower base, partition and door				2.000		73.50		73.50	97.50	120
	Install shower base, partition and door				3.556	795	194		989	1,125	1,300
	Cleanup				.500		23		23	30.50	37.50
	Total				6.056	795	290.50		1,085.50	1,253	1,457.50

C1033 210 Metal Lockers

	System Description	Freq. (Years)	Crew	Unit	Labor Hours	2014 Bare Costs				Total In-House	Total w/O&P
						Material	Labor	Equipment	Total		
1010	**Replace metal lockers, single tier**	20	1 SHEE	Ea.							
	Remove lockers				.533		19.55		19.55	26	32
	Install 12" x 18" x 72" single enameled locker				.400	267	22		289	320	370
	Cleanup				.500		23		23	30.50	37.50
	Total				1.433	267	64.55		331.55	376.50	439.50

C20 STAIRS   C2013   Stair Construction

C2013 110   Concrete Steps

						2014 Bare Costs				
	System Description	Freq. (Years)	Crew	Unit	Labor Hours	Material	Labor	Equipment	Total	Total In-House	Total w/O&P
0010	**Repair concrete steps** Repair spalls in steps	15	1 CEFI	S.F.	.104	11.70	4.58		16.28	18.70	22
	Total				.104	11.70	4.58		16.28	**18.70**	**22**
0020	**Replace concrete steps** Remove concrete Install new concrete steps	100	2 CEFI	S.F.	.520 .520	4.49	19.25 23.50	3.04 .36	22.29 28.35	28.50 36.50	35 44.50
	Total				1.040	4.49	42.75	3.40	50.64	**65**	**79.50**

C2013 115   Masonry Steps, Painted

						2014 Bare Costs				
	System Description	Freq. (Years)	Crew	Unit	Labor Hours	Material	Labor	Equipment	Total	Total In-House	Total w/O&P
0010	**Repair masonry steps, painted** Remove damaged masonry Replace with new masonry Refinish surface	20	1 BRIC	S.F.	.083 .232 .014	4.42 .20	3.05 9.60 .54		3.05 14.02 .74	4.06 17.45 .92	5 21 1.12
	Total				.329	4.62	13.19		17.81	**22.43**	**27.12**
0040	**Replace masonry steps, painted** Remove masonry steps Clean work area Install new masonry steps	50	2 BRIC	S.F.	4.000 .015 .891	83	147 .56 37.50	.15	147 .71 120.50	195 .90 141	241 1.08 165
	Total				4.907	83	185.06	.15	268.21	**336.90**	**407.08**

C2013 130   Metal Steps

						2014 Bare Costs				
	System Description	Freq. (Years)	Crew	Unit	Labor Hours	Material	Labor	Equipment	Total	Total In-House	Total w/O&P
0010	**Repair metal steps** Remove damaged tread Replace with cast iron tread, 1/2" thick	15	1 SSWK	S.F.	.347 .120	49.17	17.70 6.14		17.70 55.31	27 63.50	32.50 72.50
	Total				.467	49.17	23.84		73.01	**90.50**	**105**

C20 STAIRS C2013 Stair Construction

C2013 130 Metal Steps

						2014 Bare Costs				Total In-House	Total w/O&P
	System Description	Freq. (Years)	Crew	Unit	Labor Hours	Material	Labor	Equipment	Total		
0030	**Refinish metal steps**	9	1 PORD	S.F.							
	Prepare steps for painting				.026		1.03		1.03	1.33	1.65
	Wipe surface				.001		.05		.05	.06	.07
	Refinish surface, roller + brush, 1 coat				.025	.38	.99		1.37	1.70	2.07
	Total				.052	.38	2.07		2.45	3.09	3.79
0040	**Replace metal steps**	50	2 SSWK	S.F.							
	Remove steps				.264		9.74		9.74	12.85	15.85
	Install pre-erected steel steps				.163	118	8.40	.72	127.12	143	164
	Prepare steps for painting				.026		1.03		1.03	1.33	1.65
	Wipe surface				.001		.05		.05	.06	.07
	Refinish surface				.025	.38	.99		1.37	1.70	2.07
	Total				.479	118.38	20.21	.72	139.31	158.94	183.64

C2013 435 Metal Interior Stair Railing

						2014 Bare Costs				Total In-House	Total w/O&P
	System Description	Freq. (Years)	Crew	Unit	Labor Hours	Material	Labor	Equipment	Total		
0030	**Refinish metal stair railing**	7	1 PORD	S.F.							
	Prepare surface				.019		.76		.76	.98	1.21
	Refinish surface, roller + brush, 1 coat				.015	.11	.61		.72	.91	1.12
	Total				.035	.11	1.37		1.48	1.89	2.33
0040	**Replace metal stair railing**	45	2 SSWK	L.F.							
	Remove railing				.098		5.05	.43	5.48	8.10	9.75
	Install new railing				.195	20	10.10	.87	30.97	38	44.50
	Total				.293	20	15.15	1.30	36.45	46.10	54.25

C2013 437 Iron Interior Stair Railing

						2014 Bare Costs				Total In-House	Total w/O&P
	System Description	Freq. (Years)	Crew	Unit	Labor Hours	Material	Labor	Equipment	Total		
0030	**Refinish wrought iron stair railing**	7	1 PORD	L.F.							
	Prepare surface				.019		.76		.76	.98	1.21
	Refinish surface, roller + brush, 1 coat				.015	.11	.61		.72	.91	1.12
	Total				.035	.11	1.37		1.48	1.89	2.33

C20 STAIRS   C2013 Stair Construction

C2013 437 Iron Interior Stair Railing

	System Description	Freq. (Years)	Crew	Unit	Labor Hours	2014 Bare Costs				Total In-House	Total w/O&P
						Material	Labor	Equipment	Total		
0040	Replace wrought iron stair railing	45	2 SSWK	L.F.	.087		3.18		3.18	4.23	5.20
	Remove railing				1.300	28	66.50		94.50	132	157
	Install new railing										
	Total				1.387	28	69.68		97.68	136.23	162.20

C2013 440 Wood Steps

	System Description	Freq. (Years)	Crew	Unit	Labor Hours	2014 Bare Costs				Total In-House	Total w/O&P
						Material	Labor	Equipment	Total		
0030	Refinish wood steps	3	1 PORD	S.F.	.026		1.03		1.03	1.33	1.65
	Prepare surface				.026	.07	1.03		1.10	1.41	1.74
	Refinish surface, roller + brush, 1 coat										
	Total				.052	.07	2.06		2.13	2.74	3.39
0040	Replace wood steps	40	1 CARP	S.F.	.172		6.29		6.29	8.40	10.40
	Remove steps				.325	81	14.90		95.90	109	126
	Replace steps										
	Total				.497	81	21.19		102.19	117.40	136.40

C2013 445 Wood Interior Stair Railing

	System Description	Freq. (Years)	Crew	Unit	Labor Hours	2014 Bare Costs				Total In-House	Total w/O&P
						Material	Labor	Equipment	Total		
0030	Refinish wood stair railing	7	1 PORD	L.F.	.026		1.03		1.03	1.33	1.65
	Prepare railings for refinishing				.026	.07	1.03		1.10	1.41	1.74
	Refinish surface, roller + brush, 1 coat										
	Total				.052	.07	2.06		2.13	2.74	3.39
0040	Replace wood stair railing	40	1 CARP	L.F.	.087		3.18		3.18	4.23	5.20
	Remove railing				.130	2.50	5.95		8.45	10.70	12.95
	Replace railing										
	Total				.217	2.50	9.13		11.63	14.93	18.15

C20 STAIRS

C2023 Stair Finishes

C2023 110 Carpeted Steps

	System Description	Freq. (Years)	Crew	Unit	Labor Hours	2014 Bare Costs				Total In-House	Total w/O&P
						Material	Labor	Equipment	Total		
0020	**Replace carpeted steps**	8	2 TILF	S.F.							
	Remove old carpet				.010		.38		.38	.51	.63
	Install new carpet				.069	14.69	2.89		17.58	19.80	23
	Additional labor				.229	29.37	9.60		38.97	44.50	52
	Total				.308	44.06	12.87		56.93	64.81	75.63

C2023 120 Rubber Steps

	System Description	Freq. (Years)	Crew	Unit	Labor Hours	2014 Bare Costs				Total In-House	Total w/O&P
						Material	Labor	Equipment	Total		
0020	**Replace rubber steps**	18	2 TILF	L.F.							
	Remove rubber covering				.015		.54		.54	.73	.89
	Prepare floor for installation				.063		2.66		2.66	3.36	4.19
	Install new rubber tread				.090	14.10	3.79		17.89	20.50	23.50
	Install new rubber riser				.042	6.65	1.75		8.40	9.50	11.05
	Total				.210	20.75	8.74		29.49	34.09	39.63

C2023 130 Terrazzo Steps

	System Description	Freq. (Years)	Crew	Unit	Labor Hours	2014 Bare Costs				Total In-House	Total w/O&P
						Material	Labor	Equipment	Total		
0020	**Replace terrazzo steps**	50	2 MSTZ	S.F.							
	Remove old terrazzo tile				.052		1.91		1.91	2.54	3.13
	Clean area				.020		.91		.91	1.22	1.50
	Replace with new tread				.743	2.93	28.50	13.50	44.93	52.50	63
	Total				.815	2.93	31.32	13.50	47.75	56.26	67.63

C-30 INTERIOR FINISHES | C3013 | Wall Finishes

C3013 106 | Vinyl Wall Covering

	System Description	Freq. (Years)	Crew	Unit	Labor Hours	2014 Bare Costs				Total In-House	Total w/O&P
						Material	Labor	Equipment	Total		
0010	**Repair med. wt. vinyl wall covering - (2% of walls)**	1	1 PORD	C.S.F.							
	Remove damaged vinyl wall covering				1.156		46		46	60	74
	Install vinyl wall covering				2.391	90	95		185	222	266
	Total				3.547	90	141		231	282	**340**
0020	**Replace medium weight vinyl wall covering**	15	1 PORD	C.S.F.							
	Set up and secure scaffold				.750		34.50		34.50	46	56.50
	Remove vinyl wall covering				1.156		46		46	60	74
	Install vinyl wall covering				2.391	90	95		185	222	266
	Remove scaffold				.750		34.50		34.50	46	56.50
	Total				5.047	90	210		300	374	**453**

C3013 202 | Wallpaper

	System Description	Freq. (Years)	Crew	Unit	Labor Hours	2014 Bare Costs				Total In-House	Total w/O&P
						Material	Labor	Equipment	Total		
0010	**Repair wallpaper - (2% of walls)**	8	1 PORD	S.Y.							
	Remove damaged wallpaper				.104		4.14		4.14	5.40	6.65
	Repair / replace wallpaper				.175	8.82	6.93		15.75	18.70	22
	Total				.279	8.82	11.07		19.89	24.10	**28.65**
0020	**Replace wallpaper**	20	1 PORD	S.Y.							
	Set up and secure scaffold				.130		5.95		5.95	7.95	9.80
	Remove damaged wallpaper				.104		4.14		4.14	5.40	6.65
	Replace wallpaper				.175	8.82	6.93		15.75	18.70	22
	Remove scaffold				.130		5.95		5.95	7.95	9.80
	Total				.539	8.82	22.97		31.79	40	**48.25**

C30 INTERIOR FINISHES

C3013 Wall Finishes

C3013 206 Fabric Interior Wall Finish

						2014 Bare Costs				
	System Description	Freq. (Years)	Crew	Unit	Labor Hours	Material	Labor	Equipment	Total	Total In-House	Total w/O&P
0010	**Repair fabric wall finish**	9	1 PORD	S.Y.							
	Set up, secure and take down ladder				.130		5.15		5.15	6.70	8.30
	Remove fabric wall covering				.104		4.14		4.14	5.40	6.65
	Repair / replace fabric				.234	5.94	9.27		15.21	18.65	22.50
	Total				.468	5.94	18.56		24.50	**30.75**	**37.45**
0020	**Replace fabric wall finish**	50	1 PORD	S.Y.							
	Set up and secure scaffold				1.350		62.10		62.10	82.50	102
	Remove fabric wall covering				.104		4.14		4.14	5.40	6.65
	Replace fabric				.234	5.94	9.27		15.21	18.65	22.50
	Remove scaffold				1.350		62.10		62.10	82.50	102
	Total				3.038	5.94	137.61		143.55	**189.05**	**233.15**

C3013 208 Cork Tile

						2014 Bare Costs				
	System Description	Freq. (Years)	Crew	Unit	Labor Hours	Material	Labor	Equipment	Total	Total In-House	Total w/O&P
0010	**Repair 12″ x 12″ x 3/16″ cork tile wall - (2% of walls)**	8	1 CARP	C.S.F.							
	Remove old tiles				.620		23		23	30	37
	Replace with new tile				4.426	505	176		681	785	915
	Total				5.046	505	199		704	**815**	**952**
0020	**Replace 12″ x 12″ x 3/16″ cork tile wall**	16	2 CARP	C.S.F.							
	Set up and secure scaffold				.750		34.50		34.50	46	56.50
	Remove old tiles				.620		23		23	30	37
	Replace with new tile				4.426	505	176		681	785	915
	Remove scaffold				.750		34.50		34.50	46	56.50
	Total				6.546	505	268		773	**907**	**1,065**

C30 INTERIOR FINISHES C3013 Wall Finishes

C3013 210 Acoustical Tile

	System Description	Freq. (Years)	Crew	Unit	Labor Hours	2014 Bare Costs				Total In-House	Total w/O&P
						Material	Labor	Equipment	Total		
0010	**Repair acoustical tile - (2% of walls)**	25	1 CARP	C.S.F.							
	Remove damaged tile				1.067		39		39	52	64
	Install new tile				2.737	237	125		362	430	500
	Total				3.804	237	164		401	**482**	**564**
0030	**Refinish acoustical tile**	10	1 PORD	C.S.F.							
	Wipe surface				.894	6	35		41	53	64.50
	Prepare surface				.894	6	35		41	53	64.50
	Refinish surface				1.529	25	60		85	106	128
	Total				3.317	37	130		167	**212**	**257**
0040	**Replace acoustical tile**	60	2 CARP	C.S.F.							
	Set up and secure scaffold				.750		34.50		34.50	46	56.50
	Remove old tiles				1.067		39		39	52	64
	Install new tile				2.737	237	125		362	430	500
	Remove scaffold				.750		34.50		34.50	46	56.50
	Total				5.304	237	233		470	**574**	**677**

C3013 212 Stucco

	System Description	Freq. (Years)	Crew	Unit	Labor Hours	2014 Bare Costs				Total In-House	Total w/O&P
						Material	Labor	Equipment	Total		
0010	**Repair stucco wall - (2% of walls)**	20	1 BRIC	S.Y.							
	Prepare surface				.023		.90		.90	1.17	1.44
	Repair stucco				.532	3.60	21.33	1.89	26.82	33.50	41
	Place and remove mask and drops				.067		2.70		2.70	3.42	4.14
	Total				.621	3.60	24.93	1.89	30.42	**38.09**	**46.58**
0030	**Refinish stucco wall**	4	1 PORD	S.Y.							
	Wash surface				.023		.90		.90	1.17	1.44
	Prepare surface				.033		1.35		1.35	1.71	2.07
	Refinish surface, brushwork, 2 coats				.138	2.25	5.40		7.65	9.55	11.55
	Place and remove mask and drops				.033		1.35		1.35	1.71	2.07
	Total				.227	2.25	9		11.25	**14.14**	**17.13**

C30 INTERIOR FINISHES

C3013 Wall Finishes

C3013 212 Stucco

	System Description	Freq. (Years)	Crew	Unit	Labor Hours	2014 Bare Costs				Total In-House	Total w/O&P
						Material	Labor	Equipment	Total		
0040	**Replace stucco wall**	75	2 BRIC	S.Y.							
	Set up and secure scaffold				.130		5.95		5.95	7.95	9.80
	Remove stucco				.279		12.78		12.78	17.10	21
	Replace stucco				1.200	6.55	48	3.59	58.14	73	89
	Remove scaffold				.130		5.95		5.95	7.95	9.80
	Total				1.739	6.55	72.68	3.59	82.82	106	129.60
4120	**Spray refinish stucco wall**	5	1 PORD	S.Y.							
	Wash surface				.023		.90		.90	1.17	1.44
	Prepare surface				.033		1.35		1.35	1.71	2.07
	Refinish surface, spray, 2 coats				.049	2.52	1.98		4.50	5.30	6.30
	Place and remove mask and drops				.033		1.35		1.35	1.71	2.07
	Total				.138	2.52	5.58		8.10	9.89	11.88

C3013 213 Plaster

	System Description	Freq. (Years)	Crew	Unit	Labor Hours	2014 Bare Costs				Total In-House	Total w/O&P
						Material	Labor	Equipment	Total		
0010	**Repair plaster wall - (2% of walls)**	13	1 PLAS	S.Y.							
	Remove damage				.468		17.15		17.15	23	28
	Replace two coat plaster finish, incl. lath				.752	6.05	30	2.25	38.30	48	58.50
	Total				1.220	6.05	47.15	2.25	55.45	71	86.50
0030	**Refinish plaster wall**	4	1 PORD	S.Y.							
	Prepare surface				.033		1.35		1.35	1.71	2.07
	Paint / seal surface, brushwork, 1 coat				.063	.54	2.52		3.06	3.87	4.64
	Place and remove mask and drops				.033		1.35		1.35	1.71	2.07
	Total				.129	.54	5.22		5.76	7.29	8.78
0040	**Replace plaster wall**	75	2 PLAS	S.Y.							
	Set up, secure and take down ladder				.130		5.95		5.95	7.95	9.80
	Remove material				.468		17.15		17.15	23	28
	Replace two coat plaster wall, incl. lath				.752	6.05	30	2.25	38.30	48	58.50
	Total				1.350	6.05	53.10	2.25	61.40	78.95	96.30

C-30 INTERIOR FINISHES   C3013   Wall Finishes

C3013 214   Drywall

						2014 Bare Costs				Total In-House	Total w/O&P
	System Description	Freq. (Years)	Crew	Unit	Labor Hours	Material	Labor	Equipment	Total		
0010	**Repair 5/8" drywall - (2% of walls)**	20	1 CARP	S.F.							
	Remove damage				.008		.29		.29	.39	.48
	Replace 5/8" drywall, taped and finished				.022	.32	.99		1.31	1.67	2.02
	Total				.030	.32	1.28		1.60	**2.06**	**2.50**
0030	**Refinish drywall**	4	1 PORD	S.F.							
	Place and remove mask and drops				.004		.15		.15	.19	.23
	Prepare surface				.004		.15		.15	.19	.23
	Paint surface, brushwork, 1 coat				.007	.06	.28		.34	.43	.52
	Total				.014	.06	.58		.64	**.81**	**.98**
0040	**Replace 5/8" drywall**	75	2 CARP	S.F.							
	Set up, secure and take down ladder				.014		.66		.66	.88	1.09
	Remove drywall				.008		.29		.29	.39	.48
	Replace 5/8" drywall, taped and finished				.022	.32	.99		1.31	1.67	2.02
	Total				.044	.32	1.94		2.26	**2.94**	**3.59**
0050	**Office painting, 10' x 12', 10' high walls**	5	1 PORD	Ea.							
	Spread drop cloths				.004		.15		.15	.19	.23
	Prepare drywall partitions				1.628		66		66	83.50	101
	Clean drywall partitions				.220	2.02	8.07		10.09	13	15.70
	Paint drywall partitions, roller + brush, 1 coat				3.062	26.40	123.20		149.60	189	227
	Remove drop cloths				.004		.15		.15	.19	.23
	Total				4.918	28.42	197.57		225.99	**285.88**	**344.16**
0060	**Office painting, 10' x 15', 10' high walls**	5	1 PORD	Ea.							
	Spread drop cloths				.005		.19		.19	.24	.29
	Prepare drywall partitions				1.850		75		75	95	115
	Clean drywall partitions				.250	2.29	9.18		11.47	14.75	17.85
	Paint drywall partitions, roller + brush, 1 coat				3.480	30	140		170	215	258
	Remove drop cloths				.005		.19		.19	.24	.29
	Total				5.589	32.29	224.56		256.85	**325.23**	**391.43**

C30 INTERIOR FINISHES | C3013 Wall Finishes

C3013 214 Drywall

	System Description	Freq. (Years)	Crew	Unit	Labor Hours	2014 Bare Costs				Total In-House	Total w/O&P
						Material	Labor	Equipment	Total		
2030	**Refinish drywall, 12' to 24' high**	5	2 PORD	S.F.							
	Set up and secure scaffold				.015		.69		.69	.92	1.13
	Place and remove mask and drops				.004		.15		.15	.19	.23
	Prepare surface				.004		.15		.15	.19	.23
	Paint surface, roller + brushwork, 1 coat				.007	.06	.28		.34	.43	.52
	Remove scaffold				.015		.69		.69	.92	1.13
	Total				.044	.06	1.96		2.02	**2.65**	**3.24**
3030	**Refinish drywall, over 24' high**	5	2 PORD	S.F.							
	Set up and secure scaffold				.023		1.04		1.04	1.37	1.70
	Place and remove mask and drops				.004		.15		.15	.19	.23
	Prepare surface				.004		.15		.15	.19	.23
	Paint surface, roller + brushwork, 1 coat				.007	.06	.28		.34	.43	.52
	Remove scaffold				.023		1.04		1.04	1.37	1.70
	Total				.059	.06	2.66		2.72	**3.55**	**4.38**

C3013 215 Fiberglass Panels, Rigid

	System Description	Freq. (Years)	Crew	Unit	Labor Hours	2014 Bare Costs				Total In-House	Total w/O&P
						Material	Labor	Equipment	Total		
0010	**Repair glass cloth fiberglass panels - (2% of walls)**	9	1 CARP	C.S.F.							
	Remove damaged fiberglass panels				2.600		95		95	127	157
	Install new fiberglass panel				6.710	760	308		1,068	1,250	1,450
	Total				9.310	760	403		1,163	**1,377**	**1,607**
0040	**Replace glass cloth fiberglass panels**	35	2 CARP	C.S.F.							
	Set up and secure scaffold				.750		34.50		34.50	46	56.50
	Remove old panels				2.600		95		95	127	157
	Remove old furring				.516		19		19	25	31
	Install new furring				2.101	42	96		138	174	211
	Install new glass cloth fiberglass panels				6.710	760	308		1,068	1,250	1,450
	Remove scaffold				.750		34.50		34.50	46	56.50
	Total				13.427	802	587		1,389	**1,668**	**1,962**

C30 INTERIOR FINISHES   C3013   Wall Finishes

C3013 220   Tile

	System Description	Freq. (Years)	Crew	Unit	Labor Hours	2014 Bare Costs				Total In-House	Total w/O&P
						Material	Labor	Equipment	Total		
0010	**Repair 4" x 4" thin set ceramic tile - (2% of walls)**	10	1 TILF	C.S.F.							
	Remove damaged tiles				2.600		95		95	127	157
	Replace tile / grout				10.947	210	412		622	750	915
	Total				13.547	210	507		717	877	1,072
0020	**Replace 4" x 4" thin set ceramic tile**	75	1 TILF	C.S.F.							
	Set up and secure scaffold				.750		34.50		34.50	46	56.50
	Remove tiles				2.600		95		95	127	157
	Replace tile / grout				10.947	210	412		622	750	915
	Remove scaffold				.750		34.50		34.50	46	56.50
	Total				15.047	210	576		786	969	1,185

C3013 230   Plywood Paneling

	System Description	Freq. (Years)	Crew	Unit	Labor Hours	2014 Bare Costs				Total In-House	Total w/O&P
						Material	Labor	Equipment	Total		
0010	**Repair plywood paneling - (2% of walls)**	10	1 CARP	C.S.F.							
	Remove damaged paneling				1.156		42		42	56	70
	Replace paneling				4.952	83	227		310	395	475
	Total				6.108	83	269		352	451	545
0020	**Refinish plywood paneling**	10	1 PORD	C.S.F.							
	Place mask and drops				.500		20		20	26	32
	Sand & paint, brushwork, 1 coat				.990	6	39		45	58	70.50
	Remove mask and drops				.500		20		20	26	32
	Total				1.990	6	79		85	110	134.50
0040	**Replace plywood paneling**	30	2 CARP	C.S.F.							
	Set up and secure scaffold				.750		34.50		34.50	46	56.50
	Remove damaged paneling				1.156		42		42	56	70
	Replace paneling				4.952	83	227		310	395	475
	Remove scaffold				.750		34.50		34.50	46	56.50
	Total				7.608	83	338		421	543	658

C30 INTERIOR FINISHES

C3013 Wall Finishes

C3013 234 Wainscot

	System Description	Freq. (Years)	Crew	Unit	Labor Hours	2014 Bare Costs Material	2014 Bare Costs Labor	2014 Bare Costs Equipment	2014 Bare Costs Total	Total In-House	Total w/O&P
0010	**Repair wainscot - (2% of walls)**	10	1 CARP	C.S.F.							
	Remove damaged wainscot				1.733		64		64	85	104
	Replace wainscot				16.000	1,565	735		2,300	2,700	3,150
	Total				17.733	1,565	799		2,364	2,785	3,254
0020	**Refinish wainscot**	4	1 PORD	C.S.F.							
	Place mask and drops				.500		20		20	26	32
	Sand & paint, brushwork, 1 coat				.990	6	39		45	58	70.50
	Remove mask and drops				.500		20		20	26	32
	Total				1.990	6	79		85	110	134.50
0040	**Replace wainscot**	40	2 CARP	C.S.F.							
	Remove wainscot				1.733		64		64	85	104
	Replace wainscot				16.000	1,565	735		2,300	2,700	3,150
	Total				17.733	1,565	799		2,364	2,785	3,254

C3013 240 Stainless Steel Interior Finish

	System Description	Freq. (Years)	Crew	Unit	Labor Hours	2014 Bare Costs Material	2014 Bare Costs Labor	2014 Bare Costs Equipment	2014 Bare Costs Total	Total In-House	Total w/O&P
0020	**Replace stainless steel wall**	75	2 SSWK	C.S.F.							
	Set up, secure and take down ladder				1.500		69		69	91.50	113
	Remove existing stainless steel				2.342		86		86	114	141
	Replace stainless steel				13.419	425	525		950	1,250	1,475
	Total				17.261	425	680		1,105	1,455.50	1,729

C3013 410 Fireplace Mantles, Wood

	System Description	Freq. (Years)	Crew	Unit	Labor Hours	2014 Bare Costs Material	2014 Bare Costs Labor	2014 Bare Costs Equipment	2014 Bare Costs Total	Total In-House	Total w/O&P
0030	**Refinish wood surface - fireplaces**	7	1 PORD	S.F.							
	Prepare surface				.003		.13		.13	.17	.21
	Varnish, 3 coats, brushwork, sanding included				.025	.28	.97		1.25	1.57	1.91
	Total				.028	.28	1.10		1.38	1.74	2.12

C30 INTERIOR FINISHES — C3023 Floor Finishes

C3023 112 Concrete, Finished

	System Description	Freq. (Years)	Crew	Unit	Labor Hours	2014 Bare Costs				Total In-House	Total w/O&P
						Material	Labor	Equipment	Total		
0020	**Refinish concrete floor** Add topping to existing floor 1"	25	2 CEFI	C.S.F.	6.933	52	275	49	376	465	560
	Total				6.933	52	275	49	376	**465**	**560**

C3023 405 Epoxy Flooring

	System Description	Freq. (Years)	Crew	Unit	Labor Hours	2014 Bare Costs				Total In-House	Total w/O&P
						Material	Labor	Equipment	Total		
0020	**Replace epoxy flooring** Strip existing flooring Repair and seal floor	15	2 CEFI	C.S.F.	3.200 12.735	 397	147 487	 18	147 902	196 1,100	241 1,300
	Total				15.935	397	634	18	1,049	**1,296**	**1,541**

C3023 410 Vinyl Tile

	System Description	Freq. (Years)	Crew	Unit	Labor Hours	2014 Bare Costs				Total In-House	Total w/O&P
						Material	Labor	Equipment	Total		
0020	**Replace vinyl tile flooring** Remove damaged floor tile Prepare surface Install new tiles	18	1 TILF	S.Y.	.234 .645 .187	 7.74	8.55 24.30 7.83		8.55 24.30 15.57	11.45 30.50 18.45	14.15 38.50 22
	Total				1.067	7.74	40.68		48.42	**60.40**	**74.65**

C3023 412 Vinyl Sheet

	System Description	Freq. (Years)	Crew	Unit	Labor Hours	2014 Bare Costs				Total In-House	Total w/O&P
						Material	Labor	Equipment	Total		
0020	**Replace vinyl sheet flooring** Remove damaged floor tile Prepare surface Install new vinyl sheet	18	1 TILF	S.Y.	.234 .645 .407	 36.90	8.55 24.30 17.10		8.55 24.30 54	11.45 30.50 62	14.15 38.50 73
	Total				1.286	36.90	49.95		86.85	**103.95**	**125.65**

C30 INTERIOR FINISHES   C3023   Floor Finishes

C3023 414   Rubber Tile

	System Description	Freq. (Years)	Crew	Unit	Labor Hours	2014 Bare Costs				Total In-House	Total w/O&P
						Material	Labor	Equipment	Total		
0020	**Replace rubber tile floor**	18	1 TILF	S.Y.							
	Remove damaged floor tile				.234		8.55		8.55	11.45	14.15
	Prepare surface				.645		24.30		24.30	30.50	38.50
	Install new tiles				1.040	110.25	43.65		153.90	176	207
	Total				1.920	110.25	76.50		186.75	217.95	259.65

C3023 418   Rubber / Vinyl Trim

	System Description	Freq. (Years)	Crew	Unit	Labor Hours	2014 Bare Costs				Total In-House	Total w/O&P
						Material	Labor	Equipment	Total		
0010	**Replace rubber cove base**	9	1 TILF	L.F.							
	Remove damaged base				.017		.64		.64	.85	1.04
	Clean up debris				.001		.01		.01	.01	.01
	Install new base				.025	1.26	1.07		2.33	2.74	3.26
	Total				.044	1.26	1.72		2.98	3.60	4.31

C3023 420   Ceramic Tile

	System Description	Freq. (Years)	Crew	Unit	Labor Hours	2014 Bare Costs				Total In-House	Total w/O&P
						Material	Labor	Equipment	Total		
0010	**Ceramic tile floor repairs - (2% of floors)**	15	1 TILF	C.S.F.							
	Regrout ceramic tile floors				16.640	15	625		640	810	1,000
	Total				16.640	15	625		640	810	1,000
0020	**Replace 2" x 2" thin set ceramic tile floor**	50	1 TILF	C.S.F.							
	Remove damaged floor tile				2.600		95		95	127	157
	Prepare surface				7.172		270		270	340	425
	Install new tiles				10.947	580	412		992	1,150	1,375
	Total				20.719	580	777		1,357	1,617	1,957

C30 INTERIOR FINISHES C3023 Floor Finishes

C3023 428 Ceramic Trim

	System Description	Freq. (Years)	Crew	Unit	Labor Hours	2014 Bare Costs				Total In-House	Total w/O&P
						Material	Labor	Equipment	Total		
0020	**Replace ceramic trim**	50	2 TILF	L.F.							
	Remove ceramic tile trim				.015		.55		.55	.73	.90
	Clean up debris				.010		.46		.46	.60	.74
	Install new ceramic tile trim				.229	3.68	8.60		12.28	14.90	18.15
	Total				.253	3.68	9.61		13.29	16.23	19.79

C3023 430 Terrazzo

	System Description	Freq. (Years)	Crew	Unit	Labor Hours	2014 Bare Costs				Total In-House	Total w/O&P
						Material	Labor	Equipment	Total		
0100	**Terrazzo floor repairs - (2% of floors)**	15	1 MSTZ	S.F.							
	Chip existing terrazzo				.089		3.27		3.27	4.35	5.35
	Place terrazzo				.166	3.08	6.35	3.02	12.45	14.45	17.15
	Remove debris				.004		.16		.16	.21	.26
	Total				.260	3.08	9.78	3.02	15.88	19.01	22.76
0200	**Replace terrazzo floor**	75	2 MSTZ	C.S.F.							
	Break-up existing terrazzo				5.943		218		218	290	360
	Remove debris				.433		16		16	21.50	26
	Load into truck				.650		24.25	14	38.25	46.50	55.50
	Place terrazzo				16.640	308	635	302	1,245	1,450	1,725
	Total				23.666	308	893.25	316	1,517.25	1,808	2,166.50

C3023 438 Terrazzo Trim

	System Description	Freq. (Years)	Crew	Unit	Labor Hours	2014 Bare Costs				Total In-House	Total w/O&P
						Material	Labor	Equipment	Total		
0020	**Replace precast terrazzo trim**	75	2 MSTZ	L.F.							
	Remove old terrazzo trim				.023		.85		.85	1.13	1.39
	Clean area				.010		.46		.46	.60	.74
	Install new precast terrazzo trim				.416	16.85	17.35		34.20	40.50	48.50
	Total				.449	16.85	18.66		35.51	42.23	50.63

C30 INTERIOR FINISHES C3023 Floor Finishes

C3023 440 Quarry Tile

System Description	Freq. (Years)	Crew	Unit	Labor Hours	2014 Bare Costs Material	2014 Bare Costs Labor	2014 Bare Costs Equipment	Total	Total In-House	Total w/O&P
0010	Quarry tile floor repairs - (2% of floors)	15	1 TILF	S.F.							
	Regrout quarry tile floor				.166	.15	6.25		6.40	8.10	10.05
	Total				.166	.15	6.25		6.40	8.10	10.05
0020	Replace quarry tile floor	50	2 TILF	S.F.							
	Remove damaged floor tile				.026		.95		.95	1.27	1.57
	Prepare surface				.072		2.70		2.70	3.41	4.25
	Install new tiles				.109	5.80	4.12		9.92	11.60	13.75
	Total				.207	5.80	7.77		13.57	16.28	19.57

C3023 450 Brick

System Description	Freq. (Years)	Crew	Unit	Labor Hours	Material	Labor	Equipment	Total	Total In-House	Total w/O&P
0020	Replace thickset brick floor	60	2 BRIC	S.F.							
	Remove brick				.035		1.27		1.27	1.69	2.09
	Prepare 2" thick mortar bed				.260	.83	10.75		11.58	15	18.50
	Install new brick				.219	8.25	8.25		16.50	19.50	23.50
	Total				.514	9.08	20.27		29.35	36.19	44.09

C3023 460 Marble

System Description	Freq. (Years)	Crew	Unit	Labor Hours	Material	Labor	Equipment	Total	Total In-House	Total w/O&P
0020	Replace thinset marble floor	50	2 TILF	S.F.							
	Remove marble				.035		1.27		1.27	1.69	2.09
	Prepare floor				.260	.83	10.75		11.58	15	18.50
	Replace marble flooring				.347	9	13.05		22.05	26.50	32
	Total				.641	9.83	25.07		34.90	43.19	52.59

C30 INTERIOR FINISHES | C3023 | Floor Finishes

C3023 470 | Plywood

	System Description	Freq. (Years)	Crew	Unit	Labor Hours	2014 Bare Costs				Total In-House	Total w/O&P
						Material	Labor	Equipment	Total		
0020	**Replace plywood floor**	40	2 CARP	C.S.F.							
	Remove existing flooring				1.733		79		79	106	131
	Install new plywood flooring 3/4"				1.600	140	73		213	252	296
	Total				3.333	140	152		292	358	427

C3023 472 | Wood Parquet

	System Description	Freq. (Years)	Crew	Unit	Labor Hours	2014 Bare Costs				Total In-House	Total w/O&P
						Material	Labor	Equipment	Total		
0020	**Sand and refinish parquet floor**	10	1 CARP	S.F.							
	Sand floor				.035	.88	1.29		2.17	2.69	3.22
	Refinish existing floor, brushwork, 2 coats				.026	1.31	.95		2.26	2.71	3.21
	Total				.061	2.19	2.24		4.43	5.40	6.43
0030	**Replace 5/16" oak parquet floor**	40	2 CARP	S.F.							
	Remove flooring				.023		1.06		1.06	1.41	1.74
	Repair underlayment				.014	.97	.64		1.61	1.92	2.26
	Install new parquet flooring				.065	4.14	2.98		7.12	8.50	10.10
	Total				.102	5.11	4.68		9.79	11.83	14.10

C3023 474 | Maple Strip

	System Description	Freq. (Years)	Crew	Unit	Labor Hours	2014 Bare Costs				Total In-House	Total w/O&P
						Material	Labor	Equipment	Total		
0020	**Sand and refinish maple strip floor**	10	1 CARP	S.F.							
	Sand floor				.026	1.31	.95		2.26	2.71	3.21
	Refinish maple strip floor, brushwork, 2 coats				.035	.88	1.29		2.17	2.69	3.22
	Total				.061	2.19	2.24		4.43	5.40	6.43
0030	**Replace maple floor**	40	2 CARP	S.F.							
	Remove flooring				.023		1.06		1.06	1.41	1.74
	Repair underlayment				.014	.97	.64		1.61	1.92	2.26
	Install new maple flooring				.061	5.75	2.81		8.56	10.05	11.80
	Total				.098	6.72	4.51		11.23	13.38	15.80

C30 INTERIOR FINISHES

C3023 Floor Finishes

C3023 478 Wood Trim

						2014 Bare Costs			Total In-House	Total w/O&P
	System Description	Freq. (Years)	Crew	Unit	Labor Hours	Material	Labor	Equipment	Total		
0010	**Repair 1" x 3" wood trim**	13	1 CARP	L.F.							
	Remove wood				.021		.76		.76	1.02	1.25
	Install new trim				.173	4.56	7.96		12.52	15.60	18.75
	Total				.194	4.56	8.72		13.28	**16.62**	**20**
0030	**Refinish 1" x 3" wood trim**	7	1 PORD	L.F.							
	Prepare surface				.003		.13		.13	.17	.21
	Varnish, 3 coats, brushwork, sanding included				.025	.28	.97		1.25	1.57	1.91
	Total				.028	.28	1.10		1.38	**1.74**	**2.12**
0040	**Replace 1" x 3" wood trim**	75	2 CARP	L.F.							
	Remove wood				.021		.76		.76	1.02	1.25
	Install new trim				.173	4.56	7.96		12.52	15.60	18.75
	Prepare surface				.003		.13		.13	.17	.21
	Varnish, 3 coats, brushwork, sanding included				.025	.28	.97		1.25	1.57	1.91
	Total				.222	4.84	9.82		14.66	**18.36**	**22.12**

C3023 510 Carpet

						2014 Bare Costs			Total In-House	Total w/O&P
	System Description	Freq. (Years)	Crew	Unit	Labor Hours	Material	Labor	Equipment	Total		
0020	**Replace carpet**	8	2 TILF	S.Y.							
	Remove damaged carpet				.094		3.42		3.42	4.59	5.65
	Install new carpet				.138	38	5.80		43.80	49	56.50
	Total				.232	38	9.22		47.22	**53.59**	**62.15**

C30 INTERIOR FINISHES

C3033 Ceiling Finishes

C3033 105 Plaster

0010 Repair plaster ceiling - (2% of ceilings)

	Freq. (Years)	Crew	Unit	Labor Hours	2014 Bare Costs				Total In-House	Total w/O&P
					Material	Labor	Equipment	Total		
Set up, secure and take down ladder	12	1 PLAS	S.Y.	.130		5.95		5.95	7.95	9.80
Remove damaged ceiling				.267		9.81		9.81	13.05	16.10
Replace plaster, 2 coats on lath				.752	5.65	30	2.25	37.90	47.50	58
Total				1.149	5.65	45.76	2.25	53.66	68.50	83.90

0030 Refinish plaster ceiling

	Freq. (Years)	Crew	Unit	Labor Hours	Material	Labor	Equipment	Total	Total In-House	Total w/O&P
Set up, secure and take down ladder	10	1 PORD	S.Y.	.130		5.95		5.95	7.95	9.80
Place mask and drops				.033		1.35		1.35	1.71	2.07
Sand & paint				.089	.54	3.51		4.05	5.20	6.35
Total				.252	.54	10.81		11.35	14.86	18.22

0040 Replace plaster ceiling

	Freq. (Years)	Crew	Unit	Labor Hours	Material	Labor	Equipment	Total	Total In-House	Total w/O&P
Set up and secure scaffold	75	2 PLAS	S.Y.	.130		5.95		5.95	7.95	9.80
Remove ceiling				.267		9.81		9.81	13.05	16.10
Replace plaster, 2 coats on lath				.752	5.65	30	2.25	37.90	47.50	58
Remove scaffold				.130		5.95		5.95	7.95	9.80
Total				1.279	5.65	51.71	2.25	59.61	76.45	93.70

C3033 107 Gypsum Wall Board

0010 Repair gypsum board ceiling - (2% of ceilings)

	Freq. (Years)	Crew	Unit	Labor Hours	2014 Bare Costs				Total In-House	Total w/O&P
					Material	Labor	Equipment	Total		
Set up, secure and take down ladder	20	1 CARP	C.S.F.	1.300		59.50		59.50	79.50	98
Remove damaged gypsum board				2.737		100		100	134	165
Replace 5/8" gypsum board, taped and finished				3.382	41	155		196	252	305
Total				7.419	41	314.50		355.50	465.50	568

0020 Refinish gypsum board ceiling, up to 12' high

	Freq. (Years)	Crew	Unit	Labor Hours	Material	Labor	Equipment	Total	Total In-House	Total w/O&P
Set up, secure and take down ladder	20	1 PORD	C.S.F.	1.300		59.50		59.50	79.50	98
Wash surface				.250		10		10	13	16
Sand & paint				.990	6	39		45	58	70.50
Place and remove mask and drops				.370		15		15	19	23
Total				2.910	6	123.50		129.50	169.50	207.50

C30 INTERIOR FINISHES   C3033   Ceiling Finishes

C3033 107   Gypsum Wall Board

	System Description	Freq. (Years)	Crew	Unit	Labor Hours	2014 Bare Costs				Total In-House	Total w/O&P
						Material	Labor	Equipment	Total		
0040	**Replace gypsum board ceiling, up to 12' high**	40	2 CARP	C.S.F.							
	Set up and secure scaffold				.750		34.50		34.50	46	56.50
	Remove damaged gypsum board				2.737		100		100	134	165
	Replace 5/8" gypsum board ceiling, taped and finished				3.382	41	155		196	252	305
	Remove scaffold				.750		34.50		34.50	46	56.50
	Total				7.619	41	324		365	478	583
2020	**Refinish gypsum board ceiling, 12' to 24' high**	5	2 PORD	C.S.F.							
	Set up and secure scaffold				1.500		69		69	91.50	113
	Place and remove mask and drops				.370		15		15	19	23
	Prepare surface				.370		15		15	19	23
	Paint surface, roller + brushwork, 1 coat				.696	6	28		34	43	51.50
	Remove scaffold				1.500		69		69	91.50	113
	Total				4.436	6	196		202	264	323.50
3020	**Refinish gypsum board ceiling, over 24' high**	5	2 PORD	C.S.F.							
	Set up and secure scaffold				2.250		103.50		103.50	137	170
	Place and remove mask and drops				.370		15		15	19	23
	Prepare surface				.370		15		15	19	23
	Paint surface, roller + brushwork, 1 coat				.696	6	28		34	43	51.50
	Remove scaffold				2.250		103.50		103.50	137	170
	Total				5.936	6	265		271	355	437.50

C3033 108   Acoustic Tile

	System Description	Freq. (Years)	Crew	Unit	Labor Hours	2014 Bare Costs				Total In-House	Total w/O&P
						Material	Labor	Equipment	Total		
0010	**Acoustic tile repairs - (2% of ceilings)**	9	1 CARP	C.S.F.							
	Set up, secure and take down ladder				1.300		59.50		59.50	79.50	98
	Remove damaged tile				1.067		39		39	52	64
	Install new tile				2.737	237	125		362	430	500
	Total				5.104	237	223.50		460.50	561.50	662

C30 INTERIOR FINISHES

C3033 Ceiling Finishes

C3033 108 Acoustic Tile

		System Description	Freq. (Years)	Crew	Unit	Labor Hours	2014 Bare Costs				Total In-House	Total w/O&P
							Material	Labor	Equipment	Total		
0020		**Replace acoustic tile ceiling, non fire-rated**	20	1 CARP	C.S.F.							
		Set up, secure and take down scaffold				.750		34.50		34.50	46	56.50
		Remove old ceiling tiles				.500		18		18	24	30
		Remove old ceiling grid				.625		23		23	31	38
		Install new ceiling grid				1.000	70	46		116	138	163
		Install new ceiling tiles				1.280	125	59		184	216	252
		Sweep and clean debris				.520		26.50		26.50	34.50	43
		Total				4.675	195	207		402	489.50	582.50
0030		**Refinish acoustic tile ceiling/grid (unoccupied area)**	5	2 PORD	C.S.F.							
		Protect exposed surfaces and mask light fixtures				.800		31.50		31.50	41	51
		Refinish acoustic tiles, roller + brush, 1 coat				.267	.01	10.50		10.51	13.70	16.90
		Total				1.067	.01	42		42.01	54.70	67.90
0040		**Refinish acoustic tile ceiling/grid (occupied area)**	5	2 PORD	C.S.F.							
		Protect exposed surfaces and mask light fixtures				3.200		127		127	164	203
		Refinish acoustic tiles, roller + brush, 1 coat				.320	.01	12.70		12.71	16.40	20.50
		Total				3.520	.01	139.70		139.71	180.40	223.50
2030		**Refinish acoustic tile ceiling, 12' to 24' high**	5	2 PORD	C.S.F.							
		Set up and secure scaffold				1.500		69		69	91.50	113
		Place and remove mask and drops				.100		2		2	2.60	3
		Paint surface, roller + brushwork, 1 coat				.267	.01	10.50		10.51	13.70	16.90
		Remove scaffold				1.500		69		69	91.50	113
		Total				3.367	.01	150.50		150.51	199.30	245.90
3030		**Refinish acoustic tile ceiling, over 24' high**	5	2 PORD	C.S.F.							
		Set up and secure scaffold				2.250		103.50		103.50	137	170
		Place and remove mask and drops				.100		2		2	2.60	3
		Paint surface, roller + brushwork, 1 coat				.267	.01	10.50		10.51	13.70	16.90
		Remove scaffold				2.250		103.50		103.50	137	170
		Total				4.867	.01	219.50		219.51	290.30	359.90

C30 INTERIOR FINISHES

C3033 Ceiling Finishes

C3033 109 Acoustic Tile, Fire-Rated

	System Description	Freq. (Years)	Crew	Unit	Labor Hours	2014 Bare Costs - Material	2014 Bare Costs - Labor	2014 Bare Costs - Equipment	2014 Bare Costs - Total	Total In-House	Total w/O&P
0010	**Replace acoustic tile ceiling, fire-rated**	20	1 CARP	C.S.F.							
	Set up, secure and take down scaffold				.750		34.50		34.50	46	56.50
	Remove old ceiling tiles				.500		18		18	24	30
	Remove old grid system				.625		23		23	31	38
	Install new grid system				1.200	84	55.20		139.20	166	195
	Install new ceiling tiles				1.185	95	54		149	177	208
	Sweep & clean debris				.520		26.50		26.50	34.50	43
	Total				4.780	179	211.20		390.20	478.50	570.50

C3033 120 Wood

	System Description	Freq. (Years)	Crew	Unit	Labor Hours	2014 Bare Costs - Material	2014 Bare Costs - Labor	2014 Bare Costs - Equipment	2014 Bare Costs - Total	Total In-House	Total w/O&P
0010	**Repair wood ceiling - (2% of ceilings)**	10	1 CARP	C.S.F.							
	Set up, secure and take down ladder				1.300		59.50		59.50	79.50	98
	Remove damaged ceiling				1.455		53		53	71	88
	Replace wood				5.200	149	238		387	480	580
	Total				7.955	149	350.50		499.50	630.50	766
0030	**Refinish wood ceiling**	6	1 PORD	C.S.F.							
	Set up and secure scaffold				.375		17.25		17.25	23	28.50
	Wash surface				.286		11		11	15	18
	Sand & paint, brushwork, 1 coat				.990		39	6	45	58	70.50
	Place and remove mask and drops				.370		15		15	19	23
	Remove scaffold				.375		17.25		17.25	23	28.50
	Total				2.396		99.50	6	105.50	138	168.50
0040	**Replace wood ceiling**	50	2 CARP	C.S.F.							
	Set up and secure scaffold				.750		34.50		34.50	46	56.50
	Remove damaged ceiling				1.455		53		53	71	88
	Replace wood				5.200	149	238		387	480	580
	Remove scaffold				.750		34.50		34.50	46	56.50
	Total				8.155	149	360		509	643	781

D20 PLUMBING D2013 Plumbing Fixtures

D2013 110 Tankless Water Closet

	System Description	Freq. (Years)	Crew	Unit	Labor Hours	2014 Bare Costs				Total In-House	Total w/O&P
						Material	Labor	Equipment	Total		
0010	**Replace flush valve diaphragm**	10	1 PLUM	Ea.							
	Turn valve off and on				.010		.60		.60	.75	.93
	Remove / replace diaphragm				.300	3.41	17.25		20.66	25.50	31.50
	Check for leaks				.013		.75		.75	.94	1.17
	Total				.323	3.41	18.60		22.01	27.19	33.60
0015	**Rebuild flush valve**	20	1 PLUM	Ea.							
	Turn valve off and on				.010		.60		.60	.75	.93
	Remove flush valve				.100		5.75		5.75	7.20	9
	Rebuild valve				.350	59.50	20		79.50	90.50	106
	Reinstall valve				1.000		57.50		57.50	72	90
	Check operation				.024		1.38		1.38	1.73	2.16
	Total				1.484	59.50	85.23		144.73	172.18	208.09
0020	**Unplug clogged line**	5	1 PLUM	Ea.							
	Turn valve off and on				.010		.60		.60	.75	.93
	Remove water closet				2.078		120		120	150	187
	Unplug line				.200		11.50		11.50	14.40	17.95
	Reinstall water closet				.510		29.50		29.50	36.50	46
	Check operation				.167		9.60		9.60	12	15
	Total				2.965		171.20		171.20	213.65	266.88
0030	**Replace tankless water closet**	35	2 PLUM	Ea.							
	Turn valve off and on				.010		.60		.60	.75	.93
	Remove flush valve				.100		5.75		5.75	7.20	9
	Remove water closet				.460		26.50		26.50	33	41.50
	Install wall-hung water closet				2.759	755	143		898	1,000	1,175
	Check operation				.150		8.65		8.65	10.80	13.45
	Total				3.479	755	184.50		939.50	1,051.75	1,239.88
0040	**Replace tankless flush valve**	25	1 PLUM	Ea.							
	Turn valve off and on				.010		.60		.60	.75	.93
	Remove flush valve				.100		5.75		5.75	7.20	9
	Install water closet flush valve				1.000	148	57.50		205.50	235	275
	Check operation				.024		1.38		1.38	1.73	2.16
	Total				1.134	148	65.23		213.23	244.68	287.09

D20 PLUMBING D2013 Plumbing Fixtures

D2013 110 Tankless Water Closet

	System Description	Freq. (Years)	Crew	Unit	Labor Hours	2014 Bare Costs				Total In-House	Total w/O&P
						Material	Labor	Equipment	Total		
0050	**Replace wax ring gasket**	5	1 PLUM	Ea.							
	Turn valve off and on				.010		.60		.60	.75	.93
	Remove floor mounted tankless bowl				.667		38.50		38.50	48	60
	Install new wax gasket				.083	1.38	4.80		6.18	7.50	9.25
	Install floor mounted tankless bowl				1.000		57.50		57.50	72	90
	Check operation				.150		8.65		8.65	10.80	13.45
	Total				1.910	1.38	110.05		111.43	**139.05**	**173.63**
0080	**Replace stainless steel detention water closet flush valve actuator**	20	1 SKWK	Ea.							
	Turn valve off and on				.010		.60		.60	.75	.93
	Remove flush valve actuator				.100		5.75		5.75	7.20	9
	Remove flush valve				.100		5.75		5.75	7.20	9
	Detention equipment, replace toilet actuator				1.000	295	47.50		342.50	385	445
	Check operation				.024		1.38		1.38	1.73	2.16
	Total				1.234	295	60.98		355.98	**401.88**	**466.09**
0090	**Replace stainless steel detention water closet flush valve**	20	1 SKWK	Ea.							
	Turn valve off and on				.010		.60		.60	.75	.93
	Remove flush valve				.100		5.75		5.75	7.20	9
	Detention equipment, replace water closet valve				1.000	320	47.50		367.50	415	475
	Check operation				.024		1.38		1.38	1.73	2.16
	Total				1.134	320	55.23		375.23	**424.68**	**487.09**

D2013 130 Flush-Tank Water Closet

	System Description	Freq. (Years)	Crew	Unit	Labor Hours	2014 Bare Costs				Total In-House	Total w/O&P
						Material	Labor	Equipment	Total		
0010	**Unplug clogged line**	5	1 PLUM	Ea.							
	Turn valve off and on				.010		.60		.60	.75	.93
	Remove water closet				2.078		120		120	150	187
	Unplug line				.200		11.50		11.50	14.40	17.95
	Reinstall water closet				.510		29.50		29.50	36.50	46
	Check operation				.167		9.60		9.60	12	15
	Total				2.965		171.20		171.20	**213.65**	**266.88**

D20 PLUMBING D2013 Plumbing Fixtures

D2013 130 Flush-Tank Water Closet

	System Description	Freq. (Years)	Crew	Unit	Labor Hours	2014 Bare Costs				Total In-House	Total w/O&P
						Material	Labor	Equipment	Total		
0020	**Replace washer / diaphragm in ball cock**	5	1 PLUM	Ea.							
	Turn valve on and off				.010		.60		.60	.75	.93
	Remove and replace washer / diaphragm				.180	1.74	10.35		12.09	14.85	18.35
	Check operation				.024		1.38		1.38	1.73	2.16
	Total				.214	1.74	12.33		14.07	17.33	**21.44**
0030	**Replace valve and ball cock assembly**	15	1 PLUM	Ea.							
	Turn valve off and on				.010		.60		.60	.75	.93
	Remove flush valve and ball cock assembly				.300		17.25		17.25	21.50	27
	Install flush valve and ball cock assembly				.610	13.75	35		48.75	59	72
	Check operation				.167		9.60		9.60	12	15
	Total				1.087	13.75	62.45		76.20	93.25	**114.93**
0040	**Install gasket between tank and bowl**	20	1 PLUM	Ea.							
	Turn valve off and on				.010		.60		.60	.75	.93
	Remove tank and gasket				.200		11.50		11.50	14.40	17.95
	Install new gasket and reinstall tank				.300	2.38	17.25		19.63	24	30
	Inspect connection				.009		.52		.52	.65	.82
	Total				.520	2.38	29.87		32.25	39.80	**49.70**
0050	**Replace two piece water closet**	35	2 PLUM	Ea.							
	Turn water off and on				.010		.60		.60	.75	.93
	Remove floor-mounted water closet				1.000		57.50		57.50	72	90
	Install new bowl, tank and seat				3.019	233	156		389	450	535
	Check for leaks				.013		.75		.75	.94	1.17
	Total				4.042	233	214.85		447.85	523.69	**627.10**
0060	**Replace one piece water closet**	35	2 PLUM	Ea.							
	Turn water off and on				.010		.60		.60	.75	.93
	Remove floor-mounted water closet				1.000		57.50		57.50	72	90
	Install floor-mounted water closet with seat				3.019	1,100	156		1,256	1,400	1,625
	Check for leaks				.013		.75		.75	.94	1.17
	Total				4.042	1,100	214.85		1,314.85	1,473.69	**1,717.10**

D20 PLUMBING   D2013 Plumbing Fixtures

D2013 210   Urinal

							2014 Bare Costs				
	System Description	Freq. (Years)	Crew	Unit	Labor Hours	Material	Labor	Equipment	Total	Total In-House	Total w/O&P
0010	**Replace flush valve diaphragm**	7	1 PLUM	Ea.							
	Turn valve off and on				.010		.60		.60	.75	.93
	Remove / replace diaphragm				.300	3.41	17.25		20.66	25.50	31.50
	Check for leaks				.013		.75		.75	.94	1.17
	Total				.323	3.41	18.60		22.01	27.19	**33.60**
0015	**Rebuild flush valve**	20	1 PLUM	Ea.							
	Turn valve off and on				.010		.60		.60	.75	.93
	Remove flush valve				.100		5.75		5.75	7.20	9
	Rebuild valve				.350	59.50	20		79.50	90.50	106
	Reinstall valve				1.000		57.50		57.50	72	90
	Check operation				.024		1.38		1.38	1.73	2.16
	Total				1.484	59.50	85.23		144.73	172.18	**208.09**
0020	**Unplug line**	5	1 PLUM	Ea.							
	Turn valve off and on				.010		.60		.60	.75	.93
	Remove and reinstall bowl				1.600		92		92	115	144
	Unplug line				.200		11.50		11.50	14.40	17.95
	Check operation				.167		9.60		9.60	12	15
	Total				1.977		113.70		113.70	142.15	**177.88**
0030	**Replace wall-hung urinal**	35	2 PLUM	Ea.							
	Turn valve off and on				.010		.60		.60	.75	.93
	Remove urinal				2.667		153		153	192	239
	Fit up urinal to wall				.196		11.30		11.30	14.10	17.60
	Install flushometer pipe				.096		5.50		5.50	6.90	8.60
	Install urinal				5.333	315	276		591	690	825
	Check operation				.334		19.20		19.20	24	30
	Total				8.636	315	465.60		780.60	927.75	**1,121.13**

D20 PLUMBING | D2013 Plumbing Fixtures

D2013 310 Lavatory, Iron, Enamel

						2014 Bare Costs				Total In-House	Total w/O&P
	System Description	Freq. (Years)	Crew	Unit	Labor Hours	Material	Labor	Equipment	Total		
0010	**Replace washer in spud connection**	7	1 PLUM	Ea.							
	Turn valve off and on				.022		1.27		1.27	1.59	1.98
	Remove gasket or washer				.017		.97		.97	1.22	1.52
	Install new gasket / washer set				.026	3.85	1.50		5.35	6.10	7.15
	Install spud connection				.026		1.49		1.49	1.87	2.33
	Inspect connection				.009		.52		.52	.65	.82
	Total				.100	3.85	5.75		9.60	11.43	13.80
0020	**Replace washer in faucet**	2	1 PLUM	Ea.							
	Turn valves on and off				.022		1.27		1.27	1.59	1.98
	Remove valve stem and washer				.091		5.25		5.25	6.55	8.15
	Install washer				.026	.40	1.50		1.90	2.32	2.84
	Install stem				.017		.97		.97	1.22	1.52
	Test faucet				.010		.60		.60	.75	.93
	Total				.166	.40	9.59		9.99	12.43	15.42
0040	**Replace faucets**	10	1 PLUM	Ea.							
	Turn valves on and off				.022		1.27		1.27	1.59	1.98
	Remove faucets and tubing				.381		22		22	27.50	34
	Install detention sink faucets and tubing				1.040	44.50	60		104.50	124	149
	Test faucets				.010		.60		.60	.75	.93
	Total				1.453	44.50	83.87		128.37	153.84	185.91
0050	**Clean out strainer and P trap**	2	1 PLUM	Ea.							
	Remove strainer and P trap				.157		9.05		9.05	11.30	14.10
	Clean and reinstall strainer and trap				.320		18.40		18.40	23	28.50
	Total				.477		27.45		27.45	34.30	42.60
0060	**Replace lavatory**	40	2 PLUM	Ea.							
	Shut off water (hot and cold)				.022		1.27		1.27	1.59	1.98
	Disconnect and remove wall-hung lavatory				.800		46		46	57.50	72
	Install new wall-hung lavatory with trim, (18" x 15")				2.807	400	162		562	640	750
	Total				3.629	400	209.27		609.27	699.09	823.98

D20 PLUMBING   D2013   Plumbing Fixtures

D2013 330   Lavatory, Vitreous China

	System Description	Freq. (Years)	Crew	Unit	Labor Hours	2014 Bare Costs				Total In-House	Total w/O&P
						Material	Labor	Equipment	Total		
0010	**Replace washer in spud connection**	7	1 PLUM	Ea.							
	Turn valve off and on				.022		1.27		1.27	1.59	1.98
	Loosen locknuts				.011		.63		.63	.79	.99
	Remove spud connection				.017		.97		.97	1.22	1.52
	Remove gasket or washer				.017		.97		.97	1.22	1.52
	Clean spud seat				.021		1.20		1.20	1.50	1.87
	Install new gasket / washer set				.026	3.85	1.50		5.35	6.10	7.15
	Install spud connection				.026		1.49		1.49	1.87	2.33
	Tighten locknut				.014		.82		.82	1.03	1.29
	Inspect connection				.009		.52		.52	.65	.82
	Total				.163	3.85	9.37		13.22	15.97	19.47
0020	**Replace washer in faucet**	2	1 PLUM	Ea.							
	Turn valves on and off				.022		1.27		1.27	1.59	1.98
	Remove packing nut and install				.023		1.35		1.35	1.68	2.10
	Remove valve stem and cap				.017		.97		.97	1.22	1.52
	Remove seat washer and install				.023		1.35		1.35	1.68	2.10
	Pry out washer				.017		.97		.97	1.22	1.52
	Install washer				.026	.40	1.50		1.90	2.32	2.84
	Install stem				.017		.97		.97	1.22	1.52
	Tighten main gland valves				.012		.67		.67	.84	1.05
	Test faucet				.010		.60		.60	.75	.93
	Total				.168	.40	9.65		10.05	12.52	15.56
0040	**Replace faucets**	10	1 PLUM	Ea.							
	Turn valves on and off				.022		1.27		1.27	1.59	1.98
	Remove faucets and tubing				.381		22		22	27.50	34
	Install faucets and tubing				1.040	44.50	60		104.50	124	149
	Test faucets				.010		.60		.60	.75	.93
	Total				1.453	44.50	83.87		128.37	153.84	185.91
0050	**Clean out strainer and P trap**	2	1 PLUM	Ea.							
	Remove strainer and P trap				.157		9.05		9.05	11.30	14.10
	Clean and reinstall strainer and trap				.320		18.40		18.40	23	28.50
	Total				.477		27.45		27.45	34.30	42.60

D20 PLUMBING — Plumbing Fixtures

D2013 330 — Lavatory, Vitreous China

	System Description	Freq. (Years)	Crew	Unit	Labor Hours	2014 Bare Costs — Material	2014 Bare Costs — Labor	2014 Bare Costs — Equipment	2014 Bare Costs — Total	Total In-House	Total w/O&P
0060	**Replace lavatory**	35	2 PLUM	Ea.							
	Shut off water (hot and cold)				.022		1.27		1.27	1.59	1.98
	Disconnect and remove wall-hung lavatory				1.485		85.50		85.50	107	133
	Install new lavatory (to 20" x 17")				2.971	187	154		341	400	475
	Install faucets and tubing				1.040	44.50	60		104.50	124	149
	Total				5.518	231.50	300.77		532.27	632.59	758.98

D2013 350 — Lavatory, Enameled Steel

	System Description	Freq. (Years)	Crew	Unit	Labor Hours	2014 Bare Costs — Material	2014 Bare Costs — Labor	2014 Bare Costs — Equipment	2014 Bare Costs — Total	Total In-House	Total w/O&P
0010	**Replace washer in spud connection**	7	1 PLUM	Ea.							
	Turn valve off and on				.022		1.27		1.27	1.59	1.98
	Loosen locknut				.014		.82		.82	1.03	1.29
	Remove spud connection				.017		.97		.97	1.22	1.52
	Remove gasket or washer				.017		.97		.97	1.22	1.52
	Clean spud seat				.021		1.20		1.20	1.50	1.87
	Install new gasket / washer set				.026	3.85	1.50		5.35	6.10	7.15
	Install spud connection				.026		1.49		1.49	1.87	2.33
	Tighten locknut				.014		.82		.82	1.03	1.29
	Inspect connection				.009		.52		.52	.65	.82
	Total				.166	3.85	9.56		13.41	16.21	19.77
0020	**Replace washer in faucet**	2	1 PLUM	Ea.							
	Turn valves on and off				.022		1.27		1.27	1.59	1.98
	Remove valve stem and washer				.091		5.25		5.25	6.55	8.15
	Install washer				.026	.40	1.50		1.90	2.32	2.84
	Install stem				.017		.97		.97	1.22	1.52
	Test faucet				.010		.60		.60	.75	.93
	Total				.166	.40	9.59		9.99	12.43	15.42
0040	**Replace faucets**	10	1 PLUM	Ea.							
	Turn valves on and off				.022		1.27		1.27	1.59	1.98
	Remove faucets and tubing				.381		22		22	27.50	34
	Install faucets and tubing				1.040	44.50	60		104.50	124	149
	Test faucets				.010		.60		.60	.75	.93
	Total				1.453	44.50	83.87		128.37	153.84	185.91

D20 PLUMBING D2013 Plumbing Fixtures

D2013 350 Lavatory, Enameled Steel

	System Description	Freq. (Years)	Crew	Unit	Labor Hours	2014 Bare Costs				Total In-House	Total w/O&P
						Material	Labor	Equipment	Total		
0050	**Clean out strainer and P trap**	2	1 PLUM	Ea.							
	Remove strainer and P trap				.157		9.05		9.05	11.30	14.10
	Clean and reinstall strainer and trap				.320		18.40		18.40	23	28.50
	Total				.477		27.45		27.45	34.30	42.60
0060	**Replace lavatory**	35	2 PLUM	Ea.							
	Shut off water (hot and cold)				.022		1.27		1.27	1.59	1.98
	Disconnect and remove wall-hung lavatory				1.793		103		103	129	161
	Install new lavatory (to 20" x 17")				3.586	153	186		339	400	480
	Install faucets and tubing				1.040	44.50	60		104.50	124	149
	Total				6.441	197.50	350.27		547.77	654.59	791.98

D2013 410 Sink, Iron Enamel

	System Description	Freq. (Years)	Crew	Unit	Labor Hours	2014 Bare Costs				Total In-House	Total w/O&P
						Material	Labor	Equipment	Total		
0010	**Replace faucet washer**	2	1 PLUM	Ea.							
	Turn valves on and off				.022		1.27		1.27	1.59	1.98
	Remove valve stem and washer				.091		5.25		5.25	6.55	8.15
	Install washer				.026	.40	1.50		1.90	2.32	2.84
	Install stem				.017		.97		.97	1.22	1.52
	Test faucet				.010		.60		.60	.75	.93
	Total				.166	.40	9.59		9.99	12.43	15.42
0020	**Clean trap**	3	1 PLUM	Ea.							
	Drain trap				.021		1.21		1.21	1.51	1.89
	Remove / install trap				.043		2.49		2.49	3.11	3.88
	Inspect trap				.031		1.80		1.80	2.25	2.81
	Clean trap				.021		1.20		1.20	1.50	1.87
	Total				.116		6.70		6.70	8.37	10.45
0030	**Replace faucets**	10	1 PLUM	Ea.							
	Turn valves on and off				.022		1.27		1.27	1.59	1.98
	Remove faucets and tubing				.381		22		22	27.50	34
	Install faucets and tubing				1.040	44.50	60		104.50	124	149
	Test faucets				.010		.60		.60	.75	.93
	Total				1.453	44.50	83.87		128.37	153.84	185.91

D20 PLUMBING | D2013 Plumbing Fixtures

D2013 410 Sink, Iron Enamel

	System Description	Freq. (Years)	Crew	Unit	Labor Hours	2014 Bare Costs				Total In-House	Total w/O&P
						Material	Labor	Equipment	Total		
0040	**Unstop sink**		1 PLUM	Ea.							
	Unstop plugged drain	2			.571		33		33	41	51.50
	Total				.571		33		33	**41**	**51.50**
0060	**Replace kitchen sink, P.E.C.I.**		2 PLUM	Ea.							
	Remove PE sink	35			1.625		84		84	105	131
	Install countertop porcelain enameled kitchen sink, 24" x 21"				2.857	275	148		423	490	575
	Total				4.482	275	232		507	**595**	**706**
0070	**Replace laundry sink, P.E.C.I.**		2 PLUM	Ea.							
	Remove PE sink	35			1.625		84		84	105	131
	Install free-standing porcelain enameled laundry sink, 24" x 21"				2.667	580	138		718	810	940
	Total				4.292	580	222		802	**915**	**1,071**

D2013 420 Sink, Enameled Steel

	System Description	Freq. (Years)	Crew	Unit	Labor Hours	2014 Bare Costs				Total In-House	Total w/O&P
						Material	Labor	Equipment	Total		
0010	**Replace faucet washer**		1 PLUM	Ea.							
	Turn valves on and off	2			.022		1.27		1.27	1.59	1.98
	Remove valve stem and washer				.091		5.25		5.25	6.55	8.15
	Install washer				.026	.40	1.50		1.90	2.32	2.84
	Install stem				.017		.97		.97	1.22	1.52
	Test faucet				.010		.60		.60	.75	.93
	Total				.166	.40	9.59		9.99	**12.43**	**15.42**
0020	**Clean trap**		1 PLUM	Ea.							
	Drain trap	3			.021		1.21		1.21	1.51	1.89
	Remove / install trap				.043		2.49		2.49	3.11	3.88
	Inspect trap				.031		1.80		1.80	2.25	2.81
	Clean trap				.021		1.20		1.20	1.50	1.87
	Total				.116		6.70		6.70	**8.37**	**10.45**

D20 PLUMBING

D2013 Plumbing Fixtures

D2013 420 Sink, Enameled Steel

	System Description	Freq. (Years)	Crew	Unit	Labor Hours	2014 Bare Costs				Total In-House	Total w/O&P
						Material	Labor	Equipment	Total		
0030	**Replace faucets**	10	1 PLUM	Ea.							
	Turn valves on and off				.022		1.27		1.27	1.59	1.98
	Remove faucets and tubing				.381		22		22	27.50	34
	Install faucets and tubing				1.040	44.50	60		104.50	124	149
	Test faucets				.010		.60		.60	.75	.93
	Total				1.453	44.50	83.87		128.37	153.84	185.91
0040	**Unstop sink**	2	1 PLUM	Ea.							
	Unstop plugged drain				.571		33		33	41	51.50
	Total				.571		33		33	41	51.50
0060	**Replace sink, enameled steel**	35	2 PLUM	Ea.							
	Remove sink				1.793		93		93	116	145
	Install countertop enameled steel sink, 24" x 21"				3.714	505	192		697	795	930
	Total				5.507	505	285		790	911	1,075

D2013 430 Sink, Stainless Steel

	System Description	Freq. (Years)	Crew	Unit	Labor Hours	2014 Bare Costs				Total In-House	Total w/O&P
						Material	Labor	Equipment	Total		
0010	**Replace faucet washer**	2	1 PLUM	Ea.							
	Turn valves on and off				.022		1.27		1.27	1.59	1.98
	Remove valve stem and washer				.091		5.25		5.25	6.55	8.15
	Install washer				.026	.40	1.50		1.90	2.32	2.84
	Install stem				.017		.97		.97	1.22	1.52
	Test faucet				.010		.60		.60	.75	.93
	Total				.166	.40	9.59		9.99	12.43	15.42
0020	**Clean trap**	3	1 PLUM	Ea.							
	Drain trap				.021		1.21		1.21	1.51	1.89
	Remove / install trap				.043		2.49		2.49	3.11	3.88
	Inspect trap				.031		1.80		1.80	2.25	2.81
	Clean trap				.021		1.20		1.20	1.50	1.87
	Total				.116		6.70		6.70	8.37	10.45

D20 PLUMBING

D2013 Plumbing Fixtures

D2013 430 Sink, Stainless Steel

	System Description	Freq. (Years)	Crew	Unit	Labor Hours	2014 Bare Costs				Total In-House	Total w/O&P
						Material	Labor	Equipment	Total		
0030	**Replace faucets**	10	1 PLUM	Ea.							
	Turn valves on and off				.022		1.27		1.27	1.59	1.98
	Remove faucets and tubing				.381		22		22	27.50	34
	Install faucets and tubing				1.040	44.50	60		104.50	124	149
	Test faucets				.010		.60		.60	.75	.93
	Total				1.453	44.50	83.87		128.37	153.84	185.91
0040	**Unstop sink**	2	1 PLUM	Ea.							
	Unstop plugged drain				.571		33		33	41	51.50
	Total				.571		33		33	41	51.50
0060	**Replace sink, stainless steel**	40	2 PLUM	Ea.							
	Remove sink				1.625		84		84	105	131
	Install countertop stainless steel sink, 25" x 22"				3.714	660	192		852	965	1,125
	Total				5.339	660	276		936	1,070	1,256
0070	**Replace stainless steel detention sink faucets**	15	1 SKWK	Ea.							
	Turn valves on and off				.022		1.27		1.27	1.59	1.98
	Remove faucets and tubing				.381		22		22	27.50	34
	Detention equipment, replace fuacet valve and tubing				1.000	620	47.50		667.50	745	850
	Test faucets				.010		.60		.60	.75	.93
	Total				1.413	620	71.37		691.37	774.84	886.91

D2013 440 Sink, Plastic

	System Description	Freq. (Years)	Crew	Unit	Labor Hours	2014 Bare Costs				Total In-House	Total w/O&P
						Material	Labor	Equipment	Total		
0010	**Replace faucet washer**	2	1 PLUM	Ea.							
	Turn valves on and off				.022		1.27		1.27	1.59	1.98
	Remove valve stem and washer				.091		5.25		5.25	6.55	8.15
	Install washer				.026	.40	1.50		1.90	2.32	2.84
	Install stem				.017		.97		.97	1.22	1.52
	Test faucet				.010		.60		.60	.75	.93
	Total				.166	.40	9.59		9.99	12.43	15.42

D20 PLUMBING D2013 Plumbing Fixtures

D2013 440 Sink, Plastic

	System Description	Freq. (Years)	Crew	Unit	Labor Hours	2014 Bare Costs				Total In-House	Total w/O&P
						Material	Labor	Equipment	Total		
0020	**Clean trap**	3	1 PLUM	Ea.							
	Drain trap				.021		1.21		1.21	1.51	1.89
	Remove / install trap				.043		2.49		2.49	3.11	3.88
	Inspect trap				.031		1.80		1.80	2.25	2.81
	Clean trap				.021		1.20		1.20	1.50	1.87
	Total				.116		6.70		6.70	**8.37**	**10.45**
0030	**Replace faucets**	10	1 PLUM	Ea.							
	Turn valves on and off				.022		1.27		1.27	1.59	1.98
	Remove faucets and tubing				.381		22		22	27.50	34
	Install faucets and tubing				1.040	44.50	60		104.50	124	149
	Test faucets				.010		.60		.60	.75	.93
	Total				1.453	44.50	83.87		128.37	**153.84**	**185.91**
0040	**Unstop, sink**	2	1 PLUM	Ea.							
	Unstop plugged drain				.571		33		33	41	51.50
	Total				.571		33		33	**41**	**51.50**
0060	**Replace sink and fittings, polyethylene**	15	2 PLUM	Ea.							
	Remove sink				3.467		180		180	224	280
	Install bench mounted polyethylene sink, 47" x 24"				6.932	1,200	360		1,560	1,775	2,050
	Total				10.399	1,200	540		1,740	**1,999**	**2,330**
0080	**Replace lab sink and fittings, polyethylene**	15	2 PLUM	Ea.							
	Remove sink				3.467		180		180	224	280
	Install bench mounted PE lab sink, 70" x 24"				5.333	1,875	276		2,151	2,400	2,775
	Total				8.800	1,875	456		2,331	**2,624**	**3,055**

D20 PLUMBING

D2013 Plumbing Fixtures

D2013 444 Laboratory Sink

	System Description	Freq. (Years)	Crew	Unit	Labor Hours	2014 Bare Costs				Total In-House	Total w/O&P
						Material	Labor	Equipment	Total		
0010	**Replace faucet washer**	2	1 PLUM	Ea.							
	Turn valves on and off				.022		1.27		1.27	1.59	1.98
	Remove valve stem and washer				.091		5.25		5.25	6.55	8.15
	Install washer				.026	.40	1.50		1.90	2.32	2.84
	Install stem				.017		.97		.97	1.22	1.52
	Test faucet				.010		.60		.60	.75	.93
	Total				.166	.40	9.59		9.99	12.43	15.42
0030	**Replace faucets**	10	1 PLUM	Ea.							
	Turn valves on and off				.022		1.27		1.27	1.59	1.98
	Remove faucets and tubing				.381		22		22	27.50	34
	Install faucets and tubing				1.040	44.50	60		104.50	124	149
	Test faucets				.010		.60		.60	.75	.93
	Total				1.453	44.50	83.87		128.37	153.84	185.91
0060	**Replace sink and fittings, polyethylene, lab.**	15	2 PLUM	Ea.							
	Remove lab. sink				3.467		180		180	224	280
	Install bench mounted polyethylene lab. sink, 47" x 24"				5.333	1,875	276		2,151	2,400	2,775
	Total				8.800	1,875	456		2,331	2,624	3,055

D2013 450 Laundry Sink, Plastic

	System Description	Freq. (Years)	Crew	Unit	Labor Hours	2014 Bare Costs				Total In-House	Total w/O&P
						Material	Labor	Equipment	Total		
0020	**Replace washer in faucet**	2	1 PLUM	Ea.							
	Turn valves on and off				.022		1.27		1.27	1.59	1.98
	Remove valve stem and washer				.091		5.25		5.25	6.55	8.15
	Install washer				.026	.40	1.50		1.90	2.32	2.84
	Install stem				.017		.97		.97	1.22	1.52
	Test faucet				.010		.60		.60	.75	.93
	Total				.166	.40	9.59		9.99	12.43	15.42

D20 PLUMBING D2013 Plumbing Fixtures

D2013 450 Laundry Sink, Plastic

	System Description	Freq. (Years)	Crew	Unit	Labor Hours	2014 Bare Costs				Total In-House	Total w/O&P
						Material	Labor	Equipment	Total		
0040	**Replace faucets**	10	1 PLUM	Ea.							
	Turn valves on and off				.022		1.27		1.27	1.59	1.98
	Remove faucets and tubing				.381		22		22	27.50	34
	Install faucets and tubing				1.040	44.50	60		104.50	124	149
	Test faucets				.010		.60		.60	.75	.93
	Total				1.453	44.50	83.87		128.37	153.84	**185.91**
0050	**Clean out strainer and P trap**	2	1 PLUM	Ea.							
	Remove strainer and P trap				.157		9.05		9.05	11.30	14.10
	Clean and reinstall strainer and trap				.320		18.40		18.40	23	28.50
	Total				.477		27.45		27.45	34.30	**42.60**
0060	**Replace laundry sink**	20	2 PLUM	Ea.							
	Shut off water (hot and cold)				.022		1.27		1.27	1.59	1.98
	Disconnect and remove laundry sink				1.600		83		83	104	129
	Install sink (23" x 18", single)				3.200	135	166		301	355	430
	Install faucets and tubing				1.040	44.50	60		104.50	124	149
	Total				5.862	179.50	310.27		489.77	584.59	**709.98**

D2013 455 Laundry Sink, Stone

	System Description	Freq. (Years)	Crew	Unit	Labor Hours	2014 Bare Costs				Total In-House	Total w/O&P
						Material	Labor	Equipment	Total		
0060	**Replace laundry sink, molded stone**	40	2 PLUM	Ea.							
	Remove molded stone sink				1.625		84		84	105	131
	Install laundry sink, w/trim, molded stone, 22"x23",sgl compt				2.667	134	138		272	320	385
	Total				4.292	134	222		356	425	**516**

D20 PLUMBING

D2013 Plumbing Fixtures

D2013 460 Service/Utility Sink

	System Description	Freq. (Years)	Crew	Unit	Labor Hours	2014 Bare Costs				Total In-House	Total w/O&P
						Material	Labor	Equipment	Total		
0010	**Replace faucet washer**	2	1 PLUM	Ea.							
	Turn valves on and off				.022		1.27		1.27	1.59	1.98
	Remove valve stem and washer				.091		5.25		5.25	6.55	8.15
	Install washer				.026	.40	1.50		1.90	2.32	2.84
	Install stem				.017		.97		.97	1.22	1.52
	Test faucet				.010		.60		.60	.75	.93
	Total				.166	.40	9.59		9.99	**12.43**	**15.42**
0020	**Clean trap**	3	1 PLUM	Ea.							
	Drain trap				.021		1.21		1.21	1.51	1.89
	Remove / install trap				.043		2.49		2.49	3.11	3.88
	Inspect trap				.031		1.80		1.80	2.25	2.81
	Clean trap				.021		1.20		1.20	1.50	1.87
	Total				.116		6.70		6.70	**8.37**	**10.45**
0030	**Replace faucets**	10	1 PLUM	Ea.							
	Turn valves on and off				.022		1.27		1.27	1.59	1.98
	Remove faucets and tubing				.381		22		22	27.50	34
	Install faucets and tubing				1.040	44.50	60		104.50	124	149
	Test faucets				.010		.60		.60	.75	.93
	Total				1.453	44.50	83.87		128.37	**153.84**	**185.91**
0040	**Unstop sink**	2	1 PLUM	Ea.							
	Unstop plugged drain				.571		33		33	41	51.50
	Total				.571		33		33	**41**	**51.50**
0060	**Replace service sink, P.E.C.I.**	35	1 PLUM	Ea.							
	Remove sink				1.625		84		84	105	131
	Install service/utility sink, 22" x 18"				5.212	750	270		1,020	1,150	1,350
	Total				6.837	750	354		1,104	**1,255**	**1,481**

D20 PLUMBING | D2013 Plumbing Fixtures

D2013 470 Group Washfountain

	System Description	Freq. (Years)	Crew	Unit	Labor Hours	2014 Bare Costs Material	2014 Bare Costs Labor	2014 Bare Costs Equipment	2014 Bare Costs Total	Total In-House	Total w/O&P
0060	**Replace group wash fountain, 54" diameter**	20	Q-2	Ea.							
	Shut off water (hot and cold)				.022		1.27		1.27	1.59	1.98
	Disconnect and remove group wash fountain, 54" dia.				4.800		258		258	320	400
	Install group wash fountain, terrazzo, 54" diameter				9.600	9,025	515		9,540	10,600	12,100
	Install faucets and tubing				1.040	44.50	60		104.50	124	149
	Total				15.462	9,069.50	834.27		9,903.77	11,045.59	12,650.98

D2013 510 Bathtub, Cast Iron Enamel

	System Description	Freq. (Years)	Crew	Unit	Labor Hours	2014 Bare Costs Material	2014 Bare Costs Labor	2014 Bare Costs Equipment	2014 Bare Costs Total	Total In-House	Total w/O&P
0010	**Inspect / clean shower head**	3	1 PLUM	Ea.							
	Turn water off and on				.010		.60		.60	.75	.93
	Remove shower head				.217		12.45		12.45	15.60	19.45
	Check shower head				.009		.52		.52	.65	.82
	Install shower head				.433		25		25	31	39
	Total				.669		38.57		38.57	48	60.20
0020	**Replace mixing valve barrel**	2	1 PLUM	Ea.							
	Turn valves on and off				.022		1.27		1.27	1.59	1.98
	Remove mixing valve cover / barrel				.650		37.50		37.50	47	58.50
	Install new mixing valve barrel / replace cover				.650	78.50	37.50		116	133	157
	Test valve				.010		.60		.60	.75	.93
	Total				1.332	78.50	76.87		155.37	182.34	218.41
0030	**Replace mixing valve**	10	1 PLUM	Ea.							
	Turn valves off and on				.022		1.27		1.27	1.59	1.98
	Remove mixing valve				.867		50		50	62.50	78
	Install new mixing valve				1.733	153	100		253	293	345
	Test valve				.010		.60		.60	.75	.93
	Total				2.633	153	151.87		304.87	357.84	425.91
0070	**Replace tub**	40	2 PLUM	Ea.							
	Remove cast iron tub				1.891		98		98	122	153
	Install cast iron enameled bathtub				3.782	1,100	196		1,296	1,450	1,675
	Total				5.673	1,100	294		1,394	1,572	1,828

D20 PLUMBING

D2013 Plumbing Fixtures

D2013 530 Bathtub, Enameled Steel

	System Description	Freq. (Years)	Crew	Unit	Labor Hours	2014 Bare Costs				Total In-House	Total w/O&P
						Material	Labor	Equipment	Total		
0010	**Inspect / clean shower head**	3	1 PLUM	Ea.							
	Turn water off and on				.010		.60		.60	.75	.93
	Remove shower head				.217		12.45		12.45	15.60	19.45
	Check shower head				.009		.52		.52	.65	.82
	Install shower head				.433		25		25	31	39
	Total				.669		38.57		38.57	48	60.20
0020	**Replace mixing valve barrel**	2	1 PLUM	Ea.							
	Turn valves on and off				.022		1.27		1.27	1.59	1.98
	Remove mixing valve cover / barrel				.650		37.50		37.50	47	58.50
	Install new mixing valve barrel / replace cover				.650	78.50	37.50		116	133	157
	Test valve				.010		.60		.60	.75	.93
	Total				1.332	78.50	76.87		155.37	182.34	218.41
0030	**Replace mixing valve**	10	1 PLUM	Ea.							
	Turn valves off and on				.022		1.27		1.27	1.59	1.98
	Remove mixing valve				.867		50		50	62.50	78
	Install new mixing valve				1.733	153	100		253	293	345
	Test valve				.010		.60		.60	.75	.93
	Total				2.633	153	151.87		304.87	357.84	425.91
0070	**Replace tub**	35	2 PLUM	Ea.							
	Remove tub				1.891		98		98	122	153
	Install enameled steel bathtub				3.782	495	196		691	790	925
	Total				5.673	495	294		789	912	1,078

D2013 550 Bathtub, Fiberglass

	System Description	Freq. (Years)	Crew	Unit	Labor Hours	2014 Bare Costs				Total In-House	Total w/O&P
						Material	Labor	Equipment	Total		
0010	**Inspect / clean shower head**	3	1 PLUM	Ea.							
	Turn water off and on				.010		.60		.60	.75	.93
	Remove shower head				.217		12.45		12.45	15.60	19.45
	Check shower head				.009		.52		.52	.65	.82
	Install shower head				.433		25		25	31	39
	Total				.669		38.57		38.57	48	60.20

D20 PLUMBING		D2013	Plumbing Fixtures									
D2013 550			**Bathtub, Fiberglass**									

	System Description	Freq. (Years)	Crew	Unit	Labor Hours	2014 Bare Costs				Total In-House	Total w/O&P
						Material	Labor	Equipment	Total		
0020	**Replace mixing valve barrel**	2	1 PLUM	Ea.							
	Turn valves on and off				.022		1.27		1.27	1.59	1.98
	Remove mixing valve cover / barrel				.650		37.50		37.50	47	58.50
	Install new mixing valve barrel / replace cover				.650	78.50	37.50		116	133	157
	Test valve				.010		.60		.60	.75	.93
	Total				1.332	78.50	76.87		155.37	182.34	218.41
0030	**Replace mixing valve**	10	1 PLUM	Ea.							
	Turn valves off and on				.022		1.27		1.27	1.59	1.98
	Remove mixing valve				.867		50		50	62.50	78
	Install new mixing valve				1.733	153	100		253	293	345
	Test valve				.010		.60		.60	.75	.93
	Total				2.633	153	151.87		304.87	357.84	425.91
0070	**Replace tub**	20	2 PLUM	Ea.							
	Remove bathtub				2.600		135		135	168	210
	Install fiberglass bathtub with shower wall surround				5.200	815	269		1,084	1,225	1,450
	Total				7.800	815	404		1,219	1,393	1,660

D2013 710 **Shower, Terrazzo**

	System Description	Freq. (Years)	Crew	Unit	Labor Hours	2014 Bare Costs				Total In-House	Total w/O&P
						Material	Labor	Equipment	Total		
0010	**Inspect / clean shower head**	3	1 PLUM	Ea.							
	Turn water off and on				.010		.60		.60	.75	.93
	Remove shower head				.217		12.45		12.45	15.60	19.45
	Check shower head				.009		.52		.52	.65	.82
	Install shower head				.433		25		25	31	39
	Total				.669		38.57		38.57	48	60.20
0020	**Replace mixing valve barrel**	2	1 PLUM	Ea.							
	Turn valves on and off				.022		1.27		1.27	1.59	1.98
	Remove mixing valve cover / barrel				.650		37.50		37.50	47	58.50
	Install new mixing valve barrel / replace cover				.650	78.50	37.50		116	133	157
	Test valve				.010		.60		.60	.75	.93
	Total				1.332	78.50	76.87		155.37	182.34	218.41

D20 PLUMBING

D2013 Plumbing Fixtures

D2013 710 Shower, Terrazzo

	System Description	Freq. (Years)	Crew	Unit	Labor Hours	2014 Bare Costs				Total In-House	Total w/O&P
						Material	Labor	Equipment	Total		
0030	**Replace mixing valve**	10	1 PLUM	Ea.							
	Turn valves off and on				.022		1.27		1.27	1.59	1.98
	Remove mixing valve				.867		50		50	62.50	78
	Install new mixing valve				1.733	153	100		253	293	345
	Test valve				.010		.60		.60	.75	.93
	Total				2.633	153	151.87		304.87	357.84	425.91
0060	**Replace terrazzo shower surface**	30	2 PLUM	Ea.							
	Remove old terrazzo surface				3.004		109.85		109.85	147	181
	Install precast terrazzo tiles				9.014	304.85	377		681.85	815	975
	Install curb				2.080	112.50	87		199.50	234	277
	Total				14.098	417.35	573.85		991.20	1,196	1,433

D2013 730 Shower, Enameled Steel

	System Description	Freq. (Years)	Crew	Unit	Labor Hours	2014 Bare Costs				Total In-House	Total w/O&P
						Material	Labor	Equipment	Total		
0010	**Inspect / clean shower head**	3	1 PLUM	Ea.							
	Turn water off and on				.010		.60		.60	.75	.93
	Remove shower head				.217		12.45		12.45	15.60	19.45
	Check shower head				.009		.52		.52	.65	.82
	Install shower head				.433		25		25	31	39
	Total				.669		38.57		38.57	48	60.20
0020	**Replace mixing valve barrel**	2	1 PLUM	Ea.							
	Turn valves on and off				.022		1.27		1.27	1.59	1.98
	Remove mixing valve cover / barrel				.650		37.50		37.50	47	58.50
	Install new mixing valve barrel / replace cover				.650	78.50	37.50		116	133	157
	Test valve				.010		.60		.60	.75	.93
	Total				1.332	78.50	76.87		155.37	182.34	218.41
0030	**Replace mixing valve**	10	1 PLUM	Ea.							
	Turn valves off and on				.022		1.27		1.27	1.59	1.98
	Remove mixing valve				.867		50		50	62.50	78
	Install new mixing valve				1.733	153	100		253	293	345
	Test valve				.010		.60		.60	.75	.93
	Total				2.633	153	151.87		304.87	357.84	425.91

D20 PLUMBING D2013 Plumbing Fixtures

D2013 730 Shower, Enameled Steel

	System Description	Freq. (Years)	Crew	Unit	Labor Hours	2014 Bare Costs				Total In-House	Total w/O&P
						Material	Labor	Equipment	Total		
0060	**Replace shower**	35	2 PLUM	Ea.							
	Remove shower				5.200		269		269	335	420
	Install shower				4.522	1,150	234		1,384	1,550	1,800
	Total				9.722	1,150	503		1,653	1,885	2,220

D2013 750 Shower, Fiberglass

	System Description	Freq. (Years)	Crew	Unit	Labor Hours	2014 Bare Costs				Total In-House	Total w/O&P
						Material	Labor	Equipment	Total		
0010	**Inspect / clean shower head**	3	1 PLUM	Ea.							
	Turn water off and on				.010		.60		.60	.75	.93
	Remove shower head				.217		12.45		12.45	15.60	19.45
	Check shower head				.009		.52		.52	.65	.82
	Install shower head				.433		25		25	31	39
	Total				.669		38.57		38.57	48	60.20
0020	**Replace mixing valve barrel**	2	1 PLUM	Ea.							
	Turn valves on and off				.022		1.27		1.27	1.59	1.98
	Remove mixing valve cover / barrel				.650		37.50		37.50	47	58.50
	Install new mixing valve barrel / replace cover				.650	78.50	37.50		116	133	157
	Test valve				.010		.60		.60	.75	.93
	Total				1.332	78.50	76.87		155.37	182.34	218.41
0030	**Replace mixing valve**	10	1 PLUM	Ea.							
	Turn valves off and on				.022		1.27		1.27	1.59	1.98
	Remove mixing valve				.867		50		50	62.50	78
	Install new mixing valve				1.733	153	100		253	293	345
	Test valve				.010		.60		.60	.75	.93
	Total				2.633	153	151.87		304.87	357.84	425.91
0060	**Replace shower and fittings**	20	2 PLUM	Ea.							
	Remove shower				4.334		224		224	281	350
	Install shower, fiberglass				3.302	505	171		676	770	900
	Total				7.635	505	395		900	1,051	1,250

D20 PLUMBING — D2013 Plumbing Fixtures

D2013 770 Shower, Misc.

	System Description	Freq. (Years)	Crew	Unit	Labor Hours	2014 Bare Costs — Material	2014 Bare Costs — Labor	2014 Bare Costs — Equipment	2014 Bare Costs — Total	Total In-House	Total w/O&P
0010	**Inspect / clean shower head**	3	1 PLUM	Ea.							
	Turn water off and on				.010		.60		.60	.75	.93
	Remove shower head				.217		12.45		12.45	15.60	19.45
	Check shower head				.009		.52		.52	.65	.82
	Install shower head				.433		25		25	31	39
	Total				.669		38.57		38.57	48	60.20
0020	**Replace mixing valve barrel**	2	1 PLUM	Ea.							
	Turn valves on and off				.022		1.27		1.27	1.59	1.98
	Remove mixing valve cover / barrel				.650		37.50		37.50	47	58.50
	Install new mixing valve barrel / replace cover				.650	78.50	37.50		116	133	157
	Test valve				.010		.60		.60	.75	.93
	Total				1.332	78.50	76.87		155.37	182.34	218.41
0030	**Replace mixing valve**	10	1 PLUM	Ea.							
	Turn valves on and off				.022		1.27		1.27	1.59	1.98
	Remove mixing valve				.867		50		50	62.50	78
	Install new mixing valve				1.733	153	100		253	293	345
	Test valve				.010		.60		.60	.75	.93
	Total				2.633	153	151.87		304.87	357.84	425.91
0060	**Replace shower and fittings, aluminum**	25	2 PLUM	Ea.							
	Remove shower				4.334		224		224	281	350
	Install shower				3.302	505	171		676	770	900
	Total				7.635	505	395		900	1,051	1,250
0070	**Replace shower head with water conserving head**	10	1 PLUM	Ea.							
	Remove shower head				.217		12.45		12.45	15.60	19.45
	Install new water conserving head				.400	85	23		108	122	142
	Turn water on and off				.010		.60		.60	.75	.93
	Total				.627	85	36.05		121.05	138.35	162.38
0120	**Replace shower, C.M.U.**	20	D-8	Ea.							
	Remove C.M.U. shower enclosure				3.595		133.28	20.72	154	198	242
	Install C.M.U. shower enclosure				6.772	141.68	285.60		427.28	530	640
	Install built-in head, arm and valve				2.000	80	115		195	232	280
	Total				12.367	221.68	533.88	20.72	776.28	960	1,162

D20 PLUMBING | D2013 Plumbing Fixtures

D2013 770 Shower, Misc.

	System Description	Freq. (Years)	Crew	Unit	Labor Hours	2014 Bare Costs				Total In-House	Total w/O&P
						Material	Labor	Equipment	Total		
0200	**Replace shower, glazed C.M.U.**	25	D-8	Ea.							
	Remove C.M.U. shower enclosure				4.697		174		201.50	259	315
	Install glazed C.M.U. shower enclosure				9.393	534.80	394.80		929.60	1,100	1,325
	Install built-in head, arm and valve				2.000	80	115		195	232	280
	Total				16.090	614.80	683.80	27.50	1,326.10	1,591	1,920
0280	**Replace shower surface, ceramic tile**	30	D-7	Ea.							
	Remove ceramic tile shower enclosure surface				3.066		112		112	150	185
	Install ceramic tile in shower enclosure				6.130	117.60	230.72		348.32	420	510
	Install built-in head, arm and valve				2.000	80	115		195	232	280
	Total				11.197	197.60	457.72		655.32	802	975

D2013 810 Drinking Fountain

	System Description	Freq. (Years)	Crew	Unit	Labor Hours	2014 Bare Costs				Total In-House	Total w/O&P
						Material	Labor	Equipment	Total		
0010	**Check / minor repairs**	1	1 PLUM	Ea.							
	Check unit, make minor adjustments				.667		38.50		38.50	48	60
	Total				.667		38.50		38.50	48	60
0020	**Repair internal leaks**	4	1 PLUM	Ea.							
	Open unit, repair leaks, close				.615		35.50		35.50	44.50	55.50
	Total				.615		35.50		35.50	44.50	55.50
0030	**Correct water pressure**	2	1 PLUM	Ea.							
	Adjust water pressure, close				.571		33		33	41	51.50
	Total				.571		33		33	41	51.50
0040	**Replace refrigerant**	2	1 PLUM	Ea.							
	Add freon to unit				.133	32	7.80		39.80	45	52
	Total				.133	32	7.80		39.80	45	52
0050	**Repair drain leak**	4	1 PLUM	Ea.							
	Remove old pipe				.052		2.99		2.99	3.74	4.67
	Install new pipe				.186	12.60	10.70		23.30	27	32.50
	Total				.238	12.60	13.69		26.29	30.74	37.17

D20 PLUMBING D2013 Plumbing Fixtures

D2013 810 Drinking Fountain

	System Description	Freq. (Years)	Crew	Unit	Labor Hours	2014 Bare Costs				Total In-House	Total w/O&P
						Material	Labor	Equipment	Total		
0070	**Replace fountain**	10	2 PLUM	Ea.							
	Remove old water cooler				2.600		135		135	168	210
	Install new wall mounted water cooler				5.200	680	269		949	1,075	1,275
	Total				7.800	680	404		1,084	1,243	1,485

D2013 910 Emergency Shower Station

	System Description	Freq. (Years)	Crew	Unit	Labor Hours	2014 Bare Costs				Total In-House	Total w/O&P
						Material	Labor	Equipment	Total		
0020	**Inspect and clean shower head**	3	1 PLUM	Ea.							
	Turn water off and on				.010		.60		.60	.75	.93
	Remove shower head				.217		12.45		12.45	15.60	19.45
	Check shower head				.009		.52		.52	.65	.82
	Install shower head				.433		25		25	31	39
	Total				.669		38.57		38.57	48	60.20
0030	**Replace shower**	25	2 PLUM	Ea.							
	Remove emergency shower				2.600		135		135	168	210
	Install emergency shower				5.200	380	269		649	755	895
	Total				7.800	380	404		784	923	1,105

D2013 920 Emergency Eye Wash

	System Description	Freq. (Years)	Crew	Unit	Labor Hours	2014 Bare Costs				Total In-House	Total w/O&P
						Material	Labor	Equipment	Total		
0020	**Inspect and clean spray heads**	3	1 PLUM	Ea.							
	Turn water off and on				.010		.60		.60	.75	.93
	Remove spray head				.217		12.45		12.45	15.60	19.45
	Check shower head				.009		.52		.52	.65	.82
	Install spray head				.433		25		25	31	39
	Total				.669		38.57		38.57	48	60.20
0030	**Replace eye wash station**	25	2 PLUM	Ea.							
	Remove old eye wash station				2.600		135		135	168	210
	Install eye wash station				5.200	435	269		704	815	965
	Total				7.800	435	404		839	983	1,175

D20 PLUMBING D2023 Domestic Water Distribution

D2023 110 Pipe & Fittings, Copper

	System Description	Freq. (Years)	Crew	Unit	Labor Hours	2014 Bare Costs				Total In-House	Total w/O&P
						Material	Labor	Equipment	Total		
0010	**Resolder joint**	10	1 PLUM	Ea.							
	Measure, cut & ream both ends				.053		3.07		3.07	3.84	4.79
	Solder fitting (3/4" coupling)				.495	4.70	28.50		33.20	41	50.50
	Total				.549	4.70	31.57		36.27	**44.84**	**55.29**
0020	**Replace 3/4" copper pipe and fittings**	20	2 PLUM	L.F.							
	Remove old pipe				.071		4.09		4.09	5.10	6.40
	Install 3/4" copper tube with couplings and hangers				.141	7.80	8.10		15.90	18.70	22.50
	Total				.212	7.80	12.19		19.99	**23.80**	**28.90**
0030	**Replace 1" copper pipe and fittings**	25	2 PLUM	L.F.							
	Remove old pipe				.073		4.18		4.18	5.20	6.50
	Install 1" copper tube with couplings and hangers				.158	10.60	9.05		19.65	23	27.50
	Total				.230	10.60	13.23		23.83	**28.20**	**34**
0050	**Replace 1-1/2" copper pipe and fittings**	25	2 PLUM	L.F.							
	Remove old pipe				.104		6		6	7.50	9.35
	Install 1 1/2" copper tubing, couplings, hangers				.208	17.15	11.95		29.10	34	40
	Total				.312	17.15	17.95		35.10	**41.50**	**49.35**
0060	**Replace 2" copper pipe and fittings**	25	2 PLUM	L.F.							
	Remove old pipe				.130		7.50		7.50	9.35	11.65
	Install 2" copper tube with couplings and hangers				.260	26.50	14.95		41.45	48	56.50
	Total				.390	26.50	22.45		48.95	**57.35**	**68.15**
0070	**Replace 4" copper pipe and fittings**	25	2 PLUM	L.F.							
	Remove old pipe				.274		14.20		14.20	17.75	22
	Install 4" copper tube with couplings and hangers				.547	101	28.50		129.50	147	170
	Total				.821	101	42.70		143.70	**164.75**	**192**
0080	**Replace 8" copper pipe and fittings**	25	2 PLUM	L.F.							
	Remove old pipe				.694		37.50		37.50	46.50	58
	Install 8" copper tube with couplings and hangers				.918	705	49.50		754.50	835	960
	Total				1.612	705	87		792	**881.50**	**1,018**

D20 PLUMBING D2023 Domestic Water Distribution

D2023 120 Pipe & Fittings, Steel/Iron, Threaded

	System Description	Freq. (Years)	Crew	Unit	Labor Hours	2014 Bare Costs				Total In-House	Total w/O&P
						Material	Labor	Equipment	Total		
0030	**Replace 3/4" threaded steel pipe and fittings**	75	2 PLUM	L.F.							
	Remove old pipe				.086		4.95		4.95	6.20	7.70
	Install 3/4" diameter steel pipe, couplings, hangers				.131	3.17	7.55		10.72	12.95	15.75
	Total				.217	3.17	12.50		15.67	**19.15**	**23.45**
0040	**Replace 1-1/2" threaded steel pipe and fittings**	75	2 PLUM	L.F.							
	Remove old pipe				.130		6.75		6.75	8.40	10.50
	Install 1-1/2" diameter steel pipe, couplings, hangers				.260	6.70	13.45		20.15	24	29.50
	Total				.390	6.70	20.20		26.90	**32.40**	**40**
0050	**Replace 2" threaded steel pipe and fittings**	75	2 PLUM	L.F.							
	Remove old pipe				.163		8.45		8.45	10.55	13.20
	Install 2" diameter steel pipe, couplings, hangers				.325	8.95	16.85		25.80	31	37.50
	Total				.488	8.95	25.30		34.25	**41.55**	**50.70**
0060	**Replace 4" threaded steel pipe and fittings**	75	2 PLUM	L.F.							
	Remove old pipe				.289		14.95		14.95	18.70	23.50
	Install 4" diameter steel pipe, couplings, hangers				.578	27	30		57	67	80.50
	Total				.867	27	44.95		71.95	**85.70**	**104**

D2023 130 Solar Piping: Pipe & Fittings, PVC

	System Description	Freq. (Years)	Crew	Unit	Labor Hours	2014 Bare Costs				Total In-House	Total w/O&P
						Material	Labor	Equipment	Total		
0010	**Reglue joint, install 1" tee**	10	1 PLUM	Ea.							
	Cut exist pipe, install tee, 1"				.867	1.29	50		51.29	64	79.50
	Inspect joints				.126		7.25		7.25	9.05	11.30
	Total				.993	1.29	57.25		58.54	**73.05**	**90.80**
0110	**Reglue joint, install 1-1/4" tee**	10	1 PLUM	Ea.							
	Cut exist pipe, install tee, 1-1/4"				.945	2.01	54.50		56.51	70	87.50
	Inspect joints				.126		7.25		7.25	9.05	11.30
	Total				1.071	2.01	61.75		63.76	**79.05**	**98.80**

D20 PLUMBING D2023 Domestic Water Distribution

D2023 130 Solar Piping: Pipe & Fittings, PVC

	System Description	Freq. (Years)	Crew	Unit	Labor Hours	2014 Bare Costs				Total In-House	Total w/O&P
						Material	Labor	Equipment	Total		
0210	**Reglue joint, install 1-1/2" tee**	10	1 PLUM	Ea.							
	Cut exist pipe, install tee, 1-1/2"				1.040	2.45	60		62.45	77.50	96.50
	Inspect joints				.126		7.25		7.25	9.05	11.30
	Total				1.166	2.45	67.25		69.70	86.55	**107.80**
0310	**Reglue joint, install 2" tee**	10	Q-1	Ea.							
	Cut exist pipe, install tee, 2"				1.224	3.56	63.50		67.06	83	103
	Inspect joints				.126		7.25		7.25	9.05	11.30
	Total				1.350	3.56	70.75		74.31	92.05	**114.30**
1020	**Install 10' section PVC 1" diameter**	20	1 PLUM	Ea.							
	Remove broken pipe				1.131		65		65	81.50	102
	Install 10' new PVC pipe 1" diameter				2.939	51.48	169		220.48	267	330
	Inspect joints				.084		4.83		4.83	6.05	7.55
	Total				4.154	51.48	238.83		290.31	354.55	**439.55**
1120	**Install 10' PVC 1-1/4" diameter**	20	1 PLUM	Ea.							
	Remove broken pipe				1.239		71.50		71.50	89	111
	Install 10' new PVC pipe 1-1/4" diameter				3.219	64.22	185.25		249.47	305	365
	Inspect joints				.084		4.83		4.83	6.05	7.55
	Total				4.542	64.22	261.58		325.80	400.05	**483.55**
1220	**Install 10' PVC 1-1/2" diameter**	20	1 PLUM	Ea.							
	Remove broken pipe				1.445		83		83	104	130
	Install 10' new PVC pipe 1-1/2" diameter				3.756	61.88	216.45		278.33	340	415
	Inspect joints				.084		4.83		4.83	6.05	7.55
	Total				5.284	61.88	304.28		366.16	450.05	**552.55**
1320	**Install 10' section PVC 2" diameter**	20	Q-1	Ea.							
	Remove broken pipe				1.763		91.50		91.50	114	143
	Install 10' new PVC pipe 2" diameter				4.583	77.35	237.25		314.60	385	465
	Inspect joints				.084		4.83		4.83	6.05	7.55
	Total				6.430	77.35	333.58		410.93	505.05	**615.55**

D20 PLUMBING   D2023 Domestic Water Distribution

D2023 130   Solar Piping: Pipe & Fittings, PVC

	System Description	Freq. (Years)	Crew	Unit	Labor Hours	2014 Bare Costs				Total In-House	Total w/O&P
						Material	Labor	Equipment	Total		
2030	**Replace 1000' PVC pipe 1" diameter**	30	1 PLUM	M.L.F.							
	Remove broken pipe				113.080		6,500		6,500	8,150	10,200
	Install 1000' new PVC pipe 1" diameter				226.080	3,960	13,000		16,960	20,500	25,500
	Inspect joints				8.399		485		485	605	755
	Total				347.559	3,960	19,985		23,945	29,255	36,455
2130	**Replace 1000' PVC pipe 1-1/4" diameter**	30	1 PLUM	M.L.F.							
	Remove broken pipe				123.850		7,150		7,150	8,900	11,100
	Install 1000' new PVC pipe 1-1/4" diameter				247.620	4,940	14,250		19,190	23,500	28,200
	Inspect joints				8.399		485		485	605	755
	Total				379.869	4,940	21,885		26,825	33,005	40,055
2230	**Replace 1000' PVC pipe 1-1/2" diameter**	30	1 PLUM	M.L.F.							
	Remove broken pipe				144.490		8,300		8,300	10,400	13,000
	Install 1000' new PVC pipe 1-1/2" diameter				288.890	4,760	16,650		21,410	26,000	32,000
	Inspect joints				8.399		485		485	605	755
	Total				441.779	4,760	25,435		30,195	37,005	45,755
2330	**Replace 1000' PVC pipe 2" diameter**	30	Q-1	M.L.F.							
	Remove broken pipe				176.320		9,150		9,150	11,400	14,300
	Install 1000' new PVC pipe 2" diameter				352.540	5,950	18,250		24,200	29,500	35,900
	Inspect joints				8.399		485		485	605	755
	Total				537.259	5,950	27,885		33,835	41,505	50,955

D2023 150   Valve, Non-Drain, Less Than 1-1/2"

	System Description	Freq. (Years)	Crew	Unit	Labor Hours	2014 Bare Costs				Total In-House	Total w/O&P
						Material	Labor	Equipment	Total		
0020	**Replace old valve**	10	1 PLUM	Ea.							
	Remove old valve				.400		23		23	29	36
	Install new valve				.800	315	46		361	405	465
	Total				1.200	315	69		384	434	501

D20 PLUMBING D2023 Domestic Water Distribution

D2023 152 Valve, Non-Drain, 2"

	System Description	Freq. (Years)	Crew	Unit	Labor Hours	2014 Bare Costs				Total In-House	Total w/O&P
						Material	Labor	Equipment	Total		
0020	**Replace old valve**	10	1 PLUM	Ea.							
	Remove old valve (2")				.800		41.50		41.50	52	64.50
	Install new valve (to 2")				1.600	1,675	83		1,758	1,950	2,225
	Total				2.400	1,675	124.50		1,799.50	2,002	2,289.50

D2023 154 Valve, Non-Drain, 3"

	System Description	Freq. (Years)	Crew	Unit	Labor Hours	2014 Bare Costs				Total In-House	Total w/O&P
						Material	Labor	Equipment	Total		
0020	**Replace old valve**	10	1 PLUM	Ea.							
	Remove old valve				.800		41.50		41.50	52	64.50
	Install new valve				1.600	1,675	83		1,758	1,950	2,225
	Total				2.400	1,675	124.50		1,799.50	2,002	2,289.50

D2023 156 Valve, Non-Drain, 4" and Larger

	System Description	Freq. (Years)	Crew	Unit	Labor Hours	2014 Bare Costs				Total In-House	Total w/O&P
						Material	Labor	Equipment	Total		
0020	**Replace old valve, 4"**	10	2 PLUM	Ea.							
	Remove old flanged gate valve, 4"				5.200		279		279	350	435
	Install new flanged gate valve, 4"				7.038	1,125	380		1,505	1,700	2,000
	Total				12.239	1,125	659		1,784	2,050	2,435
0030	**Replace old valve, 6"**	10	2 PLUM	Ea.							
	Remove old flanged gate valve, 6"				4.615		248		248	310	385
	Install new flanged gate valve, 6"				10.399	1,900	560		2,460	2,800	3,250
	Total				15.014	1,900	808		2,708	3,110	3,635
0040	**Replace old valve, 8"**	10	2 PLUM	Ea.							
	Remove old flanged gate valve, 8"				3.871		208		208	260	325
	Install new flanged gate valve, 8"				12.371	3,275	665		3,940	4,425	5,125
	Total				16.242	3,275	873		4,148	4,685	5,450

D20 PLUMBING D2023 Domestic Water Distribution

D2023 156 Valve, Non-Drain, 4" and Larger

	System Description	Freq. (Years)	Crew	Unit	Labor Hours	2014 Bare Costs				Total In-House	Total w/O&P
						Material	Labor	Equipment	Total		
0050	**Replace old valve, 10"**	10	3 PLUM	Ea.							
	Remove old flanged gate valve, 10"				3.380		182		182	227	283
	Install new flanged gate valve, 10"				14.201	5,775	765		6,540	7,300	8,425
	Total				17.581	5,775	947		6,722	7,527	8,708

D2023 160 Insulation, Pipe

	System Description	Freq. (Years)	Crew	Unit	Labor Hours	2014 Bare Costs				Total In-House	Total w/O&P
						Material	Labor	Equipment	Total		
0020	**Remove old insulation & replace with new (1/2" pipe size, 1" wall)**	15	1 PLUM	L.F.							
	Remove pipe insulation				.025		1.42		1.42	1.78	2.22
	Install fiberglass pipe insulation (1" wall, 1/2" pipe size)				.087	.89	3.99		4.88	6.15	7.50
	Total				.111	.89	5.41		6.30	7.93	9.72
0030	**Remove old insulation & replace with new (3/4" pipe size, 1" wall)**	15	1 PLUM	L.F.							
	Remove pipe insulation				.025		1.42		1.42	1.78	2.22
	Install fiberglass pipe insulation (1" wall, 3/4" pipe size)				.090	.97	4.16		5.13	6.45	7.90
	Total				.115	.97	5.58		6.55	8.23	10.12
0040	**Remove old insulation & replace with new (1-1/2" pipe size, 1" wall)**	15	1 PLUM	L.F.							
	Remove pipe insulation				.025		1.42		1.42	1.78	2.22
	Install fiberglass pipe insulation (1" wall, 1-1/2" pipe size)				.099	1.22	4.56		5.78	7.25	8.85
	Total				.124	1.22	5.98		7.20	9.03	11.07
0050	**Remove old insulation & replace with new (1/2" pipe size, 3/4" wall)**	15	1 PLUM	L.F.							
	Remove pipe insulation				.025		1.42		1.42	1.78	2.22
	Install flexible rubber tubing (3/4" wall, 1/2" pipe size)				.117	1.10	6		7.10	8.95	11
	Total				.142	1.10	7.42		8.52	10.73	13.22
0060	**Remove old insulation & replace with new (3/4" pipe size, 3/4" wall)**	15	1 PLUM	L.F.							
	Remove pipe insulation				.025		1.42		1.42	1.78	2.22
	Install flexible rubber tubing (3/4" wall, 3/4" pipe size)				.117	1.35	6		7.35	9.25	11.30
	Total				.142	1.35	7.42		8.77	11.03	13.52

D20 PLUMBING D2023 Domestic Water Distribution

D2023 160 Insulation, Pipe

	System Description	Freq. (Years)	Crew	Unit	Labor Hours	2014 Bare Costs				Total In-House	Total w/O&P
						Material	Labor	Equipment	Total		
0070	Remove old insulation & replace w/new (1-1/2" pipe size, 3/4" wall)	15	1 PLUM	L.F.							
	Remove pipe insulation				.025		1.42		1.42	1.78	2.22
	Install flexible rubber tubing (3/4" wall, 1-1/2" pipe size)				.120	2.32	6.10		8.42	10.50	12.70
	Total				.144	2.32	7.52		9.84	12.28	14.92

D2023 210 Water Heater, Gas / Oil, 30 Gallon

	System Description	Freq. (Years)	Crew	Unit	Labor Hours	2014 Bare Costs				Total In-House	Total w/O&P
						Material	Labor	Equipment	Total		
0010	Overhaul heater	5	1 PLUM	Ea.							
	Turn valves off and on				.022		1.27		1.27	1.59	1.98
	Remove thermostat, check, and insulate				.182		10.45		10.45	13.10	16.35
	Remove burner, clean and install				.229		13.15		13.15	16.45	20.50
	Clean pilot and thermocouple				.296		17.05		17.05	21.50	26.50
	Remove burner door and install				.033		1.87		1.87	2.34	2.92
	Check after maintenance				.216		12.45		12.45	15.55	19.40
	Drain and fill tank				.615		35.50		35.50	44.50	55.50
	Total				1.593		91.74		91.74	115.03	143.15
0020	Clean and service	1	1 PLUM	Ea.							
	Check / adjust fuel burner				2.667		153		153	192	239
	Total				2.667		153		153	192	239
0030	Replace water heater	10	2 PLUM	Ea.							
	Remove old heater				2.600		150		150	187	233
	Install 30 gallon gas water heater				5.202	895	299		1,194	1,350	1,575
	Check after installation				.216		12.45		12.45	15.55	19.40
	Total				8.018	895	461.45		1,356.45	1,552.55	1,827.40

D20 PLUMBING D2023 Domestic Water Distribution

D2023 212 Water Heater, Gas / Oil, 70 Gallon

	System Description	Freq. (Years)	Crew	Unit	Labor Hours	2014 Bare Costs				Total In-House	Total w/O&P
						Material	Labor	Equipment	Total		
0010	**Overhaul heater**	5	1 PLUM	Ea.							
	Remove hot water heater and insulate				.667		38.50		38.50	48	60
	Remove thermostat, check and install				.182		10.45		10.45	13.10	16.35
	Remove burner, clean and install				.229		13.15		13.15	16.45	20.50
	Clean pilot and thermocouple				.296		17.05		17.05	21.50	26.50
	Remove burner door and install				.033		1.87		1.87	2.34	2.92
	Drain and refill tank				.615		35.50		35.50	44.50	55.50
	Turn valve off and on				.211		12.10		12.10	15.15	18.90
	Total				2.232		128.62		128.62	161.04	200.67
0020	**Clean & service**	1	1 PLUM	Ea.							
	Check / adjust fuel burner				2.667		153		153	192	239
	Total				2.667		153		153	192	239
0030	**Replace hot water heater**	12	2 PLUM	Ea.							
	Remove old heater				2.600		150		150	187	233
	Install water heater, oil fired, 70 gallon				6.932	1,975	400		2,375	2,675	3,100
	Total				9.532	1,975	550		2,525	2,862	3,333

D2023 214 Water Heater, Gas / Oil, 1150 GPH

	System Description	Freq. (Years)	Crew	Unit	Labor Hours	2014 Bare Costs				Total In-House	Total w/O&P
						Material	Labor	Equipment	Total		
0010	**Minor repairs, adjustments**	2	2 PLUM	Ea.							
	Reset, adjust controls				1.333		76.50		76.50	96	120
	Total				1.333		76.50		76.50	96	120
0020	**Clean & service**	2	2 PLUM	Ea.							
	Drain tank, clean, and service				8.000		460		460	575	720
	Total				8.000		460		460	575	720
0030	**Replace hot water heater**	20	2 PLUM	Ea.							
	Remove water heater				26.016		1,350		1,350	1,675	2,100
	Install water heater, gas fired, 1150 GPH				51.948	26,500	2,700		29,200	32,500	37,300
	Total				77.964	26,500	4,050		30,550	34,175	39,400

D20 PLUMBING | D2023 | Domestic Water Distribution

D2023 220 | Water Heater, Electric, 120 Gallon

	System Description	Freq. (Years)	Crew	Unit	Labor Hours	2014 Bare Costs				Total In-House	Total w/O&P
						Material	Labor	Equipment	Total		
0010	**Drain and flush**	7	1 PLUM	Ea.							
	Drain, clean and readjust				4.000		230		230	288	360
	Total				4.000		230		230	**288**	**360**
0020	**Check operation**	3	1 PLUM	Ea.							
	Check for relief valve				.010		.60		.60	.75	.93
	Check for proper water temperature				.003		.15		.15	.19	.23
	Fill out maintenance report				.022		1.27		1.27	1.59	1.98
	Total				.035		2.02		2.02	**2.53**	**3.14**
0030	**Replace water heater**	15	2 PLUM	Ea.							
	Remove old heater				4.334		249		249	310	390
	Install water heater				8.667	7,750	500		8,250	9,150	10,500
	Total				13.001	7,750	749		8,499	**9,460**	**10,890**

D2023 222 | Water Heater, Electric, 300 Gallon

	System Description	Freq. (Years)	Crew	Unit	Labor Hours	2014 Bare Costs				Total In-House	Total w/O&P
						Material	Labor	Equipment	Total		
0010	**Drain and flush**	7	1 PLUM	Ea.							
	Drain, clean and adjust				4.000		230		230	288	360
	Total				4.000		230		230	**288**	**360**
0020	**Check operation**	3	1 PLUM	Ea.							
	Check relief valve				.010		.60		.60	.75	.93
	Check for proper water temperature				.003		.15		.15	.19	.23
	Fill out maintenance report				.022		1.27		1.27	1.59	1.98
	Total				.035		2.02		2.02	**2.53**	**3.14**
0030	**Replace water heater**	15	2 PLUM	Ea.							
	Remove old heater				8.000		415		415	520	645
	Install water heater				16.000	37,500	830		38,330	42,300	48,200
	Total				24.000	37,500	1,245		38,745	**42,820**	**48,845**

D20 PLUMBING | D2023 | Domestic Water Distribution

D2023 224 | Water Heater, Electric, 1000 Gallon

	System Description	Freq. (Years)	Crew	Unit	Labor Hours	2014 Bare Costs				Total In-House	Total w/O&P
						Material	Labor	Equipment	Total		
0010	**Drain and flush**	7	1 PLUM	Ea.							
	Drain, clean and adjust				4.000		230		230	288	360
	Total				4.000		230		230	**288**	**360**
0020	**Check operation**	3	1 PLUM	Ea.							
	Check relief valve				.010		.60		.60	.75	.93
	Check for proper water temperature				.003		.15		.15	.19	.23
	Fill out maintenance report				.022		1.27		1.27	1.59	1.98
	Total				.035		2.02		2.02	**2.53**	**3.14**
0030	**Replace water heater**	15	2 PLUM	Ea.							
	Remove heater				22.284		1,200		1,200	1,500	1,875
	Install water heater				44.610	49,200	2,400		51,600	57,000	65,500
	Total				66.894	49,200	3,600		52,800	**58,500**	**67,375**

D2023 226 | Water Heater, Electric, 2000 Gallon

	System Description	Freq. (Years)	Crew	Unit	Labor Hours	2014 Bare Costs				Total In-House	Total w/O&P
						Material	Labor	Equipment	Total		
0010	**Drain and flush**	7	1 PLUM	Ea.							
	Drain, clean and adjust				4.000		230		230	288	360
	Total				4.000		230		230	**288**	**360**
0020	**Check operation**	3	1 PLUM	Ea.							
	Check relief valve				.010		.60		.60	.75	.93
	Check for proper water temperature				.003		.15		.15	.19	.23
	Fill out maintenance report				.022		1.27		1.27	1.59	1.98
	Total				.035		2.02		2.02	**2.53**	**3.14**
0030	**Replace water heater**	15	2 PLUM	Ea.							
	Remove heater				31.209		1,675		1,675	2,100	2,625
	Install water heater				62.338	92,000	3,350		95,350	105,500	120,000
	Total				93.547	92,000	5,025		97,025	**107,600**	**122,625**

D20 PLUMBING   D2023 Domestic Water Distribution

D2023 230 Steam Converter, Domestic Hot Water

	System Description	Freq. (Years)	Crew	Unit	Labor Hours	2014 Bare Costs				Total In-House	Total w/O&P
						Material	Labor	Equipment	Total		
0020	**Inspect for leaks**	1	1 PLUM	Ea.							
	Check relief valve operation				.038		2.17		2.17	2.72	3.39
	Check for proper water temperature				.054		3.13		3.13	3.92	4.89
	Total				.092		5.30		5.30	6.64	8.28
0030	**Replace steam converter**	20	2 PLUM	Ea.							
	Remove heat exchanger				1.733		100		100	125	156
	Install new heat exchanger				3.467	2,150	183		2,333	2,600	2,975
	Total				5.200	2,150	283		2,433	2,725	3,131

D2023 240 Storage Tank, Domestic Hot Water

	System Description	Freq. (Years)	Crew	Unit	Labor Hours	2014 Bare Costs				Total In-House	Total w/O&P
						Material	Labor	Equipment	Total		
0020	**Replace storage tank, glass lined, P.E., 80 gal.**	50	2 PLUM	Ea.							
	Remove hot water storage tank				.578		33.50		33.50	41.50	52
	Install new glass lined, P.E., 80 Gal. hot water storage tank				1.156	4,725	66.50		4,791.50	5,275	6,000
	Total				1.733	4,725	100		4,825	5,316.50	6,052

D2023 245 Solar Storage Tank, 1000 Gallon

	System Description	Freq. (Years)	Crew	Unit	Labor Hours	2014 Bare Costs				Total In-House	Total w/O&P
						Material	Labor	Equipment	Total		
0010	**Replace 1000 gallon solar storage tank**	20	Q-9	Ea.							
	Remove tank				8.000		395		395	500	620
	Install 1000 gallon solar storage tank				16.000	7,275	790		8,065	9,000	10,300
	Total				24.000	7,275	1,185		8,460	9,500	10,920

| D20 PLUMBING | | D2023 | | Domestic Water Distribution | | | | | | |

D2023 250 Expansion Chamber

	System Description	Freq. (Years)	Crew	Unit	Labor Hours	2014 Bare Costs				Total In-House	Total w/O&P
						Material	Labor	Equipment	Total		
0010	Refill	5	1 PLUM	Ea.							
	Refill air chamber				.039		2.25		2.25	2.81	3.50
	Total				.039		2.25		2.25	2.81	3.50
0020	Remove old chamber, install new	10	1 PLUM	Ea.							
	Remove / install expansion device, 15 gal.				.941	640	49.50		689.50	765	880
	Total				.941	640	49.50		689.50	765	880

D2023 260 Circulation Pump, 1/12 HP

	System Description	Freq. (Years)	Crew	Unit	Labor Hours	2014 Bare Costs				Total In-House	Total w/O&P
						Material	Labor	Equipment	Total		
0020	Pump maintenance, 1/12 H.P.	0.50	1 PLUM	Ea.							
	Check for leaks				.010		.60		.60	.75	.93
	Check pump operation				.012		.67		.67	.84	1.05
	Lubricate pump and motor				.047		2.69		2.69	3.37	4.20
	Check suction or discharge pressure				.004		.22		.22	.28	.35
	Check packing glands, tighten				.014		.82		.82	1.03	1.29
	Fill out inspection report				.022		1.27		1.27	1.59	1.98
	Total				.109		6.27		6.27	7.86	9.80
0030	Replace pump / motor assembly	10	2 PLUM	Ea.							
	Remove flanged connection pump				1.733		90		90	112	140
	Install new flanged connection pump (1/12 H.P.)				3.467	545	180		725	825	960
	Total				5.200	545	270		815	937	1,100

D20 PLUMBING

D2023 Domestic Water Distribution

D2023 261 Circulation Pump, 1/8 HP

	System Description	Freq. (Years)	Crew	Unit	Labor Hours	2014 Bare Costs				Total In-House	Total w/O&P
						Material	Labor	Equipment	Total		
0020	**Pump maintenance, 1/8 H.P.**	1	1 PLUM	Ea.							
	Check for leaks				.010		.60		.60	.75	.93
	Check pump operation				.012		.67		.67	.84	1.05
	Lubricate pump and motor				.047		2.69		2.69	3.37	4.20
	Check suction or discharge pressure				.004		.22		.22	.28	.35
	Check packing glands, tighten				.014		.82		.82	1.03	1.29
	Fill out inspection report				.022		1.27		1.27	1.59	1.98
	Total				.109		6.27		6.27	7.86	9.80
0030	**Replace pump / motor assembly**	20	2 PLUM	Ea.							
	Remove flanged connection pump				1.733		90		90	112	140
	Install new flanged connection pump (1/8 H.P.)				3.467	940	180		1,120	1,250	1,450
	Total				5.200	940	270		1,210	1,362	1,590

D2023 262 Circulation Pump, 1/6 HP

	System Description	Freq. (Years)	Crew	Unit	Labor Hours	2014 Bare Costs				Total In-House	Total w/O&P
						Material	Labor	Equipment	Total		
0020	**Pump maintenance, 1/6 H.P.**	1	1 PLUM	Ea.							
	Check for leaks				.010		.60		.60	.75	.93
	Check pump operation				.012		.67		.67	.84	1.05
	Lubricate pump and motor				.047		2.69		2.69	3.37	4.20
	Check suction or discharge pressure				.004		.22		.22	.28	.35
	Check packing glands, tighten				.014		.82		.82	1.03	1.29
	Fill out inspection report				.022		1.27		1.27	1.59	1.98
	Total				.109		6.27		6.27	7.86	9.80
0030	**Replace pump / motor assembly**	20	2 PLUM	Ea.							
	Remove flanged connection pump				2.080		120		120	150	187
	Install new flanged connection pump (1/6 H.P.)				4.160	1,350	216		1,566	1,750	2,025
	Total				6.240	1,350	336		1,686	1,900	2,212

D20 PLUMBING		D2023		Domestic Water Distribution								

D2023 264 Circulation Pump, 1/2 HP

	System Description	Freq. (Years)	Crew	Unit	Labor Hours	2014 Bare Costs				Total In-House	Total w/O&P
						Material	Labor	Equipment	Total		
0020	**Pump maintenance, 1/2 H.P.**	1	1 PLUM	Ea.							
	Check for leaks				.010		.60		.60	.75	.93
	Check pump operation				.012		.67		.67	.84	1.05
	Check alignment				.053		3.07		3.07	3.84	4.79
	Lubricate pump and motor				.047		2.69		2.69	3.37	4.20
	Check suction or discharge pressure				.004		.22		.22	.28	.35
	Check packing glands, tighten				.014		.82		.82	1.03	1.29
	Fill out inspection report				.022		1.27		1.27	1.59	1.98
	Total				.163		9.34		9.34	11.70	14.59
0030	**Replace pump / motor assembly, partial**	20	2 PLUM	Ea.							
	Remove flanged connection pump				2.600		135		135	168	210
	Install new flanged connection pump (1/2 H.P.)				5.200	2,325	269		2,594	2,900	3,325
	Total				7.800	2,325	404		2,729	3,068	3,535

D2023 266 Circulation Pump, Bronze 1 HP

	System Description	Freq. (Years)	Crew	Unit	Labor Hours	2014 Bare Costs				Total In-House	Total w/O&P
						Material	Labor	Equipment	Total		
0020	**Pump maintenance, 1 H.P.**	0.50	1 PLUM	Ea.							
	Check for leaks				.010		.60		.60	.75	.93
	Check pump operation				.012		.67		.67	.84	1.05
	Lubricate pump and motor				.047		2.69		2.69	3.37	4.20
	Check suction or discharge pressure				.004		.22		.22	.28	.35
	Check packing glands, tighten				.014		.82		.82	1.03	1.29
	Fill out inspection report				.022		1.27		1.27	1.59	1.98
	Total				.109		6.27		6.27	7.86	9.80
0040	**Replace pump / motor assembly**	20	2 PLUM	Ea.							
	Remove pump / motor				2.600		135		135	168	210
	Install new flanged connection pump, bronze (1 H.P.)				5.200	4,100	269		4,369	4,850	5,550
	Total				7.800	4,100	404		4,504	5,018	5,760

D20 PLUMBING   D2023 Domestic Water Distribution

D2023 267   Circulation Pump, C.I. 1-1/2 HP

	System Description	Freq. (Years)	Crew	Unit	Labor Hours	2014 Bare Costs				Total In-House	Total w/O&P
						Material	Labor	Equipment	Total		
0020	**Pump maintenance, 1-1/2 H.P.**	0.50	1 PLUM	Ea.							
	Check for leaks				.010		.60		.60	.75	.93
	Check pump operation				.012		.67		.67	.84	1.05
	Check alignment				.053		3.07		3.07	3.84	4.79
	Lubricate pump and motor				.047		2.69		2.69	3.37	4.20
	Check suction / discharge pressure				.004		.22		.22	.28	.35
	Check packing glands, tighten				.014		.82		.82	1.03	1.29
	Fill out inspection report				.022		1.27		1.27	1.59	1.98
	Total				.163		9.34		9.34	11.70	14.59
0040	**Replace pump / motor assembly**	20	2 PLUM	Ea.							
	Remove flanged connection pump				2.600		135		135	168	210
	Install new flanged connection pump, C.I. (1-1/2 H.P.)				5.200	1,900	269		2,169	2,425	2,800
	Total				7.800	1,900	404		2,304	2,593	3,010

D2023 268   Circulation Pump, 3 HP

	System Description	Freq. (Years)	Crew	Unit	Labor Hours	2014 Bare Costs				Total In-House	Total w/O&P
						Material	Labor	Equipment	Total		
0020	**Pump maintenance, 3 H.P.**	0.50	Q-1	Ea.							
	Check for leaks				.012		.67		.67	.83	1.03
	Check pump operation				.012		.67		.67	.84	1.05
	Check alignment				.059		3.41		3.41	4.27	5.30
	Lubricate pump and motor				.052		2.99		2.99	3.74	4.67
	Check suction / discharge pressure				.004		.22		.22	.28	.35
	Check packing glands, tighten				.016		.91		.91	1.14	1.43
	Fill out inspection report				.022		1.27		1.27	1.59	1.98
	Total				.176		10.14		10.14	12.69	15.81
0040	**Replace pump / motor assembly**	20	Q-1	Ea.							
	Remove flanged connection pump				4.400		227.70		227.70	285	360
	Install new flanged connection pump, 3 HP				8.889	7,350	460		7,810	8,650	9,900
	Total				13.289	7,350	687.70		8,037.70	8,935	10,260

D20 PLUMBING D2023 Domestic Water Distribution

D2023 310 Hose Bibb

					2014 Bare Costs				Total In-House	Total w/O&P
	System Description	Freq. (Years)	Crew	Unit	Labor Hours	Material	Labor	Equipment	Total		
0020	**Replace old valve with new**	10	1 PLUM	Ea.							
	Remove hose bibb				.217		12.45		12.45	15.60	19.45
	Install new hose bibb				.433	9.70	25		34.70	42	51
	Total				.650	9.70	37.45		47.15	**57.60**	**70.45**

D2023 320 Water Meter

					2014 Bare Costs				Total In-House	Total w/O&P
	System Description	Freq. (Years)	Crew	Unit	Labor Hours	Material	Labor	Equipment	Total		
0010	**Overhaul**	13	1 PLUM	Ea.							
	Open meter, replace worn parts				.300	3.41	17.25		20.66	25.50	31.50
	Total				.300	3.41	17.25		20.66	**25.50**	**31.50**
0020	**Remove old 5/8" meter, install new**	25	1 PLUM	Ea.							
	Remove 5/8" water meter				.325		18.70		18.70	23.50	29
	Install 5/8" threaded water meter				.650	48.50	37.50		86	100	119
	Total				.975	48.50	56.20		104.70	**123.50**	**148**
0025	**Remove old 3/4" meter, install new**	25	1 PLUM	Ea.							
	Remove 3/4" water meter				.371		21.50		21.50	26.50	33.50
	Install 3/4" threaded water meter				.743	88.50	43		131.50	151	177
	Total				1.114	88.50	64.50		153	**177.50**	**210.50**
0030	**Remove old 1" meter, install new**	25	1 PLUM	Ea.							
	Remove 1" water meter				.433		25		25	31	39
	Install 1" threaded water meter				.867	134	50		184	210	246
	Total				1.300	134	75		209	**241**	**285**
0035	**Remove old 1-1/2" meter, install new**	25	1 PLUM	Ea.							
	Remove 1-1/2" water meter				.650		37.50		37.50	47	58.50
	Install 1-1/2" threaded water meter				1.300	330	75		405	455	530
	Total				1.950	330	112.50		442.50	**502**	**588.50**
0040	**Remove old 2" meter, install new**	25	1 PLUM	Ea.							
	Remove 2" water meter				.867		50		50	62.50	78
	Install 2" threaded water meter				1.733	445	100		545	615	710
	Total				2.600	445	150		595	**677.50**	**788**

D20 PLUMBING D2023 Domestic Water Distribution

D2023 320 Water Meter

	System Description	Freq. (Years)	Crew	Unit	Labor Hours	2014 Bare Costs				Total In-House	Total w/O&P
						Material	Labor	Equipment	Total		
0045	**Remove old 3" meter, install new**	25	Q-1	Ea.							
	Remove 3" water meter				3.467		180		180	224	280
	Install 3" flanged water meter				6.932	3,025	360		3,385	3,775	4,350
	Total				10.399	3,025	540		3,565	**3,999**	**4,630**
0050	**Remove old 4" meter, install new**	25	Q-1	Ea.							
	Remove 4" water meter				6.932		360		360	450	560
	Install 4" flanged water meter				13.865	4,850	720		5,570	6,225	7,200
	Total				20.797	4,850	1,080		5,930	**6,675**	**7,760**
0055	**Remove old 6" meter, install new**	25	Q-1	Ea.							
	Remove 6" water meter				10.403		540		540	675	840
	Install 6" flanged water meter				20.806	7,725	1,075		8,800	9,850	11,300
	Total				31.209	7,725	1,615		9,340	**10,525**	**12,140**
0060	**Remove old 8" meter, install new**	25	Q-1	Ea.							
	Remove 8" water meter				12.998		675		675	840	1,050
	Install 8" flanged water meter				26.016	12,100	1,350		13,450	15,000	17,200
	Total				39.014	12,100	2,025		14,125	**15,840**	**18,250**

D2023 370 Water Softener

	System Description	Freq. (Years)	Crew	Unit	Labor Hours	2014 Bare Costs				Total In-House	Total w/O&P
						Material	Labor	Equipment	Total		
0030	**Replace softener**	15	2 PLUM	Ea.							
	Remove old unit				2.600		150		150	187	233
	Install new water softener				5.200	650	299		949	1,100	1,275
	Total				7.800	650	449		1,099	**1,287**	**1,508**

D20 PLUMBING | D2033 Sanitary Waste

D2033 110 Pipe & Fittings, Cast Iron

	System Description	Freq. (Years)	Crew	Unit	Labor Hours	2014 Bare Costs				Total In-House	Total w/O&P
						Material	Labor	Equipment	Total		
0020	**Unclog main drain**	10	1 PLUM	Ea.							
	Unclog main drain using auger				.630		36.50		36.50	45.50	56.50
	Total				.630		36.50		36.50	**45.50**	**56.50**
0030	**Replace cast iron pipe & fittings, 4"**	40	2 PLUM	L.F.							
	Remove old pipe				.189		9.80		9.80	12.25	15.30
	Install new 4" cast iron pipe				.378	17.45	19.60		37.05	43.50	52.50
	Total				.567	17.45	29.40		46.85	**55.75**	**67.80**

D2033 130 Pipe & Fittings, PVC

	System Description	Freq. (Years)	Crew	Unit	Labor Hours	2014 Bare Costs				Total In-House	Total w/O&P
						Material	Labor	Equipment	Total		
0010	**Unclog floor drain**	20	1 PLUM	Ea.							
	Unclog plugged floor drain				.650		37.50		37.50	47	58.50
	Total				.650		37.50		37.50	**47**	**58.50**
0020	**Unclog 4" - 12" diameter PVC main drain per L.F.**	10	1 PLUM	L.F.							
	Unclog main drain using auger				.052		2.97		2.97	3.71	4.63
	Total				.052		2.97		2.97	**3.71**	**4.63**
0040	**Repair joint**	10	1 PLUM	Ea.							
	Remove plastic fitting (2 connections)				.650		33.50		33.50	42	52.50
	Connect fitting (2 connections)				1.300	8.65	67.50		76.15	93.50	116
	Inspect joints				.163		9.40		9.40	11.75	14.65
	Total				2.113	8.65	110.40		119.05	**147.25**	**183.15**
0060	**Replace pipe, 1-1/2"**	30	2 PLUM	L.F.							
	Remove broken pipe				.145		7.80		7.80	9.75	12.15
	Install new pipe, couplings, hangers				.289	4.76	16.65		21.41	26	32
	Inspect joints				.327		18.80		18.80	23.50	29.50
	Total				.760	4.76	43.25		48.01	**59.25**	**73.65**

D20 PLUMBING D2033 Sanitary Waste

D2033 130 Pipe & Fittings, PVC

						2014 Bare Costs				Total In-House	Total w/O&P
	System Description	Freq. (Years)	Crew	Unit	Labor Hours	Material	Labor	Equipment	Total		
0080	**Replace pipe, 2"**	30	2 PLUM	L.F.							
	Remove broken pipe				.176		9.15		9.15	11.40	14.25
	Install new pipe, couplings, hangers				.353	5.95	18.25		24.20	29.50	36
	Inspect joints				.327		18.80		18.80	23.50	29.50
	Total				.855	5.95	46.20		52.15	**64.40**	**79.75**
0100	**Replace pipe, 4"**	30	2 PLUM	L.F.							
	Remove broken pipe				.217		11.20		11.20	14.05	17.50
	Install new pipe, couplings, hangers				.433	14.20	22.50		36.70	43.50	53
	Inspect joints				.327		18.80		18.80	23.50	29.50
	Total				.977	14.20	52.50		66.70	**81.05**	**100**
0120	**Replace pipe, 6"**	30	2 PLUM	L.F.							
	Remove broken pipe				.267		13.80		13.80	17.25	21.50
	New pipe, couplings, hangers				.533	27	30.50		57.50	68	82
	Inspect joints				.327		18.80		18.80	23.50	29.50
	Total				1.126	27	63.10		90.10	**108.75**	**133**
0140	**Replace pipe, 8"**	30	2 PLUM	L.F.							
	Remove broken pipe				.325		17.45		17.45	22	27
	Install new pipe, couplings, hangers				.650	42	35		77	90	107
	Inspect joints				.327		18.80		18.80	23.50	29.50
	Total				1.302	42	71.25		113.25	**135.50**	**163.50**

D2033 305 Pipe & Fittings

						2014 Bare Costs				Total In-House	Total w/O&P
	System Description	Freq. (Years)	Crew	Unit	Labor Hours	Material	Labor	Equipment	Total		
3010	**Unclog floor drain**	10	1 PLUM	Ea.							
	Unclog plugged floor drain				4.000		234		234	293	365
	Solvent					28.50			28.50	31.50	35.50
	Total				4.000	28.50	234		262.50	**324.50**	**400.50**
3020	**Unclog 4" - 12" diameter main drain per L.F.**	10	2 PLUM	L.F.							
	Unclog main drain using auger				.052		2.97		2.97	3.71	4.63
	Total				.052		2.97		2.97	**3.71**	**4.63**

D20 PLUMBING

D2033 Sanitary Waste

D2033 310 Floor Drain W/O Bucket

	System Description	Freq. (Years)	Crew	Unit	Labor Hours	2014 Bare Costs				Total In-House	Total w/O&P
						Material	Labor	Equipment	Total		
0010	**Clean drain**	4	1 PLUM	Ea.							
	Remove debris from drain				1.590		91.50		91.50	114	143
	Total				1.590		91.50		91.50	**114**	**143**
0030	**Replace floor drain**	40	1 PLUM	Ea.							
	Remove floor drain				1.040		60		60	75	93.50
	Install floor drain				2.080	900	108		1,008	1,125	1,300
	Total				3.120	900	168		1,068	**1,200**	**1,393.50**

D2033 330 Floor Drain With Bucket

	System Description	Freq. (Years)	Crew	Unit	Labor Hours	2014 Bare Costs				Total In-House	Total w/O&P
						Material	Labor	Equipment	Total		
0010	**Clean out bucket**	5	1 PLUM	Ea.							
	Remove debris from bucket				4.000		230		230	288	360
	Total				4.000		230		230	**288**	**360**
0030	**Replace floor drain**	40	1 PLUM	Ea.							
	Remove floor drain				1.040		60		60	75	93.50
	Install floor drain				2.080	900	108		1,008	1,125	1,300
	Total				3.120	900	168		1,068	**1,200**	**1,393.50**

D20 PLUMBING   D2043   Rain Water Drainage

D2043 110   Distribution: Gutters, Pipe

	System Description	Freq. (Years)	Crew	Unit	Labor Hours	2014 Bare Costs				Total In-House	Total w/O&P
						Material	Labor	Equipment	Total		
1010	**General maintenance & repair**	1	1 PLUM	M.L.F.							
	Inspect rainwater pipes				.104		6		6	7.50	9.35
	Clean out pipes				4.000		230		230	288	360
	Total				4.104		236		236	**295.50**	**369.35**
1020	**Replace pipe or gutter**	20	1 PLUM	L.F.	.650						
	Remove 4" PVC roof drain system				.217		11.20		11.20	14.05	17.50
	Install 4" roof drain system				.433	7.30	22.50		29.80	36	44
	Total				.650	7.30	33.70		41	**50.05**	**61.50**

D2043 210   Drain: Roof, Scupper, Area

	System Description	Freq. (Years)	Crew	Unit	Labor Hours	2014 Bare Costs				Total In-House	Total w/O&P
						Material	Labor	Equipment	Total		
1010	**General maintenance & repair**	1	1 PLUM	Ea.							
	Inspect drain				.104		6		6	7.50	9.35
	Clean drain				.400		23		23	29	36
	Total				.504		29		29	**36.50**	**45.35**
1020	**Replace drain**	40	1 PLUM	Ea.	.800		41.50		41.50	52	64.50
	Remove drain				1.600	365	83		448	505	585
	Install roof drain										
	Total				2.400	365	124.50		489.50	**557**	**649.50**

D20 PLUMBING

D2043 Rain Water Drainage

D2043 310 Rainwater Sump Pump

	System Description	Freq. (Years)	Crew	Unit	Labor Hours	2014 Bare Costs				Total In-House	Total w/O&P
						Material	Labor	Equipment	Total		
1020	**Pump maintenance, 87 GPM**	1	1 PLUM	Ea.							
	Check for leaks				.010		.60		.60	.75	.93
	Check pump operation				.012		.67		.67	.84	1.05
	Check alignment				.053		3.07		3.07	3.84	4.79
	Lubricate pump and motor				.047		2.69		2.69	3.37	4.20
	Check suction or discharge pressure				.004		.22		.22	.28	.35
	Check packing glands, tighten				.014		.82		.82	1.03	1.29
	Fill out inspection report				.022		1.27		1.27	1.59	1.98
	Total				.163		9.34		9.34	11.70	14.59
1030	**Replace sump pump / motor assembly**	20	2 PLUM	Ea.							
	Remove sump pump assembly				1.300		75		75	93.50	117
	Install new cast iron sump pump (87 GPM)				2.078	269	120		389	445	525
	Total				3.378	269	195		464	538.50	642

D20 PLUMBING | D2093 Other Plumbing Systems

D2093 910 Pipe & Fittings, Industrial Gas

	System Description	Freq. (Years)	Crew	Unit	Labor Hours	2014 Bare Costs				Total In-House	Total w/O&P
						Material	Labor	Equipment	Total		
1010	**General maintenance**	2	1 PLUM	M.L.F.							
	General maintenance				.500		29		29	36	45
	Total				.500		29		29	**36**	**45**
1030	**Replace pipe and fittings**	75	2 PLUM	L.F.							
	Shut off valve				1.067		61.50		61.50	77	96
	Remove old pipe				.162		8.40		8.40	10.50	13.10
	Install 2" black steel pipe, couplings, and hangers				.325	8.95	16.85		25.80	31	37.50
	Check for leaks				.889		51		51	64	80
	Total				2.443	8.95	137.75		146.70	182.50	226.60

D2093 920 Pipe & Fittings, Anesthesia

	System Description	Freq. (Years)	Crew	Unit	Labor Hours	2014 Bare Costs				Total In-House	Total w/O&P
						Material	Labor	Equipment	Total		
1010	**Resolder joint**	12	1 PLUM	Ea.							
	Measure, cut & clean both ends				.053		3.07		3.07	3.84	4.79
	Attach fitting				.578	9.60	33.50		43.10	52	64
	Total				.631	9.60	36.57		46.17	**55.84**	**68.79**
1030	**Replace pipe and fittings**	25	2 PLUM	L.F.							
	Remove old pipe				.069		3.99		3.99	4.99	6.25
	Install new 3/4" copper tubing, couplings, hangers				.141	7.80	8.10		15.90	18.70	22.50
	Total				.210	7.80	12.09		19.89	23.69	28.75

D2093 930 Pipe & Fittings, Oxygen

	System Description	Freq. (Years)	Crew	Unit	Labor Hours	2014 Bare Costs				Total In-House	Total w/O&P
						Material	Labor	Equipment	Total		
1010	**Resolder joint**	12	1 PLUM	Ea.							
	Measure, cut & clean both ends				.289		16.65		16.65	21	26
	Attach fitting				.578	9.60	33.50		43.10	52	64
	Total				.867	9.60	50.15		59.75	**73**	**90**

D20 PLUMBING   D2093 Other Plumbing Systems

D2093 930 Pipe & Fittings, Oxygen

	System Description	Freq. (Years)	Crew	Unit	Labor Hours	2014 Bare Costs				Total In-House	Total w/O&P
						Material	Labor	Equipment	Total		
1030	**Replace pipe and fittings**	25	2 PLUM	L.F.							
	Remove old pipe				.069		3.99		3.99	4.99	6.25
	Install new 3/4" copper tubing, couplings, hangers				.141	7.80	8.10		15.90	18.70	22.50
	Total				.210	7.80	12.09		19.89	23.69	28.75

D2093 940 Pipe & Fittings, Compressed Air

	System Description	Freq. (Years)	Crew	Unit	Labor Hours	2014 Bare Costs				Total In-House	Total w/O&P
						Material	Labor	Equipment	Total		
1010	**General maintenance**	2	1 PLUM	M.L.F.							
	General maintenance				.500		29		29	36	45
	Total				.500		29		29	**36**	**45**
1030	**Replace pipe and fittings**	75	2 PLUM	L.F.							
	Shut off valve				1.067		61.50		61.50	77	96
	Remove old pipe				.052		2.99		2.99	3.74	4.67
	Install 2" black steel pipe, couplings, and hangers				.325	8.95	16.85		25.80	31	37.50
	Check for leaks				.889		51		51	64	80
	Total				2.333	8.95	132.34		141.29	**175.74**	**218.17**

D20 PLUMBING D2093 Other Plumbing Systems

D2093 946 Compressed Air Systems, Compressors

	System Description	Freq. (Years)	Crew	Unit	Labor Hours	2014 Bare Costs Material	2014 Bare Costs Labor	2014 Bare Costs Equipment	Total	Total In-House	Total w/O&P
1004	**Check and adjust 3/4 H.P. compressor**	1	1 SPRI	Ea.							
	Check compressor oil level				.022		1.27		1.27	1.59	1.98
	Check V-belt tension				.038		2.17		2.17	2.72	3.39
	Drain moisture from air tank				.033		1.87		1.87	2.34	2.92
	Clean air intake filter on compressor				.178		10.25		10.25	12.80	15.95
	Clean reusable oil filter				.182		10.45		10.45	13.10	16.35
	Check pressure relief valve				.108		6.20		6.20	7.80	9.70
	Clean cooling fins				.105		6.05		6.05	7.55	9.45
	Check motor				.060		3.44		3.44	4.30	5.35
	Check control circuits				.182		10.45		10.45	13.10	16.35
	Operation checks				.222		12.80		12.80	16	19.95
	Fill out maintenance report				.022		1.27		1.27	1.59	1.98
	Total				1.151		66.22		66.22	82.89	103.37
1006	**Replace 3/4 H.P. compressor**	25	1 SPRI	Ea.							
	Remove air compressor				4.457		243		243	305	380
	Install new 3/4 H.P. compressor				6.250	735	345		1,080	1,250	1,450
	Total				10.707	735	588		1,323	1,555	1,830
1010	**Check and adjust 2 H.P. compressor**	1	1 PLUM	Ea.							
	Check compressor oil level				.022		1.27		1.27	1.59	1.98
	Check V-belt tension				.038		2.17		2.17	2.72	3.39
	Drain moisture from air tank				.033		1.87		1.87	2.34	2.92
	Clean air intake filter on compressor				.178		10.25		10.25	12.80	15.95
	Clean reusable oil filter				.182		10.45		10.45	13.10	16.35
	Check pressure relief valve				.108		6.20		6.20	7.80	9.70
	Clean cooling fins				.105		6.05		6.05	7.55	9.45
	Check motor				.060		3.44		3.44	4.30	5.35
	Check control circuits				.182		10.45		10.45	13.10	16.35
	Operation checks				.222		12.80		12.80	16	19.95
	Fill out maintenance report				.022		1.27		1.27	1.59	1.98
	Total				1.151		66.22		66.22	82.89	103.37
1030	**Replace 2 H.P. compressor**	25	2 PLUM	Ea.							
	Remove air compressor				4.457		243		243	305	380
	Install new 2 H.P. compressor				8.915	3,775	485		4,260	4,750	5,475
	Total				13.372	3,775	728		4,503	5,055	5,855

D20 PLUMBING D2093 Other Plumbing Systems

D2093 946 Compressed Air Systems, Compressors

	System Description	Freq. (Years)	Crew	Unit	Labor Hours	2014 Bare Costs				Total In-House	Total w/O&P
						Material	Labor	Equipment	Total		
3010	**Check and adjust 10 H.P. compressor**	1	1 PLUM	Ea.							
	Check compressor oil level				.022		1.27		1.27	1.59	1.98
	Check tension on V-belt				.038		2.17		2.17	2.72	3.39
	Drain moisture from air tank				.033		1.87		1.87	2.34	2.92
	Clean air intake filter on compressor				.178		10.25		10.25	12.80	15.95
	Clean reusable oil filter				.182		10.45		10.45	13.10	16.35
	Check operation of pressure valve				.108		6.20		6.20	7.80	9.70
	Clean cooling fins				.105		6.05		6.05	7.55	9.45
	Check motor				.060		3.44		3.44	4.30	5.35
	Check compressor circuits				.182		10.45		10.45	13.10	16.35
	Perform operation				.222		12.80		12.80	16	19.95
	Fill out maintenance report				.022		1.27		1.27	1.59	1.98
	Total				1.151		66.22		66.22	82.89	103.37
3030	**Replace 10 H.P. compressor**	25	2 PLUM	Ea.							
	Remove compressor				17.335		915		915	1,150	1,425
	Install 2-stage compressor 10 H.P.				34.632	6,825	1,825		8,650	9,775	11,400
	Total				51.967	6,825	2,740		9,565	10,925	12,825
4010	**Check and adjust 25 H.P. compressor**	1	1 PLUM	Ea.							
	Check compressor oil level				.022		1.27		1.27	1.59	1.98
	Check tension on V-belt				.038		2.17		2.17	2.72	3.39
	Drain moisture from air tank				.033		1.87		1.87	2.34	2.92
	Clean air intake filter on compressor				.178		10.25		10.25	12.80	15.95
	Clean reusable oil filter				.182		10.45		10.45	13.10	16.35
	Check operation of pressure valve				.108		6.20		6.20	7.80	9.70
	Clean cooling fins				.105		6.05		6.05	7.55	9.45
	Check motor				.060		3.44		3.44	4.30	5.35
	Check compressor circuits				.182		10.45		10.45	13.10	16.35
	Perform operation				.222		12.80		12.80	16	19.95
	Fill out maintenance report				.022		1.27		1.27	1.59	1.98
	Total				1.151		66.22		66.22	82.89	103.37
4030	**Replace 25 H.P. compressor**	25	2 PLUM	Ea.							
	Remove compressor				26.002		1,425		1,425	1,775	2,225
	Install 2-stage compressor 25 H.P.				51.948	14,100	2,825		16,925	19,100	22,100
	Total				77.950	14,100	4,250		18,350	20,875	24,325

D20 PLUMBING D2093 Other Plumbing Systems

D2093 946 Compressed Air Systems, Compressors

	System Description	Freq. (Years)	Crew	Unit	Labor Hours	2014 Bare Costs				Total In-House	Total w/O&P
						Material	Labor	Equipment	Total		
5030	**Check operation**										
	Check operation of unit	1	1 STPI	Ea.	.320		18.70		18.70	23.50	29
	Total				.320		18.70		18.70	**23.50**	**29**

D30 HVAC | D3013 Energy Supply

D3013 110 Fuel Oil Storage Tank, 275 Gallon

	System Description	Freq. (Years)	Crew	Unit	Labor Hours	2014 Bare Costs				Total In-House	Total w/O&P
						Material	Labor	Equipment	Total		
0010	Replace 275 gallon fuel storage tank	30	Q-5	Ea.							
	Remove 275 gallon tank				2.081		110		110	137	171
	Install 275 gallon tank				4.156	490	219		709	815	955
	Total				6.236	490	329		819	952	1,126

D3013 150 Fuel Level Meter

	System Description	Freq. (Years)	Crew	Unit	Labor Hours	2014 Bare Costs				Total In-House	Total w/O&P
						Material	Labor	Equipment	Total		
0010	Preventive maintenance	5	1 STPI	Ea.							
	Check fuel level meter calibration				.500		29.50		29.50	36.50	45.50
	Total				.500		29.50		29.50	36.50	45.50
0020	Replace remote tank fuel gauge	20	1 STPI	Ea.							
	Remove remote read gauge				2.081		122		122	152	190
	Replace remote read gauge 5" pointer travel				4.167	3,650	244		3,894	4,325	4,950
	Total				6.247	3,650	366		4,016	4,477	5,140

D3013 160 Oil Filter

	System Description	Freq. (Years)	Crew	Unit	Labor Hours	2014 Bare Costs				Total In-House	Total w/O&P
						Material	Labor	Equipment	Total		
0010	Preventive maintenance	1	1 STPI	Ea.							
	Replace filter element				.052	3	3.05		6.05	7.10	8.50
	Total				.052	3	3.05		6.05	7.10	8.50
0020	Replace filter housing	30	1 STPI	Ea.							
	Replace filter housing				.520	30	30.50		60.50	71	85
	Total				.520	30	30.50		60.50	71	85

D30 HVAC   D3013 Energy Supply

D3013 170   Fuel Oil Storage: Pipe & Fittings, Copper

					2014 Bare Costs					
System Description	Freq. (Years)	Crew	Unit	Labor Hours	Material	Labor	Equipment	Total	Total In-House	Total w/O&P
0010 **Remake flare type joint**										
Remake flare type joint	10	1 STPI	M.L.F.	.286		16.70		16.70	21	26
Total				.286		16.70		16.70	**21**	**26**
0020 **Install 10' sect. 3/8" type L copper per M.L.F.**										
Remove section 3/8" diameter copper	20	1 PLUM	Ea.	.619		35.60		35.60	44.50	55.50
Install 10' section new 3/8" diameter type L				1.238	31.30	71		102.30	124	150
Total				1.857	31.30	106.60		137.90	**168.50**	**205.50**
0030 **Install 10' sect. 1/2" type L copper per M.L.F.**										
Remove section 1/2" diameter copper	20	1 PLUM	Ea.	.642		37		37	46	57.50
Install 10' section new 1/2" diameter type L				1.284	33.70	74		107.70	130	158
Total				1.926	33.70	111		144.70	**176**	**215.50**
0040 **Install 10' sect. 5/8" type L copper per M.L.F.**										
Remove section 5/8" diameter copper	20	1 PLUM	Ea.	.658		37.90		37.90	47.50	59
Install 10' section new 5/8" diameter type L				1.316	53.50	76		129.50	154	185
Total				1.975	53.50	113.90		167.40	**201.50**	**244**
0050 **Install 10' sect. 3/4" type L copper per M.L.F.**										
Remove section 3/4" diameter copper	20	1 PLUM	Ea.	.684		39.40		39.40	49	61.50
Install 10' section new 3/4" diameter type L				1.369	51.50	79		130.50	155	187
Total				2.053	51.50	118.40		169.90	**204**	**248.50**
0060 **Install 10' section 1" type L copper per M.L.F.**										
Remove section 1" diameter copper	20	1 PLUM	Ea.	.765		44		44	55	68.50
Install 10' section new 1" diameter type L				1.529	78.50	88		166.50	197	236
Total				2.294	78.50	132		210.50	**252**	**304.50**
0130 **Replace 1000' type L 3/8" copper**										
Remove section 3/8" diameter copper	25	1 PLUM	M.L.F.	61.920		3,560		3,560	4,450	5,550
Install 1000' section new 3/8" diameter type L				123.800	3,130	7,100		10,230	12,400	15,000
Total				185.720	3,130	10,660		13,790	**16,850**	**20,550**
0140 **Replace 1000' type L 1/2" copper**										
Remove section 1/2" diameter copper	25	1 PLUM	M.L.F.	64.220		3,700		3,700	4,625	5,750
Install 1000' section new 1/2" diameter type L				128.390	3,370	7,400		10,770	13,000	15,800
Total				192.610	3,370	11,100		14,470	**17,625**	**21,550**

D30 HVAC | D3013 Energy Supply

D3013 170 Fuel Oil Storage: Pipe & Fittings, Copper

	System Description	Freq. (Years)	Crew	Unit	Labor Hours	2014 Bare Costs				Total In-House	Total w/O&P
						Material	Labor	Equipment	Total		
0150	**Replace 1000' type L 5/8" copper**	25	1 PLUM	M.L.F.							
	Remove section 5/8" diameter copper				65.840		3,790		3,790	4,750	5,900
	Install 1000' section new 5/8" diameter type L				131.640	5,350	7,600		12,950	15,400	18,500
	Total				197.480	5,350	11,390		16,740	20,150	24,400
0160	**Replace 1000' type L 3/4" copper**	25	1 PLUM	M.L.F.							
	Remove section 3/4" diameter copper				68.440		3,940		3,940	4,925	6,150
	Install 1000' section new 3/4" diameter type L				136.850	5,150	7,900		13,050	15,500	18,700
	Total				205.290	5,150	11,840		16,990	20,425	24,850
0170	**Replace 1000' type L 1" copper**	25	1 PLUM	M.L.F.							
	Remove section 1" diameter copper				76.480		4,400		4,400	5,500	6,850
	Install 1000' section new 1" diameter type L				152.930	7,850	8,800		16,650	19,700	23,600
	Total				229.410	7,850	13,200		21,050	25,200	30,450

D3013 210 Natural Gas: Pipe & Fittings, Steel/Iron

	System Description	Freq. (Years)	Crew	Unit	Labor Hours	2014 Bare Costs				Total In-House	Total w/O&P
						Material	Labor	Equipment	Total		
0010	**Install new 2" gasket, 1 per M.L.F.**	30	1 STPI	Ea.							
	Disconnect joint, remove gasket & clean faces				.803		47.18		47.18	58.50	73
	Install gasket and make up bolts				.800	4.85	47		51.85	64	79
	Total				1.603	4.85	94.18		99.03	122.50	152
0020	**Install new 3" gasket, 1 per M.L.F.**	30	1 STPI	Ea.							
	Disconnect joint, remove gasket & clean faces				.945		55.50		55.50	69	86
	Install gasket and make up bolts				.945	5.80	55.50		61.30	75.50	93.50
	Total				1.890	5.80	111		116.80	144.50	179.50
0030	**Install new 4" gasket, 1 per M.L.F.**	30	1 STPI	Ea.							
	Disconnect joint, remove gasket & clean faces				1.300		76		76	95	119
	Install gasket and make up bolts				1.300	10.40	75		85.40	105	130
	Total				2.600	10.40	151		161.40	200	249

D30 HVAC

D3013 Energy Supply

D3013 210 Natural Gas: Pipe & Fittings, Steel/Iron

	System Description	Freq. (Years)	Crew	Unit	Labor Hours	2014 Bare Costs				Total In-House	Total w/O&P
						Material	Labor	Equipment	Total		
0040	**Install new 6" gasket, 1 per M.L.F.**	30	1 STPI	Ea.							
	Disconnect joint, remove gasket & clean faces				1.733		101		101	127	158
	Install gasket and make up bolts				1.733	18.10	101		119.10	147	181
	Total				3.466	18.10	202		220.10	274	339
0110	**Replace 10' of buried 2" diam. st. pipe/M.L.F.**	12	Q-4	Ea.							
	Check break or leak				.016		.94		.94	1.17	1.46
	Turn valve on and off				.010		.59		.59	.73	.91
	Excavate by hand				5.771		211.09		211.09	282	345
	Remove broken 2" diameter section				.743		40.70	1.30	42	52	65
	Install 10' of pipe coated & wrapped				1.143	82.50	62.50	2	147	171	203
	Check for leaks, additional section				.900		53		53	66	82
	Backfill by hand				2.101		76.66		76.66	102	127
	Total				10.684	82.50	445.48	3.30	531.28	674.90	824.37
0120	**Replace 10' of buried 3" diam. st. pipe/M.L.F.**	12	Q-4	Ea.							
	Check break or leak				.016		.94		.94	1.17	1.46
	Turn valve on and off				.010		.59		.59	.73	.91
	Excavate by hand				5.771		211.09		211.09	282	345
	Remove broken 3" diameter section				.800		43.80	1.40	45.20	56	70
	Install 10' of pipe coated & wrapped				1.231	136.50	67.50	2.10	206.10	235	278
	Check for leaks, additional section				.900		53		53	66	82
	Backfill by hand				2.101		76.66		76.66	102	127
	Total				10.829	136.50	453.58	3.50	593.58	742.90	904.37
0130	**Replace 10' of buried 4" diam. st. pipe/M.L.F.**	12	B-35	Ea.							
	Check break or leak				.016		.94		.94	1.17	1.46
	Turn valve on and off				.010		.59		.59	.73	.91
	Excavate by hand				5.771		211.09		211.09	282	345
	Remove broken 4" diameter section				1.224		56	18.30	74.30	90.50	110
	Install 10' of pipe coated & wrapped				1.882	177.50	86	28.10	291.60	335	390
	Check for leaks, additional section				.900		53		53	66	82
	Backfill by hand				2.101		76.66		76.66	102	127
	Total				11.905	177.50	484.28	46.40	708.18	877.40	1,056.37

D30 HVAC

D3013 Energy Supply

D3013 210 Natural Gas: Pipe & Fittings, Steel/Iron

System Description	Freq. (Years)	Crew	Unit	Labor Hours	2014 Bare Costs Material	2014 Bare Costs Labor	2014 Bare Costs Equipment	2014 Bare Costs Total	Total In-House	Total w/O&P
0140 Replace 10' of buried 6" diam. st. pipe/M.L.F.	12	B-35	Ea.							
Check break or leak				.016		.94		.94	1.17	1.46
Turn valve on and off				.010		.59		.59	.73	.91
Excavate by hand				5.771		211.09		211.09	282	345
Remove broken 6" diameter section				1.734		.79	25.90	104.90	129	156
Install 10' of pipe coated & wrapped				2.667	315	122	39.80	476.80	545	635
Check for leaks, additional section				.900		53		53	66	82
Backfill by hand				2.101		76.66		76.66	102	127
Total				13.199	315	543.28	65.70	923.98	1,125.90	1,347.37
0210 Replace 1000 L.F. of buried 2" diam. st. pipe	75	Q-4	M.L.F.							
Check break or leak				1.600		93.50		93.50	117	146
Turn valve on and off				1.000		58.50		58.50	73	91.50
Excavate with 1/2 C.Y. backhoe				23.111		988.89	573.33	1,562.22	1,850	2,225
Remove broken 2" diameter section				74.250		4,070	130	4,200	5,200	6,500
Install 1000' of pipe coated & wrapped				114.290	8,250	6,250	200	14,700	17,100	20,300
Check for leaks, additional section				9.000		530		530	660	820
Backfill with 1 C.Y. bucket				8.667		388.89	213.33	602.22	715	860
Total				231.918	8,250	12,379.78	1,116.66	21,746.44	25,715	30,942.50
0220 Replace 1000 L.F. of buried 3" diam. st. pipe	75	Q-4	M.L.F.							
Check break or leak				1.600		93.50		93.50	117	146
Turn valve on and off				1.000		58.50		58.50	73	91.50
Excavate with 1/2 C.Y. backhoe				23.111		988.89	573.33	1,562.22	1,850	2,225
Remove broken 3" diameter section				80.000		4,380	140	4,520	5,600	7,000
Install 1000' of pipe coated & wrapped				123.080	13,650	6,750	210	20,610	23,500	27,800
Check for leaks, additional section				9.000		530		530	660	820
Backfill with 1 C.Y. bucket				8.667		388.89	213.33	602.22	715	860
Total				246.458	13,650	13,189.78	1,136.66	27,976.44	32,515	38,942.50
0230 Replace 1000 L.F. of buried 4" diam. st. pipe	75	B-35	M.L.F.							
Check break or leak				1.600		93.50		93.50	117	146
Turn valve on and off				1.000		58.50		58.50	73	91.50
Excavate with 1/2 C.Y. backhoe				23.111		988.89	573.33	1,562.22	1,850	2,225
Remove broken 4" diameter section				122.390		5,600	1,830	7,430	9,050	11,000
Install 1000' of pipe coated & wrapped				188.240	17,750	8,600	2,810	29,160	33,500	39,100
Check for leaks, additional section				9.000		530		530	660	820
Backfill with 1 C.Y. bucket				8.667		388.89	213.33	602.22	715	860
Total				354.008	17,750	16,259.78	5,426.66	39,436.44	45,965	54,242.50

D30 HVAC

D3013 Energy Supply

D3013 210 Natural Gas: Pipe & Fittings, Steel/Iron

	System Description	Freq. (Years)	Crew	Unit	Labor Hours	2014 Bare Costs				Total In-House	Total w/O&P
						Material	Labor	Equipment	Total		
0240	**Replace 1000 L.F. of buried 6″ diam. st. pipe**	75	B-35	M.L.F.							
	Check break or leak				1.600		93.50		93.50	117	146
	Turn valve on and off				1.000		58.50		58.50	73	91.50
	Excavate with 1/2 C.Y. backhoe				23.111		988.89	573.33	1,562.22	1,850	2,225
	Remove broken 6″ diameter section				173.390		7,900	2,590	10,490	12,900	15,600
	Install 1000′ of pipe coated & wrapped				266.670	31,500	12,200	3,980	47,680	54,500	63,500
	Check for leaks, additional section				9.000		530		530	660	820
	Backfill with 1 C.Y. bucket				8.667		388.89	213.33	602.22	715	860
	Total				483.438	31,500	22,159.78	7,356.66	61,016.44	70,815	83,242.50
0410	**Replace 10′ of hung 2″ diam. st. pipe/M.L.F.**	12	Q-1	Ea.							
	Check break or leak				.016		.94		.94	1.17	1.46
	Turn valve on and off				1.000		59		59	73	91
	Remove broken 2″ diameter section				1.626		84		84	106	132
	Install 10′ of black 2″ diameter steel pipe				3.250	89.50	168.50		258	310	375
	Check for leaks, additional section				.900		53		53	66	82
	Total				6.792	89.50	365.44		454.94	556.17	681.46
0420	**Replace 10′ of hung 3″ diam. st. pipe/M.L.F.**	12	Q-15	Ea.							
	Check break or leak				.016		.94		.94	1.17	1.46
	Turn valve on and off				.010		.59		.59	.73	.91
	Remove broken 3″ diameter section				2.420		125.50	8.40	133.90	165	205
	Install 10′ of black 3″ diameter steel pipe				4.837	147	250	16.70	413.70	490	590
	Check for leaks, additional section				.900		53		53	66	82
	Total				8.182	147	430.03	25.10	602.13	722.90	879.37
0430	**Replace 10′ of hung 4″ diam. st. pipe/M.L.F.**	12	Q-15	Ea.							
	Check break or leak				.016		.94		.94	1.17	1.46
	Turn valve on and off				.010		.59		.59	.73	.91
	Remove broken 4″ diameter section				2.812		145.50	9.70	155.20	192	236
	Install 10′ of black 4″ diameter steel pipe				5.622	215	290	19.50	524.50	620	745
	Check for leaks, additional section				.900		53		53	66	82
	Total				9.359	215	490.03	29.20	734.23	879.90	1,065.37

D30 HVAC — D3013 Energy Supply

D3013 210 — Natural Gas: Pipe & Fittings, Steel/Iron

	System Description	Freq. (Years)	Crew	Unit	Labor Hours	2014 Bare Costs — Material	2014 Bare Costs — Labor	2014 Bare Costs — Equipment	2014 Bare Costs — Total	Total In-House	Total w/O&P
0440	**Replace 10' of hung 6" diam. st. pipe/M.L.F.**	12	Q-16	Ea.							
	Check break or leak				.016		.94		.94	1.17	1.46
	Turn valve on and off				.010		.59		.59	.73	.91
	Remove broken 6" diameter section				4.335		235	10	245	300	375
	Install 10' of black 6" diameter steel pipe				8.667	435	465	19.90	919.90	1,075	1,300
	Check for leaks, additional section				.900		53		53	66	82
	Total				13.928	435	754.53	29.90	1,219.43	1,442.90	1,759.37
0510	**Replace 1000 L.F. of hung 2" diam. steel pipe**	75	Q-1	M.L.F.							
	Check break or leak				1.600		93.50		93.50	117	146
	Turn valve on and off				100.000		58.50		58.50	73	91.50
	Remove broken 2" diameter section				162.550		5,850		5,850	7,300	9,150
	Install 1000' of black 2" diameter steel pipe				325.000	8,950	16,850		25,800	31,000	37,700
	Check for leaks, additional section				9.000		530		530	660	820
	Total				598.150	8,950	31,723.50		40,673.50	49,677	61,016
0520	**Replace 1000 L.F. of hung 3" diam. steel pipe**	75	Q-15	M.L.F.							
	Check break or leak				1.600		93.50		93.50	117	146
	Turn valve on and off				1.000		58.50		58.50	73	91.50
	Remove broken 3" diameter section				241.950		12,550	840	13,390	16,500	20,500
	Install 1000' of black 3" diameter steel pipe				483.680	14,700	25,000	1,670	41,370	49,000	59,000
	Check for leaks, additional section				9.000		530		530	660	820
	Total				737.230	14,700	38,232	2,510	55,442	66,350	80,557.50
0530	**Replace 1000 L.F. of hung 4" diam. steel pipe**	75	Q-15	M.L.F.							
	Check break or leak				1.600		93.50		93.50	117	146
	Turn valve on and off				1.000		58.50		58.50	73	91.50
	Remove broken 4" diameter section				281.150		14,550	970	15,520	19,200	23,600
	Install 1000' of black 4" diameter steel pipe				562.190	21,500	29,000	1,950	52,450	62,000	74,500
	Check for leaks, additional section				9.000		530		530	660	820
	Total				854.940	21,500	44,232	2,920	68,652	82,050	99,157.50

D30 HVAC   D3013 Energy Supply

D3013 210   Natural Gas: Pipe & Fittings, Steel/Iron

	System Description	Freq. (Years)	Crew	Unit	Labor Hours	2014 Bare Costs				Total In-House	Total w/O&P
						Material	Labor	Equipment	Total		
0540	**Replace 1000 L.F. of hung 6" diam. steel pipe**	75	Q-16	M.L.F.							
	Check break or leak				1.600		93.50		93.50	117	146
	Turn valve on and off				1.000		58.50		58.50	73	91.50
	Remove broken 6" diameter section				433.450		23,500	1,000	24,500	30,000	37,600
	Install 1000' of black 6" diameter steel pipe				866.740	43,500	46,500	1,990	91,990	108,000	129,000
	Check for leaks, additional section				9.000		530		530	660	820
	Total				1311.790	43,500	70,682	2,990	117,172	138,850	167,657.50

D3013 240   Natural Gas: Pressure Reducing Valve

	System Description	Freq. (Years)	Crew	Unit	Labor Hours	2014 Bare Costs				Total In-House	Total w/O&P
						Material	Labor	Equipment	Total		
0010	**Check gas pressure**	5	1 STPI	Ea.							
	Check pressure				.130		7.60		7.60	9.50	11.85
	Total				.130		7.60		7.60	9.50	11.85
0110	**Replace pressure regulator 1/2" diam. pipe**	14	1 STPI	Ea.							
	Turn valve on and off				.010		.59		.59	.73	.91
	Remove 1/2" diameter regulator				.217		12.70		12.70	15.85	19.80
	Replace 1/2" diameter pipe size regulator				.433	47.50	25.50		73	84	99
	Total				.660	47.50	38.79		86.29	100.58	119.71
0120	**Replace pressure regulator 1" diam. pipe**	14	1 STPI	Ea.							
	Turn valve on and off				.010		.59		.59	.73	.91
	Remove 1" diameter regulator				.274		16		16	20	25
	Replace 1" diameter pipe size regulator				.547	97.50	32		129.50	147	172
	Total				.831	97.50	48.59		146.09	167.73	197.91
0130	**Replace pressure regulator 1-1/2" diam. pipe**	14	1 STPI	Ea.							
	Turn valve on and off				.010		.59		.59	.73	.91
	Remove 1-1/2" diameter regulator				.400		23.50		23.50	29.50	36.50
	Replace 1-1/2" diameter pipe size regulator				.800	620	47		667	740	850
	Total				1.210	620	71.09		691.09	770.23	887.41

D30 HVAC

D3013 Energy Supply

D3013 240 Natural Gas: Pressure Reducing Valve

	System Description	Freq. (Years)	Crew	Unit	Labor Hours	Material	2014 Bare Costs Labor	2014 Bare Costs Equipment	2014 Bare Costs Total	Total In-House	Total w/O&P
0140	**Replace pressure regulator 2" diam. pipe**	14	1 STPI	Ea.							
	Turn valve on and off				.010		.59		.59	.73	.91
	Remove 2" diameter regulator				.473		27.50		27.50	34.50	43
	Replace 2" diameter pipe size regulator				.946	620	55.50		675.50	750	860
	Total				1.428	620	83.59		703.59	785.23	903.91

D3013 260 LPG Distribution: Pipe & Fittings, Steel/Iron

	System Description	Freq. (Years)	Crew	Unit	Labor Hours	Material	2014 Bare Costs Labor	2014 Bare Costs Equipment	2014 Bare Costs Total	Total In-House	Total w/O&P
0120	**Replace 10' st. pipe 1/2" diam. per M.L.F.**	12	1 PLUM	Ea.							
	Check break or leak				.016		.94		.94	1.17	1.46
	Turn valve on and off				.010		.59		.59	.73	.91
	Remove broken 1/2" diameter section				.826		47.50		47.50	59.50	74
	Install section of pipe 1/2" diameter				1.651	27.30	95		122.30	149	182
	Check for leaks, additional section				.900		53		53	66	82
	Total				3.402	27.30	197.03		224.33	276.40	340.37
0220	**Replace 10' st. pipe 3/4" diam. per M.L.F.**	12	1 PLUM	Ea.							
	Check break or leak				.016		.94		.94	1.17	1.46
	Turn valve on and off				.010		.59		.59	.73	.91
	Remove broken 3/4" diameter section				.853		49.10		49.10	61.50	76.50
	Install section of pipe 3/4" diameter				1.312	31.70	75.50		107.20	130	158
	Check for leaks, additional section				.900		53		53	66	82
	Total				3.090	31.70	179.13		210.83	259.40	318.87
0320	**Replace 10' st. pipe 1" diam. per M.L.F.**	12	1 PLUM	Ea.							
	Check break or leak				.016		.94		.94	1.17	1.46
	Turn valve on and off				.010		.59		.59	.73	.91
	Remove broken 1" diameter section				.982		56.50		56.50	70.50	88
	Install section of pipe 1" diameter				1.962	46.40	113		159.40	192	234
	Check for leaks, additional section				.900		53		53	66	82
	Total				3.870	46.40	224.03		270.43	330.40	406.37

D30 HVAC

D3013 Energy Supply

D3013 260 LPG Distribution: Pipe & Fittings, Steel/Iron

	System Description	Freq. (Years)	Crew	Unit	Labor Hours	2014 Bare Costs					Total In-House	Total w/O&P
						Material	Labor	Equipment	Total		
0420	**Replace 10' st. pipe 1-1/4" diam. M.L.F.**	12	Q-1	Ea.							
	Check break or leak				.016		.94		.94	1.17	1.46
	Turn valve on and off				.010		.59		.59	.73	.91
	Remove broken 1-1/4" diameter section				1.169		60.50		60.50	75.50	94.50
	Install section of pipe 1-1/4" diameter				2.337	57.50	121		178.50	215	261
	Check for leaks, additional section				.900		53		53	66	82
	Total				4.432	57.50	236.03		293.53	358.40	439.87
0520	**Replace 10' st. pipe 1-1/2" diam. M.L.F.**	12	Q-1	Ea.							
	Check break or leak				.016		.94		.94	1.17	1.46
	Turn valve on and off				.010		.59		.59	.73	.91
	Remove broken 1-1/2" diameter section				1.300		67.50		67.50	84	105
	Install section of pipe 1-1/2" diameter				2.600	67	134.50		201.50	240	294
	Check for leaks, additional section				.900		53		53	66	82
	Total				4.826	67	256.53		323.53	391.90	483.37
0620	**Replace 10' section st. pipe 2" diam. M.L.F.**	12	Q-1	Ea.							
	Check break or leak				.016		.94		.94	1.17	1.46
	Turn valve on and off				.010		.59		.59	.73	.91
	Remove broken 2" diameter section				1.626		84		84	106	132
	Install section of pipe 2" diameter				3.250	89.50	168.50		258	310	375
	Check for leaks, additional section				.900		53		53	66	82
	Total				5.802	89.50	307.03		396.53	483.90	591.37
1130	**Replace 1000' of 1/2" diameter steel pipe**	75	Q-1	M.L.F.							
	Check break or leak				1.600		93.50		93.50	117	146
	Turn valve on and off				1.000		58.50		58.50	73	91.50
	Remove broken 1/2" diameter section				82.560		4,750		4,750	5,950	7,400
	Install section of pipe 1/2" diameter				165.080	2,730	9,500		12,230	14,900	18,200
	Check for leaks, additional section				9.000		530		530	660	820
	Total				259.240	2,730	14,932		17,662	21,700	26,657.50

D30 HVAC | D3013 Energy Supply

D3013 260 LPG Distribution: Pipe & Fittings, Steel/Iron

	System Description	Freq. (Years)	Crew	Unit	Labor Hours	2014 Bare Costs Material	2014 Bare Costs Labor	2014 Bare Costs Equipment	2014 Bare Costs Total	Total In-House	Total w/O&P
1230	**Replace 1000' of 3/4" diameter steel pipe**	75	Q-1	M.L.F.							
	Check break or leak				1.600		93.50		93.50	117	146
	Turn valve on and off				1.000		58.50		58.50	73	91.50
	Remove broken 3/4" diameter section				85.270		4,910		4,910	6,150	7,650
	Install section of pipe 3/4" diameter				131.150	3,170	7,550		10,720	13,000	15,800
	Check for leaks, additional section				9.000		530		530	660	820
	Total				228.020	3,170	13,142		16,312	**20,000**	**24,507.50**
1330	**Replace 1000' of 1" diameter steel pipe**	75	Q-1	M.L.F.							
	Check break or leak				1.600		93.50		93.50	117	146
	Turn valve on and off				1.000		58.50		58.50	73	91.50
	Remove broken 1" diameter section				98.150		5,650		5,650	7,050	8,800
	Install section of pipe 1" diameter				196.220	4,640	11,300		15,940	19,200	23,400
	Check for leaks, additional section				9.000		530		530	660	820
	Total				305.970	4,640	17,632		22,272	**27,100**	**33,257.50**
1430	**Replace 1000' of 1-1/4" diameter steel pipe**	75	Q-1	M.L.F.							
	Check break or leak				1.600		93.50		93.50	117	146
	Turn valve on and off				1.000		58.50		58.50	73	91.50
	Remove broken 1-1/4" diameter section				116.890		6,050		6,050	7,550	9,450
	Install section of pipe 1-1/4" diameter				233.710	5,750	12,100		17,850	21,500	26,100
	Check for leaks, additional section				9.000		530		530	660	820
	Total				362.200	5,750	18,832		24,582	**29,900**	**36,607.50**
1530	**Replace 1000' of 1-1/2" diameter steel pipe**	75	Q-1	M.L.F.							
	Check break or leak				1.600		93.50		93.50	117	146
	Turn valve on and off				1.000		58.50		58.50	73	91.50
	Remove broken 1-1/2" diameter section				130.040		6,750		6,750	8,400	10,500
	Install section of pipe 1-1/2" diameter				260.000	6,700	13,450		20,150	24,000	29,400
	Check for leaks, additional section				9.000		530		530	660	820
	Total				401.640	6,700	20,882		27,582	**33,250**	**40,957.50**

D30 HVAC

D3013 Energy Supply

D3013 260 LPG Distribution: Pipe & Fittings, Steel/Iron

	System Description	Freq. (Years)	Crew	Unit	Labor Hours	Material	Labor	Equipment	Total	Total In-House	Total w/O&P
							2014 Bare Costs				
1630	Replace 1000' of 2" diameter steel pipe	75	Q-1	M.L.F.							
	Check break or leak				1.600		93.50		93.50	117	146
	Turn valve on and off				1.000		58.50		58.50	73	91.50
	Remove broken 2" diameter section				162.550		8,400		8,400	10,600	13,200
	Install section of pipe 2" diameter				325.000	8,950	16,850		25,800	31,000	37,700
	Check for leaks, additional section				9.000		530		530	660	820
	Total				499.150	8,950	25,932		34,882	42,450	51,957.50

D3013 601 Solar Panel, 3' x 8'

	System Description	Freq. (Years)	Crew	Unit	Labor Hours	Material	Labor	Equipment	Total	Total In-House	Total w/O&P
							2014 Bare Costs				
0010	Replace solar panel 3' x 8'	15	Q-1	Ea.							
	Remove panel				1.891		98		98	122	153
	Install 3' x 8' solar panel aluminum frame				3.783	915	196		1,111	1,250	1,450
	Total				5.674	915	294		1,209	1,372	1,603

D30 HVAC D3023 Heat Generating Systems

D3023 180 Boiler, Gas

		System Description	Freq. (Years)	Crew	Unit	Labor Hours	2014 Bare Costs				Total In-House	Total w/O&P
							Material	Labor	Equipment	Total		
1010		**Repair boiler, gas, 250 MBH**	7	1 STPI	Ea.							
		Remove / replace burner blower				.600		35		35	44	55
		Remove burner blower bearing				1.000		58.50		58.50	73	91.50
		Replace burner blower bearing				2.000	55.50	117		172.50	207	252
		Remove burner blower motor				.976		57		57	71.50	89
		Replace burner blower motor				1.951	192	114		306	355	420
		Remove burner fireye				.195		11.40		11.40	14.25	17.80
		Replace burner fireye				.300	208	17.55		225.55	251	288
		Remove burner gas regulator				.274		16		16	20	25
		Replace burner gas regulator				.547	97.50	32		129.50	147	172
		Remove burner auto gas valve				.274		16		16	20	25
		Replace burner auto gas valve				.421	172	24.50		196.50	220	254
		Remove burner solenoid valve				.279		15.65		15.65	19.50	24.50
		Replace burner solenoid valve				.557	595	31.50		626.50	695	790
		Repair controls				.500		29.50		29.50	36.50	45.50
		Total				9.874	1,320	575.60		1,895.60	2,173.75	2,549.30
1060		**Replace boiler, gas, 250 MBH**	30	Q-7	Ea.							
		Remove boiler				21.888		1,225		1,225	1,525	1,900
		Replace boiler, 250 MBH				43.776	3,550	2,450		6,000	6,950	8,250
		Total				65.663	3,550	3,675		7,225	8,475	10,150
2010		**Repair boiler, gas, 2000 MBH**	7	Q-5	Ea.							
		Remove / replace burner blower				.600		35		35	44	55
		Remove burner blower bearing				1.000		58.50		58.50	73	91.50
		Replace burner blower bearing				2.000	55.50	117		172.50	207	252
		Remove burner blower motor				.976		57		57	71.50	89
		Replace burner blower motor				1.951	198	104		302	345	410
		Remove burner fireye				.195		11.40		11.40	14.25	17.80
		Replace burner fireye				.300	208	17.55		225.55	251	288
		Remove burner gas regulator				.727		38.50		38.50	48	59.50
		Replace burner gas regulator				1.455	1,200	76.50		1,276.50	1,425	1,625
		Remove burner auto gas valve				.800		41.50		41.50	52	64.50
		Replace burner auto gas valve				1.231	370	64		434	485	560
		Remove burner solenoid valve				2.600		146		146	182	227
		Replace burner solenoid valve				5.199	1,275	292		1,567	1,775	2,050
		Repair controls				.500		29.50		29.50	36.50	45.50
		Total				19.534	3,306.50	1,088.45		4,394.95	5,009.25	5,834.80

D30 HVAC D3023 Heat Generating Systems

D3023 180 Boiler, Gas

	System Description	Freq. (Years)	Crew	Unit	Labor Hours	2014 Bare Costs Material	Labor	Equipment	Total	Total In-House	Total w/O&P
2070	**Replace boiler, gas, 2000 MBH**	30	Q-7	Ea.							
	Remove boiler				57.762		3,225		3,225	4,025	5,025
	Replace boiler, 2000 MBH				109.000	19,600	6,100		25,700	29,200	34,000
	Total				166.762	19,600	9,325		28,925	33,225	39,025
3010	**Repair boiler, gas, 10,000 MBH**	7	Q-5	Ea.							
	Remove / replace burner blower				1.000		58.50		58.50	73	91.50
	Remove burner blower bearing				1.000		58.50		58.50	73	91.50
	Replace burner blower bearing				2.000	55.50	117		172.50	207	252
	Remove burner motor				1.000		58.50		58.50	73	91.50
	Replace burner motor				1.951	207	104		311	355	420
	Remove burner fireye				.195		11.40		11.40	14.25	17.80
	Replace burner fireye				.300	208	17.55		225.55	251	288
	Remove burner gas regulator				1.300		68.50		68.50	85.50	107
	Replace burner gas regulator				2.600	2,150	137		2,287	2,525	2,900
	Remove burner auto gas valve				5.195		279		279	350	435
	Replace burner auto gas valve				8.000	5,875	430		6,305	7,000	8,025
	Remove burner solenoid valve				2.600		146		146	182	227
	Replace burner solenoid valve				10.399	3,225	585		3,810	4,275	4,950
	Repair controls				.500		29.50		29.50	36.50	45.50
	Total				38.040	11,720.50	2,100.45		13,820.95	15,500.25	17,941.80
3070	**Replace boiler, gas, 10,000 MBH**	30	Q-7	Ea.							
	Remove boiler				364.000		20,300		20,300	25,300	31,600
	Replace boiler, gas, 10,000 MBH				722.400	154,700	40,180		194,880	220,500	256,000
	Total				1086.400	154,700	60,480		215,180	245,800	287,600

D30 HVAC | D3023 Heat Generating Systems

D3023 182 Boiler, Coal

	System Description	Freq. (Years)	Crew	Unit	Labor Hours	Material	2014 Bare Costs Labor	Equipment	Total	Total In-House	Total w/O&P
1010	**Repair boiler, coal, 4600 MBH**	20	4 STPI	Ea.							
	Remove coal feeder				2.703		158.60		158.60	198	247
	Replace coal feeder				4.054	1,560	236.60		1,796.60	2,025	2,325
	Remove vibrating grates				33.592		1,965.60		1,965.60	2,450	3,075
	Replace vibrating grates				62.712	19,864	3,671.20		23,535.20	26,400	30,600
	Remove motor				.832		48.88		48.88	61	76
	Replace motor				.450	228.80	24.02		252.82	281	325
	Repair controls				5.200		304.20		304.20	380	475
	Total				109.543	21,652.80	6,409.10		28,061.90	31,795	37,123
1050	**Replace boiler, coal, 4600 MBH**	30	5 STPI	Ea.							
	Remove boiler				122.304		7,176		7,176	8,950	11,200
	Replace boiler, coal, 4600 MBH				244.608	100,984	14,300		115,284	129,000	148,500
	Total				366.912	100,984	21,476		122,460	137,950	159,700

D3023 184 Boiler, Oil

	System Description	Freq. (Years)	Crew	Unit	Labor Hours	Material	2014 Bare Costs Labor	Equipment	Total	Total In-House	Total w/O&P
1010	**Repair boiler, oil, 250 MBH**	7	Q-5	Ea.							
	Remove / replace burner blower				.600		35		35	44	55
	Remove burner blower bearing				1.000		58.50		58.50	73	91.50
	Replace burner blower bearing				2.000	55.50	117		172.50	207	252
	Remove burner blower motor				.976		57		57	71.50	89
	Replace burner blower motor				1.951	192	114		306	355	420
	Remove burner fireye				.195		11.40		11.40	14.25	17.80
	Replace burner fireye				.300	208	17.55		225.55	251	288
	Remove burner ignition transformer				.250		14.65		14.65	18.30	23
	Replace burner ignition transformer				.500	80.50	29.50		110	125	146
	Remove burner ignition electrode				.200		11.70		11.70	14.65	18.25
	Replace burner ignition electrode				.400	16	23.50		39.50	47	56.50
	Remove burner oil pump				.350		20.50		20.50	25.50	32
	Replace burner oil pump				.700	74.50	41		115.50	133	157
	Remove burner nozzle				.167		9.75		9.75	12.20	15.20
	Replace burner nozzle				.300	13.50	17.55		31.05	37	44.50
	Repair controls				.600		35		35	44	55
	Total				10.489	640	613.60		1,253.60	1,472.40	1,760.75

D30 HVAC — D3023 Heat Generating Systems

D3023 184 Boiler, Oil

	System Description	Freq. (Years)	Crew	Unit	Labor Hours	2014 Bare Costs				Total In-House	Total w/O&P
						Material	Labor	Equipment	Total		
1060	**Replace boiler, oil, 250 MBH**	30	Q-7	Ea.							
	Remove boiler				24.465		1,375		1,375	1,700	2,125
	Replace boiler, oil, 250 MBH				48.930	3,525	2,725		6,250	7,275	8,650
	Total				73.394	3,525	4,100		7,625	8,975	10,775
1066	**Replace boiler, oil, 300 MBH**	30	Q-7	Ea.							
	Remove boiler				29.091		1,625		1,625	2,025	2,525
	Replace boiler, oil, 300 MBH				59.480	4,250	3,325		7,575	8,825	10,500
	Total				88.570	4,250	4,950		9,200	10,850	13,025
1070	**Replace boiler, oil, 420 MBH**	30	Q-7	Ea.							
	Remove boiler				35.955		2,000		2,000	2,500	3,125
	Replace boiler, oil, 420 MBH				71.749	4,750	4,000		8,750	10,200	12,200
	Total				107.704	4,750	6,000		10,750	12,700	15,325
1080	**Replace boiler, oil, 530 MBH**	30	Q-7	Ea.							
	Remove boiler				42.440		2,375		2,375	2,950	3,700
	Replace boiler, oil, 530 MBH				84.881	6,025	4,725		10,750	12,500	14,900
	Total				127.321	6,025	7,100		13,125	15,450	18,600
2010	**Repair boiler, oil, 2000 MBH**	7	1 STPI	Ea.							
	Remove / replace burner blower				.600		35		35	44	55
	Remove burner blower bearing				1.000		58.50		58.50	73	91.50
	Replace burner blower bearing				2.000	55.50	117		172.50	207	252
	Remove burner blower motor				.976		52		52	64.50	80.50
	Replace burner blower motor				1.951	198	104		302	345	410
	Remove burner fireye				.195		11.40		11.40	14.25	17.80
	Replace burner fireye				.300	208	17.55		225.55	251	288
	Remove burner ignition transformer				.286		16.70		16.70	21	26
	Replace burner ignition transformer				.571	80.50	33.50		114	130	153
	Remove burner ignition electrode				.200		11.70		11.70	14.65	18.25
	Replace burner ignition electrode				.400	16	23.50		39.50	47	56.50
	Remove burner oil pump				.400		23.50		23.50	29.50	36.50
	Replace burner oil pump				.800	77.50	47		124.50	144	170
	Remove burner nozzle				.167		9.75		9.75	12.20	15.20
	Replace burner nozzle				.333	13.50	19.50		33	39	47.50
	Repair controls				.600		35		35	44	55
	Total				10.779	649	615.60		1,264.60	1,480.10	1,772.75

D30 HVAC		D3023			Heat Generating Systems						
D3023		**184**		**Boiler, Oil**							

	System Description	Freq. (Years)	Crew	Unit	Labor Hours	2014 Bare Costs				Total In-House	Total w/O&P
						Material	Labor	Equipment	Total		
2060	**Replace boiler, oil, 2000 MBH**	30	Q-7	Ea.							
	Remove boiler				57.762		3,225		3,225	4,025	5,025
	Replace boiler, oil, 2000 MBH				148.000	17,400	8,300		25,700	29,400	34,700
	Total				205.762	17,400	11,525		28,925	33,425	39,725
2070	**Replace boiler, oil, 3000 MBH**	30	Q-7	Ea.							
	Remove boiler				110.000		6,150		6,150	7,650	9,600
	Replace boiler, oil, 3000 MBH				220.000	22,800	12,300		35,100	40,400	47,700
	Total				330.000	22,800	18,450		41,250	48,050	57,300
3010	**Repair boiler, oil, 10,000 MBH**	7	1 STPI	Ea.							
	Remove / replace burner blower				.600		35		35	44	55
	Remove burner blower bearing				1.000		58.50		58.50	73	91.50
	Replace burner blower bearing				2.000	55.50	117		172.50	207	252
	Remove burner blower motor				.800		47		47	58.50	73
	Replace burner blower motor				1.951	207	104		311	355	420
	Remove burner fireye				.195		11.40		11.40	14.25	17.80
	Replace burner fireye				.300	208	17.55		225.55	251	288
	Remove burner ignition transformer				.333		19.50		19.50	24.50	30.50
	Replace burner ignition transformer				.667	80.50	39		119.50	137	162
	Remove burner ignition electrode				.200		11.70		11.70	14.65	18.25
	Replace burner ignition electrode				.444	16	26		42	50	60.50
	Remove burner oil pump				.500		29.50		29.50	36.50	45.50
	Replace burner oil pump				1.000	102	58.50		160.50	185	219
	Remove burner nozzle				.167		9.75		9.75	12.20	15.20
	Replace burner nozzle				.400	13.50	23.50		37	44	53.50
	Repair controls				.600		35		35	44	55
	Total				11.158	682.50	642.90		1,325.40	1,550.60	1,856.75
3060	**Replace boiler, oil, 10,000 MBH**	30	Q-7	Ea.							
	Remove boiler				373.100		20,807.50		20,807.50	26,000	32,400
	Replace boiler, oil, 10,000 MBH				520.905	139,912.50	29,130.50		169,043	190,000	220,000
	Total				894.005	139,912.50	49,938		189,850.50	216,000	252,400

D30 HVAC — Heat Generating Systems

D3023 186 Boiler, Gas/Oil

D3023

	System Description	Freq. (Years)	Crew	Unit	Labor Hours	2014 Bare Costs				Total In-House	Total w/O&P
						Material	Labor	Equipment	Total		
1010	**Repair boiler, gas/oil, 2000 MBH**	7	Q-5	Ea.							
	Remove/replace burner blower				.600		35		35	44	55
	Remove burner blower bearing				1.000		58.50		58.50	73	91.50
	Replace burner blower bearing				2.000	55.50	117		172.50	207	252
	Remove burner blower motor				.976		57		57	71.50	89
	Replace burner blower motor				1.951	198	104		302	345	410
	Remove burner fireye				.195		11.40		11.40	14.25	17.80
	Replace burner fireye				.300	208	17.55		225.55	251	288
	Remove burner ignition transformer				.333		19.50		19.50	24.50	30.50
	Replace burner ignition transformer				.571	80.50	33.50		114	130	153
	Remove burner ignition electrode				.200		11.70		11.70	14.65	18.25
	Replace burner ignition electrode				.400	16	23.50		39.50	47	56.50
	Remove burner oil pump				.500		29.50		29.50	36.50	45.50
	Replace burner oil pump				.800	77.50	47		124.50	144	170
	Remove burner nozzle				.167		9.75		9.75	12.20	15.20
	Replace burner nozzle				.333	13.50	19.50		33	39	47.50
	Remove burner gas regulator				.727		38.50		38.50	48	59.50
	Replace burner gas regulator				1.455	1,200	76.50		1,276.50	1,425	1,625
	Remove burner auto gas valve				.800		41.50		41.50	52	64.50
	Replace burner auto gas valve				1.231	370	64		434	485	560
	Remove burner solenoid valve				2.600		146		146	182	227
	Replace burner solenoid valve				5.199	1,275	292		1,567	1,775	2,050
	Repair controls				.600		35		35	44	55
	Total				22.939	3,494	1,287.90		4,781.90	5,464.60	6,380.75
1050	**Replace boiler, gas/oil, 2000 MBH**	30	Q-7	Ea.							
	Remove boiler				57.762		3,225		3,225	4,025	5,025
	Replace boiler, gas/oil, 2000 MBH				115.000	30,100	6,425		36,525	41,100	47,600
	Total				172.762	30,100	9,650		39,750	45,125	52,625

D30 HVAC D3023 Heat Generating Systems

D3023 186 Boiler, Gas/Oil

	System Description	Freq. (Years)	Crew	Unit	Labor Hours	2014 Bare Costs				Total In-House	Total w/O&P
						Material	Labor	Equipment	Total		
2010	**Repair boiler, gas/oil, 20,000 MBH**	7	Q-5	Ea.							
	Remove/replace burner blower				.600		35		35	44	55
	Remove burner blower bearing				1.000		58.50		58.50	73	91.50
	Replace burner blower bearing				2.000	55.50	117		172.50	207	252
	Remove burner blower motor				.976		52		52	64.50	80.50
	Replace burner blower motor				1.951	248	104		352	400	470
	Remove burner fireye				.195		11.40		11.40	14.25	17.80
	Replace burner fireye				.300	208	17.55		225.55	251	288
	Remove burner ignition transformer				.333		19.50		19.50	24.50	30.50
	Replace burner ignition transformer				.667	80.50	39		119.50	137	162
	Remove burner ignition electrode				.267		15.60		15.60	19.50	24.50
	Replace burner ignition electrode				.444	16	26		42	50	60.50
	Remove burner oil pump				.500		29.50		29.50	36.50	45.50
	Replace burner oil pump				2.602	162	137		299	350	415
	Remove burner nozzle				.167		9.75		9.75	12.20	15.20
	Replace burner nozzle				.571	13.50	33.50		47	56.50	69
	Remove burner gas regulator				1.300		68.50		68.50	85.50	107
	Replace burner gas regulator				2.600	2,150	137		2,287	2,525	2,900
	Remove burner auto gas valve				5.195		279		279	350	435
	Replace burner auto gas valve				8.000	5,875	430		6,305	7,000	8,025
	Remove burner solenoid valve				2.600		146		146	182	227
	Replace burner solenoid valve				10.399	3,225	585		3,810	4,275	4,950
	Repair controls				.600		35		35	44	55
	Total				43.266	12,033.50	2,385.80		14,419.30	16,201.45	18,776
2050	**Replace boiler, gas/oil, 20,000 MBH**	30	Q-7	Ea.							
	Remove boiler				1527.360		85,100		85,100	106,500	132,500
	Replace boiler, gas/oil, 20,000 MBH				3156.840	278,980	176,120		455,100	527,000	623,500
	Total				4684.200	278,980	261,220		540,200	633,500	756,000

D3023 198 Blowoff System

	System Description	Freq. (Years)	Crew	Unit	Labor Hours	2014 Bare Costs				Total In-House	Total w/O&P
						Material	Labor	Equipment	Total		
1010	**Repair boiler blowoff system**	10	1 STPI	Ea.							
	Repair leak				1.000		58.50		58.50	73	91.50
	Total				1.000		58.50		58.50	73	91.50

D30 HVAC   D3023   Heat Generating Systems

D3023 198   Blowoff System

	System Description	Freq. (Years)	Crew	Unit	Labor Hours	2014 Bare Costs				Total In-House	Total w/O&P
						Material	Labor	Equipment	Total		
1020	Replace boiler blowoff system	15	Q-5	Ea.							
	Remove boiler blowoff				2.773		146		146	183	228
	Replace boiler blowoff				5.556	4,850	293		5,143	5,700	6,525
	Total				8.329	4,850	439		5,289	5,883	6,753

D3023 292   Chemical Feed System

	System Description	Freq. (Years)	Crew	Unit	Labor Hours	2014 Bare Costs				Total In-House	Total w/O&P
						Material	Labor	Equipment	Total		
1010	Repair chemical feed	15	1 STPI	Ea.							
	Repair controls				.040		2.34		2.34	2.93	3.65
	Remove / replace agitator motor				1.951	192	114		306	355	420
	Remove / replace pump seals / bearings				2.078	32	120		152	185	227
	Remove / replace pump motor				1.951	198	104		302	345	410
	Total				6.020	422	340.34		762.34	887.93	1,060.65
1030	Replace chemical feed	15	2 STPI	Ea.							
	Remove / replace feeder				2.500	580	144		724	820	950
	Total				2.500	580	144		724	820	950

D3023 294   Feed Water Supply

	System Description	Freq. (Years)	Crew	Unit	Labor Hours	2014 Bare Costs				Total In-House	Total w/O&P
						Material	Labor	Equipment	Total		
1010	Repair feed water supply pump	15	1 STPI	Ea.							
	Repair controls				.600		35		35	44	55
	Remove / replace pump seals / bearings				1.600		93.50		93.50	117	146
	Remove pump motor				2.078		122		122	152	190
	Replace pump motor				4.167	2,250	244		2,494	2,775	3,200
	Remove / replace pump coupling				1.000		58.50		58.50	73	91.50
	Total				9.445	2,250	553		2,803	3,161	3,682.50
1030	Replace feed water pump	15	Q-2	Ea.							
	Remove pump				11.111		595		595	745	930
	Replace pump				22.222	19,200	1,200		20,400	22,600	25,900
	Total				33.333	19,200	1,795		20,995	23,345	26,830

D30 HVAC | D3023 | Heat Generating Systems

D3023 296 Deaerator

	System Description	Freq. (Years)	Crew	Unit	Labor Hours	2014 Bare Costs				Total In-House	Total w/O&P
						Material	Labor	Equipment	Total		
1010	**Repair deaerator**	10	1 STPI	Ea.							
	Repair controls				1.000		58.50		58.50	73	91.50
	Total				1.000		58.50		58.50	73	91.50
1030	**Replace deaerator**	20	4 STPI	Ea.							
	Remove unit				7.500		410		410	510	640
	Replace unit				179.000	30,000	10,500		40,500	46,100	54,000
	Total				186.500	30,000	10,910		40,910	46,610	54,640

D3023 298 Separators for 9000 Ton Chilled Water System

	System Description	Freq. (Years)	Crew	Unit	Labor Hours	2014 Bare Costs				Total In-House	Total w/O&P
						Material	Labor	Equipment	Total		
1010	**Clean separator strainer**	10	Q-6	Ea.							
	Clean separator strainer				10.000		585		585	730	915
	Total				10.000		585		585	730	915
1030	**Replace separator**	20	Q-6	Ea.							
	Remove unit				22.500		1,230		1,230	1,525	1,925
	Replace separator				42.857	51,900	2,340		54,240	60,000	68,500
	Total				65.357	51,900	3,570		55,470	61,525	70,425

D3023 310 Metal Flue / Chimney

	System Description	Freq. (Years)	Crew	Unit	Labor Hours	2014 Bare Costs				Total In-House	Total w/O&P
						Material	Labor	Equipment	Total		
0010	**Replace metal flue, all fuel SS, 6" diameter**	15	Q-9	L.F.							
	Replace flue, all fuel stainless steel, 6" diameter				.387	81.20	19.07		100.27	113	131
	Total				.387	81.20	19.07		100.27	113	131
0020	**Replace metal flue, all fuel SS, 10" diameter**	15	Q-9	L.F.							
	Replace flue, all fuel stainless steel, 10" diameter				.483	95.70	23.78		119.48	136	157
	Total				.483	95.70	23.78		119.48	136	157

D30 HVAC   D3023   Heat Generating Systems

D3023 310   Metal Flue / Chimney

	System Description	Freq. (Years)	Crew	Unit	Labor Hours	2014 Bare Costs				Total In-House	Total w/O&P
						Material	Labor	Equipment	Total		
0030	Replace metal flue, all fuel SS, 20" diameter	15	Q-10	L.F.							
	Replace flue, all fuel stainless steel, 20" diameter				.967	178.35	49.30		227.65	258	300
	Total				.967	178.35	49.30		227.65	258	300
0040	Replace metal flue, all fuel SS, 32" diameter	15	Q-10	L.F.							
	Replace flue, all fuel stainless steel, 32" diameter				1.289	297.25	65.98		363.23	410	475
	Total				1.289	297.25	65.98		363.23	410	475
0050	Replace metal flue, all fuel SS, 48" diameter	15	Q-10	L.F.							
	Replace flue, all fuel stainless steel, 48" diameter				1.832	442.25	93.53		535.78	600	700
	Total				1.832	442.25	93.53		535.78	600	700
2010	Replace metal flue, gas vent, galvanized, 48" diameter	10	Q-10	L.F.							
	Replace flue, gas vent, galvanized, 48" diameter				1.895	1,485	96.75		1,581.75	1,775	2,025
	Total				1.895	1,485	96.75		1,581.75	1,775	2,025

D3023 388   Pneumatic Coal Spreader

	System Description	Freq. (Years)	Crew	Unit	Labor Hours	2014 Bare Costs				Total In-House	Total w/O&P
						Material	Labor	Equipment	Total		
1010	Repair spreader, pneumatic coal	10	2 STPI	Ea.							
	Repair controls				1.000		58.50		58.50	73	91.50
	Remove conveyor bearing				1.111		65		65	81.50	101
	Replace conveyor bearing				2.222	62.50	130		192.50	231	281
	Remove conveyor motor				2.597		152		152	190	237
	Replace conveyor motor				5.195	2,925	277		3,202	3,550	4,075
	Remove / replace blower				1.200		70		70	88	110
	Remove blower bearings				1.000		58.50		58.50	73	91.50
	Replace blower bearings				2.000	55.50	117		172.50	207	252
	Remove blower motor				2.000		117		117	146	183
	Replace blower motor				4.000	1,525	213		1,738	1,950	2,225
	Total				22.326	4,568	1,258		5,826	6,589.50	7,647
1060	Replace coal spreader	12	4 STPI	Ea.							
	Remove spreader				65.041		3,800		3,800	4,750	5,925
	Replace spreader				130.000	7,100	7,600		14,700	17,300	20,800
	Total				195.041	7,100	11,400		18,500	22,050	26,725

D30 HVAC   D3023   Heat Generating Systems

D3023 390   Fuel Oil Equipment

	System Description	Freq. (Years)	Crew	Unit	Labor Hours	2014 Bare Costs				Total In-House	Total w/O&P
						Material	Labor	Equipment	Total		
1010	**Repair fuel oil equipment, pump**	10	1 STPI	Ea.							
	Remove pump seals / bearings				.600		35		35	44	55
	Replace pump seals / bearings				1.500	29.50	88		117.50	142	174
	Remove / replace pump coupling				.900		52.50		52.50	66	82
	Remove / replace impeller / shaft				.900		52.50		52.50	66	82
	Remove / replace motor				.650		38		38	47.50	59.50
	Remove strainers				.289		16.90		16.90	21	26.50
	Replace strainers				.578	12.95	34		46.95	56.50	68.50
	Total				5.417	42.45	316.90		359.35	443	547.50
1030	**Replace fuel oil 25 GPH pump / motor set**	15	Q-5	Ea.							
	Remove equipment				1.733		91.50		91.50	114	142
	Replace equipment				3.468	1,050	183		1,233	1,375	1,600
	Total				5.201	1,050	274.50		1,324.50	1,489	1,742
1040	**Replace fuel oil 45 GPH pump / motor set**	15	Q-5	Ea.							
	Remove equipment				1.733		91.50		91.50	114	142
	Replace equipment				3.468	1,050	183		1,233	1,375	1,600
	Total				5.201	1,050	274.50		1,324.50	1,489	1,742
1050	**Replace fuel oil 90 GPH pump / motor set**	15	Q-5	Ea.							
	Remove equipment				2.081		110		110	137	171
	Replace equipment				4.161	1,050	219		1,269	1,425	1,650
	Total				6.242	1,050	329		1,379	1,562	1,821
1060	**Replace fuel oil 160 GPH pump / motor set**	15	Q-5	Ea.							
	Remove equipment				2.597		137		137	171	213
	Replace equipment				5.195	1,100	274		1,374	1,550	1,800
	Total				7.792	1,100	411		1,511	1,721	2,013

D30 HVAC

D3033 Cooling Generating Systems

D3033 115 Cooling Tower

	System Description	Freq. (Years)	Crew	Unit	Labor Hours	2014 Bare Costs				Total In-House	Total w/O&P
						Material	Labor	Equipment	Total		
1010	**Repair cooling tower, 50 ton**	10	2 STPI	Ea.							
	Repair controls				1.000		58.50		58.50	73	91.50
	Remove bearings				1.200		70		70	88	110
	Replace bearings				2.000	55.50	117		172.50	207	252
	Remove fan motor				1.155		67.50		67.50	84.50	105
	Replace fan motor				2.312	500	123		623	705	815
	Remove / replace float valve				.500	19.85	26.50		46.35	55	66
	Total				8.167	575.35	462.50		1,037.85	1,212.50	1,439.50
1030	**Replace cooling tower, 50 ton**	15	Q-6	Ea.							
	Remove cooling tower				10.397		567		567	710	885
	Replace cooling tower, 50 ton				20.779	12,660	1,134		13,794	15,400	17,600
	Total				31.177	12,660	1,701		14,361	16,110	18,485
2010	**Repair cooling tower, 100 ton**	10	2 STPI	Ea.							
	Repair controls				1.000		58.50		58.50	73	91.50
	Remove bearings				4.800		280		280	350	440
	Replace bearings				8.000	222	468		690	830	1,000
	Remove fan motors				4.621		270		270	340	420
	Replace fan motors				9.249	2,000	492		2,492	2,825	3,275
	Remove / replace float valve				.575	22.83	30.48		53.31	63.50	75.50
	Total				28.245	2,244.83	1,598.98		3,843.81	4,481.50	5,302
2030	**Replace cooling tower, 100 ton**	15	Q-6	Ea.							
	Remove cooling tower				14.286		780		780	975	1,225
	Replace cooling tower, 100 ton				28.571	17,700	1,560		19,260	21,400	24,600
	Total				42.857	17,700	2,340		20,040	22,375	25,825
3010	**Repair cooling tower, 300 ton**	10	2 STPI	Ea.							
	Repair controls				1.000		58.50		58.50	73	91.50
	Remove bearings				7.200		420		420	530	660
	Replace bearings				13.333	375	780		1,155	1,375	1,675
	Remove fan motors				6.931		405		405	505	630
	Replace fan motors				13.865	3,600	738		4,338	4,900	5,650
	Remove / replace float valve				1.250	49.63	66.25		115.88	138	165
	Total				43.580	4,024.63	2,467.75		6,492.38	7,521	8,871.50

D30 HVAC | D3033 Cooling Generating Systems

D3033 115 Cooling Tower

	System Description	Freq. (Years)	Crew	Unit	Labor Hours	2014 Bare Costs — Material	2014 Bare Costs — Labor	2014 Bare Costs — Equipment	2014 Bare Costs — Total	Total In-House	Total w/O&P
3030	**Replace cooling tower, 300 ton**	15	Q-6	Ea.							
	Remove cooling tower				36.273		1,980		1,980	2,475	3,100
	Replace cooling tower, 300 ton				72.582	30,900	3,960		34,860	39,000	44,800
	Total				108.855	30,900	5,940		36,840	41,475	47,900
4010	**Repair cooling tower, 1000 ton**	10	2 STPI	Ea.							
	Repair controls				1.000		58.50		58.50	73	91.50
	Remove bearings				12.000		700		700	880	1,100
	Replace bearings				23.529	1,760	1,380		3,140	3,650	4,350
	Remove fan motors				11.552		675		675	845	1,050
	Replace fan motors				25.974	8,550	1,390		9,940	11,000	12,800
	Remove / replace float valve				2.350	93.30	124.55		217.85	259	310
	Total				76.406	10,403.30	4,328.05		14,731.35	16,707	19,701.50
4030	**Replace cooling tower, 1000 ton**	15	Q-6	Ea.							
	Remove cooling tower				104.350		5,700		5,700	7,100	8,900
	Replace cooling tower, 1000 ton				208.700	77,500	11,400		88,900	99,500	114,500
	Total				313.050	77,500	17,100		94,600	106,600	123,400

D3033 130 Chiller, Water Cooled, Reciprocating

	System Description	Freq. (Years)	Crew	Unit	Labor Hours	2014 Bare Costs — Material	2014 Bare Costs — Labor	2014 Bare Costs — Equipment	2014 Bare Costs — Total	Total In-House	Total w/O&P
1010	**Repair water cooled chiller, 20 ton**	10	Q-6	Ea.							
	Repair controls				1.000		58.50		58.50	73	91.50
	Remove compressor				24.390		1,325		1,325	1,675	2,075
	Replace compressor				48.780	27,600	2,675		30,275	33,700	38,700
	Remove / replace evaporator tube				1.200	171	65.50		236.50	270	315
	Remove / replace condenser tube				1.200	171	65.50		236.50	270	315
	Replace refrigerant				2.667	640	156		796	900	1,050
	Total				79.237	28,582	4,345.50		32,927.50	36,888	42,546.50
1030	**Replace chiller, water cooled, 20 ton**	20	Q-7	Ea.							
	Remove chiller				50.633		2,825		2,825	3,525	4,400
	Replace chiller, water cooled, 20 ton				101.000	19,900	5,650		25,550	28,900	33,700
	Total				151.633	19,900	8,475		28,375	32,425	38,100

D30 HVAC D3033 Cooling Generating Systems

D3033 130 Chiller, Water Cooled, Reciprocating

	System Description	Freq. (Years)	Crew	Unit	Labor Hours	2014 Bare Costs Material	2014 Bare Costs Labor	2014 Bare Costs Equipment	2014 Bare Costs Total	Total In-House	Total w/O&P
2010	**Repair water cooled chiller, 50 ton**	10	Q-6	Ea.							
	Repair controls				1.000		58.50		58.50	73	91.50
	Remove compressor				60.976		3,343.75		3,343.75	4,150	5,200
	Replace compressor				121.951	69,000	6,687.50		75,687.50	84,500	96,500
	Remove / replace evaporator tube				1.200	171	65.50		236.50	270	315
	Remove / replace condenser tube				1.200	171	65.50		236.50	270	315
	Replace refrigerant				4.000	960	234		1,194	1,350	1,575
	Total				190.327	70,302	10,454.75		80,756.75	90,613	103,996.50
2030	**Replace chiller, water cooled 50 ton**	20	Q-7	Ea.							
	Remove chiller				74.074		4,125		4,125	5,150	6,425
	Replace chiller, water cooled, 50 ton				148.000	36,500	8,250		44,750	50,500	58,500
	Total				222.074	36,500	12,375		48,875	55,650	64,925
3010	**Repair water cooled chiller, 100 ton**	10	Q-6	Ea.							
	Repair controls				1.000		58.50		58.50	73	91.50
	Remove compressor				155.844		8,500		8,500	10,700	13,300
	Replace compressor				310.000	62,600	17,000		79,600	90,000	105,000
	Remove / replace evaporator tube				4.798	655	267		922	1,050	1,225
	Remove / replace condenser tube				4.798	655	267		922	1,050	1,225
	Replace refrigerant				8.000	1,920	468		2,388	2,700	3,125
	Total				484.439	65,830	26,560.50		92,390.50	105,573	123,966.50
3030	**Replace chiller, water cooled, 100 ton**	20	Q-7	Ea.							
	Remove chiller				116.000		6,500		6,500	8,075	10,100
	Replace chiller, water cooled, 100 ton				233.000	62,000	13,000		75,000	84,500	98,000
	Total				349.000	62,000	19,500		81,500	92,575	108,100
5010	**Repair water cooled chiller, 200 ton**	10	Q-7	Ea.							
	Repair controls				1.000		58.50		58.50	73	91.50
	Remove compressor				152.099		8,470		8,470	10,600	13,200
	Replace compressor				303.380	60,830	16,940		77,770	88,000	102,500
	Remove / replace evaporator tube				9.610	1,300	535		1,835	2,100	2,450
	Remove / replace condenser tube				9.610	1,300	535		1,835	2,100	2,450
	Replace refrigerant				13.333	3,200	780		3,980	4,500	5,225
	Total				489.031	66,630	27,318.50		93,948.50	107,373	125,916.50

D30 HVAC		D3033		Cooling Generating Systems							

D3033 130 Chiller, Water Cooled, Reciprocating

	System Description	Freq. (Years)	Crew	Unit	Labor Hours	2014 Bare Costs				Total In-House	Total w/O&P
						Material	Labor	Equipment	Total		
5030	Replace chiller, water cooled, 200 ton	20	Q-7	Ea.							
	Remove chiller				163.000		9,100		9,100	11,400	14,200
	Replace chiller, water cooled, 200 ton				326.000	95,000	18,200		113,200	127,000	147,000
	Total				489.000	95,000	27,300		122,300	138,400	161,200

D3033 135 Chiller, Air Cooled, Reciprocating

	System Description	Freq. (Years)	Crew	Unit	Labor Hours	2014 Bare Costs				Total In-House	Total w/O&P
						Material	Labor	Equipment	Total		
1010	Repair recip. chiller, air cooled, 20 ton	10	2 STPI	Ea.							
	Repair controls				.800		47		47	58.50	73
	Remove fan bearing				3.600		210		210	264	330
	Replace fan bearing				6.000	166.50	351		517.50	620	755
	Remove fan motor				2.927		171		171	215	267
	Replace fan motor				5.854	744	312		1,056	1,200	1,425
	Remove compressor				24.490		1,325		1,325	1,675	2,075
	Replace compressor				48.780	27,600	2,675		30,275	33,700	38,700
	Replace refrigerant				2.667	640	156		796	900	1,050
	Total				95.117	29,150.50	5,247		34,397.50	38,632.50	44,675
1030	Replace chiller, 20 ton	20	Q-7	Ea.							
	Remove chiller				42.857		2,350		2,350	2,925	3,650
	Replace chiller, air cooled, reciprocating, 20 ton				118.000	28,200	6,625		34,825	39,200	45,600
	Total				160.857	28,200	8,975		37,175	42,125	49,250
2010	Repair recip. chiller, air cooled, 50 ton	10	Q-6	Ea.							
	Repair controls				1.000		58.50		58.50	73	91.50
	Remove fan bearing				3.600		210		210	264	330
	Replace fan bearing				6.667	187.50	390		577.50	695	845
	Remove fan motor				3.096		165		165	204	255
	Replace fan motor				6.936	1,500	369		1,869	2,125	2,450
	Remove compressor				60.976		3,312.50		3,312.50	4,200	5,200
	Replace compressor				121.951	69,000	6,687.50		75,687.50	84,500	96,500
	Remove / replace compressor motor				7.201		384		384	475	595
	Replace refrigerant				4.000	960	234		1,194	1,350	1,575
	Total				215.427	71,647.50	11,810.50		83,458	93,886	107,841.50

D30 HVAC D3033 Cooling Generating Systems

D3033 135 Chiller, Air Cooled, Reciprocating

	System Description	Freq. (Years)	Crew	Unit	Labor Hours	2014 Bare Costs				Total In-House	Total w/O&P
						Material	Labor	Equipment	Total		
2030	**Replace chiller, 50 ton**	20	Q-7	Ea.							
	Remove chiller				73.395		4,100		4,100	5,100	6,375
	Replace chiller, air cooled, reciprocating, 50 ton				146.000	41,400	8,175		49,575	55,500	64,500
	Total				219.395	41,400	12,275		53,675	60,600	70,875
3010	**Repair chiller, air cooled, 100 ton**	10	Q-6	Ea.							
	Repair controls				1.000		58.50		58.50	73	91.50
	Remove fan bearing				4.200		246		246	305	385
	Replace fan bearing				6.667	187.50	390		577.50	695	845
	Remove fan motor				3.900		208.50		208.50	258	320
	Replace fan motor				7.792	2,565	417		2,982	3,300	3,850
	Remove compressor				194.805		10,625		10,625	13,300	16,600
	Replace compressor				191.250	43,250	10,718.75		53,968.75	61,000	71,000
	Remove / replace compressor motor				9.359		500		500	620	770
	Replace refrigerant				6.667	1,600	390		1,990	2,250	2,600
	Total				425.640	47,602.50	23,553.75		71,156.25	81,801	96,461.50
3030	**Replace chiller, 100 ton**	20	Q-7	Ea.							
	Remove chiller				84.211		4,700		4,700	5,875	7,325
	Replace chiller, air cooled, reciprocating, 100 ton				168.000	75,000	9,375		84,375	94,000	108,500
	Total				252.211	75,000	14,075		89,075	99,875	115,825

D3033 137 Chiller, Water Cooled, Scroll

	System Description	Freq. (Years)	Crew	Unit	Labor Hours	2014 Bare Costs				Total In-House	Total w/O&P
						Material	Labor	Equipment	Total		
4010	**Repair water cooled chiller, 5 ton**	10	1 STPI	Ea.							
	Repair controls				.500		29.50		29.50	36.50	45.50
	Remove fan bearing				1.200		70		70	88	109
	Replace fan bearing				2.000	55.50	117		172.50	207	252
	Remove compressor				2.477		145		145	181	226
	Replace compressor				4.954	485	290		775	895	1,050
	Replace refrigerant				.667	160	39		199	225	261
	Total				11.797	700.50	690.50		1,391	1,632.50	1,943.50

D30 HVAC | D3033 Cooling Generating Systems

D3033 137 Chiller, Water Cooled, Scroll

	System Description	Freq. (Years)	Crew	Unit	Labor Hours	2014 Bare Costs				Total In-House	Total w/O&P
						Material	Labor	Equipment	Total		
4030	**Replace chiller, water cooled, 5 ton**	20	Q-5	Ea.							
	Remove chiller				18.182		955		955	1,200	1,500
	Replace chiller, water cooled, 5 ton				36.364	3,500	1,925		5,425	6,250	7,350
	Total				54.545	3,500	2,880		6,380	7,450	8,850
5010	**Repair water cooled chiller, 10 ton**	10	2 STPI	Ea.							
	Repair controls				1.000		58.50		58.50	73	91.50
	Remove fan bearing				3.600		210		210	264	330
	Replace fan bearing				6.000	166.50	351		517.50	620	755
	Remove compressor				4.954		290		290	360	450
	Replace compressor				9.907	970	580		1,550	1,800	2,125
	Replace refrigerant				1.333	320	78		398	450	520
	Total				26.794	1,456.50	1,567.50		3,024	3,567	4,271.50
5030	**Replace chiller, water cooled, 10 ton**	20	Q-6	Ea.							
	Remove chiller				43.478		2,375		2,375	2,975	3,700
	Replace chiller, water cooled, 10 ton				86.957	6,625	4,750		11,375	13,200	15,700
	Total				130.435	6,625	7,125		13,750	16,175	19,400
6010	**Repair water cooled chiller, 15 ton**	10	Q-5	Ea.							
	Repair controls				1.000		58.50		58.50	73	91.50
	Remove fan bearing				24.000		1,140		1,140	1,500	1,850
	Replace fan bearing				6.000	166.50	351		517.50	620	755
	Remove compressor				6.932		366		366	455	570
	Replace compressor				13.865	2,200	730		2,930	3,350	3,900
	Replace refrigerant				2.000	480	117		597	675	780
	Total				53.797	2,846.50	2,762.50		5,609	6,673	7,946.50
6030	**Replace chiller, water cooled, 15 ton**	20	Q-6	Ea.							
	Remove chiller				42.857		2,350		2,350	2,925	3,650
	Replace chiller, water cooled, 15 ton				85.714	19,400	4,675		24,075	27,200	31,600
	Total				128.571	19,400	7,025		26,425	30,125	35,250

D30 HVAC

D3033 Cooling Generating Systems

D3033 140 Chiller, Hermetic Centrifugal

	System Description	Freq. (Years)	Crew	Unit	Labor Hours	2014 Bare Costs				Total In-House	Total w/O&P
						Material	Labor	Equipment	Total		
1010	**Repair centrifugal chiller, 100 ton**	10	Q-7	Ea.							
	Repair controls				1.000		58.50		58.50	73	91.50
	Remove compressor				24.390		1,325		1,325	1,675	2,075
	Replace compressor				197.000	39,500	11,000		50,500	57,000	66,500
	Remove / replace condenser tube				4.800	684	262		946	1,075	1,275
	Remove / replace evaporator tube				4.800	684	262		946	1,075	1,275
	Remove / replace refrigerant filter				1.600	268	89		357	405	475
	Remove refrigerant				.440		25.70		25.70	32	40
	Replace refrigerant				8.000	1,920	468		2,388	2,700	3,125
	Total				242.030	43,056	13,490.20		56,546.20	64,035	74,856.50
1030	**Replace centrifugal chiller, 100 ton**	20	Q-7	Ea.							
	Remove chiller				163.000		9,100		9,100	11,400	14,200
	Replace chiller, centrifugal, 100 ton				326.000	82,000	18,200		100,200	113,000	131,000
	Total				489.000	82,000	27,300		109,300	124,400	145,200
2010	**Repair centrifugal chiller, 300 ton**	10	Q-7	Ea.							
	Repair controls				1.000		58.50		58.50	73	91.50
	Remove compressor				56.341		3,060.75		3,060.75	3,875	4,800
	Replace compressor				455.070	91,245	25,410		116,655	131,500	154,000
	Remove / replace condenser tube				12.000	1,710	655		2,365	2,700	3,150
	Remove / replace evaporator tube				12.000	1,710	655		2,365	2,700	3,150
	Remove / replace refrigerant filter				1.200	570	67	2.08	639.08	715	820
	Remove refrigerant				1.319		76.50		76.50	96	120
	Replace refrigerant				21.333	5,120	1,248		6,368	7,200	8,350
	Total				560.263	100,355	31,230.75	2.08	131,587.83	148,859	174,481.50
2030	**Replace centrifugal chiller, 300 ton**	20	Q-7	Ea.							
	Remove chiller				173.000		9,675		9,675	12,000	15,100
	Replace chiller, 300 ton				347.000	133,000	19,400		152,400	170,500	196,500
	Total				520.000	133,000	29,075		162,075	182,500	211,600

D30 HVAC | D3033 Cooling Generating Systems

D3033 140 Chiller, Hermetic Centrifugal

	System Description	Freq. (Years)	Crew	Unit	Labor Hours	Material	Labor	Equipment	Total	Total In-House	Total w/O&P
3010	**Repair centrifugal chiller, 1000 ton**	10	Q-7	Ea.							
	Repair controls				1.000		58.50		58.50	73	91.50
	Remove compressor				187.561		10,189.25		10,189.25	12,900	16,000
	Replace compressor				1514.930	303,755	84,590		388,345	438,500	512,000
	Remove / replace condenser tube				16.800	2,394	980		3,374	3,875	4,525
	Remove / replace evaporator tube				12.600	2,394	735		3,129	3,550	4,150
	Remove / replace refrigerant filter				1.600	570	93.50		663.50	745	860
	Remove refrigerant				3.516		204		204	256	320
	Replace refrigerant				29.333	7,040	1,716		8,756	9,900	11,500
	Total				1767.340	316,153	98,566.25		414,719.25	**469,799**	**549,446.50**
3030	**Replace centrifugal chiller, 1000 ton**	20	Q-7	Ea.							
	Remove chiller				242.000		13,500		13,500	16,900	21,100
	Replace chiller, 1000 ton				484.000	447,500	27,000		474,500	526,000	601,500
	Total				726.000	447,500	40,500		488,000	**542,900**	**622,600**
6010	**Repair centrifugal chiller, 9000 ton**	10	Q-7	Ea.							
	Repair controls				1.000		58.50		58.50	73	91.50
	Remove compressor				2025.974		110,500		110,500	138,500	172,500
	Replace compressor				5122.000	1,027,000	286,000		1,313,000	1,482,000	1,731,000
	Remove / replace condenser tube				151.204	21,546	8,820		30,366	34,800	40,800
	Remove / replace evaporator tube				113.399	21,546	6,615		28,161	32,000	37,300
	Remove / replace refrigerant filter				14.400	5,130	841.50		5,971.50	6,700	7,725
	Remove refrigerant				31.648		1,836		1,836	2,300	2,875
	Replace refrigerant				263.993	63,360	15,444		78,804	89,000	103,500
	Total				7723.618	1,138,582	430,115		1,568,697	**1,785,373**	**2,095,791.50**
6030	**Replace centrifugal chiller, 9000 ton**	20	Q-7	Ea.							
	Remove chiller				1452.000		81,000		81,000	101,500	126,500
	Replace chiller, 9000 ton				3306.000	3,921,000	184,200		4,105,200	4,542,000	5,188,500
	Total				4758.000	3,921,000	265,200		4,186,200	**4,643,500**	**5,315,000**

D30 HVAC

D3033 Cooling Generating Systems

D3033 142 Chiller, Open Centrifugal

	System Description	Freq. (Years)	Crew	Unit	Labor Hours	2014 Bare Costs				Total In-House	Total w/O&P
						Material	Labor	Equipment	Total		
1010	**Repair open centrifugal chiller, 300 ton**	10	Q-7	Ea.							
	Repair controls				1.000		58.50		58.50	73	91.50
	Remove compressor				56.341		3,060.75		3,060.75	3,875	4,800
	Replace compressor				455.070	91,245	25,410		116,655	131,500	154,000
	Remove / replace condenser tube				12.000	1,710	655		2,365	2,700	3,150
	Remove / replace evaporator tube				12.000	1,710	655		2,365	2,700	3,150
	Remove / replace refrigerant filter				1.200	570	67	2.08	639.08	715	820
	Remove refrigerant				20.000		950		950	1,250	1,550
	Replace refrigerant				21.333	5,120	1,248		6,368	7,200	8,350
	Total				578.944	100,355	32,104.25	2.08	132,461.33	150,013	175,911.50
1030	**Replace open centrifugal chiller, 300 ton**	20	Q-7	Ea.							
	Remove chiller				91.429		5,100		5,100	6,375	7,950
	Replace chiller, 300 ton				907.440	170,050	50,540	1,572.25	222,162.25	252,000	293,000
	Total				998.869	170,050	55,640	1,572.25	227,262.25	258,375	300,950
2010	**Repair open centrifugal chiller, 1000 ton**	10	Q-7	Ea.							
	Repair controls				1.000		58.50		58.50	73	91.50
	Remove compressor				187.561		10,189.25		10,189.25	12,900	16,000
	Replace compressor				1514.930	303,755	84,590		388,345	438,500	512,000
	Remove / replace condenser tube				16.800	2,394	917		3,311	3,775	4,425
	Remove / replace evaporator tube				16.800	2,394	917		3,311	3,775	4,425
	Remove / replace refrigerant filter				1.600	1,150	89	2.77	1,241.77	1,375	1,575
	Remove refrigerant				35.204		2,056		2,056	2,550	3,200
	Replace refrigerant				34.666	8,320	2,028		10,348	11,700	13,600
	Total				1808.560	318,013	100,844.75	2.77	418,860.52	474,648	555,316.50
2030	**Replace open centrifugal chiller, 1000 ton**	20	Q-7	Ea.							
	Remove chiller				313.000		17,500		17,500	21,800	27,300
	Replace chiller, 1000 ton				3104.400	581,750	172,900	5,378.75	760,028.75	861,500	1,003,000
	Total				3417.400	581,750	190,400	5,378.75	777,528.75	883,300	1,030,300

D30 HVAC D3033 Cooling Generating Systems

D3033 145 Chiller, Absorption

	System Description	Freq. (Years)	Crew	Unit	Labor Hours	2014 Bare Costs				Total In-House	Total w/O&P
						Material	Labor	Equipment	Total		
1010	**Repair chiller, absorption, 100 ton**	10	2 STPI	Ea.							
	Repair controls				1.000		58.50		58.50	73	91.50
	Remove / replace concentrator tube				.900	171	47.50		218.50	247	288
	Remove / replace condenser tube				6.000	855	327.50		1,182.50	1,350	1,575
	Remove / replace evaporator tube				6.000	855	327.50		1,182.50	1,350	1,575
	Remove / replace absorber tube				.900	179	47.50		226.50	256	298
	Remove / replace gang pump				1.600	7,150	84		7,234	7,975	9,075
	Remove gang pump motor				3.704		198		198	244	305
	Replace gang pump motor				7.407	5,200	395		5,595	6,200	7,100
	Add refrigerant				8.000	256	468		724	865	1,050
	Total				35.511	14,666	1,953.50		16,619.50	18,560	21,357.50
1030	**Replace chiller, absorption, 100 ton**	20	Q-7	Ea.							
	Remove chiller				155.000		8,650		8,650	10,800	13,500
	Replace chiller				310.000	125,000	17,300		142,300	159,000	183,500
	Total				465.000	125,000	25,950		150,950	169,800	197,000
2010	**Repair chiller, absorption, 350 ton**	10	2 STPI	Ea.							
	Repair controls				1.000		58.50		58.50	73	91.50
	Remove / replace concentrator tube				1.800	342	95		437	495	575
	Remove / replace condenser tube				7.200	1,026	393		1,419	1,625	1,900
	Remove / replace evaporator tube				7.200	1,026	393		1,419	1,625	1,900
	Remove / replace absorber tube				1.800	358	95		453	510	595
	Remove / replace gang pump				1.600	7,150	84		7,234	7,975	9,075
	Remove gang pump motor				3.704		198		198	244	305
	Replace gang pump motor				7.407	5,200	395		5,595	6,200	7,100
	Add refrigerant				8.000	256	468		724	865	1,050
	Total				39.711	15,358	2,179.50		17,537.50	19,612	22,591.50
2030	**Replace chiller, absorption, 350 ton**	20	Q-7	Ea.							
	Remove chiller				200.000		11,100		11,100	13,900	17,400
	Replace chiller				400.000	335,000	22,300		357,300	396,500	453,500
	Total				600.000	335,000	33,400		368,400	410,400	470,900

D30 HVAC | D3033 Cooling Generating Systems

D3033 145 Chiller, Absorption

	System Description	Freq. (Years)	Crew	Unit	Labor Hours	2014 Bare Costs Material	2014 Bare Costs Labor	2014 Bare Costs Equipment	2014 Bare Costs Total	Total In-House	Total w/O&P
3010	**Repair chiller, absorption, 950 ton**	10	4 STPI	Ea.							
	Repair controls				1.000		58.50		58.50	73	91.50
	Remove / replace concentrator tube				3.600	684	190		874	990	1,150
	Remove / replace condenser tube				12.000	1,710	655		2,365	2,700	3,150
	Remove / replace evaporator tube				12.000	1,710	655		2,365	2,700	3,150
	Remove / replace absorber tube				3.600	716	190		906	1,025	1,200
	Remove / replace gang pump				1.600	7,150	84		7,234	7,975	9,075
	Remove gang pump motor				3.704		198		198	244	305
	Replace gang pump motor				8.696	6,375	465		6,840	7,575	8,700
	Add refrigerant				8.000	256	468		724	865	1,050
	Total				54.198	18,601	2,963.50		21,564.50	24,147	27,871.50
3030	**Replace chiller, absorption, 950 ton**	20	Q-7	Ea.							
	Remove chiller				266.000		14,900		14,900	18,500	23,200
	Replace chiller				533.000	686,000	29,700		715,700	791,500	904,000
	Total				799.000	686,000	44,600		730,600	810,000	927,200

D3033 210 Air Cooled Condenser

	System Description	Freq. (Years)	Crew	Unit	Labor Hours	2014 Bare Costs Material	2014 Bare Costs Labor	2014 Bare Costs Equipment	2014 Bare Costs Total	Total In-House	Total w/O&P
1010	**Repair condenser, air cooled, 5 ton**	10	1 STPI	Ea.							
	Repair controls				.500		29.50		29.50	36.50	45.50
	Remove fan motor				1.800		105		105	132	164
	Replace fan motor				2.311	279	123		402	460	540
	Total				4.611	279	257.50		536.50	628.50	749.50
1030	**Replace condenser, air cooled, 5 ton**	15	Q-5	Ea.							
	Remove condenser				5.195		274		274	340	425
	Replace condenser, air cooled, 5 ton				10.390	1,950	545		2,495	2,825	3,300
	Total				15.584	1,950	819		2,769	3,165	3,725
2010	**Repair condenser, air cooled, 20 ton**	10	2 STPI	Ea.							
	Repair controls				1.000		58.50		58.50	73	91.50
	Remove fan motor				2.310		135		135	169	210
	Replace fan motor				4.622	670	246		916	1,050	1,225
	Total				7.932	670	439.50		1,109.50	1,292	1,526.50

D30 HVAC | D3033 Cooling Generating Systems

D3033 210 Air Cooled Condenser

	System Description	Freq. (Years)	Crew	Unit	Labor Hours	2014 Bare Costs				Total In-House	Total w/O&P
						Material	Labor	Equipment	Total		
2030	**Replace condenser, air cooled, 20 ton**	15	Q-5	Ea.							
	Remove condenser				10.390		545		545	685	855
	Replace condenser, air cooled, 20 ton				20.779	4,875	1,100		5,975	6,725	7,800
	Total				31.169	4,875	1,645		6,520	7,410	8,655
3010	**Repair condenser, air cooled, 50 ton**	10	2 STPI	Ea.							
	Repair controls				1.000		58.50		58.50	73	91.50
	Remove fan motor				3.466		202.50		202.50	254	315
	Replace fan motor				6.936	1,500	369		1,869	2,125	2,450
	Total				11.402	1,500	630		2,130	2,452	2,856.50
3030	**Replace condenser, air cooled, 50 ton**	15	Q-6	Ea.							
	Remove condenser				38.961		2,125		2,125	2,650	3,325
	Replace condenser, air cooled, 50 ton				77.922	14,600	4,250		18,850	21,400	24,900
	Total				116.883	14,600	6,375		20,975	24,050	28,225
4010	**Repair condenser, air cooled, 100 ton**	10	2 STPI	Ea.							
	Repair controls				1.000		58.50		58.50	73	91.50
	Remove fan motor				6.931		405		405	505	630
	Replace fan motor				13.873	3,000	738		3,738	4,225	4,900
	Total				21.804	3,000	1,201.50		4,201.50	4,803	5,621.50
4030	**Replace condenser, air cooled, 100 ton**	15	Q-7	Ea.							
	Remove condenser				69.264		3,850		3,850	4,825	6,025
	Replace condenser, air cooled, 100 ton				138.000	29,700	7,725		37,425	42,300	49,100
	Total				207.264	29,700	11,575		41,275	47,125	55,125

D30 HVAC — D3033 Cooling Generating Systems

D3033 260 Evaporative Condenser

	System Description	Freq. (Years)	Crew	Unit	Labor Hours	2014 Bare Costs				Total In-House	Total w/O&P
						Material	Labor	Equipment	Total		
1010	**Repair evaporative condenser, 20 ton**	10	2 STPI	Ea.							
	Repair controls				1.000		58.50		58.50	73	91.50
	Remove fan motors				2.310		135		135	169	210
	Replace fan motors				4.622	670	246		916	1,050	1,225
	Remove pump and motor				20.779		1,075		1,075	1,350	1,675
	Replace pump and motor				41.558	3,675	2,150		5,825	6,725	7,950
	Remove / replace pump nozzles				5.000	47.50	265		312.50	380	470
	Remove / replace float valve				.500	19.85	26.50		46.35	55	66
	Total				75.770	4,412.35	3,956		8,368.35	9,802	11,687.50
1030	**Replace evaporative condenser, 20 ton**	15	Q-5	Ea.							
	Remove condenser				22.130		1,175		1,175	1,450	1,825
	Replace evaporative condenser, 20 ton				44.199	10,200	2,325		12,525	14,100	16,400
	Total				66.329	10,200	3,500		13,700	15,550	18,225
2010	**Repair evaporative condenser, 100 ton**	10	2 STPI	Ea.							
	Repair controls				1.000		58.50		58.50	73	91.50
	Remove fan motors				6.931		405		405	505	630
	Replace fan motors				13.873	3,000	738		3,738	4,225	4,900
	Remove pump and motor				20.779		1,075		1,075	1,350	1,675
	Replace pump and motor				41.558	3,675	2,150		5,825	6,725	7,950
	Remove / replace nozzles				25.000	237.50	1,325		1,562.50	1,900	2,350
	Remove / replace float valve				.500	19.85	26.50		46.35	55	66
	Total				109.642	6,932.35	5,778		12,710.35	14,833	17,662.50
2030	**Replace evaporative condenser, 100 ton**	15	Q-7	Ea.							
	Remove condenser				57.762		3,225		3,225	4,025	5,025
	Replace evaporative condenser, 100 ton				115.000	19,300	6,425		25,725	29,200	34,100
	Total				172.762	19,300	9,650		28,950	33,225	39,125

D30 HVAC		D3033	Cooling Generating Systems								
D3033 260		**Evaporative Condenser**									

							2014 Bare Costs				
	System Description	Freq. (Years)	Crew	Unit	Labor Hours	Material	Labor	Equipment	Total	Total In-House	Total w/O&P
3010	**Repair evaporative condenser, 300 ton**	10	2 STPI	Ea.							
	Repair controls				1.000		58.50		58.50	73	91.50
	Remove fan motors				11.552		675		675	845	1,050
	Replace fan motors				23.121	5,000	1,230		6,230	7,050	8,150
	Remove pump and motor				20.779		1,075		1,075	1,350	1,675
	Replace pump and motor				24.000	4,450	1,300		5,750	6,500	7,575
	Remove / replace nozzles				35.000	332.50	1,855		2,187.50	2,650	3,275
	Remove / replace float valve				1.000	39.70	53		92.70	110	132
	Total				116.453	9,822.20	6,246.50		16,068.70	18,578	21,948.50
3030	**Replace evaporative condenser, 300 ton**	15	Q-7	Ea.							
	Remove condenser				152.471		8,505		8,505	10,600	13,200
	Replace evaporative condenser, 300 ton				304.560	48,762	17,010		65,772	75,000	87,500
	Total				457.031	48,762	25,515		74,277	85,600	100,700

D30 HVAC D3043 Distribution Systems

D3043 120 Fan Coil

	System Description	Freq. (Years)	Crew	Unit	Labor Hours	Material	Labor	Equipment	Total	Total In-House	Total w/O&P
							2014 Bare Costs				
1010	**Repair fan coil unit, 1 ton**	10	1 STPI	Ea.							
	Remove fan motor				1.200		70		70	88	109
	Replace fan motor				1.951	198	104		302	345	410
	Total				3.151	198	174		372	433	519
1030	**Replace fan coil unit, 1 ton**	15	Q-5	Ea.							
	Remove fan coil unit				1.733		91.50		91.50	114	142
	Replace fan coil unit, 1 ton				3.467	935	183		1,118	1,250	1,450
	Total				5.200	935	274.50		1,209.50	1,364	1,592
2010	**Repair fan coil unit, 3 ton**	10	1 STPI	Ea.							
	Remove fan motor				1.200		70		70	88	109
	Replace fan motor				1.951	248	104		352	400	470
	Total				3.151	248	174		422	488	579
2030	**Replace fan coil unit, 3 ton**	15	Q-5	Ea.							
	Remove fan coil unit				2.597		137		137	171	213
	Replace fan coil unit, 3 ton				5.195	1,875	274		2,149	2,400	2,775
	Total				7.792	1,875	411		2,286	2,571	2,988
3010	**Repair fan coil unit, 5 ton**	10	1 STPI	Ea.							
	Remove fan motor				1.200		70		70	88	109
	Replace fan motor				2.311	279	123		402	460	540
	Total				3.511	279	193		472	548	649
3030	**Replace fan coil unit, 5 ton**	15	Q-5	Ea.							
	Remove fan coil unit				2.680		141		141	176	220
	Replace fan coil unit, 5 ton				5.371	2,585	282		2,867	3,200	3,675
	Total				8.052	2,585	423		3,008	3,376	3,895
4010	**Repair fan coil unit, 10 ton**	10	1 STPI	Ea.							
	Remove fan motor				1.400		82		82	102	128
	Replace fan motor				2.312	500	123		623	705	815
	Total				3.712	500	205		705	807	943

D30 HVAC / D3043 Distribution Systems

D3043 120 Fan Coil

							2014 Bare Costs				Total In-House	Total w/O&P
	System Description	Freq. (Years)	Crew	Unit	Labor Hours	Material	Labor	Equipment	Total			
4030	**Replace fan coil unit, 10 ton**	15	Q-6	Ea.								
	Remove fan coil unit				11.538		630		630		790	985
	Replace fan coil unit, 10 ton				23.077	3,950	1,250		5,200		5,925	6,925
	Total				34.615	3,950	1,880		5,830		6,715	7,910
5010	**Repair fan coil unit, 20 ton**	10	1 STPI	Ea.								
	Remove fan motor				1.600		93.50		93.50		117	146
	Replace fan motor				2.311	600	123		723		815	940
	Total				3.911	600	216.50		816.50		932	1,086
5030	**Replace fan coil unit, 20 ton**	15	Q-6	Ea.								
	Remove fan coil unit				19.512		1,075		1,075		1,325	1,650
	Replace fan coil unit, 20 ton				39.024	5,825	2,125		7,950		9,075	10,600
	Total				58.537	5,825	3,200		9,025		10,400	12,250
6010	**Repair fan coil unit, 30 ton**	10	1 STPI	Ea.								
	Remove fan motor				1.800		105		105		132	164
	Replace fan motor				2.597	855	139		994		1,100	1,275
	Total				4.398	855	244		1,099		1,232	1,439
6030	**Replace fan coil unit, 30 ton**	15	Q-6	Ea.								
	Remove fan coil unit				25.974		1,425		1,425		1,775	2,200
	Replace fan coil unit, 30 ton				51.948	9,825	2,825		12,650		14,400	16,700
	Total				77.922	9,825	4,250		14,075		16,175	18,900

D30 HVAC

D3043 Distribution Systems

D3043 122 Fan Coil, DX Air Conditioner, Cooling Only

	System Description	Freq. (Years)	Crew	Unit	Labor Hours	Material	Labor	Equipment	Total	Total In-House	Total w/O&P
1010	**Repair fan coil, DX 1-1/2 ton, cooling only**	10	1 STPI	Ea.							
	Repair controls				.300		17.55		17.55	22	27.50
	Remove / replace supply fan				.650		38		38	47.50	59.50
	Remove supply fan bearing				1.000		58.50		58.50	73	91.50
	Replace supply fan bearing				2.000	55.50	117		172.50	207	252
	Remove supply fan motor				1.156		61.50		61.50	76.50	95.50
	Replace supply fan motor				1.951	192	114		306	355	420
	Remove compressor				2.477		145		145	181	226
	Replace compressor				3.714	259	217		476	555	665
	Replace refrigerant				.267	64	15.60		79.60	90	104
	Total				13.515	570.50	784.15		1,354.65	1,607	1,941
1040	**Replace fan coil, DX 1-1/2 ton, no heat**	15	Q-5	Ea.							
	Remove fan coil unit				2.080		110		110	137	171
	Replace fan coil unit, 1-1/2 ton, no heat				4.160	755	219		974	1,100	1,275
	Total				6.240	755	329		1,084	1,237	1,446
2010	**Repair fan coil, DX 2 ton, cooling only**	10	1 STPI	Ea.							
	Repair controls				.300		17.55		17.55	22	27.50
	Remove / replace supply fan				.650		38		38	47.50	59.50
	Remove supply fan bearing				1.000		58.50		58.50	73	91.50
	Replace supply fan bearing				2.000	55.50	117		172.50	207	252
	Remove supply fan motor				1.156		61.50		61.50	76.50	95.50
	Replace supply fan motor				1.951	198	104		302	345	410
	Remove compressor				2.477		145		145	181	226
	Replace compressor				4.000	266	234		500	585	700
	Replace refrigerant				.400	96	23.40		119.40	135	156
	Total				13.934	615.50	798.95		1,414.45	1,672	2,018
2040	**Replace fan coil, DX 2 ton, no heat**	15	Q-5	Ea.							
	Remove fan coil unit				2.167		114		114	143	178
	Replace fan coil unit, DX 2 ton, no heat				4.334	810	228		1,038	1,175	1,375
	Total				6.500	810	342		1,152	1,318	1,553

D30 HVAC D3043 Distribution Systems

D3043 122 Fan Coil, DX Air Conditioner, Cooling Only

	System Description	Freq. (Years)	Crew	Unit	Labor Hours	2014 Bare Costs				Total In-House	Total w/O&P
						Material	Labor	Equipment	Total		
3010	**Repair fan coil, DX 2-1/2 ton**	10	1 STPI	Ea.							
	Repair controls				.300		17.55		17.55	22	27.50
	Remove / replace supply fan				.650		38		38	47.50	59.50
	Remove supply fan bearing				1.000		58.50		58.50	73	91.50
	Replace supply fan bearing				2.000	55.50	117		172.50	207	252
	Remove supply fan motor				1.156		61.50		61.50	76.50	95.50
	Replace supply fan motor				1.951	207	104		311	355	420
	Remove compressor				2.477		145		145	181	226
	Replace compressor				4.334	315	254		569	665	790
	Replace refrigerant				.533	128	31.20		159.20	180	209
	Total				14.401	705.50	826.75		1,532.25	1,807	2,171
3040	**Replace fan coil, DX 2-1/2 ton, no heat**	15	Q-5	Ea.							
	Remove fan coil unit				2.363		124		124	156	194
	Replace fan coil, DX 2-1/2 ton, no heat				4.727	835	249		1,084	1,225	1,425
	Total				7.090	835	373		1,208	1,381	1,619
4010	**Repair fan coil, DX 3 ton, cooling only**	10	1 STPI	Ea.							
	Repair controls				.300		17.55		17.55	22	27.50
	Remove / replace supply fan				.650		38		38	47.50	59.50
	Remove supply fan bearing				1.000		58.50		58.50	73	91.50
	Replace supply fan bearing				2.000	55.50	117		172.50	207	252
	Remove supply fan motor				1.156		61.50		61.50	76.50	95.50
	Replace supply fan motor				1.951	207	104		311	355	420
	Remove compressor				2.477		145		145	181	226
	Replace compressor				3.844	293.25	225.25		518.50	605	720
	Replace refrigerant				.533	128	31.20		159.20	180	209
	Total				13.911	683.75	798		1,481.75	1,747	2,101
4040	**Replace fan coil, DX 3 ton, no heat**	15	Q-5	Ea.							
	Remove fan coil unit				2.737		144		144	180	225
	Replace fan coil, DX 3 ton, no heat				5.474	1,050	288		1,338	1,525	1,775
	Total				8.211	1,050	432		1,482	1,705	2,000

D30 HVAC D3043 Distribution Systems

D3043 122 Fan Coil, DX Air Conditioner, Cooling Only

	System Description	Freq. (Years)	Crew	Unit	Labor Hours	2014 Bare Costs Material	2014 Bare Costs Labor	2014 Bare Costs Equipment	2014 Bare Costs Total	Total In-House	Total w/O&P
5050	**Repair fan coil, DX 5 ton, cooling only**	10	1 STPI	Ea.							
	Repair controls				.300		17.55		17.55	22	27.50
	Remove / replace supply fan				.650		38		38	47.50	59.50
	Remove supply fan bearing				1.000		58.50		58.50	73	91.50
	Replace supply fan bearing				2.000	55.50	117		172.50	207	252
	Remove supply fan motor				1.156		61.50		61.50	76.50	95.50
	Replace supply fan motor				1.951	207	104		311	355	420
	Remove compressor				2.477		145		145	181	226
	Replace compressor				4.954	485	290		775	895	1,050
	Replace refrigerant				.533	128	31.20		159.20	180	209
	Total				15.021	875.50	862.75		1,738.25	2,037	2,431
6060	**Replace fan coil, DX 5 ton, no heat**	15	Q-5	Ea.							
	Remove fan coil unit				3.466		183		183	228	285
	Replace fan coil, DX 5 ton, no heat				6.932	1,450	365		1,815	2,050	2,375
	Total				10.399	1,450	548		1,998	2,278	2,660
7070	**Repair fan coil, DX 10 ton, cooling only**	10	Q-6	Ea.							
	Repair controls				.300		17.55		17.55	22	27.50
	Remove / replace supply fan				.650		38		38	47.50	59.50
	Remove supply fan bearing				1.000		58.50		58.50	73	91.50
	Replace supply fan bearing				2.000	55.50	117		172.50	207	252
	Remove supply fan motor				1.156		61.50		61.50	76.50	95.50
	Replace supply fan motor				2.311	335	123		458	520	610
	Remove compressor				14.513		787.25		787.25	990	1,250
	Replace compressor				28.714	17,822	1,574.50		19,396.50	21,600	24,700
	Replace refrigerant				.800	192	46.80		238.80	270	315
	Total				51.444	18,404.50	2,824.10		21,228.60	23,806	27,401
8080	**Replace fan coil, DX 10 ton, no heat**	15	Q-6	Ea.							
	Remove fan coil unit				6.000		330		330	410	510
	Replace fan coil, DX 10 ton, no heat				12.000	3,175	655		3,830	4,300	5,000
	Total				18.000	3,175	985		4,160	4,710	5,510

D30 HVAC		D3043	Distribution Systems								
D3043 122		Fan Coil, DX Air Conditioner, Cooling Only									

	System Description	Freq. (Years)	Crew	Unit	Labor Hours	2014 Bare Costs				Total In-House	Total w/O&P
						Material	Labor	Equipment	Total		
9090	**Repair fan coil, DX 20 ton, cooling only**	10	Q-6	Ea.							
	Repair controls				.300		17.55		17.55	22	27.50
	Remove / replace supply fan				.650		38		38	47.50	59.50
	Remove supply fan bearing				1.000		58.50		58.50	73	91.50
	Replace supply fan bearing				2.000	55.50	117		172.50	207	252
	Remove supply fan motor				1.156		61.50		61.50	76.50	95.50
	Replace supply fan motor				2.476	700	132		832	935	1,075
	Remove compressor				21.929		1,192.50		1,192.50	1,500	1,875
	Replace compressor				43.902	24,840	2,407.50		27,247.50	30,300	34,800
	Replace refrigerant				1.333	320	78		398	450	520
	Total				74.747	25,915.50	4,102.55		30,018.05	33,611	38,796
9590	**Replace fan coil, DX 20 ton, no heat**	15	Q-6	Ea.							
	Remove fan coil unit				22.284		1,225		1,225	1,525	1,900
	Replace fan coil, DX 20 ton, no heat				44.610	6,250	2,425		8,675	9,925	11,600
	Total				66.894	6,250	3,650		9,900	11,450	13,500
9600	**Repair fan coil, DX 30 ton, cooling only**	10	Q-6	Ea.							
	Repair controls				.300		17.55		17.55	22	27.50
	Remove / replace supply fan				.650		38		38	47.50	59.50
	Remove supply fan bearing				1.000		58.50		58.50	73	91.50
	Replace supply fan bearing				2.000	55.50	117		172.50	207	252
	Remove supply fan motor				1.156		61.50		61.50	76.50	95.50
	Replace supply fan motor				2.476	700	132		832	935	1,075
	Remove compressor				24.490		1,325		1,325	1,675	2,075
	Replace compressor				48.780	27,600	2,675		30,275	33,700	38,700
	Replace refrigerant				1.733	416	101.40		517.40	585	680
	Total				82.586	28,771.50	4,525.95		33,297.45	37,321	43,056
9610	**Replace fan coil, DX 30 ton, no heat**	15	Q-6	Ea.							
	Remove fan coil unit				25.974		1,425		1,425	1,775	2,200
	Replace fan coil, DX 20 ton, no heat				48.000	7,800	2,625		10,425	11,900	13,900
	Total				73.974	7,800	4,050		11,850	13,675	16,100

D30 HVAC

D3043 Distribution Systems

D3043 124 Fan Coil, DX Air Conditioner W/ Heat

	System Description	Freq. (Years)	Crew	Unit	Labor Hours	Material	Labor	Equipment	Total	Total In-House	Total w/O&P
1010	**Replace fan coil, DX 1-1/2 ton, with heat**	15	Q-5	Ea.							
	Remove fan coil unit				2.312		122		122	152	190
	Replace fan coil unit, 1-1/2 ton, with heat				6.474	1,057	340.20		1,397.20	1,575	1,850
	Total				8.786	1,057	462.20		1,519.20	1,727	2,040
2010	**Replace fan coil, DX 2 ton, with heat**	15	Q-5	Ea.							
	Remove fan coil unit				2.407		127		127	158	198
	Replace fan coil unit, DX 2 ton, with heat				6.741	1,134	355.60		1,489.60	1,675	1,975
	Total				9.148	1,134	482.60		1,616.60	1,833	2,173
3010	**Replace fan coil, DX 2-1/2 ton, with heat**	15	Q-5	Ea.							
	Remove fan coil unit				2.626		138		138	173	216
	Replace fan coil unit, DX 2-1/2 ton, with heat				7.354	1,169	387.80		1,556.80	1,775	2,075
	Total				9.979	1,169	525.80		1,694.80	1,948	2,291
4010	**Replace fan coil, DX 3 ton, with heat**	15	Q-5	Ea.							
	Remove fan coil unit				3.042		160		160	200	250
	Replace fan coil unit, DX 3 ton, with heat				8.514	1,470	448		1,918	2,175	2,550
	Total				11.556	1,470	608		2,078	2,375	2,800
5010	**Replace fan coil, DX 5 ton, with heat**	10	Q-5	Ea.							
	Remove fan coil unit				3.852		203		203	254	315
	Replace fan coil unit, DX 5 ton, with heat				10.785	2,030	567		2,597	2,950	3,425
	Total				14.637	2,030	770		2,800	3,204	3,740
6010	**Replace fan coil, DX 10 ton, with heat**	10	Q-6	Ea.							
	Remove fan coil unit				6.667		365		365	455	570
	Replace fan coil unit, DX 10 ton, with heat				18.667	4,445	1,022		5,467	6,150	7,125
	Total				25.333	4,445	1,387		5,832	6,605	7,695
7010	**Replace fan coil, DX 20 ton, with heat**	10	Q-6	Ea.							
	Remove fan coil unit				24.793		1,350		1,350	1,700	2,100
	Replace fan coil unit, DX 20 ton, with heat				69.421	8,750	3,780		12,530	14,400	16,900
	Total				94.215	8,750	5,130		13,880	16,100	19,000

D30 HVAC — Distribution Systems

D3043 128 — Unit Ventilator

	System Description	Freq. (Years)	Crew	Unit	Labor Hours	2014 Bare Costs				Total In-House	Total w/O&P
						Material	Labor	Equipment	Total		
1010	**Repair unit ventilator, 750 CFM, 2 ton**	10	1 STPI	Ea.							
	Repair controls				.200		11.70		11.70	14.65	18.25
	Remove / replace fan				.650		38		38	47.50	59.50
	Remove fan motor				1.156		61.50		61.50	76.50	95.50
	Replace fan motor				1.951	198	104		302	345	410
	Total				3.957	198	215.20		413.20	**483.65**	**583.25**
1030	**Replace unit vent, 750 CFM, heat/cool coils**	15	Q-6	Ea.							
	Remove unit ventilator				7.792		425		425	530	665
	Replace unit, 750 CFM, with heat / cool coils				15.605	4,050	850		4,900	5,525	6,400
	Total				23.397	4,050	1,275		5,325	**6,055**	**7,065**
2010	**Repair unit ventilator, 1250 CFM, 3 ton**	10	1 STPI	Ea.							
	Repair controls				.200		11.70		11.70	14.65	18.25
	Remove / replace fan				.650		38		38	47.50	59.50
	Remove fan motor				1.156		61.50		61.50	76.50	95.50
	Replace fan motor				1.951	248	104		352	400	470
	Total				3.957	248	215.20		463.20	**538.65**	**643.25**
2030	**Replace unit vent, 1250 CFM, heat/cool coils**	15	Q-6	Ea.							
	Remove unit ventilator				11.142		610		610	760	950
	Replace unit, 1250 CFM, with heat / cool coils				22.284	4,950	1,225		6,175	6,975	8,100
	Total				33.426	4,950	1,835		6,785	**7,735**	**9,050**
2040	**Repair unit ventilator, 2000 CFM, 5 ton**	10	1 STPI	Ea.							
	Repair controls				.300		17.55		17.55	22	27.50
	Remove / replace fan				.650		38		38	47.50	59.50
	Remove fan motor				1.156		61.50		61.50	76.50	95.50
	Replace fan motor				2.311	279	123		402	460	540
	Total				4.417	279	240.05		519.05	**606**	**722.50**
2050	**Replace unit vent, 2000 CFM, heat/cool coils**	15	Q-6	Ea.							
	Remove unit ventilator				31.209		1,700		1,700	2,125	2,650
	Replace unit, 2000 CFM, with heat / cool coils				62.338	6,225	3,400		9,625	11,100	13,100
	Total				93.547	6,225	5,100		11,325	**13,225**	**15,750**

D30 HVAC   D3043 Distribution Systems

D3043 140 Duct Heater

							2014 Bare Costs				Total In-House	Total w/O&P
	System Description	Freq. (Years)	Crew	Unit	Labor Hours	Material	Labor	Equipment	Total			
0020	**Maintenance and inspection**											
	Inspect and clean duct heater	0.50	1 ELEC	Ea.	1.143		61		61		75.50	91.50
	Total				1.143		61		61		75.50	94.50
0030	**Replace duct heater**											
	Remove duct heater	15	1 ELEC	Ea.	.667		35.50		35.50		44	55
	Duct heater 12 KW				2.000	1,850	107		1,957		2,175	2,475
	Total				2.667	1,850	142.50		1,992.50		2,219	2,530

D3043 210 Draft Fan

							2014 Bare Costs				Total In-House	Total w/O&P
	System Description	Freq. (Years)	Crew	Unit	Labor Hours	Material	Labor	Equipment	Total			
1010	**Repair fan, induced draft, 2000 CFM**											
	Remove bearings	10	1 STPI	Ea.	1.000		58.50		58.50		73	91.50
	Replace bearings				2.000	55.50	117		172.50		207	252
	Total				3.000	55.50	175.50		231		280	343.50
1030	**Replace fan, induced draft, 2000 CFM**											
	Remove fan	20	Q-9	Ea.	3.151		155		155		197	245
	Replace fan, induced draft, 2000 CFM				6.304	2,650	310		2,960		3,300	3,800
	Total				9.456	2,650	465		3,115		3,497	4,045
2010	**Repair fan, induced draft, 6700 CFM**											
	Remove bearings	10	1 STPI	Ea.	1.000		58.50		58.50		73	91.50
	Replace bearings				2.000	55.50	117		172.50		207	252
	Total				3.000	55.50	175.50		231		280	343.50
2030	**Replace fan, induced draft, 6700 CFM**											
	Remove fan	20	Q-9	Ea.	4.522		223		223		282	350
	Replace fan, induced draft, 6700 CFM				9.045	3,250	445		3,695		4,150	4,775
	Total				13.567	3,250	668		3,918		4,432	5,125
3010	**Repair fan, induced draft, 17,700 CFM**											
	Remove bearings	10	1 STPI	Ea.	1.000		58.50		58.50		73	91.50
	Replace bearings				2.222	62.50	130		192.50		231	281
	Total				3.222	62.50	188.50		251		304	372.50

D30 HVAC		D3043		Distribution Systems							
D3043 210		**Draft Fan**									

						2014 Bare Costs				Total In-House	Total w/O&P
	System Description	Freq. (Years)	Crew	Unit	Labor Hours	Material	Labor	Equipment	Total		
3030	Replace fan, induced draft, 17,700 CFM	20	Q-9	Ea.							
	Remove fan				12.998		640		640	810	1,000
	Replace fan, induced draft, 17,700 CFM				26.016	8,375	1,275		9,650	10,800	12,500
	Total				39.014	8,375	1,915		10,290	11,610	13,500

D3043 220		**Exhaust Fan**									

						2014 Bare Costs				Total In-House	Total w/O&P
	System Description	Freq. (Years)	Crew	Unit	Labor Hours	Material	Labor	Equipment	Total		
1010	Replace fan & motor, propeller exh., 375 CFM	15	Q-20	Ea.							
	Remove fan and motor				1.300		65		65	82	102
	Replace fan & motor, propeller exhaust, 375 CFM				2.600	183	130		313	365	435
	Total				3.900	183	195		378	447	537
1030	Replace fan & motor, propeller exh., 1000 CFM	15	Q-20	Ea.							
	Remove fan and motor				1.625		81.50		81.50	103	128
	Replace fan & motor, propeller exhaust, 1000 CFM				3.250	305	163		468	540	635
	Total				4.875	305	244.50		549.50	643	763
1040	Replace fan & motor, propeller exh., 4700 CFM	15	Q-20	Ea.							
	Remove fan and motor				2.600		130		130	164	204
	Replace fan & motor, propeller exhaust, 4700 CFM				5.200	1,175	260		1,435	1,625	1,875
	Total				7.800	1,175	390		1,565	1,789	2,079
2030	Replace roof mounted exhaust fan, 800 CFM	20	Q-20	Ea.							
	Remove fan & motor				2.600		130		130	164	204
	Replace fan & motor, roof mounted exhaust, 800 CFM				5.200	865	260		1,125	1,275	1,500
	Total				7.800	865	390		1,255	1,439	1,704
2040	Replace roof mounted exhaust fan, 2000 CFM	20	Q-20	Ea.							
	Remove fan & motor				3.250		163		163	205	255
	Replace fan & motor, roof mounted exhaust, 2000 CFM				6.500	1,675	325		2,000	2,250	2,600
	Total				9.750	1,675	488		2,163	2,455	2,855

D30 HVAC D3043 Distribution Systems

D3043 220 Exhaust Fan

							2014 Bare Costs				Total In-House	Total w/O&P
	System Description	Freq. (Years)	Crew	Unit	Labor Hours	Material	Labor	Equipment	Total			
2050	**Replace roof mounted exhaust fan, 8500 CFM**	20	Q-20	Ea.								
	Remove fan & motor				4.333		217		217	273	340	
	Replace fan & motor, roof mounted exhaust, 8500 CFM				8.666	2,725	435		3,160	3,550	4,075	
	Total				12.998	2,725	652		3,377	**3,823**	**4,415**	
2060	**Replace roof mounted exhaust fan, 20,300 CFM**	20	Q-20	Ea.								
	Remove fan & motor				13.004		650		650	820	1,025	
	Replace fan & motor, roof mounted exhaust, 20,300 CFM				26.008	7,625	1,300		8,925	10,000	11,600	
	Total				39.012	7,625	1,950		9,575	**10,820**	**12,625**	
3010	**Replace utility set, belt drive, 800 CFM**	10	Q-20	Ea.								
	Remove fan and motor				2.167		108		108	137	170	
	Replace utility set, belt drive, 800 CFM				4.334	955	217		1,172	1,325	1,525	
	Total				6.500	955	325		1,280	**1,462**	**1,695**	
3020	**Replace utility set, belt drive, 3600 CFM**	10	Q-20	Ea.								
	Remove fan and motor				3.250		163		163	205	255	
	Replace utility set, belt drive, 3600 CFM				6.500	1,975	325		2,300	2,575	2,975	
	Total				9.750	1,975	488		2,463	**2,780**	**3,230**	
3030	**Replace utility set, belt drive, 11,000 CFM**	10	Q-20	Ea.								
	Remove fan and motor				6.500		325		325	410	510	
	Replace utility set, belt drive, 11,000 CFM				13.004	5,200	650		5,850	6,550	7,525	
	Total				19.504	5,200	975		6,175	**6,960**	**8,035**	
3040	**Replace utility set, belt drive, 20,000 CFM**	10	Q-20	Ea.								
	Remove fan and motor				16.247		815		815	1,025	1,275	
	Replace utility set, belt drive, 20,000 CFM				32.520	6,950	1,625		8,575	9,700	11,200	
	Total				48.767	6,950	2,440		9,390	**10,725**	**12,475**	
4010	**Replace axial flow fan, 3800 CFM**	10	Q-20	Ea.								
	Remove fan and motor				3.823		191		191	241	300	
	Replace axial flow fan, 3800 CFM				7.648	5,025	385		5,410	6,000	6,875	
	Total				11.472	5,025	576		5,601	**6,241**	**7,175**	

D30 HVAC D3043 Distribution Systems

D3043 220 Exhaust Fan

	System Description	Freq. (Years)	Crew	Unit	Labor Hours	2014 Bare Costs				Total In-House	Total w/O&P
						Material	Labor	Equipment	Total		
4020	**Replace axial flow fan, 6400 CFM**	10	Q-20	Ea.							
	Remove fan and motor				4.643		232		232	293	365
	Replace axial flow fan, 6400 CFM				9.285	5,625	465		6,090	6,775	7,750
	Total				13.928	5,625	697		6,322	7,068	**8,115**
4030	**Replace axial flow fan, 15,600 CFM**	10	Q-20	Ea.							
	Remove fan and motor				8.123		405		405	515	640
	Replace axial flow fan, 15,600 CFM				16.247	8,825	815		9,640	10,700	12,300
	Total				24.370	8,825	1,220		10,045	11,215	**12,940**
4040	**Replace axial flow fan, 28,000 CFM**	10	Q-20	Ea.							
	Remove fan and motor				32.520		1,625		1,625	2,050	2,550
	Replace axial flow fan, 28,000 CFM				64.935	15,100	3,250		18,350	20,700	24,000
	Total				97.455	15,100	4,875		19,975	22,750	**26,550**

D3043 250 Fireplaces, Clay Flue

	System Description	Freq. (Years)	Crew	Unit	Labor Hours	2014 Bare Costs				Total In-House	Total w/O&P
						Material	Labor	Equipment	Total		
0010	**Replace baked clay flue, architectural**	75	2 BRIC	L.F.							
	Set up and secure scaffold				.022		1.01		1.01	1.32	1.64
	Masonry demolition				.083		3.05		3.05	4.06	5
	Remove damaged flue				.002		.10		.10	.13	.16
	Install new flue lining				.224	10.20	9.25		19.45	23.50	28
	Install new chimney				1.518	50	62.50		112.50	137	165
	Remove scaffold				.022		1.01		1.01	1.32	1.64
	Total				1.872	60.20	76.92		137.12	167.33	**201.44**

D3043 252 Fireplaces, Metal Flue

	System Description	Freq. (Years)	Crew	Unit	Labor Hours	2014 Bare Costs				Total In-House	Total w/O&P
						Material	Labor	Equipment	Total		
0020	**Replace metal pipe flue, architectural**	50	2 SSWK	L.F.							
	Remove damaged pipe				.008		.43		.43	.65	.79
	Install new vent chimney, prefabricated metal				.306	7.20	15.05		22.25	27	33
	Total				.314	7.20	15.48		22.68	27.65	**33.79**

D30 HVAC D3043 Distribution Systems

D3043 310 Steam Converter, Commercial

	System Description	Freq. (Years)	Crew	Unit	Labor Hours	2014 Bare Costs				Total In-House	Total w/O&P
						Material	Labor	Equipment	Total		
0010	**Repair steam converter**	5	1 STPI	Ea.							
	Repair heat exchanger				5.944		350		350	435	540
	Total				5.944		350		350	**435**	**540**
0020	**Inspect for leaks**	2	1 STPI	Ea.							
	Check relief valve operation				.038		2.21		2.21	2.76	3.44
	Check for proper water temperature				.055		3.19		3.19	3.99	4.98
	Total				.092		5.40		5.40	**6.75**	**8.42**
0030	**Replace steam converter**	30	Q-5	Ea.							
	Remove exchanger				2.081		110		110	137	171
	Install new exchanger 10 GPM of 40F to 180F				4.160	3,250	219		3,469	3,850	4,400
	Total				6.241	3,250	329		3,579	**3,987**	**4,571**

D3043 320 Flash Tank, 24 Gallon

	System Description	Freq. (Years)	Crew	Unit	Labor Hours	2014 Bare Costs				Total In-House	Total w/O&P
						Material	Labor	Equipment	Total		
0010	**Repair flash tank**	5	1 STPI	Ea.							
	Repair flash tank				5.944		350		350	435	540
	Total				5.944		350		350	**435**	**540**
0030	**Replace flash tank 24 gal**	15	Q-5	Ea.							
	Remove old tank				.743		39		39	49	61
	Install new tank 24 gallon				1.486	715	78		793	885	1,025
	Total				2.229	715	117		832	**934**	**1,086**

D3043 330 Steam Regulator Valve

	System Description	Freq. (Years)	Crew	Unit	Labor Hours	2014 Bare Costs				Total In-House	Total w/O&P
						Material	Labor	Equipment	Total		
0010	**Replace steam regulator valve 1-1/2" diameter**	6	1 STPI	Ea.							
	Remove steam valve				.400		23.50		23.50	29.50	36.50
	Install steam regulator valve 1-1/2" diameter				.800	3,025	47		3,072	3,375	3,850
	Total				1.200	3,025	70.50		3,095.50	**3,404.50**	**3,886.50**

D30 HVAC | D3043 Distribution Systems

D3043 330 | Steam Regulator Valve

	System Description	Freq. (Years)	Crew	Unit	Labor Hours	2014 Bare Costs				Total In-House	Total w/O&P
						Material	Labor	Equipment	Total		
0110	**Replace steam regulator valve 2" diameter**	6	1 STPI	Ea.							
	Remove steam valve				.473		27.50		27.50	34.50	43
	Install steam regulator valve 2" diameter				.945	3,725	55.50		3,780.50	4,175	4,750
	Total				1.418	3,725	83		3,808	**4,209.50**	**4,793**
0210	**Replace steam regulator valve 2-1/2" diameter**	6	Q-5	Ea.							
	Remove steam valve				.867		45.50		45.50	57	71
	Install steam regulator valve 2-1/2" diameter				1.733	4,650	91.50		4,741.50	5,225	5,950
	Total				2.600	4,650	137		4,787	**5,282**	**6,021**
0310	**Replace steam regulator valve 3" diameter**	6	Q-5	Ea.							
	Remove steam valve				.946		50		50	62.50	77.50
	Install steam regulator valve 3" diameter flanged				1.891	5,800	99.50		5,899.50	6,500	7,400
	Total				2.836	5,800	149.50		5,949.50	**6,562.50**	**7,477.50**

D3043 340 | Condensate Meter

	System Description	Freq. (Years)	Crew	Unit	Labor Hours	2014 Bare Costs				Total In-House	Total w/O&P
						Material	Labor	Equipment	Total		
0010	**Repair meter**	15	1 STPI	Ea.							
	Replace drum				1.538	780	88.50		868.50	970	1,125
	Replace bearings				2.597	128	149		277	330	395
	Total				4.136	908	237.50		1,145.50	**1,300**	**1,520**
0030	**Replace condensate meter 500 lb./hr.**	30	1 STPI	Ea.							
	Remove meter				.372		21.50		21.50	27	34
	Replace meter 500#/hr				.743	3,275	43.50		3,318.50	3,650	4,150
	Total				1.114	3,275	65		3,340	**3,677**	**4,184**
0130	**Replace condensate meter 1500 lb./hr.**	30	1 STPI	Ea.							
	Remove meter				.743		43.50		43.50	54.50	68
	Replace meter 1500#/hr				1.486	3,575	87		3,662	4,050	4,600
	Total				2.229	3,575	130.50		3,705.50	**4,104.50**	**4,668**

D30 HVAC — D3043 Distribution Systems

D3043 350 Steam Traps

	System Description	Freq. (Years)	Crew	Unit	Labor Hours	2014 Bare Costs				Total In-House	Total w/O&P
						Material	Labor	Equipment	Total		
1030	**Replace steam trap, 15 PSIG, 3/4" threaded**	7	1 STPI	Ea.							29.50
	Remove old CI body float & thermostatic trap				.325		19		19	24	215
	Install new CI body float & thermostatic trap				.650	124	38		162	184	
	Total				.975	124	57		181	**208**	**244.50**
1040	**Replace steam trap, 15 PSIG, 1" threaded**	7	1 STPI	Ea.							31.50
	Remove old CI body float & thermostatic trap				.347		20.50		20.50	25.50	251
	Install new CI body float & thermostatic trap				.693	150	40.50		190.50	216	
	Total				1.040	150	61		211	**241.50**	**282.50**
1050	**Replace steam trap, 15 PSIG, 1-1/4" threaded**	7	1 STPI	Ea.							36.50
	Remove old CI body float & thermostatic trap				.400		23.50		23.50	29.50	299
	Install new CI body float & thermostatic trap				.800	181	47		228	258	
	Total				1.200	181	70.50		251.50	**287.50**	**335.50**
1060	**Replace steam trap, 15 PSIG, 1-1/2" threaded**	7	1 STPI	Ea.							52.50
	Remove old CI body float & thermostatic trap				.578		34		34	42.50	435
	Install new CI body float & thermostatic trap				1.156	263	67.50		330.50	375	
	Total				1.733	263	101.50		364.50	**417.50**	**487.50**
1070	**Replace steam trap, 15 PSIG, 2" threaded**	7	1 STPI	Ea.							79
	Remove old CI body float & thermostatic trap				.867		50.50		50.50	63.50	765
	Install new CI body float & thermostatic trap				1.733	485	101		586	660	
	Total				2.600	485	151.50		636.50	**723.50**	**844**

D3043 410 Radiator Valve

	System Description	Freq. (Years)	Crew	Unit	Labor Hours	2014 Bare Costs				Total In-House	Total w/O&P
						Material	Labor	Equipment	Total		
0010	**Replace radiator valve 1/2" angle union**	50	1 STPI	Ea.							19.80
	Remove radiator valve				.217		12.70		12.70	15.85	101
	Install radiator valve 1/2" angle union				.433	49	25.50		74.50	85.50	
	Total				.650	49	38.20		87.20	**101.35**	**120.80**

D30 HVAC D3043 Distribution Systems

D3043 410 Radiator Valve

	System Description	Freq. (Years)	Crew	Unit	Labor Hours	2014 Bare Costs				Total In-House	Total w/O&P
						Material	Labor	Equipment	Total		
0020	**Replace radiator valve 3/4" angle union**	50	1 STPI	Ea.							
	Remove radiator valve				.260		15.20		15.20	19	23.50
	Install radiator valve 3/4" angle union				.520	53	30.50		83.50	96.50	114
	Total				.780	53	45.70		98.70	**115.50**	**137.50**
0030	**Replace radiator valve 1" angle union**	50	1 STPI	Ea.							
	Remove radiator valve				.274		16		16	20	25
	Install radiator valve 1" angle union				.547	61	32		93	107	126
	Total				.821	61	48		109	**127**	**151**
0040	**Replace radiator valve 1-1/4" angle union**	50	1 STPI	Ea.							
	Remove radiator valve				.347		20.50		20.50	25.50	31.50
	Install radiator valve 1-1/4" angle union				.693	78.50	40.50		119	137	162
	Total				1.040	78.50	61		139.50	**162.50**	**193.50**

D3043 420 Cast Iron Radiator, 10' Section

	System Description	Freq. (Years)	Crew	Unit	Labor Hours	2014 Bare Costs				Total In-House	Total w/O&P
						Material	Labor	Equipment	Total		
0010	**Replace C.I. radiator 4 tube 25"H 10' section**	50	Q-5	Section							
	Remove radiator				1.084		57		57	71.50	89
	Install C.I. radiator 4 tube 25" high 10' section				2.167	330	114		444	505	590
	Total				3.250	330	171		501	**576.50**	**679**

D3043 430 Baseboard Radiation, 10' Section

	System Description	Freq. (Years)	Crew	Unit	Labor Hours	2014 Bare Costs				Total In-House	Total w/O&P
						Material	Labor	Equipment	Total		
0010	**Replace radiator, baseboard 10'**	20	Q-5	Ea.							
	Remove radiator				2.262		119		119	149	186
	Install 10' baseboard radiator				4.522	365	240		605	700	825
	Total				6.783	365	359		724	**849**	**1,011**

D30 HVAC / D3043 Distribution Systems

D3043 440 Finned Radiator, Wall, 10' Section

	System Description	Freq. (Years)	Crew	Unit	Labor Hours	2014 Bare Costs				Total In-House	Total w/O&P
						Material	Labor	Equipment	Total		
0010	**Replace finned radiator**	20	Q-5	Ea.							
	Replace radiator				3.468		182.50		182.50	230	285
	Install 10' finned tube, 2 tier, 1-1/4" cop tube				6.933	495	365		860	1,000	1,200
	Total				10.401	495	547.50		1,042.50	1,230	1,485

D3043 450 Duct Coil, 1-Row, Hot Water

	System Description	Freq. (Years)	Crew	Unit	Labor Hours	2014 Bare Costs				Total In-House	Total w/O&P
						Material	Labor	Equipment	Total		
1020	**Replace coil, hot water boost, 12" x 24"**	25	Q-5	Ea.							
	Remove coil				.858		45		45	56.50	70.50
	Replace coil, hot water duct, 12" x 24"				1.111	375	58.50		433.50	485	560
	Total				1.969	375	103.50		478.50	541.50	630.50
1030	**Replace coil, hot water boost, 24" x 24"**	25	Q-5	Ea.							
	Remove coil				1.717		90.50		90.50	113	141
	Replace coil, hot water duct, 24" x 24"				3.433	465	181		646	740	865
	Total				5.150	465	271.50		736.50	853	1,006
1040	**Replace coil, hot water boost, 24" x 36"**	25	Q-5	Ea.							
	Remove coil				2.581		136		136	170	212
	Replace coil, hot water duct, 24" x 36"				5.161	525	272		797	915	1,075
	Total				7.742	525	408		933	1,085	1,287
1050	**Replace coil, hot water boost, 36" x 36"**	25	Q-5	Ea.							
	Remove coil				3.867		204		204	255	320
	Replace coil, hot water duct, 36" x 36"				7.729	640	405		1,045	1,225	1,425
	Total				11.596	640	609		1,249	1,480	1,745

D30 HVAC D3043 Distribution Systems

D3043 510 Pipe & Fittings, Steel/Iron, Flanged

						2014 Bare Costs				Total In-House	Total w/O&P
	System Description	Freq. (Years)	Crew	Unit	Labor Hours	Material	Labor	Equipment	Total		
0010	**Install new gasket, 4" pipe size**	25	1 PLUM	Ea.							
	Disconnect joint / remove flange gasket				1.301		75		75	93.50	117
	Install new gasket				1.300	10.40	75		85.40	105	130
	Total				2.601	10.40	150		160.40	**198.50**	**247**
0030	**Replace 2" flanged steel pipe and fittings**	75	2 PLUM	L.F.							
	Remove old pipe				.231		11.95		11.95	14.95	18.65
	Install flanged 2" steel pipe with hangers				.356	13	18.40	1.23	32.63	38.50	46
	Total				.587	13	30.35	1.23	44.58	**53.45**	**64.65**
0040	**Replace 4" flanged steel pipe and fittings**	75	2 PLUM	L.F.							
	Remove old pipe				.400		20.50		20.50	26	32.50
	Install flanged 4" steel pipe with hangers				.800	30	41.50	2.77	74.27	87.50	105
	Total				1.200	30	62	2.77	94.77	**113.50**	**137.50**
0050	**Replace 6" flanged steel pipe and fittings**	75	2 PLUM	L.F.							
	Remove old pipe				.624		32.50		32.50	40.50	50.50
	Install flanged 6" steel pipe with hangers				1.248	58	67	2.87	127.87	150	181
	Total				1.872	58	99.50	2.87	160.37	**190.50**	**231.50**
0060	**Replace 8" flanged steel pipe and fittings**	75	2 PLUM	L.F.							
	Remove old pipe				.821		44		44	55	69
	Install flanged 8" steel pipe with hangers				1.642	95	88	3.78	186.78	219	261
	Total				2.463	95	132	3.78	230.78	**274**	**330**

D3043 520 Valves

						2014 Bare Costs				Total In-House	Total w/O&P
	System Description	Freq. (Years)	Crew	Unit	Labor Hours	Material	Labor	Equipment	Total		
1010	**Repack gate valve gland, 3/8" - 1-1/2"**	10	1 STPI	Ea.							
	Remove / replace gate valve packing				.219	10.60	11.50		22.10	26	31
	Total				.219	10.60	11.50		22.10	**26**	**31**

D30 HVAC D3043 Distribution Systems

D3043 520 Valves

	System Description	Freq. (Years)	Crew	Unit	Labor Hours	2014 Bare Costs				Total In-House	Total w/O&P
						Material	Labor	Equipment	Total		
1020	**Replace gate valve, partial, 3/8" - 1-1/2"**	20	1 STPI	Ea.							
	Turn valve off and on				.017		.99		.99	1.24	1.55
	Remove old				.530		31		31	39	48.50
	Install new, 3/8" - 1-1/2" size				.800	205	47		252	284	330
	Total				1.347	205	78.99		283.99	324.24	380.05
2010	**Repack gate valve gland, 2" - 3"**	10	1 STPI	Ea.							
	Remove / replace gate valve packing				.291	14.10	15.30		29.40	34.50	41.50
	Total				.291	14.10	15.30		29.40	34.50	41.50
2020	**Replace gate valve, partial, 2" - 3"**	20	Q-1	Ea.							
	Turn valve off and on				.017		.99		.99	1.24	1.55
	Remove valve parts				.440		25.50		25.50	32	40
	Replace valve parts, 2" - 3"				1.600	815	83		898	1,000	1,150
	Total				2.057	815	109.49		924.49	1,033.24	1,191.55
2060	**Repack gate valve gland, 8" - 12"**	5	1 PLUM	Ea.							
	Replace valve packing, 8" to 12" diameter				1.600	42.50	92		134.50	162	197
	Total				1.600	42.50	92		134.50	162	197
3010	**Repack drain valve gland, 3/4"**	3	1 STPI	Ea.							
	Remove / replace drain valve packing				.219	10.60	11.50		22.10	26	31
	Total				.219	10.60	11.50		22.10	26	31
3020	**Replace drain valve stem assembly, 3/4"**	8	1 STPI	Ea.							
	Turn valve off and on				.017		.99		.99	1.24	1.55
	Remove valve stem assembly				.267		15.60		15.60	19.50	24.50
	Replace valve stem assembly				.267	12.95	15.60		28.55	34	40.50
	Total				.550	12.95	32.19		45.14	54.74	66.55
3030	**Replace drain valve, 3/4"**	20	1 STPI	Ea.							
	Turn valve off and on				.017		.99		.99	1.24	1.55
	Remove old valve				.308		18		18	22.50	28
	Install new valve, 3/4"				.308	12.95	18		30.95	37	44
	Total				.632	12.95	36.99		49.94	60.74	73.55

D30 HVAC D3043 Distribution Systems

D3043 530 Circulator Pump

	System Description	Freq. (Years)	Crew	Unit	Labor Hours	2014 Bare Costs				Total In-House	Total w/O&P
						Material	Labor	Equipment	Total		
1010	**Repair circulator pump, 1/12 - 3/4 H.P.**	5	1 STPI	Ea.							
	Loosen coupling from pump shaft				.009		.53		.53	.66	.82
	Remove 4 stud bolts in motor				.035		2.05		2.05	2.56	3.19
	Remove motor				.052		3.04		3.04	3.80	4.75
	Remove impeller nut from shaft				.033		1.93		1.93	2.41	3.01
	Remove seal assembly				.019		1.11		1.11	1.39	1.73
	Inspect bearing brackets face				.005		.29		.29	.37	.46
	Clean bearing				.016		.94		.94	1.17	1.46
	Inspect shaft				.040	39.50	2.11		41.61	46	52.50
	Clean shaft				.017		.99		.99	1.24	1.55
	Install new seal and rubber				.013		.76		.76	.95	1.19
	Install seal spring and washers				.006		.35		.35	.44	.55
	Install impeller nut on shaft				.040	9.90	2.11		12.01	13.50	15.65
	Clean pump body gasket area				.037		2.16		2.16	2.71	3.38
	Install new gasket (ready made)				.010		.59		.59	.73	.91
	Install impeller				.009		.53		.53	.66	.82
	Install 4 stud bolts				.035		2.05		2.05	2.56	3.19
	Assemble motor to bearing				.024		1.40		1.40	1.76	2.19
	Install 4 stud bolts				.027		1.58		1.58	1.98	2.46
	Tighten coupling				.009		.53		.53	.66	.82
	Final tightening of bolts				.027		1.58		1.58	1.98	2.46
	Test run				.042		2.46		2.46	3.07	3.83
	Total				.505	49.40	29.09		78.49	90.60	106.92
1030	**Replace circulator pump, 1/12 - 3/4 H.P.**	15	Q-1	Ea.							
	Remove pump and motor				2.597		135		135	168	210
	Install new pump and motor, 1/12-3/4 H.P.				5.195	2,550	269		2,819	3,150	3,600
	Total				7.792	2,550	404		2,954	3,318	3,810

D30 HVAC   D3043 Distribution Systems

D3043 530 Circulator Pump

	System Description	Freq. (Years)	Crew	Unit	Labor Hours	2014 Bare Costs				Total In-House	Total w/O&P
						Material	Labor	Equipment	Total		
2010	**Repair circulator pump, 1 H.P.**	5	1 STPI	Ea.							
	Loosen coupling from pump shaft				.009		.53		.53	.66	.82
	Remove 4 stud bolts in motor				.035		2.05		2.05	2.56	3.19
	Remove motor				.052		3.04		3.04	3.80	4.75
	Remove impeller nut from shaft				.033		1.93		1.93	2.41	3.01
	Remove seal assembly				.019		1.11		1.11	1.39	1.73
	Inspect bearing brackets face				.005		.29		.29	.37	.46
	Clean bearing				.016		.94		.94	1.17	1.46
	Inspect shaft				.040	39.50	2.11		41.61	46	52.50
	Clean shaft				.017		.99		.99	1.24	1.55
	Install new seal and rubber				.013		.76		.76	.95	1.19
	Install seal spring and washers				.006		.35		.35	.44	.55
	Install impeller nut on shaft				.040	9.90	2.11		12.01	13.50	15.65
	Clean pump body gasket area				.037		2.16		2.16	2.71	3.38
	Install new gasket (ready made)				.013		.76		.76	.95	1.19
	Install impeller				.009		.53		.53	.66	.82
	Install 4 stud bolts				.035		2.05		2.05	2.56	3.19
	Assemble motor to bearing				.024		1.40		1.40	1.76	2.19
	Install 4 stud bolts				.027		1.58		1.58	1.98	2.46
	Tighten coupling				.009		.53		.53	.66	.82
	Final tightening of bolts				.027		1.58		1.58	1.98	2.46
	Test run				.042		2.46		2.46	3.07	3.83
	Total				.508	49.40	29.26		78.66	90.82	107.20
2030	**Replace circulator. pump, 1 H.P.**	15	Q-1	Ea.							
	Remove pump				2.597		135		135	168	210
	Install new pump, 1 H.P.				5.200	4,100	269		4,369	4,850	5,550
	Total				7.797	4,100	404		4,504	5,018	5,760

D3043 540 Expansion Tank

	System Description	Freq. (Years)	Crew	Unit	Labor Hours	2014 Bare Costs				Total In-House	Total w/O&P
						Material	Labor	Equipment	Total		
0010	**Refill expansion tank**	5	1 STPI	Ea.							
	Refill air chamber				.200		11.70		11.70	14.65	18.25
	Total				.200		11.70		11.70	14.65	18.25

D30 HVAC D3043 Distribution Systems

D3043 540 Expansion Tank

		Freq. (Years)	Crew	Unit	Labor Hours	2014 Bare Costs				Total In-House	Total w/O&P
	System Description					Material	Labor	Equipment	Total		
0020	**Replace expansion tank, 24 gal capacity**	50	Q-5	Ea.							
	Remove expansion tank				.743		39		39	49	61
	Install expansion tank, 24 gallon capacity				1.486	1,150	78		1,228	1,375	1,550
	Total				2.228	1,150	117		1,267	**1,424**	**1,611**
0120	**Replace expansion tank, 60 gal capacity**	50	Q-5	Ea.							
	Remove expansion tank				1.300		68.50		68.50	85.50	107
	Install expansion tank, 60 gallon capacity				2.602	1,825	137		1,962	2,175	2,500
	Total				3.901	1,825	205.50		2,030.50	**2,260.50**	**2,607**
0220	**Replace expansion tank, 175 gal capacity**	50	Q-5	Ea.							
	Remove expansion tank				2.597		137		137	171	213
	Install expansion tank, 175 gallon capacity				5.195	4,950	274		5,224	5,775	6,625
	Total				7.792	4,950	411		5,361	**5,946**	**6,838**
0320	**Replace expansion tank, 400 gal capacity**	50	Q-5	Ea.							
	Remove expansion tank				3.721		196		196	245	305
	Install expansion tank, 400 gallon capacity				7.442	10,600	390		10,990	12,100	13,900
	Total				11.163	10,600	586		11,186	**12,345**	**14,205**

D3043 550 Pipe Insulation

		Freq. (Years)	Crew	Unit	Labor Hours	2014 Bare Costs				Total In-House	Total w/O&P
	System Description					Material	Labor	Equipment	Total		
1010	**Repair damaged pipe insulation, fiberglass 1/2"**	5	Q-14	Ea.							
	Remove 2' length old insulation				.086		3.98		3.98	5.15	6.40
	Install insulation, fiberglass, 1" wall - 1/2" diam.				.173	1.78	7.98		9.76	12.30	15.05
	Total				.260	1.78	11.96		13.74	**17.45**	**21.45**
1110	**Repair damaged pipe insulation, fiberglass 3/4"**	5	Q-14	Ea.							
	Remove 2' length old insulation				.090		4.16		4.16	5.40	6.70
	Install insulation, fiberglass, 1" wall - 3/4" diam.				.181	1.94	8.32		10.26	12.90	15.85
	Total				.271	1.94	12.48		14.42	**18.30**	**22.55**

D30 HVAC D3043 Distribution Systems

D3043 550 Pipe Insulation

						2014 Bare Costs				Total	Total
	System Description	Freq. (Years)	Crew	Unit	Labor Hours	Material	Labor	Equipment	Total	In-House	w/O&P
1120	**Repair damaged pipe insulation, fiberglass 1"**	5	Q-14	Ea.							
	Remove 2' length old insulation				.095		4.36		4.36	5.65	7
	Install insulation, fiberglass, 1" wall - 1" diam.				.189	2.08	8.72		10.80	13.60	16.60
	Total				.284	2.08	13.08		15.16	**19.25**	**23.60**
1130	**Repair damaged pipe insulation, fbgs 1-1/4"**	5	Q-14	Ea.							
	Remove 2' length old insulation				.099		4.54		4.54	5.90	7.30
	Install insulation, fiberglass, 1" wall - 1-1/4" diam.				.198	2.26	9.10		11.36	14.30	17.45
	Total				.296	2.26	13.64		15.90	**20.20**	**24.75**
1140	**Repair damaged pipe insulation, fbgs 1-1/2"**	5	Q-14	Ea.							
	Remove 2' length old insulation				.099		4.54		4.54	5.90	7.30
	Install insulation, fiberglass, 1" wall - 1-1/2" diam.				.198	2.44	9.12		11.56	14.50	17.65
	Total				.297	2.44	13.66		16.10	**20.40**	**24.95**
1150	**Repair damaged pipe insulation, fiberglass 2"**	5	Q-14	Ea.							
	Remove 2' length old insulation				.104		4.78		4.78	6.20	7.70
	Install insulation, fiberglass, 1" wall - 2" diam.				.208	2.64	9.56		12.20	15.30	18.70
	Total				.312	2.64	14.34		16.98	**21.50**	**26.40**
1160	**Repair damaged pipe insulation, fiberglass 3"**	5	Q-14	Ea.							
	Remove 2' length old insulation				.116		5.34		5.34	6.90	8.60
	Install insulation, fiberglass, 1" wall - 3" diam.				.232	3.20	10.70		13.90	17.40	21
	Total				.348	3.20	16.04		19.24	**24.30**	**29.60**
1170	**Repair damaged pipe insulation, fiberglass 4"**	5	Q-14	Ea.							
	Remove 2' length old insulation				.139		6.40		6.40	8.30	10.30
	Install insulation, fiberglass, 1" wall - 4" diam.				.278	4.24	12.80		17.04	21.50	26
	Total				.417	4.24	19.20		23.44	**29.80**	**36.30**
1180	**Repair damaged pipe insulation, fiberglass 6"**	5	Q-14	Ea.							
	Remove 2' length old insulation				.173		7.96		7.96	10.30	12.80
	Install insulation, fiberglass, 1" wall - 6" diam.				.347	5.08	16		21.08	26.50	32
	Total				.520	5.08	23.96		29.04	**36.80**	**44.80**

D30 HVAC   D3043   Distribution Systems

D3043 550   Pipe Insulation

	System Description	Freq. (Years)	Crew	Unit	Labor Hours	2014 Bare Costs				Total In-House	Total w/O&P
						Material	Labor	Equipment	Total		
1220	**Replace pipe insulation, fiberglass 1/2"**	5	Q-14	M.L.F.							
	Remove 1000' length old insulation				33.330		1,530		1,530	2,000	2,450
	Install insulation, fiberglass, 1" wall – 1/2" diam.				66.670	890	3,070		3,960	4,950	6,050
	Total				100.000	890	4,600		5,490	**6,950**	**8,500**
1230	**Replace pipe insulation, fiberglass 3/4"**	5	Q-14	M.L.F.							
	Remove 1000' length old insulation				34.780		1,600		1,600	2,075	2,575
	Install insulation, fiberglass, 1" wall – 3/4" diam.				69.570	970	3,200		4,170	5,200	6,375
	Total				104.350	970	4,800		5,770	**7,275**	**8,950**
1240	**Replace pipe insulation, fiberglass 1"**	5	Q-14	M.L.F.							
	Remove 1000' length old insulation				36.360		1,670		1,670	2,175	2,700
	Install insulation, fiberglass, 1" wall – 1" diam.				72.730	1,040	3,350		4,390	5,500	6,700
	Total				109.090	1,040	5,020		6,060	**7,675**	**9,400**
1250	**Replace pipe insulation, fiberglass 1-1/4"**	5	Q-14	M.L.F.							
	Remove 1000' length old insulation				38.100		1,750		1,750	2,275	2,825
	Install insulation, fiberglass, 1" wall – 1-1/4" diam.				76.190	1,130	3,510		4,640	5,800	7,075
	Total				114.290	1,130	5,260		6,390	**8,075**	**9,900**
1260	**Replace pipe insulation, fiberglass 1-1/2"**	5	Q-14	M.L.F.							
	Remove 1000' length old insulation				38.100		1,750		1,750	2,275	2,825
	Install insulation, fiberglass, 1" wall – 1-1/2" diam.				76.190	1,220	3,510		4,730	5,900	7,175
	Total				114.290	1,220	5,260		6,480	**8,175**	**10,000**
1270	**Replace pipe insulation, fiberglass 2"**	5	Q-14	M.L.F.							
	Remove 1000' length old insulation				40.000		1,840		1,840	2,400	2,950
	Install insulation, fiberglass, 1" wall – 2" diam.				80.000	1,320	3,680		5,000	6,200	7,550
	Total				120.000	1,320	5,520		6,840	**8,600**	**10,500**
1280	**Replace pipe insulation, fiberglass 3"**	5	Q-14	M.L.F.							
	Remove 1000' length old insulation				44.440		2,050		2,050	2,650	3,300
	Install insulation, fiberglass, 1" wall – 3" diam.				88.890	1,600	4,090		5,690	7,050	8,550
	Total				133.330	1,600	6,140		7,740	**9,700**	**11,850**

D30 HVAC — D3043 Distribution Systems

D3043 550 Pipe Insulation

	System Description	Freq. (Years)	Crew	Unit	Labor Hours	Material	Labor (2014 Bare Costs)	Equipment	Total	Total In-House	Total w/O&P
1290	**Replace pipe insulation, fiberglass 4"**	5	Q-14	M.L.F.							
	Remove 1000' length old insulation				53.330		2,450		2,450	3,175	3,950
	Install insulation, fiberglass, 1" wall - 4" diam.				106.670	2,120	4,910		7,030	8,700	10,600
	Total				160.000	2,120	7,360		9,480	11,875	14,550
1300	**Replace pipe insulation, fiberglass 6"**	5	Q-14	M.L.F.							
	Remove 1000' length old insulation				66.670		3,070		3,070	3,975	4,925
	Install insulation, fiberglass, 1" wall - 6" diam.				133.330	2,540	6,150		8,690	10,800	13,000
	Total				200.000	2,540	9,220		11,760	14,775	17,925
1410	**Repair damaged pipe insulation rubber 1/2"**	5	1 ASBE	Ea.							
	Remove 2' length old insulation				.117		5.98		5.98	7.75	9.60
	Install insulation, foam rubber, 3/4" wall - 1/2" diam.				.234	2.20	12		14.20	17.90	22
	Total				.350	2.20	17.98		20.18	25.65	31.60
1420	**Repair damaged pipe insulation rubber 3/4"**	5	1 ASBE	Ea.							
	Remove 2' length old insulation				.117		5.98		5.98	7.75	9.60
	Install insulation, foam rubber, 3/4" wall - 3/4" diam.				.234	2.70	12		14.70	18.50	22.50
	Total				.350	2.70	17.98		20.68	26.25	32.10
1430	**Repair damaged pipe insulation rubber 1"**	5	1 ASBE	Ea.							
	Remove 2' length old insulation				.118		6.04		6.04	7.85	9.70
	Install insulation, foam rubber, 3/4" wall - 1" diam.				.236	3.08	12.10		15.18	19	23.50
	Total				.355	3.08	18.14		21.22	26.85	33.20
1440	**Repair damaged pipe insulation rubber 1-1/4"**	5	1 ASBE	Ea.							
	Remove 2' length old insulation				.120		6.12		6.12	7.90	9.80
	Install insulation, foam rubber, 3/4" wall - 1-1/4" diam.				.239	4.10	12.20		16.30	20.50	24.50
	Total				.359	4.10	18.32		22.42	28.40	34.30
1450	**Repair damaged pipe insulation rubber 1-1/2"**	5	1 ASBE	Ea.							
	Remove 2' length old insulation				.120		6.12		6.12	7.90	9.80
	Install insulation, foam rubber, 3/4" wall - 1-1/2" diam.				.239	4.64	12.20		16.84	21	25.50
	Total				.359	4.64	18.32		22.96	28.90	35.30

D30 HVAC   D3043   Distribution Systems

D3043 550   Pipe Insulation

	System Description	Freq. (Years)	Crew	Unit	Labor Hours	2014 Bare Costs				Total In-House	Total w/O&P
						Material	Labor	Equipment	Total		
1460	**Repair damaged pipe insulation rubber 2"**	5	1 ASBE	Ea.							
	Remove 2' length old insulation				.121		6.20		6.20	8.05	9.95
	Install insulation, foam rubber, 3/4" wall - 2" diam.				.242	5.36	12.40		17.76	22	26.50
	Total				.364	5.36	18.60		23.96	30.05	36.45
1470	**Repair damaged pipe insulation rubber 3"**	5	1 ASBE	Ea.							
	Remove 2' length old insulation				.122		6.26		6.26	8.10	10
	Install insulation, foam rubber, 3/4" wall - 3" diam.				.245	8.16	12.50		20.66	25	30.50
	Total				.367	8.16	18.76		26.92	33.10	40.50
1480	**Repair damaged pipe insulation rubber 4"**	5	1 ASBE	Ea.							
	Remove 2' length old insulation				.130		6.66		6.66	8.60	10.70
	Install insulation, foam rubber, 3/4" wall - 4" diam.				.260	10.20	13.30		23.50	28.50	34
	Total				.390	10.20	19.96		30.16	37.10	44.70
1490	**Repair damaged pipe insulation rubber 6"**	5	1 ASBE	Ea.							
	Remove 2' length old insulation				.130		6.66		6.66	8.60	10.70
	Install insulation, foam rubber, 3/4" wall - 6" diam.				.260	14.90	13.30		28.20	33.50	40
	Total				.390	14.90	19.96		34.86	42.10	50.70
1510	**Replace pipe insulation foam rubber 1/2"**	5	1 ASBE	L.F.							
	Remove 1000' length old insulation				.090		4.60		4.60	5.95	7.40
	Install insulation, foam rubber, 3/4" wall - 1/2" diam.				.180	2.20	9.20		11.40	14.30	17.55
	Total				.270	2.20	13.80		16	20.25	24.95
1520	**Replace pipe insulation foam rubber 3/4"**	5	1 ASBE	L.F.							
	Remove 1000' length old insulation				.090		4.60		4.60	5.95	7.40
	Install insulation, foam rubber, 3/4" wall - 3/4" diam.				.180	2.70	9.20		11.90	14.90	18.20
	Total				.270	2.70	13.80		16.50	20.85	25.60
1530	**Replace pipe insulation foam rubber 1"**	5	1 ASBE	L.F.							
	Remove 1000' length old insulation				.091		4.64		4.64	6	7.45
	Install insulation, foam rubber, 3/4" wall - 1" diam.				.182	3.08	9.30		12.38	15.40	18.75
	Total				.273	3.08	13.94		17.02	21.40	26.20

D30 HVAC   D3043   Distribution Systems

D3043 550   Pipe Insulation

							2014 Bare Costs				Total In-House	Total w/O&P
	System Description	Freq. (Years)	Crew	Unit	Labor Hours	Material	Labor	Equipment	Total			
1540	**Replace pipe insulation foam rubber 1-1/4"**	5	1 ASBE	L.F.								
	Remove 1000' length old insulation				.092		4.70		4.70	6.10	7.55	
	Install insulation, foam rubber, 3/4" wall - 1-1/4" diam.				.184	4.10	9.40		13.50	16.70	20	
	Total				.276	4.10	14.10		18.20	22.80	27.55	
1550	**Replace pipe insulation foam rubber 1-1/2"**	5	1 ASBE	L.F.								
	Remove 1000' length old insulation				.092		4.70		4.70	6.10	7.55	
	Install insulation, foam rubber, 3/4" wall - 1-1/2" diam.				.184	4.64	9.40		14.04	17.30	21	
	Total				.276	4.64	14.10		18.74	23.40	28.55	
1560	**Replace pipe insulation foam rubber 2"**	5	1 ASBE	L.F.								
	Remove 1000' length old insulation				.093		4.76		4.76	6.15	7.65	
	Install insulation, foam rubber, 3/4" wall - 2" diam.				.186	5.36	9.52		14.88	18.20	22	
	Total				.279	5.36	14.28		19.64	24.35	29.65	
1570	**Replace pipe insulation foam rubber 3"**	5	1 ASBE	L.F.								
	Remove 1000' length old insulation				.094		4.82		4.82	6.25	7.75	
	Install insulation, foam rubber, 3/4" wall - 3" diam.				.188	8.16	9.62		17.78	21.50	25.50	
	Total				.282	8.16	14.44		22.60	27.75	33.25	
1580	**Replace pipe insulation foam rubber 4"**	5	1 ASBE	L.F.								
	Remove 1000' length old insulation				.100		5.12		5.12	6.65	8.20	
	Install insulation, foam rubber, 3/4" wall - 4" diam.				.200	10.20	10.20		20.40	24.50	29	
	Total				.300	10.20	15.32		25.52	31.15	37.20	
1590	**Replace pipe insulation foam rubber 6"**	5	1 ASBE	L.F.								
	Remove 1000' length old insulation				.100		5.12		5.12	6.65	8.20	
	Install insulation, foam rubber, 3/4" wall - 6" diam.				.200	14.90	10.20		25.10	29.50	35	
	Total				.300	14.90	15.32		30.22	36.15	43.20	

D30 HVAC　　D3053　Terminal and Package Units

D3053 110　Unit Heater

	System Description	Freq. (Years)	Crew	Unit	Labor Hours	2014 Bare Costs				Total In-House	Total w/O&P
						Material	Labor	Equipment	Total		
1010	**Repair unit heater, 12 MBH, 2 PSI steam**	10	1 STPI	Ea.							
	Repair controls				.200		11.70		11.70	14.65	18.25
	Remove fan motor				.976		57		57	71.50	89
	Replace fan motor				1.951	198	104		302	345	410
	Total				3.127	198	172.70		370.70	**431.15**	**517.25**
1030	**Replace unit heater, 12 MBH, 2 PSI steam**	15	Q-5	Ea.							
	Remove unit heater				.867		45.50		45.50	57	71
	Replace unit heater, 12 MBH, 2 PSI steam				1.733	360	91.50		451.50	510	590
	Total				2.600	360	137		497	**567**	**661**
2010	**Repair unit heater, 36 MBH, 2 PSI steam**	10	1 STPI	Ea.							
	Repair controls				.200		11.70		11.70	14.65	18.25
	Remove fan motor				.976		57		57	71.50	89
	Replace fan motor				1.951	207	104		311	355	420
	Total				3.127	207	172.70		379.70	**441.15**	**527.25**
2030	**Replace unit heater, 36 MBH, 2 PSI steam**	15	Q-5	Ea.							
	Remove unit heater				1.300		68.50		68.50	85.50	107
	Replace unit heater, 36 MBH, 2 PSI steam				2.600	545	137		682	770	895
	Total				3.900	545	205.50		750.50	**855.50**	**1,002**
3010	**Repair unit heater, 85 MBH, 2 PSI steam**	10	1 STPI	Ea.							
	Repair controls				.200		11.70		11.70	14.65	18.25
	Remove fan motor				.976		57		57	71.50	89
	Replace fan motor				1.951	248	104		352	400	470
	Total				3.127	248	172.70		420.70	**486.15**	**577.25**
3030	**Replace unit heater, 85 MBH, 2 PSI steam**	15	Q-5	Ea.							
	Remove unit heater				1.600		84		84	105	131
	Replace unit heater, 85 MBH, 2 PSI steam				3.200	650	168		818	925	1,075
	Total				4.800	650	252		902	**1,030**	**1,206**
4010	**Repair unit heater, 250 MBH, 2 PSI steam**	10	1 STPI	Ea.							
	Repair controls				.200		11.70		11.70	14.65	18.25
	Remove fan motor				1.156		61.50		61.50	76.50	95.50
	Replace fan motor				2.311	279	123		402	460	540
	Total				3.667	279	196.20		475.20	**551.15**	**653.75**

D30 HVAC D3053 Terminal and Package Units

D3053 110 Unit Heater

	System Description	Freq. (Years)	Crew	Unit	Labor Hours	2014 Bare Costs				Total In-House	Total w/O&P
						Material	Labor	Equipment	Total		
4030	**Replace unit heater, 250 MBH, 2 PSI steam**	15	Q-5	Ea.							
	Remove unit heater				4.160		219		219	274	340
	Replace unit heater, 250 MBH, 2 PSI steam				8.320	1,475	440		1,915	2,175	2,525
	Total				12.481	1,475	659		2,134	2,449	**2,865**
5010	**Repair unit heater, 400 MBH, 2 PSI steam**	10	1 STPI	Ea.							
	Repair controls				.200		11.70		11.70	14.65	18.25
	Remove fan motor				1.156		61.50		61.50	76.50	95.50
	Replace fan motor				2.311	335	123		458	520	610
	Total				3.667	335	196.20		531.20	611.15	**723.75**
5020	**Replace unit heater, 400 MBH, 2 PSI steam**	15	Q-5	Ea.							
	Remove unit heater				6.499		340		340	430	535
	Replace unit heater, 400 MBH, 2 PSI steam				13.008	2,150	685		2,835	3,225	3,775
	Total				19.507	2,150	1,025		3,175	3,655	**4,310**

D3053 112 Infrared Heater Suspended, Commercial

	System Description	Freq. (Years)	Crew	Unit	Labor Hours	2014 Bare Costs				Total In-House	Total w/O&P
						Material	Labor	Equipment	Total		
0010	**Maintenance and repair**	1	1 ELEC	Ea.							
	Repair wiring connections				.615		33		33	40.50	51
	Total				.615		33		33	**40.50**	**51**
0020	**Maintenance and inspection**	0.50	1 ELEC	Ea.							
	Inspect and clean infrared heater				1.143		61		61	75.50	94.50
	Total				1.143		61		61	**75.50**	**94.50**
0030	**Replace infrared heater**	15	1 ELEC	Ea.							
	Remove infrared heater				.667		35.50		35.50	44	55
	Infrared heater 240 V, 1500 W				2.105	236	112		348	400	470
	Total				2.772	236	147.50		383.50	**444**	**525**

D30 HVAC   D3053 Terminal and Package Units

D3053 114 Standard Suspended Heater

						2014 Bare Costs				Total In-House	Total w/O&P
	System Description	Freq. (Years)	Crew	Unit	Labor Hours	Material	Labor	Equipment	Total		
0010	**Maintenance and repair**	2	1 ELEC	Ea.							
	Remove unit heater component				.151		8.05		8.05	9.95	12.45
	Unit heater component				.444	65.50	23.50		89	101	118
	Total				.595	65.50	31.55		97.05	**110.95**	**130.45**
0020	**Maintenance and inspection**	0.50	1 ELEC	Ea.							
	Inspect and clean unit heater				1.143		61		61	75.50	94.50
	Total				1.143		61		61	**75.50**	**94.50**
0030	**Replace heater**	15	1 ELEC	Ea.							
	Remove unit heater				.667		35.50		35.50	44	55
	Unit heater, heavy duty, 480 V, 3 KW				1.333	480	71		551	615	710
	Total				2.000	480	106.50		586.50	**659**	**765**

D3053 116 Explosionproof Industrial Heater

						2014 Bare Costs				Total In-House	Total w/O&P
	System Description	Freq. (Years)	Crew	Unit	Labor Hours	Material	Labor	Equipment	Total		
0010	**Maintenance and repair**	2	1 ELEC	Ea.							
	Remove explosionproof heater part				.200		10.65		10.65	13.20	16.50
	Explosionproof heater part				.615	83.50	33		116.50	132	155
	Total				.815	83.50	43.65		127.15	**145.20**	**171.50**
0020	**Maintenance and inspection**	0.50	1 ELEC	Ea.							
	Inspect / clean explosionproof heater				1.143		61		61	75.50	94.50
	Total				1.143		61		61	**75.50**	**94.50**
0030	**Replace heater**	15	1 ELEC	Ea.							
	Remove explosionproof heater				.800		42.50		42.50	53	66
	Explosionproof heater, 3 KW				3.810	5,325	203		5,528	6,100	6,975
	Total				4.610	5,325	245.50		5,570.50	**6,153**	**7,041**

D30 HVAC D3053 Terminal and Package Units

D3053 150 Wall Mounted/Recessed Heater, With Fan

	System Description	Freq. (Years)	Crew	Unit	Labor Hours	2014 Bare Costs				Total In-House	Total w/O&P
						Material	Labor	Equipment	Total		
0010	**Maintenance and repair**	5	1 ELEC	Ea.							
	Remove heater fan				.250		13.35		13.35	16.50	20.50
	Heater fan				.727	101	39		140	159	186
	Total				.977	101	52.35		153.35	**175.50**	**206.50**
0020	**Maintenance and inspection**	1	1 ELEC	Ea.							
	Inspect and clean heater with fan				1.143		61		61	75.50	94.50
	Total				1.143		61		61	**75.50**	**94.50**
0030	**Replace heater**	20	1 ELEC	Ea.							
	Remove heater with fan				.667		35.50		35.50	44	55
	Heater with fan, commercial 2000 W				2.667	189	142		331	385	455
	Total				3.333	189	177.50		366.50	**429**	**510**

D3053 160 Convector Suspended, Commercial

	System Description	Freq. (Years)	Crew	Unit	Labor Hours	2014 Bare Costs				Total In-House	Total w/O&P
						Material	Labor	Equipment	Total		
0010	**Maintenance and repair**	2	1 ELEC	Ea.							
	Repair wiring connections				.615		33		33	40.50	51
	Total				.615		33		33	**40.50**	**51**
0020	**Maintenance and inspection**	0.50	1 ELEC	Ea.							
	Inspect and clean convector heater				1.143		61		61	75.50	94.50
	Total				1.143		61		61	**75.50**	**94.50**
0030	**Replace heater**	15	1 ELEC	Ea.							
	Remove cabinet convector heater				.667		35.50		35.50	44	55
	Cabinet convector heater, 2' long				2.000	1,975	107		2,082	2,300	2,625
	Total				2.667	1,975	142.50		2,117.50	**2,344**	**2,680**

D30 HVAC D3053 Terminal and Package Units

D3053 170 Terminal Reheat

	System Description	Freq. (Years)	Crew	Unit	Labor Hours	2014 Bare Costs				Total In-House	Total w/O&P
						Material	Labor	Equipment	Total		
1010	**Repair terminal reheat, 12" x 24" coil**	10	1 STPI	Ea.							
	Fix leak				1.200		70		70	88	109
	Total				1.200		70		70	**88**	**109**
1040	**Replace terminal reheat, 12" x 24" coil**	15	Q-5	Ea.							
	Remove terminal reheat unit				.858		45		45	56.50	70.50
	Replace terminal reheat unit, 12" x 24" coil				1.716	810	90.50		900.50	1,000	1,150
	Total				2.574	810	135.50		945.50	**1,056.50**	**1,220.50**
2010	**Repair terminal reheat, 18" x 24" coil**	10	1 STPI	Ea.							
	Fix leak				1.400		82		82	102	128
	Total				1.400		82		82	**102**	**128**
2040	**Replace terminal reheat, 18" x 24" coil**	15	Q-5	Ea.							
	Remove terminal reheat unit				1.284		67.50		67.50	84.50	105
	Replace terminal reheat unit, 18" x 24" coil				2.568	975	135		1,110	1,250	1,425
	Total				3.852	975	202.50		1,177.50	**1,334.50**	**1,530**
3010	**Repair terminal reheat, 36" x 36" coil**	10	1 STPI	Ea.							
	Fix leak				2.200		129		129	161	201
	Total				2.200		129		129	**161**	**201**
3040	**Replace terminal reheat, 36" x 36" coil**	15	Q-5	Ea.							
	Remove terminal reheat unit				3.852		203		203	254	315
	Replace terminal reheat unit, 36" x 36" coil				7.703	1,625	405		2,030	2,300	2,675
	Total				11.555	1,625	608		2,233	**2,554**	**2,990**
4010	**Repair terminal reheat, 48" x 126" coil**	10	1 STPI	Ea.							
	Fix leak				4.000		234		234	293	365
	Total				4.000		234		234	**293**	**365**
4040	**Replace terminal reheat, 48" x 126" coil**	15	Q-5	Ea.							
	Remove terminal reheat unit				17.937		945		945	1,175	1,475
	Replace terminal reheat unit, 48" x 126" coil				35.874	4,800	1,900		6,700	7,650	8,950
	Total				53.812	4,800	2,845		7,645	**8,825**	**10,425**

D30 HVAC / D3053 Terminal and Package Units

D3053 245 Heat Pump

	System Description	Freq. (Years)	Crew	Unit	Labor Hours	2014 Bare Costs				Total In-House	Total w/O&P
						Material	Labor	Equipment	Total		
1010	**Repair heat pump, 1.5 ton, air to air split**	10	1 STPI	Ea.							
	Repair controls				.600		35		35	44	55
	Remove / replace supply fan				.650		38		38	47.50	59.50
	Remove supply fan motor				.700		41		41	51	64
	Replace supply fan motor				1.951	192	114		306	355	420
	Remove compressor				1.860		109		109	136	170
	Replace compressor				3.714	259	217		476	555	665
	Remove / replace condenser fan				1.800		105		105	132	164
	Remove condenser fan motor				1.800		105		105	132	164
	Replace condenser fan motor				1.951	198	104		302	345	410
	Replace refrigerant				.133	32	7.80		39.80	45	52
	Remove / replace heater				.500		29.50		29.50	36.50	45.50
	Total				15.661	681	905.30		1,586.30	1,879	2,269
1030	**Replace heat pump, 1.5 ton, air to air split**	20	Q-5	Ea.							
	Remove heat pump				4.336		228		228	285	355
	Replace heat pump, 1.5 ton, air to air split				8.649	2,175	455		2,630	2,950	3,425
	Total				12.985	2,175	683		2,858	3,235	3,780
2010	**Repair heat pump, 5 ton, air to air split**	10	1 STPI	Ea.							
	Repair controls				.600		35		35	44	55
	Remove / replace supply fan				.650		38		38	47.50	59.50
	Remove supply fan motor				1.156		61.50		61.50	76.50	95.50
	Replace supply fan motor				1.951	198	104		302	345	410
	Remove compressor				2.477		145		145	181	226
	Replace compressor				4.954	485	290		775	895	1,050
	Remove / replace condenser fan				1.800		105		105	132	164
	Remove condenser fan motor				.976		57		57	71.50	89
	Replace condenser fan motor				1.951	207	104		311	355	420
	Replace refrigerant				.667	160	39		199	225	261
	Remove / replace heater				.500		29.50		29.50	36.50	45.50
	Total				17.681	1,050	1,008		2,058	2,409	2,875.50
2030	**Replace heat pump, 5 ton, air to air split**	20	Q-5	Ea.							
	Remove heat pump				20.806		1,100		1,100	1,375	1,700
	Replace heat pump, 5 ton, air to air split				41.558	3,625	2,200		5,825	6,725	7,950
	Total				62.365	3,625	3,300		6,925	8,100	9,650

D30 HVAC — D3053 Terminal and Package Units

D3053 245 Heat Pump

	System Description	Freq. (Years)	Crew	Unit	Labor Hours	2014 Bare Costs				Total In-House	Total w/O&P
						Material	Labor	Equipment	Total		
3010	**Repair heat pump, 10 ton, air to air split**	10	Q-5	Ea.							
	Repair controls				.600		35		35	44	55
	Remove / replace supply fan				.650		38		38	47.50	59.50
	Remove supply fan motor				.700		41		41	51	64
	Replace supply fan motor				2.311	335	123		458	520	610
	Remove compressor				4.000		211		211	263	330
	Replace compressor				8.000	1,275	420		1,695	1,925	2,250
	Remove / replace condenser fan				1.800		105		105	132	164
	Remove condenser fan motor				1.800		105		105	132	164
	Replace condenser fan motor				2.311	279	123		402	460	540
	Replace refrigerant				1.067	256	62.40		318.40	360	415
	Remove / replace heater				.500		29.50		29.50	36.50	45.50
	Total				23.739	2,145	1,292.90		3,437.90	3,971	4,697
3030	**Replace heat pump, 10 ton, air to air split**	20	Q-6	Ea.							
	Remove heat pump				24.490		1,325		1,325	1,675	2,075
	Replace heat pump, 10 ton, air to air split				48.980	8,675	2,675		11,350	12,900	15,000
	Total				73.469	8,675	4,000		12,675	14,575	17,075
4010	**Repair heat pump, 25 ton, air to air split**	10	Q-5	Ea.							
	Repair controls				.600		35		35	44	55
	Remove / replace supply fan				1.000		58.50		58.50	73	91.50
	Remove supply fan motor				1.156		61.50		61.50	76.50	95.50
	Replace supply fan motor				2.311	600	123		723	815	940
	Remove compressor				10.000		525		525	660	820
	Replace compressor				20.000	3,187.50	1,050		4,237.50	4,825	5,625
	Remove / replace condenser fan				1.800		105		105	132	164
	Remove condenser fan motor				1.800		105		105	132	164
	Replace condenser fan motor				2.312	500	123		623	705	815
	Replace refrigerant				2.000	480	117		597	675	780
	Remove / replace heater				.500		29.50		29.50	36.50	45.50
	Total				43.479	4,767.50	2,332.50		7,100	8,174	9,595.50
4030	**Replace heat pump, 25 ton, air to air split**	20	Q-7	Ea.							
	Remove heat pump				61.538		3,425		3,425	4,275	5,350
	Replace heat pump, 25 ton, air to air split				123.000	20,400	6,850		27,250	31,000	36,200
	Total				184.538	20,400	10,275		30,675	35,275	41,550

D30 HVAC D3053 Terminal and Package Units

D3053 245 Heat Pump

		Freq. (Years)	Crew	Unit	Labor Hours	2014 Bare Costs				Total In-House	Total w/O&P
	System Description					Material	Labor	Equipment	Total		
5010	**Repair heat pump, 50 ton, air to air split**	10	Q-6	Ea.							
	Repair controls				.600		35		35	44	55
	Remove / replace supply fan				1.000		58.50		58.50	73	91.50
	Remove supply fan motor				1.156		61.50		61.50	76.50	95.50
	Replace supply fan motor				4.167	1,925	222		2,147	2,400	2,750
	Remove compressor				77.922		4,250		4,250	5,325	6,625
	Replace compressor				155.000	31,300	8,500		39,800	45,000	52,500
	Remove / replace condenser fan				2.400		140		140	176	219
	Remove condenser fan motor				1.800		105		105	132	164
	Replace condenser fan motor				4.000	1,525	213		1,738	1,950	2,225
	Replace refrigerant				4.667	1,120	273		1,393	1,575	1,825
	Remove / replace heater				1.200		70		70	88	109
	Total				253.912	35,870	13,928		49,798	56,839.50	66,659
5030	**Replace heat pump, 50 ton, air to air split**	20	Q-7	Ea.							
	Remove heat pump				96.970		5,400		5,400	6,750	8,425
	Replace heat pump, 50 ton, air to air split				207.000	51,000	11,600		62,600	70,500	82,000
	Total				303.970	51,000	17,000		68,000	77,250	90,425
6010	**Repair heat pump, thru-wall unit, 1.5 ton**	10	1 STPI	Ea.							
	Repair controls				.600		35		35	44	55
	Remove / replace supply fan				.650		38		38	47.50	59.50
	Remove supply fan motor				.700		41		41	51	64
	Replace supply fan motor				1.951	192	114		306	355	420
	Remove compressor				1.860		109		109	136	170
	Replace compressor				3.714	259	217		476	555	665
	Remove / replace condenser fan				1.800		105		105	132	164
	Remove condenser fan motor				1.800		105		105	132	164
	Replace condenser fan motor				1.951	198	104		302	345	410
	Replace refrigerant				.133	32	7.80		39.80	45	52
	Remove / replace heater				.500		29.50		29.50	36.50	45.50
	Total				15.661	681	905.30		1,586.30	1,879	2,269
6030	**Replace heat pump, thru-wall unit, 1.5 ton**	20	Q-5	Ea.							
	Remove heat pump				6.709		355		355	440	550
	Replace heat pump, thru-wall, 1.5 ton				13.423	2,750	705		3,455	3,900	4,550
	Total				20.131	2,750	1,060		3,810	4,340	5,100

D30 HVAC		D3053		Terminal and Package Units								
D3053 245		**Heat Pump**										

	System Description	Freq. (Years)	Crew	Unit	Labor Hours	2014 Bare Costs				Total In-House	Total w/O&P
						Material	Labor	Equipment	Total		
7010	Repair heat pump, thru-wall unit, 5 ton	10	1 STPI	Ea.							
	Repair controls				.600		35		35	44	55
	Remove / replace supply fan				.650		38		38	47.50	59.50
	Remove supply fan motor				1.156		61.50		61.50	76.50	95.50
	Replace supply fan motor				1.951	198	104		302	345	410
	Remove compressor				2.477		145		145	181	226
	Replace compressor				4.954	485	290		775	895	1,050
	Remove / replace condenser fan				1.800		105		105	132	164
	Remove condenser fan motor				.976		57		57	71.50	89
	Replace condenser fan motor				1.951	207	104		311	355	420
	Replace refrigerant				.667	160	39		199	225	261
	Remove / replace heater				.500		29.50		29.50	36.50	45.50
		Total			17.681	1,050	1,008		2,058	2,409	2,875.50
7030	Replace heat pump, thru-wall unit, 5 ton	20	Q-5	Ea.							
	Remove heat pump				16.000		840		840	1,050	1,325
	Replace heat pump, thru-wall, 5 ton				32.000	4,375	1,675		6,050	6,925	8,100
		Total			48.000	4,375	2,515		6,890	7,975	9,425

D3053 265		**Air Conditioner, Window, 1 Ton**										

	System Description	Freq. (Years)	Crew	Unit	Labor Hours	2014 Bare Costs				Total In-House	Total w/O&P
						Material	Labor	Equipment	Total		
1010	Repair air conditioner, window, 1 ton	8	1 STPI	Ea.							
	Find / fix leak				1.000		58.50		58.50	73	91.50
	Replace refrigerant				.067	16	3.90		19.90	22.50	26
		Total			1.067	16	62.40		78.40	95.50	117.50
1030	Replace air conditioner, window, 1 ton	10	L-2	Ea.							
	Remove air conditioner				1.300		52.50		52.50	70	86
	Replace air conditioner, 1 ton				2.000	695	80.50		775.50	870	1,000
		Total			3.300	695	133		828	940	1,086

D30 HVAC D3053 Terminal and Package Units

D3053 266 Air Conditioner, Thru-The-Wall

	System Description	Freq. (Years)	Crew	Unit	Labor Hours	2014 Bare Costs				Total In-House	Total w/O&P
						Material	Labor	Equipment	Total		
2010	Repair air conditioner, thru-the-wall, 2 ton	8	1 STPI	Ea.							
	Find / fix leak				1.000		58.50		58.50	73	91.50
	Replace refrigerant				.133	32	7.80		39.80	45	52
	Total				1.133	32	66.30		98.30	118	143.50
2030	Replace air conditioner, thru-the-wall, 2 ton	10	L-2	Ea.							
	Remove air conditioner				2.600		105		105	140	172
	Replace air conditioner, 2 ton				5.200	1,175	209		1,384	1,575	1,825
	Total				7.800	1,175	314		1,489	1,715	1,997

D3053 272 Air Conditioner, DX Package

	System Description	Freq. (Years)	Crew	Unit	Labor Hours	2014 Bare Costs				Total In-House	Total w/O&P
						Material	Labor	Equipment	Total		
1010	Repair air conditioner, DX, 5 ton	10	1 STPI	Ea.							
	Repair controls				.600		35		35	44	55
	Remove / replace supply fan				.650		38		38	47.50	59.50
	Remove supply fan bearing				1.000		58.50		58.50	73	91.50
	Replace supply fan bearing				2.000	55.50	117		172.50	207	252
	Remove supply fan motor				1.156		61.50		61.50	76.50	95.50
	Replace supply fan motor				2.311	279	123		402	460	540
	Remove compressor				2.477		145		145	181	226
	Replace compressor				4.954	485	290		775	895	1,050
	Remove / replace condenser fan				1.800		105		105	132	164
	Remove condenser fan bearing				1.200		70		70	88	109
	Replace condenser fan bearing				2.000	55.50	117		172.50	207	252
	Remove condenser fan motor				.976		57		57	71.50	89
	Replace condenser fan motor				1.951	248	104		352	400	470
	Replace refrigerant				.667	160	39		199	225	261
	Remove / replace heating coils				3.900	667.50	213.75		881.25	1,000	1,175
	Total				27.641	1,950.50	1,573.75		3,524.25	4,107.50	4,889.50
1030	Replace air conditioner, DX, 5 ton	20	Q-6	Ea.							
	Remove air conditioner				13.001		710		710	890	1,100
	Replace air conditioner, 5 ton				26.002	4,450	1,425		5,875	6,675	7,800
	Total				39.003	4,450	2,135		6,585	7,565	8,900

D30 HVAC		D3053		Terminal and Package Units								
D3053 272		Air Conditioner, DX Package										

	System Description	Freq. (Years)	Crew	Unit	Labor Hours	2014 Bare Costs				Total In-House	Total w/O&P
						Material	Labor	Equipment	Total		
2010	Repair air conditioner, DX, 20 ton	10	Q-6	Ea.							
	Repair controls				.800		47		47	58.50	73
	Remove / replace supply fan				.650		38		38	47.50	59.50
	Remove supply fan bearing				1.000		58.50		58.50	73	91.50
	Replace supply fan bearing				2.000	55.50	117		172.50	207	252
	Remove supply fan motor				1.156		61.50		61.50	76.50	95.50
	Replace supply fan motor				2.312	500	123		623	705	815
	Remove compressor				24.390		1,325		1,325	1,675	2,075
	Replace compressor				48.780	27,600	2,675		30,275	33,700	38,700
	Remove / replace condenser fan				1.800		105		105	132	164
	Remove condenser fan bearing				1.200		70		70	88	109
	Replace condenser fan bearing				2.000	55.50	117		172.50	207	252
	Remove condenser fan motor				.976		57		57	71.50	89
	Replace condenser fan motor				2.311	279	123		402	460	540
	Replace refrigerant				1.333	320	78		398	450	520
	Remove / replace heating coils				6.936	2,580	386.25		2,966.25	3,325	3,825
	Total				97.645	31,390	5,381.25		36,771.25	41,276	47,660.50
2030	Replace air conditioner, DX, 20 ton	20	Q-7	Ea.							
	Remove air conditioner				23.105		1,275		1,275	1,600	2,000
	Replace air conditioner, 20 ton				46.243	17,200	2,575		19,775	22,100	25,500
	Total				69.347	17,200	3,850		21,050	23,700	27,500

D30 HVAC   D3053 Terminal and Package Units

D3053 272   Air Conditioner, DX Package

	System Description	Freq. (Years)	Crew	Unit	Labor Hours	2014 Bare Costs				Total In-House	Total w/O&P
						Material	Labor	Equipment	Total		
3010	**Repair air conditioner, DX, 50 ton**	10	Q-6	Ea.							
	Repair controls				.800		47		47	58.50	73
	Remove / replace supply fan				.650		38		38	47.50	59.50
	Remove supply fan bearing				1.200		70		70	88	109
	Replace supply fan bearing				2.222	62.50	130		192.50	231	281
	Remove supply fan motor				1.156		61.50		61.50	76.50	95.50
	Replace supply fan motor				2.597	855	139		994	1,100	1,275
	Remove compressor				77.922		4,250		4,250	5,325	6,625
	Replace compressor				155.000	31,300	8,500		39,800	45,000	52,500
	Remove / replace condenser fan				1.800		105		105	132	164
	Remove condenser fan bearing				1.200		70		70	88	109
	Replace condenser fan bearing				2.222	62.50	130		192.50	231	281
	Remove condenser fan motor				.976		57		57	71.50	89
	Replace condenser fan motor				2.311	279	123		402	460	540
	Replace refrigerant				3.333	800	195		995	1,125	1,300
	Remove / replace heating coil				12.468	7,080	693.75		7,773.75	8,625	9,925
	Total				265.857	40,439	14,609.25		55,048.25	62,659	73,426
3030	**Replace air conditioner, DX, 50 ton**	20	Q-7	Ea.							
	Remove air conditioner				41.558		2,325		2,325	2,900	3,600
	Replace air conditioner, 50 ton				83.117	47,200	4,625		51,825	57,500	66,000
	Total				124.675	47,200	6,950		54,150	60,400	69,600

D30 HVAC

D3053 Terminal and Package Units

D3053 274 Computer Room A/C Units, Air Cooled

	System Description	Freq. (Years)	Crew	Unit	Labor Hours	2014 Bare Costs				Total In-House	Total w/O&P
						Material	Labor	Equipment	Total		
1010	**Repair computer room A/C, air cooled, 5 ton**	10	1 STPI	Ea.							
	Repair controls				.600		35		35	44	55
	Remove / replace supply fan				.650		38		38	47.50	59.50
	Remove supply fan bearing				1.000		58.50		58.50	73	91.50
	Replace supply fan bearing				2.000	55.50	117		172.50	207	252
	Remove supply fan motor				1.156		61.50		61.50	76.50	95.50
	Replace supply fan motor				2.311	279	123		402	460	540
	Replace fan belt for supply fan				.400	34	23.50		57.50	66.50	79
	Remove compressor				2.477		145		145	181	226
	Replace compressor				4.954	485	290		775	895	1,050
	Remove / replace condenser fan				1.800		105		105	132	164
	Remove condenser fan bearing				1.200		70		70	88	109
	Replace condenser fan bearing				2.000	55.50	117		172.50	207	252
	Remove condenser fan motor				.976		57		57	71.50	89
	Replace condenser fan motor				1.951	248	104		352	400	470
	Replace fan belt for condenser fan				.400	34	23.50		57.50	66.50	79
	Replace refrigerant				.667	160	39		199	225	261
	Remove / replace heating coils				3.900	667.50	213.75		881.25	1,000	1,175
	Total				28.441	2,018.50	1,620.75		3,639.25	4,240.50	5,047.50
1015	**Replace computer room A/C, air cooled, 5 ton**	20	Q-6	Ea.							
	Remove air conditioner and condenser				13.001		710		710	890	1,100
	Replace air conditioner and remote condenser, 5 ton				35.556	19,500	1,875		21,375	23,800	27,300
	Total				48.557	19,500	2,585		22,085	24,690	28,400

D30 HVAC

D3053 Terminal and Package Units

D3053 274 Computer Room A/C Units, Air Cooled

	System Description	Freq. (Years)	Crew	Unit	Labor Hours	2014 Bare Costs				Total In-House	Total w/O&P
						Material	Labor	Equipment	Total		
1020	**Repair computer room A/C, air cooled, 10 ton**	10	1 STPI	Ea.							
	Repair controls				.400		23.50		23.50	29.50	36.50
	Remove / replace supply fan				.650		38		38	47.50	59.50
	Remove supply fan bearing				1.200		70		70	88	110
	Replace supply fan bearing				2.000	55.50	117		172.50	207	252
	Remove supply fan motor				1.156		61.50		61.50	76.50	95.50
	Replace supply fan motor				2.311	335	123		458	520	610
	Replace fan belt for supply fan				.400	34	23.50		57.50	66.50	79
	Remove compressor				14.513		787.25		787.25	990	1,250
	Replace compressor				28.714	17,822	1,574.50		19,396.50	21,600	24,700
	Remove / replace condenser fan				1.800		105		105	132	164
	Remove condenser fan bearing				1.200		70		70	88	109
	Replace condenser fan bearing				2.000	55.50	117		172.50	207	252
	Remove condenser fan motor				.976		57		57	71.50	89
	Replace condenser fan motor				1.951	248	104		352	400	470
	Replace fan belt for condenser fan				.400	34	23.50		57.50	66.50	79
	Replace refrigerant				1.333	320	78		398	450	520
	Remove / replace heating coils				5.200	890	285		1,175	1,325	1,550
	Total				66.204	19,794	3,657.75		23,451.75	26,365	30,425.50
1025	**Replace computer room A/C, air cooled, 10 ton**	20	Q-6	Ea.							
	Remove air conditioner and condenser				33.898		1,850		1,850	2,325	2,875
	Replace air conditioner and remote condenser, 10 ton				64.000	38,000	3,375		41,375	46,000	53,000
	Total				97.898	38,000	5,225		43,225	48,325	55,875

D30 HVAC D3053 Terminal and Package Units

D3053 274 Computer Room A/C Units, Air Cooled

							2014 Bare Costs				Total In-House	Total w/O&P
	System Description	Freq. (Years)	Crew	Unit	Labor Hours	Material	Labor	Equipment	Total			
1030	**Repair computer room A/C, air cooled, 15 ton**	10	1 STPI	Ea.								
	Repair controls				.500		29.50		29.50		36.50	45.50
	Remove / replace supply fan				.650		38		38		47.50	59.50
	Remove supply fan bearing				1.200		70		70		88	110
	Replace supply fan bearing				2.000	55.50	117		172.50		207	252
	Remove supply fan motor				1.156		61.50		61.50		76.50	95.50
	Replace supply fan motor				2.312	500	123		623		705	815
	Replace fan belt for supply fan				.400	34	23.50		57.50		66.50	79
	Remove compressor				21.661		1,175		1,175		1,475	1,850
	Replace compressor				42.857	26,600	2,350		28,950		32,200	36,900
	Remove / replace condenser fan				1.800		105		105		132	164
	Remove condenser fan bearing				1.200		70		70		88	109
	Replace condenser fan bearing				2.000	55.50	117		172.50		207	252
	Remove condenser fan motor				.976		57		57		71.50	89
	Replace condenser fan motor				1.951	248	104		352		400	470
	Replace fan belt for condenser fan				.400	34	23.50		57.50		66.50	79
	Replace refrigerant				2.000	480	117		597		675	780
	Remove / replace heating coils				6.501	1,112.50	356.25		1,468.75		1,675	1,950
	Total				89.563	29,119.50	4,937.25		34,056.75		38,217	44,099.50
1035	**Replace computer room A/C, air cooled, 15 ton**	20	Q-6	Ea.								
	Remove air conditioner and condenser				50.314		2,750		2,750		3,425	4,275
	Replace air conditioner and remote condenser, 15 ton				72.727	41,800	3,825		45,625		51,000	58,000
	Total				123.042	41,800	6,575		48,375		54,425	62,275

D30 HVAC   D3053   Terminal and Package Units

D3053 274   Computer Room A/C Units, Air Cooled

	System Description	Freq. (Years)	Crew	Unit	Labor Hours	2014 Bare Costs Material	2014 Bare Costs Labor	2014 Bare Costs Equipment	2014 Bare Costs Total	Total In-House	Total w/O&P
1040	**Repair computer room A/C, air cooled, 20 ton**	10	1 STPI	Ea.							
	Repair controls				.800		47		47	58.50	73
	Remove / replace supply fan				.650		38		38	47.50	59.50
	Remove supply fan bearing				1.000		58.50		58.50	73	91.50
	Replace supply fan bearing				2.000	55.50	117		172.50	207	252
	Remove supply fan motor				1.156		61.50		61.50	76.50	95.50
	Replace supply fan motor				2.312	500	123		623	705	815
	Replace fan belt for supply fan				.400	34	23.50		57.50	66.50	79
	Remove compressor				24.390		1,325		1,325	1,675	2,075
	Replace compressor				48.780	27,600	2,675		30,275	33,700	38,700
	Remove / replace condenser fan				1.800		105		105	132	164
	Remove condenser fan bearing				1.200		70		70	88	109
	Replace condenser fan bearing				2.000	55.50	117		172.50	207	252
	Remove condenser fan motor				.976		57		57	71.50	89
	Replace condenser fan motor				2.311	279	123		402	460	540
	Replace fan belt for condenser fan				.400	34	23.50		57.50	66.50	79
	Replace refrigerant				2.667	640	156		796	900	1,050
	Remove / replace heating coils				6.936	2,580	386.25		2,966.25	3,325	3,825
	Total				99.778	31,778	5,506.25		37,284.25	41,859	48,348.50
1045	**Replace computer room A/C, air cooled, 20 ton**	20	Q-6	Ea.							
	Remove air conditioner and condenser				23.105		1,275		1,275	1,600	2,000
	Replace air conditioner and remote condenser, 20 ton				100.000	50,500	5,450		55,950	62,500	71,500
	Total				123.105	50,500	6,725		57,225	64,100	73,500

D30 HVAC   D3053 Terminal and Package Units

D3053 276 Computer Room A/C Units, Chilled Water

	System Description	Freq. (Years)	Crew	Unit	Labor Hours	2014 Bare Costs				Total In-House	Total w/O&P
						Material	Labor	Equipment	Total		
1010	**Repair computer room A/C, chilled water, 5 ton**	10	1 STPI	Ea.							
	Repair controls				.600		35		35	44	55
	Remove / replace supply fan				.650		38		38	47.50	59.50
	Remove supply fan bearing				1.000		58.50		58.50	73	91.50
	Replace supply fan bearing				2.000	55.50	117		172.50	207	252
	Remove supply fan motor				1.156		61.50		61.50	76.50	95.50
	Replace supply fan motor				2.311	279	123		402	460	540
	Replace fan belt for supply fan				.400	34	23.50		57.50	66.50	79
	Remove / replace heating coils				26.002	4,450	1,425		5,875	6,675	7,800
				Total	34.119	4,818.50	1,881.50		6,700	7,649.50	8,972.50
1015	**Replace computer room A/C, chilled water, 5 ton**	20	Q-6	Ea.							
	Remove air conditioner				13.001		710		710	890	1,100
	Replace air conditioner, 5 ton				28.070	13,900	1,475		15,375	17,100	19,700
				Total	41.071	13,900	2,185		16,085	17,990	20,800
1020	**Repair computer room A/C, chilled water, 10 ton**	10	1 STPI	Ea.							
	Repair controls				.400		23.50		23.50	29.50	36.50
	Remove / replace supply fan				.650		38		38	47.50	59.50
	Remove supply fan bearing				1.200		70		70	88	110
	Replace supply fan bearing				2.000	55.50	117		172.50	207	252
	Remove supply fan motor				1.156		61.50		61.50	76.50	95.50
	Replace supply fan motor				2.311	335	123		458	520	610
	Replace fan belt for supply fan				.400	34	23.50		57.50	66.50	79
	Remove / replace heating coils				26.002	4,450	1,425		5,875	6,675	7,800
				Total	34.119	4,874.50	1,881.50		6,756	7,710	9,042.50
1025	**Replace computer room A/C, chilled water, 10 ton**	20	Q-6	Ea.							
	Remove air conditioner				33.898		1,850		1,850	2,325	2,875
	Replace air conditioner, 10 ton				42.105	14,300	2,225		16,525	18,500	21,300
				Total	76.004	14,300	4,075		18,375	20,825	24,175

D30 HVAC

D3053 Terminal and Package Units

D3053 276 Computer Room A/C Units, Chilled Water

	System Description	Freq. (Years)	Crew	Unit	Labor Hours	2014 Bare Costs				Total In-House	Total w/O&P
						Material	Labor	Equipment	Total		
1030	**Repair computer room A/C, chilled water, 15 ton**	10	1 STPI	Ea.							
	Repair controls				.500		29.50		29.50	36.50	45.50
	Remove / replace supply fan				.650		38		38	47.50	59.50
	Remove supply fan bearing				1.200		70		70	88	110
	Replace supply fan bearing				2.000	55.50	117		172.50	207	252
	Remove supply fan motor				1.156		61.50		61.50	76.50	95.50
	Replace supply fan motor				2.312	500	123		623	705	815
	Replace fan belt for supply fan				.400	34	23.50		57.50	66.50	79
	Remove / replace heating coils				26.002	4,450	1,425		5,875	6,675	7,800
	Total				34.220	5,039.50	1,887.50		6,927	7,902	9,256.50
1035	**Replace computer room A/C, chilled water, 15 ton**	20	Q-6	Ea.							
	Remove air conditioner				50.314		2,750		2,750	3,425	4,275
	Replace air conditioner, 15 ton				44.444	14,900	2,350		17,250	19,300	22,300
	Total				94.759	14,900	5,100		20,000	22,725	26,575
1040	**Repair computer room A/C, chilled water, 20 ton**	10	1 STPI	Ea.							
	Repair controls				.500		29.50		29.50	36.50	45.50
	Remove / replace supply fan				.650		38		38	47.50	59.50
	Remove supply fan bearing				1.200		70		70	88	110
	Replace supply fan bearing				2.000	55.50	117		172.50	207	252
	Remove supply fan motor				1.156		61.50		61.50	76.50	95.50
	Replace supply fan motor				2.312	500	123		623	705	815
	Replace fan belt for supply fan				.400	34	23.50		57.50	66.50	79
	Remove / replace heating coils				46.243	17,200	2,575		19,775	22,100	25,500
	Total				54.461	17,789.50	3,037.50		20,827	23,327	26,956.50
1045	**Replace computer room A/C, chilled water, 20 ton**	20	Q-6	Ea.							
	Remove air conditioner				23.105		1,275		1,275	1,600	2,000
	Replace air conditioner, 20 ton				47.059	15,900	2,475		18,375	20,600	23,800
	Total				70.164	15,900	3,750		19,650	22,200	25,800

D30 HVAC D3053 Terminal and Package Units

D3053 278 Multi-Zone Air Conditioner

						2014 Bare Costs				Total In-House	Total w/O&P
	System Description	Freq. (Years)	Crew	Unit	Labor Hours	Material	Labor	Equipment	Total		
1010	**Repair multi-zone rooftop unit, 15 ton**	10	Q-6	Ea.							
	Repair controls				1.000		58.50		58.50	73	91.50
	Remove fan bearings				2.400		140		140	176	220
	Replace fan bearings				4.000	111	234		345	415	505
	Remove fan motor				1.156		61.50		61.50	76.50	95.50
	Replace fan motor				2.312	500	123		623	705	815
	Remove compressor				17.778		935		935	1,175	1,450
	Replace compressor				42.857	26,600	2,350		28,950	32,200	36,900
	Replace refrigerant				1.333	320	78		398	450	520
	Remove / replace heater igniter				.500	29	29.50		58.50	68.50	82
	Total				73.336	27,560	4,009.50		31,569.50	35,339	40,679
1040	**Replace multi-zone rooftop unit, 15 ton**	15	Q-7	Ea.							
	Remove multi-zone rooftop unit				94.675		5,275		5,275	6,600	8,225
	Replace multi-zone rooftop, 15 ton				70.022	65,000	3,900		68,900	76,500	87,500
	Total				164.696	65,000	9,175		74,175	83,100	95,725
2010	**Repair multi - zone rooftop unit, 25 ton**	10	Q-6	Ea.							
	Repair controls				1.000		58.50		58.50	73	91.50
	Remove fan bearings				2.400		140		140	176	220
	Replace fan bearings				4.000	111	234		345	415	505
	Remove fan motor				1.156		61.50		61.50	76.50	95.50
	Replace fan motor				2.311	600	123		723	815	940
	Remove compressor				22.222		1,168.75		1,168.75	1,475	1,825
	Replace compressor				60.976	34,500	3,343.75		37,843.75	42,100	48,300
	Replace refrigerant				2.000	480	117		597	675	780
	Remove / replace heater igniter				1.000	58	59		117	137	164
	Total				97.065	35,749	5,305.50		41,054.50	45,942.50	52,921
2040	**Replace multi-zone rooftop unit, 25 ton**	15	Q-7	Ea.							
	Remove multi-zone rooftop unit				115.000		6,425		6,425	8,000	10,000
	Replace multi-zone rooftop, 25 ton				95.808	83,500	5,325		88,825	98,500	112,500
	Total				210.808	83,500	11,750		95,250	106,500	122,500

D30 HVAC

D3053 Terminal and Package Units

D3053 278 Multi-Zone Air Conditioner

	System Description	Freq. (Years)	Crew	Unit	Labor Hours	2014 Bare Costs				Total In-House	Total w/O&P
						Material	Labor	Equipment	Total		
3010	**Repair multi-zone rooftop unit, 40 ton**	10	Q-6	Ea.							
	Repair controls				1.000		58.50		58.50	73	91.50
	Remove fan bearings				2.400		140		140	176	220
	Replace fan bearings				4.444	125	260		385	460	560
	Remove fan motor				2.000		107		107	132	165
	Replace fan motor				4.000	1,525	213		1,738	1,950	2,225
	Remove compressor				35.451		1,925		1,925	2,425	3,025
	Replace compressor				71.006	29,700	3,875		33,575	37,500	43,200
	Replace refrigerant				2.667	640	156		796	900	1,050
	Remove / replace heater igniter				1.000	58	59		117	137	164
	Total				123.967	32,048	6,793.50		38,841.50	43,753	50,700.50
3040	**Replace multi-zone rooftop unit, 40 ton**	15	Q-7	Ea.							
	Remove multi-zone				172.000		9,625		9,625	12,000	15,000
	Replace multi-zone rooftop, 40 ton				151.000	122,500	8,450		130,950	145,500	166,500
	Total				323.000	122,500	18,075		140,575	157,500	181,500
4010	**Repair multi-zone rooftop unit, 70 ton**	10	Q-7	Ea.							
	Repair controls				1.000		58.50		58.50	73	91.50
	Remove fan bearings				2.400		140		140	176	220
	Replace fan bearings				4.444	125	260		385	460	560
	Remove fan motor				2.601		139		139	172	215
	Replace fan motor				5.195	2,925	277		3,202	3,550	4,075
	Remove compressor				76.923		4,275		4,275	5,350	6,675
	Replace compressor				153.000	34,600	8,575		43,175	48,700	56,500
	Replace refrigerant				4.667	1,120	273		1,393	1,575	1,825
	Remove / replace heater igniter				1.000	58	59		117	137	164
	Total				251.230	38,828	14,056.50		52,884.50	60,193	70,325.50
4040	**Replace multi-zone rooftop unit, 70 ton**	15	Q-7	Ea.							
	Remove multi-zone rooftop unit				231.000		12,900		12,900	16,100	20,100
	Replace multi-zone rooftop, 70 ton				264.000	164,500	14,700		179,200	199,500	228,500
	Total				495.000	164,500	27,600		192,100	215,600	248,600

D30 HVAC | D3053 Terminal and Package Units

D3053 278 Multi-Zone Air Conditioner

	System Description	Freq. (Years)	Crew	Unit	Labor Hours	2014 Bare Costs				Total In-House	Total w/O&P
						Material	Labor	Equipment	Total		
5010	**Repair multi-zone rooftop unit, 105 ton**	10	Q-7	Ea.							
	Repair controls				1.000		58.50		58.50	73	91.50
	Remove fan bearings				4.800		280		280	350	440
	Replace fan bearings				9.412	704	552		1,256	1,450	1,750
	Remove fan motor				4.334		231		231	286	360
	Replace fan motor				8.696	3,600	465		4,065	4,525	5,225
	Remove compressor				79.012		4,400		4,400	5,500	6,850
	Replace compressor				157.600	31,600	8,800		40,400	45,600	53,500
	Replace refrigerant				6.667	1,600	390		1,990	2,250	2,600
	Remove / replace heater igniter				1.500	87	88.50		175.50	206	245
	Total				273.020	37,591	15,265		52,856	60,240	71,061.50
5040	**Replace multi-zone rooftop unit, 105 ton**	15	Q-7	Ea.							
	Remove multi-zone rooftop unit				347.000		19,400		19,400	24,200	30,200
	Replace multi-zone rooftop, 105 ton				390.000	217,500	21,700		239,200	266,500	306,000
	Total				737.000	217,500	41,100		258,600	290,700	336,200

D3053 280 Single Zone Air Conditioner

	System Description	Freq. (Years)	Crew	Unit	Labor Hours	2014 Bare Costs				Total In-House	Total w/O&P
						Material	Labor	Equipment	Total		
1001	**Repair single zone rooftop unit, 3 ton**	10	2 STPI	Ea.							
	Repair controls				.300		17.55		17.55	22	27.50
	Remove fan bearings				2.400		140		140	176	220
	Replace fan bearings				4.000	111	234		345	415	505
	Remove fan motor				1.156		61.50		61.50	76.50	95.50
	Replace fan motor				1.951	207	104		311	355	420
	Remove compressor				2.477		145		145	181	226
	Replace compressor				4.522	345	265		610	710	845
	Replace refrigerant				.533	128	31.20		159.20	180	209
	Remove/replace heater igniter				.500	29	29.50		58.50	68.50	82
	Total				17.840	820	1,027.75		1,847.75	2,184	2,630
1002	**Replace single zone rooftop unit, 3 ton**	15	Q-5	Ea.							
	Remove single zone				8.000		420		420	525	655
	Replace single zone rooftop, 3 ton				30.476	3,850	1,600		5,450	6,250	7,325
	Total				38.476	3,850	2,020		5,870	6,775	7,980

D30 HVAC D3053 Terminal and Package Units

D3053 280 Single Zone Air Conditioner

	System Description	Freq. (Years)	Crew	Unit	Labor Hours	2014 Bare Costs Material	2014 Bare Costs Labor	2014 Bare Costs Equipment	2014 Bare Costs Total	Total In-House	Total w/O&P
1003	**Repair single zone rooftop unit, 5 ton**	10	2 STPI	Ea.							
	Repair controls				.300		17.55		17.55	22	27.50
	Remove fan bearings				2.400		140		140	176	220
	Replace fan bearings				4.000	111	234		345	415	505
	Remove fan motor				1.156		61.50		61.50	76.50	95.50
	Replace fan motor				1.951	207	104		311	355	420
	Remove compressor				2.477		145		145	181	226
	Replace compressor				4.954	485	290		775	895	1,050
	Replace refrigerant				.533	128	31.20		159.20	180	209
	Remove/replace heater igniter				.500	29	29.50		58.50	68.50	82
	Total				18.271	960	1,052.75		2,012.75	2,369	2,835
1004	**Replace single zone rooftop unit, 5 ton**	15	Q-5	Ea.							
	Remove single zone				18.561		975		975	1,225	1,525
	Replace single zone rooftop, 5 ton				38.005	5,275	2,000		7,275	8,300	9,725
	Total				56.566	5,275	2,975		8,250	9,525	11,250
1005	**Repair single zone rooftop unit, 7.5 ton**	10	2 STPI	Ea.							
	Repair controls				.400		23.50		23.50	29.50	36.50
	Remove fan bearings				2.400		140		140	176	220
	Replace fan bearings				4.000	111	234		345	415	505
	Remove fan motor				1.156		61.50		61.50	76.50	95.50
	Replace fan motor				2.311	279	123		402	460	540
	Remove compressor				3.467		183		183	228	285
	Replace compressor				6.932	1,100	365		1,465	1,675	1,950
	Replace refrigerant				.667	160	39		199	225	261
	Remove/replace heater igniter				.500	29	29.50		58.50	68.50	82
	Total				21.833	1,679	1,198.50		2,877.50	3,353.50	3,975
1006	**Replace single zone rooftop unit, 7.5 ton**	15	Q-5	Ea.							
	Remove single zone				26.016		1,375		1,375	1,725	2,125
	Replace single zone rooftop, 7.5 ton				43.011	8,125	2,275		10,400	11,800	13,700
	Total				69.027	8,125	3,650		11,775	13,525	15,825

D30 HVAC D3053 Terminal and Package Units

D3053 280 Single Zone Air Conditioner

							2014 Bare Costs				Total In-House	Total w/O&P
	System Description	Freq. (Years)	Crew	Unit	Labor Hours	Material	Labor	Equipment	Total			
1007	**Repair single zone rooftop unit, 10 ton**	10	2 STPI	Ea.								
	Repair controls				.400		23.50		23.50	29.50	36.50
	Remove fan bearings				2.400		140		140	176	220
	Replace fan bearings				4.000	111	234		345	415	505
	Remove fan motor				1.156		61.50		61.50	76.50	95.50
	Replace fan motor				2.311	335	123		458	520	610
	Remove compressor				14.513		787.25		787.25	990	1,250
	Replace compressor				28.714	17,822	1,574.50		19,396.50	21,600	24,700
	Replace refrigerant				.800	192	46.80		238.80	270	315
	Remove/replace heater igniter				.500	29	29.50		58.50	68.50	82
	Total				54.794	18,489	3,020.05		21,509.05	**24,145.50**	27,814
1008	**Replace single zone rooftop unit, 10 ton**	15	Q-6	Ea.								
	Remove single zone				33.898		1,850		1,850	2,325	2,875
	Replace single zone rooftop, 10 ton				48.000	11,600	2,625		14,225	16,000	18,600
	Total				81.898	11,600	4,475		16,075	**18,325**	21,475
1010	**Repair single zone rooftop unit, 15 ton**	10	2 STPI	Ea.								
	Repair controls				.500		29.50		29.50	36.50	45.50
	Remove fan bearings				2.400		140		140	176	220
	Replace fan bearings				4.000	111	234		345	415	505
	Remove fan motor				1.156		61.50		61.50	76.50	95.50
	Replace fan motor				2.312	500	123		623	705	815
	Remove compressor				21.661		1,175		1,175	1,475	1,850
	Replace compressor				42.857	26,600	2,350		28,950	32,200	36,900
	Replace refrigerant				1.333	320	78		398	450	520
	Remove / replace heater igniter				.500	29	29.50		58.50	68.50	82
	Total				76.719	27,560	4,220.50		31,780.50	**35,602.50**	41,033
1040	**Replace single zone rooftop unit, 15 ton**	15	Q-6	Ea.								
	Remove single zone rooftop unit				50.314		2,750		2,750	3,425	4,275
	Replace single zone rooftop, 15 ton				56.075	17,000	3,050		20,050	22,500	26,000
	Total				106.389	17,000	5,800		22,800	**25,925**	30,275

D30 HVAC / D3053 Terminal and Package Units

D3053 280 Single Zone Air Conditioner

	System Description	Freq. (Years)	Crew	Unit	Labor Hours	2014 Bare Costs				Total In-House	Total w/O&P
						Material	Labor	Equipment	Total		
2010	**Repair single zone rooftop unit, 25 ton**	10	Q-6	Ea.							
	Repair controls				.500		29.50		29.50	36.50	45.50
	Remove fan bearings				2.400		140		140	176	220
	Replace fan bearings				4.000	111	234		345	415	505
	Remove fan motor				1.156		61.50		61.50	76.50	95.50
	Replace fan motor				2.311	600	123		723	815	940
	Remove compressor				22.222		1,168.75		1,168.75	1,475	1,825
	Replace compressor				60.976	34,500	3,343.75		37,843.75	42,100	48,300
	Replace refrigerant				2.000	480	117		597	675	780
	Remove / replace heater igniter				1.000	58	59		117	137	164
	Total				96.565	35,749	5,276.50		41,025.50	45,906	52,875
2040	**Replace single zone rooftop unit, 25 ton**	15	Q-7	Ea.							
	Remove single zone rooftop unit				76.923		4,275		4,275	5,350	6,675
	Replace single zone rooftop, 25 ton				76.739	27,700	4,275		31,975	35,800	41,300
	Total				153.662	27,700	8,550		36,250	41,150	47,975
3010	**Repair single zone rooftop unit, 60 ton**	10	Q-6	Ea.							
	Repair controls				.500		29.50		29.50	36.50	45.50
	Remove fan bearings				2.400		140		140	176	220
	Replace fan bearings				4.444	125	260		385	460	560
	Remove fan motor				2.000		107		107	132	165
	Replace fan motor				4.167	1,925	222		2,147	2,400	2,750
	Remove compressor				93.507		5,100		5,100	6,400	7,950
	Replace compressor				186.000	37,560	10,200		47,760	54,000	63,000
	Replace refrigerant				4.667	1,120	273		1,393	1,575	1,825
	Remove / replace heater igniter				1.000	58	59		117	137	164
	Total				298.684	40,788	16,390.50		57,178.50	65,316.50	76,679.50
3040	**Replace single zone rooftop unit, 60 ton**	15	Q-7	Ea.							
	Remove single zone rooftop unit				188.000		10,500		10,500	13,100	16,400
	Replace single zone rooftop, 60 ton				181.000	51,500	10,100		61,600	69,500	80,000
	Total				369.000	51,500	20,600		72,100	82,600	96,400

D30 HVAC | D3053 Terminal and Package Units

D3053 280 Single Zone Air Conditioner

	System Description	Freq. (Years)	Crew	Unit	Labor Hours	2014 Bare Costs				Total In-House	Total w/O&P
						Material	Labor	Equipment	Total		
4010	**Repair single zone rooftop unit, 100 ton**	10	Q-7	Ea.							
	Repair controls				.500		29.50		29.50	36.50	45.50
	Remove fan bearings				4.800		280		280	350	440
	Replace fan bearings				9.412	704	552		1,256	1,450	1,750
	Remove fan motor				4.334		231		231	286	360
	Replace fan motor				7.407	4,125	395		4,520	5,025	5,775
	Remove compressor				98.765		5,500		5,500	6,875	8,575
	Replace compressor				197.000	39,500	11,000		50,500	57,000	66,500
	Replace refrigerant				6.667	1,600	390		1,990	2,250	2,600
	Remove / replace heater igniter				1.500	87	88.50		175.50	206	245
	Total				330.385	46,016	18,466		64,482	73,478.50	86,290.50
4040	**Replace single zone rooftop unit, 100 ton**	15	Q-7	Ea.							
	Remove single zone rooftop unit				296.000		16,500		16,500	20,600	25,700
	Replace single zone rooftop, 100 ton				304.000	104,000	17,000		121,000	135,500	156,500
	Total				600.000	104,000	33,500		137,500	156,100	182,200

D3053 282 Multi-Zone Variable Volume

	System Description	Freq. (Years)	Crew	Unit	Labor Hours	2014 Bare Costs				Total In-House	Total w/O&P
						Material	Labor	Equipment	Total		
1010	**Repair multi-zone variable volume, 50 ton**	10	Q-6	Ea.							
	Repair controls				1.300		76		76	95	119
	Remove fan bearings				2.400		140		140	176	220
	Replace fan bearings				4.444	125	260		385	460	560
	Remove fan motor				2.000		107		107	132	165
	Replace fan motor				4.000	1,525	213		1,738	1,950	2,225
	Remove compressor				77.922		4,250		4,250	5,325	6,625
	Replace compressor				155.000	31,300	8,500		39,800	45,000	52,500
	Replace refrigerant				3.333	800	195		995	1,125	1,300
	Remove / replace heater igniter				1.000	58	59		117	137	164
	Total				251.400	33,808	13,800		47,608	54,400	63,878
1040	**Replace multi-zone variable volume, 50 ton**	15	Q-7	Ea.							
	Remove multi-zone variable volume unit				160.000		8,900		8,900	11,100	13,900
	Replace multi-zone variable volume unit, 50 ton				219.000	113,000	12,200		125,200	139,500	160,500
	Total				379.000	113,000	21,100		134,100	150,600	174,400

D30 HVAC D3053 Terminal and Package Units

D3053 282 Multi-Zone Variable Volume

	System Description	Freq. (Years)	Crew	Unit	Labor Hours	2014 Bare Costs Material	2014 Bare Costs Labor	2014 Bare Costs Equipment	2014 Bare Costs Total	Total In-House	Total w/O&P
2010	**Repair multi-zone variable volume, 70 ton**	10	Q-7	Ea.							
	Repair controls				1.300		76		76	95	119
	Remove fan bearings				2.400		140		140	176	220
	Replace fan bearings				4.444	125	260		385	460	560
	Remove fan motor				2.601		139		139	172	215
	Replace fan motor				5.195	2,925	277		3,202	3,550	4,075
	Remove compressor				76.923		4,275		4,275	5,350	6,675
	Replace compressor				153.000	34,600	8,575		43,175	48,700	56,500
	Replace refrigerant				4.667	1,120	273		1,393	1,575	1,825
	Remove / replace heater igniter				1.000	58	59		117	137	164
	Total				251.530	38,828	14,074		52,902	60,215	70,353
2040	**Replace multi-zone variable volume, 70 ton**	15	Q-7	Ea.							
	Remove multi-zone variable volume unit				231.000		12,900		12,900	16,100	20,100
	Replace multi-zone variable volume unit, 70 ton				304.000	158,000	17,000		175,000	195,000	224,000
	Total				535.000	158,000	29,900		187,900	211,100	244,100
3010	**Repair multi-zone variable volume, 90 ton**	10	Q-7	Ea.							
	Repair controls				1.300		76		76	95	119
	Remove fan bearings				2.400		140		140	176	220
	Replace fan bearings				4.444	125	260		385	460	560
	Remove fan motor				2.601		139		139	172	215
	Replace fan motor				6.504	2,775	345		3,120	3,475	4,000
	Remove compressor				92.308		5,130		5,130	6,425	8,000
	Replace compressor				183.600	41,520	10,290		51,810	58,500	68,000
	Replace refrigerant				6.667	1,600	390		1,990	2,250	2,600
	Remove / replace heater igniter				1.500	87	88.50		175.50	206	245
	Total				301.323	46,107	16,858.50		62,965.50	71,759	83,959
3040	**Replace multi-zone variable volume, 90 ton**	15	Q-7	Ea.							
	Remove multi-zone variable volume unit				296.000		16,500		16,500	20,600	25,700
	Replace multi-zone variable volume unit, 90 ton				395.000	181,500	22,000		203,500	227,000	261,000
	Total				691.000	181,500	38,500		220,000	247,600	286,700

D30 HVAC   D3053   Terminal and Package Units

D3053 282   Multi-Zone Variable Volume

	System Description	Freq. (Years)	Crew	Unit	Labor Hours	2014 Bare Costs				Total In-House	Total w/O&P
						Material	Labor	Equipment	Total		
4010	**Repair multi-zone variable volume, 105 ton**	10	Q-7	Ea.							
	Repair controls				1.300		76		76	95	119
	Remove fan bearings				4.800		280		280	350	440
	Replace fan bearings				9.412	704	552		1,256	1,450	1,750
	Remove fan motor				4.334		231		231	286	360
	Replace fan motor				8.696	3,600	465		4,065	4,525	5,225
	Remove compressor				79.012		4,400		4,400	5,500	6,850
	Replace compressor				157.600	31,600	8,800		40,400	45,600	53,500
	Replace refrigerant				7.333	1,760	429		2,189	2,475	2,875
	Remove / replace heater igniter				1.500	87	88.50		175.50	206	245
	Total				273.987	37,751	15,321.50		53,072.50	60,487	71,364
4040	**Replace multi-zone variable volume, 105 ton**	15	Q-7	Ea.							
	Remove multi-zone variable volume unit				347.000		19,400		19,400	24,200	30,200
	Replace multi-zone variable volume unit, 105 ton				444.000	203,000	24,800		227,800	254,000	292,500
	Total				791.000	203,000	44,200		247,200	278,200	322,700
5010	**Repair multi-zone variable volume, 140 ton**	10	Q-7	Ea.							
	Repair controls				1.300		76		76	95	119
	Remove fan bearings				7.200		420		420	530	660
	Replace fan bearings				14.118	1,056	828		1,884	2,200	2,600
	Remove fan motor				4.334		231		231	286	360
	Replace fan motor				11.594	4,625	620		5,245	5,850	6,725
	Remove compressor				106.370		5,923.50		5,923.50	7,400	9,225
	Replace compressor				212.169	42,541.50	11,847		54,388.50	61,500	71,500
	Replace refrigerant				8.666	2,080	507		2,587	2,925	3,400
	Remove / replace heater igniter				2.000	116	118		234	274	325
	Total				367.751	50,418.50	20,570.50		70,989	81,060	94,914
5040	**Replace multi-zone variable volume, 140 ton**	15	Q-7	Ea.							
	Remove multi-zone variable volume unit				516.000		28,700		28,700	35,900	44,800
	Replace multi-zone variable volume unit, 140 ton				592.000	248,500	33,000		281,500	314,500	362,000
	Total				1108.000	248,500	61,700		310,200	350,400	406,800

D30 HVAC D3053 Terminal and Package Units

D3053 284 Single Zone Variable Volume

	System Description	Freq. (Years)	Crew	Unit	Labor Hours	2014 Bare Costs				Total In-House	Total w/O&P
						Material	Labor	Equipment	Total		
1010	**Repair single zone variable volume, 20 ton**	10	Q-6	Ea.							
	Repair controls				1.000		58.50		58.50	73	91.50
	Remove fan bearings				2.400		140		140	176	220
	Replace fan bearings				4.000	111	234		345	415	505
	Remove fan motor				1.156		61.50		61.50	76.50	95.50
	Replace fan motor				2.312	500	123		623	705	815
	Remove compressor				17.778		935		935	1,175	1,450
	Replace compressor				48.780	27,600	2,675		30,275	33,700	38,700
	Replace refrigerant				1.333	320	78		398	450	520
	Remove / replace heater igniter				1.000	58	59		117	137	164
	Total				79.760	28,589	4,364		32,953	36,907.50	42,561
1040	**Replace single zone variable volume, 20 ton**	15	Q-7	Ea.							
	Remove single zone variable volume unit				71.545		3,987.50		3,987.50	4,975	6,225
	Replace single zone variable volume unit, 20 ton				64.945	29,260	3,630		32,890	36,700	42,200
	Total				136.489	29,260	7,617.50		36,877.50	41,675	48,425
2010	**Repair single zone variable volume, 30 ton**	10	Q-6	Ea.							
	Repair controls				1.000		58.50		58.50	73	91.50
	Remove fan bearings				2.400		140		140	176	220
	Replace fan bearings				4.000	111	234		345	415	505
	Remove fan motor				1.156		61.50		61.50	76.50	95.50
	Replace fan motor				2.476	700	132		832	935	1,075
	Remove compressor				26.667		1,402.50		1,402.50	1,775	2,175
	Replace compressor				73.171	41,400	4,012.50		45,412.50	50,500	58,000
	Replace refrigerant				2.667	640	156		796	900	1,050
	Remove / replace heater igniter				1.000	58	59		117	137	164
	Total				114.536	42,909	6,256		49,165	54,987.50	63,376
2040	**Replace single zone variable volume, 30 ton**	15	Q-7	Ea.							
	Remove single zone variable volume unit				104.142		5,802.50		5,802.50	7,250	9,050
	Replace single zone variable volume unit, 30 ton				92.147	30,470	5,142.50		35,612.50	39,900	46,100
	Total				196.289	30,470	10,945		41,415	47,150	55,150

D30 HVAC D3053 Terminal and Package Units

D3053 284 Single Zone Variable Volume

	System Description	Freq. (Years)	Crew	Unit	Labor Hours	2014 Bare Costs				Total In-House	Total w/O&P
						Material	Labor	Equipment	Total		
3010	**Repair single zone variable volume, 40 ton**	10	Q-6	Ea.							
	Repair controls				1.000		58.50		58.50	73	91.50
	Remove fan bearings				2.400		140		140	176	220
	Replace fan bearings				4.000	111	234		345	415	505
	Remove fan motor				1.156		61.50		61.50	76.50	95.50
	Replace fan motor				4.000	1,525	213		1,738	1,950	2,225
	Remove compressor				35.451		1,925		1,925	2,425	3,025
	Replace compressor				71.006	29,700	3,875		33,575	37,500	43,200
	Replace refrigerant				3.333	800	195		995	1,125	1,300
	Remove / replace heater igniter				1.000	58	59		117	137	164
	Total				123.346	32,194	6,761		38,955	43,877.50	50,826
3040	**Replace single zone variable volume, 40 ton**	15	Q-7	Ea.							
	Remove single zone variable volume unit				143.000		7,975		7,975	9,950	12,400
	Replace single zone variable volume unit, 40 ton				122.100	40,590	6,847.50		47,437.50	53,000	61,500
	Total				265.100	40,590	14,822.50		55,412.50	62,950	73,900
4010	**Repair single zone variable volume, 60 ton**	10	Q-6	Ea.							
	Repair controls				.500		29.50		29.50	36.50	45.50
	Remove fan bearings				2.400		140		140	176	220
	Replace fan bearings				4.444	125	260		385	460	560
	Remove fan motor				2.000		107		107	132	165
	Replace fan motor				4.167	1,925	222		2,147	2,400	2,750
	Remove compressor				93.507		5,100		5,100	6,400	7,950
	Replace compressor				186.000	37,560	10,200		47,760	54,000	63,000
	Replace refrigerant				4.000	960	234		1,194	1,350	1,575
	Remove / replace heater igniter				1.000	58	59		117	137	164
	Total				298.018	40,628	16,351.50		56,979.50	65,091.50	76,429.50
4040	**Replace single zone variable volume, 60 ton**	15	Q-7	Ea.							
	Remove single zone variable volume unit				189.200		10,587.50		10,587.50	13,200	16,500
	Replace single zone variable volume unit, 60 ton				184.800	47,850	10,312.50		58,162.50	65,500	76,000
	Total				374.000	47,850	20,900		68,750	78,700	92,500

D30 HVAC

D3053 Terminal and Package Units

D3053 286 Central Station Air Conditioning Air Handling Unit

	System Description	Freq. (Years)	Crew	Unit	Labor Hours	2014 Bare Costs				Total In-House	Total w/O&P
						Material	Labor	Equipment	Total		
1010	**Repair central station A.H.U., 1300 CFM**	10	1 STPI	Ea.							
	Repair controls				.300		17.55		17.55	22	27.50
	Remove blower motor				1.200		70		70	88	109
	Replace blower motor				1.951	248	104		352	400	470
	Total				3.451	248	191.55		439.55	**510**	**606.50**
1040	**Replace central station A.H.U., 1300 CFM**	15	Q-5	Ea.							
	Remove central station A.H.U.				8.667		455		455	570	710
	Replace central station A.H.U., 1300 CFM				17.335	6,950	915		7,865	8,775	10,100
	Total				26.002	6,950	1,370		8,320	**9,345**	**10,810**
2010	**Repair central station A.H.U., 1900 CFM**	10	1 STPI	Ea.							
	Repair controls				.300		17.55		17.55	22	27.50
	Remove blower motor				1.200		70		70	88	109
	Replace blower motor				2.311	279	123		402	460	540
	Total				3.811	279	210.55		489.55	**570**	**676.50**
2040	**Replace central station A.H.U., 1900 CFM**	15	Q-5	Ea.							
	Remove central station A.H.U.				9.456		500		500	625	775
	Replace central station A.H.U., 1900 CFM				18.913	10,700	995		11,695	13,000	14,900
	Total				28.369	10,700	1,495		12,195	**13,625**	**15,675**
3010	**Repair central station A.H.U., 5400 CFM**	10	1 STPI	Ea.							
	Repair controls				.300		17.55		17.55	22	27.50
	Remove blower motor				1.400		82		82	102	128
	Replace blower motor				2.312	500	123		623	705	815
	Total				4.012	500	222.55		722.55	**829**	**970.50**
3040	**Replace central station A.H.U., 5400 CFM**	15	Q-6	Ea.							
	Remove central station A.H.U.				19.496		1,075		1,075	1,325	1,650
	Replace central station A.H.U., 5400 CFM				39.024	17,100	2,125		19,225	21,500	24,700
	Total				58.521	17,100	3,200		20,300	**22,825**	**26,350**
4010	**Repair central station A.H.U., 8000 CFM**	10	1 STPI	Ea.							
	Repair controls				.300		17.55		17.55	22	27.50
	Remove blower motor				1.600		93.50		93.50	117	146
	Replace blower motor				2.311	600	123		723	815	940
	Total				4.211	600	234.05		834.05	**954**	**1,113.50**

D30 HVAC   D3053   Terminal and Package Units

D3053 286   Central Station Air Conditioning Air Handling Unit

	System Description	Freq. (Years)	Crew	Unit	Labor Hours	2014 Bare Costs				Total In-House	Total w/O&P
						Material	Labor	Equipment	Total		
4040	**Replace central station A.H.U., 8000 CFM**	15	Q-6	Ea.							
	Remove central station A.H.U.				25.974		1,425		1,425	1,775	2,200
	Replace central station A.H.U., 8000 CFM				51.948	27,400	2,825		30,225	33,700	38,700
	Total				77.922	27,400	4,250		31,650	35,475	40,900
5010	**Repair central station A.H.U., 16,000 CFM**	10	1 STPI	Ea.							
	Repair controls				.300		17.55		17.55	22	27.50
	Remove blower motor				1.800		105		105	132	164
	Replace blower motor				2.597	855	139		994	1,100	1,275
	Total				4.698	855	261.55		1,116.55	1,254	1,466.50
5040	**Replace central station A.H.U., 16,000 CFM**	15	Q-6	Ea.							
	Remove central station A.H.U.				41.026		2,250		2,250	2,800	3,500
	Replace central station A.H.U., 16,000 CFM				82.192	52,000	4,500		56,500	63,000	72,000
	Total				123.217	52,000	6,750		58,750	65,800	75,500
6010	**Repair central station A.H.U., 33,500 CFM**	10	1 STPI	Ea.							
	Repair controls				.300		17.55		17.55	22	27.50
	Remove blower motor				2.601		139		139	172	215
	Replace blower motor				5.195	2,925	277		3,202	3,550	4,075
	Total				8.096	2,925	433.55		3,358.55	3,744	4,317.50
6040	**Replace central station A.H.U., 33,500 CFM**	15	Q-6	Ea.							
	Remove central station A.H.U.				82.192		4,500		4,500	5,600	7,000
	Replace central station A.H.U., 33,500 CFM				164.000	109,500	8,975		118,475	131,500	151,000
	Total				246.192	109,500	13,475		122,975	137,100	158,000
7010	**Repair central station A.H.U., 63,000 CFM**	10	1 STPI	Ea.							
	Repair controls				.300		17.55		17.55	22	27.50
	Remove blower motor				4.334		231		231	286	360
	Replace blower motor				8.696	3,600	465		4,065	4,525	5,225
	Total				13.329	3,600	713.55		4,313.55	4,833	5,612.50
7040	**Replace central station A.H.U., 63,000 CFM**	15	Q-7	Ea.							
	Remove central station A.H.U.				160.000		8,900		8,900	11,100	13,900
	Replace central station A.H.U., 63,000 CFM				320.000	207,000	17,800		224,800	250,000	286,500
	Total				480.000	207,000	26,700		233,700	261,100	300,400

D30 HVAC D3053 Terminal and Package Units

D3053 310 Residential Furnace, Gas

	System Description	Freq. (Years)	Crew	Unit	Labor Hours	Material	Labor	Equipment	Total	Total In-House	Total w/O&P
							2014 Bare Costs				
1010	**Repair furnace, gas, 25 MBH**	10	Q-1	Ea.							
	Repair controls				.500		29.50		29.50	36.50	45.50
	Remove fan motor				.976		57		57	71.50	89
	Replace fan motor				1.951	198	104		302	345	410
	Remove auto vent damper				.274		14.95		14.95	18.95	23.50
	Replace auto vent damper				.547	71.50	30		101.50	117	136
	Remove burner				1.000		52		52	65	81
	Replace burner				2.000	257	104		361	410	485
	Total				7.248	526.50	391.45		917.95	1,063.95	1,270
1030	**Replace furnace, gas, 25 MBH**	15	Q-9	Ea.							
	Remove furnace				2.000		98.50		98.50	125	155
	Replace furnace				4.000	535	197		732	840	980
	Total				6.000	535	295.50		830.50	965	1,135
2010	**Repair furnace, gas, 100 MBH**	10	Q-1	Ea.							
	Repair controls				.500		29.50		29.50	36.50	45.50
	Remove fan motor				.976		57		57	71.50	89
	Replace fan motor				1.951	207	104		311	355	420
	Remove auto vent damper				.289		15.80		15.80	20	25
	Replace auto vent damper				.578	92.50	31.50		124	142	166
	Remove burner				1.200		62		62	77.50	97
	Replace burner				2.400	580	124		704	795	920
	Total				7.894	879.50	423.80		1,303.30	1,497.50	1,762.50
2030	**Replace furnace, gas, 100 MBH**	15	Q-9	Ea.							
	Remove furnace				3.250		160		160	203	252
	Replace furnace				6.499	625	320		945	1,100	1,275
	Total				9.749	625	480		1,105	1,303	1,527

D30 HVAC D3053 Terminal and Package Units

D3053 310 Residential Furnace, Gas

	System Description	Freq. (Years)	Crew	Unit	Labor Hours	2014 Bare Costs				Total In-House	Total w/O&P
						Material	Labor	Equipment	Total		
3010	**Repair furnace, gas, 200 MBH**	10	Q-1	Ea.							
	Repair controls				.500		29.50		29.50	36.50	45.50
	Remove fan motor				.976		57		57	71.50	89
	Replace fan motor				1.951	248	104		352	400	470
	Remove auto vent damper				.325		17.80		17.80	22.50	28
	Replace auto vent damper				.650	104	35.50		139.50	159	186
	Remove burner				3.077		159		159	199	249
	Replace burner				10.403	1,200	540		1,740	2,000	2,350
	Total				17.882	1,552	942.80		2,494.80	2,888.50	3,417.50
3030	**Replace furnace, gas, 200 MBH**	15	Q-9	Ea.							
	Remove furnace				4.000		197		197	250	310
	Replace furnace				8.000	2,800	395		3,195	3,575	4,125
	Total				12.000	2,800	592		3,392	3,825	4,435

D3053 320 Residential Furnace, Oil

	System Description	Freq. (Years)	Crew	Unit	Labor Hours	2014 Bare Costs				Total In-House	Total w/O&P
						Material	Labor	Equipment	Total		
1010	**Repair furnace, oil, 55 MBH**	10	Q-1	Ea.							
	Repair controls				.500		29.50		29.50	36.50	45.50
	Remove fan motor				.976		57		57	71.50	89
	Replace fan motor				1.951	198	104		302	345	410
	Remove auto vent damper				.473		23.50		23.50	29.50	36.50
	Replace auto vent damper				.945	155	46.50		201.50	229	267
	Remove burner				4.334		224		224	281	350
	Replace burner				8.667	289	450		739	880	1,050
	Total				17.846	642	934.50		1,576.50	1,872.50	2,248
1030	**Replace furnace, oil, 55 MBH**	15	Q-9	Ea.							
	Remove furnace				2.889		142		142	180	224
	Replace furnace, oil, 55 mbh				5.778	1,550	284		1,834	2,075	2,400
	Total				8.667	1,550	426		1,976	2,255	2,624

D30 HVAC / D3053 Terminal and Package Units

D3053 320 Residential Furnace, Oil

	System Description	Freq. (Years)	Crew	Unit	Labor Hours	2014 Bare Costs — Material	2014 Bare Costs — Labor	2014 Bare Costs — Equipment	2014 Bare Costs — Total	Total In-House	Total w/O&P
2010	**Repair furnace, oil, 100 MBH**	10	Q-1	Ea.							
	Repair controls				.500		29.50		29.50	36.50	45.50
	Remove fan motor				.976		57		57	71.50	89
	Replace fan motor				1.951	207	104		311	355	420
	Remove auto vent damper				.495		24.50		24.50	31	38.50
	Replace auto vent damper				.990	71.50	49		120.50	140	166
	Remove burner				4.334		224		224	281	350
	Replace burner				8.667	289	450		739	880	1,050
	Total				17.914	567.50	938		1,505.50	1,795	2,159
2030	**Replace furnace, oil, 100 MBH**	15	Q-1	Ea.							
	Remove furnace				3.059		151		151	191	237
	Replace furnace, oil, 100 mbh				6.119	1,875	300		2,175	2,450	2,825
	Total				9.177	1,875	451		2,326	2,641	3,062
3010	**Repair furnace, oil, 200 MBH**	10	Q-1	Ea.							
	Repair controls				.500		29.50		29.50	36.50	45.50
	Remove fan motor				.976		57		57	71.50	89
	Replace fan motor				1.951	248	104		352	400	470
	Remove auto vent damper				.520		25.50		25.50	32.50	40.50
	Replace auto vent damper				1.040	92.50	51		143.50	167	196
	Remove burner				4.334		224		224	281	350
	Replace burner				8.667	289	450		739	880	1,050
	Total				17.988	629.50	941		1,570.50	1,868.50	2,241
3030	**Replace furnace, oil, 200 MBH**	15	Q-1	Ea.							
	Remove furnace				4.000		197		197	250	310
	Replace furnace, oil, 200 mbh				8.000	2,450	395		2,845	3,200	3,675
	Total				12.000	2,450	592		3,042	3,450	3,985

D30 HVAC

D3053 Terminal and Package Units

D3053 330 Residential Furnace, Electric

	System Description	Freq. (Years)	Crew	Unit	Labor Hours	2014 Bare Costs				Total In-House	Total w/O&P
						Material	Labor	Equipment	Total		
1010	**Repair furnace, electric, 25 MBH**	10	Q-20	Ea.							
	Repair controls				.500		29.50		29.50	36.50	45.50
	Remove fan motor				.976		57		57	71.50	89
	Replace fan motor				1.951	198	104		302	345	410
	Remove heat element				.867		43.50		43.50	54.50	68
	Replace heat element				1.733	1,325	87		1,412	1,575	1,800
	Total				6.027	1,523	321		1,844	2,082.50	2,412.50
1030	**Replace furnace, electric, 25 MBH**	15	Q-20	Ea.							
	Remove furnace				2.826		141		141	178	222
	Replace furnace, electric, 25 mbh				5.653	600	283		883	1,025	1,200
	Total				8.479	600	424		1,024	1,203	1,422
2010	**Repair furnace, electric, 50 MBH**	10	Q-20	Ea.							
	Repair controls				.500		29.50		29.50	36.50	45.50
	Remove fan motor				.976		57		57	71.50	89
	Replace fan motor				1.951	198	104		302	345	410
	Remove heat element				.929		46.50		46.50	58.50	73
	Replace heat element				1.857	1,350	93		1,443	1,600	1,825
	Total				6.212	1,548	330		1,878	2,111.50	2,442.50
2030	**Replace furnace, electric, 50 MBH**	15	Q-20	Ea.							
	Remove furnace				3.095		155		155	195	243
	Replace furnace, electric, 50 mbh				6.190	630	310		940	1,075	1,275
	Total				9.285	630	465		1,095	1,270	1,518
3010	**Repair furnace, electric, 85 MBH**	10	Q-20	Ea.							
	Repair controls				.500		29.50		29.50	36.50	45.50
	Remove fan motor				.976		57		57	71.50	89
	Replace fan motor				1.951	207	104		311	355	420
	Remove heat element				1.083		54		54	68.50	85
	Replace heat element				2.167	2,450	108		2,558	2,825	3,225
	Total				6.677	2,657	352.50		3,009.50	3,356.50	3,864.50
3030	**Replace furnace, electric, 85 MBH**	15	Q-20	Ea.							
	Remove furnace				4.000		197		197	250	310
	Replace furnace, electric, 85 mbh				6.842	660	340		1,000	1,150	1,375
	Total				10.842	660	537		1,197	1,400	1,685

D30 HVAC   D3053   Terminal and Package Units

D3053 410   Electric Baseboard Heating Units

	System Description	Freq. (Years)	Crew	Unit	Labor Hours	2014 Bare Costs				Total In-House	Total w/O&P
						Material	Labor	Equipment	Total		
0010	**Maintenance and repair**	2	1 ELEC	Ea.	.615		33		33	40.50	51
	Repair wiring connections										
	Total				.615		33		33	**40.50**	**51**
0020	**Maintenance and inspection**	0.50	1 ELEC	Ea.	1.143		61		61	75.50	94.50
	Inspect and clean baseboard heater										
	Total				1.143		61		61	**75.50**	**94.50**
0030	**Replace baseboard heater**	20	1 ELEC	Ea.	.667		35.50		35.50	44	55
	Remove baseboard heater										
	Baseboard heater, 5' long				1.860	51	99.50		150.50	179	218
	Total				2.527	51	135		186	223	273

D3053 420   Cast Iron Radiator, 10' section, 1 side

	System Description	Freq. (Years)	Crew	Unit	Labor Hours	2014 Bare Costs				Total In-House	Total w/O&P
						Material	Labor	Equipment	Total		
0020	**Refinish C.I. radiator, 1 side**	5	1 PORD	Ea.	.296		11.80		11.80	15.20	18.80
	Prepare surface										
	Prime and paint 1 coat, brushwork				.471	3.80	18.60		22.40	28	34.50
	Total				.767	3.80	30.40		34.20	**43.20**	**53.30**

D3053 710   Fireplaces, Firebrick

	System Description	Freq. (Years)	Crew	Unit	Labor Hours	2014 Bare Costs				Total In-House	Total w/O&P
						Material	Labor	Equipment	Total		
0020	**Replace firebrick**	75	2 BRIC	S.F.	1.155		42.40		42.40	56.50	70
	Masonry demolition, fireplace, brick, 30" x 24"										
	Install new firebrick				.208	8.25	8.55		16.80	20.50	24.50
	Total				1.363	8.25	50.95		59.20	77	**94.50**

D40 FIRE PROTECTION   D4013   Sprinklers

D4013 110   Backflow Preventer

						2014 Bare Costs				Total In-House	Total w/O&P
	System Description	Freq. (Years)	Crew	Unit	Labor Hours	Material	Labor	Equipment	Total		
3010	**Rebuild 4" diam. reduced pressure backflow preventer**	10	1 PLUM	Ea.							
	Notify proper authorities prior to closing valves				.250		14.40		14.40	18	22.50
	Close valves at 4" diameter BFP and open by-pass				.167		9.60		9.60	12	14.95
	Rebuild 4" diameter reduced pressure BFP				2.000	340	115		455	520	605
	Test backflow preventer				.500		29		29	36	45
	Total				2.917	340	168		508	586	687.45
3020	**Rebuild 6" diam. reduced pressure backflow preventer**	10	1 PLUM	Ea.							
	Notify proper authorities prior to closing valves				.250		14.40		14.40	18	22.50
	Close valves at 6" diameter BFP and open by-pass				.182		10.45		10.45	13.10	16.35
	Rebuild 6" diameter reduced pressure BFP				3.008	355	173		528	605	715
	Test backflow preventer				.500		29		29	36	45
	Total				3.939	355	226.85		581.85	672.10	798.85
3030	**Rebuild 8" diam. reduced pressure backflow preventer**	10	1 PLUM	Ea.							
	Notify proper authorities prior to closing valves				.250		14.40		14.40	18	22.50
	Close valves at 8" diameter BFP and open by-pass				.200		11.50		11.50	14.40	17.95
	Rebuild 8" diameter reduced pressure BFP				4.000	440	230		670	770	910
	Test backflow preventer				.500		29		29	36	45
	Total				4.950	440	284.90		724.90	838.40	995.45
3040	**Rebuild 10" diam. reduced pressure backflow preventer**	10	1 PLUM	Ea.							
	Notify proper authorities prior to closing valves				.250		14.40		14.40	18	22.50
	Close valves at 10" diameter BFP and open by-pass				.222		12.80		12.80	16	19.95
	Rebuild 10" diameter reduced pressure BFP				5.000	515	288		803	925	1,100
	Test backflow preventer				.500		29		29	36	45
	Total				5.972	515	344.20		859.20	995	1,187.45

D40 FIRE PROTECTION

D4013 Sprinklers

D4013 310 Sprinkler System, Fire Supression

	System Description	Freq. (Years)	Crew	Unit	Labor Hours	2014 Bare Costs				Total In-House	Total w/O&P
						Material	Labor	Equipment	Total		
1020	**Inspect sprinkler system**	1	1 PLUM	Ea.							
	Inspect water pressure gauges				.044		2.50		2.50	3.20	4
	Test fire alarm (wet or dry)				.030		1.70		1.70	2.20	2.70
	Motor, exposed, inspect				.333		19.20		19.20	24	30
	Inspect, visual				.060		3.40		3.40	4.30	5.40
	Total				.467		26.80		26.80	33.70	42.10
1030	**Replace sprinkler head**	20	1 PLUM	Ea.							
	Remove old sprinkler head				.325		18		18	22.50	28
	Install new head				.650	11.20	36		47.20	57.50	70
	Total				.975	11.20	54		65.20	80	98
1040	**Rebuild double check 3" backflow preventer**	1	1 PLUM	Ea.							
	Rebuild 3" backflow preventer				3.910	375	225		600	695	820
	Test system				1.300		75		75	93.50	117
	Total				5.210	375	300		675	788.50	937
1050	**Rebuild double check 4" backflow preventer**	1	1 PLUM	Ea.							
	Rebuild 4" backflow preventer				4.561	510	262		772	890	1,050
	Test system				1.300		75		75	93.50	117
	Total				5.861	510	337		847	983.50	1,167
1060	**Rebuild double check 6" backflow preventer**	1	1 PLUM	Ea.							
	Rebuild 6" backflow preventer				5.202	605	299		904	1,050	1,225
	Test system				1.300		75		75	93.50	117
	Total				6.502	605	374		979	1,143.50	1,342
1070	**Rebuild reduced pressure backflow preventer**	1	1 PLUM	Ea.							
	Rebuild 3" backflow preventer				3.910	375	225		600	695	820
	Test system				1.300		75		75	93.50	117
	Total				5.210	375	300		675	788.50	937

D40 FIRE PROTECTION   D4013   Sprinklers

D4013 410   Fire Pump

	System Description	Freq. (Years)	Crew	Unit	Labor Hours	2014 Bare Costs				Total In-House	Total w/O&P
						Material	Labor	Equipment	Total		
1030	**Replace fire pump / electric motor assembly 100 H.P.**	25	Q-13	Ea.							
	Remove 100 H.P. electric-drive fire pump				31.373		1,650		1,650	2,075	2,575
	Install new electric-drive fire pump and controls, 100 H.P.				64.000	23,500	3,375		26,875	30,100	34,700
	Total				95.373	23,500	5,025		28,525	32,175	37,275

D50 ELECTRICAL   D5013   Electrical Service/Distribution

D5013 110   Primary Transformer, Liquid Filled

	System Description	Freq. (Years)	Crew	Unit	Labor Hours	2014 Bare Costs				Total In-House	Total w/O&P
						Material	Labor	Equipment	Total		
0010	**Repair 500 kVA transformer**	10	1 ELEC	Ea.							
	Remove transformer component				1.000		53.50		53.50	66	82.50
	Transformer component				2.963	2,225	158		2,383	2,650	3,025
	Total				3.963	2,225	211.50		2,436.50	2,716	3,107.50
0020	**Maintenance and inspection**	0.50	1 ELEC	Ea.							
	Test energized transformer				.500		26.50		26.50	33	41.50
	Total				.500		26.50		26.50	**33**	**41.50**
0030	**Replace transformer**	30	R-3	Ea.							
	Remove transformer				13.793		730	96.50	826.50	1,000	1,225
	Transformer 500 kVA, 5 KV primary, 277/480 V secondary				50.000	20,200	2,650	350	23,200	25,900	29,800
	Total				63.793	20,200	3,380	446.50	24,026.50	26,900	31,025

D5013 120   Primary Transformer, Dry

	System Description	Freq. (Years)	Crew	Unit	Labor Hours	2014 Bare Costs				Total In-House	Total w/O&P
						Material	Labor	Equipment	Total		
0010	**Repair 15 KV primary transformer**	15	1 ELEC	Ea.							
	Repair transformer terminal				2.000	48.50	107		155.50	185	226
	Total				2.000	48.50	107		155.50	**185**	**226**
0020	**Maintenance and inspection**	0.50	1 ELEC	Ea.							
	Test energized transformer				.500		26.50		26.50	33	41.50
	Total				.500		26.50		26.50	**33**	**41.50**
0030	**Replace transformer**	30	R-3	Ea.							
	Remove transformer				13.793		730	96.50	826.50	1,000	1,225
	Transformer 500 kVA, 15 KV primary, 277/480 V secondary				57.143	47,700	3,025	400	51,125	56,500	65,000
	Total				70.936	47,700	3,755	496.50	51,951.50	57,500	66,225

D50 ELECTRICAL D5013 Electrical Service/Distribution

D5013 210 Switchgear, Mainframe

	System Description	Freq. (Years)	Crew	Unit	Labor Hours	2014 Bare Costs				Total In-House	Total w/O&P
						Material	Labor	Equipment	Total		
0010	**Repair switchgear 1200 A**	5	1 ELEC	Ea.							
	Remove relay				1.702		91		91	112	141
	Relay				5.000	1,150	267		1,417	1,600	1,850
	Total				6.702	1,150	358		1,508	1,712	**1,991**
0020	**Maintenance and inspection**	1	1 ELEC	Ea.							
	Check & retighten switchgear connections				.800		42.50		42.50	53	66
	Total				.800		42.50		42.50	**53**	**66**
0030	**Replace switchgear 1200 A**	20	3 ELEC	Ea.							
	Remove switchgear 1200 A				6.000		320		320	395	495
	Switchgear 1200 A				21.818	1,725	1,175		2,900	3,350	3,950
	Test switchgear				2.000		107		107	132	165
	Total				29.818	1,725	1,602		3,327	3,877	**4,610**

D5013 216 Fuses

	System Description	Freq. (Years)	Crew	Unit	Labor Hours	2014 Bare Costs				Total In-House	Total w/O&P
						Material	Labor	Equipment	Total		
0010	**Replace fuse**	25	1 ELEC	Ea.							
	Remove fuse				.151		8.05		8.05	9.95	12.45
	Fuse 600 A				.400	355	21.50		376.50	415	475
	Total				.551	355	29.55		384.55	424.95	**487.45**

D5013 220 Switchgear, Indoor, Less Than 600 V

	System Description	Freq. (Years)	Crew	Unit	Labor Hours	2014 Bare Costs				Total In-House	Total w/O&P
						Material	Labor	Equipment	Total		
0010	**Repair switchgear, - (5% of total C.B.)**	10	1 ELEC	Ea.							
	Remove circuit breaker				.400		21.50		21.50	26.50	33
	Circuit breaker				1.143	226	61		287	325	375
	Total				1.543	226	82.50		308.50	351.50	**408**
0020	**Maintenance and inspection**	3	1 ELEC	Ea.							
	Inspect and clean panel				.500		26.50		26.50	33	41.50
	Total				.500		26.50		26.50	**33**	**41.50**

D50 ELECTRICAL

D5013 Electrical Service/Distribution

D5013 220 Switchgear, Indoor, Less Than 600 V

	System Description	Freq. (Years)	Crew	Unit	Labor Hours	Material	2014 Bare Costs Labor	Equipment	Total	Total In-House	Total w/O&P
0030	**Replace switchgear, 225 A**	30	1 ELEC	Ea.							
	Remove panelboard, 4 wire, 120/208 V, 225 A				4.000		213		213	264	330
	Panelboard, 4 wire, 120/208 V, 225 A main circuit breaker				11.940	2,400	635		3,035	3,425	3,975
	Test indoor switchgear				1.509		80.50		80.50	99.50	125
	Total				17.450	2,400	928.50		3,328.50	3,788.50	4,430
0200	**Replace switchgear, 400 A**	30	2 ELEC	Ea.							
	Turn power off, later turn on				.200		10.70		10.70	13.20	16.50
	Remove panelboard, 4 wire, 120/208 V, 100 A				3.333		178		178	220	275
	Remove panelboard, 4 wire, 120/208 V, 400 A				8.333		445		445	550	690
	Panelboard, 4 wire, 120/208 V, 100 A main circuit breaker				17.021	1,175	910		2,085	2,425	2,875
	Panelboard, 4 wire, 120/208 V, 400 A main circuit breaker				33.333	3,225	1,775		5,000	5,750	6,775
	Test circuits				.145		7.76		7.76	9.60	12
	Total				62.367	4,400	3,326.46		7,726.46	8,967.80	10,643.50
0240	**Replace switchgear, 600 A**	30	2 ELEC	Ea.							
	Turn power off, later turn on				.300		16.05		16.05	19.80	25
	Remove panelboard, 4 wire, 120/208 V, 200 A				6.667		355		355	440	550
	Remove switchboard, incoming section, 600 A				4.848		259		259	320	400
	Remove switchboard, distribution section, 600 A				4.000		213		213	264	330
	Panelboard, 4 wire, 120/208 V, 225 A main circuit breaker				28.571	2,400	1,525		3,925	4,525	5,350
	Switchboard, incoming section, 600 A				16.000	3,975	855		4,830	5,425	6,300
	Switchboard, distribution section, 600 A				16.000	1,600	855		2,455	2,825	3,325
	Test circuits				.218		11.64		11.64	14.40	18
	Total				76.605	7,975	4,089.69		12,064.69	13,833.20	16,298
0280	**Replace switchgear, 800 A**	30	2 ELEC	Ea.							
	Turn power off, later turn on				.400		21.40		21.40	26.50	33
	Remove panelboard, 4 wire, 120/208 V, 200 A (2 panelboards)				13.333		710		710	880	1,100
	Remove switchboard, incoming section, 800 A				5.517		294		294	365	455
	Remove switchboard, distribution section, 800 A				4.444		237		237	293	365
	Panelboard, 4 wire, 120/208 V, 225 A main lugs (2 panelboards)				47.059	2,700	2,500		5,200	6,100	7,275
	Switchboard, incoming section, 800 A				18.182	3,975	970		4,945	5,575	6,475
	Switchboard, distribution section, 800 A				18.182	2,100	970		3,070	3,500	4,125
	Test circuits				.291		15.52		15.52	19.20	24
	Total				107.408	8,775	5,717.92		14,492.92	16,758.70	19,852

D50 ELECTRICAL D5013 Electrical Service/Distribution

D5013 220 Switchgear, Indoor, Less Than 600 V

	System Description	Freq. (Years)	Crew	Unit	Labor Hours	2014 Bare Costs				Total In-House	Total w/O&P
						Material	Labor	Equipment	Total		
0320	**Replace switchgear, 1200 A**	30	3 ELEC	Ea.							
	Turn power off, later turn on				.600		32.10		32.10	39.50	49.50
	Remove panelboard, 4 wire, 120/208 V, 200 A (4 panelboards)				26.667		1,420		1,420	1,750	2,200
	Remove switchboard, incoming section, 1200 A				6.667		355		355	440	550
	Remove switchboard, distribution section, 1200 A				5.161		275		275	340	425
	Panelboard, 4 wire, 120/208 V, 225 A main lugs (4 panelboards)				94.118	5,400	5,000		10,400	12,200	14,600
	Switchboard, incoming section, 1200 A				22.222	4,775	1,175		5,950	6,725	7,800
	Switchboard, distribution section, 1200 A				22.222	3,100	1,175		4,275	4,875	5,700
	Test circuits				.436		23.28		23.28	29	36
	Total				178.093	13,275	9,455.38		22,730.38	26,398.50	31,360.50
0360	**Replace switchgear, 1600 A**	30	3 ELEC	Ea.							
	Turn power off, later turn on				.800		42.80		42.80	53	66
	Remove panelboard, 4 wire, 120/208 V, 200 A (6 panelboards)				40.000		2,130		2,130	2,650	3,300
	Remove switchboard, incoming section, 1600 A				7.273		390		390	480	600
	Remove switchboard, distribution section, 1600 A				5.517		294		294	365	455
	Panelboard, 4 wire, 120/208 V, 225 A main lugs (6 panelboards)				141.176	8,100	7,500		15,600	18,300	21,800
	Switchboard, incoming section, 1600 A				24.242	4,775	1,300		6,075	6,850	7,975
	Switchboard, distribution section, 1600 A				24.242	3,550	1,300		4,850	5,500	6,450
	Test circuits				.582		31.04		31.04	38.50	48
	Total				243.833	16,425	12,987.84		29,412.84	34,236.50	40,694
0400	**Replace switchgear, 2000 A**	30	3 ELEC	Ea.							
	Turn power off, later turn on				1.000		53.50		53.50	66	82.50
	Remove panelboard, 4 wire, 120/208 V, 200 A (8 panelboards)				53.333		2,840		2,840	3,525	4,400
	Remove switchboard, incoming section, 2000 A				7.619		405		405	505	630
	Remove switchboard, distribution section, 2000 A				5.926		315		315	390	490
	Panelboard, 4 wire, 120/208 V, 225 A main lugs (8 panelboards)				188.235	10,800	10,000		20,800	24,400	29,100
	Switchboard, incoming section, 2000 A				25.806	5,150	1,375		6,525	7,375	8,575
	Switchboard, distribution section, 2000 A				25.806	4,150	1,375		5,525	6,275	7,325
	Test circuits				.727		38.80		38.80	48	60
	Total				308.454	20,100	16,402.30		36,502.30	42,584	50,662.50

D50 ELECTRICAL D5013 Electrical Service/Distribution

D5013 224 Switchgear, Indoor, 600 V

	System Description	Freq. (Years)	Crew	Unit	Labor Hours	2014 Bare Costs				Total In-House	Total w/O&P
						Material	Labor	Equipment	Total		
0010	Maintenance and repair, - (5% of total fuses)	10	1 ELEC	Ea.							
	Remove fuse				.151		8.05		8.05	9.95	12.45
	Fuse				.400	355	21.50		376.50	415	475
	Total				.551	355	29.55		384.55	424.95	487.45
0020	Maintenance and inspection	3	1 ELEC	Ea.							
	Inspect and clean panel				.500		26.50		26.50	33	41.50
	Total				.500		26.50		26.50	33	41.50
0030	Replace switchgear	30	2 ELEC	Ea.							
	Remove indoor switchgear				4.000		213		213	264	330
	Switchgear				14.035	1,325	750		2,075	2,375	2,800
	Test indoor switchgear				2.000		107		107	132	165
	Total				20.035	1,325	1,070		2,395	2,771	3,295

D5013 230 Meters

	System Description	Freq. (Years)	Crew	Unit	Labor Hours	2014 Bare Costs				Total In-House	Total w/O&P
						Material	Labor	Equipment	Total		
0010	Repair switchboard meter	10	1 ELEC	Ea.							
	Repair meter				5.000	1,075	267		1,342	1,525	1,750
	Total				5.000	1,075	267		1,342	1,525	1,750
0020	Replace switchboard meter	20	1 ELEC	Ea.							
	Remove meter				.500		26.50		26.50	33	41.50
	Meter				1.600	5,375	85.50		5,460.50	6,025	6,850
	Total				2.100	5,375	112		5,487	6,058	6,891.50

D5013 240 Inverter

	System Description	Freq. (Years)	Crew	Unit	Labor Hours	2014 Bare Costs				Total In-House	Total w/O&P
						Material	Labor	Equipment	Total		
0010	Maintenance and repair	1	2 ELEC	Ea.							
	Repair inverter component				5.000	244	267		511	600	720
	Total				5.000	244	267		511	600	720

D50 ELECTRICAL — D5013 Electrical Service/Distribution

D5013 240 Inverter

	System Description	Freq. (Years)	Crew	Unit	Labor Hours	2014 Bare Costs				Total In-House	Total w/O&P
						Material	Labor	Equipment	Total		
0020	Maintenance and inspection	0.25	1 ELEC	Ea.							
	Check connections, controls & relays				1.600		85.50		85.50	106	132
	Total				1.600		85.50		85.50	**106**	**132**
0030	Replace 1 kVA inverter	20	2 ELEC	Ea.							
	Remove inverter				.500		26.50		26.50	33	41.50
	Inverter				1.600	2,450	85.50		2,535.50	2,800	3,200
	Total				2.100	2,450	112		2,562	2,833	3,241.50

D5013 250 Rectifier, Up To 600 V

	System Description	Freq. (Years)	Crew	Unit	Labor Hours	2014 Bare Costs				Total In-House	Total w/O&P
						Material	Labor	Equipment	Total		
0010	Maintenance and repair	2	1 ELEC	Ea.							
	Repair rectifier component				4.000	455	213		668	765	900
	Total				4.000	455	213		668	**765**	**900**
0020	Maintenance and inspection	0.33	1 ELEC	Ea.							
	Check connections, controls & relays				1.600		85.50		85.50	106	132
	Total				1.600		85.50		85.50	**106**	**132**
0030	Replace rectifier	20	2 ELEC	Ea.							
	Remove rectifier				.500		26.50		26.50	33	41.50
	Rectifier				1.600	1,375	85.50		1,460.50	1,625	1,850
	Total				2.100	1,375	112		1,487	1,658	1,891.50

D5013 264 Motor Starter, Up To 600 V

	System Description	Freq. (Years)	Crew	Unit	Labor Hours	2014 Bare Costs				Total In-House	Total w/O&P
						Material	Labor	Equipment	Total		
0010	Maintenance and repair	5	1 ELEC	Ea.							
	Remove motor starter coil				.500		26.50		26.50	33	41.50
	Motor starter coil				1.600	34.50	85.50		120	144	175
	Total				2.100	34.50	112		146.50	177	216.50

D50 ELECTRICAL   D5013   Electrical Service/Distribution

D5013 264   Motor Starter, Up To 600 V

	System Description	Freq. (Years)	Crew	Unit	Labor Hours	2014 Bare Costs				Total In-House	Total w/O&P
						Material	Labor	Equipment	Total		
0020	**Maintenance and inspection**										
	Inspect & clean motor starter	0.50	1 ELEC	Ea.	.667		35.50		35.50	44	55
	Total				.667		35.50		35.50	**44**	**55**
0030	**Replace starter**										
	Remove motor starter	18	1 ELEC	Ea.	.899		48		48	59.50	74
	Magnetic motor starter 5 H.P., size 0				3.478	236	186		422	490	580
	Cut, form, align & connect wires				.151		8.05		8.05	9.95	12.45
	Check operation of motor starter				.018		.98		.98	1.21	1.51
	Total				4.546	236	243.03		479.03	560.66	667.96

D5013 266   Motor Starter, 600 V

	System Description	Freq. (Years)	Crew	Unit	Labor Hours	2014 Bare Costs				Total In-House	Total w/O&P
						Material	Labor	Equipment	Total		
0010	**Maintenance and repair**										
	Remove motor starter coil	3	1 ELEC	Ea.	1.702		91		91	112	141
	Motor starter coil				5.000	102	267		369	440	545
	Total				6.702	102	358		460	**552**	**686**
0020	**Maintenance and inspection**										
	Inspect & clean motor starter	0.25	1 ELEC	Ea.	.667		35.50		35.50	44	55
	Total				.667		35.50		35.50	**44**	**55**
0030	**Replace starter**										
	Remove motor starter	18	2 ELEC	Ea.	3.333		178		178	220	275
	Motor starter FVNR 600 V, 3 P, 100 H.P. motor				13.333	2,375	710		3,085	3,500	4,075
	Cut, form, align & connect wires				.170		9.10		9.10	11.25	14.05
	Check operation of motor starter				.018		.98		.98	1.21	1.51
	Total				16.855	2,375	898.08		3,273.08	3,732.46	4,365.56

D50 ELECTRICAL | D5013 Electrical Service/Distribution

D5013 272 Secondary Transformer, Liquid Filled

	System Description	Freq. (Years)	Crew	Unit	Labor Hours	2014 Bare Costs				Total In-House	Total w/O&P
						Material	Labor	Equipment	Total		
0010	**Maintenance and repair**	25	1 ELEC	Ea.							
	Remove transformer terminal				.348		18.55		18.55	23	28.50
	Transformer terminal				1.000	36.50	53.50		90	106	128
	Total				1.348	36.50	72.05		108.55	**129**	**156.50**
0020	**Maintenance and inspection**	0.50	1 ELEC	Ea.							
	Check connections & fluid				.667		35.50		35.50	44	55
	Total				.667		35.50		35.50	**44**	**55**

D5013 274 Secondary Transformer, Dry

	System Description	Freq. (Years)	Crew	Unit	Labor Hours	2014 Bare Costs				Total In-House	Total w/O&P
						Material	Labor	Equipment	Total		
0010	**Maintenance and repair**	10	1 ELEC	Ea.							
	Remove transformer terminal				.500		26.50		26.50	33	41.50
	Transformer terminal				1.000	109	53.50		162.50	186	219
	Total				1.500	109	80		189	**219**	**260.50**
0020	**Maintenance and inspection**	0.50	1 ELEC	Ea.							
	Inspect connections & tighten				1.000		53.50		53.50	66	82.50
	Total				1.000		53.50		53.50	**66**	**82.50**
0030	**Replace 15 kVA transformer**	30	2 ELEC	Ea.							
	Remove transformer 15 KVA, incl. supp., wire & conduit terminations				3.810		203		203	251	315
	Transformer 15 KVA, 240/480 V primary, 120/208 V secondary				14.545	1,075	775		1,850	2,150	2,550
	Total				18.355	1,075	978		2,053	**2,401**	**2,865**
0040	**Replace 112.5 kVA transformer**	30	R-3	Ea.							
	Remove transformer 112.5 KVA, incl. support, wire, conduit termination				6.897		365	48.50	413.50	500	620
	Transformer 112.5 KVA, 240/480 V primary, 120/208 V secondary				22.222	2,825	1,175	156	4,156	4,725	5,525
	Total				29.119	2,825	1,540	204.50	4,569.50	**5,225**	**6,145**
0050	**Replace 500 kVA transformer**	30	R-3	Ea.							
	Remove transformer 500 KVA, incl. support, wire & conduit terminations				14.286		755	100	855	1,050	1,275
	Transformer 500 KVA, 240/480 V primary, 120/208 V secondary				44.444	10,500	2,350	310	13,160	14,800	17,100
	Total				58.730	10,500	3,105	410	14,015	**15,850**	**18,375**

D50 ELECTRICAL   D5013   Electrical Service/Distribution

D5013 280   Lighting Panel, Indoor

	System Description	Freq. (Years)	Crew	Unit	Labor Hours	2014 Bare Costs				Total In-House	Total w/O&P
						Material	Labor	Equipment	Total		
0020	**Maintenance and inspection**										
	Inspect and clean panels	3	1 ELEC	Ea.	.500		26.50		26.50	33	41.50
	Total				.500		26.50		26.50	**33**	**41.50**
0030	**Replace load center, 100 A**	20	2 ELEC	Ea.							
	Turn power off, later turn on				.100		5.35		5.35	6.60	8.25
	Remove load center, 3 wire, 120/240 V				3.077		164		164	203	254
	Load centers, 3 Wire, 120/240V, 100 A				6.667	325	355		680	800	955
	Test circuits				.073		3.88		3.88	4.80	6
	Total				9.916	325	528.23		853.23	**1,014.40**	**1,223.25**

D50 ELECTRICAL D5023 Lighting & Branch Wiring

D5023 110 Wireway

0030 Replace 8" x 8" wireway

System Description	Freq. (Years)	Crew	Unit	Labor Hours	2014 Bare Costs				Total In-House	Total w/O&P
					Material	Labor	Equipment	Total		
Remove and reinstall supply box	20	2 ELEC	L.F.	.050		2.65		2.65	3.30	4.15
Remove wireway, fittings & support				.100		5.35		5.35	6.60	8.25
Wireway 8" x 8", fittings & support				.400	28	21.50		49.50	57	68
Total				.550	28	29.50		57.50	66.90	80.40

D5023 112 Conduit EMT

0010 Replace 1" EMT conduit

System Description	Freq. (Years)	Crew	Unit	Labor Hours	2014 Bare Costs				Total In-House	Total w/O&P
					Material	Labor	Equipment	Total		
Remove & reinstall cover plate	50	2 ELEC	M.L.F.	1.290		69		69	85	107
Remove conduit w/ fittings & hangers				20.300		1,080		1,080	1,350	1,675
Remove junction box				1.000		53.50		53.50	66	82.50
1" EMT with couplings				38.650	1,120	2,060		3,180	3,775	4,600
Outlet box square				4.000	21	215		236	285	355
Box connector				.889	35.90	47.40		83.30	98	118
Beam clamp				5.000	195	288		483	575	695
Conduit hanger				5.926	450	316		766	885	1,050
Total				77.055	1,821.90	4,128.90		5,950.80	7,119	8,682.50

D5023 120 Cable, Non-Metallic (NM) Sheathed

0010 Replace NM cable

System Description	Freq. (Years)	Crew	Unit	Labor Hours	2014 Bare Costs				Total In-House	Total w/O&P
					Material	Labor	Equipment	Total		
Shut off power, later turn on	50	2 ELEC	M.L.F.	.100		5.35		5.35	6.60	8.25
Remove & reinstall cover plate				1.290		69		69	85	107
Remove junction box				1.000		53.50		53.50	66	82.50
Remove NM cable #12, 2 wire				12.720		680		680	840	1,050
NM cable #12, 2 wire				32.000	385	1,710		2,095	2,550	3,125
Staples, for Romex or BX cable				4.000	6	214		220	270	340
NM #12-2 wire cable connectors				.602	14.10	32.10		46.20	55	67.50
Outlet box square 4" for Romex or BX				4.000	89.50	215		304.50	365	440
Total				55.712	494.60	2,978.95		3,473.55	4,237.60	5,220.25

D50 ELECTRICAL | D5023 Lighting & Branch Wiring

D5023 122 Cable, Service

					2014 Bare Costs				Total In-House	Total w/O&P
System Description	Freq. (Years)	Crew	Unit	Labor Hours	Material	Labor	Equipment	Total		
0010 Replace service cable	50	2 ELEC	M.L.F.							
Shut off power, later turn on				.100		5.35		5.35	6.60	8.25
Remove & reinstall cover plate				1.290		69		69	85	107
Remove junction box				1.000		53.50		53.50	66	82.50
Remove wire #6 XLPE				2.000		106.50		106.50	132	165
600 V, copper type XLPE-USE				12.308	855	655		1,510	1,750	2,100
Staples, for service cable				5.000	8	266		274	340	420
SER cable connectors				3.333	34.10	178		212.10	260	320
Outlet box square 4"				4.000	21	215		236	285	355
Total				29.031	918.10	1,548.35		2,466.45	2,924.60	3,557.75

D5023 124 Cable, Armored

					2014 Bare Costs				Total In-House	Total w/O&P
System Description	Freq. (Years)	Crew	Unit	Labor Hours	Material	Labor	Equipment	Total		
0010 Replace armored cable	60	2 ELEC	M.L.F.							
Shut off power, later turn on				.100		5.35		5.35	6.60	8.25
Remove & reinstall cover plate				1.290		69		69	85	107
Remove junction box				1.000		53.50		53.50	66	82.50
Remove armored cable				13.220		710		710	870	1,100
BX #12, 2 wire				34.783	545	1,860		2,405	2,900	3,550
Staples, for Romex or BX cable				4.000	6	214		220	270	340
Cable connector				2.000	9	106.50		115.50	142	176
Outlet box square 4" for Romex or BX				4.000	89.50	215		304.50	365	440
Total				60.393	649.50	3,233.35		3,882.85	4,704.60	5,803.75

D50 ELECTRICAL — D5023 Lighting & Branch Wiring

D5023 126 Branch Wiring, With Junction Box

	System Description	Freq. (Years)	Crew	Unit	Labor Hours	2014 Bare Costs Material	2014 Bare Costs Labor	2014 Bare Costs Equipment	2014 Bare Costs Total	Total In-House	Total w/O&P
0010	Replace branch wiring	50	2 ELEC	M.L.F.							
	Shut off power, later turn on				.100		5.35		5.35	6.60	8.25
	Remove & reinstall cover plate				1.290		69		69	85	107
	Remove junction box				1.000		53.50		53.50	66	82.50
	Remove #12 wire				1.455		77.50		77.50	96	120
	Conductor THW-THHN, copper, stranded, #12				7.273	157	390		547	655	795
	Outlet box square 4"				4.000	21	215		236	285	355
	Wire connectors				1.000	2.40	53.40		55.80	68.50	85.50
	Total				16.117	180.40	863.75		1,044.15	1,262.10	1,553.25

D5023 128 Branch Wiring, 600 V

	System Description	Freq. (Years)	Crew	Unit	Labor Hours	2014 Bare Costs Material	2014 Bare Costs Labor	2014 Bare Costs Equipment	2014 Bare Costs Total	Total In-House	Total w/O&P
0010	Replace branch wiring	50	2 ELEC	M.L.F.							
	Shut off power, later turn on				.100		5.35		5.35	6.60	8.25
	Remove #12 wire				1.455		77.50		77.50	96	120
	Conductor THW-THHN, copper, stranded, #12				7.273	157	390		547	655	795
	Cut, form and align 4 wires				.410		22		22	27	34
	Splice, solder, insulate 2 wires				2.667		142		142	176	220
	Total				11.904	157	636.85		793.85	960.60	1,177.25

D5023 130 Circuit Breaker

	System Description	Freq. (Years)	Crew	Unit	Labor Hours	2014 Bare Costs Material	2014 Bare Costs Labor	2014 Bare Costs Equipment	2014 Bare Costs Total	Total In-House	Total w/O&P
0010	Maintenance and repair breaker, molded case, 480 V, 1 pole	20	1 ELEC	Ea.							
	Tighten connections				.889		47.50		47.50	58.50	73.50
	Total				.889		47.50		47.50	58.50	73.50
0020	Maintenance and inspection C.B., molded case, 480 V, 1 pole	0.50	1 ELEC	Ea.							
	Inspect, tighten C.B. wiring				.400		21.50		21.50	26.50	33
	Total				.400		21.50		21.50	26.50	33

D50 ELECTRICAL D5023 Lighting & Branch Wiring

D5023 130 Circuit Breaker

		Freq. (Years)	Crew	Unit	Labor Hours	2014 Bare Costs				Total In-House	Total w/O&P
	System Description					Material	Labor	Equipment	Total		
0030	**Replace C.B. molded case, 480 V, 1 pole**	50	1 ELEC	Ea.							
	Remove circuit breaker				.348		18.55		18.55	23	28.50
	Circuit breaker molded case, 480 V, 70 to 100 A, 1 P				1.143	226	61		287	325	375
	Cut, form, align & connect wires				.151		8.05		8.05	9.95	12.45
	Check operation of breaker				.018		.98		.98	1.21	1.51
	Total				1.660	226	88.58		314.58	**359.16**	**417.46**
1010	**Maintenance and repair breaker, molded case, 480 V, 2 pole**	20	1 ELEC	Ea.							
	Tighten connections				.889		47.50		47.50	58.50	73.50
	Total				.889		47.50		47.50	**58.50**	**73.50**
1020	**Maintenance and inspection C.B., molded case, 480 V, 2 pole**	0.50	1 ELEC	Ea.							
	Inspect, tighten C.B. wiring				.400		21.50		21.50	26.50	33
	Total				.400		21.50		21.50	**26.50**	**33**
1030	**Replace C.B. molded case, 480 V, 2 pole**	50	1 ELEC	Ea.							
	Remove circuit breaker				.500		26.50		26.50	33	41.50
	Circuit breaker molded case, 480 V, 70 to 100 A, 2 P				1.600	570	85.50		655.50	735	845
	Cut, form, align & connect wires				.151		8.05		8.05	9.95	12.45
	Check operation of breaker				.018		.98		.98	1.21	1.51
	Total				2.269	570	121.03		691.03	**779.16**	**900.46**
1040	**Maintenance and repair breaker, molded case, 480 V, 3 pole**	20	1 ELEC	Ea.							
	Tighten connections				.889		47.50		47.50	58.50	73.50
	Total				.889		47.50		47.50	**58.50**	**73.50**
1050	**Maintenance and inspection C.B., molded case, 480 V, 3 pole**	0.50	1 ELEC	Ea.							
	Inspect & tighten C.B. wiring				.400		21.50		21.50	26.50	33
	Total				.400		21.50		21.50	**26.50**	**33**
1060	**Replace C.B. molded case, 480 V, 3 pole**	50	1 ELEC	Ea.							
	Remove circuit breaker				.500		26.50		26.50	33	41.50
	Circuit breaker molded case, 480 V, 70 to 100 A, 3 P				2.000	670	107		777	870	1,000
	Cut, form, align & connect wires				.151		8.05		8.05	9.95	12.45
	Check operation of breaker				.018		.98		.98	1.21	1.51
	Total				2.669	670	142.53		812.53	**914.16**	**1,055.46**

D50 ELECTRICAL — D5023 Lighting & Branch Wiring

D5023 130 Circuit Breaker

	System Description	Freq. (Years)	Crew	Unit	Labor Hours	2014 Bare Costs — Material	2014 Bare Costs — Labor	2014 Bare Costs — Equipment	2014 Bare Costs — Total	Total In-House	Total w/O&P
2010	**Repair failed breaker, molded case, 600 V, 2 pole**	10	1 ELEC	Ea.							
	Remove trip switch				.500		26.50		26.50	33	41.50
	Trip switch				1.600	206	85.50		291.50	330	390
	Total				2.100	206	112		318	363	431.50
2020	**Maintenance and inspection C.B., molded case, 600 V, 2 pole**	0.33	1 ELEC	Ea.							
	Inspect & tighten C.B. wiring				.400		21.50		21.50	26.50	33
	Total				.400		21.50		21.50	**26.50**	**33**
2030	**Replace C.B. molded case, 600 V, 2 pole**	50	1 ELEC	Ea.							
	Remove circuit breaker				.500		26.50		26.50	33	41.50
	Circuit breaker molded case, 600 V, 70 to 100 A, 2 P				1.600	645	85.50		730.50	815	940
	Cut, form, align & connect wires				.170		9.10		9.10	11.25	14.05
	Check operation of breakers				.018		.98		.98	1.21	1.51
	Total				2.289	645	122.08		767.08	860.46	997.06
3010	**Repair failed breaker, molded case, 600 V, 3 pole**	10	1 ELEC	Ea.							
	Remove trip switch				.500		26.50		26.50	33	41.50
	Trip switch				1.600	206	85.50		291.50	330	390
	Total				2.100	206	112		318	363	431.50
3020	**Maintenance and inspection C.B., molded case, 600 V, 3 pole**	0.33	1 ELEC	Ea.							
	Inspect & tighten C.B. wiring				.400		21.50		21.50	26.50	33
	Total				.400		21.50		21.50	**26.50**	**33**
3030	**Replace C.B. molded case, 600 V, 3 pole**	50	1 ELEC	Ea.							
	Remove circuit breaker				.889		47.50		47.50	58.50	73.50
	Circuit breaker molded case, 600 V, 125 to 400 A, 3 P				3.478	2,775	186		2,961	3,275	3,750
	Cut, form, align & connect wires				.170		9.10		9.10	11.25	14.05
	Check operation of breakers				.018		.98		.98	1.21	1.51
	Total				4.556	2,775	243.58		3,018.58	3,345.96	3,839.06
3040	**Maintenance and repair breaker, enclosed, 240 V, 1 pole**	25	1 ELEC	Ea.							
	Tighten connections				.889		47.50		47.50	58.50	73.50
	Total				.889		47.50		47.50	**58.50**	**73.50**

D50 ELECTRICAL   D5023   Lighting & Branch Wiring

D5023 130   Circuit Breaker

	System Description	Freq. (Years)	Crew	Unit	Labor Hours	Material	Labor	Equipment	Total	Total In-House	Total w/O&P
3050	**Maintenance and inspection C.B., enclosed, 240 V, 1 pole**	1	1 ELEC	Ea.							
	Inspect & tighten C.B. wiring				.400		21.50		21.50	26.50	33
	Total				.400		21.50		21.50	**26.50**	**33**
3060	**Replace C.B. enclosed, 240 V, 1 pole**	50	1 ELEC	Ea.							
	Remove circuit breaker				.500		26.50		26.50	33	41.50
	Circuit breaker enclosed, 240 V, 15 to 60 A, 1 P				2.105	240	112		352	405	475
	Cut, form, align & connect wires				.151		8.05		8.05	9.95	12.45
	Check operation of breaker				.018		.98		.98	1.21	1.51
	Total				2.775	240	147.53		387.53	449.16	530.46
4010	**Maintenance and repair breaker, enclosed, 240 V, 2 pole**	25	1 ELEC	Ea.							
	Tighten connections				.889		47.50		47.50	58.50	73.50
	Total				.889		47.50		47.50	**58.50**	**73.50**
4020	**Maintenance and inspection C.B., enclosed, 240 V, 2 pole**	1	1 ELEC	Ea.							
	Inspect & tighten C.B. wiring				.400		21.50		21.50	26.50	33
	Total				.400		21.50		21.50	**26.50**	**33**
4030	**Replace C.B. enclosed, 240 V, 2 pole**	50	1 ELEC	Ea.							
	Remove circuit breaker				.500		26.50		26.50	33	41.50
	Circuit breaker enclosed, 240 V, 15 to 60 A, 2 P				2.286	360	122		482	545	640
	Cut, form, align & connect wires				.151		8.05		8.05	9.95	12.45
	Check operation of breaker				.018		.98		.98	1.21	1.51
	Total				2.955	360	157.53		517.53	589.16	695.46
4040	**Maintenance and repair breaker, enclosed, 240 V, 3 pole**	25	1 ELEC	Ea.							
	Tighten connections				.889		47.50		47.50	58.50	73.50
	Total				.889		47.50		47.50	**58.50**	**73.50**
4050	**Maintenance and inspection C.B., enclosed, 240 V, 3 pole**	1	1 ELEC	Ea.							
	Inspect & tighten C.B. wiring				.400		21.50		21.50	26.50	33
	Total				.400		21.50		21.50	**26.50**	**33**

D50 ELECTRICAL D5023 Lighting & Branch Wiring

D5023 130 Circuit Breaker

	System Description	Freq. (Years)	Crew	Unit	Labor Hours	2014 Bare Costs Material	2014 Bare Costs Labor	2014 Bare Costs Equipment	2014 Bare Costs Total	Total In-House	Total w/O&P
4060	**Replace C.B. enclosed, 240 V, 3 pole**	50	1 ELEC	Ea.							
	Remove circuit breaker				.500		26.50		26.50	33	41.50
	Circuit breaker enclosed, 240 V, 15 to 60 A, 3 P				2.500	480	133		613	695	805
	Cut, form, align & connect wires				.151		8.05		8.05	9.95	12.45
	Check operation of breaker				.018		.98		.98	1.21	1.51
	Total				3.169	480	168.53		648.53	739.16	860.46
5010	**Repair failed breaker, enclosed, 600 V, 2 pole**	4	1 ELEC	Ea.							
	Repair circuit breaker component				2.105	365	112		477	540	630
	Total				2.105	365	112		477	540	630
5020	**Maintenance and inspection C.B., enclosed, 600 V, 2 pole**	0.33	1 ELEC	Ea.							
	Inspect & tighten C.B. wiring				.400		21.50		21.50	26.50	33
	Total				.400		21.50		21.50	**26.50**	**33**
5030	**Replace C.B. enclosed, 600 V, 2 pole**	50	1 ELEC	Ea.							
	Remove circuit breaker				.800		42.50		42.50	53	66
	Circuit breaker enclosed, 600 V, 100 A, 2 P				3.200	645	171		816	920	1,075
	Cut, form, align & connect wires				.170		9.10		9.10	11.25	14.05
	Check operation of breaker				.018		.98		.98	1.21	1.51
	Total				4.189	645	223.58		868.58	985.46	1,156.56
6010	**Repair failed breaker, enclosed, 600 V, 3 pole**	4	1 ELEC	Ea.							
	Repair circuit breaker component				2.105	365	112		477	540	630
	Total				2.105	365	112		477	**540**	**630**
6020	**Maintenance and inspection C.B., enclosed, 600 V, 3 pole**	0.33	1 ELEC	Ea.							
	Inspect & tighten C.B. wiring				.400		21.50		21.50	26.50	33
	Total				.400		21.50		21.50	**26.50**	**33**
6030	**Replace C.B. enclosed, 600 V, 3 pole**	50	1 ELEC	Ea.							
	Remove circuit breaker				.889		47.50		47.50	58.50	73.50
	Circuit breaker enclosed, 600 V, 100 A, 3 P				3.478	700	186		886	1,000	1,150
	Cut, form, align & connect wires				.170		9.10		9.10	11.25	14.05
	Check operation of breaker				.018		.98		.98	1.21	1.51
	Total				4.556	700	243.58		943.58	**1,070.96**	**1,239.06**

D50 ELECTRICAL   D5023   Lighting & Branch Wiring

D5023 130 Circuit Breaker

	System Description	Freq. (Years)	Crew	Unit	Labor Hours	2014 Bare Costs Material	2014 Bare Costs Labor	2014 Bare Costs Equipment	2014 Bare Costs Total	Total In-House	Total w/O&P
6830	**Replace C.B. enclosed, 600 V, 3 pole 800 A**	50	2 ELEC	Ea.							
	Remove circuit breaker				4.444		237		237	293	365
	Circuit breaker enclosed, 600 V, 800 A, 3 P				17.021	5,200	910		6,110	6,850	7,900
	Cut, form, align & connect wires				.340		18.20		18.20	22.50	28
	Check operation of breaker				.037		1.96		1.96	2.42	3.02
	Total				21.843	5,200	1,167.16		6,367.16	7,167.92	8,296.02
6930	**Replace C.B. enclosed, 600 V, 3 pole 1000 A**	50	2 ELEC	Ea.							
	Remove circuit breaker				5.161		275		275	340	425
	Circuit breaker enclosed, 600 V, 1000 A, 3 P				19.048	6,575	1,025		7,600	8,500	9,800
	Cut, form, align & connect wires				.516		27.58		27.58	34	42.50
	Check operation of breaker				.055		2.97		2.97	3.67	4.58
	Total				24.780	6,575	1,330.55		7,905.55	8,877.67	10,272.08
7130	**Replace C.B. enclosed, 600 V, 3 pole 1600 A**	50	2 ELEC	Ea.							
	Remove circuit breaker				5.714		305		305	375	470
	Circuit breaker enclosed, 600 V, 1600 A, 3 P				22.222	15,500	1,175		16,675	18,500	21,200
	Cut, form, align & connect wires				.681		36.40		36.40	45	56
	Check operation of breaker				.073		3.92		3.92	4.84	6.05
	Total				28.691	15,500	1,520.32		17,020.32	18,924.84	21,732.05
7230	**Replace C.B. enclosed, 600 V, 3 pole 2000 A**	50	2 ELEC	Ea.							
	Remove circuit breaker				6.400		340		340	420	530
	Circuit breaker enclosed, 600 V, 2000 A, 3 P				25.000	16,800	1,325		18,125	20,100	23,100
	Cut, form, align & connect wires				1.418		75.83		75.83	94	117
	Check operation of breaker				.153		8.17		8.17	10.10	12.60
	Total				32.971	16,800	1,749		18,549	20,624.10	23,759.60

D5023 132 Safety Switch, Heavy Duty

	System Description	Freq. (Years)	Crew	Unit	Labor Hours	2014 Bare Costs Material	2014 Bare Costs Labor	2014 Bare Costs Equipment	2014 Bare Costs Total	Total In-House	Total w/O&P
0010	**Maintenance and repair safety switch, H.D., 3 pole**	8	1 ELEC	Ea.							
	Resecure lugs				.500		26.50		26.50	33	41.50
	Total				.500		26.50		26.50	33	41.50

D50 ELECTRICAL D5023 Lighting & Branch Wiring

D5023 132 Safety Switch, Heavy Duty

	System Description	Freq. (Years)	Crew	Unit	Labor Hours	Material	2014 Bare Costs Labor	Equipment	Total	Total In-House	Total w/O&P
0020	**Maintenance and inspection safety switch, 3 pole**	1	1 ELEC	Ea.							
	Inspect & clean safety switch				.500		26.50		26.50	33	41.50
	Total				.500		26.50		26.50	**33**	**41.50**
0030	**Replace safety switch, H.D. 30 A**	25	2 ELEC	Ea.							
	Remove safety switch				.650		34.50		34.50	43	53.50
	Fused safety switch, heavy duty, 600 V, 3 P, 30 A				2.500	236	133		369	425	500
	Fuses				.600	49.20	31.95		81.15	93	111
	Check operation of switch				.018		.98		.98	1.21	1.51
	Total				3.769	285.20	200.43		485.63	**562.21**	**666.01**
0130	**Replace safety switch, H.D. 100 A**	25	2 ELEC	Ea.							
	Remove safety switch				1.096		58.50		58.50	72.50	90.50
	Fused safety switch, heavy duty, 600 V, 3 P, 100 A				4.211	520	225		745	850	1,000
	Fuses				.667	174	35.55		209.55	236	273
	Check operation of switch				.018		.98		.98	1.21	1.51
	Total				5.991	694	320.03		1,014.03	**1,159.71**	**1,365.01**
0230	**Replace safety switch, H.D. 200 A**	25	2 ELEC	Ea.							
	Remove safety switch				1.600		85.50		85.50	106	132
	Fused safety switch, heavy duty, 600 V, 3 P, 200 A				6.154	755	330		1,085	1,225	1,450
	Fuses				.800	351	42.75		393.75	440	505
	Check operation of switch				.018		.98		.98	1.21	1.51
	Total				8.572	1,106	459.23		1,565.23	**1,772.21**	**2,088.51**
0430	**Replace safety switch, H.D. 400 A**	25	2 ELEC	Ea.							
	Remove safety switch				2.353		126		126	155	194
	Fused safety switch, heavy duty, 600 V, 3 P, 400 A				8.889	1,950	475		2,425	2,725	3,175
	Fuses				1.000	699	53.40		752.40	835	955
	Check operation of switch				.037		1.96		1.96	2.42	3.02
	Total				12.278	2,649	656.36		3,305.36	**3,717.42**	**4,327.02**
0630	**Replace safety switch, H.D. 600 A**	25	2 ELEC	Ea.							
	Remove safety switch				3.478		186		186	230	287
	Fused safety switch, heavy duty, 600 V, 3 P, 600 A				13.333	3,300	710		4,010	4,500	5,225
	Fuses				1.200	1,005	64.50		1,069.50	1,175	1,350
	Check operation of switch				.037		1.96		1.96	2.42	3.02
	Total				18.048	4,305	962.46		5,267.46	**5,907.42**	**6,865.02**

D50 ELECTRICAL — Lighting & Branch Wiring

D5023 140 Safety Switch, General Duty

	System Description	Freq. (Years)	Crew	Unit	Labor Hours	2014 Bare Costs Material	2014 Bare Costs Labor	2014 Bare Costs Equipment	2014 Bare Costs Total	Total In-House	Total w/O&P
2010	**Maintenance and repair safety switch gen., 2 pole**	8	1 ELEC	Ea.							
	Resecure lugs				.500		26.50		26.50	33	41.50
	Total				.500		26.50		26.50	**33**	**41.50**
2020	**Maintenance and inspection safety switch, 2 pole**	1	1 ELEC	Ea.							
	Inspect & clean safety switch				.500		26.50		26.50	33	41.50
	Total				.500		26.50		26.50	**33**	**41.50**
2030	**Replace safety switch 240 V, 2 pole**	25	2 ELEC	Ea.							
	Remove safety switch				.650		34.50		34.50	43	53.50
	Fused safety switch, general duty, 240 V, 2 P, 30 A				2.299	104	123		227	266	320
	Fuse, dual element, time delay, 250 V, 30 A				.320	16.50	17.10		33.60	39.50	47
	Cut, form, align & connect wires				.151		8.05		8.05	9.95	12.45
	Check operation of switch				.018		.98		.98	1.21	1.51
	Total				3.439	120.50	183.63		304.13	**359.66**	**434.46**
3010	**Maintenance and repair safety switch gen., 3 pole**	8	1 ELEC	Ea.							
	Resecure lugs				.500		26.50		26.50	33	41.50
	Total				.500		26.50		26.50	**33**	**41.50**
3020	**Maintenance and inspection safety switch, 3 pole**	1	1 ELEC	Ea.							
	Inspect & clean safety switch				.500		26.50		26.50	33	41.50
	Total				.500		26.50		26.50	**33**	**41.50**
3030	**Replace safety switch 240 V, 3 pole**	25	2 ELEC	Ea.							
	Remove safety switch				.650		34.50		34.50	43	53.50
	Fused safety switch, general duty, 240 V, 3 P, 30 A				2.500	139	133		272	320	380
	Fuse, dual element, time delay, 250 V, 30 A				.480	24.75	25.65		50.40	59	70.50
	Cut, form, align & connect wires				.151		8.05		8.05	9.95	12.45
	Check operation of switch				.018		.98		.98	1.21	1.51
	Total				3.800	163.75	202.18		365.93	**433.16**	**517.96**
4010	**Replace low voltage cartridge**	50	1 ELEC	Ea.							
	Remove cartridge fuse cutouts				.100		5.35		5.35	6.60	8.25
	Cartridge fuse cutouts 3 P				.296	16	15.80		31.80	37	44.50
	Total				.396	16	21.15		37.15	**43.60**	**52.75**

D50 ELECTRICAL — D5023 Lighting & Branch Wiring

D5023 140 Safety Switch, General Duty

	System Description	Freq. (Years)	Crew	Unit	Labor Hours	2014 Bare Costs Material	2014 Bare Costs Labor	2014 Bare Costs Equipment	2014 Bare Costs Total	Total In-House	Total w/O&P
5010	**Replace plug fuse**	25	1 ELEC	Ea.							
	Remove plug fuse				.080		4.27		4.27	5.30	6.60
	Fuse, plug, 120 V, 15 to 30 A				.160	5.70	8.55		14.25	16.85	20.50
	Total				.240	5.70	12.82		18.52	22.15	27.10

D5023 150 Receptacles And Plugs

	System Description	Freq. (Years)	Crew	Unit	Labor Hours	2014 Bare Costs Material	2014 Bare Costs Labor	2014 Bare Costs Equipment	2014 Bare Costs Total	Total In-House	Total w/O&P
0010	**Maintenance and repair**	20	1 ELEC	Ea.							
	Repair receptacle component (cover)				.500	2.53	26.50		29.03	36	44.50
	Total				.500	2.53	26.50		29.03	36	44.50
0020	**Replace receptacle/plug**	20	1 ELEC	Ea.							
	Turn power off, later turn on				.100		5.35		5.35	6.60	8.25
	Remove & install cover plates				.170		9.10		9.10	11.25	14.05
	Remove receptacle				.040		2.13		2.13	2.64	3.30
	Wiring devices, receptacle				.381	8.35	20.50		28.85	34.50	42
	Cut, form, align & connect wires				.026		1.39		1.39	1.71	2.14
	Test for operation				.015		.78		.78	.96	1.21
	Total				.732	8.35	39.25		47.60	57.66	70.95

D5023 154 4-Pin Receptacle

	System Description	Freq. (Years)	Crew	Unit	Labor Hours	2014 Bare Costs Material	2014 Bare Costs Labor	2014 Bare Costs Equipment	2014 Bare Costs Total	Total In-House	Total w/O&P
0010	**Repair 4-pin receptacle cover**	10	1 ELEC	Ea.							
	Repair receptacle(cover)				.615	6.40	33		39.40	47.50	59
	Total				.615	6.40	33		39.40	47.50	59
0020	**Replace 4 - pin receptacle**	20	1 ELEC	Ea.							
	Remove 4-pin receptacle				.200		10.65		10.65	13.20	16.50
	Receptacle locking, 4-pin				.615	60.50	33		93.50	107	127
	Total				.815	60.50	43.65		104.15	120.20	143.50

D50 ELECTRICAL · D5023 Lighting & Branch Wiring

D5023 210 Contactors And Relays

	System Description	Freq. (Years)	Crew	Unit	Labor Hours	Material	Labor	Equipment	Total	Total In-House	Total w/O&P
0010	**Maintenance and repair**	3	1 ELEC	Ea.							
	Remove contactor coil				.602		32		32	39.50	49.50
	Contactor coil				2.000	32.50	107		139.50	168	206
	Total				2.602	32.50	139		171.50	207.50	**255.50**
0020	**Maintenance and inspection**	0.50	1 ELEC	Ea.							
	Inspect and clean contacts				.250		13.35		13.35	16.50	20.50
	Total				.250		13.35		13.35	16.50	**20.50**
0030	**Replace contactor**	18	1 ELEC	Ea.							
	Remove magnetic contactor				.800		42.50		42.50	53	66
	Magnetic contactor 600 V, 3 P, 25 H.P.				2.667	420	142		562	640	745
	Cut, form, align & connect wires				.151		8.05		8.05	9.95	12.45
	Check operation of contactor				.018		.98		.98	1.21	1.51
	Total				3.636	420	193.53		613.53	704.16	**824.96**

D5023 220 Wiring Devices, Switches

	System Description	Freq. (Years)	Crew	Unit	Labor Hours	Material	Labor	Equipment	Total	Total In-House	Total w/O&P
0010	**Maintenance and repair**	10	1 ELEC	Ea.							
	Repair switch component (cover)				.500	2.53	26.50		29.03	36	44.50
	Total				.500	2.53	26.50		29.03	36	**44.50**
0020	**Replace switch**	15	1 ELEC	Ea.							
	Turn power off, later turn on				.100		5.35		5.35	6.60	8.25
	Remove & install cover plates				.170		9.10		9.10	11.25	14.05
	Remove switch				.040		2.13		2.13	2.64	3.30
	Wiring devices, switch				.381	9.65	20.50		30.15	36	43.50
	Cut, form, align & connect wires				.026		1.39		1.39	1.71	2.14
	Test for operation				.015		.78		.78	.96	1.21
	Total				.732	9.65	39.25		48.90	59.16	**72.45**

D50 ELECTRICAL D5023 Lighting & Branch Wiring

D5023 222 Switch, Pull Cord

	System Description	Freq. (Years)	Crew	Unit	Labor Hours	2014 Bare Costs				Total In-House	Total w/O&P
						Material	Labor	Equipment	Total		
0010	**Maintenance and repair**		1 ELEC	Ea.							
	Repair switch component (cover)	5			.500	2.78	26.50		29.28	36	45
	Total				.500	2.78	26.50		29.28	**36**	**45**
0020	**Replace switch**		1 ELEC	Ea.							
	Turn power off, later turn on	15			.100		5.35		5.35	6.60	8.25
	Remove & install cover plates				.170		9.10		9.10	11.25	14.05
	Remove switch, pull cord				.040		2.13		2.13	2.64	3.30
	Switch, pull cord				.381	10.10	20.50		30.60	36.50	44
	Cut, form, align & connect wires				.026		1.39		1.39	1.71	2.14
	Test for operation				.015		.78		.78	.96	1.21
	Total				.732	10.10	39.25		49.35	**59.66**	**72.95**

D5023 230 Light Dimming Panel

	System Description	Freq. (Years)	Crew	Unit	Labor Hours	2014 Bare Costs				Total In-House	Total w/O&P
						Material	Labor	Equipment	Total		
0010	**Minor repairs to light dimming panel**		1 ELEC	Ea.							
	Check and test components	5			.400		21.50		21.50	26.50	33
	Remove dimming panel component				.222		11.85		11.85	14.65	18.35
	Dimming panel component				.667	84.50	35.50		120	137	161
	Total				1.289	84.50	68.85		153.35	**178.15**	**212.35**
0020	**Maintenance and inspection**		1 ELEC	Ea.							
	Check light dimming panel	1			.533		28.50		28.50	35	44
	Total				.533		28.50		28.50	**35**	**44**
0030	**Replace light dimming panel**		1 ELEC	Ea.							
	Turn power off, later turn on	15			.100		5.35		5.35	6.60	8.25
	Remove dimming panel				1.600		85.50		85.50	106	132
	Dimming panel				8.000	630	425		1,055	1,225	1,450
	Test circuits				.073		3.88		3.88	4.80	6
	Total				9.773	630	519.73		1,149.73	**1,342.40**	**1,596.25**

D50 ELECTRICAL   D5023   Lighting & Branch Wiring

D5023 240   Incandescent Lighting Fixtures

						2014 Bare Costs				Total	Total
	System Description	Freq. (Years)	Crew	Unit	Labor Hours	Material	Labor	Equipment	Total	In-House	w/O&P
0010	**Maintenance and repair**	10	1 ELEC	Ea.							
	Remove glass globe				.100		5.35		5.35	6.60	8.25
	Glass globe				.296	29	15.80		44.80	51.50	61
	Total				.396	29	21.15		50.15	58.10	69.25
0020	**Replace lamp**	5	1 ELEC	Ea.							
	Remove incand. lighting fixture lamp				.022		1.17		1.17	1.45	1.81
	Incand. lighting fixture lamp				.065	2.56	3.47		6.03	7.10	8.55
	Total				.087	2.56	4.64		7.20	8.55	10.36
0030	**Replace lighting fixture**	20	1 ELEC	Ea.							
	Turn circuit off and on for fixture				.018		.98		.98	1.21	1.51
	Remove incand. lighting fixture				.258		13.75		13.75	17.05	21.50
	Incand. lighting fixture, 75 W				.800	55.50	42.50		98	114	135
	Total				1.076	55.50	57.23		112.73	132.26	158.01
0120	**Replace lamp, 200 W**	5	1 ELEC	Ea.							
	Remove incand. lighting fixture lamp				.022		1.19		1.19	1.47	1.83
	Incand. lighting fixture lamp, 200 W				.065	3.63	3.47		7.10	8.30	9.90
	Total				.087	3.63	4.66		8.29	9.77	11.73
0220	**Replace lamp for explosion proof fixture**	5	1 ELEC	Ea.							
	Remove incand. lighting fixture lamp				.044		2.38		2.38	2.94	3.66
	Incand. lighting fixture lamp for explosion proof fixture				.130	7.26	6.94		14.20	16.60	19.80
	Total				.175	7.26	9.32		16.58	19.54	23.46
0230	**Replace incan. lighting fixt., explosion proof, ceiling mtd, 200 W**	20	1 ELEC	Ea.							
	Turn branch circuit off and on				.018		.98		.98	1.21	1.51
	Remove incandescent lighting fixture, explosion proof				.920		49		49	60.50	76
	Incan. lighting fixture, explosion proof, ceiling mtd, 200 W				2.000	1,075	107		1,182	1,325	1,500
	Total				2.938	1,075	156.98		1,231.98	1,386.71	1,577.51
1230	**Replace incan. lighting fixt., high hat can type, ceiling mtd., 200 W**	20	1 ELEC	Ea.							
	Turn branch circuit off and on				.018		.98		.98	1.21	1.51
	Remove incandescent lighting fixture, can type				.258		13.75		13.75	17.05	21.50
	Incan. lighting fixture, high hat can type, ceiling mtd, 200 W				1.194	96	63.50		159.50	184	219
	Total				1.470	96	78.23		174.23	202.26	242.01

D50 ELECTRICAL   D5023 Lighting & Branch Wiring

D5023 250   Quartz Fixture

	System Description	Freq. (Years)	Crew	Unit	Labor Hours	2014 Bare Costs				Total In-House	Total w/O&P
						Material	Labor	Equipment	Total		
0010	**Maintenance and repair**										
	Repair glass lens	10	1 ELEC	Ea.	.400		21.50		52.50	60.50	72
	Total				.400		21.50		52.50	**60.50**	**72**
0020	**Replace 1500 W quartz lamp**										
	Remove quartz lamp	10	1 ELEC	Ea.	.087		4.64		4.64	5.75	7.20
	Quartz lamp				.258	39.50	13.75		53.25	60.50	71
	Total				.345	39.50	18.39		57.89	**66.25**	**78.20**
0030	**Replace fixture**										
	Turn branch circuit off and on	20	1 ELEC	Ea.	.018		.98		.98	1.21	1.51
	Remove exterior fixt., quartz, 1500 W				.593		31.50		31.50	39	49
	Exterior fixt. quartz, 1500 W				1.905	102	102		204	238	285
	Total				2.516	102	134.48		236.48	**278.21**	**335.51**

D5023 260   Fluorescent Lighting Fixture

	System Description	Freq. (Years)	Crew	Unit	Labor Hours	2014 Bare Costs				Total In-House	Total w/O&P
						Material	Labor	Equipment	Total		
0010	**Replace fluor. ballast**										
	Remove indoor fluor. ballast	10	1 ELEC	Ea.	.333		17.80		17.80	22	27.50
	Fluorescent, electronic ballast				.667	37	35.50		72.50	84.50	101
	Test fixture				.018		.98		.98	1.21	1.51
	Total				1.018	37	54.28		91.28	**107.71**	**130.01**
0020	**Replace lamps (2 lamps), 4', 34 W energy saver**										
	Remove fluor. lamps in fixture	10	1 ELEC	Ea.	.078		4.16		4.16	5.15	6.45
	Fluor. lamp, 4' energy saver				.232	9.38	12.40		21.78	25.50	31
	Total				.310	9.38	16.56		25.94	**30.65**	**37.45**
0030	**Replace fixture, lay-in, recess mtd, 2' x 4', two 40 W**										
	Turn branch circuit off and on	20	1 ELEC	Ea.	.018		.98		.98	1.21	1.51
	Remove fluor. lighting fixture				.485		26		26	32	40
	Fluor. 2' x 4', recess mounted, two 40 W				1.509	50	80.50		130.50	155	188
	Total				2.013	50	107.48		157.48	**188.21**	**229.51**

D50 ELECTRICAL D5023 Lighting & Branch Wiring

D5023 260 Fluorescent Lighting Fixture

	System Description	Freq. (Years)	Crew	Unit	Labor Hours	2014 Bare Costs				Total In-House	Total w/O&P
						Material	Labor	Equipment	Total		
0040	**Replace fixture, lay-in, recess mtd., 2' x 4', four 32 W**	20	1 ELEC	Ea.							
	Turn branch circuit off and on				.018		.98		.98	1.21	1.51
	Remove fluor. lighting fixture				.533		28.50		28.50	35	44
	Fluor, lay-in, recess mtd., 2' x 4', four 32 W				1.702	64.50	91		155.50	183	222
	Total				2.254	64.50	120.48		184.98	**219.21**	**267.51**
0120	**Replace lamps (2 lamps), 8', 60 W energy saver**	10	1 ELEC	Ea.							
	Remove fluor. lamps in fixture				.078		4.16		4.16	5.15	6.45
	Fluor. lamps, 8' energy saver				.200	9	10.70		19.70	23	28
	Total				.278	9	14.86		23.86	**28.15**	**34.45**
0130	**Replace fixture, strip, surface mtd., 8', two 75 W**	20	2 ELEC	Ea.							
	Turn branch circuit off and on				.018		.98		.98	1.21	1.51
	Remove fluor. lighting fixture				.400		21.50		21.50	26.50	33
	Fluor. strip, surface mounted, 8', two 75 W				1.290	57.50	69		126.50	148	179
	Total				1.709	57.50	91.48		148.98	**175.71**	**213.51**
0140	**Replace fixture, strip, pendent mtd, 8', two 75 W**	20	2 ELEC	Ea.							
	Turn branch circuit off and on				.018		.98		.98	1.21	1.51
	Remove fluor. lighting fixture				.593		31.50		31.50	39	49
	Fluor, strip, pendent mtd, 8', two 75 W				1.818	97	97		194	227	271
	Total				2.429	97	129.48		226.48	**267.21**	**321.51**

D5023 270 Metal Halide Fixture

	System Description	Freq. (Years)	Crew	Unit	Labor Hours	2014 Bare Costs				Total In-House	Total w/O&P
						Material	Labor	Equipment	Total		
2010	**Replace M.H. ballast, 175 W**	10	1 ELEC	Ea.							
	Remove metal halide fixt. ballast				.333		17.80		17.80	22	27.50
	Metal halide fixt. 175 W, ballast				.667	98.50	35.50		134	152	178
	Test fixture				.018		.98		.98	1.21	1.51
	Total				1.018	98.50	54.28		152.78	**175.21**	**207.01**
2020	**Replace lamp, 175 W**	5	1 ELEC	Ea.							
	Remove metal halide fixt. lamp				.116		6.20		6.20	7.65	9.55
	Metal halide lamp, 175 W				.348	33	18.55		51.55	59.50	70
	Total				.464	33	24.75		57.75	**67.15**	**79.55**

D50 ELECTRICAL   D5023   Lighting & Branch Wiring

D5023 270   Metal Halide Fixture

	Freq. (Years)	Crew	Unit	Labor Hours	2014 Bare Costs				Total In-House	Total w/O&P
System Description					Material	Labor	Equipment	Total		
2030	20	1 ELEC	Ea.							
Replace fixture, 175 W										
Turn branch circuit off and on				.018		.98		.98	1.21	1.51
Remove metal halide fixt.				.800		42.50		42.50	53	66
Metal halide, recessed, square 175 W				2.353	385	126		511	580	675
Total				3.171	385	169.48		554.48	634.21	742.51
2210	10	1 ELEC	Ea.							
Replace M.H. ballast, 400 W										
Remove metal halide fixt. ballast				.333		17.80		17.80	22	27.50
Metal halide fixt. 400 W, ballast				.762	159	40.50		199.50	225	262
Test fixture				.018		.98		.98	1.21	1.51
Total				1.114	159	59.28		218.28	248.21	291.01
2220	5	1 ELEC	Ea.							
Replace lamp, 400 W										
Remove metal halide fixt. lamp				.118		6.30		6.30	7.75	9.70
Metal halide lamp, 400 W				.348	61.50	18.55		80.05	90.50	105
Total				.465	61.50	24.85		86.35	98.25	114.70
2230	20	2 ELEC	Ea.							
Replace fixture, 400 W										
Turn branch circuit off and on				.018		.98		.98	1.21	1.51
Remove metal halide fixt.				1.067		57		57	70.50	88
Metal halide, surface mtd, high bay, 400 W				3.478	405	186		591	675	795
Total				4.563	405	243.98		648.98	746.71	884.51
2310	10	1 ELEC	Ea.							
Replace M.H. ballast, 1000 W										
Remove metal halide fixt. ballast				.333		17.80		17.80	22	27.50
Metal halide fixt. 1000 W, ballast				.889	272	47.50		319.50	360	415
Test fixture				.018		.98		.98	1.21	1.51
Total				1.241	272	66.28		338.28	383.21	444.01
2320	5	1 ELEC	Ea.							
Replace lamp, 1000 W										
Remove metal halide fixt. lamp				.118		6.30		6.30	7.75	9.70
Metal halide lamp, 1000 W				.523	85.50	28		113.50	129	150
Total				.641	85.50	34.30		119.80	136.75	159.70

D50 ELECTRICAL D5023 Lighting & Branch Wiring

D5023 270 Metal Halide Fixture

	System Description	Freq. (Years)	Crew	Unit	Labor Hours	2014 Bare Costs				Total In-House	Total w/O&P
						Material	Labor	Equipment	Total		
2330	**Replace fixture, 1000 W**	20	2 ELEC	Ea.							
	Turn branch circuit off and on				.018		.98		.98	1.21	1.51
	Remove metal halide fixt.				1.333		71		71	88	110
	Metal halide, surface mtd, high bay, 1000 W				4.000	585	213		798	910	1,050
	Total				5.352	585	284.98		869.98	999.21	1,161.51

D5023 280 H.P. Sodium Fixture, 250 W

	System Description	Freq. (Years)	Crew	Unit	Labor Hours	2014 Bare Costs				Total In-House	Total w/O&P
						Material	Labor	Equipment	Total		
1010	**Replace H.P. sodium ballast**	10	2 ELEC	Ea.							
	Remove H.P. Sodium fixt. ballast				.333		17.80		17.80	22	27.50
	H.P. Sodium fixt., 250 W, ballast				.667	268	35.50		303.50	340	390
	Test fixture				.018		.98		.98	1.21	1.51
	Total				1.018	268	54.28		322.28	363.21	419.01
1020	**Replace lamp**	10	2 ELEC	Ea.							
	Remove H.P. Sodium fixt. lamp				.116		6.20		6.20	7.65	9.55
	H.P. Sodium fixt. lamp, 250 W				.348	64	18.55		82.55	93.50	109
	Total				.464	64	24.75		88.75	101.15	118.55
1030	**Replace fixture**	20	2 ELEC	Ea.							
	Turn branch circuit off and on				.018		.98		.98	1.21	1.51
	Remove H.P. Sodium fixt.				.889		47.50		47.50	58.50	73.50
	H.P. Sodium fixt, recessed, square 250 W				2.667	670	142		812	915	1,050
	Total				3.574	670	190.48		860.48	974.71	1,125.01

D50 ELECTRICAL

D5033 Communications and Security

D5033 310 Telephone Cable

		System Description	Freq. (Years)	Crew	Unit	Labor Hours	2014 Bare Costs				Total In-House	Total w/O&P
							Material	Labor	Equipment	Total		
0010		**Repair cable, 22 A.W.G., 4 pair**	8	1 ELEC	M.L.F.							
		Cut telephone cable				.080		4.27		4.27	5.30	6.60
		Splice and insulate telephone cable				.471	7.95	25		32.95	40	49
		Total				.551	7.95	29.27		37.22	**45.30**	**55.60**
0020		**Replace telephone cable, #22-4 conductor**	50	2 ELEC	M.L.F.							
		Turn power off, later turn on				.100		5.35		5.35	6.60	8.25
		Remove & reinstall cover plates				1.290		69		69	85	107
		Remove telephone cable				2.581		138		138	170	213
		Telephone twisted, #22-4 conductor				10.000	186.50	535		721.50	865	1,050
		Total				13.971	186.50	747.35		933.85	**1,126.60**	**1,378.25**
1020		**Replace telephone jack**	20	1 ELEC	Ea.							
		Remove telephone jack				.059		3.16		3.16	3.91	4.89
		Telephone jack				.250	4.84	13.35		18.19	22	26.50
		Total				.309	4.84	16.51		21.35	**25.91**	**31.39**

D5033 410 Master Clock Control

		System Description	Freq. (Years)	Crew	Unit	Labor Hours	2014 Bare Costs				Total In-House	Total w/O&P
							Material	Labor	Equipment	Total		
0010		**Maintenance and repair**	10	1 ELEC	Ea.							
		Remove time control clock motor				.296		15.80		15.80	19.55	24.50
		Time control clock motor				1.000	66	53.50		119.50	139	165
		Total				1.296	66	69.30		135.30	**158.55**	**189.50**
0020		**Check operation**	1	1 ELEC	Ea.							
		Check time control clock operation				.533		28.50		28.50	35	44
		Total				.533		28.50		28.50	**35**	**44**
0030		**Replace time control clock**	15	1 ELEC	Ea.							
		Remove time control clock				.296		15.80		15.80	19.55	24.50
		Time control clock				1.333	128	71		199	229	270
		Total				1.630	128	86.80		214.80	**248.55**	**294.50**

D50 ELECTRICAL D5033 Communications and Security

D5033 410 Master Clock Control

	System Description	Freq. (Years)	Crew	Unit	Labor Hours	2014 Bare Costs				Total In-House	Total w/O&P
						Material	Labor	Equipment	Total		
1120	**Maintenance and inspection**	1	2 ELEC	Ea.							
	Check operation of program bell				.320		17.05		17.05	21	26.50
	Total				.320		17.05		17.05	**21**	**26.50**
1130	**Replace program bell**	15	2 ELEC	Ea.							
	Remove program bell				.667		35.50		35.50	44	55
	Program bell				1.000	86	53.50		139.50	161	190
	Total				1.667	86	89		175	**205**	**245**

D5033 510 TV Cable Outlet

	System Description	Freq. (Years)	Crew	Unit	Labor Hours	2014 Bare Costs				Total In-House	Total w/O&P
						Material	Labor	Equipment	Total		
0010	**Maintenance and repair**	10	1 ELEC	Ea.							
	Repair TV cable outlet (cover)				.615	2.78	33		35.78	43.50	54.50
	Total				.615	2.78	33		35.78	**43.50**	**54.50**
0020	**Replace TV cable outlet**	20	1 ELEC	Ea.							
	Remove TV cable outlet				.200		10.65		10.65	13.20	16.50
	TV cable outlet				.615	6.55	33		39.55	48	59
	Total				.815	6.55	43.65		50.20	**61.20**	**75.50**

D5033 610 Door Bell

	System Description	Freq. (Years)	Crew	Unit	Labor Hours	2014 Bare Costs				Total In-House	Total w/O&P
						Material	Labor	Equipment	Total		
0010	**Maintenance and repair**	10	1 ELEC	Ea.							
	Remove door bell transformer				.200		10.65		10.65	13.20	16.50
	Door bell transformer				.667	21	35.50		56.50	67	81.50
	Total				.867	21	46.15		67.15	**80.20**	**98**
0020	**Maintenance and inspection**	1	1 ELEC	Ea.							
	Check operation of door bell				.320		17.05		17.05	21	26.50
	Total				.320		17.05		17.05	**21**	**26.50**

D50 ELECTRICAL | D5033 Communications and Security

D5033 610 Door Bell

	System Description	Freq. (Years)	Crew	Unit	Labor Hours	2014 Bare Costs				Total In-House	Total w/O&P
						Material	Labor	Equipment	Total		
0030	**Replace door bell**	15	1 ELEC	Ea.							
	Remove door bell system				.667		35.50		35.50	44	55
	Door bell system				2.667	113	142		255	300	360
	Total				3.333	113	177.50		290.50	344	415

D5033 620 Sound System Components

	System Description	Freq. (Years)	Crew	Unit	Labor Hours	2014 Bare Costs				Total In-House	Total w/O&P
						Material	Labor	Equipment	Total		
0030	**Replace speaker**	20	1 ELEC	Ea.							
	Turn power off, later turn on				.100		5.35		5.35	6.60	8.25
	Remove speaker				.333		17.80		17.80	22	27.50
	Speaker				1.000	116	53.50		169.50	194	228
	Test for operation				.015		.78		.78	.96	1.21
	Total				1.448	116	77.43		193.43	223.56	264.96
0120	**Inspect monitor panel**	0.50	1 ELEC	Ea.							
	Check monitor panel				.533		28.50		28.50	35	44
	Total				.533		28.50		28.50	35	44
0130	**Replace monitor panel**	15	1 ELEC	Ea.							
	Turn power off, later turn on				.100		5.35		5.35	6.60	8.25
	Remove monitor panel				1.667		89		89	110	138
	Monitor panel				2.000	385	107		492	555	645
	Test circuits				.073		3.88		3.88	4.80	6
	Total				3.839	385	205.23		590.23	676.40	797.25
0220	**Inspect volume control**	1	1 ELEC	Ea.							
	Check operation of volume control				.320		17.05		17.05	21	26.50
	Total				.320		17.05		17.05	21	26.50
0230	**Replace volume control**	15	2 ELEC	Ea.							
	Remove volume control				.333		17.80		17.80	22	27.50
	Volume control				1.000	87.50	53.50		141	162	192
	Total				1.333	87.50	71.30		158.80	184	219.50

D50 ELECTRICAL D5033 Communications and Security

D5033 620 Sound System Components

	System Description	Freq. (Years)	Crew	Unit	Labor Hours	2014 Bare Costs				Total In-House	Total w/O&P
						Material	Labor	Equipment	Total		
0330	**Replace amplifier**	15	2 ELEC	Ea.							
	Turn power off, later turn on				.100		5.35		5.35	6.60	8.25
	Remove amplifier				1.667		89		89	110	138
	Amplifier				8.000	1,250	425		1,675	1,900	2,225
	Test for operation				.015		.78		.78	.96	1.21
	Total				9.781	1,250	520.13		1,770.13	2,017.56	2,372.46

D5033 630 Intercom System Components

	System Description	Freq. (Years)	Crew	Unit	Labor Hours	2014 Bare Costs				Total In-House	Total w/O&P
						Material	Labor	Equipment	Total		
0120	**Inspect intercom master station**	0.50	1 ELEC	Ea.							
	Check intercom master station				2.133		114		114	140	176
	Total				2.133		114		114	140	176
0130	**Replace intercom master station**	15	2 ELEC	Ea.							
	Turn power off, later turn on				.100		5.35		5.35	6.60	8.25
	Remove intercom master station				1.333		71		71	88	110
	Intercom master station				8.000	2,300	425		2,725	3,050	3,525
	Test circuits				.073		3.88		3.88	4.80	6
	Total				9.506	2,300	505.23		2,805.23	3,149.40	3,649.25
0220	**Inspect intercom remote station**	1	1 ELEC	Ea.							
	Check operation of intercom remote station				.320		17.05		17.05	21	26.50
	Total				.320		17.05		17.05	21	26.50
0230	**Replace intercom remote station**	15	1 ELEC	Ea.							
	Remove intercom remote station				.333		17.80		17.80	22	27.50
	Intercom remote station				1.000	162	53.50		215.50	244	285
	Total				1.333	162	71.30		233.30	266	312.50

D50 ELECTRICAL D5033 Communications and Security

D5033 640 Security System Components

	System Description	Freq. (Years)	Crew	Unit	Labor Hours	2014 Bare Costs Material	2014 Bare Costs Labor	2014 Bare Costs Equipment	2014 Bare Costs Total	Total In-House	Total w/O&P
0020	**Inspect camera and monitor**	0.50	1 ELEC	Ea.							
	Check camera and monitor				.808		43.18		43.18	53	66.50
	Total				.808		43.18		43.18	**53**	**66.50**
0030	**Replace camera and monitor**	12	2 ELEC	Ea.							
	Turn power off, later turn on				.100		5.35		5.35	6.60	8.25
	Remove closed circuit TV, one camera & one monitor				2.051		109		109	135	169
	Closed circuit TV, one camera & one monitor				6.154	1,350	330		1,680	1,900	2,200
	Test for operation				.015		.78		.78	.96	1.21
	Total				8.320	1,350	445.13		1,795.13	**2,042.56**	**2,378.46**
0120	**Inspect camera**	0.50	1 ELEC	Ea.							
	Check camera				.533		28.50		28.50	35	44
	Total				.533		28.50		28.50	**35**	**44**
0130	**Replace camera**	12	1 ELEC	Ea.							
	Turn power off, later turn on				.100		5.35		5.35	6.60	8.25
	Remove closed circuit TV, one camera				1.000		53.50		53.50	66	82.50
	Closed circuit TV, one camera				2.963	750	158		908	1,025	1,175
	Test for operation				.015		.78		.78	.96	1.21
	Total				4.078	750	217.63		967.63	**1,098.56**	**1,266.96**

D5033 710 Smoke Detector

	System Description	Freq. (Years)	Crew	Unit	Labor Hours	2014 Bare Costs Material	2014 Bare Costs Labor	2014 Bare Costs Equipment	2014 Bare Costs Total	Total In-House	Total w/O&P
0010	**Repair smoke detector**	10	1 ELEC	Ea.							
	Remove smoke detector lamp				.151		8.05		8.05	9.95	12.45
	Smoke detector lamp				.444	3.83	23.50		27.33	33.50	41.50
	Total				.595	3.83	31.55		35.38	**43.45**	**53.95**
0020	**Check operation**	1	1 ELEC	Ea.							
	Check smoke detector operation				.200		10.65		10.65	13.20	16.50
	Total				.200		10.65		10.65	**13.20**	**16.50**

D50 ELECTRICAL D5033 Communications and Security

D5033 710 Smoke Detector

	System Description	Freq. (Years)	Crew	Unit	Labor Hours	2014 Bare Costs				Total In-House	Total w/O&P
						Material	Labor	Equipment	Total		
0030	**Replace smoke detector**	15	1 ELEC	Ea.							
	Turn power off, later turn on				.100		5.35		5.35	6.60	8.25
	Remove smoke detector				.348		18.55		18.55	23	28.50
	Smoke detector, ceiling type				1.290	109	69		178	205	243
	Test for operation				.015		.78		.78	.96	1.21
	Total				1.753	109	93.68		202.68	235.56	280.96

D5033 712 Heat Detector

	System Description	Freq. (Years)	Crew	Unit	Labor Hours	2014 Bare Costs				Total In-House	Total w/O&P
						Material	Labor	Equipment	Total		
0010	**Repair heat detector**	10	1 ELEC	Ea.							
	Remove heat detector element				.151		8.05		8.05	9.95	12.45
	Heat detector element				.444	8.90	23.50		32.40	39	47.50
	Total				.595	8.90	31.55		40.45	48.95	59.95
0020	**Check operation**	1	1 ELEC	Ea.							
	Check heat detector operation				.200		10.65		10.65	13.20	16.50
	Total				.200		10.65		10.65	13.20	16.50
0030	**Replace heat detector**	15	1 ELEC	Ea.							
	Turn power off, later turn on				.100		5.35		5.35	6.60	8.25
	Remove heat detector				.348		18.55		18.55	23	28.50
	Heat detector				1.000	50.50	53.50		104	122	146
	Test for operation				.015		.78		.78	.96	1.21
	Total				1.462	50.50	78.18		128.68	152.56	183.96

D5033 720 Manual Pull Station

	System Description	Freq. (Years)	Crew	Unit	Labor Hours	2014 Bare Costs				Total In-House	Total w/O&P
						Material	Labor	Equipment	Total		
0010	**Check and repair manual pull station**	10	1 ELEC	Ea.							
	Remove pull station component				.200		10.65		10.65	13.20	16.50
	Manual pull station component				.667	17.15	35.50		52.65	63	76.50
	Total				.867	17.15	46.15		63.30	76.20	93

D50 ELECTRICAL | D5033 Communications and Security

D5033 720 Manual Pull Station

		Freq. (Years)	Crew	Unit	Labor Hours	2014 Bare Costs				Total In-House	Total w/O&P
	System Description					Material	Labor	Equipment	Total		
0020	**Replace manual pull station**	15	1 ELEC	Ea.							
	Turn power off, later turn on				.100		5.35		5.35	6.60	8.25
	Remove manual pull station				.348		18.55		18.55	23	28.50
	Manual pull station				1.000	82.50	53.50		136	157	186
	Test for operation				.015		.78		.78	.96	1.21
	Total				1.462	82.50	78.18		160.68	187.56	223.96

D5033 760 Fire Alarm Control Panel

		Freq. (Years)	Crew	Unit	Labor Hours	2014 Bare Costs				Total In-House	Total w/O&P
	System Description					Material	Labor	Equipment	Total		
0010	**Minor repairs to fire alarm**	5	1 ELEC	Ea.							
	Check and test components				.400		21.50		21.50	26.50	33
	Remove relay				.222		11.85		11.85	14.65	18.35
	Relay				.667	41	35.50		76.50	89	106
	Total				1.289	41	68.85		109.85	130.15	157.35
0020	**Maintenance and inspection**	0.50	1 ELEC	Ea.							
	Check control panel				.533		28.50		28.50	35	44
	Total				.533		28.50		28.50	35	44
0030	**Replace fire alarm panel**	15	1 ELEC	Ea.							
	Turn power off, later turn on				.100		5.35		5.35	6.60	8.25
	Remove fire alarm control panel				2.000		107		107	132	165
	Fire alarm control panel				16.000	775	855		1,630	1,900	2,300
	Test circuits				.073		3.88		3.88	4.80	6
	Total				18.173	775	971.23		1,746.23	2,043.40	2,479.25

D50 ELECTRICAL D5033 Communications and Security

D5033 766 Annunciation Panel

						2014 Bare Costs				Total	Total
	System Description	Freq. (Years)	Crew	Unit	Labor Hours	Material	Labor	Equipment	Total	In-House	w/O&P
0010	**Minor repairs to annunciation panel**	5	1 ELEC	Ea.							
	Check and test components				.400		21.50		21.50	26.50	33
	Remove relay				.222		11.85		11.85	14.65	18.35
	Relay				.667	41	35.50		76.50	89	106
	Total				1.289	41	68.85		109.85	130.15	157.35
0020	**Maintenance and inspection**	0.50	1 ELEC	Ea.							
	Check annunciator panel				.533		28.50		28.50	35	44
	Total				.533		28.50		28.50	35	44
0030	**Replace annunciation panel**	15	1 ELEC	Ea.							
	Turn power off, later turn on				.100		5.35		5.35	6.60	8.25
	Remove annunciation panel				1.600		85.50		85.50	106	132
	Annunciation panel				6.154	330	330		660	770	925
	Test circuits				.073		3.88		3.88	4.80	6
	Total				7.927	330	424.73		754.73	887.40	1,071.25

D5033 770 Fire Alarm Bell

						2014 Bare Costs				Total	Total
	System Description	Freq. (Years)	Crew	Unit	Labor Hours	Material	Labor	Equipment	Total	In-House	w/O&P
0010	**Replace fire alarm bell, 6"**	20	1 ELEC	Ea.							
	Turn power off, later turn on				.100		5.35		5.35	6.60	8.25
	Remove fire alarm bell				.348		18.55		18.55	23	28.50
	Fire alarm bell				1.000	78	53.50		131.50	152	180
	Test for operation				.015		.78		.78	.96	1.21
	Total				1.462	78	78.18		156.18	182.56	217.96

D50 ELECTRICAL D5093 Other Electrical Systems

D5093 110 Electrical Service Ground

	System Description	Freq. (Years)	Crew	Unit	Labor Hours	2014 Bare Costs				Total In-House	Total w/O&P
						Material	Labor	Equipment	Total		
0010	**Maintenance and repair**	25	1 ELEC	M.L.F.							
	Repair service ground (1 termination)				1.053	5.35	56		61.35	75.50	93.50
	Total				1.053	5.35	56		61.35	75.50	93.50
0020	**Replace electrical service ground**	50	1 ELEC	M.L.F.							
	Remove ground wire #6				2.000		106.50		106.50	132	165
	Excavate hole for ground rod				2.222		118.50		118.50	147	184
	Ground rod 5/8", 10'				17.391	245	930		1,175	1,425	1,750
	Ground clamp				2.500	53.50	133.50		187	225	272
	Backfill over top of ground rod				1.053		56		56	69.50	87
	Insulated copper wire #6				12.308	650	655		1,305	1,525	1,825
	Total				37.474	948.50	1,999.50		2,948	3,523.50	4,283

D5093 120 Building Structure Ground

	System Description	Freq. (Years)	Crew	Unit	Labor Hours	2014 Bare Costs				Total In-House	Total w/O&P
						Material	Labor	Equipment	Total		
0010	**Maintenance and repair**	7	1 ELEC	M.L.F.							
	Repair bldg structure ground (1 termination)				1.053	5.35	56		61.35	75.50	93.50
	Total				1.053	5.35	56		61.35	75.50	93.50
0020	**Replace building structure ground**	50	1 ELEC	M.L.F.							
	Remove ground wire #4				3.077		164		164	205	255
	Excavate hole for ground rod				2.222		118.50		118.50	147	184
	Ground rod 5/8", 10'				17.391	245	930		1,175	1,425	1,750
	Ground clamp				2.500	53.50	133.50		187	225	272
	Backfill over top of ground rod				1.053		56		56	69.50	87
	Insulated copper wire #4				15.094	1,030	805		1,835	2,125	2,550
	Total				41.337	1,328.50	2,207		3,535.50	4,196.50	5,098

D50 ELECTRICAL   D5093 Other Electrical Systems

D5093 130 Lightning Protection System

	System Description	Freq. (Years)	Crew	Unit	Labor Hours	2014 Bare Costs				Total In-House	Total w/O&P
						Material	Labor	Equipment	Total		
1010	**Maintenance and repair of general wiring**	1	1 ELEC	M.L.F.							
	Repair lightning protection (1 termination)				1.053	13.95	56		69.95	85	104
	Total				1.053	13.95	56		69.95	85	104
1020	**Replace lightning protection general wiring**	25	2 ELEC	M.L.F.							
	Remove lightning protection cable				4.000		213		213	264	330
	Excavate hole for ground rod				2.222		118.50		118.50	147	184
	Ground rod 5/8", 10'				17.391	245	930		1,175	1,425	1,750
	Ground clamp				2.500	53.50	133.50		187	225	272
	Backfill over top of ground rod				1.053		56		56	69.50	87
	Lightning protection cable				25.000	3,000	1,330		4,330	4,950	5,800
	Air terminals, copper 3/8" x 10"				10.000	245	535		780	930	1,125
	Total				62.166	3,543.50	3,316		6,859.50	8,010.50	9,548

D5093 140 Lightning Ground Rod

	System Description	Freq. (Years)	Crew	Unit	Labor Hours	2014 Bare Costs				Total In-House	Total w/O&P
						Material	Labor	Equipment	Total		
1010	**Maintenance and repair**	1	1 ELEC	Ea.							
	Repair ground rod connector				1.053	5.35	56		61.35	75.50	93.50
	Total				1.053	5.35	56		61.35	75.50	93.50
1020	**Replace lightning ground rod**	25	2 ELEC	Ea.							
	Excavate hole for ground rod				.333		17.80		17.80	22	27.50
	Ground rod 5/8", 10'				1.739	24.50	93		117.50	142	175
	Ground clamp				.250	5.35	13.35		18.70	22.50	27
	Backfill over top of ground rod				.157		8.35		8.35	10.35	12.95
	Total				2.479	29.85	132.50		162.35	196.85	242.45

D5093 150 Computer Ground System

	System Description	Freq. (Years)	Crew	Unit	Labor Hours	2014 Bare Costs				Total In-House	Total w/O&P
						Material	Labor	Equipment	Total		
0010	**Maintenance and repair**	4	1 ELEC	Ea.							
	Test bonding each connection				.267		14.25		14.25	17.60	22
	Total				.267		14.25		14.25	17.60	22

D50 ELECTRICAL D5093 Other Electrical Systems

D5093 150 Computer Ground System

	System Description	Freq. (Years)	Crew	Unit	Labor Hours	2014 Bare Costs				Total In-House	Total w/O&P
						Material	Labor	Equipment	Total		
0020	Replace computer ground system	50	1 ELEC	M.L.F.							
	Remove ground wire #6				2.000		106.50		106.50	132	165
	Insulated ground wire #6				12.308	650	655		1,305	1,525	1,825
	Grounding connection				2.500	53.50	133.50		187	225	272
	Total				16.808	703.50	895		1,598.50	1,882	2,262

D5093 190 Special Ground System

	System Description	Freq. (Years)	Crew	Unit	Labor Hours	2014 Bare Costs				Total In-House	Total w/O&P
						Material	Labor	Equipment	Total		
0010	Maintenance and repair	4	1 ELEC	Ea.							
	Test bonding each connection				.267		14.25		14.25	17.60	22
	Total				.267		14.25		14.25	17.60	22
0020	Replace special ground system	50	2 ELEC	M.L.F.							
	Remove ground wire #6				2.000		106.50		106.50	132	165
	Insulated ground wire #6				12.308	650	655		1,305	1,525	1,825
	Grounding connection				2.500	53.50	133.50		187	225	272
	Total				16.808	703.50	895		1,598.50	1,882	2,262

D5093 210 Generator, Gasoline, 175 KW

	System Description	Freq. (Years)	Crew	Unit	Labor Hours	2014 Bare Costs				Total In-House	Total w/O&P
						Material	Labor	Equipment	Total		
0010	Maintenance and inspection	0.08	1 ELEC	Ea.							
	Run test gasoline generator				.800		42.50		42.50	53	66
	Total				.800		42.50		42.50	53	66
0020	Replace generator component	25	2 ELEC	Ea.							
	Remove gasoline generator				36.364		1,950		1,950	2,400	3,000
	Gasoline generator, 175 KW				106.000	70,500	5,700		76,200	84,500	97,000
	Total				142.364	70,500	7,650		78,150	86,900	100,000

D50 ELECTRICAL   D5093 Other Electrical Systems

D5093 220 Generator, Diesel, 750 KW

	System Description	Freq. (Years)	Crew	Unit	Labor Hours	2014 Bare Costs				Total In-House	Total w/O&P
						Material	Labor	Equipment	Total		
0010	**Maintenance and inspection**										
	Run test diesel generator	0.08	1 ELEC	Ea.	.800		42.50		42.50	53	66
	Total				.800		42.50		42.50	**53**	**66**
0020	**Replace diesel generator component**										
	Remove diesel generator	25	4 ELEC	Ea.	57.143		3,050		3,050	3,775	4,725
	Diesel generator, 750 KW				160.000	138,500	8,525		147,025	163,000	186,500
	Total				217.143	138,500	11,575		150,075	166,775	191,225

D5093 230 Transfer Switch

	System Description	Freq. (Years)	Crew	Unit	Labor Hours	2014 Bare Costs				Total In-House	Total w/O&P
						Material	Labor	Equipment	Total		
0010	**Maintenance and repair**										
	Remove transfer switch coil	5	1 ELEC	Ea.	.667		35.50		35.50	44	55
	Transfer switch coil				2.000	97	107		204	239	286
	Total				2.667	97	142.50		239.50	**283**	**341**
0020	**Maintenance and inspection**										
	Inspect and clean transfer switch	0.50	1 ELEC	Ea.	.500		26.50		26.50	33	41.50
	Total				.500		26.50		26.50	**33**	**41.50**
0030	**Replace transfer switch**										
	Remove transfer switch	18	1 ELEC	Ea.	.667		35.50		35.50	44	55
	Transfer switch, 480 V, 3 P, 30 A				3.478	2,625	186		2,811	3,125	3,575
	Cut, form, align & connect wires				.151		8.05		8.05	9.95	12.45
	Check operation of switch				.018		.98		.98	1.21	1.51
	Total				4.314	2,625	230.53		2,855.53	3,180.16	3,643.96

D50 ELECTRICAL D5093 Other Electrical Systems

D5093 240 Emergency Lighting Fixture

	System Description	Freq. (Years)	Crew	Unit	Labor Hours	2014 Bare Costs				Total In-House	Total w/O&P
						Material	Labor	Equipment	Total		
0020	**Replace lamp**	2	1 ELEC	Ea.							
	Remove emergency remote lamp				.100		5.35		5.35	6.60	8.25
	Emergency remote lamp				.296	31.50	15.80		47.30	54	64
	Total				.396	31.50	21.15		52.65	60.60	72.25
0030	**Replace fixture**	20	1 ELEC	Ea.							
	Turn branch circuit off and on				.018		.98		.98	1.21	1.51
	Remove emergency light units				.667		35.50		35.50	44	55
	Emergency light units, battery operated				2.000	660	107		767	860	990
	Total				2.685	660	143.48		803.48	905.21	1,046.51

D5093 250 Exit Light

	System Description	Freq. (Years)	Crew	Unit	Labor Hours	2014 Bare Costs				Total In-House	Total w/O&P
						Material	Labor	Equipment	Total		
0010	**Maintenance and repair**	20	1 ELEC	Ea.							
	Remove exit light face plate				.100		5.35		5.35	6.60	8.25
	Exit light face plate				.296	18.25	15.80		34.05	39.50	47.50
	Total				.396	18.25	21.15		39.40	46.10	55.75
0020	**Replace lamp**	5	1 ELEC	Ea.							
	Remove exit light lamp				.022		1.17		1.17	1.45	1.81
	Exit light lamp				.065	5.40	3.47		8.87	10.25	12.10
	Total				.087	5.40	4.64		10.04	11.70	13.91
0030	**Replace lighting fixture**	20	1 ELEC	Ea.							
	Turn circuit off and on for fixture				.018		.98		.98	1.21	1.51
	Remove exit light				.333		17.80		17.80	22	27.50
	Exit light				1.000	39	53.50		92.50	109	131
	Total				1.352	39	72.28		111.28	132.21	160.01

D50 ELECTRICAL   D5093 Other Electrical Systems

D5093 255   Exit Light L.E.D.

	System Description	Freq. (Years)	Crew	Unit	Labor Hours	2014 Bare Costs				Total In-House	Total w/O&P
						Material	Labor	Equipment	Total		
0020	**Replace lamp with exit light L.E.D. retrofit kits**	15	1 ELEC	Ea.							
	Remove exit light lamp				.022		1.17		1.17	1.45	1.81
	Exit light L.E.D. retrofit kits				.133	50.50	7.10		57.60	64.50	74
	Total				.155	50.50	8.27		58.77	**65.95**	**75.81**
0030	**Replace lighting fixture with exit light L.E.D. std.**	20	1 ELEC	Ea.							
	Turn circuit off and on for fixture				.018		.98		.98	1.21	1.51
	Remove exit light				.333		17.80		17.80	22	27.50
	Exit light, L.E.D. standard, double face				1.194	76	63.50		139.50	162	194
	Total				1.546	76	82.28		158.28	**185.21**	**223.01**
0040	**Replace lighting fixture with exit light L.E.D. w/battery unit**	20	1 ELEC	Ea.							
	Turn circuit off and on for fixture				.018		.98		.98	1.21	1.51
	Remove exit light				.333		17.80		17.80	22	27.50
	Exit light, L.E.D. with battery unit, double face				2.000	142	107		249	288	345
	Total				2.352	142	125.78		267.78	**311.21**	**374.01**

D5093 260   Battery, Wet

	System Description	Freq. (Years)	Crew	Unit	Labor Hours	2014 Bare Costs				Total In-House	Total w/O&P
						Material	Labor	Equipment	Total		
0010	**Maintenance and inspection**	0.02	1 ELEC	Ea.							
	Clean & maintain wet cell battery				.500		26.50		26.50	33	41.50
	Total				.500		26.50		26.50	**33**	**41.50**
0020	**Replace battery**	10	1 ELEC	Ea.							
	Remove wet cell battery				.100		5.35		5.35	6.60	8.25
	Wet cell battery				.296	515	15.80		530.80	585	670
	Total				.396	515	21.15		536.15	**591.60**	**678.25**

D50 ELECTRICAL D5093 Other Electrical Systems

D5093 265 Battery, Dry

	System Description	Freq. (Years)	Crew	Unit	Labor Hours	2014 Bare Costs				Total In-House	Total w/O&P
						Material	Labor	Equipment	Total		
0010	**Maintenance and inspection** Clean & maintain dry cell battery	0.08	1 ELEC	Ea.	.500		26.50		26.50	33	41.50
	Total				.500		26.50		26.50	**33**	**41.50**
0020	**Replace battery** Remove dry cell battery Dry cell battery	5	1 ELEC	Ea.	.100 .296	182	5.35 15.80		5.35 197.80	6.60 220	8.25 252
	Total				.396	182	21.15		203.15	**226.60**	**260.25**

D5093 270 Battery Charger

	System Description	Freq. (Years)	Crew	Unit	Labor Hours	2014 Bare Costs				Total In-House	Total w/O&P
						Material	Labor	Equipment	Total		
0010	**Maintenance and repair** Repair battery charger component	2	1 ELEC	Ea.	3.077	44.50	164		208.50	252	310
	Total				3.077	44.50	164		208.50	**252**	**310**
0020	**Maintenance and inspection** Check battery charger	0.25	1 ELEC	Ea.	1.600		85.50		85.50	106	132
	Total				1.600		85.50		85.50	**106**	**132**
0030	**Replace charger** Remove battery charger Battery charger	20	1 ELEC	Ea.	.615 1.600	865	33 85.50		33 950.50	40.50 1,050	51 1,225
	Total				2.215	865	118.50		983.50	**1,090.50**	**1,276**

D5093 280 UPS Battery

	System Description	Freq. (Years)	Crew	Unit	Labor Hours	2014 Bare Costs				Total In-House	Total w/O&P
						Material	Labor	Equipment	Total		
0020	**Maintenance and inspection** UPS battery electrical test	0.17	1 ELEC	Ea.	.800		42.50		42.50	53	66
	Total				.800		42.50		42.50	**53**	**66**

D50 ELECTRICAL   D5093   Other Electrical Systems

D5093 280   UPS Battery

						2014 Bare Costs				Total	Total
	System Description	Freq. (Years)	Crew	Unit	Labor Hours	Material	Labor	Equipment	Total	In-House	w/O&P
0030	**Replace motor generator UPS battery**	15	1 ELEC	Ea.							
	Remove UPS battery				1.333		71		71	88	110
	UPS battery				4.000	435	213		648	745	875
	Total				5.333	435	284		719	833	985

D5093 920   Communications Components

						2014 Bare Costs				Total	Total
	System Description	Freq. (Years)	Crew	Unit	Labor Hours	Material	Labor	Equipment	Total	In-House	w/O&P
0020	**Maintenance and repair voice/data outlet**	10	1 ELEC	Ea.							
	Repair voice/data outlet				.615	2.78	33		35.78	43.50	54.50
	Total				.615	2.78	33		35.78	43.50	54.50
0030	**Replace voice/data outlet**	20	1 ELEC	Ea.							
	Remove voice/data outlet				.057		3.05		3.05	3.77	4.72
	Voice/data outlet, single opening, excludes voice/data device				.167	7.85	8.90		16.75	19.65	23.50
	Total				.224	7.85	11.95		19.80	23.42	28.22
0120	**Maintenance and inspection patch panel**	0.50	1 ELEC	Ea.							
	Check patch panel				1.067		57		57	70	88
	Total				1.067		57		57	70	88
0130	**Replace patch panel**	15	1 ELEC	Ea.							
	Remove patch panel				1.010		53.41		53.41	67	83
	Patch panel				4.000	375	213		588	675	800
	Test circuits				.996		53.15		53.15	66	82
	Total				6.006	375	319.56		694.56	808	965

E10 EQUIPMENT   E1023 Institutional Equipment

E1023 710 Laboratory Equipment

		Freq. (Years)	Crew	Unit	Labor Hours	2014 Bare Costs				Total In-House	Total w/O&P
	System Description					Material	Labor	Equipment	Total		
0020	**Replace glove box gloves**	5	1 SKWK	Ea.							
	Laboratory equipment, remove bacteriological glove box gloves				.500		23.50		23.50	31	38.50
	Lab equipment, replace fiberglass, bacteriological glove box gloves				1.000	189	47.50		236.50	270	315
	Laboratory equipment, test replacement gloves				1.000		47.50		47.50	62.50	77
	Total				2.500	189	118.50		307.50	363.50	430.50
0080	**Replace fume hood sash**	20	2 SKWK	Ea.							
	Laboratory equipment, remove fume hood sash				1.000		47.50		47.50	62.50	77
	Laboratory equipment, replace fume hood sash				2.000	800	94.50		894.50	1,000	1,150
	Total				3.000	800	142		942	**1,062.50**	**1,227**

E10 EQUIPMENT

E1033 Vehicular Equipment

E1033 110 Automotive Equipment

	System Description	Freq. (Years)	Crew	Unit	Labor Hours	2014 Bare Costs				Total In-House	Total w/O&P
						Material	Labor	Equipment	Total		
0020	**Remove and replace vehicle lift hydraulic pump** Automotive hoist, two posts, adj. frame, remove and replace pump	15	1 SKWK	Ea.	4.000	3,000	189		3,189	3,550	4,050
	Total				4.000	3,000	189		3,189	**3,550**	**4,050**
0100	**Remove and replace compressor, electric, 5 H.P.** Automotive compressors, remove and replace 5 H.P. motor	10	1 ELEC	Ea.	2.000	885	107		992	1,100	1,275
	Total				2.000	885	107		992	**1,100**	**1,275**

E10 EQUIPMENT E1093 Other Equipment

E1093 310 Loading Dock Equipment

	System Description	Freq. (Years)	Crew	Unit	Labor Hours	2014 Bare Costs				Total In-House	Total w/O&P
						Material	Labor	Equipment	Total		
0020	**Remove and replace hydraulic dock leveler lift cylinder**	15	2 SKWK	Ea.							
	Loading dock, dock levelers, 10 ton cap., replace lift cylinder				4.000	5,925	189		6,114	6,775	7,725
	Total				4.000	5,925	189		6,114	**6,775**	**7,725**
0040	**Remove and replace hydraulic dock leveler hydraulic pump**	20	1 SKWK	Ea.							
	Loading dock, dock levelers, 10 ton cap, replace lift cylinder				3.077	1,425	146		1,571	1,750	2,025
	Total				3.077	1,425	146		1,571	**1,750**	**2,025**

E1093 315 Dishwasher

	System Description	Freq. (Years)	Crew	Unit	Labor Hours	2014 Bare Costs				Total In-House	Total w/O&P
						Material	Labor	Equipment	Total		
0020	**Replace commercial dishwasher, 10 to 12 racks per hour**	10	2 PLUM	Ea.							
	Cleaning and disposal, dishwasher, commercial, rack type				12.000		675		675	840	1,050
	Install new dishwasher				6.499	3,475	335		3,810	4,250	4,875
	Total				18.499	3,475	1,010		4,485	**5,090**	**5,925**
0040	**Remove and replace dishwasher pump**	15	1 PLUM	Ea.							
	Laboratory equip., glassware washer, remove and replace pump				1.778	720	102		822	920	1,050
	Total				1.778	720	102		822	**920**	**1,050**
0080	**Replace commercial dishwasher, to 375 racks per hour**	20	L-4	Ea.							
	Commercial dishwasher, to 275 racks per hour, selective demo				76.000		1,025		1,025	1,375	1,700
	Dishwasher, automatic, 235 to 275 racks/hr.				24.000	31,100	3,375		34,475	38,400	44,100
	Total				60.000						
					84.000	31,100	4,400		35,500	**39,775**	**45,800**

E1093 316 Waste Disposal, Residential

	System Description	Freq. (Years)	Crew	Unit	Labor Hours	2014 Bare Costs				Total In-House	Total w/O&P
						Material	Labor	Equipment	Total		
0010	**Unstop disposal**	1	1 PLUM	Ea.							
	Unstop clogged unit				.889		51		51	64	80
	Total				.889		51		51	**64**	**80**

E10 EQUIPMENT   E1093   Other Equipment

E1093 316   Waste Disposal, Residential

	System Description	Freq. (Years)	Crew	Unit	Labor Hours	2014 Bare Costs				Total In-House	Total w/O&P
						Material	Labor	Equipment	Total		
0020	Replace waste disposal unit, residential	8	1 PLUM	Ea.							
	Remove unit				1.040		57.50		57.50	71.50	89.50
	Install new disposal				2.080	217	115		332	380	450
	Total				3.120	217	172.50		389.50	451.50	539.50

E1093 320   Waste Handling Equipment

	System Description	Freq. (Years)	Crew	Unit	Labor Hours	2014 Bare Costs				Total In-House	Total w/O&P
						Material	Labor	Equipment	Total		
0020	Remove and replace waste compactor hydraulic cylinder	15	2 SKWK	Ea.							
	Replace ram, 5 C.Y. capacity				4.000	5,075	189		5,264	5,825	6,650
	Total				4.000	5,075	189		5,264	5,825	6,650
0040	Remove and replace waste compactor hydraulic pump	20	1 SKWK	Ea.							
	Replace hydraulic pump, 5 C.Y. capacity				3.200	1,025	151		1,176	1,325	1,525
	Total				3.200	1,025	151		1,176	1,325	1,525

E1093 510   Darkroom Dryer

	System Description	Freq. (Years)	Crew	Unit	Labor Hours	2014 Bare Costs				Total In-House	Total w/O&P
						Material	Labor	Equipment	Total		
0020	Remove and replace darkroom dryer fan 48" x 25" x 68" high	10	1 SKWK	Ea.	1.600	176	75.50		251.50	293	345
	Total				1.600	176	75.50		251.50	293	345
0040	Remove and replace darkroom dryer heating element Darkroom equip., dryers, dehumidifier, replace compressor	15	1 SKWK	Ea.	1.333	85	63		148	177	209
	Total				1.333	85	63		148	177	209

E10 EQUIPMENT E1093 Other Equipment

E1093 610 Dust Collector

						2014 Bare Costs				Total In-House	Total w/O&P
	System Description	Freq. (Years)	Crew	Unit	Labor Hours	Material	Labor	Equipment	Total		
0100	Remove and replace 20" dia dust collector bag	5	1 SKWK	Ea.	1.600	435	87.50		522.50	590	680
	Dust collector bag, 20" diameter Total				1.600	435	87.50		522.50	590	680

E1093 910 Pump Systems

						2014 Bare Costs				Total In-House	Total w/O&P
	System Description	Freq. (Years)	Crew	Unit	Labor Hours	Material	Labor	Equipment	Total		
0500	Remove and replace 50 HP pump motor	25	2 ELEC	Ea.							
	Electrical dml, motors, 230/460 V, 60 Hz, 50 HP				1.667		89		89	110	138
	Install 50 HP motor, 230/460 V, totly encld, 1800 RPM				8.000	3,100	425		3,525	3,950	4,525
	Total				9.667	3,100	514		3,614	4,060	4,663

G10 SITE PREPARATION

G1023 Site Demolition and Relocations

G1023 210 Underground Storage Tank Removal

	System Description	Freq. (Years)	Crew	Unit	Labor Hours	2014 Bare Costs				Total In-House	Total w/O&P
						Material	Labor	Equipment	Total		
1010	**Remove 500 gal. underground storage tank(non-leaking)**	20	B-34P	Ea.							
	Saw cut asphalt				2.700	7	113	88.50	208.50	242	287
	Disconnect and remove piping				1.500		86.40		86.40	108	135
	Transfer liquids, 10% of volume				.250		14.50		14.50	18	22.50
	Insert solid carbon dioxide, 1.5 lbs./100 gal.				.150	8.10	5.48		13.58	16.30	19.15
	Cut accessway to inside of tank				1.501		55		55	73.50	90.50
	Remove sludge, wash and wipe tank				1.000		57.50		57.50	72	90
	Properly dispose of sludge/water					34.25			34.25	38	43
	Excavate, pull, & load tank, backfill hole				16.000		685	365	1,050	1,250	1,500
	Select structural backfill					168			168	184	210
	Haul tank to certified dump, 100 miles round trip				8.000		375	270	645	750	890
	Total				31.101	217.35	1,391.88	723.50	2,332.73	2,751.80	3,287.15
1020	**Remove 3000 gal. underground storage tank(non-leaking)**	20	B-34R	Ea.							
	Saw cut asphalt				5.400	14	226	177	417	485	575
	Disconnect and remove piping				2.500		144		144	180	225
	Transfer liquids, 10% of volume				1.500		87		87	108	135
	Insert solid carbon dioxide, 1.5 lbs./100 gal.				.900	48.60	32.85		81.45	97.50	115
	Cut accessway to inside of tank				1.501		55		55	73.50	90.50
	Remove sludge, wash and wipe tank				1.199		69		69	86.50	108
	Properly dispose of sludge/water					308.25			308.25	340	385
	Excavate, pull, & load tank, backfill hole				32.000		1,375	795	2,170	2,575	3,075
	Select structural backfill					630			630	690	790
	Haul tank to certified dump, 100 miles round trip				8.000		375	335	710	815	960
	Total				53.000	1,000.85	2,363.85	1,307	4,671.70	5,450.50	6,458.50
1030	**Remove 5000 gal. underground storage tank(non-leaking)**	20	B-34R	Ea.							
	Saw cut asphalt				5.940	15.40	248.60	194.70	458.70	530	630
	Disconnect and remove piping				2.500		144		144	180	225
	Transfer liquids, 10% of volume				2.500		145		145	180	225
	Insert solid carbon dioxide, 1.5 lbs./100 gal.				1.500	81	54.75		135.75	163	191
	Cut accessway to inside of tank				1.501		55		55	73.50	90.50
	Remove sludge, wash and wipe tank				1.301		75		75	93.50	117
	Properly dispose of sludge/water					513.75			513.75	565	640
	Excavate, pull, & load tank, backfill hole				32.000		1,375	795	2,170	2,575	3,075
	Select structural backfill					1,050			1,050	1,150	1,325
	Haul tank to certified dump, 100 miles round trip				8.000		375	335	710	815	960
	Total				55.242	1,660.15	2,472.35	1,324.70	5,457.20	6,325	7,478.50

G10 SITE PREPARATION G1023 Site Demolition and Relocations

G1023 210 Underground Storage Tank Removal

							2014 Bare Costs				Total In-House	Total w/O&P
	System Description	Freq. (Years)	Crew	Unit	Labor Hours	Material	Labor	Equipment	Total			
1040	**Remove 8000 gal. underground storage tank(non-leaking)**	20	B-34R	Ea.								
	Saw cut asphalt				6.480		271.20	212.40	500.40		580	690
	Disconnect and remove piping				2.500		144		144		180	225
	Transfer liquids, 10% of volume				4.000		232		232		288	360
	Insert solid carbon dioxide, 1.5 lbs./100 gal.				2.400	129.60	87.60		217.20		260	305
	Cut accessway to inside of tank				1.501		55		55		73.50	90.50
	Remove sludge, wash and wipe tank				1.501		86.50		86.50		108	135
	Properly dispose of sludge/water					822			822		905	1,025
	Excavate, pull, & load tank, backfill hole				32.000		1,400	2,350	3,750		4,150	4,825
	Select structural backfill					1,470			1,470		1,600	1,850
	Haul tank to certified dump, 100 miles round trip				8.000		300	960	1,260		1,350	1,550
	Total				58.382	2,438.40	2,576.30	3,522.40	8,537.10		9,494.50	11,055.50
1050	**Remove 10000 gal. ugnd. storage tank(non-leaking)**	20	B-34S	Ea.								
	Saw cut asphalt				10.800	28	452	354	834		970	1,150
	Disconnect and remove piping				3.750		216		216		270	335
	Transfer liquids, 10% of volume				5.000		290		290		360	450
	Insert solid carbon dioxide, 1.5 lbs./100 gal.				3.000	162	109.50		271.50		325	385
	Cut accessway to inside of tank				1.501		55		55		73.50	90.50
	Remove sludge, wash and wipe tank				1.751		101		101		126	157
	Properly dispose of sludge/water					1,027.50			1,027.50		1,125	1,275
	Excavate, pull, & load tank, backfill hole				32.000		1,400	2,350	3,750		4,150	4,825
	Select structural backfill					1,890			1,890		2,075	2,375
	Haul tank to certified dump, 100 miles round trip				8.000		300	960	1,260		1,350	1,550
	Total				65.801	3,107.50	2,923.50	3,664	9,695		10,824.50	12,592.50
1060	**Remove 12000 gal. ugnd. storage tank(non-leaking)**	20	B-34S	Ea.								
	Saw cut asphalt				12.150	31.50	508.50	398.25	938.25		1,100	1,300
	Disconnect and remove piping				5.000		288		288		360	450
	Transfer liquids, 10% of volume				6.000		348		348		430	540
	Insert solid carbon dioxide, 1.5 lbs./100 gal.				3.600	194.40	131.40		325.80		390	460
	Cut accessway to inside of tank				1.501		55		55		73.50	90.50
	Remove sludge, wash and wipe tank				1.900		109		109		137	171
	Properly dispose of sludge/water					1,233			1,233		1,350	1,550
	Excavate, pull, & load tank, backfill hole				32.000		1,400	2,350	3,750		4,150	4,825
	Select structural backfill					2,100			2,100		2,300	2,625
	Haul tank to certified dump, 100 miles round trip				8.000		300	960	1,260		1,350	1,550
	Total				70.151	3,558.90	3,139.90	3,708.25	10,407.05		11,640.50	13,561.50

G20 SITE IMPROVEMENTS G2023 Parking Lots

G2023 210 Parking Lot Repairs

		Freq. (Years)	Crew	Unit	Labor Hours	2014 Bare Costs				Total In-House	Total w/O&P
	System Description					Material	Labor	Equipment	Total		
1010	**Parking lot repair and sealcoating**	5	B-1	M.S.F.							
	Thoroughly clean surface				.312		11.10	3.33	14.43	18.85	22
	Patch holes				.240	25.11	8.91	1.35	35.37	41	47.50
	Fill cracks				.715	75	26.50	10.50	112	128	149
	Install 2 coat petroleum resistant emulsion				3.330	143.19	124.32		267.51	325	385
	Restripe lot				.291	18.48	10.76	3.92	33.16	38.50	45
	Total				4.887	261.78	181.59	19.10	462.47	551.35	648.50
1020	**Parking lot repair and resurface**	10	B-25B	M.S.F.							
	Thoroughly clean surface				.312		11.10	3.33	14.43	18.85	22
	Patch holes				.240	25.11	8.91	1.35	35.37	41	47.50
	Fill cracks				.715	75	26.50	10.50	112	128	149
	Emulsion tack coat, .05 gal. per S.Y.				.710	31.08	31.08	41.07	103.23	115	133
	Install 1" thick asphaltic concrete wearing course				1.008	407.37	41.07	31.08	479.52	535	610
	Restripe lot				.291	18.48	10.76	3.92	33.16	38.50	45
	Total				3.276	557.04	129.42	91.25	777.71	876.35	1,006.50

G2023 310 General

		Freq. (Years)	Crew	Unit	Labor Hours	2014 Bare Costs				Total In-House	Total w/O&P
	System Description					Material	Labor	Equipment	Total		
1000	**Remove and replace steel guard rail**	7	B-80	L.F.							
	Remove old metal guide/guard rail and posts				.213		8.45	5.05	13.50	16.15	19.30
	Install metal guide/guard rail including posts 6'3"				.107	21.50	4.24	2.53	28.27	31.50	36.50
	Total				.320	21.50	12.69	7.58	41.77	47.65	55.80
1100	**Raise MH or catch basin frame and cover**	10	D-1	Ea.							
	Demolish pavement around frame				4.800		186	23.50	209.50	269	330
	Raise MH or catch basin frame with brick				1.332	9.16	54.95		64.11	82.50	101
	Total				6.132	9.16	240.95	23.50	273.61	351.50	431

G20 SITE IMPROVEMENTS G2033 Pedestrian Paving

G2033 130 Asphalt Sidewalk & Curb

	System Description	Freq. (Years)	Crew	Unit	Labor Hours	2014 Bare Costs				Total In-House	Total w/O&P
						Material	Labor	Equipment	Total		
0100	**Remove & replace asphalt sidewalk, 4' wide**	15	B-37	L.F.							
	Remove existing asphalt pavement				.080		3.16	1.96	5.12	6.10	7.25
	Bank run gravel, 3" avg.				.002	.74	.09	.22	1.05	1.15	1.31
	Vibratory plate compaction				.003		.10	.01	.11	.14	.17
	Fine grade gravel base				.026		1.11	1.15	2.26	2.60	3.05
	Place asphalt sidewalk, 2-1/2" thick				.027	2.56	1.03	.09	3.68	4.26	4.98
	Total				.138	3.30	5.49	3.43	12.22	14.25	16.76
1000	**Remove and replace asphalt curb or berm**	10	B-37	L.F.							
	Demolish existing curb or berm				.096		3.72	.47	4.19	5.40	6.55
	Remove broken curb				.032		1.26	.78	2.04	2.44	2.91
	Asphalt curb, 8" wide, 6" high machine formed				.042	.05	1.54	.38	1.97	2.50	3.02
	Total				.170	.05	6.52	1.63	8.20	10.34	12.48

G2033 140 Concrete Sidewalk & Curb

	System Description	Freq. (Years)	Crew	Unit	Labor Hours	2014 Bare Costs				Total In-House	Total w/O&P
						Material	Labor	Equipment	Total		
0100	**Remove and replace concrete sidewalk, 4' wide**	25	B-37	L.F.							
	Remove existing concrete pavement				.107		4.21	2.61	6.82	8.15	9.70
	Bank run gravel, 3" avg.				.002	.74	.09	.22	1.05	1.15	1.31
	Vibratory plate compaction				.003		.10	.01	.11	.14	.17
	Fine grade gravel				.026		1.11	1.15	2.26	2.60	3.05
	Broom finished concrete sidewalk, 4" thick				.209	6.60	8.80		15.40	18.80	22.50
	Total				.346	7.34	14.31	3.99	25.64	30.84	36.73
2000	**Remove and replace concrete curb or berm**	25	B-37	L.F.							
	Demolish existing curb or berm				.240		9.30	1.17	10.47	13.45	16.45
	Remove broken curb				.080		3.16	1.96	5.12	6.10	7.25
	Concrete curb, 6" x 18", steel forms				.091	.83	4.04		4.87	6.30	7.70
	Total				.411	.83	16.50	3.13	20.46	25.85	31.40

G20 SITE IMPROVEMENTS G2033 Pedestrian Paving

G2033 150 Patios

	System Description	Freq. (Years)	Crew	Unit	Labor Hours	2014 Bare Costs				Total In-House	Total w/O&P
						Material	Labor	Equipment	Total		
1030	**Refinish Concrete Patio**	3	1 PORD	S.F.							
	Prepare surface				.003		.12	.03	.15	.19	.23
	Paint 1 coat, roller & brush				.008	.08	.32		.40	.51	.62
	Total				.012	.08	.44	.03	.55	**.70**	**.85**
2030	**Refinish Masonry Patio**	3	1 PORD	S.F.							
	Prepare surface				.003		.12	.03	.15	.19	.23
	Apply 1 coat of sealer, roller & brush				.005	.25	.19		.44	.52	.61
	Total				.008	.25	.31	.03	.59	**.71**	**.84**
3030	**Refinish Stone Patio**	3	1 PORD	S.F.							
	Prepare surface				.003		.12	.03	.15	.19	.23
	Apply 1 coat of sealer, roller & brush				.005	.25	.19		.44	.52	.61
	Total				.008	.25	.31	.03	.59	**.71**	**.84**
4030	**Refinish Wood Patio**	3	1 PORD	S.F.							
	Prepare surface				.015		.59		.59	.76	.94
	Primer + 1 top coat, roller & brush				.022	.13	.88		1.01	1.28	1.57
	Total				.037	.13	1.47		1.60	**2.04**	**2.51**

G2033 250 Handicap Ramp

	System Description	Freq. (Years)	Crew	Unit	Labor Hours	2014 Bare Costs				Total In-House	Total w/O&P
						Material	Labor	Equipment	Total		
1030	**Refinish Metal Handicap Ramp**	3	1 PORD	S.F.							
	Prepare surface				.015		.59		.59	.76	.94
	Prime & paint 1 coat, roller & brush				.015	.28	.58		.86	1.05	1.27
	Total				.029	.28	1.17		1.45	**1.81**	**2.21**
2030	**Refinish Wood Handicap Ramp**	3	1 PORD	S.F.							
	Prepare surface				.015		.59		.59	.76	.94
	Prime & paint 1 coat, roller & brush				.022	.13	.88		1.01	1.28	1.57
	Total				.037	.13	1.47		1.60	**2.04**	**2.51**

G20 SITE IMPROVEMENTS — G2043 Site Development

G2043 105 Chain Link Fence and Gate Repairs

	System Description	Freq. (Years)	Crew	Unit	Labor Hours	Material	2014 Bare Costs Labor	Equipment	Total	Total In-House	Total w/O&P
1010	**Minor chain link fence repairs (per 10 L.F.)**	1	2 CLAB	Ea.							
	Straighten bent 2" line post				.167		6.10		6.10	8.15	10.05
	Re-tie fence fabric				.167	1.72	6.10		7.82	10	12.15
	Refasten loose barbed wire arm				.083		3.05		3.05	4.07	5
	Refasten loose barbed wire, 3 strands				.083		3.05		3.05	4.07	5
	Total				.500	1.72	18.30		20.02	26.29	32.20
1110	**Replace bent 1-5/8" top rail (per 20 L.F.)**	2	2 CLAB	Ea.							
	Remove fabric ties, top rail & couplings				.333		12.20		12.20	16.30	20
	Install new top rail, couplings & fabric ties				.667	37.84	24.42		62.26	73.50	87.50
	Total				1.000	37.84	36.62		74.46	89.80	107.50
1120	**Replace broken barbed wire arm**	2	2 CLAB	Ea.							
	Remove fabric ties, top rail & couplings				.333		12.20		12.20	16.30	20
	Release barbed wires & remove old arm				.067		2.44		2.44	3.26	4.01
	Install new arm & refasten barbed wires				.167	.52	6.12		6.64	8.75	10.70
	Install old top rail, couplings & fabric ties				.667		24.40		24.40	32.50	40
	Total				1.234	.52	45.16		45.68	60.81	74.71
1130	**Replace barbed wire, 3 strands (per 100 L.F.)**	5	2 CLAB	Ea.							
	Remove old barbed wire				3.002		110		110	147	181
	Install new barbed wire				3.000	103.20	110.10		213.30	261	310
	Total				6.002	103.20	220.10		323.30	408	491
1140	**Replace fence fabric, 6' high, 9 ga. (per 10 L.F.)**	10	2 CLAB	Ea.							
	Remove ties from top rail, bottom tension wire & posts				.167		6.10		6.10	8.15	10.05
	Unstitch & remove section of fabric				.667		24.50		24.50	32.50	40
	Hang & stretch new fabric, stitch into existing fabric				1.333	68.80	48.80		117.60	140	166
	Re-tie fence fabric				.167	1.72	6.10		7.82	10	12.15
	Total				2.333	70.52	85.50		156.02	190.65	228.20
1150	**Replace double swing gates, 6' high, 20' opng.**	5	2 CLAB	Opng.							
	Remove 2 old gates (6'x10' ea.) & 4 hinges				1.111		40.50		40.50	54.50	67
	Supply 4 new gate hinges					52			52	57	65
	Install 2 new gates (6'x10' ea.), 4 hinges, adjust alignment				2.412	477.60	88.36		565.96	640	740
	Remove & replace gate latch				.914	12.80	33.60		46.40	59	71
	Remove & replace gate cane bolt				.883	12.80	32.40		45.20	57	69
	Total				5.320	555.20	194.86		750.06	867.50	1,012

G20 SITE IMPROVEMENTS — G2043 Site Development

G2043 105 Chain Link Fence and Gate Repairs

	System Description	Freq. (Years)	Crew	Unit	Labor Hours	Material	Labor	Equipment	Total	Total In-House	Total w/O&P
							2014 Bare Costs				
1160	**Replace 6'x18' cantilever slide gate**	5	2 CLAB	Opng.							
	Remove old gate & 4 rollers				.926		34		34	45	55.50
	Supply 4 new cantilever gate rollers				.003	290	.03		290.03	320	365
	Install new 6'x18' cantilever slide gate, 4 rollers, adjust				2.148	1,442.81	78.98		1,521.79	1,700	1,925
	Remove & replace gate latch				.914	12.80	33.60		46.40	59	71
	Total				3.991	1,745.61	146.61		1,892.22	2,124	2,416.50
1210	**Replace 2" line post**	5	B-55	Ea.							
	Remove fabric ties, top rail & couplings				.333		12.20		12.20	16.30	20
	Release barbed wires & remove old arm				.067		2.44		2.44	3.26	4.01
	Unstitch & roll up section of fabric				1.000		36.50	46.50	83	95	112
	Remove 2" post & concrete base from ground				.750		27.50	35	62.50	71.50	83.50
	Install new 2" line post with concrete in old hole				1.500	60.20	54.95	70	185.15	208	242
	Install old top rail, couplings & new fabric ties				.667		24.40		24.40	32.50	40
	Unroll & stretch fabric, stitch into existing fabric				2.000		73	93.50	166.50	190	223
	Re-tie fence fabric				.333	3.44	12.20		15.64	20	24.50
	Total				6.650	63.64	243.19	245	551.83	636.56	749.01
1220	**Replace 3" corner post**	10	B-55	Ea.							
	Remove fabric ties, retainer bars & ring fittings				.500		18.30	23.50	41.80	47.50	55.50
	Roll back fabric				.250		9.15	11.65	20.80	24	28
	Remove top rail & ring fittings				.500		18.30	23.50	41.80	47.50	55.50
	Remove barbed wires & ring fittings				.750		27.50	35	62.50	71.50	83.50
	Remove diagonal & horizontal braces, ring fittings				.500		18.30	23.50	41.80	47.50	55.50
	Remove 3" post & concrete base from ground				1.000		36.50	46.50	83	95	112
	Install new 3" post with concrete in old hole				2.000	90	73	93.50	256.50	289	335
	Install old diagonal & horizontal braces, ring fittings				.500		18.30	23.50	41.80	47.50	55.50
	Install old barbed wires & ring fittings				.750		27.50	35	62.50	71.50	83.50
	Install old top rail & ring fittings				.500		18.30	23.50	41.80	47.50	55.50
	Hang & stretch fabric, insert retainer bars & ring fittings				1.500		55	70	125	143	167
	Re-tie fence fabric				.333	3.44	12.20		15.64	20	24.50
	Total				9.083	93.44	332.35	409.15	834.94	951.50	1,111

G20 SITE IMPROVEMENTS

G2043 Site Development

G2043 105 Chain Link Fence and Gate Repairs

	System Description	Freq. (Years)	Crew	Unit	Labor Hours	2014 Bare Costs				Total In-House	Total w/O&P
						Material	Labor	Equipment	Total		
1230	**Replace 3" gate post**	5	B-55	Ea.							
	Remove gate & hinges				.833		30.50	39	69.50	79.50	93
	Remove fabric ties, retainer bars & ring fittings				.500		18.30	23.50	41.80	47.50	55.50
	Roll back fabric				.250		9.15	11.65	20.80	24	28
	Remove top rail & ring fittings				.500		18.30	23.50	41.80	47.50	55.50
	Remove barbed wires & ring fittings				.750		27.50	35	62.50	71.50	83.50
	Remove diagonal & horizontal braces, ring fittings				.500		18.30	23.50	41.80	47.50	55.50
	Remove 3" post & concrete base from ground				1.000		36.50	46.50	83	95	112
	Install new 3" post with concrete in old hole				2.000	90	73	93.50	256.50	289	335
	Install old diagonal & horizontal braces, ring fittings				.500		18.30	23.50	41.80	47.50	55.50
	Install old barbed wires & ring fittings				.750		27.50	35	62.50	71.50	83.50
	Install old top rail & ring fittings				.500		18.30	23.50	41.80	47.50	55.50
	Hang & stretch fabric, insert retainer bars & ring fittings				1.500		55	70	125	143	167
	Re-tie fence fabric				.333	3.44	12.20		15.64	20	24.50
	Install old hinges & gate				1.807		66	84.50	150.50	172	201
	Total				11.724	93.44	428.85	532.65	1,054.94	1,203	1,405

G2043 110 Wood Fence

	System Description	Freq. (Years)	Crew	Unit	Labor Hours	2014 Bare Costs				Total In-House	Total w/O&P
						Material	Labor	Equipment	Total		
1020	**Refinish Wood Fence, 6' High**	3	1 PORD	L.F.							
	Prepare surface				.089		3.54		3.54	4.56	5.65
	Prime & 1 top coat, roller & brush				.046	.48	1.80		2.28	2.88	3.54
	Total				.135	.48	5.34		5.82	7.44	9.19

G2043 710 Bleachers, Exterior

	System Description	Freq. (Years)	Crew	Unit	Labor Hours	2014 Bare Costs				Total In-House	Total w/O&P
						Material	Labor	Equipment	Total		
1020	**Refinish Wood Bleachers**	3	1 PORD	S.F.							
	Prepare surface				.015		.59		.59	.76	.94
	Prime & 1 top coat, roller & brush				.022	.13	.88		1.01	1.28	1.57
	Total				.037	.13	1.47		1.60	2.04	2.51

G20 SITE IMPROVEMENTS

G2043 Site Development

G2043 750 Tennis Court Resurfacing

	System Description	Freq. (Years)	Crew	Unit	Labor Hours	2014 Bare Costs				Total In-House	Total w/O&P
						Material	Labor	Equipment	Total		
1010	**Resurface asphalt tennis court**	7	2 SKWK	Ea.							
	Remove/reinstall tennis net				1.000		36.50		36.50	49	60
	Thoroughly clean surface				7.760		288		288	375	465
	Install 3 coat acrylic emulsion sealcoat				16.000	4,800	584		5,384	6,075	6,950
	Total				24.760	4,800	908.50		5,708.50	6,499	7,475
1020	**Resurface cushioned asphalt tennis court**	7	2 SKWK	Ea.							
	Remove/reinstall tennis net				1.000		36.50		36.50	49	60
	Thoroughly clean surface				7.760		288		288	375	465
	Two cushion coats				16.000	2,175	585		2,760	3,175	3,675
	Install 3 coat acrylic emulsion sealcoat				16.000	4,800	584		5,384	6,075	6,950
	Total				40.760	6,975	1,493.50		8,468.50	9,674	11,150
2010	**Clay tennis court preparation**	1	2 SKWK	Ea.							
	Remove/reinstall tennis net				1.000		36.50		36.50	49	60
	Remove and reinstall tapes				6.015		220		220	294	360
	Remove weeds and deteriorated surface material				2.000		73.50		73.50	97.50	120
	Check court for level				1.000		36.50		36.50	49	60
	Clay court top dressing, 1.5 tons				7.500	495	274.50		769.50	910	1,075
	Clay court, roll after dressing				1.500		67	23	90	110	133
	Total				19.015	495	708	23	1,226	1,509.50	1,808

G2043 810 Flag Pole

	System Description	Freq. (Years)	Crew	Unit	Labor Hours	2014 Bare Costs				Total In-House	Total w/O&P
						Material	Labor	Equipment	Total		
1020	**Refinish 25' Wood Flag Pole**	5	1 PORD	Ea.							
	Set up, secure and take down ladder				.295		10.20		10.20	13.20	16.35
	Prepare surface				.258		14.75		14.75	19	23.50
	Prime & 1 top coat, roller & brush				.370	3.25	22		25.25	32	39.50
					.556						
	Total				1.184	3.25	46.95		50.20	64.20	79.35

G30 SITE MECH. UTILITIES G3013 Water Supply

G3013 400 Ground Level Water Storage Tank

	System Description	Freq. (Years)	Crew	Unit	Labor Hours	2014 Bare Costs				Total In-House	Total w/O&P
						Material	Labor	Equipment	Total		
0100	**Prep & paint 100k gal. ground level water stor tank, 30' dia x 19' tall**	10	E-11	Ea.							
	Scaffolding, outer ring				72.000		3,312		3,312	4,400	5,425
	Scaffolding planks				35.666		1,637.10		1,637.10	2,200	2,675
	Sandblast exterior, near white blast (SSPC-SP10), loose scale, fine rust				177.775	2,350	7,325	1,325	11,000	14,500	17,300
	Epoxy primer, spray exterior				13.325	725	550		1,275	1,650	1,925
	Epoxy topcoat, 2 coats, spray exterior				28.550	1,300	1,150		2,450	3,300	3,825
	Clean up				10.000		400		400	575	700
	Total				337.316	4,375	14,374.10	1,325	20,074.10	26,625	31,850
0200	**Prep & paint 250k gal. ground level water stor tank, 40' dia x 27' tall**	10	E-11	Ea.							
	Scaffolding, outer ring				126.000		5,796		5,796	7,675	9,500
	Scaffolding planks				63.333		2,907		2,907	3,900	4,750
	Sandblast exterior, near white blast (SSPC-SP10), loose scale, fine rust				327.106	4,324	13,478	2,438	20,240	26,700	31,800
	Epoxy primer, spray exterior				24.518	1,334	1,012		2,346	3,025	3,550
	Epoxy topcoat, 2 coats, spray exterior				52.532	2,392	2,116		4,508	6,075	7,050
	Clean up				18.400		736		736	1,050	1,300
	Total				611.889	8,050	26,045	2,438	36,533	48,425	57,950
0300	**Prep & paint 500k gal. ground level water stor tank, 50' dia x 34' tall**	10	E-11	Ea.							
	Scaffolding, outer ring				192.000		8,832		8,832	11,700	14,500
	Scaffolding planks				96.666		4,437		4,437	5,950	7,250
	Sandblast exterior, near white blast (SSPC-SP10), loose scale, fine rust				519.103	6,862	21,389	3,869	32,120	42,300	50,500
	Epoxy primer, spray exterior				38.909	2,117	1,606		3,723	4,825	5,650
	Epoxy topcoat, 2 coats, spray exterior				83.366	3,796	3,358		7,154	9,625	11,200
	Clean up				29.200		1,168		1,168	1,675	2,050
	Total				959.244	12,775	40,790	3,869	57,434	76,075	91,150
0400	**Prep & paint 750k gal. ground level water stor tank, 60' dia x 36' tall**	10	E-11	Ea.							
	Scaffolding, outer ring				234.000		10,764		10,764	14,300	17,600
	Scaffolding planks				117.332		5,385.60		5,385.60	7,225	8,800
	Sandblast exterior, near white blast (SSPC-SP10), loose scale, fine rust				675.545	8,930	27,835	5,035	41,800	55,000	65,500
	Epoxy primer, spray exterior				50.635	2,755	2,090		4,845	6,275	7,350
	Epoxy topcoat, 2 coats, spray exterior				108.490	4,940	4,370		9,310	12,500	14,500
	Clean up				38.000		1,520		1,520	2,175	2,650
	Total				1224.002	16,625	51,964.60	5,035	73,624.60	97,475	116,400

G30 SITE MECH. UTILITIES G3013 Water Supply

G3013 400 Ground Level Water Storage Tank

	System Description	Freq. (Years)	Crew	Unit	Labor Hours	2014 Bare Costs				Total In-House	Total w/O&P
						Material	Labor	Equipment	Total		
0500	**Prep & paint 1M gal. ground level water stor tank, 70' dia x 35' tall**	10	E-11	Ea.							
	Scaffolding, outer ring				261.000		12,006		12,006	15,900	19,700
	Scaffolding planks				131.332		6,028.20		6,028.20	8,075	9,850
	Sandblast exterior, near white blast (SSPC-SP10), loose scale, fine rust				817.765	10,810	33,695	6,095	50,600	66,500	79,500
	Epoxy primer, spray exterior				61.295	3,335	2,530		5,865	7,600	8,875
	Epoxy topcoat, 2 coats, spray exterior				131.330	5,980	5,290		11,270	15,200	17,600
	Clean up				46.000		1,840		1,840	2,650	3,225
	Total				1448.722	20,125	61,389.20	6,095	87,609.20	115,925	138,750
0600	**Prep & paint 2M gal. ground level water stor tank, 100' dia x 34' tall**	10	E-11	Ea.							
	Scaffolding, outer ring				354.000		16,284		16,284	21,600	26,700
	Scaffolding planks				176.998		8,124.30		8,124.30	10,900	13,300
	Sandblast exterior, near white blast (SSPC-SP10), loose scale, fine rust				1315.535	17,390	54,205	9,805	81,400	107,500	127,500
	Epoxy primer, spray exterior				98.605	5,365	4,070		9,435	12,200	14,300
	Epoxy topcoat, 2 coats, spray exterior				211.270	9,620	8,510		18,130	24,400	28,300
	Clean up				74.000		2,960		2,960	4,250	5,175
	Total				2230.408	32,375	94,153.30	9,805	136,333.30	180,850	215,275
0700	**Prep & paint 4M gal. ground level water stor tank, 130' dia x 40' tall**	10	E-11	Ea.							
	Scaffolding, outer ring				531.000		24,426		24,426	32,400	40,000
	Scaffolding planks				266.664		12,240		12,240	16,400	20,000
	Sandblast exterior, near white blast (SSPC-SP10), loose scale, fine rust				2133.300	28,200	87,900	15,900	132,000	174,000	207,000
	Epoxy primer, spray exterior				159.900	8,700	6,600		15,300	19,800	23,200
	Epoxy topcoat, 2 coats, spray exterior				342.600	15,600	13,800		29,400	39,600	45,900
	Clean up				120.000		4,800		4,800	6,900	8,400
	Total				3553.464	52,500	149,766	15,900	218,166	289,100	344,500

G30 SITE MECH. UTILITIES G3013 Water Supply

G3013 405 Ground Level Water Storage Standpipe

	System Description	Freq. (Years)	Crew	Unit	Labor Hours	2014 Bare Costs				Total In-House	Total w/O&P
						Material	Labor	Equipment	Total		
0100	**Prep & paint 100k gal. grnd. level water standpipe, 24' dia x 30' tall**	10	E-11	Ea.							
	Scaffolding, outer ring				96.000		4,416		4,416	5,850	7,225
	Scaffolding planks				47.666		2,187.90		2,187.90	2,925	3,575
	Sandblast exterior, near white blast (SSPC-SP10), loose scale, fine rust				191.997	2,538	7,911	1,431	11,880	15,700	18,600
	Epoxy primer, spray exterior				14.391	783	594		1,377	1,775	2,075
	Epoxy topcoat, 2 coats, spray exterior				30.834	1,404	1,242		2,646	3,575	4,125
	Clean up				10.800		432		432	620	755
	Total				391.688	4,725	16,782.90	1,431	22,938.90	30,445	36,355
0200	**Prep & paint 250k gal. grnd. level water standpipe, 32' dia x 42' tall**	10	E-11	Ea.							
	Scaffolding, outer ring				165.000		7,590		7,590	10,100	12,400
	Scaffolding planks				82.666		3,794.40		3,794.40	5,075	6,200
	Sandblast exterior, near white blast (SSPC-SP10), loose scale, fine rust				355.550	4,700	14,650	2,650	22,000	29,000	34,500
	Epoxy primer, spray exterior				26.650	1,450	1,100		2,550	3,300	3,875
	Epoxy topcoat, 2 coats, spray exterior				57.100	2,600	2,300		4,900	6,600	7,650
	Clean up				20.000		800		800	1,150	1,400
	Total				706.966	8,750	30,234.40	2,650	41,634.40	55,225	66,025
0300	**Prep & paint 500k gal. grnd. level water standpipe, 40' dia x 53' tall**	10	E-11	Ea.							
	Scaffolding, outer ring				252.000		11,592		11,592	15,400	19,000
	Scaffolding planks				125.665		5,768.10		5,768.10	7,725	9,425
	Sandblast exterior, near white blast (SSPC-SP10), loose scale, fine rust				568.880	7,520	23,440	4,240	35,200	46,400	55,000
	Epoxy primer, spray exterior				42.640	2,320	1,760		4,080	5,275	6,175
	Epoxy topcoat, 2 coats, spray exterior				91.360	4,160	3,680		7,840	10,600	12,200
	Clean up				32.000		1,280		1,280	1,850	2,250
	Total				1112.545	14,000	47,520.10	4,240	65,760.10	87,250	104,050
0400	**Prep & paint 750k gal. grnd. level water standpipe, 46' dia x 60' tall**	10	E-11	Ea.							
	Scaffolding, outer ring				318.000		14,628		14,628	19,400	24,000
	Scaffolding planks				159.665		7,328.70		7,328.70	9,825	12,000
	Sandblast exterior, near white blast (SSPC-SP10), loose scale, fine rust				739.544	9,776	30,472	5,512	45,760	60,500	72,000
	Epoxy primer, spray exterior				55.432	3,016	2,288		5,304	6,875	8,025
	Epoxy topcoat, 2 coats, spray exterior				118.768	5,408	4,784		10,192	13,700	15,900
	Clean up				41.600		1,664		1,664	2,400	2,900
	Total				1433.009	18,200	61,164.70	5,512	84,876.70	112,700	134,825

G30 SITE MECH. UTILITIES G3013 Water Supply

G3013 405 Ground Level Water Storage Standpipe

	System Description	Freq. (Years)	Crew	Unit	Labor Hours	2014 Bare Costs				Total In-House	Total w/O&P
						Material	Labor	Equipment	Total		
0500	**Prep & paint 1M gal. grnd. level water standpipe, 50' dia x 68' tall**	10	E-11	Ea.							
	Scaffolding, outer ring				384.000		17,664		17,664	23,400	28,900
	Scaffolding planks				192.998		8,858.70		8,858.70	11,900	14,500
	Sandblast exterior, near white blast (SSPC-SP10), loose scale, fine rust				903.097	11,938	37,211	6,731	55,880	73,500	87,500
	Epoxy primer, spray exterior				67.691	3,683	2,794		6,477	8,375	9,800
	Epoxy topcoat, 2 coats, spray exterior				145.034	6,604	5,842		12,446	16,800	19,400
	Clean up				50.800		2,032		2,032	2,925	3,550
	Total				1743.620	22,225	74,401.70	6,731	103,357.70	136,900	163,650
0600	**Prep & paint 2M gal. grnd. level water standpipe, 64' dia x 83' tall**	10	E-11	Ea.							
	Scaffolding, outer ring				579.000		26,634		26,634	35,300	43,600
	Scaffolding planks				290.664		13,341.60		13,341.60	17,900	21,800
	Sandblast exterior, near white blast (SSPC-SP10), loose scale, fine rust				1422.200	18,800	58,600	10,600	88,000	116,000	138,000
	Epoxy primer, spray exterior				106.600	5,800	4,400		10,200	13,200	15,500
	Epoxy topcoat, 2 coats, spray exterior				228.400	10,400	9,200		19,600	26,400	30,600
	Clean up				80.000		3,200		3,200	4,600	5,600
	Total				2706.864	35,000	115,375.60	10,600	160,975.60	213,400	255,100

G3013 410 Fire Hydrants

	System Description	Freq. (Years)	Crew	Unit	Labor Hours	2014 Bare Costs				Total In-House	Total w/O&P
						Material	Labor	Equipment	Total		
1010	**Remove and replace fire hydrant**	25	B-21	Ea.							
	Isolate hydrant from system				1.000		36.50		36.50	49	60
	Excavate to expose base and feeder pipe				.974		41.65	22.26	63.91	76.50	91.50
	Disconnect and remove hydrant and associated piping				4.000		147		147	195	241
	Install pipe to new hydrant				3.200	177.50	143		320.50	380	450
	Install hydrant, including lower barrel and shoe				4.000	1,875	169	20	2,064	2,300	2,650
	Install thrust blocks				1.200	49.20	52.80	.35	102.35	125	148
	Backfill excavation				5.091		185.50		185.50	249	310
	Fine grade, hand				8.000		294		294	390	480
	Total				27.465	2,101.70	1,069.45	42.61	3,213.76	3,764.50	4,430.50

G30 SITE MECH. UTILITIES

G3013 Water Supply

G3013 470 Post Indicator Valve

	System Description	Freq. (Years)	Crew	Unit	Labor Hours	2014 Bare Costs				Total In-House	Total w/O&P
						Material	Labor	Equipment	Total		
1010	**Remove and replace post indicator valve**	35	B-21	Ea.							
	Isolate PIV from system				.126		7.25		7.25	9.05	11.30
	Excavate to expose base of valve				.139		5.95	3.18	9.13	10.90	13.10
	Disconnect and remove post indicator and valve				4.000		147		147	195	241
	Install new 4" diameter valve				4.000	1,400	160	61	1,621	1,800	2,075
	Install new post indicator				3.636	800	153	18.20	971.20	1,100	1,275
	Backfill excavation				.727		26.50		26.50	35.50	44
	Fine grade, hand				8.000		294		294	390	480
	Total				20.629	2,200	793.70	82.38	3,076.08	3,540.45	4,139.40

G3013 510 Elevated Water Storage Tank

	System Description	Freq. (Years)	Crew	Unit	Labor Hours	2014 Bare Costs				Total In-House	Total w/O&P
						Material	Labor	Equipment	Total		
0100	**Prep & paint 50K gal. stg tank, 30' dia x 10' tall, 50' above grnd.**	10	E-11	Ea.							
	Scaffolding, outer ring				225.000		10,350		10,350	13,700	17,000
	Scaffolding, center core under tank				80.000		3,660		3,660	4,875	6,025
	Scaffolding planks				58.333		2,677.50		2,677.50	3,600	4,375
	Sandblast exterior, near white blast (SSPC–SP10), loose scale, fine rust				167.464	2,213.70	6,900.15	1,248.15	10,362	13,700	16,300
	Epoxy primer, spray exterior				12.552	682.95	518.10		1,201.05	1,550	1,825
	Epoxy topcoat, 2 coats, spray exterior				26.894	1,224.60	1,083.30		2,307.90	3,100	3,600
	Clean up				9.420		376.80		376.80	540	660
	Total				579.663	4,121.25	25,565.85	1,248.15	30,935.25	41,065	49,785
0200	**Prep & paint 100K gal. stg tank, 30' dia x 20' tall, 50' above grnd.**	10	E-11	Ea.							
	Scaffolding, outer ring				261.000		12,006		12,006	15,900	19,700
	Scaffolding, center core under tank				80.000		3,660		3,660	4,875	6,025
	Scaffolding planks				153.332		7,038		7,038	9,425	11,500
	Sandblast exterior, near white blast (SSPC–SP10), loose scale, fine rust				227.552	3,008	9,376	1,696	14,080	18,600	22,100
	Epoxy primer, spray exterior				17.056	928	704		1,632	2,100	2,475
	Epoxy topcoat, 2 coats, spray exterior				36.544	1,664	1,472		3,136	4,225	4,900
	Clean up				12.800		512		512	735	895
	Total				788.284	5,600	34,768	1,696	42,064	55,860	67,595

G30 SITE MECH. UTILITIES G3013 Water Supply

G3013 510 Elevated Water Storage Tank

	System Description	Freq. (Years)	Crew	Unit	Labor Hours	2014 Bare Costs				Total	Total In-House	Total w/O&P
						Material	Labor	Equipment	Total			
0300	**Prep & paint 250K gal. stg tank, 40' dia x 27' tall, 50' above grnd.**	10	E-11	Ea.								
	Scaffolding, outer ring				360.000		16,560		16,560	22,000	27,100
	Scaffolding, center core under tank				134.000		6,130.50		6,130.50	8,175	10,100
	Scaffolding planks				219.998		10,098		10,098	13,500	16,500
	Sandblast exterior, near white blast (SSPC-SP10), loose scale, fine rust				419.549	5,546	17,287	3,127	25,960	34,200	40,700
	Epoxy primer, spray exterior				31.447	1,711	1,298		3,009	3,900	4,550
	Epoxy topcoat, 2 coats, spray exterior				67.378	3,068	2,714		5,782	7,800	9,025
	Clean up				23.600		944		944	1,350	1,650
	Total				1255.972	10,325	55,031.50	3,127	68,483.50	90,925	109,625
0400	**Prep & paint 500K gal. stg tank, 50' dia x 34' tall, 50' above grnd.**	10	E-11	Ea.								
	Scaffolding, outer ring				480.000		22,080		22,080	29,300	36,200
	Scaffolding, center core under tank				202.000		9,241.50		9,241.50	12,300	15,200
	Scaffolding planks				296.664		13,617		13,617	18,200	22,300
	Sandblast exterior, near white blast (SSPC-SP10), loose scale, fine rust				661.323	8,742	27,249	4,929	40,920	54,000	64,000
	Epoxy primer, spray exterior				49.569	2,697	2,046		4,743	6,150	7,175
	Epoxy topcoat, 2 coats, spray exterior				106.206	4,836	4,278		9,114	12,300	14,200
	Clean up				37.200		1,488		1,488	2,150	2,600
	Total				1832.962	16,275	79,999.50	4,929	101,203.50	134,400	161,675
0500	**Prep & paint 750K gal. stg tank, 60' dia x 36' tall, 50' above grnd.**	10	E-11	Ea.								
	Scaffolding, outer ring				570.000		26,220		26,220	34,800	42,900
	Scaffolding, center core under tank				283.000		12,947.25		12,947.25	17,300	21,300
	Scaffolding planks				366.663		16,830		16,830	22,600	27,500
	Sandblast exterior, near white blast (SSPC-SP10), loose scale, fine rust				881.764	11,656	36,332	6,572	54,560	72,000	85,500
	Epoxy primer, spray exterior				66.092	3,596	2,728		6,324	8,175	9,575
	Epoxy topcoat, 2 coats, spray exterior				141.608	6,448	5,704		12,152	16,400	19,000
	Clean up				49.600		1,984		1,984	2,850	3,475
	Total				2358.727	21,700	102,745.25	6,572	131,017.25	174,125	209,250
0600	**Prep & paint 1M gal. stg tank, 65' dia x 40' tall, 50' above grnd.**	10	E-11	Ea.								
	Scaffolding, outer ring				645.000		29,670		29,670	39,300	48,600
	Scaffolding, center core under tank				329.000		15,051.75		15,051.75	20,100	24,800
	Scaffolding planks				413.329		18,972		18,972	25,400	31,000
	Sandblast exterior, near white blast (SSPC-SP10), loose scale, fine rust				1059.539	14,006	43,657	7,897	65,560	86,500	103,000
	Epoxy primer, spray exterior				79.417	4,321	3,278		7,599	9,825	11,500
	Epoxy topcoat, 2 coats, spray exterior				170.158	7,748	6,854		14,602	19,700	22,800
	Clean up				59.600		2,384		2,384	3,425	4,175
	Total				2756.043	26,075	119,866.75	7,897	153,838.75	204,250	245,875

G30 SITE MECH. UTILITIES G3063 Fuel Distribution

G3063 410 Ground Level Fuel Storage Tank

						2014 Bare Costs				Total In-House	Total w/O&P
	System Description	Freq. (Years)	Crew	Unit	Labor Hours	Material	Labor	Equipment	Total		
0100	**Prep & paint 100k gal. ground level fuel storage tank, 24' dia x 30' tall**	25	E-11	Ea.							
	Scaffolding, outer ring				96.000		4,416		4,416	5,850	7,225
	Scaffolding planks				47.333		2,172.60		2,172.60	2,900	3,550
	Sandblast exterior, near white blast (SSPC-SP10), loose scale, fine rust				191.997	2,538	7,911	1,431	11,880	15,700	18,600
	Epoxy primer, spray exterior				14.391	783	594		1,377	1,775	2,075
	Epoxy topcoat, 2 coats, spray exterior				30.834	1,404	1,242		2,646	3,575	4,125
	Clean up				10.800		432		432	620	755
	Total				391.355	4,725	16,767.60	1,431	22,923.60	30,420	36,330

G40 SITE ELEC. UTILITIES G4013 Electrical Distribution

G4013 210 Overhead Service Cables

	System Description	Freq. (Years)	Crew	Unit	Labor Hours	2014 Bare Costs				Total In-House	Total w/O&P
						Material	Labor	Equipment	Total		
0010	**Repair cable splice**	12	1 ELEC	M.L.F.							
	Repair splice cable, polyethylene jacket				7.619	9.90	405		414.90	515	640
	Total				7.619	9.90	405		414.90	**515**	**640**
0020	**Cable inspection**	5	1 ELEC	M.L.F.							
	Check cables for damage				.667		35.50		35.50	44	55
	Total				.667		35.50		35.50	**44**	**55**
0030	**Replace service cable**	30	2 ELEC	M.L.F.							
	Remove overhead service cables				1.860		99.50		99.50	123	154
	Overhead service 3 ACSR cable				16.000	1,875	855		2,730	3,125	3,675
	Install 3 conductors at terminations				.941		50		50	62	77.50
	Install conductors to insulators and clipping				2.500		133		133	165	206
	Install and remove sag gauge				1.000		53.50		53.50	66	82.50
	Make up & install conductor jumper				2.000		107		107	132	165
	Total				24.302	1,875	1,298		3,173	**3,673**	**4,360**

G40 SITE ELEC. UTILITIES G4023 Site Lighting

G4023 210 Outdoor Pole Lights

	System Description	Freq. (Years)	Crew	Unit	Labor Hours	2014 Bare Costs				Total In-House	Total w/O&P
						Material	Labor	Equipment	Total		
1010	**Replace 400W H.P.S. lamp, pole-mounted fixture**	10	R-26	Ea.							
	Position truck, raise and lower boom bucket				.289		15.40	5.30	20.70	24.50	30
	Remove & install 400W H.P.S. lamp, pole-mounted fixture				.185	42.75	9.90	3.40	56.05	63	72.50
	Total				.474	42.75	25.30	8.70	76.75	87.50	102.50
1020	**Replace 400W H.P.S. ballast, pole-mounted fixture**	10	R-26	Ea.							
	Position truck, raise and lower boom bucket				.289		15.40	5.30	20.70	24.50	30
	Remove & install 400W H.P.S. multi-tap ballast, pole-mounted fixt.				2.974	305	159	55	519	585	690
	Test pole-mounted H.I.D. fixture				.074		3.95	1.37	5.32	6.25	7.60
	Total				3.337	305	178.35	61.67	545.02	615.75	727.60
1030	**Replace 400W H.P.S. pole-mounted fixt. w/ lamp & blst.**	20	R-26	Ea.							
	Turn branch cicuit off and on, pole light				.370		19.75	6.85	26.60	31.50	38
	Turn branch cicuit off and on, pole light				.370		19.75	6.85	26.60	31.50	38
	Position truck, raise and lower boom bucket				.289		15.40	5.30	20.70	24.50	30
	Remove & install pole-mounted 400W H.P.S. fixture w/ lamp & blst.				9.456	375	505	174	1,054	1,200	1,450
	Test pole-mounted H.I.D. fixture				.074		3.95	1.37	5.32	6.25	7.60
	Total				10.560	375	563.85	194.37	1,133.22	1,293.75	1,563.60
1040	**Replace light pole, 2 fixtures (conc. base not incl.)**	10	R-3	Ea.							
	Turn branch cicuit off and on, pole light				.370		19.75	6.85	26.60	31.50	38
	Remove parking lot light pole with fixtures				2.500		132	17.50	149.50	182	225
	Install 2 brackets arms onto aluminum light pole				2.602	480	139		619	700	815
	Install two 400W H.P.S. fixtures w/ lamps & ballasts onto light pole				9.467	750	506		1,256	1,450	1,725
	Install 30' aluminum light pole (incl. arms & fixtures) onto base				10.000	1,750	530	70	2,350	2,650	3,100
	Pull new wires in existing underground conduits (3#10 500')				15.605	322.50	832.50		1,155	1,400	1,700
	Test pole-mounted H.I.D. fixture				.074		3.95	1.37	5.32	6.25	7.60
	Total				40.618	3,302.50	2,163.20	95.72	5,561.42	6,419.75	7,610.60

G40 SITE ELEC. UTILITIES   G4023   Site Lighting

G4023 210   Outdoor Pole Lights

						2014 Bare Costs				Total w/O&P
	System Description	Freq. (Years)	Crew	Unit	Labor Hours	Material	Labor	Equipment	Total	Total In-House	
1050	**Replace concrete base for parking lot light pole**	20	B-18	Ea.							
	Sawcut asphalt paving				1.296	3.36	54.24	42.48	100.08	116	138
	Remove concrete light pole base				10.000		372	58.40	430.40	550	670
	Excavate by hand				12.987		475		475	635	780
	Install 1-1/4" PVC electrical conduit				.612	13.60	32.60		46.20	55.50	67.50
	Install 1-1/4" PVC electrical sweep				.433	44	23		67	77	91
	Install new concrete light pole base, 24" dia. x 8'				12.030	265	525		790	990	1,200
	Backfill excavation by hand				4.728		172.50		172.50	230	285
	Compact backfill				.578		21.20	2.60	23.80	31	37.50
	Install asphalt patch				.960	71.80	37.20	3.16	112.16	132	154
	Total				43.624	397.76	1,712.74	106.64	2,217.14	2,816.50	3,423

H10 GENERAL CONDITIONS   H1043   Construction Aids

H1043 200   Temporary Cranes

| | | | | | | 2014 Bare Costs | | | | Total | Total |
	System Description	Freq. (Years)	Crew	Unit	Labor Hours	Material	Labor	Equipment	Total	In-House	w/O&P
1010	**Daily use of crane, portal to portal, 12-ton**										
	12-ton truck-mounted hydraulic crane, daily use for small jobs	1	A-3H	Day	8.000		400	875	1,275	1,375	1,600
	Total				8.000		400	875	1,275	**1,375**	**1,600**
1020	**Daily use of crane, portal to portal, 25-ton**										
	25-ton truck-mounted hydraulic crane, daily use for small jobs	1	A-3I	Day	8.000		400	1,000	1,400	1,525	1,750
	Total				8.000		400	1,000	1,400	**1,525**	**1,750**
1030	**Daily use of crane, portal to portal, 40-ton**										
	40-ton truck-mounted hydraulic crane, daily use for small jobs	1	A-3J	Day	8.000		400	1,250	1,650	1,750	2,025
	Total				8.000		400	1,250	1,650	**1,750**	**2,025**
1040	**Daily use of crane, portal to portal, 55-ton**										
	55-ton truck-mounted hydraulic crane, daily use for small jobs	1	A-3K	Day	16.000		750	1,650	2,400	2,625	3,025
	Total				16.000		750	1,650	2,400	**2,625**	**3,025**
1050	**Daily use of crane, portal to portal, 80-ton**										
	80-ton truck-mounted hydraulic crane, daily use for small jobs	1	A-3L	Day	16.000		750	2,400	3,150	3,350	3,825
	Total				16.000		750	2,400	3,150	**3,350**	**3,825**
1060	**Daily use of crane, portal to portal, 100-ton**										
	100-ton truck-mounted hydraulic crane, daily use for small jobs	1	A-3M	Day	16.000		750	2,375	3,125	3,350	3,825
	Total				16.000		750	2,375	3,125	**3,350**	**3,825**

Preventive Maintenance

Table of Contents

B20 Exterior Closure
B2035 Exterior Doors
B2035 110 Door, Sliding, Electric 383
B2035 225 Door, Emergency Egress 384
B2035 300 Door, Revolving 385
B2035 400 Door, Overhead, Manual 386
B2035 410 Door, Overhead, Roll-up, Electric 386
B2035 450 Shutter, Roll-up, Electric 387

C10 Interior Construction
C1025 Interior Doors
C1025 110 Fire Doors 388

D10 Conveying
D1015 Elevators & Lifts
D1015 100 Elevator, Electric, Pass/Freight 390
D1015 110 Elevator, Hydraulic, Pass/Freight ... 393
D1015 310 Wheelchair Lift 394

D1025 Escalators & Moving Walks
D1025 100 Escalator, Electric 395

D1095 Other Conveying Systems
D1095 100 Dumbwaiter, Electric 396
D1095 110 Dumbwaiter, Hydraulic 396
D1095 200 Pneumatic Tube System 397

D20 Plumbing
D2015 Plumbing Fixtures
D2015 100 Facility Plumbing Fixture Service ... 398
D2015 800 Drinking Fountain 399

D2025 Domestic Water Distribution
D2025 120 Valve, Butterfly 400
D2025 125 Valve, Check 400
D2025 130 Valve, Cock 401
D2025 135 Valve, Diaphragm 401
D2025 140 Valve, Gate 401
D2025 145 Valve, Globe 402
D2025 150 Valve, Motor Operated 402
D2025 155 Valve, OS&Y 403
D2025 190 Water Heater, Solar 404
D2025 260 Water Heater, Gas/Oil/Steam 405
D2025 262 Valve, Pressure Relief 407
D2025 265 Valve, Pressure Regulator 407
D2025 270 Valve, Sediment Strainer 407
D2025 310 Valve, Automatic 408

D2095 Other Plumbing Systems
D2095 905 Duplex Sump 409
D2095 910 Pump, Submersible 409
D2095 930 Oxygen Monitor 410

D30 HVAC
D3025 Heat Generating Systems
D3025 110 Boiler, Electric 411
D3025 130 Boiler, Hot Water, Oil/Gas/Comb. ... 412
D3025 140 Boiler, Steam, Oil/Gas/Comb. 416
D3025 210 Deaerator Tank 420
D3025 310 Pumps 421

D3035 Cooling Generating Systems
D3035 110 Cooling Tower 423
D3035 130 Chiller, Recip., Air Cooled 425
D3035 135 Chiller, Recip., Water Cooled 426
D3035 140 Chiller, Centrif., Water Cooled 427
D3035 150 Chiller, Absorption Unit 429
D3035 160 Chiller, Screw, Water Cooled 430
D3035 170 Evaporative Cooler 431
D3035 180 Evap. Cooler, Rotating Drum 431
D3035 210 Condenser, Air Cooled 432
D3035 220 Condensing Unit, Air Cooled 434
D3035 240 Condensing Unit, Water Cooled ... 436
D3035 260 Compressor, DX Refrigeration 438
D3035 290 Fluid Cooler 439

D3045 Distribution Systems
D3045 110 Air Handling Unit 440
D3045 112 A.H.U., Computer Room 441
D3045 120 Fan Coil Unit 442
D3045 150 Air Filters, Electrostatic 442
D3045 160 VAV Boxes 443
D3045 170 Fire Dampers 443
D3045 210 Fan, Axial 444
D3045 220 Fan, Centrifugal 446
D3045 250 Hood and Blower 447
D3045 410 Pump, Centrifugal 448
D3045 420 Pump w/Oil Reservoir, Electric 449
D3045 600 Heat Exchanger, Steam 450

D3055 Terminal & Package Units
D3055 110 Space Heater 451
D3055 122 Forced Air Heater, Oil/Gas Fired ... 453
D3055 210 Package Unit, Air Cooled 454
D3055 220 Package Unit, Water Cooled 455
D3055 230 Package Unit, Computer Room ... 456
D3055 240 Package Terminal Unit 457
D3055 250 Air Conditioning, Split System ... 458
D3055 310 Heat Pump, Air Cooled 459
D3055 320 Heat Pump, Water Cooled 460

D3065 Controls & Instrumentation
D3065 100 Controls, Central System 461

D3095 Other HVAC Systems & Equipment
D3095 110 Air Compressor, Gas Engine 462
D3095 114 Air Compressor, Centrifugal 463
D3095 118 Air Compressor, Reciprocating ... 465
D3095 210 Humidification Systems 467
D3095 220 Dehumidifier, Desiccant Wheel ... 468

D40 Fire Protection
D4015 Sprinklers
D4015 100 Backflow Prevention Device 469
D4015 150 Extinguishing System, Wet Pipe ... 470
D4015 180 Extinguishing System, Deluge/Preaction 471
D4015 210 Fire Pump, Electric Motor Driven ... 472
D4015 250 Fire Pump, Motor/Engine Driven ... 473
D4015 310 Extinguishing System, Dry Pipe ... 474

D4095 Other Fire Protection Systems
D4095 100 Extinguishing System, CO_2 475
D4095 200 Extinguishing Sys., Foam Bottle ... 476
D4095 210 Extinguishing Sys., Foam, Electric ... 477
D4095 220 Extinguishing Sys., Foam, Diesel ... 478
D4095 400 Extinguishing Sys., Dry Chemical ... 479
D4095 450 Extinguishing System, FM200 480

D50 Electrical
D5015 Electrical Service & Distribution
D5015 210 Switchboard, Electrical 481
D5015 214 Switchboard, w/Air Circ. Breaker ... 481
D5015 217 Switchboard, w/Air C.B. and Tie ... 482
D5015 220 Circuit Breaker, HV Air 483
D5015 222 Circuit Breaker, HV Oil 483
D5015 230 Switch, Selector, HV Air 484
D5015 232 Switch, Selector, HV Oil 484
D5015 234 Switch, Automatic Transfer 485
D5015 236 Switch, Interrupt, HV, Fused Air ... 485
D5015 238 Switch, HV w/Aux. Fuses, Air 485
D5015 240 Transformer 486
D5015 260 Panelboard, 225 Amps and Above ... 488
D5015 280 Motor Control Center 488

D5035 Communication & Security
D5035 610 Clock Systems 489
D5035 710 Fire Alarm Annunciator System ... 490
D5035 810 Security/Intrusion Alarm Control Panel 490

D5095 Other Electrical Systems
D5095 210 Emergency Generator 491
D5095 220 Power Stabilizer 493
D5095 230 Uninterruptible Power System ... 494
D5095 240 Wet Battery System and Charger ... 495
D5095 250 Lighting System, Emergency 496

E10 Equipment
E1015 Commercial Equipment
E1015 810 Personal Computers 497

E1035 Vehicular Equipment
E1035 100 Hydraulic Lift, Automotive 498
E1035 310 Hydraulic Lift, Loading Dock 498

Table of Contents (cont.)

E1095 Other Equipment
E1095 122 Crane, Electric Bridge 499
E1095 123 Crane, Manual Bridge 501
E1095 125 Hoist, Pneumatic 502
E1095 126 Hoist/Winch, Chain/Cable 503
E1095 302 Beverage Dispensing Unit 503
E1095 304 Braising Pan 504
E1095 306 Bread Slicer 504
E1095 310 Broiler 505
E1095 312 Chopper, Electric 506
E1095 314 Coffee Maker/Urn 507
E1095 316 Cooker 508
E1095 318 Cookie Maker 509
E1095 320 Deep Fat Fryer 510
E1095 322 Dishwasher, Electric/Steam 512
E1095 328 Dispenser, Beverage 514
E1095 332 Disposal, Garbage 514
E1095 334 Dough, Divider/Roller 515
E1095 336 Drink Cooler 516
E1095 338 Fluid Cooler 516
E1095 340 Food Saw/Slicer 517
E1095 346 Grill, Gas/Electric 518
E1095 348 Ice Machine 518
E1095 350 Kettle, Steam 519
E1095 354 Mixer, Electric 520
E1095 356 Oven, Convection/Rotary 521
E1095 366 Peeler, Vegetable 522
E1095 368 Pie Maker, Electric 523
E1095 370 Proofer, Automatic 523
E1095 380 Refrigerator 524
E1095 382 Refrigerator/Freezer 526
E1095 386 Steam Table 527
E1095 388 Steamer, Vegetable 527
E1095 390 Toaster, Rotary 528
E1095 930 Pump, Vacuum 528

F10 Special Construction
F1045 Special Facilities
F1045 110 Swimming Pool 529

G20 Site Improvements
G2015 Roadways
G2015 610 Traffic Signal Light 530

G2045 Site Development
G2045 150 Chain Link Fence Gates 531

G30 Site Mechanical Utilities
G3015 Water Supply
G3015 108 Storage Tank 533
G3015 116 Water Flow Meter 534
G3015 118 Reservoir Controls 534
G3015 126 Pumps 535
G3015 310 Pump, Turbine Well 540
G3015 320 Pump, Vertical Lift 540
G3015 410 Fire Hydrant 541
G3015 420 Valve, Post Indicator 541
G3015 612 Water Treatment, Filtration Plant 542
G3015 630 Water Treatment, De-Ionization 542
G3015 632 Reverse Osmosis System 543
G3015 634 Water Ion Exchange System 543
G3015 650 Water Softener 544
G3015 660 Fluoride Saturator 544
G3015 670 Water Treatment, Chlorine 545

G3025 Sanitary Sewer
G3025 410 Pump, Sewage Ejector 546
G3025 412 Ejector Pump 547
G3025 420 Pump, Sewage Lift 548
G3025 530 Aerator, Floating 549
G3025 532 Barminutor 549
G3025 534 Blower, Aerator 550
G3025 536 Clarifier 550
G3025 538 Comminutor 551
G3025 540 Filter, Trickling 551
G3025 542 Grit Drive 552
G3025 544 Mixer, Sewage 552

G3065 Fuel Distribution
G3065 310 Fuel Oil Tank 553

G40 Site Electrical Utilities
G4095 Other Site Electrical Utilities
G4095 110 Cathodic Protection System 554

How to Use the Preventive Maintenance Cost Tables

The following is a detailed explanation of a sample Preventive Maintenance Cost Table. The Preventive Maintenance Tables are separated into two parts: 1) the components and related labor-hours and frequencies of a typical system and 2) the costs for systems on an annual or annualized basis. Next to each bold number that follows is the described item with the appropriate component of the sample entry following in parentheses.

1 System/Line Numbers (D3025 130 4950)

Each Maintenance & Repair Assembly has been assigned a unique identification number based on the UNIFORMAT II classification system.

D30 HVAC		D3025 130 **1**	Boiler, Hot Water, Oil/Gas/Comb.					
PM Components			Labor-hrs. **3**	W	M	Q	S	A
PM System D3025 130 4950						**4**		
Boiler, hot water; oil, gas or combination fired, over 1000 MBH								
1	Check combustion chamber for air or gas leaks.		.117					√
2	Inspect and clean oil burner gun and ignition assembly where applicable.		.987					√
3	Inspect fuel system for leaks and change fuel filter element, where applicable.		.147					√
4	Check fuel lines and connections for damage.		.035		√	√	√	√
5	Check for proper operational response of burner to thermostat controls.	**2**	.199			√	√	√
6	Check and lubricate burner and blower motors.		.120			√	√	√
7	Check main flame failure protection and main flame detection scanner on boiler equipped with spark ignition (oil burner).		.186		√	√	√	√
8	Check electrical wiring to burner controls and blower.		.120					√
9	Clean firebox (sweep and vacuum).		.819					√
10	Check operation of mercury control switches (i.e., steam pressure, hot water temperature limit, atomizing or combustion air proving, etc.).		.215		√	√	√	√
11	Check operation and condition of safety pressure relief valve.		.046		√	√	√	√
12	Check operation of boiler low water cut off devices.		.085		√	√	√	√
13	Check hot water pressure gauges.		.109		√	√	√	√
14	Inspect and clean water column sight glass (or replace).		.191		√	√	√	√
15	Clean fire side of water jacket.		.433					√
16	Check condition of flue pipe, damper and exhaust stack.		.221			√	√	√
17	Check boiler operation through complete cycle, up to 30 minutes.		.887					√
18	Check fuel level with gauge pole, add as required.		.098		√	√	√	√
19	Clean area around boiler.		.182		√	√	√	√
20	Fill out maintenance checklist and report deficiencies.		.022		√	√	√	√
	Total labor-hours/period			**5**	1.169	1.709	1.709	5.219
	Total labor-hours/year				9.352	3.418	1.709	5.219

			Cost Each					
		7	2014 Bare Costs				**9** Total	**10** Total
Description **6**		Labor-hrs.	Material	Labor	Equip.	Total	In-House	w/O&P
4900	Boiler, hot water, O/G/C, over 1000 MBH, annually	5.219	120	305	**8**	425	515	625
4950	Annualized	19.698	125	1,150		1,275	1,600	1,950

② PM Components

The individual preventive maintenance operations required to be performed annually are listed separately to show what has been included in the development of the total system price. The cost table below the PM Components listing contains prices for the system on an annual or annualized basis.

③ Labor-hrs.

The "Labor-hrs." figure represents the number of labor-hours required to perform the individual operations one time.

④ Frequency

The columns marked with "W" for weekly, "M" for monthly, "Q" for quarterly, "S" for semi-annually, and "A" for annually indicate the recommended frequency for performing a task. For example, a check in the "M" column indicates the recommended frequency of the operation as monthly. Note that, as a result, checks also appear in the "Q," "S," and "A" columns, as they would be part of a monthly frequency.

⑤ Total Labor-hour

"Total Labor-hours" are provided on both a per-period and per-year basis. The per-period totals are derived by summing the hours for each operation recommended to be performed (indicated with a check mark) in a given period. In the example given here:

M	= .035 + .186 + .215 + .046 + .085 + .109 + .191 + .068 + .182 + .022	= 1.169
Similarly		
Q	= .035 + .199 + .120 + ... + .098 + .182 + .022	= 1.709
S	= .035 + .199 + .120 + ... + .098 + .182 + .022	= 1.709
A	= .117 + .987 + .147 + .035 + ... + .887 + .098 + .182 + .022	= 5.219

The per-year totals are derived by multiplying the per-period total by the net frequency for each time period. The net frequency per period is based on the following table using a 50-week work year:

Total Labor-hours		Labor-hours		
Annual	=	Annual	x	1
Semi-Annual	=	Semi-Annual	x	1
Quarterly	=	Quarterly	x	2
Monthly	=	Monthly	x	8
Weekly	=	Weekly	x	38
				50

Based on these values, the total labor-hours/year are calculated for the given example as follows:

M	=	(1.169)	x	8	=	9.352
Q	=	(1.709)	x	2	=	3.418
S	=	(1.709)	x	1	=	1.709
A	=	(5.219)	x	1	=	5.219
						19.698 hrs.

⑥ System Description (Boiler, Hot Water, O/G/C, etc.)

The "System Description" corresponds to that given at the beginning of the PM Components list. A description of the cost basis, annual or annualized, is given as well. Annual costs are based on performing each operation listed in the PM Components one time in a given year. Annualized costs are based on performing each operation at the recommended periodic frequencies in a given year.

⑦ Labor-hours: Annual (5.219)/ Annualized (19.698)

Annual labor-hours are based on performing each task once during a given year and are equal to the total labor-hours/period under the "A" column in the PM Components list. Annualized labor-hours are based on the recommended periodic frequencies and are equivalent to the sum of all the values in the total labor-hours/year row at the bottom of the PM Components.

8 Bare Costs: Annualized Material (125.00)

This figure for the system material cost is the "bare" material cost with no overhead and profit allowances included. *Costs shown reflect national average material prices for the current year and generally include delivery to the facility. Small purchases may require an additional delivery charge. No sales taxes are included.*

Labor (1,150.00)

The labor costs are derived by multiplying bare labor-hour costs by labor-hour units. (The hourly labor costs are listed in the Crew Listings, or for a single trade, the wage rate is found on the inside back cover of the book.)

Equip. (Equipment) (0.00)

The unit equipment cost is derived by multiplying the bare equipment hourly cost by the labor-hour units. Equipment costs for each crew are listed in the description of each crew.

Total (1,275.00)

The total of the bare costs is the arithmetic total of the three previous columns: mat., labor, equip.

Material	+	Labor	+	Equip.	=	Total
$125.00	+	$1,150.00	+	$0.00	=	$1,275.00

9 Total In-House Costs Annualized (1,600)

"Total In-House Costs" include markups to bare costs for the direct overhead requirements of in-house maintenance staff. The figure in this column is the sum of three components: the bare material cost plus 10% (for purchasing and handling small quantities); the bare labor cost plus workers' compensation and average fixed overhead (per the labor rate table on the inside back cover of the book or, if a crew is listed, from the Crew Listings); and the bare equipment cost.

10 Total Costs Including O&P Annualized (1,950)

"Total Costs Including O&P" include suggested markups to bare costs for an outside subcontractor's overhead and profit requirements. The figure in this column is the sum of three components: the bare material cost plus 25% for profit; the bare lobar cost plus total overhead and profit (per the labor rate table on the inside back cover of this book or, if a crew is listed, from the Crew Listings); and the bare equipment cost plus 10% for profit.

B20 EXTERIOR CLOSURE — B2035 110 Door, Sliding, Electric

PM Components	Labor-hrs.	W	M	Q	S	A
PM System B2035 110 1950						
Door, sliding, electric						
1 Check with operating or area personnel for deficiencies.	.035				✓	✓
2 Check for proper operation, binding or misalignment; adjust as necessary.	.062				✓	✓
3 Check and lubricate door guides, pulleys and hinges.	.142				✓	✓
4 Inspect and lubricate motor gearbox, drive chain (or belt), and motor; adjust as necessary.	.200				✓	✓
5 Check operation of limit switch; adjust as necessary.	.224				✓	✓
6 Check electrical operator, wiring, connections and contacts; adjust as necessary.	.473				✓	✓
7 Clean area around door.	.066				✓	✓
8 Fill out maintenance checklist and report deficiencies.	.022				✓	✓
Total labor-hours/period					1.224	1.224
Total labor-hours/year					1.224	1.224

		Cost Each					
		2014 Bare Costs				Total	Total
Description	Labor-hrs.	Material	Labor	Equip.	Total	In-House	w/O&P
1900 Door, sliding, electric, annually	1.224	14.90	56		70.90	91	111
1950 Annualized	2.448	20.50	112		132.50	172	210

PM Components	Labor-hrs.	W	M	Q	S	A
PM System B2035 110 2950						
Hanger doors, sliding						
1 Check with door operating personnel for deficiencies.	.027			✓	✓	✓
2 Remove debris from door track.	.040			✓	✓	✓
3 Operate door.	.050			✓	✓	✓
4 Check alignment of hanger door, door guides and lubricate.	.650					✓
5 Inspect and lubricate wheel drive chain.	.030			✓	✓	✓
6 Adjust brake shoes and check for wear.	.150					✓
7 Inspect and lubricate drive wheels, guides, stops and rollers.	.200			✓	✓	✓
8 Fill out maintenance report.	.017			✓	✓	✓
Total labor-hours/period				.364	.364	1.164
Total labor-hours/year				.728	.364	1.164

		Cost Each					
		2014 Bare Costs				Total	Total
Description	Labor-hrs.	Material	Labor	Equip.	Total	In-House	w/O&P
2900 Hanger doors, sliding, annually	1.164	26	59.50		85.50	119	142
2950 Annualized	2.256	47	115		162	228	271

B20 EXTERIOR CLOSURE B2035 225 Door, Emergency Egress

PM Components	Labor-hrs.	W	M	Q	S	A
PM System B2035 225 1950						
Door, emergency egress, swinging						
1 Remove obstructions that restrict full movement/swing of door.	.013			√	√	√
2 Check swing of door; door must latch on normal closing.	.013			√	√	√
3 Test operation of panic hardware and local alarm battery.	.013			√	√	√
4 Lubricate hardware.	.013			√	√	√
5 Fill out maintenance checklist and report deficiencies.	.013			√	√	√
Total labor-hours/period				.065	.065	.065
Total labor-hours/year				.130	.065	.065

				Cost Each				
				2014 Bare Costs			Total	Total
Description		Labor-hrs.	Material	Labor	Equip.	Total	In-House	w/O&P
1900	Door, emergency egress, swinging, annually	.065	14.90	2.98		17.88	20.50	23.50
1950	Annualized	.260	20.50	11.90		32.40	38.50	45

B20 EXTERIOR CLOSURE B2035 300 Door, Revolving

PM Components	Labor-hrs.	W	M	Q	S	A
PM System B2035 300 1950						
Door, revolving, manual						
1 Check with operating or area personnel for deficiencies.	.043				✓	✓
2 Check for proper operation, binding or misalignment; adjust as necessary.	.043				✓	✓
3 Check for proper emergency breakaway operation of wings.	.043				✓	✓
4 Check top, bottom, edge and shaft weatherstripping.	.217				✓	✓
5 Check and lubricate pivots and bearings.	.217				✓	✓
6 Clean area around door.	.065				✓	✓
7 Fill out maintenance checklist and report deficiencies.	.022				✓	✓
Total labor-hours/period					.650	.650
Total labor-hours/year					.650	.650

			Cost Each				
			2014 Bare Costs			Total	Total
Description	Labor-hrs.	Material	Labor	Equip.	Total	In-House	w/O&P
1900 Door, revolving, manual, annually	.650	14.90	30		44.90	56	67.50
1950 Annualized	1.300	20.50	59.50		80	102	124

PM Components	Labor-hrs.	W	M	Q	S	A
PM System B2035 300 2950						
Door, revolving, electric						
1 Check with operating or area personnel for deficiencies.	.043				✓	✓
2 Check for proper operation, binding or misalignment; adjust as necessary.	.043				✓	✓
3 Check for proper emergency breakaway operation of wings.	.043				✓	✓
4 Check top, bottom, edge and shaft weatherstripping.	.217				✓	✓
5 Check elec. operator, safety devices and activators for proper operation.	.433				✓	✓
6 Check and lubricate pivots and bearings and operator gearbox.	.325				✓	✓
7 Clean area around door.	.065				✓	✓
8 Fill out maintenance checklist and report deficiencies.	.022				✓	✓
Total labor-hours/period					1.191	1.191
Total labor-hours/year					1.191	1.191

			Cost Each				
			2014 Bare Costs			Total	Total
Description	Labor-hrs.	Material	Labor	Equip.	Total	In-House	w/O&P
2900 Door, revolving, electric, annually	1.192	43	54.50		97.50	120	144
2950 Annualized	2.382	46	109		155	196	238

B20 EXTERIOR CLOSURE — B2035 400 — Door, Overhead, Manual

PM Components	Labor-hrs.	W	M	Q	S	A
PM System B2035 400 1950						
Door, overhead, manual, up to 24' high x 25' wide						
1 Check with operating or area personnel for deficiencies.	.035				√	√
2 Check for proper operation, binding or misalignment; adjust as necessary.	.135				√	√
3 Check and lubricate door guides, pulleys and hinges.	.880				√	√
4 Clean area around door.	.026				√	√
5 Fill out maintenance checklist and report deficiencies.	.022				√	√
Total labor-hours/period					1.098	1.098
Total labor-hours/year					1.098	1.098

			Cost Each				
			2014 Bare Costs			Total	Total
Description		Labor-hrs.	Material	Labor	Equip.	Total	In-House	w/O&P
1900	Dr., overhead, manual, up to 24'H x 25'W, annually	1.098	14.90	50.50		65.40	83.50	102
1950	Annualized	2.196	20.50	101		121.50	157	191

B20 EXTERIOR CLOSURE — B2035 410 — Door, Overhead, Roll-up, Electric

PM Components	Labor-hrs.	W	M	Q	S	A
PM System B2035 410 1950						
Door, overhead, roll-up, electric, up to 24' high x 25' wide						
1 Check with operating or area personnel for deficiencies.	.035				√	√
2 Check for proper operation, binding or misalignment; adjust as necessary.	.135				√	√
3 Check and lubricate door guides, pulleys and hinges.	.880				√	√
4 Inspect and lubricate motor gearbox, drive chain (or belt), and motor; adjust as necessary.	.200				√	√
5 Check operation of limit switch; adjust as necessary.	.224				√	√
6 Check electrical operator, wiring, connections and contacts; adjust as necessary.	.473				√	√
7 Clean area around door.	.066				√	√
8 Fill out maintenance checklist and report deficiencies.	.022				√	√
Total labor-hours/period					2.035	2.035
Total labor-hours/year					2.035	2.035

			Cost Each				
			2014 Bare Costs			Total	Total
Description		Labor-hrs.	Material	Labor	Equip.	Total	In-House	w/O&P
1900	Dr., overhead, electric, to 24'H x 25'W, annually	2.035	43	93.50		136.50	173	207
1950	Annualized	4.070	46	187		233	299	365

B20 EXTERIOR CLOSURE B2035 450 | Shutter, Roll-up, Electric

PM Components	Labor-hrs.	W	M	Q	S	A
PM System B2035 450 1950						
Shutter, roll-up, electric						
1 Check with operating or area personnel for deficiencies.	.035					√
2 Check for proper operation, binding or misalignment; adjust as necessary.	.135					√
3 Check and lubricate door guides, pulleys and hinges.	.880					√
4 Inspect and lubricate motor gearbox, drive chain (or belt), and motor; adjust as necessary.	.200					√
5 Check operation of limit switch; adjust as necessary.	.224					√
6 Check electrical operator, wiring, connections and contacts; adjust as necessary.	.473					√
7 Clean area around door.	.074					√
8 Fill out maintenance checklist and report deficiencies.	.022					√
Total labor-hours/period						2.043
Total labor-hours/year						2.043

			Cost Each				
			2014 Bare Costs			Total	Total
Description	Labor-hrs.	Material	Labor	Equip.	Total	In-House	w/O&P
1900 Shutter, roll up, electric, annually	2.043	23.50	93.50		117	151	183
1950 Annualized	2.043	29.50	93.50		123	158	191

C10 INTERIOR CONSTR. C1025 110 | Fire Doors

PM Components	Labor-hrs.	W	M	Q	S	A
PM System C1025 110 1950						
Fire Door, Swinging						
1 Remove fusible link hold open devices.	.026			√	√	√
2 Remove obstructions that retard full movement/swing of door.	.013			√	√	√
3 Check swing of door. Door must latch on normal closing.	.013			√	√	√
4 Test operation of panic hardware.	.007			√	√	√
5 Check operation of special devices such as smoke detectors or magnetic door releases.	.013			√	√	√
6 Lubricate hardware.	.013			√	√	√
7 Fill out maintenance checklist and report deficiencies.	.013			√	√	√
Total labor-hours/period				.098	.098	.098
Total labor-hours/year				.196	.098	.098

			Cost Each					
			2014 Bare Costs				Total	Total
Description		Labor-hrs.	Material	Labor	Equip.	Total	In-House	w/O&P
1900	Fire doors, swinging, annually	.098	3.55	4.64		8.19	10.05	12
1950	Annualized	.392	7.10	18.55		25.65	32.50	39

PM Components	Labor-hrs.	W	M	Q	S	A
PM System C1025 110 2950						
Fire Door, Sliding						
1 Clean track.	.013			√	√	√
2 Lubricate pulleys.	.013			√	√	√
3 Inspect cable or chain for wear or damage and proper threading through pulleys.	.078			√	√	√
4 Replace fusible links or other heat activated devices that have been painted. Check operation of other heat activated devices.	.130			√	√	√
5 Permanently remove obstructions that prohibit movement of door.	.013			√	√	√
6 Check operation of door by disconnecting or lifting counterweight or other appropriate means.	.130			√	√	√
7 Check fit of door in binders and fit against stay roll.	.013			√	√	√
8 Check door for breaks in covering and examine for dry rot.	.013			√	√	√
9 Fill out maintenance checklist and report deficiencies.	.013			√	√	√
Total labor-hours/period				.416	.416	.416
Total labor-hours/year				.832	.416	.416

			Cost Each					
			2014 Bare Costs				Total	Total
Description		Labor-hrs.	Material	Labor	Equip.	Total	In-House	w/O&P
2900	Fire doors, sliding, annually	.416	4.35	19.70		24.05	31	37.50
2950	Annualized	1.664	8.70	78.50		87.20	113	139

C10 INTERIOR CONSTR.   C1025 110 | Fire Doors

PM Components	Labor-hrs.	W	M	Q	S	A
PM System C1025 110 3950						
Fire Door, Rollup						
1 Check with door operating personnel for deficiencies.	.027			√	√	√
2 Drop test heat activated or fusible link roll up fire door.	.070			√	√	√
3 Adjust rate of descent governor to desired closing speed.	.200					√
4 Inspect, clean and lubricate door gear assembly.	.200			√	√	√
5 Check alignment of overhead door guides, clean and lubricate.	.100			√	√	√
6 Check and adjust door counter balance assembly.	.250					√
7 Check electrical wiring and contacts for wear.	.200					√
8 Check and adjust limit switches.	.170					√
9 Inspect general condition of door for need of paint and repairs.	.005			√	√	√
10 Replace missing or tighten loose nuts and bolts.	.040			√	√	√
11 After maintenance clean around door and ensure access.	.050			√	√	√
12 Fill out maintenance report.	.017			√	√	√
Total labor-hours/period				.509	.509	1.329
Total labor-hours/year				1.018	.509	1.329

			Cost Each				
			2014 Bare Costs			Total	Total
Description		Labor-hrs.	Material	Labor	Equip.	Total	In-House	w/O&P
3900	Fire doors, roll-up, annually	1.329	4.70	68		72.70	109	131
3950	Annualized	2.856	18.80	146		164.80	243	292

D10 CONVEYING — D1015 100 | Elevator, Electric, Pass/Freight

PM Components	Labor-hrs.	W	M	Q	S	A
PM System D1015 100 1950						
Elevator, cable, electric, passenger/freight						
1 Ride car, check for any unusual noise or operation.	.411		✓	✓	✓	✓
2 Inspect machine room equipment.	.286		✓	✓	✓	✓
3 Motor room:						
A) visually inspect controllers and starters.	.012		✓	✓	✓	✓
B) visually inspect selector.	.012		✓	✓	✓	✓
C) lubricate and adjust tension of selector.	.113		✓	✓	✓	✓
D) inspect sleeve bearings of hoist motor.	.012		✓	✓	✓	✓
E) inspect brushes and commutator of hoist motor.	.012		✓	✓	✓	✓
F) inspect sleeve bearings of motor generator.	.012		✓	✓	✓	✓
G) inspect brushes and commutator of motor generator.	.014		✓	✓	✓	✓
H) inspect sleeve bearings of exciter.			✓	✓	✓	✓
I) inspect brushes and commutator of exciter.	.022		✓	✓	✓	✓
J) inspect sleeve bearings of regulator dampening motors and tach generators.	.022		✓	✓	✓	✓
K) inspect brushes and commutator of regulator dampening motors and tach generators.	.044		✓	✓	✓	✓
L) inspect and grease sleeve bearings of geared machines.	.047		✓	✓	✓	✓
M) inspect oil level in worm and gear of geared machines.	.044		✓	✓	✓	✓
N) visually inspect the brakes.	.022		✓	✓	✓	✓
O) inspect and grease, if necessary, sleeve bearings of drive deflectors and secondary sheaves.	.186		✓	✓	✓	✓
P) visually inspect and adjust contacts of controllers and starters.	.338			✓	✓	✓
Q) inspect and adjust main operating contactor and switches of controllers and starters.	.338			✓	✓	✓
R) inspect selector, adjusting contacts, brushes and cams if needed.	.103			✓	✓	✓
S) inspect selsyn and advancer, motor brushes and commutators of selector.	.022			✓	✓	✓
T) clean hoist motor brush rigging and commutator.	.068			✓	✓	✓
U) clean motor generator and tighten brush rigging and clean commutator.	.068			✓	✓	✓
V) clean exciter brush rigging and commutator.	.066			✓	✓	✓
W) clean brush rigging and commutator of regulator dampening motors and tach generators.	.099			✓	✓	✓
X) inspect machine worm and gear thrust and backlash and spider bolts; replace oil.	.047			✓	✓	✓
Y) clean and lubricate brakes, pins and linkage.	.099			✓	✓	✓
Z) inspect governors; lubricate sleeve bearings.	.099			✓	✓	✓
Aa) inspect controllers and starters; adjust setting of overloads.	.169					✓
Ab) tighten connections on controllers and starters; clean fuses and holders.	.169					✓
Ac) set controllers and starter timers.	.169					✓
Ad) clean controllers and starters.	.169					✓
Ae) clean and inspect selector components.	.068					✓
Af) inspect group operation controllers; clean and lubricate contacts, timers, steppers of operation controller.	.099					✓
Ag) change oil and/or grease roller bearings of the hoist motor.	.094					✓
Ah) vacuum and blowout the hoist motor.	.094					✓
Ai) tighten all connections on the hoist motor.	.022					✓

D10 CONVEYING — D1015 100 Elevator, Electric, Pass/Freight

	PM Components	Labor-hrs.	W	M	Q	S	A
	Aj) change oil and/or grease generator's roller bearings.	.069					✓
	Ak) vacuum and blowout motor generator.	.094					✓
	Al) inspect all connections on the motor generator and tighten if necessary.	.022					✓
	Am) inspect roller bearings of the exciter and change oil and/or grease.	.069					✓
	An) vacuum and blowout exciter.	.094					✓
	Ao) change oil and/or grease roller bearings of regulator dampening motors and tach generators.	.094					✓
	Ap) vacuum/blowout regulator dampening motors and tach generators.	.044					✓
	Aq) inspect all regulator dampening motors and tach generator connections.	.047					✓
	Ar) change oil and/or grease roller bearings of the geared machines.	.094					✓
	As) clean and lubricate brake cores.	.099					✓
	At) inspect drive deflectors and secondary sheaves; change oil and grease roller bearings if necessary.	.170					✓
	Au) inspect drive deflectors and secondary sheaves' cable grooves for cracks.	.012					✓
	Av) inspect governors; change oil and grease roller bearings, if necessary.	.094					✓
4	Hatch:						
	A) lubricate rails of hoistway.	.733		✓	✓	✓	✓
	B) inspect car top and grease sleeve bearings if needed.	.151		✓	✓	✓	✓
	C) inspect counterweight and grease sleeve bearings if needed.	.068		✓	✓	✓	✓
	D) visually inspect cables, chains, hoist, compensating governor for wear and equalization.	.047		✓	✓	✓	✓
	E) lubricate cables, chain hoist and compensating governor if necessary.	.186		✓	✓	✓	✓
	F) inspect door operator; clean and lubricate chain and belt tension.	.177			✓	✓	✓
	G) inspect door, clean and adjust safety edge, light ray and cables.	.052			✓	✓	✓
	H) clean and adjust proximity devices on door.	.103			✓	✓	✓
	I) inspect and lubricate governor tape and compensating sheave in pit.	.068			✓	✓	✓
	J) clean and lubricate compensating tie down in pit.	.047			✓	✓	✓
	K) inspect pit's run-by on counter weight.	.142			✓	✓	✓
	L) inspect governor and compensating sheaves or chains for clearance to pit floor.	.022			✓	✓	✓
	M) inspect/lubricate overhead hatch switches and cams.	.096					✓
	N) clean and lubricate rollers, cables, sight guards, chains, motors, closures, limit and zone switches.	.068					✓
	O) inspect door gibs and fastening.	.022					✓
5	Inspect car top; change oil and grease roller bearings, if necessary.	.044					✓
6	Check safety switches, indicators, leveling devices, selector tape, switches and hatches.	.044					✓
7	Check and lubricate fan motor.	.072					✓
8	Inspect counterweight; change oil and grease roller bearings, if necessary.	.094					✓
9	Check safety linkage safety and cwt guides.	.098					✓
10	Inspect compensation hatch.	.047					✓
11	Inspect chains, hoist and compensating governor.	.094					✓
12	Inspect traveling cables for tracking and wear.	.070					✓

D10 CONVEYING — D1015 100 Elevator, Electric, Pass/Freight

PM Components	Labor-hrs.	W	M	Q	S	A
13 Inspect door operator commutator and brushes.	.012					✓
14 Inspect/clean/lubricate door operator clutch, retiring cam, door gib, rollers, tracks, upthrusts and relating cables.	.103					✓
15 Under car:						
A) inspect/clean/lubricate safety devices, linkages, car guides, selector tape, switches and hitches.	.020					✓
B) inspect cable hitch and loops; load weighing devices.	.027					✓
16 Inspect/lubricate pit hatch switches, cams and pit oil buffer.	.083					✓
17 In car:						
A) inspect and clean fixtures and signal in operating panel and car position and direction indicator.	.103		✓	✓	✓	✓
B) check operation of emergency lights and bells.	.228		✓	✓	✓	✓
C) check handrails, ceiling panels and hang on panels for tightness.	.027		✓	✓	✓	✓
D) check for tripping hazards.	.044		✓	✓	✓	✓
18 Hallway corridor:						
A) inspect hall buttons, signal lamps, lanterns and hall position indicator.	.046		✓	✓	✓	✓
B) inspect starter station, key operation and lamps.	.069		✓	✓	✓	✓
19 Clean equipment and surrounding area.	.066		✓	✓	✓	✓
20 Fill out maintenance checklist.	.022		✓	✓	✓	✓
Total labor-hours/period			3.061	5.019	5.019	8.068
Total labor-hours/year			24.488	10.038	5.019	8.068

		Cost Each					
		2014 Bare Costs				Total	Total
Description	Labor-hrs.	Material	Labor	Equip.	Total	In-House	w/O&P
1900 Elevator, cable, electric, passenger / freight, annually	8.068	1,550	600		2,150	2,450	2,875
1950 Annualized	47.613	3,075	3,525		6,600	7,750	9,300

D10 CONVEYING | D1015 110 | Elevator, Hydraulic, Pass/Freight

PM Components	Labor-hrs.	W	M	Q	S	A
PM System D1015 110 1950						
Elevator, hydraulic, passenger/freight						
1 Ride car, checking for any unusual noise or operation.	.052		✓	✓	✓	✓
2 In car:						
A) inspect and clean fixtures and signal in operating panel and car position and direction indicator.	.103		✓	✓	✓	✓
B) check operation of emergency lights and bell.	.020		✓	✓	✓	✓
C) check handrails, ceiling panels and hang on panels for tightness.	.027		✓	✓	✓	✓
D) check for tripping hazards.	.012		✓	✓	✓	✓
3 Inspect and lubricate rails of hoistway.	.094		✓	✓	✓	✓
4 Hallway corridor:						
A) inspect hall buttons, signal lamps, lanterns and hall position indicator.	.046		✓	✓	✓	✓
B) inspect starter station, key operation and lamps.	.069		✓	✓	✓	✓
5 Motor room:						
A) inspect machine room equipment.	.077		✓	✓	✓	✓
B) lockout and log record.	.022		✓	✓	✓	✓
C) inspect tank oil level.	.012		✓	✓	✓	✓
D) inspect and adjust controller contacts; main operating contactors and switches.	.196			✓	✓	✓
E) inspect pump and valve unit for leaks.	.051			✓	✓	✓
F) inspect and adjust controller overloads; set timers.	.049					✓
G) tighten connections and clean controller fuses and holders.	.020					✓
H) inspect and lubricate pump motor bearings.	.099					✓
6 Hatch:						
A) check hoistway car rails, brackets and fish plates.	.103					✓
B) inspect/lubricate overhead hatch switches and cams.	.070					✓
C) inspect/clean/lubricate hatch doors locks, rollers, tracks, upthrusts, relating cables, racks, sight guards and closers, motors, gear boxes, limit and zone switches.	.096					✓
D) inspect door gibs and fastening.	.022					✓
E) inspect/clean/lubricate cab top guides, steading devices, safety switches, inductors, leveling devices, selector tape, switches, hitches and fan motor.	.068					✓
F) check traveling cables for wear.	.070					✓
G) inspect/clean/lubricate door operator roller tracks, upthrusts, related cables, clutch, retiring cam and door gib.	.078					✓
H) inspect door operator; clean and lubricate chain and belt tension.	.052			✓	✓	✓
I) inspect door, clean and adjust safety edge, light ray and cables.	.103			✓	✓	✓
J) clean and adjust proximity devices on door.	.068			✓	✓	✓
7 Inspect pit gland packing.	.010			✓	✓	✓
8 Inspect/clean/lubricate under car guides, selector tape, traveling cable, switches and platen plate assembly.	.068					✓
9 Clean equipment and surrounding area.	.074		✓	✓	✓	✓
10 Fill out maintenance checklist.	.022		✓	✓	✓	✓
Total labor-hours/period			.630	1.110	1.110	1.853
Total labor-hours/year			5.040	2.220	1.110	1.853

		Cost Each					
			2014 Bare Costs			Total	Total
Description	Labor-hrs.	Material	Labor	Equip.	Total	In-House	w/O&P
1900 Elevator, hydraulic, passenger / freight, annually	1.853	1,350	137		1,487	1,650	1,900
1950 Annualized	10.223	1,700	760		2,460	2,800	3,300

D10 CONVEYING   D1015 310 | Wheelchair Lift

PM Components	Labor-hrs.	W	M	Q	S	A
PM System D1015 310 1950						
Wheelchair Lift						
1 Check with operating personnel for any deficiencies.	.033				✓	✓
2 Check voltage and amperage under load.	.091				✓	✓
3 Check bolts securing drive cabinet and base, tighten accordingly.	.033				✓	✓
4 Check belt tension.	.033				✓	✓
5 Check lift nut assembly.	.026				✓	✓
6 Check cam rollers.	.026				✓	✓
7 Check wear pads for excessive wear.	.026				✓	✓
8 Inspect motor and shaft pulleys.	.039				✓	✓
9 Check Acme screw alignment.	.033				✓	✓
10 Inspect and lubricate upper and lower bearings.	.039				✓	✓
11 Check fastening of cable harness.	.026				✓	✓
12 Check alignment of platform and doors.	.033				✓	✓
13 Check door interlock switch for proper operation.	.039				✓	✓
14 Check operation of final limit switch and emergency stop/alarm.	.033				✓	✓
15 Check call/send controls at each station and on platform.	.039				✓	✓
16 Lube hinge of flip up ramp.	.033				✓	✓
17 Fill out maintenance checklist and report deficiencies.	.022				✓	✓
Total labor-hours/period					.604	.604
Total labor-hours/year					.604	.604

			Cost Each				
			2014 Bare Costs			Total	Total
Description		Labor-hrs.	Material	Labor	Equip.	Total	In-House	w/O&P
1900	Wheelchair lift, annually	.604		32		32	40	50
1950	Annualized	1.208		64.50		64.50	79.50	99.50

D10 CONVEYING — D1025 100 Escalator, Electric

PM Components	Labor-hrs.	W	M	Q	S	A
PM System D1025 100 1950						
Escalator						
1 Ride escalator; check operation for smoothness, unusual vibration or noise, condition of handrails, etc.	.104	√	√	√	√	√
2 Deenergize, tag, and lockout the electrical circuit.	.017	√	√	√	√	√
3 Inspect comb plates at both ends of escalator for broken teeth and check for proper clearance between combs and step teeth and check for broken step treads.	.039	√	√	√	√	√
4 Check clearance between steps and skirt panel; look for loose trim, screws or bolts, that could snag or damage clothing or cause injury; check operation of handrail brushes.	.087	√	√	√	√	√
5 Clean escalator machine space.	.113	√	√	√	√	√
6 Lubricate step rollers, step chain, drive gears or chains, handrail drive chains, etc., according to manufacturers' instructions; observe gears and chains for signs of wear, misalignment, etc; adjust as required.	.267	√	√	√	√	√
7 Check motor for signs of overheating.	.004	√	√	√	√	√
8 Inspect controller for loose leads, burned contacts, etc.; repair as required.	.022	√	√	√	√	√
9 Clean handrails as required.	.217	√	√	√	√	√
10 Check escalator lighting; replace bulbs as required.	.099	√	√	√	√	√
11 Operate each emergency stop button and note that the escalator stops; if the escalator has the capabilities of running in both directions, stop buttons should function properly for each direction of travel; observe the stopping distance.	.013	√	√	√	√	√
12 Clean surrounding area.	.066	√	√	√	√	√
13 Fill out maintenance checklist and report deficiencies.	.022	√	√	√	√	√
Total labor-hours/period		1.070	1.070	1.070	1.070	1.070
Total labor-hours/year		40.660	8.560	2.140	1.070	1.070

			Cost Each				
			2014 Bare Costs			Total	Total
Description		Labor-hrs.	Material	Labor	Equip.	Total	In-House	w/O&P
1900	Escalator, annually	1.070	2,275	79.50		2,354.50	2,600	2,975
1950	Annualized	53.501	2,825	3,975		6,800	8,000	9,650

D10 CONVEYING — D1095 100 Dumbwaiter, Electric

PM Components	Labor-hrs.	W	M	Q	S	A
PM System D1095 100 1950						
Dumbwaiter, electric						
1 Check with operating or area personnel for any obvious deficiencies.	.086					✓
2 Operate dumbwaiter through complete cycle and check operation for unusual noises or problems.	.052					✓
3 Inspect and clean operating panel, starter station, key operation and lamps.	.072					✓
4 Check operation of emergency lights and alarms.	.052					✓
5 Inspect, clean and lubricate motor.	.072					✓
6 Check operation of indicator lights and buttons.	.052					✓
7 Inspect and lubricate rails of guides.	.148					✓
8 Clean equipment and surrounding area.	.077					✓
9 Fill out maintenance checklist and report deficiencies.	.022					✓
Total labor-hours/period						.633
Total labor-hours/year						.633

		Cost Each					
		2014 Bare Costs				Total	Total
Description	Labor-hrs.	Material	Labor	Equip.	Total	In-House	w/O&P
1900 Dumbwaiter, electric, annually	.633	45.50	47		92.50	108	129
1950 Annualized	.633	90.50	47		137.50	158	186

D10 CONVEYING — D1095 110 Dumbwaiter, Hydraulic

PM Components	Labor-hrs.	W	M	Q	S	A
PM System D1095 110 1950						
Dumbwaiter, hydraulic						
1 Check with operating or area personnel for any obvious deficiencies.	.086				✓	✓
2 Operate dumbwaiter through complete cycle and check operation for unusual noises or problems.	.052				✓	✓
3 Inspect and clean operating panel, starter station, key operation and lamps.	.072				✓	✓
4 Check operation of emergency lights and alarms.	.052				✓	✓
5 Inspect pump and valve unit for leaks.	.072				✓	✓
6 Inspect tank oil level; add as required.	.039				✓	✓
7 Inspect and lubricate rails of guides.	.148				✓	✓
8 Inspect jack seals.	.051				✓	✓
9 Clean equipment and surrounding area.	.077				✓	✓
10 Fill out maintenance checklist and report deficiencies.	.022				✓	✓
Total labor-hours/period						.671	.671
Total labor-hours/year						.671	.671

		Cost Each					
		2014 Bare Costs				Total	Total
Description	Labor-hrs.	Material	Labor	Equip.	Total	In-House	w/O&P
1900 Dumbwaiter, hydraulic, annually	.671	64	50		114	132	157
1950 Annualized	1.342	128	99.50		227.50	264	315

D10 CONVEYING

D1095 200 Pneumatic Tube System

PM Components	Labor-hrs.	W	M	Q	S	A
PM System D1095 200 1950						
Pneumatic Tube System						
1 Check with operating personnel for any deficiencies.	.039			√	√	√
2 Clean air intake screen.	.039			√	√	√
3 Check seal at doors.	.130			√	√	√
4 Inspect check valves	.026			√	√	√
5 Check blower motor and lubricate.	.065			√	√	√
6 Inspect carrier ends, latches and air washers	.026			√	√	√
7 Fill out maintenance checklist and report deficiencies.	.026			√	√	√
Total labor-hours/period				.351	.351	.351
Total labor-hours/year				.702	.351	.351

			Cost Each				
			2014 Bare Costs			Total	Total
Description		Labor-hrs.	Material	Labor	Equip.	Total	In-House	w/O&P
1900	Pneumatic Tube System, annually	.351		18.75		18.75	23	29
1950	Annualized	1.404		75		75	92.50	116

D20 PLUMBING — D2015 100 Facility Plumbing Fixture Service

PM Components	Labor-hrs.	W	M	Q	S	A
PM System D2015 100 1950						
Urinals (Time is per fixture)						
1 Urinals - Flush and adjust water flow if required.	.020			√	√	√
2 Inspect for missing or damaged parts/caps and replace.	.020			√	√	√
3 Fill out maintenance checklist and report deficiencies.	.017			√	√	√
Total labor-hours/period				.057	.057	.057
Total labor-hours/year				.114	.057	.057

			Cost Each				
			2014 Bare Costs			Total	Total
Description	Labor-hrs.	Material	Labor	Equip.	Total	In-House	w/O&P
1900 Urinals, annually	.057		3.28		3.28	4.10	5.10
1950 Annualized	.228		13.10		13.10	16.40	20.50

PM Components	Labor-hrs.	W	M	Q	S	A
PM System D2015 100 2950						
Toilet, vacuum breaker type (Time is per fixture)						
1 Toilet (vacuum breaker type) - Flush and adjust water flow if required.	.020			√	√	√
2 Inspect for missing or damaged parts/caps, seat supports, and replace.	.020			√	√	√
3 Fill out maintenance checklist and report deficiencies.	.017			√	√	√
Total labor-hours/period				.057	.057	.057
Total labor-hours/year				.114	.057	.057

			Cost Each				
			2014 Bare Costs			Total	Total
Description	Labor-hrs.	Material	Labor	Equip.	Total	In-House	w/O&P
2900 Toilet (vacuum breaker type), annually	.057		3.28		3.28	4.10	5.10
2950 Annualized	.228		13.10		13.10	16.40	20.50

PM Components	Labor-hrs.	W	M	Q	S	A
PM System D2015 100 3950						
Toilet, tank type (Time is per fixture)						
1 Toilet - Clean flapper seat if leaking, adjust fill level if required.	.050			√	√	√
2 Inspect for missing or damaged parts/caps, seat supports, and replace.	.030			√	√	√
3 Fill out maintenance checklist and report deficiencies.	.017			√	√	√
Total labor-hours/period				.097	.097	.097
Total labor-hours/year				.194	.097	.097

			Cost Each				
			2014 Bare Costs			Total	Total
Description	Labor-hrs.	Material	Labor	Equip.	Total	In-House	w/O&P
3900 Toilet (tank type), annually	.097		5.60		5.60	7	8.70
3950 Annualized	.388		22.50		22.50	28	35

D20 PLUMBING   D2015 100   Facility Plumbing Fixture Service

PM Components	Labor-hrs.	W	M	Q	S	A
PM System D2015 100 4950 **Lavatories (Time is per fixture)**						
1 Lavatories - Operate faucets, replace washers/"O" Rings as necessary.	.040			✓	✓	✓
2 Observe drain flow, clean trap if flow is obstructed.	.030			✓	✓	✓
3 Fill out maintenance checklist and report deficiencies.	.017			✓	✓	✓
Total labor-hours/period				.087	.087	.087
Total labor-hours/year				.174	.087	.087

			Cost Each				
			2014 Bare Costs			Total	Total
Description		Labor-hrs.	Material	Labor	Equip.	Total	In-House	w/O&P
4900 Lavatories, annually	.087	6.80	4.12		10.92	12.90	15.20
4950 Annualized	.348	6.80	16.45		23.25	29	35.50

PM Components	Labor-hrs.	W	M	Q	S	A
PM System D2015 100 5950 **Showers (Time is per fixture)**						
1 Showers - Check for damaged, or missing shower heads/handles and replace if required.	.040			✓	✓	✓
2 Fill out maintenance checklist and report deficiencies.	.017			✓	✓	✓
Total labor-hours/period				.057	.057	.057
Total labor-hours/year				.114	.057	.057

			Cost Each				
			2014 Bare Costs			Total	Total
Description	Labor-hrs.	Material	Labor	Equip.	Total	In-House	w/O&P
5900 Showers, annually	.057	8.90	2.70		11.60	13.35	15.50
5950 Annualized	.228	8.90	10.80		19.70	24	28.50

D20 PLUMBING   D2015 800   Drinking Fountain

PM Components	Labor-hrs.	W	M	Q	S	A
PM System D2015 800 1950 **Drinking fountain**						
1 Check unit for proper operation, excessive noise or vibration.	.035					✓
2 Clean condenser coils and fan as required.	.222					✓
3 Check for water leaks in supply line and drain.	.077					✓
4 Check water flow; adjust as necessary.	.039					✓
5 Check drinking water temperature to ensure the unit is operating properly.	.039					✓
6 Check electrical connections and cord; tighten and repair as necessary.	.120					✓
7 Clean area around fountain.	.066					✓
8 Fill out maintenance checklist and report deficiencies.	.022					✓
Total labor-hours/period						.620
Total labor-hours/year						.620

			Cost Each				
			2014 Bare Costs			Total	Total
Description	Labor-hrs.	Material	Labor	Equip.	Total	In-House	w/O&P
1900 Drinking fountain, annually	.620	17.50	29.50		47	58.50	70
1950 Annualized	.620	17.50	29.50		47	58.50	70

D20 PLUMBING — D2025 120 Valve, Butterfly

PM Components	Labor-hrs.	W	M	Q	S	A
PM System D2025 120 1950 **Valve, butterfly, above 4"**						
1 Lubricate valves that have grease fittings.	.022					✓
2 Open and close valve using handle, wrench, or hand wheel to check operation.	.049					✓
3 Check for leaks.	.007					✓
4 Clean valve exterior and area around valve.	.066					✓
5 Fill out maintenance report and report deficiencies.	.022					✓
Total labor-hours/period						.166
Total labor-hours/year						.166

		Cost Each					
		2014 Bare Costs				Total	Total
Description	Labor-hrs.	Material	Labor	Equip.	Total	In-House	w/O&P
1900 Valve, butterfly, above 4", annually	.166	7.90	7.85		15.75	19.05	22.50
1950 Annualized	.166	7.90	7.85		15.75	19.05	22.50

PM Components	Labor-hrs.	W	M	Q	S	A
PM System D2025 120 2950 **Valve, butterfly, auto, above 4"**						
1 Lubricate valve actuator stem and valve stem, where possible.	.022					✓
2 Check automatic valve for proper operation.	.062					✓
3 Check packing gland for leaks; adjust as required.	.160					✓
4 Check pneumatic operator and tubing for air leaks.	.047					✓
5 Clean valve exterior and area around valve.	.034					✓
6 Fill out maintenance report and report deficiencies.	.022					✓
Total labor-hours/period						.347
Total labor-hours/year						.347

		Cost Each					
		2014 Bare Costs				Total	Total
Description	Labor-hrs.	Material	Labor	Equip.	Total	In-House	w/O&P
2900 Valve, butterfly, auto, above 4", annually	.347	7.90	16.40		24.30	30.50	36.50
2950 Annualized	.347	7.90	16.40		24.30	30.50	36.50

D20 PLUMBING — D2025 125 Valve, Check

PM Components	Labor-hrs.	W	M	Q	S	A
PM System D2025 125 1950 **Valve, check, above 4"**						
1 Inspect valve for leaks; repair as necessary.	.009					✓
2 Adjust system pressure/flow to verify that valve is opening and closing properly, if applicable.	.160					✓
3 Clean valve and area around valve.	.066					✓
4 Fill out maintenance report and report deficiencies.	.022					✓
Total labor-hours/period						.257
Total labor-hours/year						.257

		Cost Each					
		2014 Bare Costs				Total	Total
Description	Labor-hrs.	Material	Labor	Equip.	Total	In-House	w/O&P
1900 Valve, check, above 4", annually	.257	7.90	12.15		20.05	24.50	29.50
1950 Annualized	.257	7.90	12.15		20.05	24.50	29.50

D20 PLUMBING — D2025 130 Valve, Cock

PM Components	Labor-hrs.	W	M	Q	S	A
PM System D2025 130 1950						
Valve, ball, above 4"						
1 Lubricate valves that have grease fittings.	.022					✓
2 Open and close valve using handle, wrench, or hand wheel to check operation.	.049					✓
3 Check for leaks.	.007					✓
4 Clean valve and area around valve.	.066					✓
5 Fill out maintenance report and report deficiencies.	.022					✓
Total labor-hours/period						.166
Total labor-hours/year						.166

		Cost Each					
			2014 Bare Costs			Total	Total
Description	Labor-hrs.	Material	Labor	Equip.	Total	In-House	w/O&P
1900 Valve, ball, above 4", annually	.166	7.90	7.85		15.75	19.05	22.50
1950 Annualized	.166	7.90	7.85		15.75	19.05	22.50

D20 PLUMBING — D2025 135 Valve, Diaphragm

PM Components	Labor-hrs.	W	M	Q	S	A
PM System D2025 135 1950						
Valve, diaphragm, above 4"						
1 Lubricate valve stem, close and open valve to check operation.	.022					✓
2 Check valve for proper operation and leaks; tighten packing and flange bolts as required.	.044					✓
3 Clean valve exterior and area around valve.	.034					✓
4 Fill out maintenance report and report deficiencies.	.022					✓
Total labor-hours/period						.122
Total labor-hours/year						.122

		Cost Each					
			2014 Bare Costs			Total	Total
Description	Labor-hrs.	Material	Labor	Equip.	Total	In-House	w/O&P
1900 Valve, diaphragm, above 4", annually	.122	7.90	5.75		13.65	16.30	19.30
1950 Annualized	.122	7.90	5.75		13.65	16.30	19.30

D20 PLUMBING — D2025 140 Valve, Gate

PM Components	Labor-hrs.	W	M	Q	S	A
PM System D2025 140 1950						
Valve, gate, above 4"						
1 Lubricate valve stem; close and open valve to check operation.	.049					✓
2 Check packing gland for leaks; tighten packing and flange bolts as req.	.022					✓
3 Clean valve exterior and area around valve.	.066					✓
4 Fill out maintenance report and report deficiencies.	.022					✓
Total labor-hours/period						.159
Total labor-hours/year						.159

		Cost Each					
			2014 Bare Costs			Total	Total
Description	Labor-hrs.	Material	Labor	Equip.	Total	In-House	w/O&P
1900 Valve, gate, above 4", annually	.159	7.90	7.50		15.40	18.60	22
1950 Annualized	.159	7.90	7.50		15.40	18.60	22

D20 PLUMBING — D2025 145 Valve, Globe

PM Components	Labor-hrs.	W	M	Q	S	A
PM System D2025 145 1950						
Valve, globe, above 4"						
1 Lubricate stem, close and open valve to check operation.	.049					✓
2 Check packing gland for leaks; tighten packing and flange bolts as req.	.022					✓
3 Clean valve exterior.	.066					✓
4 Fill out maintenance report and report deficiencies.	.022					✓
Total labor-hours/period						.159
Total labor-hours/year						.159

			Cost Each				
			2014 Bare Costs			Total	Total
Description	Labor-hrs.	Material	Labor	Equip.	Total	In-House	w/O&P
1900 Valve, globe, above 4", annually	.159	7.90	7.50		15.40	18.60	22
1950 Annualized	.159	7.90	7.50		15.40	18.60	22

PM Components	Labor-hrs.	W	M	Q	S	A
PM System D2025 145 2950						
Valve, globe, auto, above 4"						
1 Lubricate valve actuator stem and valve stem, where possible.	.022					✓
2 Check automatic valve for proper operation.	.062					✓
3 Check packing gland for leaks; adjust as required.	.160					✓
4 Check penumatic operator and tubing for air leaks.	.025					✓
5 Clean valve exterior and area around valve.	.034					✓
6 Fill out maintenance report and report deficiencies.	.022					✓
Total labor-hours/period						.325
Total labor-hours/year						.325

			Cost Each				
			2014 Bare Costs			Total	Total
Description	Labor-hrs.	Material	Labor	Equip.	Total	In-House	w/O&P
2900 Valve, globe, auto, above 4", annually	.325	7.90	15.35		23.25	29	35
2950 Annualized	.325	7.90	15.35		23.25	29	35

D20 PLUMBING — D2025 150 Valve, Motor Operated

PM Components	Labor-hrs.	W	M	Q	S	A
PM System D2025 150 1950						
Valve, motor operated, above 4"						
1 Lubricate valve actuator stem and valve stem, where possible.	.091				✓	✓
2 Check motor and valve for proper operation, including limit switch; adjust as required.	.022				✓	✓
3 Check packing gland for leaks; adjust as required.	.113				✓	✓
4 Check electrical wiring, connections and contacts; repair as necessary.	.120				✓	✓
5 Inspect and lubricate motor gearbox as required.	.099				✓	✓
6 Clean valve exterior and area around valve.	.034				✓	✓
7 Fill out maintenance report and report deficiencies.	.022				✓	✓
Total labor-hours/period						.501	.501
Total labor-hours/year						.501	.501

			Cost Each				
			2014 Bare Costs			Total	Total
Description	Labor-hrs.	Material	Labor	Equip.	Total	In-House	w/O&P
1900 Valve, motor operated, above 4", annually	.501	7.90	23.50		31.40	40	48.50
1950 Annualized	1.002	15.85	47.50		63.35	80	97

D20 PLUMBING   D2025 155 Valve, OS&Y

PM Components	Labor-hrs.	W	M	Q	S	A
PM System D2025 155 1950						
Valve, OS&Y, above 4"						
1 Lubricate stem; open and close valve to check operation.	.049					√
2 Check packing gland for leaks; tighten packing and flange bolts as required.	.022					√
3 Clean valve exterior and area around valve.	.066					√
4 Fill out maintenance report and report deficiencies.	.022					√
Total labor-hours/period						.159
Total labor-hours/year						.159

			Cost Each					
				2014 Bare Costs			Total	Total
	Description	Labor-hrs.	Material	Labor	Equip.	Total	In-House	w/O&P
1900	Valve, OS&Y, above 4", annually	.159	7.90	7.50		15.40	18.60	22
1950	Annualized	.159	7.90	7.50		15.40	18.60	22

D20 PLUMBING — D2025 190 Water Heater, Solar

PM Components	Labor-hrs.	W	M	Q	S	A
PM System D2025 190 1950						
Solar, closed loop hot water heating system, up to 6 panels						
1 Check with operating or area personnel for deficiencies.	.035				✓	✓
2 Inspect interior piping and connections for leaks and damaged insulation; tighten connections and repair damaged insulation as necessary.	.125				✓	✓
3 Check zone and circulating pump motors for excessive overheating; lubricate motor bearings.	.077				✓	✓
4 Check pressure and air relief valves for proper operation.	.030				✓	✓
5 Check control panel and differential thermostat for proper operation.	.094				✓	✓
6 Clean sight glasses, controls, pumps, and flow indicators on tanks.	.127				✓	✓
7 Check system pressure on closed loop for loss of fluid.	.046				✓	✓
8 Check fluid level on drain-back systems; add fluid as necessary.	.029				✓	✓
9 Test glycol strength in closed systems, as applicable; if required, drain system and replace with new fluid mixture.	.222					✓
10 Check heat exchanger for exterior leaks.	.077				✓	✓
11 Clean strainers and traps.	.181					✓
12 Check storage and expansion tanks; for leaks and deteriorated insulation.	.077					✓
13 Inspect all collector piping for leaks and damaged insulation; tighten connections and repair as required.	.133				✓	✓
14 Inspect collector glazing for cracks and seals for tightness; tighten or replace seals as necessary.	.124				✓	✓
15 Wash/clean glazing on collector panels.	.585					✓
16 Inspect ferrule around pipe flashing where solar piping runs through roof; repair as necessary.	.086				✓	✓
17 Check collector mounting brackets and bolts; tighten as required.	.094				✓	✓
18 Clean area.	.066				✓	✓
19 Fill out maintenance checklist and report deficiencies.	.022				✓	✓
Total labor-hours/period					1.165	2.230
Total labor-hours/year					1.165	2.230

		Cost Each					
		2014 Bare Costs				Total	Total
Description	Labor-hrs.	Material	Labor	Equip.	Total	In-House	w/O&P
1900 Wtr. htng. sys., solar clsd. lp., up to 6 panels, annually	2.230	238	130		368	425	500
1950 Annualized	3.395	240	199		439	510	610

D20 PLUMBING — D2025 260 Water Heater, Gas/Oil/Steam

PM Components	Labor-hrs.	W	M	Q	S	A
PM System D2025 260 1950						
Water heater, gas, to 120 gallon						
1 Check with operating or area personnel for deficiencies.	.035				✓	✓
2 Check for water leaks to tank and piping. check for fuel system leaks.	.077				✓	✓
3 Check gas burner and pilot for proper flame; adjust if required.	.118				✓	✓
4 Check operation and condition of pressure relief valve.	.010				✓	✓
5 Check automatic controls for proper operation (temperature regulators, thermostatic devices, automatic fuel shut off valve, etc.).	.094				✓	✓
6 Check draft diverter and clear openings, if clogged.	.027				✓	✓
7 Check electrical wiring for fraying and loose connections on oil burner.	.072				✓	✓
8 Check for proper water temperature setting; adjust as required.	.029				✓	✓
9 Check condition of flue pipe, and chimney.	.147				✓	✓
10 Drain sediment from tank.	.325					✓
11 Clean up area around unit.	.066				✓	✓
12 Fill out maintenance checklist and report deficiencies.	.022				✓	✓
Total labor-hours/period					.697	1.022
Total labor-hours/year					.697	1.022

		Cost Each					
			2014 Bare Costs			Total	Total
	Description	Labor-hrs.	Material	Labor	Equip.	Total	In-House	w/O&P
1900	Water heater, gas, to 120 gal., annually	1.022	64	60		124	145	174
1950	Annualized	1.719	64	101		165	197	237

PM Components	Labor-hrs.	W	M	Q	S	A
PM System D2025 260 2950						
Water heater, oil fired, to 100 gallon						
1 Check with operating or area personnel for deficiencies.	.035				✓	✓
2 Check for water leaks to tank and piping; check for fuel system leaks.	.077				✓	✓
3 Check burner flame and pilot on oil burner; adjust if required.	.073				✓	✓
4 Check operation and condition of pressure relief valve.	.010				✓	✓
5 Check automatic controls for proper operation (temperature regulators, thermostatic devices, automatic fuel shut off valve, etc.).	.094				✓	✓
6 Check fuel filter element on oil burner.	.068				✓	✓
7 Check fuel level in tank; check tank, fill pipe and fuel lines and connections for damage.	.022				✓	✓
8 Inspect, clean, and adjust electrodes and nozzles on oil burners; inspect fire box and flame detection scanner.	.254				✓	✓
9 Check electrical wiring for fraying and loose connections on oil burner.	.072				✓	✓
10 Check for proper water temperature setting; adjust as required.	.029				✓	✓
11 Clean fire box.	.577					✓
12 Check for proper draft adjustment; adjust draft meter if necessary.	.005				✓	✓
13 Check condition of flue pipe, damper and chimney.	.147				✓	✓
14 Drain sediment from tank.	.325					✓
15 Clean up area around unit.	.066				✓	✓
16 Fill out maintenance checklist and report deficiencies.	.022				✓	✓
Total labor-hours/period					.974	1.876
Total labor-hours/year					.974	1.876

		Cost Each					
			2014 Bare Costs			Total	Total
	Description	Labor-hrs.	Material	Labor	Equip.	Total	In-House	w/O&P
2900	Water heater, oil fired, to 100 gal., annually	1.876	72	110		182	216	261
2950	Annualized	2.850	80	167		247	296	360

D20 PLUMBING — D2025 260 Water Heater, Gas/Oil/Steam

PM Components	Labor-hrs.	W	M	Q	S	A
PM System D2025 260 3950						
Water heater, steam coil, to 2500 gallon						
1 Check with operating or area personnel for deficiencies.	.035				✓	✓
2 Check for water leaks to tank and piping; check steam lines for leaks.	.077				✓	✓
3 Check operation and condition of pressure relief valve.	.010				✓	✓
4 Check steam modulating valve and steam condensate trap for proper operation.	.068				✓	✓
5 Check electrical wiring connections on controls and switches.	.094				✓	✓
6 Check for proper water temperature setting; adjust as required.	.029				✓	✓
7 Clean, test and inspect sight gauges, valves and drains.	.040				✓	✓
8 Check automatic controls for proper operation including temperature regulators and thermostatic devices.	.094				✓	✓
9 Check insulation on heater; repair as necessary.	.077				✓	✓
10 Drain sediment from tank.	.325					✓
11 Clean up area around unit.	.066				✓	✓
12 Fill out maintenance checklist and report deficiencies.	.022				✓	✓
Total labor-hours/period					.612	.937
Total labor-hours/year					.612	.937

			Cost Each				
			2014 Bare Costs			Total	Total
Description	Labor-hrs.	Material	Labor	Equip.	Total	In-House	w/O&P
3900 Water heater, steam coil, to 2500 gal., annually	.937	32	55		87	104	126
3950 Annualized	1.549	32	90.50		122.50	149	181

D20 PLUMBING — D2025 262 | Valve, Pressure Relief

PM Components	Labor-hrs.	W	M	Q	S	A
PM System D2025 262 1950						
Valve, pressure relief, above 4″						
1 Inspect valve for leaks, tighten fittings as necessary.	.022					✓
2 Manually operate to check operation.	.038					✓
3 Clean valve and area around valve.	.066					✓
4 Fill out maintenance report and report deficiencies.	.022					✓
Total labor-hours/period						.148
Total labor-hours/year						.148

	Description	Labor-hrs.	Material	Labor	Equip.	Total	Total In-House	Total w/O&P
1900	Valve, pressure relief, above 4″, annually	.148		7		7	9.25	11.40
1950	Annualized	.148		7		7	9.25	11.40

D20 PLUMBING — D2025 265 | Valve, Pressure Regulator

PM Components	Labor-hrs.	W	M	Q	S	A
PM System D2025 265 1950						
Valve, pressure regulator, above 4″						
1 Inspect valve for leaks; repair as necessary.	.160					✓
2 Manually operate to check operation; adjust as required.	.049					✓
3 Check pressure mechanism for proper opening and closing action.	.062					✓
4 Clean valve and area around regulator.	.066					✓
5 Fill out maintenance report and report deficiencies.	.022					✓
Total labor-hours/period						.359
Total labor-hours/year						.359

	Description	Labor-hrs.	Material	Labor	Equip.	Total	Total In-House	Total w/O&P
1900	Valve, pressure regular, above 4″, annually	.359		17		17	22.50	27.50
1950	Annualized	.359		17		17	22.50	27.50

D20 PLUMBING — D2025 270 | Valve, Sediment Strainer

PM Components	Labor-hrs.	W	M	Q	S	A
PM System D2025 270 1950						
Valve, sediment strainer, above 4″						
1 Inspect valve for leaks, tighten fittings as necessary.	.160					✓
2 Open valve drain to remove collected sediment.	.065					✓
3 Clean valve exterior and around strainer.	.066					✓
4 Fill out maintenance checklist and report deficiencies.	.022					✓
Total labor-hours/period						.313
Total labor-hours/year						.313

	Description	Labor-hrs.	Material	Labor	Equip.	Total	Total In-House	Total w/O&P
1900	Valve, sediment strainer, above 4″, annually	.313		14.80		14.80	19.55	24
1950	Annualized	.313		14.80		14.80	19.55	24

D20 PLUMBING — D2025 310 Valve, Automatic

PM Components	Labor-hrs.	W	M	Q	S	A
PM System D2025 310 1950						
Valve, automatic, above 4"						
1 Lubricate valve actuator stem and valve stem, where possible.	.022					✓
2 Check automatic valve for proper operation.	.062					✓
3 Check packing gland for leaks; adjust as required.	.022					✓
4 Check pneumatic operator and tubing for air leaks.	.025					✓
5 Clean valve exterior and area around valve.	.034					✓
6 Fill out maintenance report and report deficiencies.	.022					✓
Total labor-hours/period						.187
Total labor-hours/year						.187

			Cost Each				
			2014 Bare Costs			Total	Total
Description	Labor-hrs.	Material	Labor	Equip.	Total	In-House	w/O&P
1900 Valve, automatic, above 4", annually	.187	7.90	8.85		16.75	20.50	24.50
1950 Annualized	.187	7.90	8.85		16.75	20.50	24.50

PM Components	Labor-hrs.	W	M	Q	S	A
PM System D2025 310 2950						
Valve, auto diaphragm, above 4"						
1 Lubricate valve actuator mechanism and valve stem.	.022					✓
2 Check valve for proper operation and leaks; tighten packing and flange bolts as required.	.044					✓
3 Check pneumatic tubing and operator mechanism for proper alignment and damage; adjust as necessary and soap solution test for leaks after maintenance.	.053					✓
4 Clean valve exterior and area around valve.	.034					✓
5 Fill out maintenance report and report deficiencies.	.022					✓
Total labor-hours/period						.175
Total labor-hours/year						.175

			Cost Each				
			2014 Bare Costs			Total	Total
Description	Labor-hrs.	Material	Labor	Equip.	Total	In-House	w/O&P
2900 Valve, auto diaphragm, above 4", annually	.175	7.90	8.30		16.20	19.60	23.50
2950 Annualized	.175	7.90	8.30		16.20	19.60	23.50

D20 PLUMBING — D2095 905 Duplex Sump

PM Components	Labor-hrs.	W	M	Q	S	A
PM System D2095 905 1950						
Duplex sump pump						
1 Check electrical cords, plugs and connections.	.239				✓	✓
2 Activate float switches and check pumps for proper operation.	.078				✓	✓
3 Lubricate pumps as required.	.094				✓	✓
4 Inspect packing and tighten as required.	.062				✓	✓
5 Check pumps for misalignment and bearings for overheating.	.260				✓	✓
6 Clean out trash from sump bottom.	.072				✓	✓
7 Fill out maintenance checklist and report deficiencies.	.022				✓	✓
Total labor-hours/period					.827	.827
Total labor-hours/year					.827	.827

			Cost Each				
			2014 Bare Costs			Total	Total
Description	Labor-hrs.	Material	Labor	Equip.	Total	In-House	w/O&P
1900 Duplex sump pump, annually	.827	16.75	47.50		64.25	78	95.50
1950 Annualized	1.654	33.50	95		128.50	156	191

D20 PLUMBING — D2095 910 Pump, Submersible

PM Components	Labor-hrs.	W	M	Q	S	A
PM System D2095 910 1950						
Submersible pump, 1 H.P. and over						
1 Check with operating personnel for any deficiencies.	.035				✓	✓
2 Remove pump from pit.	.585				✓	✓
3 Clean out trash from pump intake.	.338				✓	✓
4 Check electrical plug, cord and connection.	.120				✓	✓
5 Inspect pump body for corrosion; prime and paint as necessary.	.053				✓	✓
6 Check pump and motor operation for excessive vibration, noise and overheating.	.022				✓	✓
7 Lubricate pump and motor, where applicable.	.099				✓	✓
8 Return pump to pit; reset and check float switch for proper operation.	.585				✓	✓
9 Clean area.	.066				✓	✓
10 Fill out maintenance checklist and report deficiencies.	.022				✓	✓
Total labor-hours/period					1.925	1.925
Total labor-hours/year					1.925	1.925

			Cost Each				
			2014 Bare Costs			Total	Total
Description	Labor-hrs.	Material	Labor	Equip.	Total	In-House	w/O&P
1900 Submersible pump, 1 H.P. and over, annually	1.925	16.75	111		127.75	157	194
1950 Annualized	3.850	33.50	222		255.50	315	385

D20 PLUMBING — D2095 930 Oxygen Monitor

PM Components	Labor-hrs.	W	M	Q	S	A
PM System D2095 930 1950						
Oxygen monitor.						
1 Check alarm set points.	.167			✓	✓	✓
2 Check alarm lights for replacement.	.167			✓	✓	✓
3 Check audible annunciator for proper operation.	.167			✓	✓	✓
4 Check battery backup for proper operation.	.167			✓	✓	✓
5 Check and test oxygen sensor for proper operation.	.276			✓	✓	✓
6 Replace oxygen sensor annually.	1.000					✓
7 Clean area.	.034			✓	✓	✓
8 Fill out maintenance checklist and report deficiencies.	.022			✓	✓	✓
Total labor-hours/period				1.000	1.000	2.000
Total labor-hours/year				2.000	1.000	2.000

	Description	Labor-hrs.	Cost Each				Total In-House	Total w/O&P
			2014 Bare Costs					
			Material	Labor	Equip.	Total		
1900	Oxygen monitor annually	2.000	205	94.50		299.50	350	410
1950	Annualized	5.000	205	237		442	540	640

| D30 HVAC | D3025 110 | Boiler, Electric | | | | | |

PM Components	Labor-hrs.	W	M	Q	S	A
PM System D3025 110 1950						
Boiler, electric, to 1500 gallon						
1 Check with operating or area personnel for deficiencies.	.035			✓	✓	✓
2 Check hot water pressure gauges.	.073			✓	✓	✓
3 Check operation and condition of pressure relief valve.	.030			✓	✓	✓
4 Check for proper operation of primary controls for resistance-type or electrode-type heating elements; check and adjust thermostat.	.133			✓	✓	✓
5 Check electrical wiring to heating elements, blower, motors, overcurrent protective devices, grounding system and other electrical components as required.	.238			✓	✓	✓
6 Check over temperature and over-pressure limit controls for proper operation.	.143			✓	✓	✓
7 Check furnace operation through complete cycle or up to 10 minutes.	.217			✓	✓	✓
8 Clean area around boiler.	.066			✓	✓	✓
9 Fill out maintenance checklist and report deficiencies.	.022			✓	✓	✓
Total labor-hours/period				.957	.957	.957
Total labor-hours/year				1.914	.957	.957

| | Description | Labor-hrs. | Cost Each | | | | Total In-House | Total w/O&P |
| | | | 2014 Bare Costs | | | | | |
			Material	Labor	Equip.	Total		
1900	Boiler, electric, to 1500 gal., annually	.957	69.50	56		125.50	146	174
1950	Annualized	3.828	69.50	224		293.50	355	435

D30 HVAC — D3025 130 Boiler, Hot Water, Oil/Gas/Comb.

PM Components	Labor-hrs.	W	M	Q	S	A
PM System D3025 130 1950						
Boiler, hot water; oil, gas or combination fired, up to 120 MBH						
1 Check combustion chamber for air or gas leaks.	.077					✓
2 Inspect and clean oil burner gun and ignition assembly, where applicable.	.658					✓
3 Inspect fuel system for leaks and change fuel filter element, where applicable.	.098					✓
4 Check fuel lines and connections for damage.	.023		✓	✓	✓	✓
5 Check for proper operational response of burner to thermostat controls.	.133			✓	✓	✓
6 Check and lubricate burner and blower motors.	.079			✓	✓	✓
7 Check main flame failure protection and main flame detection scanner on boiler equipped with spark ignition (oil burner).	.124		✓	✓	✓	✓
8 Check electrical wiring to burner controls and blower.	.079					✓
9 Clean firebox (sweep and vacuum).	.577					✓
10 Check operation of mercury control switches (i.e., steam pressure, hot water temperature limit, atomizing or combustion air proving, etc.).	.143		✓	✓	✓	✓
11 Check operation and condition of safety pressure relief valve.	.030		✓	✓	✓	✓
12 Check operation of boiler low water cut off devices.	.056		✓	✓	✓	✓
13 Check hot water pressure gauges.	.073		✓	✓	✓	✓
14 Inspect and clean water column sight glass (or replace).	.127		✓	✓	✓	✓
15 Clean fire side of water jacket boiler.	.433					✓
16 Check condition of flue pipe, damper and exhaust stack.	.147			✓	✓	✓
17 Check boiler operation through complete cycle, up to 30 minutes.	.650					✓
18 Check fuel level with gauge pole, add as required.	.046		✓	✓	✓	✓
19 Clean area around boiler.	.066		✓	✓	✓	✓
20 Fill out maintenance checklist and report deficiencies.	.022		✓	✓	✓	✓
Total labor-hours/period			.710	1.069	1.069	3.641
Total labor-hours/year			5.680	2.138	1.069	3.641

			Cost Each				
			2014 Bare Costs			Total	Total
Description		Labor-hrs.	Material	Labor	Equip.	Total	In-House	w/O&P
1900	Boiler, hot water, O/G/C, up to 120 MBH, annually	3.641	78.50	213		291.50	355	430
1950	Annualized	12.528	83.50	735		818.50	1,000	1,250

D30 HVAC

D3025 130 Boiler, Hot Water, Oil/Gas/Comb.

PM Components	Labor-hrs.	W	M	Q	S	A
PM System D3025 130 2950						
Boiler, hot water; oil, gas or combination fired, 120 to 500 MBH						
1 Check combustion chamber for air or gas leaks.	.077					√
2 Inspect and clean oil burner gun and ignition assembly where applicable.	.835					√
3 Inspect fuel system for leaks and change fuel filter element, where applicable.	.125					√
4 Check fuel lines and connections for damage.	.023		√	√	√	√
5 Check for proper operational response of burner to thermostat controls.	.169		√	√	√	√
6 Check and lubricate burner and blower motors.	.099			√	√	√
7 Check main flame failure protection and main flame detection scanner on boiler equipped with spark ignition (oil burner).	.155		√	√	√	√
8 Check electrical wiring to burner controls and blower.	.100					√
9 Clean firebox (sweep and vacuum).	.793					√
10 Check operation of mercury control switches (i.e., steam pressure, hot water temperature limit, atomizing or combustion air proving, etc.).	.185		√	√	√	√
11 Check operation and condition of safety pressure relief valve.	.038		√	√	√	√
12 Check operation of boiler low water cut off devices.	.070		√	√	√	√
13 Check hot water pressure gauges.	.073		√	√	√	√
14 Inspect and clean water column sight glass (or replace).	.160		√	√	√	√
15 Clean fire side of water jacket	.433					√
16 Check condition of flue pipe, damper and exhaust stack.	.183			√	√	√
17 Check boiler operation through complete cycle, up to 30 minutes.	.806					√
18 Check fuel level with gauge pole, add as required.	.046		√	√	√	√
19 Clean area around boiler.	.137		√	√	√	√
20 Fill out maintenance checklist and report deficiencies.	.022		√	√	√	√
Total labor-hours/period			.909	1.360	1.360	4.529
Total labor-hours/year			7.272	2.720	1.360	4.529

	Description	Labor-hrs.	Cost Each					
			2014 Bare Costs				Total In-House	Total w/O&P
			Material	Labor	Equip.	Total		
2900	Boiler, hot water, O/G/C, 120 to 500 MBH, annually	4.529	99	265		364	440	540
2950	Annualized	15.881	104	930		1,034	1,275	1,575

D30 HVAC — D3025 130 Boiler, Hot Water, Oil/Gas/Comb.

PM Components	Labor-hrs.	W	M	Q	S	A
PM System D3025 130 3950						
Boiler, hot water; oil, gas or combination fired, 500 to 1000 MBH						
1 Check combustion chamber for air or gas leaks.	.086					✓
2 Inspect and clean oil burner gun and ignition assembly where applicable.	.910					✓
3 Inspect fuel system for leaks and change fuel filter element, where applicable.	.140					✓
4 Check fuel lines and connections for damage.	.026		✓	✓	✓	✓
5 Check for proper operational response of burner to thermostat controls.	.185			✓	✓	✓
6 Check and lubricate burner and blower motors.	.109			✓	✓	✓
7 Check main flame failure protection and main flame detection scanner on boiler equipped with spark ignition (oil burner).	.169		✓	✓	✓	✓
8 Check electrical wiring to burner controls and blower.	.111					✓
9 Clean firebox (sweep and vacuum).	.875					✓
10 Check operation of mercury control switches (i.e., steam pressure, hot water temperature limit, atomizing or combustion air proving, etc.).	.203		✓	✓	✓	✓
11 Check operation and condition of safety pressure relief valve.	.042		✓	✓	✓	✓
12 Check operation of boiler low water cut off devices.	.077		✓	✓	✓	✓
13 Check hot water pressure gauges.	.081		✓	✓	✓	✓
14 Inspect and clean water column sight glass (or replace).	.176		✓	✓	✓	✓
15 Clean fire side of water jacket	.433					✓
16 Check condition of flue pipe, damper and exhaust stack.	.202			✓	✓	✓
17 Check boiler operation through complete cycle, up to 30 minutes.	.887					✓
18 Check fuel level with gauge pole, add as required.	.049		✓	✓	✓	✓
19 Clean area around boiler.	.151		✓	✓	✓	✓
20 Fill out maintenance checklist and report deficiencies.	.022		✓	✓	✓	✓
Total labor-hours/period			.996	1.492	1.492	4.934
Total labor-hours/year			7.968	2.984	1.492	4.934

		Cost Each					
		2014 Bare Costs				Total	Total
Description	Labor-hrs.	Material	Labor	Equip.	Total	In-House	w/O&P
3900 Boiler, hot water, O/G/C, 500 to 1000 MBH, annually	4.934	106	289		395	480	585
3950 Annualized	17.378	111	1,025		1,136	1,400	1,725

D30 HVAC		D3025 130	Boiler, Hot Water, Oil/Gas/Comb.					

PM Components	Labor-hrs.	W	M	Q	S	A
PM System D3025 130 4950						
Boiler, hot water; oil, gas or combination fired, over 1000 MBH						
1 Check combustion chamber for air or gas leaks.	.117					√
2 Inspect and clean oil burner gun and ignition assembly where applicable.	.987					√
3 Inspect fuel system for leaks and change fuel filter element, where applicable.	.147					√
4 Check fuel lines and connections for damage.	.035		√	√	√	√
5 Check for proper operational response of burner to thermostat controls.	.199		√	√	√	√
6 Check and lubricate burner and blower motors.	.120			√	√	√
7 Check main flame failure protection and main flame detection scanner on boiler equipped with spark ignition (oil burner).	.186		√	√	√	√
8 Check electrical wiring to burner controls and blower.	.120					√
9 Clean firebox (sweep and vacuum).	.819					√
10 Check operation of mercury control switches (i.e., steam pressure, hot water temperature limit, atomizing or combustion air proving, etc.).	.215		√	√	√	√
11 Check operation and condition of safety pressure relief valve.	.046		√	√	√	√
12 Check operation of boiler low water cut off devices.	.085		√	√	√	√
13 Check hot water pressure gauges.	.109		√	√	√	√
14 Inspect and clean water column sight glass (or replace).	.191		√	√	√	√
15 Clean fire side of water jacket	.433					√
16 Check condition of flue pipe, damper and exhaust stack.	.221			√	√	√
17 Check boiler operation through complete cycle, up to 30 minutes.	.887					√
18 Check fuel level with gauge pole, add as required.	.098		√	√	√	√
19 Clean area around boiler.	.182		√	√	√	√
20 Fill out maintenance checklist and report deficiencies.	.022		√	√	√	√
Total labor-hours/period			1.169	1.709	1.709	5.219
Total labor-hours/year			9.352	3.418	1.709	5.219

		Cost Each					
		2014 Bare Costs				Total	Total
Description	Labor-hrs.	Material	Labor	Equip.	Total	In-House	w/O&P
4900 Boiler, hot water, O/G/C, over 1000 MBH, annually	5.219	120	305		425	515	625
4950 Annualized	19.698	125	1,150		1,275	1,600	1,950

D30 HVAC — D3025 140 Boiler, Steam, Oil/Gas/Comb.

PM Components	Labor-hrs.	W	M	Q	S	A
PM System D3025 140 1950						
Boiler, steam; oil, gas or combination fired, up to 120 MBH						
1. Inspect fuel system for leaks or damage.	.098		√	√	√	√
2. Change fuel filter element and clean strainers; repair leaks, where applicable.	.581					√
3. Check main flame failure protection, positive fuel shutoff and main flame detection scanner on boiler equipped with spark ignition (oil burner).	.124		√	√	√	√
4. Check for proper operational response of burner to thermostat controls.	.133			√	√	√
5. Inspect all gas, steam and water lines, valves, connections for leaks or damage; repair as necessary.	.195			√	√	√
6. Check feedwater system and feedwater makeup control and pump.	.056		√	√	√	√
7. Check and lubricate burner and blower motors as required.	.083			√	√	√
8. Check operation and condition of safety pressure relief valve.	.030		√	√	√	√
9. Check combustion controls, combustion blower and damper modulation control.	.133					√
10. Check all indicator lamps and water/steam pressure gauges.	.073		√	√	√	√
11. Check electrical panels and wiring to burner, blowers and other components.	.079			√	√	√
12. Clean blower air-intake dampers, if required.	.055			√	√	√
13. Check condition of flue pipe, damper and exhaust stack.	.147		√	√	√	√
14. Check boiler operation through complete cycle, up to 30 minutes.	.650			√	√	√
15. Check water column sight glass and water level system; clean or replace sight glass, if required.	.127		√	√	√	√
16. Clean firebox (sweep and vacuum).	.577					√
17. Check fuel level with gauge pole for oil burning boilers.	.046		√	√	√	√
18. Inspect and clean oil burner gun and ignition assembly where applicable.	.650					√
19. Clean area around boiler.	.066		√	√	√	√
20. Fill out maintenance checklist and report deficiencies.	.022		√	√	√	√
Total labor-hours/period			.789	1.984	1.984	3.925
Total labor-hours/year			6.312	3.968	1.984	3.925

			Cost Each				
			2014 Bare Costs			Total	Total
Description	Labor-hrs.	Material	Labor	Equip.	Total	In-House	w/O&P
1900 Boiler, steam, O/G/C, up to 120 MBH, annually	3.925	155	230		385	460	555
1950 Annualized	16.189	160	945		1,105	1,350	1,675

D30 HVAC — D3025 140 | Boiler, Steam, Oil/Gas/Comb.

PM Components	Labor-hrs.	W	M	Q	S	A
PM System D3025 140 2950						
Boiler, steam; oil, gas or combination fired, 120 to 500 MBH						
1 Inspect fuel system for leaks or damage.	.098		√	√	√	√
2 Change fuel filter element and clean strainers; repair leaks, where applicable.	.835					√
3 Check main flame failure protection, positive fuel shutoff and main flame detection scanner on boiler equipped with spark ignition (oil burner).	.125		√	√	√	√
4 Check for proper operational response of burner to thermostat controls.	.169			√	√	√
5 Inspect all gas, steam and water lines, valves, connections for leaks or damage; repair as necessary.	.195			√	√	√
6 Check feedwater system and feedwater makeup control and pump.	.069		√	√	√	√
7 Check and lubricate burner, blowers and motors as required.	.099			√	√	√
8 Check operation and condition of safety pressure relief valve.	.039		√	√	√	√
9 Check combustion controls, combustion blower and damper modulation control.	.169					√
10 Check all indicator lamps and water/steam pressure gauges.	.073		√	√	√	√
11 Check electrical panels and wiring to burner, blowers and other components.	.104			√	√	√
12 Clean blower air-intake dampers, if required.	.069		√	√	√	√
13 Check condition of flue pipe, damper and exhaust stack.	.147		√	√	√	√
14 Check boiler operation through complete cycle, up to 30 minutes.	.806			√	√	√
15 Check water column sight glass and water level system; clean or replace sight glass, if required.	.127		√	√	√	√
16 Clean firebox (sweep and vacuum).	.793					√
17 Check fuel level with gauge pole for oil burning boilers.	.046		√	√	√	√
18 Inspect and clean oil burner gun and ignition assembly where applicable.	.819					√
19 Clean area around boiler.	.137		√	√	√	√
20 Fill out maintenance checklist and report deficiencies.	.022		√	√	√	√
Total labor-hours/period			.883	2.325	2.325	4.941
Total labor-hours/year			7.064	4.650	2.325	4.941

			Cost Each				
			2014 Bare Costs			Total	Total
Description		Labor-hrs.	Material	Labor	Equip.	Total	In-House	w/O&P
2900 Boiler, steam, O/G/C, 120 to 500 MBH, annually	4.941	191	289		480	570	690
2950 Annualized	18.980	202	1,100		1,302	1,600	1,975

| D30 HVAC | | D3025 140 | Boiler, Steam, Oil/Gas/Comb. | | | | | |

PM Components	Labor-hrs.	W	M	Q	S	A
PM System D3025 140 3950						
Boiler, steam; oil, gas or combination fired, 500 to 1000 MBH						
1 Inspect fuel system for leaks or damage.	.098		√	√	√	√
2 Change fuel filter element and clean strainers; repair leaks, where applicable.	.910					√
3 Check main flame failure protection, positive fuel shutoff and main flame detection scanner on boiler equipped with spark ignition (oil burner).	.140		√	√	√	√
4 Check for proper operational response of burner to thermostat controls.	.185			√	√	√
5 Inspect all gas, steam and water lines, valves, connections for leaks or damage; repair as necessary.	.195			√	√	√
6 Check feedwater system and feedwater makeup control and pump.	.075		√	√	√	√
7 Check and lubricate burner, blowers and motors as required.	.109			√	√	√
8 Check operation and condition of safety pressure relief valve.	.052		√	√	√	√
9 Check combustion controls, combustion blower and damper modulation control.	.185					√
10 Check all indicator lamps and water/steam pressure gauges.	.091		√	√	√	√
11 Check electrical panels and wiring to burner, blowers and other components.	.117			√	√	√
12 Clean blower air-intake dampers, if required.	.069			√	√	√
13 Check condition of flue pipe, damper and exhaust stack.	.147		√	√	√	√
14 Check boiler operation through complete cycle, up to 30 minutes.	.884			√	√	√
15 Check water column sight glass and water level system; clean or replace sight glass, if required.	.127		√	√	√	√
16 Clean firebox (sweep and vacuum).	1.053					√
17 Check fuel level with gauge pole for oil burning boilers.	.046		√	√	√	√
18 Inspect and clean oil burner gun and ignition assembly where applicable.	.923					√
19 Clean area around boiler.	.151		√	√	√	√
20 Fill out maintenance checklist and report deficiencies.	.022		√	√	√	√
Total labor-hours/period			.949	2.508	2.508	5.579
Total labor-hours/year			7.592	5.016	2.508	5.579

			Cost Each				
			2014 Bare Costs			Total	Total
Description		Labor-hrs.	Material	Labor	Equip.	Total	In-House	w/O&P
3900	Boiler, steam, O/G/C, 500 to 1000 MBH, annually	5.579	205	325		530	635	765
3950	Annualized	20.695	216	1,200		1,416	1,750	2,175

D30 HVAC		D3025 140	Boiler, Steam, Oil/Gas/Comb.					

	PM Components	Labor-hrs.	W	M	Q	S	A
	PM System D3025 140 4950						
	Boiler, steam; oil, gas or combination fired, over 1000 MBH						
1	Inspect fuel system for leaks or damage.	.098		√	√	√	√
2	Change fuel filter element and clean strainers; repair leaks, where applicable.	1.027					√
3	Check main flame failure protection, positive fuel shutoff and main flame detection scanner on boiler equipped with spark ignition (oil burner).	.147		√	√	√	√
4	Check for proper operational response of burner to thermostat controls.	.199			√	√	√
5	Inspect all gas, steam and water lines, valves, connections for leaks abor damage; repair as necessary.	.195			√	√	√
6	Check feedwater system and feedwater makeup control and pump.	.091		√	√	√	√
7	Check and lubricate burner, blowers and motors as required.	.120			√	√	√
8	Check operation and condition of safety pressure relief valve.	.069		√	√	√	√
9	Check combustion controls, combustion blower and damper modulation control.	.199					√
10	Check all indicator lamps and water/steam pressure gauges.	.109		√	√	√	√
11	Check electrical panels and wiring to burner, blowers and other components.	.140			√	√	√
12	Clean blower air-intake dampers, if required.	.069		√	√	√	√
13	Check condition of flue pipe, damper and exhaust stack.	.147		√	√	√	√
14	Check boiler operation through complete cycle, up to 30 minutes.	.884			√	√	√
15	Check water column sight glass and water level system; clean or replace sight glass, if required.	.127		√	√	√	√
16	Clean firebox (sweep and vacuum).	1.144					√
17	Check fuel level with gauge pole for oil burning boilers.	.046		√	√	√	√
18	Inspect and clean oil burner gun and ignition assembly where applicable.	1.196					√
19	Clean area around boiler.	.182		√	√	√	√
20	Fill out maintenance checklist and report deficiencies.	.022		√	√	√	√
	Total labor-hours/period			1.038	2.645	2.645	6.211
	Total labor-hours/year			8.304	5.290	2.645	6.211

			Cost Each				
			2014 Bare Costs			Total	Total
	Description	Labor-hrs.	Material	Labor	Equip.	Total	In-House	w/O&P
4900	Boiler, steam, O/G/C, over 1000 MBH, annually	6.211	226	365		591	705	850
4950	Annualized	22.450	237	1,325		1,562	1,900	2,350

D30 HVAC		D3025 210	Deaerator Tank					
	PM Components		Labor-hrs.	W	M	Q	S	A
PM System D3025 210 1950								
Deaerator tank								
1	Check tank and associated piping for leaks.		.013				√	√
2	Bottom - blow deaerator tank.		.022				√	√
3	Perform sulfite test on deaerator water sample.		.013				√	√
4	Check low and high float levels for proper operation and respective water level alarms.		.337				√	√
5	Clean steam and feedwater strainers.		.143				√	√
6	Check steam pressure regulating valve operation.		.007				√	√
7	Check all indicator lights.		.066				√	√
8	Clean unit and surrounding area.		.022				√	√
9	Fill out maintenance checklist and report deficiencies.		.130				√	√
		Total labor-hours/period					.753	.753
		Total labor-hours/year					.753	.753

			Cost Each					
			2014 Bare Costs				Total	Total
Description		Labor-hrs.	Material	Labor	Equip.	Total	In-House	w/O&P
1900	Deaerator tank, annually	.753	41.50	44		85.50	101	120
1950	Annualized	1.506	41.50	88		129.50	156	189

D30 HVAC   D3025 310 Pumps

PM Components	Labor-hrs.	W	M	Q	S	A
PM System D3025 310 1950						
Pump, boiler fuel oil						
1 Check with operating or area personnel for deficiencies.	.035				✓	✓
2 Check for leaks on discharge piping and seals, etc.	.077				✓	✓
3 Check pump and motor operation for vibration, noise, overheating, etc.	.022				✓	✓
4 Check alignment and clearances of shaft and coupler.	.290				✓	✓
5 Tighten or replace loose, missing, or damaged nuts, bolts, and screws.	.005				✓	✓
6 Lubricate pump and motor as required.	.099				✓	✓
7 Clean pump, motor and surrounding area.	.066				✓	✓
8 Fill out maintenance checklist and report deficiencies.	.022				✓	✓
Total labor-hours/period					.616	.616
Total labor-hours/year					.616	.616

			Cost Each				
			2014 Bare Costs			Total	Total
Description	Labor-hrs.	Material	Labor	Equip.	Total	In-House	w/O&P
1900 Pump, boiler fuel oil, annually	.616	52.50	36		88.50	104	122
1950 Annualized	1.232	54	72		126	150	180

PM Components	Labor-hrs.	W	M	Q	S	A
PM System D3025 310 2950						
Pump, condensate return, over 1 H.P.						
1 Check for proper operation of pump.	.022				✓	✓
2 Check for leaks on suction and discharge piping, seals, packing glands, etc.; make minor adjustments as required.	.077				✓	✓
3 Check pump and motor operation for vibration, noise, overheating, etc.	.022				✓	✓
4 Check alignment pump and motor; adjust as necessary.	.290				✓	✓
5 Lubricate pump and motor.	.099				✓	✓
6 Clean exterior of pump, motor and surrounding area.	.030				✓	✓
7 Fill out maintenance checklist and report deficiencies.	.022				✓	✓
Total labor-hours/period					.562	.562
Total labor-hours/year					.562	.562

			Cost Each				
			2014 Bare Costs			Total	Total
Description	Labor-hrs.	Material	Labor	Equip.	Total	In-House	w/O&P
2900 Pump, steam condensate return, over 1 H.P., annually	.562	52.50	33		85.50	99.50	117
2950 Annualized	1.124	54	66		120	142	171

D30 HVAC		D3025 310 Pumps						
PM Components		Labor-hrs.	W	M	Q	S	A
PM System D3025 310 3950							
Pump, steam condensate return, duplex							
1	Check with operating or area personnel for deficiencies.	.035				✓	✓
2	Check for proper operation.	.022				✓	✓
3	Check for leaks on suction and discharge piping, seals, packing glands, etc.; make minor adjustments as required.	.153				✓	✓
4	Check pumps and motors operation for excessive vibration, noise and overheating.	.044				✓	✓
5	Check pump controller for proper operation.	.130				✓	✓
6	Lubricate pumps and motors.	.099				✓	✓
7	Clean condensate return unit and surrounding area.	.066				✓	✓
8	Fill out maintenance checklist and report deficiencies.	.022				✓	✓
	Total labor-hours/period					.571	.571
	Total labor-hours/year					.571	.571

			Cost Each					
			2014 Bare Costs				Total	Total
Description		Labor-hrs.	Material	Labor	Equip.	Total	In-House	w/O&P
3900	Pump, steam condensate return, duplex, annually	.571	105	33.50		138.50	158	183
3950	Annualized	1.142	108	67		175	202	239

D30 HVAC   D3035 110 Cooling Tower

PM Components	Labor-hrs.	W	M	Q	S	A
PM System D3035 110 1950						
Water cooling tower, forced draft, up to 50 tons						
1 Check with operating or area personnel for deficiencies.	.035				✓	✓
2 Check operation of unit for water leaks, noise or vibration.	.077				✓	✓
3 Clean and inspect hot water basin.	.156				✓	✓
4 Remove access panel.	.048				✓	✓
5 Check electrical wiring and connections; make appropriate adjustments.	.120				✓	✓
6 Lubricate all motor and fan bearings.	.047				✓	✓
7 Check fan blades or blowers for imbalance and tip clearance.	.055				✓	✓
8 Check belt for wear, tension and alignment; adjust as required.	.029				✓	✓
9 Drain and flush cold water sump and clean strainer.	.382				✓	✓
10 Clean inside of water tower using water hose; scrape, brush and wipe as required; heavy deposits of scale should be removed with scale removing compound.	.598				✓	✓
11 Refill with water, ck. make-up water asm. for leakage, adj. float if nec.	.211				✓	✓
12 Replace access panel.	.039				✓	✓
13 Remove, clean and reinstall conductivity and ph electrodes in chemical water treatment system.	.390				✓	✓
14 Inspect and clean around cooling tower.	.066				✓	✓
15 Fill out maintenance checklist and report deficiencies.	.022				✓	✓
Total labor-hours/period					2.275	2.275
Total labor-hours/year					2.275	2.275

		Cost Each					
		2014 Bare Costs				Total	Total
Description	Labor-hrs.	Material	Labor	Equip.	Total	In-House	w/O&P
1900 Cooling tower, forced draft, up to 50 tons, annually	2.275	27	133		160	196	242
1950 Annualized	4.550	38	266		304	375	465

PM Components	Labor-hrs.	W	M	Q	S	A
PM System D3035 110 2950						
Water cooling tower, forced draft, 51 tons through 500 tons						
1 Check with operating or area personnel for deficiencies.	.074				✓	✓
2 Check operation of unit for water leaks, noise or vibration.	.163				✓	✓
3 Clean and inspect hot water basin.	.332				✓	✓
4 Remove access panel.	.103				✓	✓
5 Check electrical wiring and connections; make appropriate adjustments.	.255				✓	✓
6 Lubricate all motor and fan bearings.	.100				✓	✓
7 Check fan blades or blowers for imbalance and tip clearance.	.247				✓	✓
8 Check belt for wear, tension and alignment; adjust as required.	.096				✓	✓
9 Drain and flush cold water sump and clean strainer.	.819				✓	✓
10 Clean inside of water tower using water hose; scrape, brush and wipe as required; heavy deposits of scale should be removed with scale removing compound.	1.271				✓	✓
11 Refill with water, chk. make-up water aseb. for leakage, adj. float if nec.	.447				✓	✓
12 Replace access panel.	.078				✓	✓
13 Remove, clean and reinstall conductivity and ph electrodes in chemical water treatment system.	.832				✓	✓
14 Inspect and clean around cooling tower.	.117				✓	✓
15 Fill out maintenance checklist and report deficiencies.	.022				✓	✓
Total labor-hours/period					4.956	4.956
Total labor-hours/year					4.956	4.956

		Cost Each					
		2014 Bare Costs				Total	Total
Description	Labor-hrs.	Material	Labor	Equip.	Total	In-House	w/O&P
2900 Cooling tower, forced draft, 50 thru 500 tons, annually	4.956	49.50	290		339.50	415	510
2950 Annualized	9.912	69	580		649	800	990

D30 HVAC		D3035 110	Cooling Tower					
	PM Components	Labor-hrs.	W	M	Q	S	A

	PM Components	Labor-hrs.	W	M	Q	S	A
	PM System D3035 110 3950 **Water cooling tower, forced draft, 500 tons through 1000 tons**						
1	Check with operating or area personnel for deficiencies.	.109				✓	✓
2	Check operation of unit for water leaks, noise or vibration.	.239				✓	✓
3	Clean and inspect hot water basin.	.488				✓	✓
4	Remove access panel.	.151				✓	✓
5	Check electrical wiring and connections; make appropriate adjustments.	.374				✓	✓
6	Lubricate all motor and fan bearings.	.147				✓	✓
7	Check fan blades or blowers for imbalance and tip clearance.	.302				✓	✓
8	Check belt for wear, tension and alignment; adjust as required.	.220				✓	✓
9	Drain and flush cold water sump and clean strainer.	1.464				✓	✓
10	Clean inside of water tower using water hose; scrape, brush and wipe as required; heavy deposits of scale should be removed with scale removing compound.	2.259				✓	✓
11	Refill with water, check make-up water assembly for leakage, adjust float if necessary.	1.578				✓	✓
12	Replace access panel.	.117				✓	✓
13	Remove, clean and reinstall conductivity and ph electrodes in chemical water treatment system.	1.222				✓	✓
14	Inspect and clean around cooling tower.	.172				✓	✓
15	Fill out maintenance checklist and report deficiencies.	.022				✓	✓
	Total labor-hours/period					8.864	8.864
	Total labor-hours/year					8.864	8.864

			Cost Each					
			2014 Bare Costs				Total	Total
Description		Labor-hrs.	Material	Labor	Equip.	Total	In-House	w/O&P
3900	Cooling tower, forced draft, 500 to 1000 tons, annually	8.864	107	520		627	765	945
3950	Annualized	17.728	153	1,025		1,178	1,475	1,825

D30 HVAC		D3035 130	Chiller, Recip., Air Cooled					

PM Components	Labor-hrs.	W	M	Q	S	A
PM System D3035 130 1950						
Chiller, reciprocating, air cooled, up to 25 tons						
1 Check unit for proper operation, excessive noise or vibration.	.033		√	√	√	√
2 Run system diagnostics test.	.325		√	√	√	√
3 Check oil level in sight glass of lead compressor only, add oil as necessary.	.042		√	√	√	√
4 Check superheat and subcooling temperatures.	.325					√
5 Check liquid line sight glass, oil and refrigerant pressures.	.036		√	√	√	√
6 Check contactors, sensors and mechanical safety limits.	.094					√
7 Check electrical wiring and connections; tighten loose connections.	.120					√
8 Clean intake side of condenser coils, fans and intake screens.	1.282					√
9 Inspect fan(s) or blower(s) for bent blades of imbalance.	.086					√
10 Lubricate shaft bearings and motor bearings as required.	.281					√
11 Inspect plumbing and valves for leaks, adjust as necessary.	.077		√	√	√	√
12 Check evaporator and condenser for corrosion.	.026		√	√	√	√
13 Clean chiller and surrounding area.	.066		√	√	√	√
14 Fill out maintenance checklist and report deficiencies.	.022		√	√	√	√
Total labor-hours/period			.627	.627	.627	2.815
Total labor-hours/year			5.016	1.254	.627	2.815

			Cost Each				
			2014 Bare Costs			Total	Total
Description	Labor-hrs.	Material	Labor	Equip.	Total	In-House	w/O&P
1900 Chiller, recip., air cooled, up to 25 tons, annually	2.815	53.50	165		218.50	265	325
1950 Annualized	9.712	53.50	570		623.50	770	950

PM Components	Labor-hrs.	W	M	Q	S	A
PM System D3035 130 2950						
Chiller, reciprocating, air cooled, over 25 tons						
1 Check unit for proper operation, excessive noise or vibration.	.033		√	√	√	√
2 Run system diagnostics test.	.455		√	√	√	√
3 Check oil level in sight glass of lead compressor only, add oil as necessary.	.042		√	√	√	√
4 Check superheat and subcooling temperatures.	.325					√
5 Check liquid line sight glass, oil and refrigerant pressures.	.036		√	√	√	√
6 Check contactors, sensors and mechanical safety limits.	.094					√
7 Check electrical wiring and connections; tighten loose connections.	.120					√
8 Clean intake side of condenser coils, fans and intake screens.	1.282					√
9 Inspect fan(s) or blower(s) for bent blades of imbalance.	.237					√
10 Lubricate shaft bearings and motor bearings as required.	.341					√
11 Inspect plumbing and valves for leaks, adjust as necessary.	.117		√	√	√	√
12 Check evaporator and condenser for corrosion.	.052		√	√	√	√
13 Clean chiller and surrounding area.	.117		√	√	√	√
14 Fill out maintenance checklist and report deficiencies.	.022		√	√	√	√
Total labor-hours/period			.874	.874	.874	3.273
Total labor-hours/year			6.992	1.748	.874	3.273

			Cost Each				
			2014 Bare Costs			Total	Total
Description	Labor-hrs.	Material	Labor	Equip.	Total	In-House	w/O&P
2900 Chiller, recip., air cooled, over 25 tons, annually	3.273	76.50	191		267.50	325	395
2950 Annualized	12.887	76.50	755		831.50	1,025	1,275

D30 HVAC D3035 135 Chiller, Recip., Water Cooled

PM Components	Labor-hrs.	W	M	Q	S	A
PM System D3035 135 1950						
Chiller, reciprocating, water cooled, up to 50 tons						
1 Check unit for proper operation, excessive noise or vibration.	.033		√	√	√	√
2 Run system diagnostics test.	.325		√	√	√	√
3 Check oil level in sight glass of lead compressor only, add oil as necessary.	.042		√	√	√	√
4 Check superheat and subcooling temperatures.	.325					√
5 Check liquid line sight glass, oil and refrigerant pressures.	.036		√	√	√	√
6 Check contactors, sensors and mechanical limits, adjust as necessary.	.094					√
7 Inspect plumbing and valves for leaks, tighten connections as necessary.	.077		√	√	√	√
8 Check condenser and evaporator for corrosion.	.026		√	√	√	√
9 Clean chiller and surrounding area.	.066		√	√	√	√
10 Fill out maintenance checklist and report deficiencies.	.022		√	√	√	√
Total labor-hours/period			.627	.627	.627	1.046
Total labor-hours/year			5.016	1.254	.627	1.046

			Cost Each				
			2014 Bare Costs			Total	Total
Description	Labor-hrs.	Material	Labor	Equip.	Total	In-House	w/O&P
1900 Chiller, recip., water cooled, up to 50 tons, annually	1.046	53.50	61		114.50	135	162
1950 Annualized	7.943	53.50	465		518.50	640	790

PM Components	Labor-hrs.	W	M	Q	S	A
PM System D3035 135 2950						
Chiller, reciprocating, water cooled, over 50 tons						
1 Check unit for proper operation, excessive noise or vibration.	.033		√	√	√	√
2 Run system diagnostics test.	.455		√	√	√	√
3 Check oil level in sight glass of lead compressor only, add oil as necessary.	.042		√	√	√	√
4 Check superheat and subcooling temperatures.	.325					√
5 Check liquid line sight glass, oil and refrigerant pressures.	.036		√	√	√	√
6 Check contactors, sensors and mechanical limits, adjust as necessary.	.094					√
7 Inspect plumbing and valves for leaks, tighten connections as necessary.	.117		√	√	√	√
8 Check condenser and evaporator for corrosion.	.052		√	√	√	√
9 Clean chiller and surrounding area.	.117		√	√	√	√
10 Fill out maintenance checklist and report deficiencies.	.022		√	√	√	√
Total labor-hours/period			.874	.874	.874	1.293
Total labor-hours/year			6.992	1.748	.874	1.293

			Cost Each				
			2014 Bare Costs			Total	Total
Description	Labor-hrs.	Material	Labor	Equip.	Total	In-House	w/O&P
2900 Chiller, recip., water cooled, over 50 tons, annually	1.293	61	75.50		136.50	162	194
2950 Annualized	10.907	61	640		701	865	1,075

D30 HVAC — D3035 140 Chiller, Centrif., Water Cooled

	PM Components	Labor-hrs.	W	M	Q	S	A
	PM System D3035 140 1950						
	Chiller, centrifugal water cooled, up to 100 tons						
1	Check unit for proper operation.	.035	✓	✓	✓	✓	✓
2	Check oil level; add oil as necessary.	.022	✓	✓	✓	✓	✓
3	Check oil temperature.	.038	✓	✓	✓	✓	✓
4	Check dehydrator or purge system; remove water if observed in sight glass.	.046	✓	✓	✓	✓	✓
5	Run system control tests.	.325		✓	✓	✓	✓
6	Check refrigerant charge/level, add as necessary.	.272		✓	✓	✓	✓
7	Check compressor for excessive noise/vibration.	.025		✓	✓	✓	✓
8	Check sensor and mechanical safety limits; replace as necessary.	.094				✓	✓
9	Clean dehydrator float valve.	.195					✓
10	Perform spectrochemical analysis of compressor oil; replace oil as necessary.	.039					✓
11	Replace oil filters and add oil as necessary.	.081					✓
12	Inspect cooler and condenser tubes for leaks; clean screens as necessary.	5.200					✓
13	Inspect utility vessel vent piping and safety relief valve; replace as necessary.	.195					✓
14	Inspect/clean the economizer (vane) gas line, damper valve and actuator arm.	.650					✓
15	Run an insulation test on the centrifugal motor.	1.300					✓
16	Clean area around equipment.	.066	✓	✓	✓	✓	✓
17	Document all maintenance and cleaning procedures.	.022	✓	✓	✓	✓	✓
	Total labor-hours/period		.229	.851	.851	.945	8.605
	Total labor-hours/year		8.702	6.808	1.702	.945	8.605

			Cost Each				
				2014 Bare Costs			Total	Total
	Description	Labor-hrs.	Material	Labor	Equip.	Total	In-House	w/O&P
1900	Chiller, centrif., water cooled, up to 100 tons, annually	8.605	248	505		753	900	1,100
1950	Annualized	26.762	252	1,575		1,827	2,225	2,775

D30 HVAC — D3035 140 Chiller, Centrif., Water Cooled

PM Components	Labor-hrs.	W	M	Q	S	A
PM System D3035 140 2950						
Chiller, centrifugal water cooled, over 100 tons						
1 Check unit for proper operation.	.035	✓	✓	✓	✓	✓
2 Check oil level; add oil as necessary.	.022	✓	✓	✓	✓	✓
3 Check oil temperature.	.038	✓	✓	✓	✓	✓
4 Check dehydrator or purge system; remove water if observed in sight glass.	.046	✓	✓	✓	✓	✓
5 Run system control tests.	.455		✓	✓	✓	✓
6 Check refrigerant charge/level, add as necessary.	.381		✓	✓	✓	✓
7 Check compressor for excessive noise/vibration.	.039		✓	✓	✓	✓
8 Check sensor and mechanical safety limits; replace as necessary.	.133				✓	✓
9 Clean dehydrator float valve.	.351					✓
10 Perform spectrochemical analysis of compressor oil; replace oil as necessary.	.039					✓
11 Replace oil filters and add oil as necessary.	.161					✓
12 Inspect cooler and condenser tubes for leaks; clean screens as necessary.	5.200					✓
13 Inspect utility vessel vent piping and safety relief valve; replace as necessary.	.247					✓
14 Inspect/clean the economizer (vane), gas line damper valve and actuator arm.	.650					✓
15 Run an insulation test on the centrifugal motor.	1.950					✓
16 Clean area around equipment.	.117	✓	✓	✓	✓	✓
17 Document all maintenance and cleaning procedures.	.022	✓	✓	✓	✓	✓
Total labor-hours/period		.280	1.155	1.155	1.288	9.886
Total labor-hours/year		10.640	9.240	2.310	1.288	9.886

		Cost Each					
		2014 Bare Costs				Total	Total
Description	Labor-hrs.	Material	Labor	Equip.	Total	In-House	w/O&P
2900 Chiller, centrif., water cooled, over 100 tons, annually	9.886	279	580		859	1,025	1,250
2950 Annualized	33.364	283	1,950		2,233	2,750	3,400

D30 HVAC   D3035 150 | Chiller, Absorption Unit

PM Components	Labor-hrs.	W	M	Q	S	A
PM System D3035 150 1950						
Chiller, absorption unit, up to 500 tons						
1 Check with operating personnel for deficiencies; check operating log sheets for indications of increased temperature trends.	.035			✓	✓	✓
2 Check unit for proper operation, excessive noise or vibration.	.026			✓	✓	✓
3 Check and clean strainers in all lines as required.	1.300					✓
4 Check pulley alignment and belts for condition, proper tension and misalignment on external purge pump system, if applicable; adjust for proper tension and or alignment.	.043			✓	✓	✓
5 Check purge pump vacuum oil level, as required; add/change oil as necessary.	.650			✓	✓	✓
6 Lubricate pump shaft bearings and motor bearings.	.281			✓	✓	✓
7 Check and service system controls, wirings and connections; tighten loose connections.	.120					✓
8 Inspect cooling and system water piping circuits for leakage.	.077			✓	✓	✓
9 Clean area around equipment.	.066			✓	✓	✓
10 Fill out maintenance checklist.	.022			✓	✓	✓
Total labor-hours/period				1.200	1.200	2.620
Total labor-hours/year				2.400	1.200	2.620

			Cost Each				
			2014 Bare Costs			Total	Total
Description		Labor-hrs.	Material	Labor	Equip.	Total	In-House	w/O&P
1900 Chiller, absorption unit, up to 500 tons, annually		2.620	86	153		239	286	345
1950 Annualized		6.220	91.50	365		456.50	555	685

PM Components	Labor-hrs.	W	M	Q	S	A
PM System D3035 150 2950						
Chiller, absorption unit, 500 to 5000 tons						
1 Check with operating personnel for deficiencies; check operating log sheets for indications of increased temperature trends.	.035			✓	✓	✓
2 Check unit for proper operation, excessive noise or vibration.	.026			✓	✓	✓
3 Check and clean strainers in all lines as required.	2.340					✓
4 Check pulley alignment and belts for condition, proper tension and misalignment on external purge pump system, if applicable; adjust for proper tension and or alignment.	.086			✓	✓	✓
5 Check purge pump vacuum oil level, as required; add/change oil as necessary.	.975			✓	✓	✓
6 Lubricate pump shaft bearings and motor bearings.	.320			✓	✓	✓
7 Check and service system controls, wirings and connections; tighten loose connections.	.139					✓
8 Inspect cooling and system water piping circuits for leakage.	.153			✓	✓	✓
9 Clean area around equipment.	.117			✓	✓	✓
10 Fill out maintenance checklist.	.022			✓	✓	✓
Total labor-hours/period				1.734	1.734	4.213
Total labor-hours/year				3.468	1.734	4.213

			Cost Each				
			2014 Bare Costs			Total	Total
Description		Labor-hrs.	Material	Labor	Equip.	Total	In-House	w/O&P
2900 Chiller, absorption unit, 500 to 5000 tons, annually		4.213	174	246		420	500	605
2950 Annualized		9.415	237	550		787	950	1,150

D30 HVAC — D3035 160 Chiller, Screw, Water Cooled

PM Components	Labor-hrs.	W	M	Q	S	A
PM System D3035 160 1950						
Chiller, screw, water cooled, up to 100 tons						
1 Check unit for proper operation, excessive noise or vibration.	.033		✓	✓	✓	✓
2 Run system diagnostics test.	.325		✓	✓	✓	✓
3 Check oil level and oil temperature; add oil as necessary.	.022		✓	✓	✓	✓
4 Check refrigerant pressures; add as necessary.	.272			✓	✓	✓
5 Replace oil filters and oil, if applicable.	.081				✓	✓
6 Check contactors, sensors and mechanical safety limits.	.094				✓	✓
7 Perform spectrochemical analysis of compressor oil.	.039				✓	✓
8 Check electrical wiring and connections; tighten loose connections.	.120				✓	✓
9 Inspect cooler and condenser tubes for leaks.	5.200					✓
10 Check evaporator and condenser for corrosion.	.026					✓
11 Clean chiller and surrounding area.	.066		✓	✓	✓	✓
12 Fill out maintenance checklist and report deficiencies.	.022		✓	✓	✓	✓
Total labor-hours/period			.740	.740	1.074	6.300
Total labor-hours/year			5.920	1.480	1.074	6.300

			Cost Each				
			2014 Bare Costs			Total	Total
Description	Labor-hrs.	Material	Labor	Equip.	Total	In-House	w/O&P
1900 Chiller, screw, water cooled, up to 100 tons, annually	6.300	248	370		618	735	885
1950 Annualized	14.774	252	865		1,117	1,350	1,675

PM Components	Labor-hrs.	W	M	Q	S	A
PM System D3035 160 2950						
Chiller, screw, water cooled, over 100 tons						
1 Check unit for proper operation, excessive noise or vibration.	.033		✓	✓	✓	✓
2 Run system diagnostics test.	.325		✓	✓	✓	✓
3 Check oil level and oil temperature; add oil as necessary.	.022		✓	✓	✓	✓
4 Check refrigerant pressures; add as necessary.	.381			✓	✓	✓
5 Replace oil filters and oil, if applicable.	.081				✓	✓
6 Check contactors, sensors and mechanical safety limits.	.094				✓	✓
7 Perform spectrochemical analysis of compressor oil.	.039				✓	✓
8 Check electrical wiring and connections; tighten loose connections.	.120				✓	✓
9 Inspect cooler and condenser tubes for leaks.	5.200					✓
10 Check evaporator and condenser for corrosion.	.052					✓
11 Clean chiller and surrounding area.	.117		✓	✓	✓	✓
12 Fill out maintenance checklist and report deficiencies.	.022		✓	✓	✓	✓
Total labor-hours/period			.900	.900	1.234	6.486
Total labor-hours/year			7.200	1.800	1.234	6.486

			Cost Each				
			2014 Bare Costs			Total	Total
Description	Labor-hrs.	Material	Labor	Equip.	Total	In-House	w/O&P
2900 Chiller, screw, water cooled, over 100 tons, annually	6.486	279	380		659	780	940
2950 Annualized	16.720	283	980		1,263	1,525	1,875

D30 HVAC — D3035 170 Evaporative Cooler

PM Components	Labor-hrs.	W	M	Q	S	A
PM System D3035 170 1950						
Evaporative cooler						
1 Check with operating or area personnel for deficiencies.	.035					✓
2 Check unit for proper operation, noise and vibration.	.016					✓
3 Clean evaporation louver panels.	.083					✓
4 Lubricate fan, bearings and motor.	.052					✓
5 Check belt(s) for excessive wear and deterioration.	.095					✓
6 Visually inspect wiring for damage or loose connections; tighten loose connections.	.120					✓
7 Clean sump.	.039					✓
8 Clean and adjust float.	.257					✓
9 Check interior surfaces and check components for loose paint/lime deposits.	.016					✓
10 Check/clean, adjust nozzles (6).	.039					✓
11 Remove/clean and reinstall evaporative pads.	.352					✓
12 Start unit and check for proper operation.	.055					✓
13 Remove debris from surrounding area.	.066					✓
14 Fill out maintenance checklist.	.022					✓
Total labor-hours/period						1.247
Total labor-hours/year						1.247

	Description	Labor-hrs.	Material	Labor	Equip.	Total	Total In-House	Total w/O&P
			\multicolumn{4}{c}{2014 Bare Costs}		
1900	Evaporative cooler, annually	1.247	30.50	73		103.50	125	152
1950	Annualized	1.247	30.50	73		103.50	125	152

D30 HVAC — D3035 180 Evaporative Cooler, Rotating Drum

PM Components	Labor-hrs.	W	M	Q	S	A
PM System D3035 180 1950						
Evaporative cooler, rotating drum						
1 Check with operating or area personnel for deficiencies.	.035					✓
2 Check unit for proper operation, noise and vibration.	.016					✓
3 Clean evaporation louver panels.	.083					✓
4 Lubricate rotary and fan, bearings and motor.	.078					✓
5 Check belt(s) for excessive wear and deterioration.	.095					✓
6 Visually inspect wiring for damage or loose connections; tighten loose connections.	.120					✓
7 Clean sump and pump intake.	.039					✓
8 Clean and adjust float.	.257					✓
9 Check interior surfaces and check components for loose paint/lime deposits.	.055					✓
10 Check/clean, adjust nozzles.	.039					✓
11 Remove/clean and reinstall evaporative pads.	.352					✓
12 Start unit and check for proper operation.	.055					✓
13 Remove debris from surrounding area.	.066					✓
14 Fill out maintenance checklist.	.022					✓
Total labor-hours/period						1.312
Total labor-hours/year						1.312

	Description	Labor-hrs.	Material	Labor	Equip.	Total	Total In-House	Total w/O&P
1900	Evaporative cooler, rotating drum, annually	1.312	38	77		115	138	168
1950	Annualized	1.312	38	77		115	138	168

D30 HVAC — D3035 210 Condenser, Air Cooled

PM Components	Labor-hrs.	W	M	Q	S	A
PM System D3035 210 1950						
Condenser, air cooled, 3 tons to 25 tons						
1 Check with operating or area personnel for deficiencies.	.027			√	√	√
2 Check unit for proper operation, excessive noise or vibration.	.025			√	√	√
3 Pressure wash coils and fans with coil cleaning solution.	.470					√
4 Check electrical wiring and connections; tighten loose connections.	.092					√
5 Lubricate shaft bearings and motor bearings.	.036			√	√	√
6 Inspect fan(s) or blower(s) for bent blades or imbalance; adjust as necessary.	.024			√	√	√
7 Check belt(s) for condition, proper tension, and misalignment; adjust for proper tension and/or alignment, if applicable.	.060			√	√	√
8 Inspect piping and valves for leaks; tighten connections as necessary.	.059			√	√	√
9 Clean area around equipment.	.051			√	√	√
10 Fill out maintenance checklist and report deficiencies.	.017			√	√	√
Total labor-hours/period				.299	.299	.861
Total labor-hours/year				.598	.299	.861

			Cost Each				
			2014 Bare Costs			Total	Total
Description	Labor-hrs.	Material	Labor	Equip.	Total	In-House	w/O&P
1900 Condenser, air cooled, 3 tons to 25 tons, annually	.861	24	50.50		74.50	89.50	109
1950 Annualized	1.758	40	103		143	173	210

PM Components	Labor-hrs.	W	M	Q	S	A
PM System D3035 210 2950						
Condenser, air cooled, 26 tons through 100 tons						
1 Check with operating or area personnel for deficiencies.	.027			√	√	√
2 Check unit for proper operation, excessive noise or vibration.	.025			√	√	√
3 Pressure wash coils and fans with coil cleaning solution.	.570					√
4 Check electrical wiring and connections; tighten loose connections.	.092					√
5 Lubricate shaft bearings and motor bearings.	.042			√	√	√
6 Inspect fan(s) or blower(s) for bent blades or imbalance; adjust as necessary.	.030			√	√	√
7 Check belt(s) for condition, proper tension, and misalignment; adjust for proper tension and/or alignment, if applicable.	.080			√	√	√
8 Inspect piping and valves for leaks; tighten connections as necessary.	.059			√	√	√
9 Clean area around equipment.	.051			√	√	√
10 Fill out maintenance checklist and report deficiencies.	.017			√	√	√
Total labor-hours/period				.331	.331	.993
Total labor-hours/year				.662	.331	.993

			Cost Each				
			2014 Bare Costs			Total	Total
Description	Labor-hrs.	Material	Labor	Equip.	Total	In-House	w/O&P
2900 Condenser, air cooled, 26 tons to 100 tons, annually	.993	57	58		115	135	162
2950 Annualized	1.986	73.50	116		189.50	226	273

D30 HVAC

D3035 210 | Condenser, Air Cooled

PM Components	Labor-hrs.	W	M	Q	S	A
PM System D3035 210 3950						
Condenser, air cooled, over 100 tons						
1 Check with operating or area personnel for deficiencies.	.027			√	√	√
2 Check unit for proper operation, excessive noise or vibration.	.031			√	√	√
3 Pressure wash coils and fans with coil cleaning solution.	.670					√
4 Check electrical wiring and connections; tighten loose connections.	.092					√
5 Lubricate shaft bearings and motor bearings.	.042			√	√	√
6 Inspect fan(s) or blower(s) for bent blades or imbalance; adjust as necessary.	.030			√	√	√
7 Check belt(s) for condition, proper tension, and misalignment; adjust for proper tension and/or alignment, if applicable.	.100			√	√	√
8 Inspect piping and valves for leaks; tighten connections as necessary.	.059			√	√	√
9 Clean area around equipment.	.051			√	√	√
10 Fill out maintenance checklist and report deficiencies.	.017			√	√	√
Total labor-hours/period				.357	.357	1.119
Total labor-hours/year				.714	.357	1.119

			Cost Each				
			2014 Bare Costs			Total	Total
Description	Labor-hrs.	Material	Labor	Equip.	Total	In-House	w/O&P
3900 Condenser, air cooled, over 100 tons, annually	1.119	89.50	65.50		155	180	214
3950 Annualized	2.190	106	128		234	277	335

D30 HVAC

D3035 220 Condensing Unit, Air Cooled

PM Components	Labor-hrs.	W	M	Q	S	A
PM System D3035 220 1950						
Condensing unit, air cooled, 3 tons to 25 tons						
1　Check with operating or area personnel for deficiencies.	.035			✓	✓	✓
2　Check unit for proper operation, excessive noise or vibration.	.033			✓	✓	✓
3　Pressure wash coils and fans with coil cleaning solution.	.611					✓
4　Check electrical wiring and connections; tighten loose connections.	.120					✓
5　Lubricate shaft bearings and motor bearings.	.047			✓	✓	✓
6　Inspect fan(s) or blower(s) for bent blades or imbalance; adjust as necessary.	.031			✓	✓	✓
7　Check belt(s) for condition, proper tension, and misalignment; adjust for proper tension and/or alignment, if applicable.	.078			✓	✓	✓
8　Inspect piping and valves for leaks; tighten connections as necessary.	.077			✓	✓	✓
9　Check refrigerant pressure; add refrigerant, if necessary.	.272					✓
10　Clean area around equipment.	.066			✓	✓	✓
11　Fill out maintenance checklist and report deficiencies.	.022			✓	✓	✓
Total labor-hours/period				.389	.389	1.392
Total labor-hours/year				.778	.389	1.392

			Cost Each				
			2014 Bare Costs			Total	Total
Description	Labor-hrs.	Material	Labor	Equip.	Total	In-House	w/O&P
1900　Condensing unit, air cooled, 3 to 25 tons, annually	1.392	86	81.50		167.50	196	235
1950　　Annualized	2.559	91.50	150		241.50	288	345

PM Components	Labor-hrs.	W	M	Q	S	A
PM System D3035 220 2950						
Condensing unit, air cooled, 26 tons through 100 tons						
1　Check with operating or area personnel for deficiencies.	.035			✓	✓	✓
2　Check unit for proper operation, excessive noise or vibration.	.033			✓	✓	✓
3　Pressure wash coils and fans with coil cleaning solution.	.741					✓
4　Check electrical wiring and connections; tighten loose connections.	.120					✓
5　Lubricate shaft bearings and motor bearings.	.055			✓	✓	✓
6　Inspect fan(s) or blower(s) for bent blades or imbalance; adjust as necessary.	.039			✓	✓	✓
7　Check belt(s) for condition, proper tension, and misalignment; adjust for proper tension and/or alignment, if applicable.	.104			✓	✓	✓
8　Inspect piping and valves for leaks; tighten connections as necessary.	.077			✓	✓	✓
9　Check refrigerant pressure; add refrigerant, if necessary.	.390					✓
10　Clean area around equipment.	.066			✓	✓	✓
11　Fill out maintenance checklist and report deficiencies.	.022			✓	✓	✓
Total labor-hours/period				.431	.431	1.682
Total labor-hours/year				.862	.431	1.682

			Cost Each				
			2014 Bare Costs			Total	Total
Description	Labor-hrs.	Material	Labor	Equip.	Total	In-House	w/O&P
2900　Condensing unit, air cooled, 26 to 100 tons, annually	1.682	117	98.50		215.50	252	299
2950　　Annualized	2.975	122	174		296	350	425

D30 HVAC

D3035 220 Condensing Unit, Air Cooled

PM Components	Labor-hrs.	W	M	Q	S	A
PM System D3035 220 3950						
Condensing unit, air cooled, over 100 tons						
1 Check with operating or area personnel for deficiencies.	.035			√	√	√
2 Pressure wash coils and fans with coil cleaning solution.	.871					√
3 Clean intake side of condenser coils, fans and intake screens.	.091			√	√	√
4 Check electrical wiring and connections; tighten loose connections.	.120					√
5 Lubricate shaft bearings and motor bearings.	.055			√	√	√
6 Inspect fan(s) or blower(s) for bent blades or imbalance; adjust as necessary.	.039			√	√	√
7 Check belt(s) for condition, proper tension, and misalignment; adjust for proper tension and/or alignment, if applicable.	.130			√	√	√
8 Inspect piping and valves for leaks; tighten connections as necessary.	.077			√	√	√
9 Check refrigerant pressure; add refrigerant, if necessary.	.455					√
10 Clean area around equipment.	.066			√	√	√
11 Fill out maintenance checklist and report deficiencies.	.022			√	√	√
Total labor-hours/period				.515	.515	1.961
Total labor-hours/year				1.030	.515	1.961

	Description	Labor-hrs.	Cost Each					
			2014 Bare Costs				Total In-House	Total w/O&P
			Material	Labor	Equip.	Total		
3900	Condensing unit, air cooled, over 100 tons, annually	1.961	147	115		262	305	365
3950	Annualized	3.506	153	205		358	425	510

D30 HVAC — D3035 240 Condensing Unit, Water Cooled

PM Components	Labor-hrs.	W	M	Q	S	A
PM System D3035 240 1950						
Condensing unit, water cooled, 3 tons to 24 tons						
1 Check with operating or area personnel for deficiencies.	.035			✓	✓	✓
2 Check unit for proper operation, excessive noise or vibration.	.033			✓	✓	✓
3 Check electrical wiring and connections; tighten loose connections.	.120					✓
4 Inspect piping and valves for leaks; tighten connections as necessary.	.077			✓	✓	✓
5 Check refrigerant pressure; add refrigerant, if necessary.	.272					✓
6 Clean area around equipment.	.066			✓	✓	✓
7 Fill out maintenance checklist and report deficiencies.	.022			✓	✓	✓
Total labor-hours/period				.233	.233	.625
Total labor-hours/year				.466	.233	.625

			Cost Each				
			2014 Bare Costs			Total	Total
Description	Labor-hrs.	Material	Labor	Equip.	Total	In-House	w/O&P
1900 Condensing unit, water cooled, 3 to 24 tons, annually	.625	30.50	36.50		67	79.50	95
1950 Annualized	1.324	30.50	77.50		108	130	159

PM Components	Labor-hrs.	W	M	Q	S	A
PM System D3035 240 2950						
Condensing unit, water cooled, 25 tons through 100 tons						
1 Check with operating or area personnel for deficiencies.	.035			✓	✓	✓
2 Check unit for proper operation, excessive noise or vibration.	.033			✓	✓	✓
3 Run system diagnostics test.	.216			✓	✓	✓
4 Check electrical wiring and connections; tighten loose connections.	.120			✓	✓	✓
5 Check oil level in sight glass of lead compressor only; add oil as necessary.	.029			✓	✓	✓
6 Check contactors, sensors and mechanical limits, adjust as necessary.	.156					✓
7 Check condenser for corrosion.	.026					
8 Inspect piping and valves for leaks; tighten connections as necessary.	.077			✓	✓	✓
9 Check refrigerant pressure; add refrigerant, if necessary.	.272					✓
10 Clean area around equipment.	.066			✓	✓	✓
11 Fill out maintenance checklist and report deficiencies.	.022			✓	✓	✓
Total labor-hours/period				.598	.598	1.052
Total labor-hours/year				1.196	.598	1.052

			Cost Each				
			2014 Bare Costs			Total	Total
Description	Labor-hrs.	Material	Labor	Equip.	Total	In-House	w/O&P
2900 Condensing unit, WC, 25 to 100 tons, annually	1.052	76.50	61.50		138	161	192
2950 Annualized	2.846	76.50	166		242.50	292	355

D30 HVAC

D3035 240 Condensing Unit, Water Cooled

PM Components	Labor-hrs.	W	M	Q	S	A
PM System D3035 240 3950						
Condensing unit, water cooled, over 100 tons						
1 Check with operating or area personnel for deficiencies.	.035			√	√	√
2 Check unit for proper operation, excessive noise or vibration.	.033			√	√	√
3 Run system diagnostics test.	.216			√	√	√
4 Check electrical wiring and connections; tighten loose connections.	.120			√	√	√
5 Check oil level in sight glass of lead compressor only; add oil as necessary.	.029			√	√	√
6 Check contactors, sensors and mechanical limits, adjust as necessary.	.156					√
7 Check condenser for corrosion.	.026					√
8 Inspect piping and valves for leaks; tighten connections as necessary.	.077			√	√	√
9 Check refrigerant pressure; add refrigerant, if necessary.	.272					√
10 Clean area around equipment.	.066			√	√	√
11 Fill out maintenance checklist and report deficiencies.	.022			√	√	√
Total labor-hours/period				.598	.598	1.052
Total labor-hours/year				1.196	.598	1.052

		Cost Each					
			2014 Bare Costs			Total	Total
	Description	Labor-hrs.	Material	Labor	Equip.	Total	In-House	w/O&P
3900	Condensing unit, WC, over 100 tons, annually	1.052	122	61.50		183.50	211	249
3950	Annualized	2.846	122	166		288	340	415

D30 HVAC — D3035 260 Compressor, DX Refrigeration

PM Components	Labor-hrs.	W	M	Q	S	A
PM System D3035 260 1950						
Compressor, DX Refrigeration, to 25 tons						
1 Remove/replace access panel/cover.	.070			✓	✓	✓
2 Check unit for proper operation, excessive noise or vibration.	.030			✓	✓	✓
3 Run systems diagnostics test, if applicable.	.060			✓	✓	✓
4 Check oil level in compressor, if possible, and add oil as required.	.030			✓	✓	✓
5 Check refrigerant pressures; add refrigerant as necessary.	.200			✓	✓	✓
6 Check contactors, sensors and mechanical limits, adjust as necessary.	.100			✓	✓	✓
7 Inspect refrigerant piping and insulation.	.050			✓	✓	✓
8 Clean compressor and surrounding area.	.050			✓	✓	✓
9 Fill out maintenance checklist and report deficiencies.	.020			✓	✓	✓
Total labor-hours/period				.610	.610	.610
Total labor-hours/year				1.220	.610	.610

				Cost Each				
				2014 Bare Costs			Total	Total
Description		Labor-hrs.	Material	Labor	Equip.	Total	In-House	w/O&P
1900	Compressor, DX refrigeration, to 25 tons, annually	.610	16	35.50		51.50	62	75.50
1950	Annualized	2.440	16	143		159	196	243

PM Components	Labor-hrs.	W	M	Q	S	A
PM System D3035 260 2950						
Compressor, DX refrigeration, 25 tons to 100 tons						
1 Check unit for proper operation, excessive noise or vibration.	.030			✓	✓	✓
2 Run system diagnostics test, if applicable.	.098			✓	✓	✓
3 Check oil level in compressor; add oil as necessary.	.033			✓	✓	✓
4 Check refrigerant pressures; add refrigerant as necessary.	.272			✓	✓	✓
5 Check contactors, sensors and mechanical limits, adjust as necessary.	.216			✓	✓	✓
6 Inspect refrigerant piping and insulation.	.077			✓	✓	✓
7 Clean compressor and surrounding area.	.066			✓	✓	✓
8 Fill out maintenance checklist and report deficiencies.	.022			✓	✓	✓
Total labor-hours/period				.814	.814	.814
Total labor-hours/year				1.628	.814	.814

				Cost Each				
				2014 Bare Costs			Total	Total
Description		Labor-hrs.	Material	Labor	Equip.	Total	In-House	w/O&P
2900	Compressor, DX refrigeration, 25 to 100 tons, annually	.814	15.15	47.50		62.65	76	93.50
2950	Annualized	3.256	15.15	190		205.15	255	315

D30 HVAC — D3035 290 Fluid Cooler

PM Components	Labor-hrs.	W	M	Q	S	A
PM System D3035 290 1950						
Fluid cooler, 2 fans (no compressor)						
1 Check with operating or area personnel for deficiencies.	.035					✓
2 Check unit for proper operation, excessive noise or vibration.	.159					✓
3 Clean intake side of condenser coils, fans and intake screens.	.473					✓
4 Check electrical wiring and connections; tighten loose connections.	.120					✓
5 Inspect fan(s) for bent blades or unbalance; adjust as necessary.	.040					✓
6 Check belts for condition, proper tension and misalignment; adjust for proper tension and/or alignment, if required.	.029					✓
7 Lubricate shaft bearings and motor bearings.	.047					✓
8 Inspect piping and valves for leaks; tighten connections as necessary.	.077					✓
9 Lubricate and check operation of dampers, if applicable.	.055					✓
10 Clean area around fluid cooler.	.066					✓
11 Fill out maintenance checklist and report deficiencies.	.022					✓
Total labor-hours/period						1.123
Total labor-hours/year						1.123

			Cost Each				
			2014 Bare Costs			Total	Total
Description	Labor-hrs.	Material	Labor	Equip.	Total	In-House	w/O&P
1900 Fluid cooler, 2 fans (no compressor), annually	1.123	15.30	65.50		80.80	99	121
1950 Annualized	1.123	15.30	65.50		80.80	99	121

D30 HVAC   D3045 110 | Air Handling Unit

PM Components	Labor-hrs.	W	M	Q	S	A
PM System D3045 110 1950						
Air handling unit, 3 tons through 24 tons						
1 Check with operating or area personnel for deficiencies.	.035			√	√	√
2 Check controls and unit for proper operation.	.033			√	√	√
3 Check for unusual noise or vibration.	.033			√	√	√
4 Check tension, condition and alignment of belts, adjust as necessary.	.029			√	√	√
5 Clean coils, evaporator drain pan, blower, motor and drain piping, as required.	.380					√
6 Lubricate shaft and motor bearings.	.047			√	√	√
7 Replace air filters.	.078			√	√	√
8 Inspect exterior piping and valves for leaks; tighten connections as required.	.077			√	√	√
9 Clean area around equipment.	.066			√	√	√
10 Fill out maintenance checklist and report deficiencies.	.022			√	√	√
Total labor-hours/period				.420	.420	.800
Total labor-hours/year				.840	.420	.800

			Cost Each				
			2014 Bare Costs			Total	Total
Description		Labor-hrs.	Material	Labor	Equip.	Total	In-House	w/O&P
1900	Air handling unit, 3 thru 24 tons, annually	.800	47	47		94	110	132
1950	Annualized	2.060	101	121		222	262	315

PM Components	Labor-hrs.	W	M	Q	S	A
PM System D3045 110 2950						
Air handling unit, 25 tons through 50 tons						
1 Check with operating or area personnel for deficiencies.	.035			√	√	√
2 Check controls and unit for proper operation.	.033			√	√	√
3 Check for unusual noise or vibration.	.033			√	√	√
4 Clean coils, evaporator drain pan, blower, motor and condensate drain piping, as required.	.380					√
5 Lubricate shaft and motor bearings.	.047			√	√	√
6 Check belts for wear, proper tension, and alignment; adjust as necessary.	.029			√	√	√
7 Inspect exterior piping and valves for leaks; tighten connections as required.	.077			√	√	√
8 Check operation and clean dampers, louvers and shutters; lubricate all pivot points and linkages.	.078					√
9 Replace air filters.	.078			√	√	√
10 Clean area around equipment.	.066			√	√	√
11 Fill out maintenance checklist and report deficiencies.	.022			√	√	√
Total labor-hours/period				.420	.420	.878
Total labor-hours/year				.840	.420	.878

			Cost Each				
			2014 Bare Costs			Total	Total
Description		Labor-hrs.	Material	Labor	Equip.	Total	In-House	w/O&P
2900	Air handling unit, 25 thru 50 tons, annually	.878	147	51.50		198.50	227	264
2950	Annualized	2.138	400	125		525	595	695

D30 HVAC		D3045 110	Air Handling Unit					
	PM Components	Labor-hrs.	W	M	Q	S	A

	PM Components	Labor-hrs.	W	M	Q	S	A
PM System D3045 110 3950							
Air handling unit, over 50 tons							
1	Check with operating or area personnel for deficiencies.	.035			✓	✓	✓
2	Check controls and unit for proper operation.	.033			✓	✓	✓
3	Check for unusual noise or vibration.	.033			✓	✓	✓
4	Clean coils, evaporator drain pan, blower, motor and drain piping, as required.	.510					✓
5	Lubricate shaft and motor bearings.	.047			✓	✓	✓
6	Check belts for wear, proper tension, and alignment; adjust as necessary.	.055			✓	✓	✓
7	Inspect exterior piping and valves for leaks; tighten connections as required.	.077			✓	✓	✓
8	Check operation and clean dampers, louvers and shutters; lubricate all pivot points and linkages.	.078					✓
9	Clean centrifugal fan.	.065					✓
10	Replace air filters.	.286					✓
11	Clean area around equipment.	.066			✓	✓	✓
12	Fill out maintenance checklist and report deficiencies	.022			✓	✓	✓
	Total labor-hours/period				.654	.654	1.307
	Total labor-hours/year				1.308	.654	1.307

			Cost Each					
				2014 Bare Costs			Total	Total
	Description	Labor-hrs.	Material	Labor	Equip.	Total	In-House	w/O&P
3900	Air handling unit, over 50 tons, annually	1.307	211	76.50		287.50	330	385
3950	Annualized	3.269	515	191		706	805	940

D30 HVAC		D3045 112	A.H.U., Computer Room					

	PM Components	Labor-hrs.	W	M	Q	S	A
PM System D3045 112 1950							
Air handling unit, computer room							
1	Check with operating or area personnel for deficiencies.	.035			✓	✓	✓
2	Run microprocessor check, if available, or check controls and unit for proper operation.	.216			✓	✓	✓
3	Check for unusual noise or vibration.	.033			✓	✓	✓
4	Clean coils, evaporator drain pan, blower, motor and drain piping, as required.	.380					✓
5	Lubricate shaft and motor bearings.	.047			✓	✓	✓
6	Check belts for wear, proper tension, and alignment; adjust as necessary.	.029			✓	✓	✓
7	Check humidity lamp, replace if necessary.	.156			✓	✓	✓
8	Inspect exterior piping and valves for leaks; tighten connections as required.	.077			✓	✓	✓
9	Replace air filters.	.078			✓	✓	✓
10	Clean area around equipment.	.066			✓	✓	✓
11	Fill out maintenance checklist and report deficiencies.	.022			✓	✓	✓
	Total labor-hours/period				.759	.759	1.139
	Total labor-hours/year				1.518	.759	1.139

			Cost Each					
				2014 Bare Costs			Total	Total
	Description	Labor-hrs.	Material	Labor	Equip.	Total	In-House	w/O&P
1900	Air handling unit, computer room, annually	1.139	39	66.50		105.50	126	153
1950	Annualized	3.416	62.50	200		262.50	320	390

D30 HVAC — D3045 120 Fan Coil Unit

PM Components	Labor-hrs.	W	M	Q	S	A
PM System D3045 120 1950						
Fan coil unit						
1 Check with operating or area personnel for deficiencies.	.035			✓	✓	✓
2 Check coil unit while operating.	.120				✓	✓
3 Remove access panel and vacuum inside of unit and coils.	.468				✓	✓
4 Check coils and piping for leaks, damage and corrosion; repair as necessary.	.077			✓	✓	✓
5 Lubricate blower shaft and fan motor bearings.	.047				✓	✓
6 Clean coil, drip pan, and drain line with solvent.	.473				✓	✓
7 Replace filters as required.	.009			✓	✓	✓
8 Replace access panel.	.023				✓	✓
9 Check operation after repairs.	.120				✓	✓
10 Clean area.	.066			✓	✓	✓
11 Fill out maintenance checklist and report deficiencies.	.022			✓	✓	✓
Total labor-hours/period				.209	1.460	1.460
Total labor-hours/year				.418	1.460	1.460

			Cost Each				
			2014 Bare Costs			Total	Total
Description	Labor-hrs.	Material	Labor	Equip.	Total	In-House	w/O&P
1900 Fan coil unit, annually	1.460	6.90	85.50		92.40	114	142
1950 Annualized	3.338	27.50	195		222.50	275	340

D30 HVAC — D3045 150 Air Filters, Electrostatic

PM Components	Labor-hrs.	W	M	Q	S	A
PM System D3045 150 1950						
Filters, Electrostatic						
1 Review manufacturers instructions.	.091			✓	✓	✓
2 Check indicators for defective tubes or broken ionizing wires.	.130			✓	✓	✓
3 De-energize power supply, tag and lockout disconnect switch.	.130			✓	✓	✓
4 Ground bus trips, top and bottom.	.065			✓	✓	✓
5 Secure filter and fan unit.	.065			✓	✓	✓
6 Wash each manifold until clean, approximately 4 minutes with hot water, 7 minutes with cold water.	.390			✓	✓	✓
7 Clean or replace dry filters as necessary.	.260			✓	✓	✓
8 Inspect for broken on hum supressors and wipe insulators with soft dry cloth.	.260			✓	✓	✓
9 Disassemble unit as required, check it thoroughly, clean and adjust.	.650			✓	✓	✓
10 Restore to service and check for shorts.	.130			✓	✓	✓
11 Fill out maintenance checklist and report deficiencies.	.022			✓	✓	✓
Total labor-hours/period				2.193	2.193	2.193
Total labor-hours/year				4.386	2.193	2.193

			Cost Each				
			2014 Bare Costs			Total	Total
Description	Labor-hrs.	Material	Labor	Equip.	Total	In-House	w/O&P
1900 Air filter, electrostatic, annually	2.193	12.45	104		116.45	151	185
1950 Annualized	8.772	25	415		440	580	705

D30 HVAC — D3045 160 VAV Boxes

PM Components	Labor-hrs.	W	M	Q	S	A
PM System D3045 160 1950						
VAV Boxes						
1 Open/close VAV control box access.	.060				✓	✓
2 Check that pneumatic tubing/electrical connections are in place and tight.	.040				✓	✓
3 Tighten arm on motor output shaft.	.040				✓	✓
4 Cycle actuator while watching for proper operation.	.200				✓	✓
Verify that blades fully open and close.	.040				✓	✓
5 Lubricate actuator linkage and damper blade pivot points.	.070				✓	✓
6 Fill out maintenance checklist and report deficiencies.	.017				✓	✓
Total labor-hours/period					.467	.467
Total labor-hours/year					.467	.467

			Cost Each				
			2014 Bare Costs			Total	Total
Description	Labor-hrs.	Material	Labor	Equip.	Total	In-House	w/O&P
1900 VAV Boxes, annually	.467	8.35	27.50		35.85	43.50	53
1950 Annualized	.934	8.35	54.50		62.85	77.50	95.50

D30 HVAC — D3045 170 Fire Dampers

PM Components	Labor-hrs.	W	M	Q	S	A
PM System D3045 170 1950						
Fire dampers						
1 Remove/replace access door.	.800					✓
2 Clean out debris/dirt blown against damper.	.200					✓
3 Remove fusible link and check that blades operate freely.	.070					✓
4 Lubricate pivot points.	.050					✓
5 Replace fusible link.	.020					✓
6 Fill out maintenance checklist and report deficiencies.	.017					✓
Total labor-hours/period						1.157
Total labor-hours/year						1.157

			Cost Each				
			2014 Bare Costs			Total	Total
Description	Labor-hrs.	Material	Labor	Equip.	Total	In-House	w/O&P
1900 Fire dampers, annually	1.157	6.50	67.50		74	92	114
1950 Annualized	1.157	6.50	67.50		74	92	114

D30 HVAC — D3045 210 Fan, Axial

PM Components	Labor-hrs.	W	M	Q	S	A
PM System D3045 210 1950						
Fan, axial, up to 5,000 CFM						
1 Start and stop fan with local switch.	.012				✓	✓
2 Check fan for noise and vibration.	.091				✓	✓
3 Check electrical wiring and connections; tighten loose connections.	.029				✓	✓
4 Check motor and fan shaft bearings for noise, vibration, overheating; lubricate as required.	.325				✓	✓
5 Clean area around fan.	.143				✓	✓
6 Fill out maintenance checklist and report deficiencies.	.022				✓	✓
Total labor-hours/period					.622	.622
Total labor-hours/year					.622	.622

			Cost Each				
			2014 Bare Costs			Total	Total
Description	Labor-hrs.	Material	Labor	Equip.	Total	In-House	w/O&P
1900 Fan, axial, up to 5,000 CFM, annually	.622	3.83	29.50		33.33	43	53
1950 Annualized	1.244	7.65	59		66.65	86	106

PM Components	Labor-hrs.	W	M	Q	S	A
PM System D3045 210 2950						
Fan, axial, 5,000 to 10,000 CFM						
1 Start and stop fan with local switch.	.012				✓	✓
2 Check motor and fan shaft bearings for noise, vibration, overheating; lubricate bearings.	.325				✓	✓
3 Check belts for wear, tension, and alignment, if applicable; adjust as required.	.057				✓	✓
4 Check fan pitch operator, lubricate; if applicable.	.029				✓	✓
5 Check electrical wiring and connections; tighten loose connections.	.057				✓	✓
6 Clean fan and surrounding fan.	.143				✓	✓
7 Fill out maintenance checklist and report deficiencies.	.022				✓	✓
Total labor-hours/period					.645	.645
Total labor-hours/year					.645	.645

			Cost Each				
			2014 Bare Costs			Total	Total
Description	Labor-hrs.	Material	Labor	Equip.	Total	In-House	w/O&P
2900 Fan, axial, 5,000 to 10,000 CFM, annually	.645	3.83	30.50		34.33	44.50	54.50
2950 Annualized	1.290	7.65	61		68.65	89	109

D30 HVAC

D3045 210 Fan, Axial

PM Components	Labor-hrs.	W	M	Q	S	A
PM System D3045 210 3950						
Fan, axial, 36″ to 48″ dia (over 10,000 CFM)						
1 Start and stop fan with local switch.	.012				✓	✓
2 Check motor and fan shaft bearings for noise, vibration, overheating; lubricate bearings.	.325				✓	✓
3 Check belts for wear, tension, and alignment, if applicable; adjust as required.	.086				✓	✓
4 Check fan pitch operator, lubricate; if applicable.	.029				✓	✓
5 Check electrical wiring and connections; tighten loose connections.	.078				✓	✓
6 Clean fan and surrounding area.	.143				✓	✓
7 Fill out maintenance checklist and report deficiencies.	.022				✓	✓
Total labor-hours/period					.695	.695
Total labor-hours/year					.695	.695

	Description	Labor-hrs.	Cost Each				Total In-House	Total w/O&P
			2014 Bare Costs					
			Material	Labor	Equip.	Total		
3900	Fan, axial, 36″ to 48″ dia (over 10,000 CFM), annually	.695	3.83	33		36.83	47.50	58.50
3950	Annualized	1.390	7.65	65.50		73.15	95	117

D30 HVAC — D3045 220 Fan, Centrifugal

PM Components	Labor-hrs.	W	M	Q	S	A
PM System D3045 220 1950						
Fan, centrifugal, up to 5,000 CFM						
1 Start and stop fan with local switch.	.012				✓	✓
2 Check motor and fan shaft bearings for noise, vibration, overheating; lubricate bearings.	.325				✓	✓
3 Check belts for wear, tension, and alignment, if applicable; adjust as required.	.057				✓	✓
4 Check blower intake dampers, lubricate; if applicable.	.029				✓	✓
5 Check electrical wiring and connections; tighten loose connections.	.029				✓	✓
6 Clean fan and surrounding area.	.066				✓	✓
7 Fill out maintenance checklist and report deficiencies.	.022				✓	✓
Total labor-hours/period					.540	.540
Total labor-hours/year					.540	.540

			Cost Each				
			2014 Bare Costs			Total	Total
Description	Labor-hrs.	Material	Labor	Equip.	Total	In-House	w/O&P
1900 Fan, centrifugal, up to 5,000 CFM, annually	.540	3.83	25.50		29.33	38	46.50
1950 Annualized	1.080	7.65	51		58.65	76	92.50

PM Components	Labor-hrs.	W	M	Q	S	A
PM System D3045 220 2950						
Fan, centrifugal, 5,000 to 10,000 CFM						
1 Start and stop fan with local switch.	.012				✓	✓
2 Check motor and fan shaft bearings for noise, vibration, overheating; lubricate bearings.	.325				✓	✓
3 Check belts for wear, tension, and alignment, if applicable; adjust as required.	.057				✓	✓
4 Check blower intake dampers, lubricate; if applicable.	.029				✓	✓
5 Check electrical wiring and connections; tighten loose connections.	.057				✓	✓
6 Clean fan and surrounding area.	.066				✓	✓
7 Fill out maintenance checklist and report deficiencies.	.022				✓	✓
Total labor-hours/period					.568	.568
Total labor-hours/year					.568	.568

			Cost Each				
			2014 Bare Costs			Total	Total
Description	Labor-hrs.	Material	Labor	Equip.	Total	In-House	w/O&P
2900 Fan, centrifugal, 5,000 to 10,000 CFM, annually	.568	3.83	27		30.83	39.50	49
2950 Annualized	1.136	7.65	53.50		61.15	79.50	97

D30 HVAC — D3045 220 Fan, Centrifugal

PM Components	Labor-hrs.	W	M	Q	S	A
PM System D3045 220 3950						
Fan, centrifugal, over 10,000 CFM						
1 Start and stop fan with local switch.	.007				✓	✓
2 Check motor and fan shaft bearings for noise, vibration, overheating; lubricate bearings.	.325				✓	✓
3 Check belts for wear, tension, and alignment, if applicable; adjust as required.	.086				✓	✓
4 Check blower intake dampers, lubricate; if applicable.	.029				✓	✓
5 Check electrical wiring and connections; tighten loose connections.	.057				✓	✓
6 Clean fan and surrounding area.	.066				✓	✓
7 Fill out maintenance checklist and report deficiencies.	.022				✓	✓
Total labor-hours/period					.592	.592
Total labor-hours/year					.592	.592

			Cost Each				
			2014 Bare Costs			Total	Total
Description	Labor-hrs.	Material	Labor	Equip.	Total	In-House	w/O&P
3900 Fan, centrifugal, over 10,000 CFM, annually	.592	3.83	28		31.83	41	50.50
3950 Annualized	1.184	7.65	56		63.65	82.50	101

D30 HVAC — D3045 250 Hood And Blower

PM Components	Labor-hrs.	W	M	Q	S	A
PM System D3045 250 1950						
Hood and blower						
1 Check with operating or area personnel for any deficiencies.	.044			✓	✓	✓
2 Check unit for proper operation, including switches, controls and thermostat; calibrate thermostat and repair components as required.	.109			✓	✓	✓
3 Check operation of spray nozzles for spray coverage and drainage, if applicable.	.022					✓
4 Inspect soap and spray solution feeder lines.	.007			✓	✓	✓
5 Check components for automatic cleaning system and automatic fire protection system - solenoid valve, line strainer, shut-off valve, detergent tank, etc.	.055			✓	✓	✓
6 Clean spray nozzles.	1.104					✓
7 Tighten or replace loose, missing or damaged nuts, bolts or screws.	.005			✓	✓	✓
8 Check operation of exhaust blower on roof; lubricate bearings; adjust tension of fan belts and clean fan blades as required.	.056			✓	✓	✓
9 Fill out maintenance checklist and report deficiencies.	.022			✓	✓	✓
Total labor-hours/period				.298	.298	1.424
Total labor-hours/year				.596	.298	1.424

			Cost Each				
			2014 Bare Costs			Total	Total
Description	Labor-hrs.	Material	Labor	Equip.	Total	In-House	w/O&P
1900 Hood and blower, annually	1.424	61.50	67.50		129	157	187
1950 Annualized	2.318	87.50	110		197.50	242	288

D30 HVAC — D3045 410 Pump, Centrifugal

PM Components	Labor-hrs.	W	M	Q	S	A
PM System D3045 410 1950						
Centrifugal pump over 1 H.P.						
1 Check for proper operation of pump.	.022				✓	✓
2 Check for leaks on suction and discharge piping, seals, packing glands, etc.; make minor adjustments as required.	.077				✓	✓
3 Check pump and motor operation for excessive vibration, noise and overheating.	.022				✓	✓
4 Check alignment of pump and motor; adjust as necessary.	.260				✓	✓
5 Lubricate pump and motor.	.099				✓	✓
6 Clean exterior of pump, motor and surrounding area.	.096				✓	✓
7 Fill out maintenance checklist and report deficiencies.	.022				✓	✓
Total labor-hours/period					.598	.598
Total labor-hours/year					.598	.598

			Cost Each				
			2014 Bare Costs			Total	Total
Description	Labor-hrs.	Material	Labor	Equip.	Total	In-House	w/O&P
1900 Centrifugal pump, over 1 H.P., annually	.598	8.35	34.50		42.85	52	64
1950 Annualized	1.196	16.75	69		85.75	104	128

PM Components	Labor-hrs.	W	M	Q	S	A
PM System D3045 410 2950						
Centrifugal pump w/reduction gear, over 1 H.P.						
1 Check with operating or area personnel for deficiencies.	.035				✓	✓
2 Clean pump exterior and check for corrosion on pump exterior and base plate.	.030				✓	✓
3 Check for leaks on suction and discharge piping, seals, packing glands, etc.	.077				✓	✓
4 Check pump, gear and motor operation for vibration, noise, overheating, etc.	.022				✓	✓
5 Check alignment and clearances of shaft reduction gear and coupler.	.260				✓	✓
6 Tighten or replace loose, missing, or damaged nuts, bolts and screws.	.005				✓	✓
7 Lubricate pump and motor as required.	.099				✓	✓
8 When available, check suction or discharge, pressure gauge readings and flow rate.	.022				✓	✓
9 Clean area around pump.	.066				✓	✓
10 Fill out maintenance checklist and report deficiencies.	.022				✓	✓
Total labor-hours/period					.638	.638
Total labor-hours/year					.638	.638

			Cost Each				
			2014 Bare Costs			Total	Total
Description	Labor-hrs.	Material	Labor	Equip.	Total	In-House	w/O&P
2900 Centrifugal pump, w/ red. gear, over 1H.P., annually	.638	8.35	36.50		44.85	55	68
2950 Annualized	1.276	16.75	73.50		90.25	110	136

D30 HVAC		D3045 420	Pump w/Oil Reservoir, Electric					
	PM Components	Labor-hrs.	W	M	Q	S	A
PM System D3045 420 1950							
Pump w/oil reservoir, over 1 H.P.							
1	Check for proper operation of pump.	.022				√	√
2	Check for leaks on suction and discharge piping, seals, packing glands, etc.; make minor adjustments as required.	.077				√	√
3	Check pump and motor operation for excessive vibration, noise and overheating.	.022				√	√
4	Check alignment of pump and motor; adjust as necessary.	.260				√	√
5	Lubricate motor; check oil level in pump reservoir.	.099				√	√
6	Clean exterior of pump and surrounding area.	.096				√	√
7	Fill out maintenance checklist and report deficiencies.	.022				√	√
	Total labor-hours/period					.598	.598
	Total labor-hours/year					.598	.598

			Cost Each					
				2014 Bare Costs			Total	Total
	Description	Labor-hrs.	Material	Labor	Equip.	Total	In-House	w/O&P
1900	Pump w/ oil reservoir, electric, annually	.598	8.35	34.50		42.85	52	64
1950	Annualized	1.196	16.75	69		85.75	104	128

D30 HVAC — D3045 600 Heat Exchanger, Steam

PM Components	Labor-hrs.	W	M	Q	S	A
PM System D3045 600 1950						
Heat exchanger, steam						
1 Check with operating or area personnel for deficiencies.	.035				✓	✓
2 Check temperature gauges for proper operating temperatures.	.091				✓	✓
3 Check steam modulating valve and steam condensate trap for proper operation.	.101				✓	✓
4 Inspect heat exchanger and adjacent piping for torn or deteriorated insulation.	.147				✓	✓
5 Clean heat exchanger and surrounding area.	.066				✓	✓
6 Fill out maintenance checklist and report deficiencies.	.022				✓	✓
Total labor-hours/period					.462	.462
Total labor-hours/year					.462	.462

			Cost Each				
			2014 Bare Costs			Total	Total
Description		Labor-hrs.	Material	Labor	Equip.	Total	In-House	w/O&P
1900 Heat exchanger, steam, annually		.462	26	27		53	62.50	74.50
1950 Annualized		.924	26	54		80	96	117

| D30 HVAC | | D3055 110 | Space Heater | | | | | |

PM Components	Labor-hrs.	W	M	Q	S	A
PM System D3055 110 1950						
Unit heater, gas radiant						
1 Check with operating or area personnel for deficiencies.	.035					✓
2 Inspect, clean and adjust control valves and thermo sensing bulbs on gas burners.	.254					✓
3 Inspect fuel system for leaks.	.016					✓
4 Check for proper operation of burner controls. check and adjust thermostat.	.133					✓
5 Check fan and motor for vibration and noise. lubricate bearings.	.056					✓
6 Check electrical wiring to blower motor.	.079					✓
7 Check condition of flue pipe, damper and stack.	.147					✓
8 Check unit heater operation through complete cycle or up to ten minutes.	.133					✓
9 Clean area around unit heater.	.133					✓
10 Fill out maintenance checklist and report deficiencies.	.022					✓
Total labor-hours/period						1.008
Total labor-hours/year						1.008

		Cost Each					
			2014 Bare Costs			Total	Total
Description	Labor-hrs.	Material	Labor	Equip.	Total	In-House	w/O&P
1900 Unit heater, gas radiant, annually	1.008	3.17	59		62.17	77	96
1950 Annualized	1.008	3.17	59		62.17	77	96

PM Components	Labor-hrs.	W	M	Q	S	A
PM System D3055 110 2950						
Unit heater, gas infrared						
1 Check with operating or area personnel for deficiencies.	.035					✓
2 Inspect fuel system for leaks around unit.	.016					✓
3 Replace primary air intake filter.	.048					✓
4 Blow out burner.	.130					✓
5 Clean spark electrode and reset gap, replace if necessary.	.254					✓
6 Clean pilot and thermo-sensor bulb, replace if necessary.	.133					✓
7 Check alignment of thermo-sensor bulb.	.133					✓
8 Check wiring and connections; tighten any loose connections.	.079					✓
9 Check and clean pilot sight glass.	.091					✓
10 Vacuum pump (fan):						
A) check motor bearings for overheating; adjust or repair as required.	.022					✓
B) lubricate motor bearings.	.099					✓
C) check motor mounting; adjust as required.	.030					✓
D) check fan blade clearance; adjust as required.	.042					✓
E) check wiring, connections, switches, etc.; tighten any loose connections.	.079					✓
11 Operate unit to ensure that it is in proper working condition.	.225					✓
12 Clean equipment and surrounding area.	.066					✓
13 Fill out maintenance checklist and report deficiencies.	.022					✓
Total labor-hours/period						1.504
Total labor-hours/year						1.504

		Cost Each					
			2014 Bare Costs			Total	Total
Description	Labor-hrs.	Material	Labor	Equip.	Total	In-House	w/O&P
2900 Unit heater, gas infrared, annually	1.504	31.50	88		119.50	145	176
2950 Annualized	1.504	31.50	88		119.50	145	176

D30 HVAC		D3055 110	Space Heater					
	PM Components	Labor-hrs.	W	M	Q	S	A

PM System D3055 110 3950
Unit heater, steam

	PM Components	Labor-hrs.	W	M	Q	S	A
1	Check with operating or area personnel for deficiencies.	.035					√
2	Inspect, clean and adjust control valves and thermostat.	.254					√
3	Inspect coils, connections, trap and steam piping for leaks; repair as necessary.	.195					√
4	Check fan and motor for vibration and noise; lubricate bearings.	.056					√
5	Check electrical wiring to motor.	.079					√
6	Check unit heater operation through complete cycle or up to ten minutes.	.133					√
7	Clean equipment and surrounding area.	.066					√
8	Fill out maintenance checklist and report deficiencies.	.022					√
	Total labor-hours/period						.840
	Total labor-hours/year						.840

			Cost Each					
			2014 Bare Costs				Total	Total
	Description	Labor-hrs.	Material	Labor	Equip.	Total	In-House	w/O&P
3900	Unit heater, steam, annually	.840	31.50	49		80.50	96	116
3950	Annualized	.840	31.50	49		80.50	96	116

D30 HVAC — D3055 122 Forced Air Heater, Oil/Gas Fired

PM Components	Labor-hrs.	W	M	Q	S	A
PM System D3055 122 1950						
Forced air heater, oil or gas fired, up to 120 MBH						
1 Check with operating or area personnel for deficiencies.	.035			✓	✓	✓
2 Inspect, clean and adjust electrodes and nozzles on oil burners or controls, valves and thermo-sensing bulbs on gas burners; lubricate oil burner motor bearings as applicable.	.254					✓
3 Inspect fuel system for leaks.	.016			✓	✓	✓
4 Change fuel filter element on oil burner, where applicable.						✓
5 Check for proper operation of burner primary controls. check and adjust thermostat.	.133			✓	✓	✓
6 Replace air filters in air handler.	.009			✓	✓	✓
7 Check blower and motor for vibration and noise. lubricate bearings.	.042			✓	✓	✓
8 Check belts for wear and proper tension, tighten if required.	.029			✓	✓	✓
9 Check electrical wiring to burner controls and blower.	.079					✓
10 Inspect and clean firebox.	.577					✓
11 Clean blower and air plenum.	.294					✓
12 Check condition of flue pipe, damper and stack.	.147					✓
13 Check furnace operation through complete cycle or up to 10 minutes.	.650			✓	✓	✓
14 Clean area around furnace.	.066			✓	✓	✓
15 Fill out maintenance checklist and report deficiencies.	.022			✓	✓	✓
Total labor-hours/period				1.149	1.149	2.353
Total labor-hours/year				2.298	1.149	2.353

			Cost Each				
			2014 Bare Costs			Total	Total
Description	Labor-hrs.	Material	Labor	Equip.	Total	In-House	w/O&P
1900 Forced air heater, oil/gas, up to 120 MBH, annually	2.353	100	138		238	282	340
1950 Annualized	5.800	121	340		461	555	680

PM Components	Labor-hrs.	W	M	Q	S	A
PM System D3055 122 2950						
Forced air heater, oil or gas fired, over 120 MBH						
1 Check with operating or area personnel for deficiencies.	.035			✓	✓	✓
2 Inspect, clean and adjust electrodes and nozzles on oil burners or controls valves and thermo-sensing bulbs on gas burners; lubricate oil burner motor bearings as applicable.	.358					✓
3 Inspect fuel system for leaks.	.156			✓	✓	✓
4 Change fuel filter element on oil burner, where applicable.	.120					✓
5 Check for proper operation of burner primary controls. check and adjust thermostat.	.133			✓	✓	✓
6 Replace air filters in air handler.	.182			✓	✓	✓
7 Check blower and motor for vibration and noise. lubricate bearings.	.047			✓	✓	✓
8 Check belts for wear and proper tension, tighten if required.	.057			✓	✓	✓
9 Check electrical wiring to burner controls and blower.	.079					✓
10 Inspect and clean firebox.	.577					✓
11 Clean blower and air plenum.	.294					✓
12 Check condition of flue pipe, damper and stack.	.147					✓
13 Check furnace operation through complete cycle or up to 10 minutes.	.650			✓	✓	✓
14 Clean area around furnace.	.066			✓	✓	✓
15 Fill out maintenance checklist and report deficiencies.	.022			✓	✓	✓
Total labor-hours/period				1.495	1.495	2.923
Total labor-hours/year				2.990	1.495	2.923

			Cost Each				
			2014 Bare Costs			Total	Total
Description	Labor-hrs.	Material	Labor	Equip.	Total	In-House	w/O&P
2900 Forced air heater, oil/gas, over 120 MBH, annually	2.923	131	171		302	360	430
2950 Annualized	7.408	161	435		596	720	875

D30 HVAC — D3055 210 | Package Unit, Air Cooled

PM Components	Labor-hrs.	W	M	Q	S	A
PM System D3055 210 1950						
Package unit, air cooled, 3 tons through 24 tons						
1 Check with operating or area personnel for deficiencies.	.035			✓	✓	✓
2 Check tension, condition, and alignment of belts; adjust as necessary.	.029			✓	✓	✓
3 Lubricate shaft and motor bearings.	.047			✓	✓	✓
4 Replace air filters.	.055			✓	✓	✓
5 Clean electrical wiring and connections; tighten loose connections.	.120					✓
6 Clean coils, evaporator drain pan, blowers, fans, motors and drain piping as required.	.385					✓
7 Perform operational check of unit; make adjustments on controls and other components as required.	.077			✓	✓	✓
8 During operation of unit, check refrigerant pressure; add refrigerant as necessary.	.135			✓	✓	✓
9 Check compressor oil level; add oil as required.	.033					✓
10 Clean area around unit.	.066			✓	✓	✓
11 Fill out maintenance checklist and report deficiencies.	.022			✓	✓	✓
Total labor-hours/period				.466	.466	1.004
Total labor-hours/year				.932	.466	1.004

Cost Each

			2014 Bare Costs			Total	Total
Description	Labor-hrs.	Material	Labor	Equip.	Total	In-House	w/O&P
1900 Package unit, air cooled, 3 thru 24 ton, annually	1.004	48	58.50		106.50	126	152
1950 Annualized	2.402	76.50	141		217.50	260	315

PM Components	Labor-hrs.	W	M	Q	S	A
PM System D3055 210 2950						
Package unit, air cooled, 25 tons through 50 tons						
1 Check with operating or area personnel for deficiencies.	.035			✓	✓	✓
2 Check tension, condition, and alignment of belts; adjust as necessary.	.029			✓	✓	✓
3 Lubricate shaft and motor bearings.	.047			✓	✓	✓
4 Replace air filters.	.078			✓	✓	✓
5 Clean electrical wiring and connections; tighten loose connections.	.120					✓
6 Clean coils, evaporator drain pan, blowers, fans, motors and drain piping as required.	.385					✓
7 Perform operational check of unit; make adjustments on controls and other components as required.	.130			✓	✓	✓
8 During operation of unit, check refrigerant pressure; add refrigerant as necessary.	.272			✓	✓	✓
9 Check compressor oil level; add oil as required.	.033					✓
10 Clean area around unit.	.066			✓	✓	✓
11 Fill out maintenance checklist and report deficiencies.	.022			✓	✓	✓
Total labor-hours/period				.679	.679	1.217
Total labor-hours/year				1.358	.679	1.217

Cost Each

			2014 Bare Costs			Total	Total
Description	Labor-hrs.	Material	Labor	Equip.	Total	In-House	w/O&P
2900 Package unit, air cooled, 25 thru 50 ton, annually	1.217	67	71		138	163	195
2950 Annualized	3.254	107	190		297	355	430

D30 HVAC		D3055 220	Package Unit, Water Cooled					
	PM Components		Labor-hrs.	W	M	Q	S	A
PM System D3055 220 1950								
Package unit, water cooled, 3 tons through 24 tons								
1	Check with operating or area personnel for deficiencies.		.035			✓	✓	✓
2	Check tension, condition, and alignment of belts; adjust as necessary.		.029			✓	✓	✓
3	Lubricate shaft and motor bearings.		.047			✓	✓	✓
4	Replace air filters.		.055			✓	✓	✓
5	Clean electrical wiring and connections; tighten loose connections.		.120					✓
6	Clean coils, evaporator drain pan, blowers, fans, motors and drain piping as required.		.385					✓
7	Perform operational check of unit; make adjustments on controls and other components as required.		.077			✓	✓	✓
8	During operation of unit, check refrigerant pressure; add refrigerant as necessary.		.135			✓	✓	✓
9	Check compressor oil level; add oil as required.		.033					✓
10	Clean area around unit.		.066			✓	✓	✓
11	Fill out maintenance checklist and report deficiencies.		.022			✓	✓	✓
	Total labor-hours/period					.466	.466	1.004
	Total labor-hours/year					.932	.466	1.004

			Cost Each					
				2014 Bare Costs			Total	Total
	Description	Labor-hrs.	Material	Labor	Equip.	Total	In-House	w/O&P
1900	Package unit, water cooled, 3 thru 24 ton, annually	1.004	48	58.50		106.50	126	152
1950	Annualized	2.402	76.50	141		217.50	260	315

	PM Components		Labor-hrs.	W	M	Q	S	A
PM System D3055 220 2950								
Package unit, water cooled, 25 tons through 50 tons								
1	Check with operating or area personnel for deficiencies.		.035			✓	✓	✓
2	Check tension, condition, and alignment of belts; adjust as necessary.		.029			✓	✓	✓
3	Lubricate shaft and motor bearings.		.047			✓	✓	✓
4	Replace air filters.		.078			✓	✓	✓
5	Clean electrical wiring and connections; tighten loose connections.		.120					✓
6	Clean coils, evaporator drain pan, blowers, fans, motors and drain piping as required.		.385					✓
7	Perform operational check of unit; make adjustments on controls and other components as required.		.130			✓	✓	✓
8	During operation of unit, check refrigerant pressure; add refrigerant as necessary.		.272			✓	✓	✓
9	Check compressor oil level; add oil as required.		.033					✓
10	Clean area around unit.		.066			✓	✓	✓
11	Fill out maintenance checklist and report deficiencies.		.022			✓	✓	✓
	Total labor-hours/period					.679	.679	1.217
	Total labor-hours/year					1.358	.679	1.217

			Cost Each					
				2014 Bare Costs			Total	Total
	Description	Labor-hrs.	Material	Labor	Equip.	Total	In-House	w/O&P
2900	Package unit, water cooled, 25 thru 50 ton, annually	1.217	67	71		138	163	195
2950	Annualized	3.254	107	190		297	355	430

D30 HVAC — D3055 230 Package Unit, Computer Room

PM Components	Labor-hrs.	W	M	Q	S	A
PM System D3055 230 1950						
Package unit, computer room						
1 Check with operating or area personnel for deficiencies.	.035			✓	✓	✓
2 Run microprocessor check, if available, or check controls and unit for proper operation.	.216			✓	✓	✓
3 Check for unusual noise or vibration.	.033			✓	✓	✓
4 Clean coils, evaporator drain pan, humidifier pan, blower, motor and drain piping as required.	.380				✓	✓
5 Replace air filters.	.078			✓	✓	✓
6 Lubricate shaft and motor bearings.	.047			✓	✓	✓
7 Check belts for wear, proper tension, and alignment; adjust as necessary.	.029			✓	✓	✓
8 Check humidity lamp, replace if necessary.	.156			✓	✓	✓
9 During operation of unit, check refrigerant pressures; add refrigerant as necessary.	.270				✓	✓
10 Inspect exterior piping and valves for leaks; tighten connections as required.	.077			✓	✓	✓
11 Clean area around unit.	.066			✓	✓	✓
12 Fill out maintenance checklist and report deficiencies.	.022			✓	✓	✓
Total labor-hours/period				.759	1.409	1.409
Total labor-hours/year				1.518	1.409	1.409

		Cost Each					
		2014 Bare Costs				Total	Total
Description	Labor-hrs.	Material	Labor	Equip.	Total	In-House	w/O&P
1900 Package unit, computer room, annually	1.409	55.50	82.50		138	164	198
1950 Annualized	4.336	61	254		315	385	470

D30 HVAC   D3055 240 Package Terminal Unit

PM Components	Labor-hrs.	W	M	Q	S	A
PM System D3055 240 1950						
Package unit, with duct gas heater						
1 Check with operating or area personnel for deficiencies.	.035			√	√	√
2 Check tension, condition and alignment of belts; adjust as necessary.	.029			√	√	√
3 Lubricate shaft and motor bearings.	.047			√	√	√
4 Replace air filters.	.078			√	√	√
5 Check electrical wiring and connections; tighten loose connections.	.120					√
6 Clean coils, evaporator drain pan, blowers, fans, motors and drain piping as required.	.385					√
7 Perform operational check of unit; make adjustments on controls and other components as required.	.077			√	√	√
8 During operation of unit, check refrigerant pressures; add refrigerant as necessary.	.272			√	√	√
9 Check compressor oil level; add oil as required.	.033					√
10 Inspect, clean and adjust control valves and thermo-sensing bulbs on gas burners.	.254					√
11 Inspect fuel system for leaks.	.016			√	√	√
12 Check for proper operation of burner primary controls. Check and adjust thermostat.	.133					√
13 Check electrical wiring to burner controls.	.079					√
14 Inspect and clean firebox.	.577					√
15 Check condition of flue pipe, damper and stack.	.147			√	√	√
16 Check heater operation through complete cycle or up to 10 minutes.	.225					√
17 Clean area around entire unit.	.066			√	√	√
18 Fill out maintenance checklist and report deficiencies.	.022			√	√	√
Total labor-hours/period				.789	.789	2.595
Total labor-hours/year				1.578	.789	2.595

		Cost Each					
		2014 Bare Costs				Total	Total
Description	Labor-hrs.	Material	Labor	Equip.	Total	In-House	w/O&P
1900 Package unit with duct gas heater, annually	2.595	46	152		198	240	295
1950 Annualized	4.962	69	290		359	440	540

457

D30 HVAC — D3055 250 Air Conditioning, Split System

PM Components	Labor-hrs.	W	M	Q	S	A
PM System D3055 250 1950						
Air conditioning split system, DX, air cooled, up to 10 tons						
1 Check with operating or area personnel for deficiencies.	.035			✓	✓	✓
2 Clean intake side of condenser coils, fans and intake screens.	.055			✓	✓	✓
3 Lubricate shaft and motor bearings.	.047			✓	✓	✓
4 Pressure wash condenser coils with coil clean solution, as required.	.611					✓
5 Replace air filters.	.078			✓	✓	
6 Clean electrical wiring and connections; tighten loose connections.	.120					✓
7 Clean evaporator coils, drain pan, blowers, fans, motors and drain piping as required.	.380					✓
8 Perform operational check of unit; make adjustments on controls and other components as required.	.033			✓	✓	✓
9 During operation of unit, check refrigerant pressure; add refrigerant as necessary.	.272			✓	✓	✓
10 Clean area around equipment.	.066			✓	✓	✓
11 Fill out maintenance checklist and report deficiencies.	.022			✓	✓	✓
Total labor-hours/period				.608	.608	1.719
Total labor-hours/year				1.216	.608	1.719

			Cost Each				
			2014 Bare Costs			Total	Total
Description	Labor-hrs.	Material	Labor	Equip.	Total	In-House	w/O&P
1900 A/C, split sys., DX, air cooled, to 10 tons, annually	1.719	49	101		150	180	218
1950 Annualized	3.543	78	207		285	345	425

PM Components	Labor-hrs.	W	M	Q	S	A
PM System D3055 250 2950						
Air conditioning split system, DX, air cooled, over 10 tons						
1 Check with operating or area personnel for deficiencies.	.035			✓	✓	✓
2 Check tension, condition, and alignment of belts; adjust as necessary.	.029			✓	✓	✓
3 Lubricate shaft and motor bearings.	.047			✓	✓	✓
4 Pressure wash condenser coils with coil clean solution, as required.	.611					✓
5 Replace air filters.	.078			✓	✓	
6 Clean electrical wiring and connections; tighten loose connections.	.120					✓
7 Clean evaporator coils, drain pan, blowers, fans, motors and drain piping as required.	.380					✓
8 Perform operational check of unit; make adjustments on controls and other components as required.	.033			✓	✓	✓
9 During operation of unit, check refrigerant pressure; add refrigerant as necessary.	.272			✓	✓	✓
10 Check compressor oil level; add oil as required.	.033			✓	✓	✓
11 Clean area around equipment.	.066			✓	✓	✓
12 Fill out maintenance checklist and report deficiencies.	.022			✓	✓	✓
Total labor-hours/period				.615	.615	1.726
Total labor-hours/year				1.230	.615	1.726

			Cost Each				
			2014 Bare Costs			Total	Total
Description	Labor-hrs.	Material	Labor	Equip.	Total	In-House	w/O&P
2900 A/C, split sys., DX, air cooled, over 10 T, annually	1.726	68.50	101		169.50	203	244
2950 Annualized	3.571	109	209		318	380	460

D30 HVAC

D3055 310 Heat Pump, Air Cooled

PM Components	Labor-hrs.	W	M	Q	S	A
PM System D3055 310 1950						
Heat pump, air cooled, up to 5 tons						
1 Check with operating or area personnel for deficiencies.	.035			√	√	√
2 Check unit for proper operation, excessive noise or vibration.	.033			√	√	√
3 Clean intake side of condenser coils, fans and intake screens.	.055			√	√	√
4 Check electrical wiring and connections; tighten loose connections.	.120					√
5 Inspect fan(s) for bent blades or unbalance; adjust and clean as necessary.	.027					√
6 Check belts for condition, proper tension and misalignment; adjust as required.	.029			√	√	√
7 Lubricate shaft bearings and motor bearings.	.047			√	√	√
8 Inspect piping and valves for leaks; tighten connections as necessary.	.077			√	√	√
9 Replace air filters.	.078			√	√	√
10 Check refrigerant pressure; add refrigerant as necessary.	.135			√	√	√
11 Clean evaporative drain pan, and drain piping as required.	.385					√
12 Cycle the reverse cycle valve to insure proper operation.	.091			√	√	√
13 Clean area around equipment.	.066			√	√	√
14 Fill out maintenance checklist and report deficiencies.	.022			√	√	√
Total labor-hours/period				.668	.668	1.200
Total labor-hours/year				1.336	.668	1.200

			Cost Each				
			2014 Bare Costs			Total	Total
Description		Labor-hrs.	Material	Labor	Equip.	Total	In-House	w/O&P
1900 Heat pump, air cooled, up to 5 ton, annually	1.200	28.50	70		98.50	119	146
1950 Annualized	3.204	46	187		233	285	350

PM Components	Labor-hrs.	W	M	Q	S	A
PM System D3055 310 2950						
Heat pump, air cooled, over 5 tons						
1 Check with operating or area personnel for deficiencies.	.035			√	√	√
2 Check unit for proper operation, excessive noise or vibration.	.033			√	√	√
3 Clean intake side of condenser coils, fans and intake screens.	.055			√	√	√
4 Check electrical wiring and connections; tighten loose connections.	.120					√
5 Inspect fan(s) for bent blades or unbalance; adjust and clean as necessary.	.055					√
6 Check belts for condition, proper tension and misalignment; adjust as required.	.057			√	√	√
7 Lubricate shaft bearings and motor bearings.	.047			√	√	√
8 Inspect piping and valves for leaks; tighten connections as necessary.	.077			√	√	√
9 Replace air filters.	.156			√	√	√
10 Check refrigerant pressure; add refrigerant as necessary.	.135			√	√	√
11 Lubricate and check operation of dampers, if applicable.	.029					√
12 Check compressor oil level and add oil, if required.	.033			√	√	√
13 Cycle the reverse cycle valve to insure proper operation.	.091			√	√	√
14 Clean evaporative drain pan, and drain piping as required.	.385					√
15 Clean area around equipment.	.066			√	√	√
16 Fill out maintenance checklist and report deficiencies.	.022			√	√	√
Total labor-hours/period				.730	.730	1.396
Total labor-hours/year				1.460	.730	1.396

			Cost Each				
			2014 Bare Costs			Total	Total
Description		Labor-hrs.	Material	Labor	Equip.	Total	In-House	w/O&P
2900 Heat pump, air cooled, over 5 ton, annually	1.396	53.50	81.50		135	161	194
2950 Annualized	3.586	76.50	210		286.50	345	420

D30 HVAC — D3055 320 Heat Pump, Water Cooled

PM Components	Labor-hrs.	W	M	Q	S	A
PM System D3055 320 1950						
Heat pump, water cooled, up to 5 tons						
1 Check with operating or area personnel for deficiencies.	.035			✓	✓	✓
2 Check unit for proper operation, excessive noise or vibration.	.033			✓	✓	✓
3 Clean intake side of evaporator coil, fans and intake screens.	.055			✓	✓	✓
4 Check electrical wiring and connections; tighten loose connections.	.120					✓
5 Inspect fan(s) for bent blades or unbalance; adjust and clean as necessary.	.027					✓
6 Check belts for condition, proper tension and misalignment; adjust as required.	.029			✓	✓	✓
7 Lubricate shaft bearings and motor bearings.	.047			✓	✓	✓
8 Inspect piping and valves for leaks; tighten connections as necessary.	.077			✓	✓	✓
9 Replace air filters.	.078			✓	✓	✓
10 Check refrigerant pressure; add refrigerant as necessary.	.135			✓	✓	✓
11 Lubricate and check operation of dampers, if applicable.	.029					
12 Clean evaporator drain pan and drain line with solvent.	.385					
13 Cycle reverse cycle valve to insure proper operation.	.091			✓	✓	
14 Backwash condenser coil to remove sediment.	.325					
15 Clean area around equipment.	.066			✓	✓	✓
16 Fill out maintenance checklist and report deficiencies.	.022			✓	✓	✓
Total labor-hours/period				.668	.668	1.554
Total labor-hours/year				1.336	.668	1.554

		Cost Each					
		2014 Bare Costs				Total	Total
Description	Labor-hrs.	Material	Labor	Equip.	Total	In-House	w/O&P
1900 Heat pump, water cooled, up to 5 ton, annually	1.554	28.50	91		119.50	146	178
1950 Annualized	3.558	46	208		254	310	385

PM Components	Labor-hrs.	W	M	Q	S	A
PM System D3055 320 2950						
Heat pump, water cooled, over 5 tons						
1 Check with operating or area personnel for deficiencies.	.035			✓	✓	✓
2 Check unit for proper operation, excessive noise or vibration.	.033			✓	✓	✓
3 Clean intake side of condenser coils, fans and intake screens.	.055			✓	✓	✓
4 Check electrical wiring and connections; tighten loose connections.	.120					✓
5 Inspect fan(s) for bent blades or unbalance; adjust and clean as necessary.	.027					✓
6 Check belts for condition, proper tension and misalignment; adjust as required.	.029			✓	✓	✓
7 Lubricate shaft bearings and motor bearings.	.047			✓	✓	✓
8 Inspect piping and valves for leaks; tighten connections as necessary.	.077			✓	✓	✓
9 Replace air filters.	.078			✓	✓	✓
10 Check refrigerant pressure; add refrigerant as necessary.	.135			✓	✓	✓
11 Lubricate and check operation of dampers, if applicable.	.029					
12 Clean evaporator drain pan and drain line with solvent.	.385					
13 Cycle reverse cycle valve to insure proper operation.	.091			✓	✓	✓
14 Clean area around equipment.	.066			✓	✓	✓
15 Fill out maintenance checklist and report deficiencies.	.022			✓	✓	✓
Total labor-hours/period				.668	.668	1.229
Total labor-hours/year				1.336	.668	1.229

		Cost Each					
		2014 Bare Costs				Total	Total
Description	Labor-hrs.	Material	Labor	Equip.	Total	In-House	w/O&P
2900 Heat pump, water cooled, over 5 ton, annually	1.229	53.50	72		125.50	149	179
2950 Annualized	3.233	76.50	189		265.50	320	390

D30 HVAC | D3065 100 | Controls, Central System

PM Components	Labor-hrs.	W	M	Q	S	A
PM System D3065 100 1950						
Controls, central system, electro/pneumatic						
1 With panel disconnected from power source, clean patrol panel compartment with a vacuum.	.468					✓
2 Inspect wiring/components for loose connections; tighten, as required.	.120					✓
3 Check set point of controls temperature, humidity or pressure.	.033					✓
4 Check unit over its range of control.	.033					✓
5 Check for correct pressure differential on all two position controllers.	.195					✓
6 Check source of the signal and its amplification on electronic controls.	.468					✓
7 Check air systems for leaks; repair as necessary.	.156					✓
8 Check relays, pilot valves and pressure regulators for proper operation; repair or replace as necessary.	.325					✓
9 Replace air filters in sensors, controllers, and thermostats as necessary.	.029					✓
10 Clean area around equipment.	.066					✓
11 Fill out maintenance checklist and report deficiencies.	.022					✓
Total labor-hours/period						1.915
Total labor-hours/year						1.915

	Description	Labor-hrs.	Cost Each				Total In-House	Total w/O&P
			2014 Bare Costs					
			Material	Labor	Equip.	Total		
1900	Controls, central system, electro/pneumatic, annually	1.915	46	112		158	191	233
1950	Annualized	1.915	46	112		158	191	233

D30 HVAC — D3095 110 Air Compressor, Gas Engine

PM Components	Labor-hrs.	W	M	Q	S	A
PM System D3095 110 1950						
Air compressor, gas engine						
1 Check with operating or area personnel for any obvious deficiencies.	.035		√	√	√	√
2 Check compressor oil level; add oil as required.	.022		√	√	√	√
3 Replace compressor oil.	.221				√	√
4 Clean air intake filter on air compressor(s); replace if necessary.	.177				√	√
5 Clean cylinder cooling fins and air cooler on compressor(s).	.156				√	√
6 Check tension, condition, and alignment of v-belts; adjust as necessary.	.030				√	√
7 Clean oil and water traps.	.178				√	√
8 Drain moisture from air storage tank and check discharge for indication of interior corrosion.	.046				√	√
9 Perform operation check of compressor; check the operation of low pressure cut-in and high pressure cut-out switches.	.221		√	√	√	√
10 Check operation of safety pressure relief valve.	.030				√	√
11 Check and tighten compressor foundation anchor bolts.	.022				√	√
12 Check radiator coolant; add if required.	.012		√	√	√	√
13 Check battery water; add if required.	.241		√	√	√	√
14 Check wiring, connections, switches, etc.; tighten loose connections.	.120				√	√
15 Check engine oil level; add if required.	.014		√	√	√	√
16 Change engine oil.	.224				√	√
17 Change engine oil filter.	.059				√	√
18 Check spark plug and reset cap.	.035				√	√
19 Check condition of engine air filter; replace if necessary.	.039				√	√
20 Test run engine for proper operation.	.413		√	√	√	√
21 Wipe dust and dirt from engine and compressor.	.109		√	√	√	√
22 Check muffler/exhaust system for corrosion.	.020				√	√
23 Clean area around equipment.	.066		√	√	√	√
24 Fill out maintenance checklist and report deficiencies.	.022		√	√	√	√
Total labor-hours/period			1.155	1.155	2.512	2.512
Total labor-hours/year			9.240	2.310	2.512	2.512

			Cost Each				
			2014 Bare Costs			Total	Total
Description		Labor-hrs.	Material	Labor	Equip.	Total	In-House	w/O&P
1900 Air compressor, gas engine powered, annually	2.512	106	147		253	300	360
1950 Annualized	16.574	257	970		1,227	1,500	1,825

D30 HVAC — D3095 114 Air Compressor, Centrifugal

PM Components	Labor-hrs.	W	M	Q	S	A
PM System D3095 114 1950						
Air compressor, centrifugal, to 40 H.P.						
1 Check compressor oil level; add oil as necessary.	.022			√	√	√
2 Perform operational check of compressor system and adjust as required.	.221			√	√	√
3 Check motor for excessive vibration, noise and overheating; lubricate.	.039			√	√	√
4 Check operation of pressure relief valve.	.030			√	√	√
5 Clean cooling fans and air cooler.	.023			√	√	√
6 Check tension, condition, and alignment of V-belts; adjust as necessary.	.030			√	√	√
7 Drain moisture from air storage tank and check low pressure cut-in; while draining, check discharge for indication of interior corrosion.	.046			√	√	√
8 Replace air intake filter, as needed.	.177			√	√	√
9 Clean oil and water trap.	.177			√	√	√
10 Clean compressor and surrounding area.	.066			√	√	√
11 Fill out maintenance checklist and report deficiencies.	.022			√	√	√
Total labor-hours/period				.853	.853	.853
Total labor-hours/year				1.706	.853	.853

			Cost Each				
			2014 Bare Costs			Total	Total
Description	Labor-hrs.	Material	Labor	Equip.	Total	In-House	w/O&P
1900 Air compressor, centrifugal, to 40 H.P., annually	.853	45.50	50		95.50	112	135
1950 Annualized	3.412	68	200		268	325	395

D30 HVAC

D3095 114 Air Compressor, Centrifugal

PM Components	Labor-hrs.	W	M	Q	S	A
PM System D3095 114 2950						
Air compressor, centrifugal, over 40 H.P.						
1 Check with operating or area personnel for any obvious deficiencies.	.035		√	√	√	√
2 Perform control system check.	.325				√	√
3 Lubricate main driver coupling if necessary.	.047				√	√
4 Lubricate prelube pump motor and pump coupling, if necessary.	.047				√	√
5 Perform operation check of air compressor, adjust as required.	.221		√	√	√	√
6 Check main driver coupling alignment; realign if necessary.	.325				√	√
7 Check compressor and motor operation for excessive vibration, noise and overheating; lubricate motor.	.039		√	√	√	√
8 Check compressor scrolls, piping, and impeller housing for all leaks or cracks.	.325		√	√	√	√
9 Check intercoolers and aftercoolers for high cooling water temperatures or cooling water leakage.	.195		√	√	√	√
10 Check oil level in compressor oil reservoir; add oil as necessary.	.022		√	√	√	√
11 Replace compressor oil and oil filters.	.282					√
12 Record oil pressure and oil temperature.	.013		√	√	√	√
13 Check compressor for oil leaks.	.026		√	√	√	√
14 Visually inspect oil mist arrestor, clean housing, lines, and replace element if saturated, if applicable.	.177				√	√
15 Check operation of pressure relief valve.	.030		√	√	√	√
16 Visually inspect discharge check valve.	.009				√	√
17 Check oil and water trap.	.022		√	√	√	√
18 Check indicating lamps or gauges for proper operation if appropriate; replace burned out lamps or repair/replace gauges.	.020					√
19 Visually check all air intake filter elements; replace if necessary.	.022		√	√	√	√
20 Replace all air intake filter elements.	.177					√
21 Lubricate motor.	.047				√	√
22 Clean compressor, motor and surrounding area.	.066		√	√	√	√
23 Fill out maintenance check and report deficiencies.	.022		√	√	√	√
Total labor-hours/period			1.038	1.038	2.015	2.494
Total labor-hours/year			8.304	2.076	2.015	2.494

		Cost Each						
			2014 Bare Costs				Total	Total
Description	Labor-hrs.	Material	Labor	Equip.	Total	In-House	w/O&P
2900 Air compressor, centrifugal, over 40 H.P., annually	2.494	91	146		237	283	340
2950 Annualized	14.889	212	870		1,082	1,325	1,625

D30 HVAC   D3095 118 | Air Compressor, Reciprocating

PM Components	Labor-hrs.	W	M	Q	S	A
PM System D3095 118 1950						
Air compressor, reciprocating, less than 5 H.P.						
1 Replace compressor oil.	.341			✓	✓	✓
2 Perform operation check of compressor system and adjust as required.	.221			✓	✓	✓
3 Check motor operation for excessive vibration, noise and overheating.	.042			✓	✓	✓
4 Lubricate motor.	.047			✓	✓	✓
5 Check operation of pressure relief valve.	.030			✓	✓	✓
6 Check tension, condition, and alignment of V-belts; adjust as necessary.	.030			✓	✓	✓
7 Drain moisture from air storage tank and check low pressure cut-in; while draining, check discharge for indication of interior corrosion.	.046			✓	✓	✓
8 Clean air intake filter on compressor.	.177			✓	✓	✓
9 Clean oil and water trap.	.177			✓	✓	✓
10 Clean exterior of compressor, motor and surrounding area.	.066			✓	✓	✓
11 Fill out maintenance checklist and report deficiencies.	.022			✓	✓	✓
Total labor-hours/period				1.199	1.199	1.199
Total labor-hours/year				2.398	1.199	1.199

			Cost Each				
			2014 Bare Costs			Total	Total
Description	Labor-hrs.	Material	Labor	Equip.	Total	In-House	w/O&P
1900 Air compressor, recip., less than 5 H.P., annually	1.199	22.50	70		92.50	113	137
1950 Annualized	4.796	91	281		372	450	555

PM Components	Labor-hrs.	W	M	Q	S	A
PM System D3095 118 2950						
Air compressor, reciprocating, 5 to 40 H.P.						
1 Replace compressor oil.	.341			✓	✓	✓
2 Perform operation check of compressor system and adjust as required.	.221			✓	✓	✓
3 Check motor operation for excessive vibration, noise and overheating; lubricate motor.	.042			✓	✓	✓
4 Check operation of pressure relief valve.	.043			✓	✓	✓
5 Clean cooling fans and air cooler on compressor.	.023			✓	✓	✓
6 Check tension, condition, and alignment of V-belts; adjust as necessary.	.030			✓	✓	✓
7 Drain moisture from air storage tank and check low pressure cut-in; while draining, check discharge for indication of interior corrosion.	.059			✓	✓	✓
8 Clean air intake filter on compressor.	.177			✓	✓	✓
9 Clean oil and water trap.	.190			✓	✓	✓
10 Clean compressor and surrounding area.	.066			✓	✓	✓
11 Fill out maintenance checklist and report deficiencies.	.022			✓	✓	✓
Total labor-hours/period				1.214	1.214	1.214
Total labor-hours/year				2.428	1.214	1.214

			Cost Each				
			2014 Bare Costs			Total	Total
Description	Labor-hrs.	Material	Labor	Equip.	Total	In-House	w/O&P
2900 Air compressor, reciprocating, 5 to 40 H.P., annually	1.214	22.50	71		93.50	115	139
2950 Annualized	4.856	91	284		375	455	560

D30 HVAC		D3095 118	Air Compressor, Reciprocating					
	PM Components	Labor-hrs.	W	M	Q	S	A
	PM System D3095 118 3950							
	Air compressor, reciprocating, over 40 H.P.							
1	Check with operating or area personnel for deficiencies.	.035			√	√	√
2	Perform operation check of compressor system and adjust as required.	.221			√	√	√
3	Replace compressor oil.	.341			√	√	√
4	Check motor(s) operation for excessive vibration, noise and overheating; lubricate motor(s).	.042			√	√	√
5	Clean cylinder cooling fins and air cooler on compressor.	.023			√	√	√
6	Check tension, condition, and alignment of V-belts; adjust as necessary.	.056			√	√	√
7	Check operation of pressure relief valve.	.030			√	√	√
8	Check low pressure cut in and high pressure cut out switches.	.120			√	√	√
9	Drain moisture from air storage tank and check low pressure cut-in; while draining, check discharge for indication of interior corrosion.	.072			√	√	√
10	Clean air intake filter on air compressor(s); replace if necessary.	.177			√	√	√
11	Clean oil and water trap.	.204			√	√	√
12	Check indicating lamps or gauges for proper operation if appropriate; replace burned out lamps or repair/replace gauges.	.020			√	√	√
13	Clean area around equipment.	.066			√	√	√
14	Fill out maintenance checklist and report deficiencies.	.022			√	√	√
	Total labor-hours/period				1.429	1.429	1.429
	Total labor-hours/year				2.858	1.429	1.429

			Cost Each					
			2014 Bare Costs				Total	Total
Description		Labor-hrs.	Material	Labor	Equip.	Total	In-House	w/O&P
3900	Air compressor, reciprocating, over 40 H.P., annually	1.429	22.50	83.50		106	129	158
3950	Annualized	5.716	91	335		426	520	635

D30 HVAC — D3095 210 Humidification Systems

PM Components	Labor-hrs.	W	M	Q	S	A
PM System D3095 210 1950						
Steam Humidification System						
1 Operate humidistat through its throttling range to verify activation and deactivation.	.117				✓	✓
2 Inspect steam trap for proper operation.	.117				✓	✓
3 Turn off steam supply.	.065				✓	✓
4 Secure electrical service before servicing humidification unit.	.065				✓	✓
5 Clean strainer.	.195				✓	✓
6 Clean and/or replace water/steam nozzles as necessary.	.455				✓	✓
7 Inspect pneumatic controller for air leaks.	.039				✓	✓
8 Inspect steam lines for leaks and corrosion and repair leaks.	.195				✓	✓
9 Fill out maintenance checklist and report deficiencies	.022				✓	✓
Total labor-hours/period					1.270	1.270
Total labor-hours/year					1.270	1.270

			Cost Each				
			2014 Bare Costs			Total	Total
	Description	Labor-hrs.	Material	Labor	Equip.	Total	In-House	w/O&P
1900	Steam humidification system, annually	1.270	23	74.50		97.50	118	145
1950	Annualized	2.540	46	149		195	236	290

PM Components	Labor-hrs.	W	M	Q	S	A
PM System D3095 210 2950						
Evaporative Pan with Heating Coil Humidification System						
1 Operate humidistat through its throttling range to verify activation and deactivation.	.117				✓	✓
2 Inspect steam trap for proper operation.	.117				✓	✓
3 Turn off water and steam supply.	.065				✓	✓
4 Secure electrical service before servicing humidification unit.	.065				✓	✓
5 Drain and flush water pans, clean drains, etc.	.260				✓	✓
6 Check condition of heating element/steam coils and clean.	.065				✓	✓
7 Inspect pneumatic controller for air leaks.	.039				✓	✓
8 Inspect steam lines for leaks and corrosion and repair leaks.	.195				✓	✓
9 Fill out maintenance checklist and report deficiencies.	.022				✓	✓
Total labor-hours/period					.945	.945
Total labor-hours/year					.945	.945

			Cost Each				
			2014 Bare Costs			Total	Total
	Description	Labor-hrs.	Material	Labor	Equip.	Total	In-House	w/O&P
2900	Evap. pan with heating coil humidif. system, annually	.945	15.30	55.50		70.80	86	105
2950	Annualized	1.890	30.50	111		141.50	172	210

D30 HVAC — D3095 220 Dehumidifier, Desiccant Wheel

PM Components	Labor-hrs.	W	M	Q	S	A
PM System D3095 220 1950						
Dehumidifier, desiccant wheel						
1 Check with operating or area personnel for deficiencies.	.035		√	√	√	√
2 Check filters for any blockage or fouling, remove and clean as required.	.036		√	√	√	√
3 Check wheel seals for tears or punctures.	.055		√	√	√	√
4 Check the control valve, thermo sensor bulb and burner, if it has heaters; or check for scaling or leaking on steam heating coils, as applicable.	.094				√	√
5 Check that the outlet air temperature is within the proper heat range.	.105		√	√	√	√
6 Clean the desiccant wheel and check for softening of wheel faces.	.031				√	√
7 Check desiccant wheel and motor for vibration and noise, adjust as required.	.109				√	√
8 Check gear reducer oil level; add as required.	.035				√	√
9 Check wheel belt(s) for wear, proper tension and alignment; adjust as required.	.029				√	√
10 Check blower and motor for excessive vibration and noise; adjust as required.	.109				√	√
11 Check blower belt(s) for wear, proper tension and alignment; adjust as required.	.029				√	√
12 Lubricate wheel, blower and motor bearings.	.047				√	√
13 Check electrical wiring and connections; make appropriate adjustments.	.120				√	√
14 Check reactivation ductwork for condensation and air leaks.	.013				√	√
15 Clean the equipment and the surrounding area.	.066		√	√	√	√
16 Fill out maintenance checklist and report deficiencies.	.022		√	√	√	√
Total labor-hours/period			.319	.319	.935	.935
Total labor-hours/year			2.552	.638	.935	.935

		Cost Each					
		2014 Bare Costs				Total	Total
Description	Labor-hrs.	Material	Labor	Equip.	Total	In-House	w/O&P
1900 Dehumidifier, desiccant wheel, annually	.935	32	54.50		86.50	104	126
1950 Annualized	5.060	99.50	296		395.50	480	585

D40 FIRE PROTECTION | D4015 100 | Backflow Prevention Device

	PM Components	Labor-hrs.	W	M	Q	S	A
	PM System D4015 100 1950						
	Backflow prevention device, up to 4"						
	NOTE: Test frequency may vary depending on local regulations and application.						
1	Test and calibrate check valve operation of backflow prevention device with test set.	.191					√
2	Bleed air from backflow preventer.	.047					√
3	Inspect for leaks under pressure.	.007					√
4	Clean backflow preventer and surrounding area.	.066					√
5	Fill out maintenance checklist and report deficiencies.	.022					√
	Total labor-hours/period						.333
	Total labor-hours/year						.333

			Cost Each					
				2014 Bare Costs			Total	Total
	Description	Labor-hrs.	Material	Labor	Equip.	Total	In-House	w/O&P
1900	Backflow prevention device, up to 4", annually	.333		19.15		19.15	24	30
1950	Annualized	.333		19.15		19.15	24	30

	PM Components	Labor-hrs.	W	M	Q	S	A
	PM System D4015 100 2950						
	Backflow prevention device, over 4"						
	NOTE: Test frequency may vary depending on local regulations and application.						
1	Test and calibrate check valve operation of backflow prevention device with test set.	.334					√
2	Bleed air from backflow preventer.	.065					√
3	Inspect for leaks under pressure.	.007					√
4	Clean backflow preventer and surrounding area.	.066					√
5	Fill out maintenance checklist and report deficiencies.	.022					√
	Total labor-hours/period						.494
	Total labor-hours/year						.494

			Cost Each					
				2014 Bare Costs			Total	Total
	Description	Labor-hrs.	Material	Labor	Equip.	Total	In-House	w/O&P
2900	Backflow prevention device, over 4", annually	.494		28.50		28.50	35.50	44.50
2950	Annualized	.494		28.50		28.50	35.50	44.50

D40 FIRE PROTECTION — D4015 150 Extinguishing System, Wet Pipe

PM Components	Labor-hrs.	W	M	Q	S	A
PM System D4015 150 1950						
Extinguishing system, wet pipe						
1 Notify proper authorities prior to testing any alarm systems.	.130		√	√	√	√
2 Open and close post indicator valve (PIV) to check operation; make minor adjustments such as lubricating valve stem, cleaning and/or replacing target windows as required.	.176					√
3 Open and close OS&Y (outside stem and yoke) cut-off valve to check operation; make minor repairs such as lubricating stems and tightening packing glands as required.	.176					√
4 Perform operational test of water flow detectors; make minor adjustments and restore system to proper operating condition.	.148					√
5 Check to ensure that alarm drain is open; clean drain line if necessary.	.081		√	√	√	√
6 Open water motor alarm test valve and ensure that outside alarm operates; lubricate alarm, make adjustments as required.	.176		√	√	√	√
7 Conduct main drain test by opening 2″ test valve; maintain a continuous record of drain tests; make minor adjustments if applicable; restore system to proper operating condition.	.333			√	√	√
8 Check general condition of sprinklers and sprinkler system; make minor adjustments as required.	.228					√
9 Check equipment gaskets, piping, packing glands, and valves for leaks; tighten flange bolts and loose connections to stop all leaks.	.013					√
10 Check condition of fire department connections; replace missing or broken covers as required.	.103					√
11 Trip test wet pipe system using test valve furthest from wet pipe valve (control valve); make minor adjustments as necessary and restore system to proper operating condition.	1.733					√
12 Inspect OS&Y and PIV cut-off valves for open position.	.066		√	√	√	√
13 Clean area around system components.	.433			√	√	√
14 Fill out maintenance checklist and report deficiencies.	.022		√	√	√	√
Total labor-hours/period			.475	1.241	1.241	3.818
Total labor-hours/year			3.800	2.482	1.241	3.818

		Cost Each					
		2014 Bare Costs				Total	Total
Description	Labor-hrs.	Material	Labor	Equip.	Total	In-House	w/O&P
1900 Extinguishing system, wet pipe, annually	3.818	86	212		298	360	440
1950 Annualized	11.341	172	630		802	975	1,200

D40 FIRE PROTECTION — D4015 180 Exting. System, Deluge/Preaction

	PM Components	Labor-hrs.	W	M	Q	S	A
PM System D4015 180 1950							
Extinguishing system, deluge/preaction							
1	Notify proper authorites prior to testing any alarm systems.	.130		√	√	√	√
2	Open and close post indicator valve (PIV) to check operation; make minor repairs such as lubricating valve stem, cleaning and/or replacing target windows as required.	.176					√
3	Open and close outside stem and yoke (OS&Y) cut-off valve to check operation; make minor repairs such as lubricating stems, tightening packing glands as required.	.176					√
4	Perform operational test of supervisory initiating devices and water flow detectors; make minor adjustments and restore system to proper operating condition.	.148					√
5	Check to ensure that alarm drain is open; clean drain if necessary.	.081		√	√	√	√
6	Open water motor alarm test valve and ensure that outside alarm operates; lubricate alarm and adjust as required.	.176		√	√	√	√
7	Visually check water pressure to ensure adequate operating pressure is available; make adjustments as required.	.004		√	√	√	√
8	Conduct main drain test by opening 2″ test valve and observing drop in water pressure on gauge; pressure drop should not exceed 20 PSI; maintain a continuous record of drain tests; make minor adjustments; restore system to proper operating condition.	.333			√	√	√
9	Check general condition of sprinklers and sprinkler system; make minor adjustments as required.	.228					√
10	Check equipment gaskets, piping, packing glands, and valves for leaks; tighten flange bolts and loose connections to stop all leaks.	.013					√
11	Check condition of fire department connections; replace missing or broken covers as required.	.103					√
12	Check and inspect pneumatic system for physical damage and proper operation; make minor adjustments as required.	.320					√
13	Trip test deluge system (control valve closed); make minor adjustments as necessary and restore system to fully operational condition.	1.956					√
14	Inspect OS&Y and PIV cut-off valves for open position.	.066		√	√	√	√
15	Clean area around system components.	.066		√	√	√	√
16	Ensure that system is restored to proper operating condition.	.109		√	√	√	√
17	Fill out maintenance checklist and report deficiencies.	.022		√	√	√	√
	Total labor-hours/period			.588	.921	.921	4.107
	Total labor-hours/year			4.704	1.842	.921	4.107

			Cost Each					
	Description	Labor-hrs.	2014 Bare Costs				Total In-House	Total w/O&P
			Material	Labor	Equip.	Total		
1900	Extinguishing system, deluge / preaction, annually	4.107	86	228		314	380	465
1950	Annualized	11.574	172	640		812	990	1,225

D40 FIRE PROTECTION — D4015 210 Fire Pump, Electric Motor Driven

PM Components	Labor-hrs.	W	M	Q	S	A
PM System D4015 210 1950						
Fire pump, electric motor driven						
1 Check control panel and wiring for loose connections; tighten connections as required.	.109			√	√	√
2 Ensure all valves relating to water system are in correct position.	.008	√	√	√	√	√
3 Open and close OS&Y (outside steam and yoke) cut-off valve to check operation; make minor repairs such as lubricating stems and tightening packing glands as required.	.176			√	√	√
4 Centrifugal pump:						
A) perform 10 minute pump test run; check for proper operation and adjust if required.	.217	√	√	√	√	√
B) check for leaks on suction and discharge piping, seals, packing glands, etc.	.077	√	√	√	√	√
C) check for excessive vibration, noise, overheating, etc.	.022	√	√	√	√	√
D) check alignment, clearances, and rotation of shaft and coupler (includes removing and reinstalling safety cover).	.160	√	√	√	√	√
E) tighten or replace loose, missing or damaged nuts, bolts, or screws.	.005	√	√	√	√	√
F) lubricate pump and motor as required.	.099			√	√	√
G) check suction or discharge pressure gauge readings and flow rate.	.078	√	√	√	√	√
H) check packing glands and tighten or repack as required; note that slight dripping is required for proper lubrication of shaft.	.113			√	√	√
5 Inspect and clean strainers after each use and flow test.	.260	√	√	√	√	√
6 Clean equipment and surrounding area.	.066	√	√	√	√	√
7 Fill out maintenance checklist and report deficiencies.	.022	√	√	√	√	√
Total labor-hours/period		.915	.915	1.412	1.412	1.412
Total labor-hours/year		34.770	7.320	2.824	1.412	1.412

		Cost Each					
		2014 Bare Costs				Total	Total
Description	Labor-hrs.	Material	Labor	Equip.	Total	In-House	w/O&P
1900 Fire pump, electric motor driven, annually	1.412	88	78		166	195	232
1950 Annualized	47.738	136	2,650		2,786	3,450	4,300

D40 FIRE PROTECTION — D4015 250 Fire Pump, Motor/Engine Driven

PM Components	Labor-hrs.	W	M	Q	S	A
PM System D4015 250 1950						
Fire pump, engine driven						
1 Ensure all valves relating to water system are in correct position.	.008	✓	✓	✓	✓	✓
2 Open and close OS&Y (outside steam and yoke) cut-off valve to check operation; make minor repairs such as lubricating stems and tightening packing glands as required.	.176			✓	✓	✓
3 Controller panel:						
A) check controller panel for proper operation in accordance with standard procedures for the type of panel installed, replace burned out bulbs and fuses; make other repairs as necessary.	.039			✓	✓	✓
B) check wiring for loose connections and fraying; tighten or replace as required.	.109			✓	✓	✓
4 Diesel engine:						
A) check oil level in crankcase with dipstick; add oil as necessary.	.014	✓	✓	✓	✓	✓
B) drain and replace engine oil.	.511					✓
C) replace engine oil filter.	.059					✓
D) check radiator coolant level; add water if low.	.012	✓	✓	✓	✓	✓
E) drain and replace radiator coolant.	.511					✓
F) check battery water level; add water if low.	.012			✓	✓	✓
G) check battery terminals for corrosion; clean if necessary.	.124			✓	✓	✓
H) check belts for tension and wear; adjust or replace if required.	.012			✓	✓	✓
I) check fuel level in tank; refill as required.	.046			✓	✓	✓
J) check electrical wiring, connections, switches, etc.; adjust or tighten as necessary.	.120			✓	✓	✓
K) perform 10 minute pump check for proper operation and adjust as required.	.217	✓	✓	✓	✓	✓
5 Centrifugal pump:						
A) check for leaks on suction and discharge piping, seals, packing glands, etc.	.077	✓	✓	✓	✓	✓
B) check pump operation; vibration, noise, overheating, etc.	.022	✓	✓	✓	✓	✓
C) check alignment, clearances, and rotation of shaft and coupler (includes removing and reinstalling safety cover).	.160	✓	✓	✓	✓	✓
D) tighten or replace loose, missing or damaged nuts, bolts, or screws.	.005	✓	✓	✓	✓	✓
E) lubricate pump as required.	.099			✓	✓	✓
F) check suction or discharge pressure gauge readings and flow rate.	.078	✓	✓	✓	✓	✓
G) check packing glands and tighten or repack as required; note that slight dripping is required for proper lubrication of shaft.	.113			✓	✓	✓
6 Inspect and clean strainers after each use and flow test.	.260	✓	✓	✓	✓	✓
7 Clean equipment and surrounding area.	.433	✓	✓	✓	✓	✓
8 Fill out maintenance checklist and report deficiencies.	.022	✓	✓	✓	✓	✓
Total labor-hours/period		1.308	1.308	2.158	2.158	3.239
Total labor-hours/year		49.704	10.464	4.316	2.158	3.239

			Cost Each				
			2014 Bare Costs			Total	Total
Description	Labor-hrs.	Material	Labor	Equip.	Total	In-House	w/O&P
1900 Fire pump, motor/engine driven, annually	3.239	192	179		371	435	520
1950 Annualized	69.881	224	3,875		4,099	5,100	6,325

D40 FIRE PROTECTION — D4015 310 Extinguishing System, Dry Pipe

PM Components	Labor-hrs.	W	M	Q	S	A
PM System D4015 310 1950						
Extinguishing system, dry pipe						
1 Notify proper authorities prior to testing any alarm systems.	.130		√	√	√	√
2 Open and close post indicator valve to check operation; make minor repairs such as lubricating valve stem, cleaning and/or replacing target windows as required.	.176					√
3 Open and close OS&Y (outside stem and yoke) cut-off valve to check operation; make minor repairs such as lubricating stems and tightening packing glands as required.	.176					√
4 Perform operational test of water flow detectors; make minor adjustments and restore system to proper operating condition.	.148					√
5 Check to ensure that alarm drain is open; clean drain line if necessary.	.081		√	√	√	√
6 Open water motor alarm test valve and ensure that outside alarm operates; lubricate alarm, make adjustments as required.	.176		√	√	√	√
7 Visually check water pressure to ensure that adequate operating pressure is available; make adjustments as required.	.003		√	√	√	√
8 Conduct main drain test by opening 2" test valve; maintain a continuous record of drain tests; make minor adjustments if applicable; restore system to proper operating condition.	.333			√		√
9 Check general condition of sprinklers and sprinkler system; make minor adjustments as required.	.228					√
10 Check equipment gaskets, piping, packing glands, and valves for leaks; tighten flange bolts and loose connections to stop all leaks.	.013					√
11 Check condition of fire department connections; replace missing or broken covers as required.	.103					√
12 Visually check system air pressure; pump up system and make minor adjustments as required.	.135		√	√	√	√
13 Trip test dry pipe system using test valve furthest from dry pipe valve (control valve partially open); make minor adjustments as necessary and restore system to fully operational condition.	1.733					√
14 Check operation of quick opening device.	.022					√
15 Inspect OS&Y and PIV cut-off valves for open position.	.066		√	√	√	√
16 Clean area around system components.	.433			√	√	√
17 Fill out maintenance checklist and report deficiencies.	.022		√	√	√	√
Total labor-hours/period			.613	1.379	1.379	3.978
Total labor-hours/year			4.904	2.758	1.379	3.978

			Cost Each				
			2014 Bare Costs			Total	Total
Description	Labor-hrs.	Material	Labor	Equip.	Total	In-House	w/O&P
1900 Extinguishing system, dry pipe, annually	3.978	133	220		353	420	510
1950 Annualized	13.019	219	720		939	1,150	1,400

D40 FIRE PROTECTION — D4095 100 Extinguishing System, CO_2

PM Components	Labor-hrs.	W	M	Q	S	A
PM System D4095 100 1950						
Extinguishing system, CO_2						
1 Check that nozzles, heads and hand hose lines are clear from obstructions, have not been damaged and in proper position.	.098		√	√	√	√
2 Check to ensure that all operating controls are properly set.	.066		√	√	√	√
3 Clean nozzles as required.	.098		√	√	√	√
4 Visually inspect the control panel for obstructions or physical damage; clean dirt and dust from interior and exterior; make sure that all cards are plugged in tightly, tighten loose connections and make other minor adjustments as necessary.	.222		√	√	√	√
5 Check battery voltages where installed, recharge or replace as required.	.012		√	√	√	√
6 Weigh CO_2 cylinders; replace any that show a weight loss greater than 10%.	.650				√	√
7 Blow out entire system to make sure that the system is not plugged or restricted.	.325					√
8 Run out hose reels and check hoses.	.217					√
9 Conduct operational test of actuating devices, both automatic, manual and alarm system. Prior to conducting tests, close valves, disable CO_2 dumping control and manually override computer shutdown feature if applicable, all in accordance with the manufacturer's specifications.	.433					√
10 Check the operation and timing of the time delay control and check operation of the abort station.	.108					√
11 Restore system to proper operating condition and notify personnel upon completion of tests.	.079					√
12 Clean up around system.	.066		√	√	√	√
13 Fill out maintenance checklist and report deficiencies.	.022		√	√	√	√
Total labor-hours/period			.584	.584	1.234	2.396
Total labor-hours/year			4.672	1.168	1.234	2.396

	Description	Labor-hrs.	Cost Each				Total In-House	Total w/O&P
			2014 Bare Costs					
			Material	Labor	Equip.	Total		
1900	Extinguishing system, CO_2, annually	2.396	70.50	133		203.50	244	295
1950	Annualized	9.470	219	525		744	895	1,100

D40 FIRE PROTECTION   D4095 200 Exting. System, Foam Bottle

PM Components	Labor-hrs.	W	M	Q	S	A
PM System D4095 200 1950						
Extinguishing system, foam bottle						
1 Check that all nozzles, heads and hand held lines are clear from obstructions, have not been damaged and are in proper position.	.195		√	√	√	√
2 Check to ensure that all operating controls are properly set.	.022		√	√	√	√
3 Observe pressure on system to ensure that proper pressure is being maintained.	.004		√	√	√	√
4 Clean area around system.	.066		√	√	√	√
5 Fill out maintenance checklist and report deficiencies.	.022		√	√	√	√
Total labor-hours/period			.309	.309	.309	.309
Total labor-hours/year			2.472	.618	.309	.309

			Cost Each				
			2014 Bare Costs			Total	Total
Description	Labor-hrs.	Material	Labor	Equip.	Total	In-House	w/O&P
1900 Extinguishing system, foam bottle, annually	.309		17.10		17.10	21.50	26.50
1950 Annualized	3.708		205		205	257	320

D40 FIRE PROTECTION | D4095 210 Exting. System, Foam, Electric

PM Components	Labor-hrs.	W	M	Q	S	A
PM System D4095 210 1950						
Extinguishing system, foam, electric pump, deluge system						
1 Check foam concentrate level in tank; add concentrate as required to maintain proper level.	.043		√	√	√	√
2 Ensure all valves relating to foam/water system are in correct position.	.008		√	√	√	√
3 Visually check proportioning devices, pumps, and foam nozzles; correct any observed deficiencies.	.386		√	√	√	√
4 Check water supply pressure; adjust as necessary.	.029		√	√	√	√
5 Open and close OS&Y (outside stem and yoke) cut-off valve to check operation; make minor repairs such as lubricating stems and tightening packing glands as required.	.059			√	√	√
6 Centrifugal pump:						
A) perform 10 minute pump test run, check for proper operation and adjust as required.	.217			√	√	√
B) test start pump without foam discharger.	.022		√	√	√	√
C) check for leaks on suction and discharge piping, seals, packing glands, etc.	.077			√	√	√
D) check for excessive vibration, noise, overheating, etc.	.022			√	√	√
E) check alignment, clearances, and rotation of shaft and coupler (includes removing and reinstalling safety cover).	.160			√	√	√
F) tighten or replace loose, missing or damaged nuts, bolts, or screws.	.005			√	√	√
G) lubricate pump and motor as required.	.099			√	√	√
H) check suction or discharge pressure gauge readings and flow rate.	.078			√	√	√
I) check packing glands and tighten or repack as required; note that slight dripping is required for proper lubrication of shaft.	.113			√	√	√
7 Inspect and clean strainers after each use and flow test.	.260		√	√	√	√
8 Clean equipment and surrounding area.	.433		√	√	√	√
9 Fill out maintenance checklist.	.022		√	√	√	√
Total labor-hours/period			1.203	2.033	2.033	2.033
Total labor-hours/year			9.624	4.066	2.033	2.033

	Description	Labor-hrs.	Material	Labor	Equip.	Total	Total In-House	Total w/O&P
			2014 Bare Costs					
1900	Extinguishing system, foam, electric pump, annually	2.033	128	113		241	282	335
1950	Annualized	17.756	176	985		1,161	1,425	1,750

D40 FIRE PROTECTION   D4095 220 Exting. System, Foam, Diesel

PM Components	Labor-hrs.	W	M	Q	S	A
PM System D4095 220 1950						
Extinguishing system, foam, diesel pump, deluge system						
1 Check foam concentrate level in tank; add concentrate as required to maintain proper level.	.043		√	√	√	√
2 Ensure all valves relating to foam/water system are in correct position.	.008		√	√	√	√
3 Visually check proportioning devices, pumps, and foam nozzles; correct any observed deficiencies.	.386		√	√	√	√
4 Check water supply pressure; adjust as necessary.	.003		√	√	√	√
5 Open and close OS&Y (outside stem and yoke) cut-off valve to check operation; make minor repairs such as lubricating stems and tightening packing glands as required.	.059			√	√	√
6 Controller panel:						
A) check controller panel for proper operation in accordance with standard procedures for the type of panel installed, replace burned out bulbs and fuses; make other repairs as necessary.	.039			√	√	√
B) check wiring for loose connections and fraying; tighten or replace as required.	.109				√	√
7 Diesel engine:						
A) check oil level in crankcase with dipstick; add oil as necessary.	.014		√	√	√	√
B) drain and replace engine oil.	.511					√
C) replace engine oil filter.	.059					√
D) check radiator coolant level; add water if low.	.012		√	√	√	√
E) drain and replace radiator coolant.	.511					√
F) check battery water level; add water if low.	.012			√	√	√
G) check battery terminals for corrosion; clean if necessary.	.124			√	√	√
H) check belts for tension and wear; adjust or replace if required.	.012			√	√	√
I) check fuel level in tank; refill as required.	.046			√	√	√
J) check electrical wiring, connections, switches, etc.; adjust or tighten as necessary.	.120			√	√	√
K) perform 10 minute pump check for proper operation and adjust as required, start pump without foam discharger.	.217		√	√	√	√
8 Centrifugal pump:						
A) check for leaks on suction and discharge piping, seals, packing glands, etc.	.077			√	√	√
B) check pump operation; vibration, noise, overheating, etc.	.022			√	√	√
C) check alignment, clearances, and rotation of shaft and coupler (includes removing and reinstalling safety cover).	.160				√	√
D) tighten or replace loose, missing or damaged nuts, bolts, or screws.	.005			√	√	√
E) lubricate pump as required.	.099			√	√	√
F) check suction or discharge pressure gauge readings and flow rate.	.078			√	√	√
G) check packing glands and tighten or repack as as required; note that slight dripping is required for proper lubrication of shaft.	.113			√	√	√
9 Inspect and clean strainers after each use and flow test.	.260		√	√	√	√
10 Clean equipment and surrounding area.	.433		√	√	√	√
11 Fill out maintenance checklist.	.022		√	√	√	√
Total labor-hours/period			1.398	2.473	2.473	3.554
Total labor-hours/year			11.184	4.946	2.473	3.554

	Description	Labor-hrs.	2014 Bare Costs				Total In-House	Total w/O&P
			Material	Labor	Equip.	Total		
1900	Extinguishing system, foam, diesel pump, annually	3.554	232	197		429	500	595
1950	Annualized	22.157	264	1,225		1,489	1,825	2,250

D40 FIRE PROTECTION D4095 400 | Exting. System, Dry Chemical

PM Components	Labor-hrs.	W	M	Q	S	A
PM System D4095 400 1950						
Extinguishing system, dry chemical						
1 Check nozzles, heads, and fusible links or heat detectors to ensure that they are clean, have adequate clearance from obstructions and have not been damaged.	.098		√	√	√	√
2 Check hand hose lines, if applicable, to ensure that they are clear, in proper position and in good condition.	.098		√	√	√	√
3 Check to ensure that all operating controls are properly set.	.066		√	√	√	√
4 Observe pressure on system to ensure that proper pressure is maintained.	.004		√	√	√	√
5 Check expellant gas cylinders of gas cartridge systems for proper pressure or weight to ensure that sufficient gas is available.	.260				√	√
6 Test all actuating and operating devices; turn valve on dry chemical bottle or cylinder to off position in accordance with manufacturers specifications, and test operation of control head by removing fusible link, activating heat detectors or pulling test control. Restore system to normal operation and make minor adjustments as applicable.	.433					√
7 Open and check dry chemical in cylinder and stored pressure systems to ensure it is free flowing and without lumps.	.325					√
8 When an alarm system is reactivated, ensure that it functions in accordance with manufacturers specifications.	.079					√
9 Clean area around system.	.066		√	√	√	√
10 Fill out maintenance checklist and report deficiencies.	.022		√	√	√	√
Total labor-hours/period			.354	.354	.614	1.451
Total labor-hours/year			2.832	.708	.614	1.451

			Cost Each					
			2014 Bare Costs				Total	Total
	Description	Labor-hrs.	Material	Labor	Equip.	Total	In-House	w/O&P
1900	Extinguishing system, dry chemical, annually	1.451	70.50	80.50		151	178	214
1950	Annualized	5.605	219	310		529	630	760

D40 FIRE PROTECTION — D4095 450 Extinguishing System, FM200

PM Components	Labor-hrs.	W	M	Q	S	A
PM System D4095 450 1950						
Extinguishing system, FM200						
1 Check that nozzles, heads and hand hose lines are clear from obstructions, have not been damaged and in proper position.	.098		√	√	√	√
2 Check to ensure that all operating controls are properly set.	.066		√	√	√	√
3 Clean nozzles as required.	.098		√	√	√	√
4 Visually inspect the control panel for obstructions or physical damage; clean dirt and dust from interior and exterior; make sure that all cards are plugged in tightly, tighten loose connections and make other minor adjustments as necessary.	.222		√	√	√	√
5 Check battery voltages where installed, recharge or replace as required.	.012		√	√	√	√
6 Weigh cylinders; replace any that show a weight loss greater than 5% or a pressure loss greater than 10%.	.650				√	√
7 Blow out entire system to make sure that the system is not plugged or restricted.	.333					√
8 Conduct operational test of actuating devices, both automatic, manual and alarm system; prior to conducting tests, close valves disable CO_2 dumping control and manually override computer shutdown feature if applicable, all in accordance with manufacturers specifications.	.436					√
9 Check the operation and timing of the time delay control and check operation of abort station.	.108					√
10 Restore system to proper operating condition and notify cognizant personnel upon completion of tests.	.079					√
11 Clean out around system.	.066		√	√	√	√
12 Fill out maintenance checklist and report deficiencies.	.022		√	√	√	√
Total labor-hours/period			.584	.584	1.234	2.190
Total labor-hours/year			4.672	1.168	1.234	2.190

			Cost Each				
			2014 Bare Costs			Total	Total
Description	Labor-hrs.	Material	Labor	Equip.	Total	In-House	w/O&P
1900 Extinguishing system, FM200, annually	2.190	47	121		168	203	248
1950 Annualized	9.264	172	515		687	830	1,025

D50 ELECTRICAL — D5015 210 Switchboard, Electrical

PM Components	Labor-hrs.	W	M	Q	S	A
PM System D5015 210 1950						
Switchboard, electrical						
1 Check with operating or area personnel for deficiencies.	.044					✓
2 Check indicating lamps for proper operation, if appropriate; replace burned out lamps.	.018					✓
3 Remove and reinstall cover.	.196					✓
4 Check for discolorations, hot spots, odors and charred insulation.	.359					✓
5 Clean switchboard exterior and surrounding area.	.066					✓
6 Fill out maintenance checklist and report deficiencies.	.022					✓
Total labor-hours/period						.705
Total labor-hours/year						.705

			Cost Each				
			2014 Bare Costs			Total	Total
Description	Labor-hrs.	Material	Labor	Equip.	Total	In-House	w/O&P
1900 Switchboard, electrical, annually	.705	8	37.50		45.50	55.50	68
1950 Annualized	.705	8	37.50		45.50	55.50	68

D50 ELECTRICAL — D5015 214 Switchboard, w/Air Circ. Breaker

PM Components	Labor-hrs.	W	M	Q	S	A
PM System D5015 214 1950						
Switchboard w/air circuit breaker						
1 Check with operating or area personnel for deficiencies.	.035					✓
2 Check indicating lamps for proper operation, if appropriate; replace burned out lamps.	.020					✓
3 Remove switchboard from service, coordinate with appropriate authority.	.038					✓
4 Remove and reinstall cover.	.051					✓
5 Check for discolorations, hot spots, odors and charred insulation.	.016					✓
6 Check the records on the circuit breaker for any full current rated trips which may indicate a potential problem.	.020					✓
7 Examine the exterior of the switchboard for any damage or moisture.	.020					✓
8 Check operation of circuit breaker tripping mechanism and relays.	.096					✓
9 Check the operation of the circuit breaker charging motor.	.020					✓
10 Check hardware for tightness.	.120					✓
11 Check interlock for proper operation.	.013					✓
12 Check that the charger is operating properly.	.022		✓	✓	✓	✓
13 Check for any signs of corrosion on the battery terminals or wires.	.020		✓	✓	✓	✓
14 Check the electrolyte level in the batteries; add if required.	.624		✓	✓	✓	✓
15 Check the specific gravity of the electrolyte in a 10% sample of the batteries.	.069			✓	✓	✓
16 Check 25% of terminal-to-cell connection resistance; rehabilitate connections as required; add anti-corrosion grease to battery terminals and connections.	.113					✓
17 Measure and record individual cell and string float voltages.	2.808					✓
18 Place switchboard back in service, notify appropriate authority.	.057					✓
19 Clean switchboard, charger and the surrounding area.	.066		✓	✓	✓	✓
20 Fill out maintenance checklist and report deficiencies.	.022		✓	✓	✓	✓
Total labor-hours/period			.823	.823	.823	4.250
Total labor-hours/year			6.584	1.646	.823	4.250

			Cost Each				
			2014 Bare Costs			Total	Total
Description	Labor-hrs.	Material	Labor	Equip.	Total	In-House	w/O&P
1900 Switchboard, with air circuit breaker, annually	4.250	22	227		249	305	380
1950 Annualized	13.303	29	710		739	910	1,125

D50 ELECTRICAL
D5015 217 Switchboard, w/Air C.B. and Tie

PM Components	Labor-hrs.	W	M	Q	S	A
PM System D5015 217 1950						
Switchboard w/air circuit breaker and tie switch						
1 Check with operating or area personnel for deficiencies.	.035					√
2 Check indicating lamps for proper operation, if appropriate; replace burned out lamps.	.020					√
3 Remove switchboard from service, coordinate with appropriate authority.	.038					√
4 Remove and reinstall cover.	.051					√
5 Check for discolorations, hot spots, odors and charred insulation.	.016					√
6 Check the records on the circuit breaker for any full current rated trips which may indicate a potential problem.	.020					√
7 Examine the exterior of the switchboard for any damage or moisture.	.020					√
8 Check operation of circuit breaker tripping mechanism and relays.	.096					√
9 Check the operation of the circuit breaker charging motor.	.020					√
10 Check hardware for tightness.	.120					√
11 Check interlock for proper operation.	.013					√
12 Operate the switching handle and check the locking mechanism.	.052					√
13 Check that the charger is operating properly.	.022		√	√	√	√
14 Check for any signs of corrosion on the battery terminals or wires.	.020		√	√	√	√
15 Check the electrolyte level in the batteries; add if required.	.624		√	√	√	√
16 Check the specific gravity of the electrolyte in a 10% sample of the batteries.	.069		√	√	√	√
17 Check 25% of terminal-to-cell connection resistance; rehabilitate connections as required; add anti-corrosion grease to battery terminals and connections.	.113					√
18 Measure and record individual cell and string float voltages.	2.808					√
19 Place switchboard back in service, notify appropriate authority.	.057					√
20 Clean switchboard, charger and the surrounding area.	.066		√	√	√	√
21 Fill out maintenance checklist and report deficiencies.	.022		√	√	√	√
Total labor-hours/period			.823	.823	.823	4.302
Total labor-hours/year			6.584	1.646	.823	4.302

			Cost Each				
			2014 Bare Costs			Total	Total
Description		Labor-hrs.	Material	Labor	Equip.	Total	In-House	w/O&P
1900 Switchboard, with air C.B. and tie switch, annually	4.302	33	230		263	320	395
1950 Annualized	13.355	47	715		762	935	1,150

D50 ELECTRICAL — D5015 220 Circuit Breaker, HV Air

PM Components	Labor-hrs.	W	M	Q	S	A
PM System D5015 220 1950						
Circuit breaker, HV air						
1 Check the records on the circuit breaker for any full current rated trips which may indicate a potential problem.	.066					✓
2 Examine the exterior of the circuit breaker for any damage or moisture.	.060					✓
3 Remove the circuit breaker from service, coordinate with appropriate authority.	.038					✓
4 Check operation of tripping mechanisms and relays.	.096					✓
5 Check the operation of the charging motor, if applicable.	.038					✓
6 Check hardware for tightness.	.005					✓
7 Check interlock for proper operation.	.039					✓
8 Clean area around switch.	.066					✓
9 Place the circuit breaker back in service, coordinate with appropriate authority.	.040					✓
10 Fill out maintenance checklist.	.022					✓
Total labor-hours/period						.470
Total labor-hours/year						.470

		Cost Each					
		2014 Bare Costs				Total	Total
Description	Labor-hrs.	Material	Labor	Equip.	Total	In-House	w/O&P
1900 Circuit breaker, high voltage air, annually	.470		25		25	31	39
1950 Annualized	.470		25		25	31	39

D50 ELECTRICAL — D5015 222 Circuit Breaker, HV Oil

PM Components	Labor-hrs.	W	M	Q	S	A
PM System D5015 222 1950						
Circuit breaker, HV oil						
1 Check the records on the circuit breaker for any full current rated trips which may indicate a potential problem.	.066					✓
2 Examine the exterior of the circuit breaker for any damage, cracks, rust or leaks; check around the gaskets and the drain valves.	.074					✓
3 Check the entrance bushings for cracks in the porcelain.	.321					✓
4 Check the condition of the ground system.	.170					✓
5 Draw oil sample from bottom of circuit breaker; have sample tested; replace removed oil to the proper level.	.042					✓
6 Clean the exterior of the circuit breaker and operating mechanism.	.094					✓
7 Clean the area around the circuit breaker.	.066					✓
8 Fill out maintenance checklist.	.022					✓
Total labor-hours/period						.855
Total labor-hours/year						.855

		Cost Each					
		2014 Bare Costs				Total	Total
Description	Labor-hrs.	Material	Labor	Equip.	Total	In-House	w/O&P
1900 Circuit breaker, high voltage oil, annually	.855	17.90	45.50		63.40	76	93
1950 Annualized	.855	17.90	45.50		63.40	76	93

D50 ELECTRICAL — D5015 230 Switch, Selector, HV Air

PM Components	Labor-hrs.	W	M	Q	S	A
PM System D5015 230 1950						
Switch, selector, HV, air						
1 Examine the exterior of the selector switch for any damage.	.060					✓
2 Check that the switching handle is locked.	.044					✓
3 Check the condition of the ground system.	.170					✓
4 Clean the exterior of the selector switch and operating mechanism.	.046					✓
5 Clean the area around the selector switch.	.066					✓
6 Fill out maintenance checklist and report deficiencies.	.022					✓
Total labor-hours/period						.408
Total labor-hours/year						.408

			Cost Each				
			2014 Bare Costs			Total	Total
Description	Labor-hrs.	Material	Labor	Equip.	Total	In-House	w/O&P
1900 Switch, selector, high voltage, air, annually	.408	11.95	22		33.95	40	48.50
1950 Annualized	.408	11.95	22		33.95	40	48.50

D50 ELECTRICAL — D5015 232 Switch, Selector, HV Oil

PM Components	Labor-hrs.	W	M	Q	S	A
PM System D5015 232 1950						
Switch, selector, HV, oil						
1 Examine the exterior of the interrupt switch for any damage, cracks, rust or leaks; check around the gaskets and the drain valves.	.060					✓
2 Check that the switching handle is locked.	.044					✓
3 Check the condition of the ground system.	.170					✓
4 Draw oil sample from bottom of selector switch; have sample tested; replace removed oil to the proper level.	.042					✓
5 Clean the exterior of the selector switch and operating mechanism.	.046					✓
6 Clean the area around the selector switch.	.066					✓
7 Fill out maintenance checklist and report deficiencies.	.022					✓
Total labor-hours/period						.450
Total labor-hours/year						.450

			Cost Each				
			2014 Bare Costs			Total	Total
Description	Labor-hrs.	Material	Labor	Equip.	Total	In-House	w/O&P
1900 Switch, selector, high voltage, oil, annually	.450	24	24		48	56	67
1950 Annualized	.450	24	24		48	56	67

D50 ELECTRICAL — D5015 234 Switch, Automatic Transfer

PM Components	Labor-hrs.	W	M	Q	S	A
PM System D5015 234 1950						
Switch, automatic transfer						
1 Check with operating or area personnel for deficiencies.	.044		✓	✓	✓	✓
2 Inspect wiring, wiring connections, and fuse blocks for looseness, charring, evidence of short circuiting, overheating and tighten all connections.	.263		✓	✓	✓	✓
3 Inspect gen. cond. of transf. switch and clean exter. and surrounding area.	.114		✓	✓	✓	✓
4 Fill out maintenance checklist and report deficiencies.	.022		✓	✓	✓	✓
Total labor-hours/period			.443	.443	.443	.443
Total labor-hours/year			3.544	.886	.443	.443

			Cost Each				
			2014 Bare Costs			Total	Total
Description	Labor-hrs.	Material	Labor	Equip.	Total	In-House	w/O&P
1900 Switch, automatic transfer, annually	.443	11.95	23.50		35.45	42.50	51.50
1950 Annualized	5.316	11.95	284		295.95	365	455

D50 ELECTRICAL — D5015 236 Switch, Interrupt, HV, Fused Air

PM Components	Labor-hrs.	W	M	Q	S	A
PM System D5015 236 1950						
Switch, interrupt, HV, fused air						
1 Examine the exterior of the interrupt switch for any damage.	.060					✓
2 Check the condition of the ground system.	.170					✓
3 Clean the exterior of the interrupt switch and operating mechanism.	.046					✓
4 Clean the area around the interrupt switch.	.066					✓
5 Fill out maintenance checklist and report deficiencies.	.022					✓
Total labor-hours/period						.364
Total labor-hours/year						.364

			Cost Each				
			2014 Bare Costs			Total	Total
Description	Labor-hrs.	Material	Labor	Equip.	Total	In-House	w/O&P
1900 Switch, interrupt, high voltage, fused air, annually	.364	11.95	19.40		31.35	37	45
1950 Annualized	.364	11.95	19.40		31.35	37	45

D50 ELECTRICAL — D5015 238 Switch, HV w/Aux Fuses, Air

PM Components	Labor-hrs.	W	M	Q	S	A
PM System D5015 238 1950						
Switch, interrupt, HV, w/auxiliary fuses, air						
1 Examine the exterior of the interrupt switch for any damage.	.060					✓
2 Check the condition of the ground system.	.170					✓
3 Clean the exterior of the interrupt switch and operating mechanism.	.046					✓
4 Clean the area around the interrupt switch.	.066					✓
5 Fill out maintenance checklist and report deficiencies.	.022					✓
Total labor-hours/period						.364
Total labor-hours/year						.364

			Cost Each				
			2014 Bare Costs			Total	Total
Description	Labor-hrs.	Material	Labor	Equip.	Total	In-House	w/O&P
1900 Switch, interrupt, HV, w/ aux fuses, air, annually	.364	11.95	19.40		31.35	37	45
1950 Annualized	.364	11.95	19.40		31.35	37	45

D50 ELECTRICAL — D5015 240 Transformer

PM Components	Labor-hrs.	W	M	Q	S	A
PM System D5015 240 1950						
Transformer, dry type 500 KVA and over						
1 Examine the exterior of the transformer for any damage.	.060					√
2 Remove the covers or open the doors.	.196					√
3 Check for signs of moisture or overheating.	.059					√
4 Check for voltage creeping over insulated surfaces, such as evidenced by tracking or carbonization.	.018					√
5 Check fans, motors and other auxiliary devices for proper operation; where applicable.	.049					√
6 Check the condition of the ground system.	.170					√
7 Replace the covers or close the doors.	.196					√
8 Fill out maintenance checklist and report deficiencies.	.022					√
Total labor-hours/period						.770
Total labor-hours/year						.770

		Cost Each						
			2014 Bare Costs				Total	Total
Description		Labor-hrs.	Material	Labor	Equip.	Total	In-House	w/O&P
1900	Transformer, dry type 500 KVA and over, annually	.770	11.95	41		52.95	64	78.50
1950	Annualized	.770	11.95	41		52.95	64	78.50

PM Components	Labor-hrs.	W	M	Q	S	A
PM System D5015 240 2950						
Transformer, oil, pad mounted						
1 Examine the exterior of the transformer for any damage, cracks, rust or leaks; check around bushings, gaskets and pressure relief device.	.014					√
2 Check the condition of the ground system.	.170					√
3 Check and record the oil level, pressure and temperature readings.	.263					√
4 Check fans, motors and other auxiliary devices for proper operation; where applicable.	.022					√
5 Draw oil sample from top of transformer and have sample tested for dielectric strength; replace removed oil.	.525					√
6 Clean transformer exterior and the surrounding area.	.066					√
7 Fill out maintenance checklist and report deficiencies.	.022					√
Total labor-hours/period						1.082
Total labor-hours/year						1.082

		Cost Each						
			2014 Bare Costs				Total	Total
Description		Labor-hrs.	Material	Labor	Equip.	Total	In-House	w/O&P
2900	Transformer, oil, pad mounted, annually	1.082	24	57.50		81.50	98	120
2950	Annualized	1.082	24	57.50		81.50	98	120

D50 ELECTRICAL — D5015 240 Transformer

PM Components	Labor-hrs.	W	M	Q	S	A
PM System D5015 240 3950						
Transformer, oil, pad mounted, PCB						
1 Check with operating or area personnel for deficiencies.	.040			√	√	√
2 Examine the exterior of the transformer for any damage, cracks, rust or leaks; check around the gaskets, the pressure relief device and the cooling fins.	.014			√	√	√
3 Check and record the oil levels, pressure and temperature readings: level ____; pressure ____; temperature ____.	.263					√
4 Check fans, motors and other auxiliary devices for proper operation; where applicable.	.051					√
5 Check the exterior bushings for leaking oil.	.321			√	√	√
6 Check the condition of the ground system.	.170					√
7 Draw oil sample from top of transformer and have sample tested; replace removed oil.	.525					√
8 Clean transformer exterior and the surrounding area.	.066					√
9 Verify the PCB warning markings are intact and clearly visible.	.022					√
10 Fill out maintenance checklist and report deficiencies. Note: the EPA PCB regulations under 40 CFR 716.30 shall apply to all inspections, samples, disposal, cleanups, decontamination, reporting and record-keeping.	.022			√	√	√
Total labor-hours/period				.397	.397	1.494
Total labor-hours/year				.794	.397	1.494

			Cost Each				
			2014 Bare Costs			Total	Total
Description	Labor-hrs.	Material	Labor	Equip.	Total	In-House	w/O&P
3900 Transformer, oil, pad mounted, PCB, annually	1.494	47.50	79.50		127	152	182
3950 Annualized	2.655	47.50	142		189.50	228	278

D50 ELECTRICAL — D5015 260 Panelboard, 225 Amps and Above

PM Components	Labor-hrs.	W	M	Q	S	A
PM System D5015 260 1950						
Panelboard, 225 amps and above						
1 Check with operating or area personnel for deficiencies.	.066					✓
2 Check for excessive heat, odors, noise, and vibration.	.239					✓
3 Clean and check general condition of panel.	.114					✓
4 Fill out maintenance checklist and report deficiencies.	.022					✓
Total labor-hours/period						.441
Total labor-hours/year						.441

			Cost Each				
			2014 Bare Costs			Total	Total
Description	Labor-hrs.	Material	Labor	Equip.	Total	In-House	w/O&P
1900 Panelboard, 225 A and above, annually	.441	28	23.50		51.50	60	71.50
1950 Annualized	.441	28	23.50		51.50	60	71.50

D50 ELECTRICAL — D5015 280 Motor Control Center

PM Components	Labor-hrs.	W	M	Q	S	A
PM System D5015 280 1950						
Motor control center, over 400 amps						
1 Check with operating or area personnel for deficiencies.	.044					✓
2 Check starter lights, replace if required.	.018					✓
3 Check for excessive heat, odors, noise, and vibration.	.239					✓
4 Clean motor control center exterior and surrounding area.	.066					✓
5 Fill out maintenance checklist and report deficiencies.	.022					✓
Total labor-hours/period						.389
Total labor-hours/year						.389

			Cost Each				
			2014 Bare Costs			Total	Total
Description	Labor-hrs.	Material	Labor	Equip.	Total	In-House	w/O&P
1900 Motor control center, over 400 A, annually	.389	28	21		49	56.50	67
1950 Annualized	.389	28	21		49	56.50	67

D50 ELECTRICAL   D5035 610 Clock Systems

PM Components	Labor-hrs.	W	M	Q	S	A
PM System D5035 610 1950						
Clocks, Central System						
1 Clean dirt and dust from interior and exterior of cabinet.	.260				√	√
2 Adjust relays and check transmission of signal.	.117				√	√
3 Tighten contacts and terminal screws.	.065				√	√
4 Burnish contacts.	.195				√	√
5 Fill out maintenance checklist and report deficiencies.	.022				√	√
Total labor-hours/period					.659	.659
Total labor-hours/year					.659	.659

			Cost Each				
			2014 Bare Costs			Total	Total
Description	Labor-hrs.	Material	Labor	Equip.	Total	In-House	w/O&P
1900 Central clock systems, annually	.659	2	35		37	45.50	57
1950 Annualized	1.318	2	70.50		72.50	89	112

D50 ELECTRICAL — D5035 710 Fire Alarm Annunciator System

PM Components	Labor-hrs.	W	M	Q	S	A
PM System D5035 710 1950						
Fire alarm annunciator system						
1 Visually inspect all alarm equipment for obstructions or physical damage, clean dirt and dust from interior and exterior of panel/pull boxes, tighten loose connections.	.222		√	√	√	√
2 Notify cogniant personnel prior to testing.	.112		√	√	√	√
3 Conduct operational test of initiating and signal transmitting devices in populated buildings by building zone/area; for those circuits which do not operate properly, check detectors, control units, and annunciators for dust on defective components; make minor adjustments as required.	.112		√	√	√	√
4 Check battery voltages where installed; replace as required.	.012		√	√	√	√
5 Conduct operational test of 10% of total number of spot type heat detectors and all smoke detectors; for those circuits which do not operate properly, check to determine if problem relates to circuit, device, or control unit, make minor adjustments as required; if detector is defective but no replacement is immediately available, remove detector, re-establish Initiating circuit and tag location until a replacement detector is installed.	2.171				√	
6 Restore system to proper operating condition and notify personnel upon completion of tests.	.079		√	√	√	√
7 Fill out maintenance checklist and report deficiencies.	.022		√	√	√	√
Total labor-hours/period			.559	.559	2.730	2.730
Total labor-hours/year			4.472	1.118	2.730	2.730

			Cost Each				
			2014 Bare Costs			Total	Total
Description		Labor-hrs.	Material	Labor	Equip.	Total	In-House	w/O&P
1900 Fire alarm annunciator system, annually	2.730	128	151		279	330	395
1950 Annualized	11.050	160	610		770	940	1,150

D50 ELECTRICAL — D5035 810 Security/Intrusion Alarm Ctl. Panel

PM Components	Labor-hrs.	W	M	Q	S	A
PM System D5035 810 1950						
Security / intrusion alarm system						
1 Check in and out with area security officer, notify operating/facility personnel, obtain necessary alarm keys, alarm codes and escort when required.	.520			√	√	√
2 Inspect alarm control panel and conduct operational test of initiating and signal transmitting devices; make minor adjustments as required.	.241			√	√	√
3 Check indicating lamps for proper operation, replace if necessary.	.018			√	√	√
4 Check battery voltages where installed, replace as required.	.012			√	√	√
5 Restore system to proper operating condition and notify personnel upon completion of tests.	.079			√	√	√
6 Clean exterior of cabinet and surrounding area.	.066			√	√	√
7 Fill out maintenance checklist and report deficiencies.	.022			√	√	√
Total labor-hours/period				.958	.958	.958
Total labor-hours/year				1.916	.958	.958

			Cost Each				
			2014 Bare Costs			Total	Total
Description		Labor-hrs.	Material	Labor	Equip.	Total	In-House	w/O&P
1900 Security, intrusion alarm system, annually	.958	31	51		82	97.50	118
1950 Annualized	3.832	41	204		245	298	365

D50 ELECTRICAL — D5095 210 | Emergency Generator

PM Components	Labor-hrs.	W	M	Q	S	A
PM System D5095 210 1950						
Emergency generator up to 15 KVA						
1 Check with the operating or area personnel for any obvious deficiencies.	.044		√	√	√	√
2 Check engine oil level; add as required.	.010		√	√	√	√
3 Change engine oil and oil filter.	.170					√
4 Check battery charge and electrolyte specific gravity, add water as required; check terminals for corrosion, clean as required.	.120		√	√	√	√
5 Check belt tension and wear; adjust as required, if applicable.	.012		√	√	√	√
6 Check engine air filter; change as required.	.042					√
7 Check spark plug or injector nozzle condition; service or replace as required.	.211					√
8 Check wiring, connections, switches, etc., adjust as required.	.036		√	√	√	√
9 Perform 30 minute generator test run; check for proper operation.	.650		√	√	√	√
10 Check fuel level; add as required.	.046		√	√	√	√
11 Wipe dust and dirt from engine and generator.	.056		√	√	√	√
12 Clean area around generator.	.066		√	√	√	√
13 Fill out maintenance checklist and report deficiencies.	.022		√	√	√	√
Total labor-hours/period			1.062	1.062	1.062	1.485
Total labor-hours/year			8.496	2.124	1.062	1.485

		Cost Each					
			2014 Bare Costs			Total	Total
Description	Labor-hrs.	Material	Labor	Equip.	Total	In-House	w/O&P
1900 Emergency diesel generator, up to 15 KVA, annually	1.485	97	79		176	206	244
1950 Annualized	13.167	125	700		825	1,000	1,225

PM Components	Labor-hrs.	W	M	Q	S	A
PM System D5095 210 2950						
Emergency diesel generator, over 15 KVA						
1 Check with the operating or area personnel for any obvious deficiencies.	.044		√	√	√	√
2 Check engine oil level; add as required.	.010		√	√	√	√
3 Change engine oil and oil filter.	.511					√
4 Check battery charge and electrolyte specific gravity, add water as required; check terminals for corrosion, clean as required.	.241		√	√	√	√
5 Check belts for wear and proper tension; adjust as necessary.	.012		√	√	√	√
6 Check that crank case heater is operating.	.038		√	√	√	√
7 Check engine air filter; change as required.	.042					√
8 Check wiring, connections, switches, etc.; adjust as required.	.036		√	√	√	√
9 Check spark plug or injector nozzle condition; service or replace as required.	.281					√
10 Perform 30 minute generator test run; check for proper operation.	.650		√	√	√	√
11 Check fuel level with gauge pole, add as required.	.046		√	√	√	√
12 Wipe dust and dirt from engine and generator.	.109		√	√	√	√
13 Clean area around generator.	.066		√	√	√	√
14 Fill out maintenance checklist and report deficiencies.	.025		√	√	√	√
Total labor-hours/period			1.277	1.277	1.277	2.111
Total labor-hours/year			10.216	2.554	1.277	2.111

		Cost Each					
			2014 Bare Costs			Total	Total
Description	Labor-hrs.	Material	Labor	Equip.	Total	In-House	w/O&P
2900 Emergency diesel generator, over 15 KVA, annually	2.111	109	113		222	259	310
2950 Annualized	16.158	137	860		997	1,225	1,500

D50 ELECTRICAL D5095 210 Emergency Generator

PM Components	Labor-hrs.	W	M	Q	S	A
PM System D5095 210 3950						
Emergency diesel generator, turbine						
1 Check with the operating or area personnel for any obvious deficiencies.	.044		√	√	√	√
2 Check turbine oil level; add oil as required.	.036		√	√	√	√
3 Change turbine oil and oil filter; check transmission oil level.	.569					√
4 Check that the crankcase heater is operating properly.	.022		√	√	√	√
5 Replace turbine air filter.	.224					√
6 Check wiring, connections, switches, etc.; adjust as required.	.060					√
7 Check starter for proper operation; lubricate as necessary.	.047		√	√	√	√
8 Check fuel nozzles, fuel regulator and ignition device condition; service or replace as required.	.813					√
9 Perform 30 minute generator test run; check for proper operation.	.650		√	√	√	√
10 Check and record transmission oil pressure and temperature, and natural gas pressure: oil press:____ oil temp:____ gas press:____.	.021					√
11 Record running time: beginning ____ hours; ending ____ hours.	.010					√
12 Check that the charger is operating properly.	.127		√	√	√	√
13 Check for any signs of corrosion on battery terminals or wires.	.117		√	√	√	√
14 Check the electrolyte level in the batteries; add if required.	.049		√	√	√	√
15 Check the specific gravity of the electrolyte in a 10% sample of the batteries.	.069					√
16 Check 25% of terminal-to-cell connection resistance; rehabilitate connections as required; add anti-corrosion grease to battery terminals and connections.	.038					√
17 Measure and record individual cell and string float voltages.	.140					√
18 Clean area around generator.	.066		√	√	√	√
19 Fill out maintenance checklist and report deficiencies.	.022		√	√	√	√
Total labor-hours/period			1.180	1.180	1.180	3.124
Total labor-hours/year			9.440	2.360	1.180	3.124

			Cost Each				
			2014 Bare Costs			Total	Total
Description		Labor-hrs.	Material	Labor	Equip.	Total	In-House	w/O&P
3900	Emergency diesel generator, turbine, annually	3.124	165	167		332	390	465
3950	Annualized	16.104	215	860		1,075	1,300	1,600

D50 ELECTRICAL D5095 220 | Power Stabilizer

PM Components	Labor-hrs.	W	M	Q	S	A
PM System D5095 220 1950						
Power stabilizer						
1 Remove the covers or open the doors.	.218					✓
2 Check for signs of moisture or overheating.	.022					✓
3 Check fans and other auxiliary devices for proper operation.	.027					✓
4 Check for voltage creeping over transformer insulated surfaces, such as evidenced by tracking or carbonization.	.018					✓
5 Check condition of the ground system.	.170					✓
6 Check wiring, connections, relays, etc; tighten and adjust as necessary.	.060					✓
7 Replace the covers or close the doors.	.022					✓
8 Clean the surrounding area.	.066					✓
9 Fill out maintenance checklist and report deficiencies.	.022					✓
Total labor-hours/period						.625
Total labor-hours/year						.625

			Cost Each				
			2014 Bare Costs			Total	Total
Description		Labor-hrs.	Material	Labor	Equip.	Total	In-House	w/O&P
1900	Power stablizer, annually	.625		33.50		33.50	41.50	51.50
1950	Annualized	.625		33.50		33.50	41.50	51.50

D50 ELECTRICAL — D5095 230 Uninterruptible Power System

PM Components	Labor-hrs.	W	M	Q	S	A
PM System D5095 230 1950						
Uninterruptible power system, up to 200 KVA						
1 Check with operating or area personnel for any obvious deficiencies.	.044		√	√	√	√
2 Check electrolyte level of batteries; add water as required; check terminals for corrosion, clean as required.	.878		√	√	√	√
3 Check 25% of the batteries for charge and electrolyte specific gravity.	.777			√	√	√
4 Check batteries for cracks or leaks.	.059		√	√	√	√
5 Check 25% of the terminal-to-cell connection resistances; rehabilitate connections as required; add anti-corrosion grease to battery terminals and connections.	.143			√	√	√
6 Measure and record individual cell and string float voltages.	.618				√	√
7 Check integrity of battery rack.	.176				√	√
8 Check battery room temperature and ventilation systems.	.010		√	√	√	√
9 Replace air filters on UPS modules.	.036		√	√	√	√
10 Check output voltage and amperages, from control panel.	.007		√	√	√	√
11 Check UPS room temperature and ventilation system.	.062		√	√	√	√
12 Notify personnel and test UPS fault alarm system.	.112		√	√	√	√
13 Clean around batteries and UPS modules.	.066		√	√	√	√
14 Fill out maintenance checklist and report deficiencies.	.022		√	√	√	√
Total labor-hours/period			1.360	3.010	3.010	3.010
Total labor-hours/year			10.880	6.020	3.010	3.010

			Cost Each				
			2014 Bare Costs			Total	Total
Description		Labor-hrs.	Material	Labor	Equip.	Total	In-House	w/O&P
1900 Uninter. power system, up to 200 KVA, annually	3.010	202	161		363	420	500
1950 Annualized	22.920	232	1,225		1,457	1,775	2,200

PM Components	Labor-hrs.	W	M	Q	S	A
PM System D5095 230 2950						
Uninterrupted power system, 200 KVA to 800 KVA						
1 Check with operating or area personnel for any obvious deficiencies.	.044		√	√	√	√
2 Check electrolyte level of batteries; add water as required; check terminals for corrosion, clean as required.	3.510		√	√	√	√
3 Check 25% of the batteries for charge and electrolyte specific gravity.	3.107			√	√	√
4 Check batteries for cracks or leaks.	.585		√	√	√	√
5 Check 25% of the terminal-to-cell connection resistances; rehabilitate connections as required; add anti-corrosion grease to battery terminals and connections.	.234			√	√	√
6 Measure and record individual cell and string float voltages.	.936			√	√	√
7 Check integrity of battery rack.	.702		√	√	√	√
8 Check battery room temperature and ventilation systems.	.010		√	√	√	√
9 Replace air filters on UPS modules.	.036		√	√	√	√
10 Check output voltage and amperages, from control panel.	.007		√	√	√	√
11 Check UPS room temperature and ventilation system.	.062		√	√	√	√
12 Notify personnel and test UPS fault alarm system.	.112		√	√	√	√
13 Clean around batteries and UPS modules.	.066		√	√	√	√
14 Fill out maintenance checklist and report deficiencies.	.022		√	√	√	√
Total labor-hours/period			5.044	9.433	9.433	9.433
Total labor-hours/year			40.352	18.866	9.433	9.433

			Cost Each				
			2014 Bare Costs			Total	Total
Description		Labor-hrs.	Material	Labor	Equip.	Total	In-House	w/O&P
2900 UPS, 200 KVA to 800 KVA, annually	9.433	246	505		751	895	1,100
2950 Annualized	78.087	277	4,175		4,452	5,450	6,800

D50 ELECTRICAL — D5095 240 Wet Battery System and Charger

PM Components	Labor-hrs.	W	M	Q	S	A
PM System D5095 240 1950						
Battery system and charger						
1 Check that the charger is operating properly.	.051		√	√	√	√
2 Check for any signs of corrosion on the battery terminals or wires.	.117		√	√	√	√
3 Check the electrolyte level in the batteries; add if required.	.062		√	√	√	√
4 Check the specific gravity of the electrolyte in a 10% sample of the batteries.	.069		√	√	√	√
5 Check 25% of terminal-to-cell connection resistance; rehabilitate connections as required; add anti-corrosion grease to battery terminals and connections.	.741			√	√	√
6 Measure and record individual cell and string float voltages.	.281			√	√	√
7 Clean area around batteries.	.066		√	√	√	√
8 Fill out maintenance checklist and report deficiencies.	.022		√	√	√	√
Total labor-hours/period			.387	1.409	1.409	1.409
Total labor-hours/year			3.096	2.818	1.409	1.409

	Description	Labor-hrs.	Cost Each					
			2014 Bare Costs				Total In-House	Total w/O&P
			Material	Labor	Equip.	Total		
1900	Battery system and charger, annually	1.409	28	75		103	124	151
1950	Annualized	8.732	34	465		499	620	765

D50 ELECTRICAL — D5095 250 Lighting System, Emergency

PM Components	Labor-hrs.	W	M	Q	S	A
PM System D5095 250 1950						
Light, emergency, hardwired system						
1 Test for proper operation, replace lamps as required.	.013				✓	✓
2 Clean exterior of cabinet and light heads.	.025				✓	✓
3 Check wiring for obvious defects.	.026				✓	✓
4 Adjust lamp heads for maximum illumination of area.	.039				✓	✓
5 Fill out maintenance checklist and report deficiencies.	.022				✓	✓
Total labor-hours/period					.125	.125
Total labor-hours/year					.125	.125

			Cost Each				
			2014 Bare Costs			Total	Total
Description		Labor-hrs.	Material	Labor	Equip.	Total	In-House	w/O&P
1900 Light, emergency, hardwired system, annually	.125		6.65		6.65	8.25	10.30
1950 Annualized	.250		13.35		13.35	16.50	20.50

PM Components	Labor-hrs.	W	M	Q	S	A
PM System D5095 250 2950						
Light, emergency, dry cell						
1 Test for proper operation, replace lamps as required.	.013				✓	✓
2 Clean exterior of emergency light cabinet and lamp heads.	.025				✓	✓
3 Check condition of batteries and change as required.	.012				✓	✓
4 Clean interior of cabinet, top of battery, and battery terminal.	.009				✓	✓
5 Inspect cabinet, relay, relay contacts, pilot light, wiring, and gen. cond.	.065				✓	✓
6 Adjust lamp heads for maximum illumination of area.	.039				✓	✓
7 Fill out maintenance checklist and report deficiencies.	.022				✓	✓
Total labor-hours/period					.185	.185
Total labor-hours/year					.185	.185

			Cost Each				
			2014 Bare Costs			Total	Total
Description	Labor-hrs.	Material	Labor	Equip.	Total	In-House	w/O&P
2900 Light, emergency, dry cell, annually	.185	34	9.85		43.85	49.50	58
2950 Annualized	.370	41	19.75		60.75	69.50	82

PM Components	Labor-hrs.	W	M	Q	S	A
PM System D5095 250 3950						
Light, emergency, wet cell						
1 Test for proper operation; correct deficiencies as required.	.013				✓	✓
2 Clean exterior of emergency light cabinet and lamp heads.	.025				✓	✓
3 Clean interior of cabinet, top of battery, and battery terminal.	.009				✓	✓
4 Check battery charge and electrolyte specific gravity; add water if nec.	.012				✓	✓
5 Inspect cabinet, relay, relay contacts, pilot light, wiring, and gen. cond.	.065				✓	✓
6 Apply anti-corrosion coating to battery terminals.	.010				✓	✓
7 Adjust lamp heads for maximum illumination of area.	.039				✓	✓
8 Fill out maintenance checklist and report deficiencies.	.022				✓	✓
Total labor-hours/period					.195	.195
Total labor-hours/year					.195	.195

			Cost Each				
			2014 Bare Costs			Total	Total
Description	Labor-hrs.	Material	Labor	Equip.	Total	In-House	w/O&P
3900 Light, emergency, wet cell, annually	.195	48	10.40		58.40	65.50	76
3950 Annualized	.390	62	21		83	94	110

E10 EQUIPMENT | E1015 810 Personal Computers

PM Components	Labor-hrs.	W	M	Q	S	A
PM System E1015 810 1950						
Personal Computers						
1 Check with operating personnel for any deficiencies.	.065				√	√
2 Check all cables for fraying, cracking, or exposed wires; ensure connections are secure.	.065				√	√
3 Remove metal casing and vacuum interior, power supply vents, keyboard, monitor, and surrounding work area.	.130				√	√
4 Check for secure connections of all interior wires and cables. Secure all peripheral cards.	.065				√	√
5 Replace PC cover and clean case, keyboard, monitor screen and housing with an anti-static spray cleaner. Apply cleaner to towel and not the components.	.130				√	√
6 Disassemble the mouse; clean the tracking ball, interior rollers and housing. Blow out the interior of the mouse with clean compressed air.	.065				√	√
7 Turn on PC and check for unusual noises.	.026				√	√
8 Clean floppy drive heads with cleaning diskette and solution.	.039				√	√
9 Defragment the hard drive using a defragmentation utility.	.234				√	√
10 Run diagnostics software to test all the PC's components.	.234				√	√
11 Fill out maintenance checklist and report deficiencies.	.029				√	√
Total labor-hours/period					1.082	1.082
Total labor-hours/year					1.082	1.082

			Cost Each				
			2014 Bare Costs			Total	Total
Description	Labor-hrs.	Material	Labor	Equip.	Total	In-House	w/O&P
1900 Personnel Computers, annually	1.082		57.50		57.50	71.50	89.50
1950 Annualized	2.164		115		115	143	179

E10 EQUIPMENT | E1035 100 | Hydraulic Lift, Automotive

PM Components	Labor-hrs.	W	M	Q	S	A
PM System E1035 100 1950						
Hydraulic lift						
1 Check for proper operation of pump.	.035				✓	✓
2 Check for leaks on suction and discharge piping, seals, packing glands, etc.; make minor adjustments as required.	.077				✓	✓
3 Check pump and motor operation for excessive vibration, noise and overheating.	.022				✓	✓
4 Check alignment of pump and motor; adjust as necessary.	.260				✓	✓
5 Lubricate pump and motor.	.099				✓	✓
6 Inspect hydraulic lift post(s) for wear or leaks.	.065				✓	✓
7 Inspect, clean and tighten valves.	.034				✓	✓
8 Inspect and clean motor contactors.	.026				✓	✓
9 Inspect and test control relays and check wiring terminals.	.034				✓	✓
10 Clean pump unit and surrounding area.	.099				✓	✓
11 Fill out maintenance checklist and report deficiencies.	.022				✓	✓
Total labor-hours/period					.773	.773
Total labor-hours/year					.773	.773

		Cost Each					
		2014 Bare Costs				Total	Total
Description	Labor-hrs.	Material	Labor	Equip.	Total	In-House	w/O&P
1900 Hydraulic lift, annually	.773	50	44.50		94.50	111	132
1950 Annualized	1.546	67	89		156	185	223

E10 EQUIPMENT | E1035 310 | Hydraulic Lift, Loading Dock

PM Components	Labor-hrs.	W	M	Q	S	A
PM System E1035 310 1950						
Hydraulic lift, loading dock						
1 Check for proper operation of pump.	.022				✓	✓
2 Check for leaks on suction and discharge piping, seals, packing glands, etc.; make minor adjustments as required.	.077				✓	✓
3 Check pump and motor operation for excessive vibration, noise and overheating.	.022				✓	✓
4 Check alignment of pump and motor; adjust as necessary.	.100				✓	✓
5 Lubricate pump and motor.	.099				✓	✓
6 Inspect hydraulic lift post(s) for wear or leaks.	.051				✓	✓
7 Inspect, clean and tighten valves.	.072				✓	✓
8 Inspect and clean motor contactors.	.148				✓	✓
9 Inspect and test control relays and check wiring terminals.	.137				✓	✓
10 Clean pump unit and surrounding area.	.066				✓	✓
11 Fill out maintenance checklist and report deficiencies.	.022				✓	✓
Total labor-hours/period					.816	.816
Total labor-hours/year					.816	.816

		Cost Each					
		2014 Bare Costs				Total	Total
Description	Labor-hrs.	Material	Labor	Equip.	Total	In-House	w/O&P
1900 Hydraulic lift, loading dock, annually	.816	8.70	39.50		48.20	59	72.50
1950 Annualized	1.632	17.35	78.50		95.85	118	145

E10 EQUIPMENT | E1095 122 | Crane, Electric Bridge

PM Components	Labor-hrs.	W	M	Q	S	A
PM System E1095 122 1950						
Crane, electric bridge, up to 5 ton						
1 Check with operating or area personnel for deficiencies.	.035					✓
2 Inspect crane for damaged structural members, alignment, worn pins, column fasteners, stops and capacity markings.	.235					✓
3 Inspect and lubricate power cable tension reel and inspect condition of cable.	.491					✓
4 Check hoist operation - travel function, up, down, speed stepping, upper and lower travel limit switches, etc.	.072					✓
5 Inspect, clean and lubricate trolley assembly and reduction gear bearings.	.112					✓
6 Inspect and clean hook assembly and chain or wire rope sheaves.	.452					✓
7 Inspect, clean and lubricate hoist motor.	.094					✓
8 Inspect hoist brake mechanism.	.445					✓
9 Lubricate rails.	.114					✓
10 Weight capacity test hoisting equipment at 125% of rated capacity in accordance with 29 CFR 1910.179.	.780					✓
11 Clean work area.	.066					✓
12 Fill out maintenance report and report deficiencies.	.022					✓
Total labor-hours/period						2.918
Total labor-hours/year						2.918

			Cost Each				
			2014 Bare Costs			Total	Total
Description	Labor-hrs.	Material	Labor	Equip.	Total	In-House	w/O&P
1900 Crane, electric bridge, up to 5 ton, annually	2.918	45.50	140		185.50	227	278
1950 Annualized	2.918	45.50	140		185.50	227	278

PM Components	Labor-hrs.	W	M	Q	S	A
PM System E1095 122 2950						
Crane, electric bridge 5 to 15 ton						
1 Check with operating or area personnel for deficiencies.	.035					
2 Inspect crane for damaged structural members, alignment, worn pins, column fasteners, stops and capacity markings.	.235			✓	✓	✓
3 Inspect and lubricate power cable tension reel and inspect condition of cable.	.491			✓	✓	✓
4 Check operation of upper and lower travel limit switches.	.072			✓	✓	✓
5 Inspect brake mechanism.	.250					✓
6 Inspect, clean and lubricate trolley assembly and reduction gear bearings.	.099			✓	✓	✓
7 Drain and replace reduction gear housing oil.	.170					✓
8 Inspect and clean hook assembly and chain or wire rope sheaves.	.099					✓
9 Inspect, clean and lubricate hoist motor.	.094			✓	✓	✓
10 Check pendent control for hoist operation - travel function, up, down, speed stepping, etc.	.088			✓	✓	✓
11 Lubricate rails.	.325			✓	✓	✓
12 Weight capacity test hoisting equipment at 125% of rated capacity in accordance with 29 CFR 1910.179.	1.040					✓
13 Clean work area.	.066			✓	✓	✓
14 Fill out maintenance report and report deficiencies.	.022			✓	✓	✓
Total labor-hours/period				1.626	1.626	3.086
Total labor-hours/year				3.252	1.626	3.086

			Cost Each				
			2014 Bare Costs			Total	Total
Description	Labor-hrs.	Material	Labor	Equip.	Total	In-House	w/O&P
2900 Crane, electric bridge, 5 to 15 ton, annually	3.086	22.50	148		170.50	212	261
2950 Annualized	7.964	68	385		453	560	685

E10 EQUIPMENT — E1095 122 Crane, Electric Bridge

PM Components	Labor-hrs.	W	M	Q	S	A
PM System E1095 122 3950						
Crane, electric bridge, over 15 tons						
1 Check with operating or area personnel for deficiencies.	.035			✓	✓	✓
2 Inspect crane for damaged structural members, alignment, worn pins, column fasteners, stops and capacity markings.	.235			✓	✓	✓
3 Inspect and lubricate power cable tension reel and inspect condition of cable.	.491			✓	✓	✓
4 Check operation of upper and lower travel limit switches.	.072			✓	✓	✓
5 Inspect brake mechanism.	.250					✓
6 Inspect, clean and lubricate trolley assembly and reduction gear bearings.	.099			✓	✓	✓
7 Drain and replace reduction gear housing oil.	.170					✓
8 Inspect and clean hook assembly and chain or wire rope sheaves.	.099			✓	✓	✓
9 Inspect, clean and lubricate hoist motor.	.094			✓	✓	✓
10 Check pendent control for hoist operation - travel function, up, down, speed stepping, etc.	.088			✓	✓	✓
11 Lubricate rails.	.325			✓	✓	✓
12 Inspect control cabinet.	.238					✓
13 Weight capacity test hoisting equipment, at 125% of rated capacity in accordance with 29 CFR 1910.179.	1.170					✓
14 Clean work area.	.066			✓	✓	✓
15 Fill out maintenance report and report deficiencies.	.022			✓	✓	✓
Total labor-hours/period				1.626	1.626	3.454
Total labor-hours/year				3.252	1.626	3.454

		Cost Each					
		2014 Bare Costs				Total	Total
Description	Labor-hrs.	Material	Labor	Equip.	Total	In-House	w/O&P
3900 Crane, electric bridge, over 15 tons, annually	3.454	34	166		200	248	305
3950 Annualized	8.332	90.50	400		490.50	605	745

E10 EQUIPMENT — E1095 123 | Crane, Manual Bridge

PM Components	Labor-hrs.	W	M	Q	S	A
PM System E1095 123 1950						
Crane, manual bridge, up to 5 tons						
1 Check with operating or area personnel for deficiencies.	.035					✓
2 Operate crane and hoist, inspect underhung hoist monorail for damaged structural members, braces, stops and capacity markings.	.075					✓
3 Inspect and lubricate trolley wheels and bearings.	.096					✓
4 Inspect, clean and lubricate hooks and chain sheaves.	.099					✓
5 Check operation of crane and hoist brakes.	.016					✓
6 Make minor adjustments as required for operational integrity; tighten bolts, etc.	.135					✓
7 Weight capacity test hoisting equipment at 125% of rated capacity in accordance with 29 CFR 1910.179.	.390					✓
8 Clean work area.	.066					✓
9 Fill out maintenance report and report deficiencies.	.022					✓
Total labor-hours/period						.934
Total labor-hours/year						.934

			Cost Each					
			2014 Bare Costs				Total	Total
	Description	Labor-hrs.	Material	Labor	Equip.	Total	In-House	w/O&P
1900	Crane, manual bridge, up to 5 ton, annually	.934	22.50	45		67.50	82	98.50
1950	Annualized	.934	22.50	45		67.50	82	98.50

PM Components	Labor-hrs.	W	M	Q	S	A
PM System E1095 123 2950						
Crane, manual bridge, 5 to 15 tons						
1 Check with operating or area personnel for deficiencies.	.035					✓
2 Operate crane and hoist, inspect underhung hoist monorail for damaged structural members, braces, stops and capacity markings.	.130					✓
3 Inspect and lubricate trolley wheels and bearings.	.096					✓
4 Inspect, clean and lubricate hooks and chain sheaves.	.099					✓
5 Check operation of crane and hoist brakes.	.016					✓
6 Make minor adjustments as required for operational integrity; tighten bolts, etc.	.135					✓
7 Weight capacity test hoisting equipment at 125% of rated capacity in accordance with 29 CFR 1910.179.	.772					✓
8 Clean work area.	.066					✓
9 Fill out maintenance report and report deficiencies.	.022					✓
Total labor-hours/period						1.371
Total labor-hours/year						1.371

			Cost Each					
			2014 Bare Costs				Total	Total
	Description	Labor-hrs.	Material	Labor	Equip.	Total	In-House	w/O&P
2900	Crane, manual bridge, 5 to 15 ton, annually	1.371	45.50	66		111.50	133	161
2950	Annualized	1.371	45.50	66		111.50	133	161

E10 EQUIPMENT		E1095 123	Crane, Manual Bridge					
	PM Components	Labor-hrs.	W	M	Q	S	A

PM System E1095 123 3950
Crane, manual bridge, over 15 tons

	PM Components	Labor-hrs.	W	M	Q	S	A
1	Check with operating or area personnel for deficiencies.	.035					✓
2	Operate crane and hoist, inspect for damaged structural members, alignment, worn pins, fasteners, stops and capacity markings.	.235					✓
3	Inspect and lubricate trolley wheels and bearings.	.096					✓
4	Inspect, clean and lubricate hooks and chain sheaves.	.099					✓
5	Check operation of crane and hoist brakes.	.016					✓
6	Make minor adjustments as required for operational integrity; tighten bolts, etc.	.135					✓
7	Weight capacity test hoisting equipment at 125% of rated capacity in accordance with 29 CFR 1910.179.	1.162					✓
8	Clean work area.	.066					✓
9	Fill out maintenance report and report deficiencies.	.022					✓
	Total labor-hours/period						1.866
	Total labor-hours/year						1.866

			Cost Each				
			2014 Bare Costs			Total	Total
Description		Labor-hrs.	Material	Labor	Equip.	Total	In-House	w/O&P
3900	Crane, manual bridge, over 15 tons, annually	1.866	68	90		158	188	226
3950	Annualized	1.866	68	90		158	188	226

E10 EQUIPMENT		E1095 125	Hoist, Pneumatic					
	PM Components	Labor-hrs.	W	M	Q	S	A

PM System E1095 125 1950
Hoist, pneumatic

	PM Components	Labor-hrs.	W	M	Q	S	A
1	Check with operating or area personnel for deficiencies.	.035					✓
2	Inspect for damaged structural members, alignment, worn pins, column fasteners, stops and capacity markings.	.235					✓
3	Inspect and lubricate cable tension reel and inspect condition of cable.	.452					✓
4	Check pendent control for hoist operation - travel function, up, down, speed stepping, upper and lower travel limit switches, etc.	.088					✓
5	Inspect, clean and lubricate trolley assembly.	.099					✓
6	Drain and replace reduction gear housing oil.	.170					✓
7	Inspect brake mechanism.	.016					✓
8	Inspect and clean hook assembly and chain or wire rope sheaves.	.099					✓
9	Inspect, clean and lubricate hoist motor.	.029					✓
10	Clean work area.	.066					✓
11	Fill out maintenance report and report deficiencies.	.022					✓
	Total labor-hours/period						1.311
	Total labor-hours/year						1.311

			Cost Each				
			2014 Bare Costs			Total	Total
Description		Labor-hrs.	Material	Labor	Equip.	Total	In-House	w/O&P
1900	Hoist, pneumatic, annually	1.311	90.50	63		153.50	179	212
1950	Annualized	1.311	90.50	63		153.50	179	212

E10 EQUIPMENT E1095 126 Hoist/Winch, Chain/Cable

PM Components	Labor-hrs.	W	M	Q	S	A
PM System E1095 126 1950						
Hoist/winch, chain fall/cable, electric						
1 Check with operating or area personnel for deficiencies.	.035					✓
2 Inspect for damaged structural members, alignment, worn pins, column fasteners, stops and capacity markings.	.235					✓
3 Inspect and lubricate power cable tension reel; inspect condition of cable.	.491					✓
4 Check hoist operation - travel function, up, down, speed stepping, upper and lower travel limit switches, etc.	.088					✓
5 Inspect, clean and lubricate trolley assembly and reduction gear bearings.	.099					✓
6 Drain and replace reduction gear housing oil.	.170					✓
7 Inspect brake mechanism.	.250					✓
8 Inspect and clean hook assembly and chain or wire rope sheaves.	.099					✓
9 Inspect, clean and lubricate hoist motor.	.094					✓
10 Clean work area.	.066					✓
11 Fill out maintenance report and report deficiencies.	.022					✓
Total labor-hours/period						1.649
Total labor-hours/year						1.649

			Cost Each					
			2014 Bare Costs				Total	Total
Description		Labor-hrs.	Material	Labor	Equip.	Total	In-House	w/O&P
1900	Hoist / winch, chain / cable, electric, annually	1.649	22.50	79.50		102	125	153
1950	Annualized	1.649	22.50	79.50		102	125	153

E10 EQUIPMENT E1095 302 Beverage Dispensing Unit

PM Components	Labor-hrs.	W	M	Q	S	A
PM System E1095 302 1950						
Beverage dispensing unit						
1 Check with operating or area personnel for deficiencies.	.044			✓	✓	✓
2 Remove access panels and clean coils, fans, fins, and other areas; vacuum, blow down with compressed air or CO_2, and/or wipe off.	.517			✓	✓	✓
3 Lubricate motor bearings.	.008			✓	✓	✓
4 Check electrical wiring and connections; tighten loose connections.	.120			✓	✓	✓
5 Visually check for refrigerant, water, and other liquid leaks.	.153			✓	✓	✓
6 Check door gaskets for proper sealing; adjust door catch as required.	.033			✓	✓	✓
7 Inspect dispensing valves; adjust as necessary.	.108			✓	✓	✓
8 Perform operational check of unit; make adjustments as necessary.	.040			✓	✓	✓
9 Clean area around equipment.	.066			✓	✓	✓
10 Fill out maintenance checklist and report deficiencies.	.022			✓	✓	✓
Total labor-hours/period				1.111	1.111	1.111
Total labor-hours/year				2.222	1.111	1.111

			Cost Each					
			2014 Bare Costs				Total	Total
Description		Labor-hrs.	Material	Labor	Equip.	Total	In-House	w/O&P
1900	Beverage dispensing unit, annually	1.111	8.75	52.50		61.25	79	96.50
1950	Annualized	4.444	17.50	210		227.50	297	360

| E10 EQUIPMENT | E1095 304 | Braising Pan |

PM Components	Labor-hrs.	W	M	Q	S	A
PM System E1095 304 1950						
Braising pan, tilting, gas/electric						
1 Check with operating or area personnel for deficiencies.	.044			√	√	√
2 Check electric transformer, heating elements, insulators, connections and wiring; tighten connections and repair as required.	.239			√	√	√
3 Check surface temperature with meter for hot spots.	.073			√	√	√
4 Check piping and valves on gas operated units for leaks.	.077			√	√	√
5 Check pilot/igniter, burner and combustion chamber for proper operation; adjust as required.	.079			√	√	√
6 Check temperature control system for proper operation; adjust as required.	.458			√	√	√
7 Lubricate tilting gear mechanism and trunnion bearings.	.047			√	√	√
8 Check nuts, bolts and screws for tightness; tighten as required.	.007			√	√	√
9 Check complete operation of unit.	.055			√	√	√
10 Fill out maintenance checklist and report deficiencies.	.022			√	√	√
Total labor-hours/period				1.101	1.101	1.101
Total labor-hours/year				2.202	1.101	1.101

			Cost Each				
			2014 Bare Costs			Total	Total
Description	Labor-hrs.	Material	Labor	Equip.	Total	In-House	w/O&P
1900 Braising pan, tilting, gas / electric, annually	1.101	164	52		216	249	290
1950 Annualized	4.404	193	208		401	485	580

| E10 EQUIPMENT | E1095 306 | Bread Slicer |

PM Components	Labor-hrs.	W	M	Q	S	A
PM System E1095 306 1950						
Bread slicer, electric						
1 Check with operating or area personnel for deficiencies.	.044			√	√	√
2 Check electric motor, controls, wiring connections, insulation and contacts for damage and/or proper operation; adjust as required.	.338			√	√	√
3 Check V-belt for wear, tension and alignment; change/adjust belt as necessary.	.029			√	√	√
4 Lubricate bushings and check oil level in gear drive mechanism; add oil if required.	.065			√	√	√
5 Clean machine of dust, grease and food particles.	.074			√	√	√
6 Check motor and motor bearings for overheating.	.039			√	√	√
7 Check for loose, missing or damaged nuts, bolts, or screws; tighten or replace as necessary.	.005			√	√	√
8 Fill out maintenance checklist and report deficiencies.	.022			√	√	√
Total labor-hours/period				.616	.616	.616
Total labor-hours/year				1.232	.616	.616

			Cost Each				
			2014 Bare Costs			Total	Total
Description	Labor-hrs.	Material	Labor	Equip.	Total	In-House	w/O&P
1900 Bread slicer, electric, annually	.616	67.50	29		96.50	114	132
1950 Annualized	2.464	96.50	117		213.50	260	310

E10 EQUIPMENT E1095 310 | Broiler

PM Components	Labor-hrs.	W	M	Q	S	A
PM System E1095 310 1950						
Broiler, conveyor type, gas						
1 Check with operating or area personnel for deficiencies.	.044			√	√	√
2 Check broiler pilot/igniters and jets for uniform flame; adjust as required.	.166			√	√	√
3 Check air filter; change as required.	.059			√	√	√
4 Check operation of chains and counterweights.	.012			√	√	√
5 Check piping and valves for leaks.	.077			√	√	√
6 Check electrical connections and wiring for defects; tighten loose connections.	.120			√	√	√
7 Check thermocouple and thermostat; calibrate thermostat as required.	.646			√	√	√
8 Tighten and replace any loose nuts, bolts, or screws.	.005			√	√	√
9 Check operation of unit.	.055			√	√	√
10 Fill out maintenance checklist and report deficiencies.	.022			√	√	√
Total labor-hours/period				1.206	1.206	1.206
Total labor-hours/year				2.412	1.206	1.206

			Cost Each				
			2014 Bare Costs			Total	Total
Description	Labor-hrs.	Material	Labor	Equip.	Total	In-House	w/O&P
1900 Broiler, conveyor type, gas, annually	1.206	173	57		230	266	310
1950 Annualized	4.824	173	228		401	490	585

PM Components	Labor-hrs.	W	M	Q	S	A
PM System E1095 310 2950						
Broiler, hot dog, electric						
1 Check with operating or area personnel for deficiencies.	.044			√	√	√
2 Check electric motor, controls, contacts, wiring connections for damage and proper operation; adjust as required.	.152			√	√	√
3 Lubricate bushings.	.007			√	√	√
4 Inspect lines and unit for mechanical defects; adjust or repair as required.	.021			√	√	√
5 Clean machine of dust and grease.	.109			√	√	√
6 Check unit for proper operation.	.082			√	√	√
7 Fill out maintenance checklist and report deficiencies.	.022			√	√	√
Total labor-hours/period				.437	.437	.437
Total labor-hours/year				.874	.437	.437

			Cost Each				
			2014 Bare Costs			Total	Total
Description	Labor-hrs.	Material	Labor	Equip.	Total	In-House	w/O&P
2900 Broiler, hot dog type, electric, annually	.437	48	20.50		68.50	80	93.50
2950 Annualized	1.748	77	82.50		159.50	194	231

E10 EQUIPMENT E1095 310 Broiler

PM Components	Labor-hrs.	W	M	Q	S	A
PM System E1095 310 3950						
Broiler/oven (salamander), gas/electric						
1 Check with operating or area personnel for deficiencies.	.044			✓	✓	✓
2 Check broiler pilot and oven for uniform flame; adjust as required.	.166			✓	✓	✓
3 Check piping and valves for leaks.	.077			✓	✓	✓
4 Check doors, shelves for warping, alignment, and seals; adjust as required.	.148			✓	✓	✓
5 Check electrical connections and wiring for defects; tighten loose connections.	.241			✓	✓	✓
6 Tighten and replace any loose nuts, bolts and/or screws.	.012			✓	✓	✓
7 Check operation of broiler arm, adjust as required.	.007			✓	✓	✓
8 Check thermocouple and thermostat; calibrate thermostat as required.	.569			✓	✓	✓
9 Fill out maintenance checklist and report deficiencies.	.022			✓	✓	✓
Total labor-hours/period				1.286	1.286	1.286
Total labor-hours/year				2.572	1.286	1.286

				Cost Each				
				2014 Bare Costs			Total	Total
Description		Labor-hrs.	Material	Labor	Equip.	Total	In-House	w/O&P
3900	Broiler / oven, (salamander), gas / electric, annually	1.286	154	61		215	250	292
3950	Annualized	5.144	154	243		397	490	590

E10 EQUIPMENT E1095 312 Chopper, Electric

PM Components	Labor-hrs.	W	M	Q	S	A
PM System E1095 312 1950						
Chopper, electric						
1 Check with operating or area personnel for deficiencies.	.044			✓	✓	✓
2 Check cutting surfaces, clutch assembly and guards for damage and proper alignment; adjust as required.	.065			✓	✓	✓
3 Check nuts, bolts, and screws for tightness; tighten or replace as required.	.005			✓	✓	✓
4 Check switch and controls for proper operation; adjust as required.	.010			✓	✓	✓
5 Clean machine of dust, grease and food particles.	.014			✓	✓	✓
6 Check operation of motor and check motor bearings for overheating; lubricate as required.	.069			✓	✓	✓
7 Check belt tension; replace belt and/or adjust if required.	.315			✓	✓	✓
8 Fill out maintenance checklist and report deficiencies.	.022			✓	✓	✓
Total labor-hours/period				.544	.544	.544
Total labor-hours/year				1.088	.544	.544

				Cost Each				
				2014 Bare Costs			Total	Total
Description		Labor-hrs.	Material	Labor	Equip.	Total	In-House	w/O&P
1900	Chopper, electric, annually	.544	83	25.50		108.50	125	146
1950	Annualized	2.176	102	103		205	248	296

E10 EQUIPMENT   E1095 314 | Coffee Maker/Urn

PM Components	Labor-hrs.	W	M	Q	S	A
PM System E1095 314 1950						
Coffee maker/urn, gas/electric/steam						
1 Check thermostat, switch, and temperature gauge; calibrate if required.	.465			√	√	√
2 Check pilots and flame on gas operated units; adjust as required.	.038			√	√	√
3 Check working pressure on steam operated unit.	.005			√	√	√
4 Check electrical connections and wiring for defects; tighten connections.	.120			√	√	√
5 Check for clogged or defective steam trap.	.100			√	√	√
6 Inspect and clean steam strainer.	.207			√	√	√
7 Examine equipment, valves, and piping for leaks.	.077			√	√	√
8 Inspect for leaks at water gauge glasses and at valves; repack valves if necessary.	.046			√	√	√
9 Tighten or replace loose, missing, or damaged nuts, bolts, or screws.	.005			√	√	√
10 Lubricate water filter valve, check rings, tighten as required.	.033			√	√	√
11 Check operation of lights.	.027			√	√	√
12 Check timer mechanism and operation of unit.	.018			√	√	√
13 Fill out maintenance checklist and report deficiencies.	.022			√	√	√
Total labor-hours/period				1.163	1.163	1.163
Total labor-hours/year				2.326	1.163	1.163

			Cost Each				
			2014 Bare Costs			Total	Total
Description	Labor-hrs.	Material	Labor	Equip.	Total	In-House	w/O&P
1900 Coffee maker / urn, gas / electric / steam, annually	1.163	52.50	55		107.50	131	155
1950 Annualized	4.652	52.50	220		272.50	350	425

E10 EQUIPMENT E1095 316 | Cooker

PM Components	Labor-hrs.	W	M	Q	S	A
PM System E1095 316 1950						
Cooker, gas						
1 Check with operating or area personnel for any deficiencies.	.044			✓	✓	✓
2 Examine doors and door gaskets; make necessary adjustments.	.021			✓	✓	✓
3 Visually inspect hinges and latches; lubricate hinges.	.091			✓	✓	✓
4 Tighten or replace any loose nuts, bolts, screws; replace broken knobs.	.007			✓	✓	✓
5 Check gas piping and valves for leaks.	.077			✓	✓	✓
6 Check pilot and thermosensor bulb.	.079			✓	✓	✓
7 Check gas burner for uniform flame.	.007			✓	✓	✓
8 Check calibration of thermostats.	.458			✓	✓	✓
9 Operate unit to ensure that it is in proper working condition.	.130			✓	✓	✓
10 Fill out maintenance checklist and report deficiencies.	.022			✓	✓	✓
Total labor-hours/period				.936	.936	.936
Total labor-hours/year				1.872	.936	.936

			Cost Each				
			2014 Bare Costs			Total	Total
Description	Labor-hrs.	Material	Labor	Equip.	Total	In-House	w/O&P
1900 Cooker, gas, annually	.936	123	44.50		167.50	194	226
1950 Annualized	3.744	123	177		300	370	445

PM Components	Labor-hrs.	W	M	Q	S	A
PM System E1095 316 2950						
Cooker, vegetable steamer						
1 Check with operating or area personnel for any deficiencies.	.044			✓	✓	✓
2 Examine doors and door gaskets; make necessary adjustments.	.021			✓	✓	✓
3 Visually inspect hinges and latches; lubricate hinges.	.091			✓	✓	✓
4 Tighten or replace any loose nuts, bolts, screws; replace broken knobs.	.005			✓	✓	✓
5 Check working pressure on steam gauge.	.005			✓	✓	✓
6 Inspect piping and valves for leaks; tighten as necessary.	.077			✓	✓	✓
7 Check operation of low water cut-off; adjust as required.	.030			✓	✓	✓
8 Check water level sight glass; adjust as required.	.007			✓	✓	✓
9 Fill out maintenance checklist and report deficiencies.	.022			✓	✓	✓
Total labor-hours/period				.302	.302	.302
Total labor-hours/year				.604	.302	.302

			Cost Each				
			2014 Bare Costs			Total	Total
Description	Labor-hrs.	Material	Labor	Equip.	Total	In-House	w/O&P
2900 Cooker, vegetable steamer, annually	.302	42	14.30		56.30	65	76
2950 Annualized	1.208	42	57		99	122	146

E10 EQUIPMENT — E1095 316 Cooker

PM Components	Labor-hrs.	W	M	Q	S	A
PM System E1095 316 3950						
Cooker, steam						
1 Check with operating or area personnel for any deficiencies.	.044			✓	✓	✓
2 Examine doors and door gaskets; make necessary adjustments.	.021			✓	✓	✓
3 Visually inspect hinges and latches; lubricate hinges.	.091			✓	✓	✓
4 Tighten or replace any loose nuts, bolts, screws; replace broken knobs.	.005			✓	✓	✓
5 Check for clogged or defective steam trap; clean trap.	.098			✓	✓	✓
6 Inspect and clean steam strainer.	.207			✓	✓	✓
7 Check working pressure on steam gauge.	.005			✓	✓	✓
8 Turn unit on, allow pressure to build, inspect piping, valve and doors for leaks; adjust as required.	.077			✓	✓	✓
9 Check operation of pressure relief valve; adjust as required.	.010			✓	✓	✓
10 Check operation of low water cut-off; adjust as required.	.030			✓	✓	✓
11 Check water level sight glass; adjust as required.	.007			✓	✓	✓
12 Check timer mechanism and operation; adjust as required.	.018			✓	✓	✓
13 Fill out maintenance checklist and report deficiencies.	.022			✓	✓	✓
Total labor-hours/period				.635	.635	.635
Total labor-hours/year				1.270	.635	.635

			Cost Each				
			2014 Bare Costs			Total	Total
Description		Labor-hrs.	Material	Labor	Equip.	Total	In-House	w/O&P
3900	Cooker, steam, annually	.635	84	30		114	132	154
3950	Annualized	2.540	84	120		204	252	300

E10 EQUIPMENT — E1095 318 Cookie Maker

PM Components	Labor-hrs.	W	M	Q	S	A
PM System E1095 318 1950						
Cookie maker, electric						
1 Check with operating or area personnel for any deficiencies.	.044			✓	✓	✓
2 Check electrical panel switches, connections, timing device for loose, frayed wiring; tighten connections as required.	.120			✓	✓	✓
3 Check operation of electric motor and bearings for overheating; lubricate motor bearings.	.086			✓	✓	✓
4 Check belt; adjust tension or replace belt as required.	.369			✓	✓	✓
5 Check conveyor and rollers by observing movement; lubricate rollers and adjust as required.	.127			✓	✓	✓
6 Tighten or replace loose, missing or damaged nuts, bolts or screws.	.012			✓	✓	✓
7 Wipe off and clean unit.	.096			✓	✓	✓
8 Fill out maintenance checklist and report deficiencies.	.022			✓	✓	✓
Total labor-hours/period				.876	.876	.876
Total labor-hours/year				1.752	.876	.876

			Cost Each				
			2014 Bare Costs			Total	Total
Description		Labor-hrs.	Material	Labor	Equip.	Total	In-House	w/O&P
1900	Cookie maker, electric, annually	.876	61.50	41.50		103	122	144
1950	Annualized	3.504	87.50	166		253.50	315	380

E10 EQUIPMENT   E1095 320 | Deep Fat Fryer

	PM Components	Labor-hrs.	W	M	Q	S	A
	PM System E1095 320 1950						
	Deep fat fryer, pressurized broaster, gas/electric						
1	Check with operating or area personnel for any deficiencies.	.044			√	√	√
2	Check piping and valves for leaks; tighten fittings as required.	.125			√	√	√
3	Check pilot and flame on gas operated unit; adjust as required.	.009			√	√	√
4	Check elements, switches, controls, contacts, and wiring on electrically heated units; repair or adjust as required.	.025			√	√	√
5	Check thermostat; calibrate if necessary.	.439			√	√	√
6	Check basket/racks for bends, breaks or defects; straighten bends or repair as necessary.	.013			√	√	√
7	Check cooking oil filter on circulating system; change as required.	.068			√	√	√
8	Check operation of unit.	.164			√	√	√
9	Check operation of motor for excessive noise and overheating.	.022			√	√	√
10	Check nuts, bolts, and screws for tightness; tighten or replace as necessary.	.005			√	√	√
11	Fill out maintenance checklist and report deficiencies.	.022			√	√	√
	Total labor-hours/period				.936	.936	.936
	Total labor-hours/year				1.872	.936	.936

			Cost Each					
				2014 Bare Costs			Total	Total
	Description	Labor-hrs.	Material	Labor	Equip.	Total	In-House	w/O&P
1900	Fryer, pressurized broaster, gas / electric, annually	.936	123	44.50		167.50	194	226
1950	Annualized	3.744	175	177		352	425	505

	PM Components	Labor-hrs.	W	M	Q	S	A
	PM System E1095 320 2950						
	Deep fat fryer, automatic conveyor belt type, gas/electric						
1	Check with operating or area personnel for any deficiencies.	.044			√	√	√
2	Check compartments, valves and piping for leaks; tighten as required.	.077			√	√	√
3	Check pilot and flame on gas operated fryers; adjust as required.	.168			√	√	√
4	Check elements, switches, controls, contacts and wiring on electrically heated units for defects; repair or adjust as necessary.	.051			√	√	√
5	Check thermostats; calibrate as necessary.	.439			√	√	√
6	Check cooking oil filter on circulating system; change as required.	.068			√	√	√
7	Check operation of motor for excessive noise and overheating; lubricate bearings.	.022			√	√	√
8	Check operation of belt hoist; lubricate and adjust as required.	.052			√	√	√
9	Check nuts, bolts and screws for tightness; replace or tighten as necessary.	.017			√	√	√
10	Clean machine of dust, grease and food particles.	.044			√	√	√
11	Clean area around unit.	.066			√	√	√
12	Fill out maintenance checklist and report deficiencies.	.022			√	√	√
	Total labor-hours/period				1.070	1.070	1.070
	Total labor-hours/year				2.140	1.070	1.070

			Cost Each					
				2014 Bare Costs			Total	Total
	Description	Labor-hrs.	Material	Labor	Equip.	Total	In-House	w/O&P
2900	Fryer, conveyor belt type, gas / electric, annually	1.070	105	50.50		155.50	183	214
2950	Annualized	4.280	105	202		307	385	460

E10 EQUIPMENT | E1095 320 | Deep Fat Fryer

PM Components	Labor-hrs.	W	M	Q	S	A
PM System E1095 320 3950						
Deep fat fryer, conventional type, gas/electric						
1 Check with operating or area personnel for any deficiencies.	.044			√	√	√
2 Clean machine thoroughly of dust, grease, and food particles.	.014			√	√	√
3 Check compartments, valves, and piping for leaks; tighten as required.	.077			√	√	√
4 Check pilot and flame on gas operated deep fat fryer; adjust as required.	.005			√	√	√
5 Check thermostat; calibrate if necessary.	.439			√	√	√
6 On electrically heated units check elements, switches, controls, contacts, and wiring for defects; repair or adjust as necessary.	.025			√	√	√
7 Check basket or rack for bends, breaks or defects, straighten bends or repair as necessary.	.013			√	√	√
8 Check nuts, bolts and screws for tightness; tighten or replace as necessary.	.005			√	√	√
9 Fill out maintenance checklist and report deficiencies.	.022			√	√	√
Total labor-hours/period				.644	.644	.644
Total labor-hours/year				1.288	.644	.644

			Cost Each				
			2014 Bare Costs			Total	Total
Description		Labor-hrs.	Material	Labor	Equip.	Total	In-House	w/O&P
3900	Fryer, conventional type, gas / electric, annually	.644	87.50	30.50		118	137	159
3950	Annualized	2.576	87.50	122		209.50	258	305

E10 EQUIPMENT   E1095 322   Dishwasher, Electric/Steam

PM Components	Labor-hrs.	W	M	Q	S	A
PM System E1095 322 1950						
Dishwasher, electric						
1 Check with operating or area personnel for any deficiencies.	.044			√	√	√
2 Check electric insulators, connections, and wiring (includes removing access panels).	.120			√	√	√
3 Check motor and bearings for excessive noise, vibration and overheating.	.039			√	√	√
4 Check dish conveyor by turning on switch and observing conveyor movement; adjust chain if necessary.	.070			√	√	√
5 Check operation of wash and rinse spray mechanism for spray coverage and drainage.	.022			√	√	√
6 Inspect soap and spray solution feeder lines; clean as necessary.	.013			√	√	√
7 Inspect water lines and fittings for leaks; tighten fittings as necessary.	.077			√	√	√
8 Lubricate conveyor drive bushings and chain drive.	.055			√	√	√
9 Check belt tension; adjust if required.	.100			√	√	√
10 Check doors for: operation of chains and counterweights, warping, alignment and water tightness; adjust if necessary.	.022			√	√	√
11 Check packing glands on wash, rinse, and drain valves; add or replace packing as required.	.147			√	√	√
12 Check lubricant in gear case; add oil if required.	.049			√	√	√
13 Inspect splash curtain for tears, clearance and water tightness; adjust if required.	.016			√	√	√
14 Check proper operation of solenoid valve and float in fill tank; adjust as required.	.096			√	√	√
15 Check pumps for leakage and obstructions; adjust as required.	.120			√	√	√
16 Check proper operation of micro-switch.	.010			√	√	√
17 Check water for proper temperature.	.004			√	√	√
18 Check operation of safety stop and clutch on conveyor belt motor; adjust cut-off switch mechanism as required.	.049			√	√	√
19 Check temperature regulator and adjust if necessary.	.017			√	√	√
20 Clean lime off thermostatic probe and heating elements.	.020			√	√	√
21 Clean area around equipment.	.066			√	√	√
22 Fill out maintenance checklist and report deficiencies.	.022			√	√	√
Total labor-hours/period				1.178	1.178	1.178
Total labor-hours/year				2.356	1.178	1.178

			Cost Each				
			2014 Bare Costs			Total	Total
Description		Labor-hrs.	Material	Labor	Equip.	Total	In-House	w/O&P
1900	Dishwasher, electric, annually	1.178	114	55.50		169.50	199	234
1950	Annualized	4.712	245	223		468	565	670

E10 EQUIPMENT | E1095 322 | Dishwasher, Electric/Steam

PM Components	Labor-hrs.	W	M	Q	S	A
PM System E1095 322 2950						
Dishwasher, steam						
1 Check with operating or area personnel for any deficiencies.	.044			✓	✓	✓
2 Check electric insulators, connections, and wiring (includes removing access panels).	.120			✓	✓	✓
3 Check motor and bearings for excessive noise, vibration and overheating.	.039			✓	✓	✓
4 Check dish conveyor by turning on switch and observing conveyor movement; adjust chain if necessary.	.070			✓	✓	✓
5 Check operation of wash and rinse spray mechanism for spray coverage and drainage.	.022			✓	✓	✓
6 Inspect soap and spray solution feeder lines; clean as necessary.	.013			✓	✓	✓
7 Inspect steam and water lines and fittings for leaks; tighten fittings as necessary.	.077			✓	✓	✓
8 Lubricate conveyor drive bushings and chain drive.	.055			✓	✓	✓
9 Check belt tension; adjust if required.	.100			✓	✓	✓
10 Check doors for: operation of chains and counterweights, warping, alignment and water tightness; adjust if necessary.	.022			✓	✓	✓
11 Tighten loose nuts, bolts and screws.	.005			✓	✓	✓
12 Check packing glands on wash, rinse, and drain valves; add or replace packing as required.	.147			✓	✓	✓
13 Check lubricant in gear case; add oil if required.	.049			✓	✓	✓
14 Inspect splash curtain for tears, clearance and water tightness; adjust if required.	.016			✓	✓	✓
15 Check proper operation of solenoid valve and float in fill tank; adjust as required.	.096			✓	✓	✓
16 Check pumps for leakage and obstructions; adjust as required.	.120			✓	✓	✓
17 Check proper operation of micro-switch.	.010			✓	✓	✓
18 Check water for proper temperature readings.	.004			✓	✓	✓
19 Check operation of safety stop and clutch on conveyor belt motor; adjust cut-off switch mechanism as required.	.049			✓	✓	✓
20 Check temperature regulator and adjust if necessary.	.017			✓	✓	✓
21 Clean lime off thermostatic probe.	.020			✓	✓	✓
22 Clean area around equipment.	.066			✓	✓	✓
23 Fill out maintenance checklist and report deficienceis.	.022			✓	✓	✓
Total labor-hours/period				1.183	1.183	1.183
Total labor-hours/year				2.366	1.183	1.183

			Cost Each				
			2014 Bare Costs			Total	Total
Description		Labor-hrs.	Material	Labor	Equip.	Total	In-House	w/O&P
2900	Dishwasher, steam, annually	1.183	114	56		170	199	234
2950	Annualized	4.732	245	224		469	565	670

E10 EQUIPMENT — E1095 328 | Dispenser, Beverage

PM Components	Labor-hrs.	W	M	Q	S	A
PM System E1095 328 1950						
Dispenser, carbonated beverage						
1 Check with operating or area personnel for any deficiencies.	.044			✓	✓	✓
2 Check all lines, connections and valves for leaks.	.077			✓	✓	✓
3 Check pressure gauges; adjust as required.	.018			✓	✓	✓
4 Check water strainer screen on carbonator; clean as required.	.159			✓	✓	✓
5 Check wiring, connectors and switches; repair as required.	.120			✓	✓	✓
6 Check nuts, bolts and screws for tightness; replace or tighten as required.	.005			✓	✓	✓
7 Wipe and clean unit.	.056			✓	✓	✓
8 Fill out maintenance checklist and report deficiencies.	.022			✓	✓	✓
Total labor-hours/period				.501	.501	.501
Total labor-hours/year				1.002	.501	.501

				Cost Each				
				2014 Bare Costs			Total	Total
Description		Labor-hrs.	Material	Labor	Equip.	Total	In-House	w/O&P
1900	Dispenser, carbonated beverage, annually	.501	70	23.50		93.50	108	126
1950	Annualized	2.004	70	95		165	202	242

E10 EQUIPMENT — E1095 332 | Disposal, Garbage

PM Components	Labor-hrs.	W	M	Q	S	A
PM System E1095 332 1950						
Disposal, garbage, electric						
1 Check with operating or area personnel for any deficiencies.	.044			✓	✓	✓
2 Check motor and drive shaft for excessive noise, vibration, overheating, etc.	.044			✓	✓	✓
3 Inspect electrical wiring.	.014			✓	✓	✓
4 Visually examine grinder; check for obstructions; adjust cutters if required.	.074			✓	✓	✓
5 Check drive belt(s) for wear and tension; adjust if required.	.014			✓	✓	✓
6 Check for leaks to supply and drain connections; tighten if required.	.103			✓	✓	✓
7 Check electrical switch for proper operation.	.014			✓	✓	✓
8 Lubricate grinder, drive, and motor.	.022			✓	✓	✓
9 Check water sprayer operation.	.030			✓	✓	✓
10 Fill out maintenance checklist and report deficiencies.	.022			✓	✓	✓
Total labor-hours/period				.381	.381	.381
Total labor-hours/year				.762	.381	.381

				Cost Each				
				2014 Bare Costs			Total	Total
Description		Labor-hrs.	Material	Labor	Equip.	Total	In-House	w/O&P
1900	Disposal, garbage, electric, annually	.381	42	18		60	70	82
1950	Annualized	1.524	42	72		114	141	170

E10 EQUIPMENT E1095 334 | Dough, Divider/Roller

PM Components	Labor-hrs.	W	M	Q	S	A
PM System E1095 334 1950						
Dough divider						
1 Check with operating or area personnel for any deficiencies.	.044			√	√	√
2 Check motor and drive shaft for excessive noise, vibration, overheating, etc.	.039			√	√	√
3 Inspect electrical wiring.	.120			√	√	√
4 Visually examine plates; check for obstructions; adjust cutters if required.	.074			√	√	√
5 Check electrical switch for proper operation.	.014			√	√	√
6 Lubricate press, drive, and motor.	.218			√	√	√
7 Fill out maintenance checklist and report deficiencies.	.022			√	√	√
Total labor-hours/period				.531	.531	.531
Total labor-hours/year				1.062	.531	.531

			Cost Each				
			2014 Bare Costs			Total	Total
Description	Labor-hrs.	Material	Labor	Equip.	Total	In-House	w/O&P
1900 Dough divider, annually	.531	87.50	25		112.50	130	150
1950 Annualized	2.124	87.50	100		187.50	230	273

PM Components	Labor-hrs.	W	M	Q	S	A
PM System E1095 334 2950						
Dough roller						
1 Check with operating or area personnel for any deficiencies.	.044			√	√	√
2 Check electric motor, switches, controls, wiring, connections and insulation for defective materials; repair as necessary.	.241			√	√	√
3 Check belt tension; adjust or replace as required.	.446			√	√	√
4 Check operation of conveyor-roller clearance; adjust as necessary.	.109			√	√	√
5 Lubricate all moving parts, motors, pivot points and chain drives.	.017			√	√	√
6 Clean unit and treat for rust and/or corrosion.	.083			√	√	√
7 Tighten or replace loose, missing or damaged nuts, bolts and screws.	.005			√	√	√
8 Fill out maintenance checklist and report deficiencies.	.022			√	√	√
Total labor-hours/period				.967	.967	.967
Total labor-hours/year				1.934	.967	.967

			Cost Each				
			2014 Bare Costs			Total	Total
Description	Labor-hrs.	Material	Labor	Equip.	Total	In-House	w/O&P
2900 Dough roller, annually	.967	131	45.50		176.50	204	238
2950 Annualized	3.868	158	183		341	415	495

| E10 EQUIPMENT | E1095 336 | Drink Cooler |

PM Components	Labor-hrs.	W	M	Q	S	A
PM System E1095 336 1950						
Drink cooler, with external condenser						
1 Check with operating or area personnel for deficiencies.	.035				✓	✓
2 Clean condenser coils, fans, and intake screens; lubricate motor.	.504				✓	✓
3 Inspect door gaskets for damage and proper fit; adjust as necessary.	.033				✓	✓
4 Check starter panels and controls for proper operation, burned or loose contacts, and loose connections.	.094				✓	✓
5 Clean coils, evaporator drain pan, blowers, fans, motors and drain piping as required; lubricate motor(s).	.473				✓	✓
6 During operation of unit, check refrigerant pressures and compressor oil level; add refrigerant and/or oil as necessary.	.066				✓	✓
7 Check operation of low pressure cut-out; adjust or replace as necessary.	.057				✓	✓
8 Inspect defrost systems for proper operation; adjust as required.	.027				✓	✓
9 Clean area around equipment.	.066				✓	✓
10 Fill out maintenance checklist and report deficiencies.	.022				✓	✓
Total labor-hours/period					1.377	1.377
Total labor-hours/year					1.377	1.377

		Cost Each					
		2014 Bare Costs				Total	Total
Description	Labor-hrs.	Material	Labor	Equip.	Total	In-House	w/O&P
1900 Drink cooler with external condenser, annually	1.377	17.50	65		82.50	105	128
1950 Annualized	2.754	35	130		165	211	256

| E10 EQUIPMENT | E1095 338 | Fluid Cooler |

PM Components	Labor-hrs.	W	M	Q	S	A
PM System E1095 338 1950						
Fluid cooler, air cooled condenser						
1 Check with operating or area personnel for deficiencies.	.035					✓
2 Check unit for proper operation, excessive noise or vibration.	.159					✓
3 Clean intake side of condenser coils, fans and intake screens.	.473					✓
4 Check electrical wiring and connections; tighten loose connections.	.120					✓
5 Inspect fan(s) for bent blades or unbalance; adjust as necessary.	.014					✓
6 Check belts for condition, proper tension and misalignment, if required.	.029					✓
7 Lubricate shaft bearings and motor bearings.	.047					✓
8 Inspect piping and valves for leaks; tighten connections as necessary.	.077					✓
9 Lubricate and check operation of dampers, if applicable.	.055					✓
10 Clean area around fluid cooler.	.066					✓
11 Fill out maintenance checklist and report deficiencies.	.022					✓
Total labor-hours/period						1.097
Total labor-hours/year						1.097

		Cost Each					
		2014 Bare Costs				Total	Total
Description	Labor-hrs.	Material	Labor	Equip.	Total	In-House	w/O&P
1900 Fluid cooler, annually	1.097	17.50	52		69.50	88	106
1950 Annualized	1.097	17.50	52		69.50	88	106

E10 EQUIPMENT   E1095 340 | Food Saw/Slicer

PM Components	Labor-hrs.	W	M	Q	S	A
PM System E1095 340 1950						
Food saw, electric						
1 Check with operating or area personnel for any deficiencies.	.044			√	√	√
2 Check operation of saw for excessive vibration, blade tension and proper adjustment of safety guards; repair or adjust as required.	.095			√	√	√
3 Check motor and bearings for overheating; lubricate motor bearings.	.039			√	√	√
4 Check switches, motor, controls and wiring for damage and proper operation; repair as required.	.120			√	√	√
5 Clean machine of dust and grease.	.044			√	√	√
6 Check nuts, bolts and screws for tightness; tighten as required.	.005			√	√	√
7 Check belt tension and alignment; adjust as required.	.029			√	√	√
8 Fill out maintenance checklist and report deficiencies.	.022			√	√	√
Total labor-hours/period				.398	.398	.398
Total labor-hours/year				.796	.398	.398

		Cost Each					
			2014 Bare Costs			Total	Total
Description	Labor-hrs.	Material	Labor	Equip.	Total	In-House	w/O&P
1900 Food saw, electric, annually	.398	61.50	18.80		80.30	92.50	107
1950 Annualized	1.592	87.50	75.50		163	197	232

PM Components	Labor-hrs.	W	M	Q	S	A
PM System E1095 340 2950						
Food slicer, electric						
1 Check with operating or area personnel for any deficiencies.	.044			√	√	√
2 Check nuts, bolts and screws for tightness; tighten or replace as required.	.005			√	√	√
3 Lubricate food slicer and add lubricant to gear case if required.	.085			√	√	√
4 Check condition of blade, blade guard, guides, and controls; adjust as required.	.047			√	√	√
5 Check operation of motor and check motor bearings for overheating; lubricate motor bearings.	.039			√	√	√
6 Check sharpening stones; install new stones as required.	.013			√	√	√
7 Clean machine of dust, grease, and food particles.	.014			√	√	√
8 Fill out maintenance checklist and report deficiencies.	.022			√	√	√
Total labor-hours/period				.269	.269	.269
Total labor-hours/year				.538	.269	.269

		Cost Each					
			2014 Bare Costs			Total	Total
Description	Labor-hrs.	Material	Labor	Equip.	Total	In-House	w/O&P
2900 Food slicer, electric, annually	.269	79	12.70		91.70	104	119
2950 Annualized	1.076	105	51		156	184	214

E10 EQUIPMENT — E1095 346 Grill, Gas/Electric

PM Components	Labor-hrs.	W	M	Q	S	A
PM System E1095 346 1950						
Grill, gas/electric						
1 Check with operating or area personnel for any deficiencies.	.044			√	√	√
2 Check nuts, bolts and screws for tightness; tighten or replace as required.	.005			√	√	√
3 On gas operated units, check piping and valves for leaks.	.077			√	√	√
4 On gas operated units, check pilot and gas burners for uniform flame; adjust as required.	.079			√	√	√
5 On electrically operated units, check switches, connections, and wiring for proper operation; adjust as required.	.135			√	√	√
6 Check calibration of thermostats; adjust as required.	.458			√	√	√
7 Fill out maintenance checklist and report deficiencies.	.022			√	√	√
Total labor-hours/period				.820	.820	.820
Total labor-hours/year				1.640	.820	.820

			Cost Each				
			2014 Bare Costs			Total	Total
Description	Labor-hrs.	Material	Labor	Equip.	Total	In-House	w/O&P
1900 Grill, gas / electric, annually	.820	87.50	39		126.50	148	172
1950 Annualized	3.280	87.50	155		242.50	300	360

E10 EQUIPMENT — E1095 348 Ice Machine

PM Components	Labor-hrs.	W	M	Q	S	A
PM System E1095 348 1950						
Ice machine, flake or cube						
1 Check with operating or area personnel for any deficiencies.	.044			√	√	√
2 Remove and install access panel.	.044			√	√	√
3 Lubricate all moving parts, pivot points and fan motor(s).	.023			√	√	√
4 Visually check for refrigerant, oil or water leaks.	.022			√	√	√
5 Open and close water valve.	.007			√	√	√
6 Replace in-line water filter.	.009			√	√	√
7 Check and clear ice machine draining system (drain vent and trap).	.198			√	√	√
8 Clean motor, compressor and condenser coil.	.403			√	√	√
9 Check and tighten any loose screw-type electrical connections.	.005			√	√	√
10 Inspect door(s) hinges, gaskets, handles; lubricate as required.	.053			√	√	√
11 Clean area around equipment.	.066			√	√	√
12 Fill out maintenance checklist and report deficiencies.	.022			√	√	√
Total labor-hours/period				.896	.896	.896
Total labor-hours/year				1.792	.896	.896

			Cost Each				
			2014 Bare Costs			Total	Total
Description	Labor-hrs.	Material	Labor	Equip.	Total	In-House	w/O&P
1900 Ice machine, flake or cube, annually	.896	26.50	42.50		69	85	102
1950 Annualized	3.584	105	170		275	340	405

E10 EQUIPMENT — E1095 350 Kettle, Steam

PM Components	Labor-hrs.	W	M	Q	S	A
PM System E1095 350 1950						
Kettle, steam, fixed or tilt						
1 Check with operating or area personnel for any deficiencies.	.044			√	√	√
2 Check piping and fittings for leaks; tighten as required.	.085			√	√	√
3 Check operation of electric water valve.	.030			√	√	√
4 Inspect cover, hinges, and seals on units to equipped; lubricate hinge.	.053			√	√	√
5 Lubricate tilting gear mechanism and trunnion bearings, if applicable.	.047			√	√	√
6 Fill out maintenance checklist and report deficiencies.	.022			√	√	√
Total labor-hours/period				.281	.281	.281
Total labor-hours/year				.562	.281	.281

				Cost Each				
				2014 Bare Costs			Total	Total
	Description	Labor-hrs.	Material	Labor	Equip.	Total	In-House	w/O&P
1900	Kettle, steam, fixed or tilt, annually	.281	44	13.30		57.30	66	76.50
1950	Annualized	1.124	44	53		97	119	142

E10 EQUIPMENT — E1095 354 | Mixer, Electric

PM Components	Labor-hrs.	W	M	Q	S	A
PM System E1095 354 1950						
Mixer, counter, electric						
1 Check with operating or area personnel for any deficiencies.	.044			√	√	√
2 Check operation of mixer at varying speeds for excessive noise and vibrations; align or adjust as required.	.033			√	√	√
3 Check belt for proper tension and alignment; adjust as required.	.068			√	√	√
4 Lubricate mixer gears.	.042			√	√	√
5 Clean machine thoroughly of dust, grease, and food particles.	.014			√	√	√
6 Check motor, switches, controls and motor bearings for overheating; lubricate motor bearings.	.039			√	√	√
7 Check anchor bolts for tightness; tighten as required.	.010			√	√	√
8 Fill out maintenance checklist and report deficiencies.	.022			√	√	√
Total labor-hours/period				.272	.272	.272
Total labor-hours/year				.544	.272	.272

			Cost Each				
			2014 Bare Costs			Total	Total
Description	Labor-hrs.	Material	Labor	Equip.	Total	In-House	w/O&P
1900 Mixer, counter, electric, annually	.272	44	12.85		56.85	65.50	76
1950 Annualized	1.088	70	51.50		121.50	145	172

PM Components	Labor-hrs.	W	M	Q	S	A
PM System E1095 354 2950						
Mixer, floor, electric						
1 Check with operating or area personnel for any deficiencies.	.044			√	√	√
2 Check operation of mixer at varying speeds for excessive noise and vibrations; align or adjust as required.	.120			√	√	√
3 Tighten loose bolts, nuts, and screws.	.005			√	√	√
4 Check oil level in transmission; add or replace oil if required.	.159			√	√	√
5 Lubricate mixer gears and bowl lift mechanism.	.042			√	√	√
6 Check switches and control for damage and proper operation; adjust as required.	.068			√	√	√
7 Check belt for proper tension and alignment; adjust as required.	.029			√	√	√
8 Clean machine of dust and grease.	.030			√	√	√
9 Check anchor bolts for tightness; tighten as required.	.010			√	√	√
10 Clean area around equipment.	.066			√	√	√
11 Fill out maintenance checklist and report deficiencies.	.022			√	√	√
Total labor-hours/period				.595	.595	.595
Total labor-hours/year				1.190	.595	.595

			Cost Each				
			2014 Bare Costs			Total	Total
Description	Labor-hrs.	Material	Labor	Equip.	Total	In-House	w/O&P
2900 Mixer, floor, electric, annually	.595	79	28		107	124	145
2950 Annualized	2.380	105	113		218	265	315

E10 EQUIPMENT — E1095 356 | Oven, Convection/Rotary

PM Components	Labor-hrs.	W	M	Q	S	A
PM System E1095 356 1950						
Oven, convection, gas/electric						
1 Check with operating or area personnel for any deficiencies.	.044		√	√	√	√
2 Check doors and seals for warping and misalignment; lubricate hinges and repair as necessary.	.027		√	√	√	√
3 Check piping and valves for leaks.	.077		√	√	√	√
4 Check nuts, bolts, and screws for tightness; replace or tighten as required.	.005		√	√	√	√
5 Check pilot and gas burner for uniform flame; adjust as required.	.083		√	√	√	√
6 Check element, switches, controls and wiring on electrically heated units for defects; repair as required.	.120		√	√	√	√
7 Check fan blades and fan motor for proper operation.	.120		√	√	√	√
8 Check operation of thermostat; calibrate as required.	.439		√	√	√	√
9 Fill out maintenance checklist and report deficiencies.	.022		√	√	√	√
Total labor-hours/period			.937	.937	.937	.937
Total labor-hours/year			7.496	1.874	.937	.937

			Cost Each				
			2014 Bare Costs			Total	Total
Description		Labor-hrs.	Material	Labor	Equip.	Total	In-House	w/O&P
1900	Oven, convection, gas / electric, annually	.937	87.50	44.50		132	156	181
1950	Annualized	11.244	87.50	530		617.50	800	975

PM Components	Labor-hrs.	W	M	Q	S	A
PM System E1095 356 2950						
Oven, rotary, electric						
1 Check with operating or area personnel for any deficiencies.	.044		√	√	√	√
2 Remove and reinstall access panel; lubricate main bearing bushings.	.064		√	√	√	√
3 Turn motor on and check operation of chain drive mechanism; lubricate chain.	.064		√	√	√	√
4 Inspect motor and wiring; sensory inspect motor and bearings for overheating; lubricate motor bearings.	.055		√	√	√	√
5 Check V-belt; adjust tension and/or pulley as required.	.029		√	√	√	√
6 Inspect shelves for defects and level if required.	.142		√	√	√	√
7 Tighten or replace any loose nuts, bolts, screws.	.005		√	√	√	√
8 Check thermostat with thermometer; adjust as required.	.439		√	√	√	√
9 Check operation of stop and reverse switch.	.008		√	√	√	√
10 Check timer mechanism and operation.	.018		√	√	√	√
11 Check lubricant in gear case; add as required.	.055		√	√	√	√
12 Check ventilator for proper operation.	.160		√	√	√	√
13 Clean area around equipment.	.066		√	√	√	√
14 Fill out maintenance checklist and report deficiencies.	.022		√	√	√	√
Total labor-hours/period			1.171	1.171	1.171	1.171
Total labor-hours/year			9.368	2.342	1.171	1.171

			Cost Each				
			2014 Bare Costs			Total	Total
Description		Labor-hrs.	Material	Labor	Equip.	Total	In-House	w/O&P
2900	Oven, rotary, electric, annually	1.171	289	55.50		344.50	390	450
2950	Annualized	14.052	385	665		1,050	1,300	1,550

E10 EQUIPMENT | E1095 356 | Oven, Convection/Rotary

PM Components	Labor-hrs.	W	M	Q	S	A
PM System E1095 356 3950						
Oven, rotary, gas						
1 Check with operating or area personnel for any deficiencies.	.044		✓	✓	✓	✓
2 Examine doors and door gaskets; make necessary adjustments.	.021		✓	✓	✓	✓
3 Visually inspect hinges and latches; lubricate hinges.	.091		✓	✓	✓	✓
4 Remove and reinstall access panel; lubricate main bearing bushings.	.025		✓	✓	✓	✓
5 Turn motor on and chk. operation of chain drive mechanism; lubricate chain.	.064		✓	✓	✓	✓
6 Inspect motor and wiring; sensory inspect motor and bearings for overheating; lubricate motor bearings.	.055		✓	✓	✓	✓
7 Check V-belt; adjust tension and/or pulley as required.	.029		✓	✓	✓	✓
8 Check pilot and thermosensor bulb.	.042		✓	✓	✓	✓
9 Check gas burner for uniform flame.	.042		✓	✓	✓	✓
10 Check piping and valves for leaks.	.077		✓	✓	✓	✓
11 Inspect shelves for defects and level if required.	.142		✓	✓	✓	✓
12 Tighten or replace any loose nuts, bolts, screws.	.005		✓	✓	✓	✓
13 Check thermostat with thermometer; calibrate if required.	.439		✓	✓	✓	✓
14 Check operation of stop and reverse switch.	.008		✓	✓	✓	✓
15 Check timer mechanism and operation.	.018		✓	✓	✓	✓
16 Check lubricant in gear case, add as required.	.055		✓	✓	✓	✓
17 Check ventilator for proper operation.	.160		✓	✓	✓	✓
18 Clean area around equipment.	.066		✓	✓	✓	✓
19 Fill out maintenance checklist and report deficiencies.	.022		✓	✓	✓	✓
Total labor-hours/period			1.405	1.405	1.405	1.405
Total labor-hours/year			11.240	2.810	1.405	1.405

			Cost Each					
				2014 Bare Costs			Total	Total
Description		Labor-hrs.	Material	Labor	Equip.	Total	In-House	w/O&P
3900	Oven, rotary, gas, annually	1.405	289	66.50		355.50	405	470
3950	Annualized	16.860	385	795		1,180	1,475	1,775

E10 EQUIPMENT | E1095 366 | Peeler, Vegetable

PM Components	Labor-hrs.	W	M	Q	S	A
PM System E1095 366 1950						
Peeler, vegetable, electric						
1 Check with operating or area personnel for any deficiencies.	.044			✓	✓	✓
2 Check vegetable peeler for proper operation, including switches and controls; lubricate motor bearings.	.156			✓	✓	✓
3 Lubricate vegetable peeler bushings, as required.	.062			✓	✓	✓
4 Check door for loose hinges and latch fittings; adjust if required.	.007			✓	✓	✓
5 Clean peeler of dust, grease, and food particles.	.014			✓	✓	✓
6 Check piping and valves for leaks.	.077			✓	✓	✓
7 Inspect abrasive disk.	.018			✓	✓	✓
8 Tighten or replace loose, missing or damaged nuts, bolts, screws.	.005			✓	✓	✓
9 Check belt for proper tension and alignment; adjust or replace as required.	.029			✓	✓	✓
10 Fill out maintenance checklist and report deficiencies.	.022			✓	✓	✓
Total labor-hours/period				.434	.434	.434
Total labor-hours/year				.868	.434	.434

			Cost Each					
				2014 Bare Costs			Total	Total
Description		Labor-hrs.	Material	Labor	Equip.	Total	In-House	w/O&P
1900	Peeler, vegetable, electric, annually	.434	44	20.50		64.50	75.50	88.50
1950	Annualized	1.736	70	82		152	185	222

E10 EQUIPMENT — E1095 368 Pie Maker, Electric

PM Components	Labor-hrs.	W	M	Q	S	A
PM System E1095 368 1950						
Pie maker, electric						
1 Check with operating or area personnel for any deficiencies.	.044			√	√	√
2 Check electrical system for loose connections and frayed wiring; tighten connections as required.	.120			√	√	√
3 Check operation of motors and bearings for overheating; lubricate motor bearings.	.169			√	√	√
4 Check V-belt for alignment and proper tension; adjust or replace belt as required.	.345			√	√	√
5 Check drain drive and sprocket for excessive wear; lubricate if applicable.	.087			√	√	√
6 Tighten or replace loose, missing or damaged nuts, bolts or screws.	.005			√	√	√
7 Wipe clean the machine.	.096			√	√	√
8 Fill out maintenance checklist and report deficiencies.	.022			√	√	√
Total labor-hours/period				.888	.888	.888
Total labor-hours/year				1.776	.888	.888

			Cost Each				
			2014 Bare Costs			Total	Total
Description		Labor-hrs.	Material	Labor	Equip.	Total	In-House	w/O&P
1900	Pie maker, electric, annually	.888	96.50	42		138.50	162	189
1950	Annualized	3.552	123	168		291	355	430

E10 EQUIPMENT — E1095 370 Proofer, Automatic

PM Components	Labor-hrs.	W	M	Q	S	A
PM System E1095 370 1950						
Proofer, automatic, electric						
1 Check with operating or area personnel for any deficiencies.	.044			√	√	√
2 Check operation of electric motor, switches, controls and bearings; adjust, repair and lubricate as required.	.280			√	√	√
3 Check belt tension; adjust as required.	.029			√	√	√
4 Check chain drive operation for proper alignment and tension; repair as required.	.129			√	√	√
5 Check action of flour sifter mechanism, connecting linkage and conveyor; lubricate bushings as required.	.290			√	√	√
6 Check roller conveyor and connecting drive mechanism.	.113			√	√	√
7 Check door latch; adjust if needed.	.014			√	√	√
8 Check nuts, bolts and screws for tightness; replace or tighten as required.	.029			√	√	√
9 Fill out maintenance checklist and report deficiencies.	.022			√	√	√
Total labor-hours/period				.950	.950	.950
Total labor-hours/year				1.900	.950	.950

			Cost Each				
			2014 Bare Costs			Total	Total
Description		Labor-hrs.	Material	Labor	Equip.	Total	In-House	w/O&P
1900	Proofer, automatic, electric, annually	.950	140	45		185	213	248
1950	Annualized	3.800	166	180		346	420	500

E10 EQUIPMENT — E1095 380 Refrigerator

PM Components	Labor-hrs.	W	M	Q	S	A
PM System E1095 380 1950						
Refrigerator, domestic						
1 Check with operating or area personnel for deficiencies.	.035					✓
2 Clean coils, fans, fan motors, drip pan and other areas with vacuum, brush or wiping as necessary.	.066					✓
3 Inspect door gaskets for damage and proper fit; adjust gaskets as required and lubricate hinges.	.022					✓
4 Check door latch and adjust as necessary.	.023					✓
5 Clean area around equipment.	.066					✓
6 Fill out maintenance checklist and report deficiencies.	.022					✓
Total labor-hours/period						.234
Total labor-hours/year						.234

		Cost Each						
			2014 Bare Costs				Total	Total
Description	Labor-hrs.	Material	Labor	Equip.	Total	In-House	w/O&P
1900 Refrigerator, domestic, annually	.234	17.50	11.05		28.55	34	40
1950 Annualized	.234	17.50	11.05		28.55	34	40

PM Components	Labor-hrs.	W	M	Q	S	A
PM System E1095 380 2950						
Refrigerator, display case						
1 Check with operating or area personnel for deficiencies.	.035					✓
2 Clean coils, fans, fan motors, drip pan and other areas with vacuum, brush or wiping as necessary.	.066					✓
3 Inspect door gaskets for damage and proper fit; adjust gaskets as required and lubricate hinges.	.044					✓
4 Check door latch and adjust as necessary.	.047					✓
5 Clean area around equipment.	.066					✓
6 Fill out maintenance checklist and report deficiencies.	.022					✓
Total labor-hours/period						.280
Total labor-hours/year						.280

		Cost Each						
			2014 Bare Costs				Total	Total
Description	Labor-hrs.	Material	Labor	Equip.	Total	In-House	w/O&P
2900 Refrigerator, display case, annually	.280	17.50	13.25		30.75	36.50	43.50
2950 Annualized	.280	17.50	13.25		30.75	36.50	43.50

E10 EQUIPMENT — E1095 380 Refrigerator

PM Components	Labor-hrs.	W	M	Q	S	A
PM System E1095 380 3950						
Refrigerator/display, walk-in w/external condenser						
1 Check with operating or area personnel for deficiencies.	.035				✓	✓
2 Clean condenser coils, fans, and intake screens; lubricate motor.	.066				✓	✓
3 Inspect door gaskets for damage and proper fit; adjust gaskets as required and lubricate hinges.	.137				✓	✓
4 Check starter panels and controls for proper operation, burned or loose contacts, and loose connections.	.182				✓	✓
5 Clean coils, evaporator drain pan, blowers, fans, motors and drain piping as required; lubricate motor(s).	.473				✓	✓
6 During operation of unit, check refrigerant pressures and compressor oil level; add refrigerant and/or oil as necessary.	.066				✓	✓
7 Check operation of low pressure cut-out; adjust or replace as required.	.047				✓	✓
8 Inspect defrost systems for proper operation; adjust as required.	.094				✓	✓
9 Clean area around equipment.	.066				✓	✓
10 Fill out maintenance checklist and report deficiencies.	.022				✓	✓
Total labor-hours/period					1.188	1.188
Total labor-hours/year					1.188	1.188

			Cost Each					
				2014 Bare Costs			Total	Total
	Description	Labor-hrs.	Material	Labor	Equip.	Total	In-House	w/O&P
3900	Refrig. display, walk-in w/ ext. condenser, annually	1.188	44	56		100	123	147
3950	Annualized	2.376	52.50	112		164.50	206	249

E10 EQUIPMENT — E1095 382 Refrigerator/Freezer

PM Components	Labor-hrs.	W	M	Q	S	A
PM System E1095 382 1950						
Refrigerator/freezer, walk-in box w/external condenser						
1 Check with operating or area personnel for deficiencies.	.035				√	√
2 Clean condenser coils, fans, and intake screens; lubricate motor.	.036				√	√
3 Inspect door gaskets for damage and proper fit; adjust gaskets as required and lubricate hinges.	.022				√	√
4 Check starter panels and controls for proper operation, burned or loose contacts, and loose connections.	.182				√	√
5 Clean coils, evaporator drain pan, blowers, fans, motors and drain piping as required; lubricate motor(s).	.114				√	√
6 During operation of unit, check refrigerant pressures and compressor oil level; add refrigerant and/or oil as necessary.	.083				√	√
7 Check operation of low pressure cut-out; adjust or replace as required.	.079				√	√
8 Inspect defrost systems for proper operation; adjust as required.	.094				√	√
9 Clean area around equipment.	.066				√	√
10 Fill out maintenance checklist and report deficiencies.	.022				√	√
Total labor-hours/period					.733	.733
Total labor-hours/year					.733	.733

			Cost Each				
			2014 Bare Costs			Total	Total
Description	Labor-hrs.	Material	Labor	Equip.	Total	In-House	w/O&P
1900 Refrig. freezer, walk-in box w/ext. condenser, annually	.733	44	34.50		78.50	94	112
1950 Annualized	1.466	52.50	69.50		122	150	179

PM Components	Labor-hrs.	W	M	Q	S	A
PM System E1095 382 2950						
Refrigerator unit/display case/freezer w/external condenser						
1 Check with operating or area personnel for deficiencies.	.035				√	√
2 Clean condenser coils, fans, and intake screens; lubricate motor.	.504				√	√
3 Inspect door gaskets for damage and proper fit; adjust gaskets as required and lubricate hinges.	.033				√	√
4 Check starter panels and controls for proper operation, burned or loose contacts, and loose connections.	.213				√	√
5 Clean coils, evaporator drain pan, blowers, fans, motors and drain piping as required; lubricate motor(s).	.473				√	√
6 During operation of unit, check refrigerant pressures and compressor oil level; add refrigerant and/or oil as necessary.	.066				√	√
7 Check operation of low pressure cut-out; adjust or replace as required.	.057				√	√
8 Inspect defrost systems for proper operation; adjust as required.	.027				√	√
9 Clean area around equipment.	.066				√	√
10 Fill out maintenance checklist and report deficiencies.	.022				√	√
Total labor-hours/period					1.496	1.496
Total labor-hours/year					1.496	1.496

			Cost Each				
			2014 Bare Costs			Total	Total
Description	Labor-hrs.	Material	Labor	Equip.	Total	In-House	w/O&P
2900 Refrig., display case, freezer w/ ext. cond., annually	1.496	17.50	71		88.50	113	137
2950 Annualized	2.992	35	142		177	226	275

E10 EQUIPMENT — E1095 386 Steam Table

PM Components	Labor-hrs.	W	M	Q	S	A
PM System E1095 386 1950						
Steam table						
1 Check with operating or area personnel for any deficiencies.	.044			✓	✓	✓
2 Inspect water compartment, steam coil, valves, and piping for leaks.	.117			✓	✓	✓
3 Check steam trap and strainer; clean as required.	.306			✓	✓	✓
4 Check operation of pressure regulating valve and gauge.	.035			✓	✓	✓
5 Check pilots and flame on gas burner units; adjust as required.	.085			✓	✓	✓
6 Check insulators, connections, and wiring, if applicable; tighten connections.	.120			✓	✓	✓
7 Check condition of covers and receptacles; adjust as required.	.025			✓	✓	✓
8 Check thermostat and temperature gauge; calibrate thermostat if necessary.	.152			✓	✓	✓
9 Fill out maintenance checklist and report deficiencies.	.022			✓	✓	✓
Total labor-hours/period				.906	.906	.906
Total labor-hours/year				1.812	.906	.906

			Cost Each				
			2014 Bare Costs			Total	Total
	Description	Labor-hrs.	Material	Labor	Equip.	Total	In-House	w/O&P
1900	Steam table, annually	.906	87.50	43		130.50	154	179
1950	Annualized	3.624	87.50	171		258.50	320	390

E10 EQUIPMENT — E1095 388 Steamer, Vegetable

PM Components	Labor-hrs.	W	M	Q	S	A
PM System E1095 388 1950						
Steamer, vegetable, direct connected units						
1 Check with operating or area personnel for any deficiencies.	.044			✓	✓	✓
2 Inspect doors, door hardware and gaskets; lubricate hinges.	.112			✓	✓	✓
3 Tighten or replace loose, missing, or damaged nuts, bolts, screws.	.005			✓	✓	✓
4 Check for clogged or defective steam and strainers; clean as required.	.304			✓	✓	✓
5 Check working pressure on steam gauge and inspect piping, valves, and doors for leaks; make necessary repairs as required.	.082			✓	✓	✓
6 Check operation of pressure relief valve, low water cut-off, timer operation and water level in sight glass.	.065			✓	✓	✓
7 Fill out maintenance checklist and report deficiencies.	.022			✓	✓	✓
Total labor-hours/period				.634	.634	.634
Total labor-hours/year				1.268	.634	.634

			Cost Each				
			2014 Bare Costs			Total	Total
	Description	Labor-hrs.	Material	Labor	Equip.	Total	In-House	w/O&P
1900	Steamer, vegetable, annually	.634	87.50	30		117.50	137	158
1950	Annualized	2.536	87.50	120		207.50	256	305

E10 EQUIPMENT | E1095 390 | Toaster, Rotary

	PM Components	Labor-hrs.	W	M	Q	S	A
PM System E1095 390 1950							
Toaster, rotary, gas/electric							
1	Check with operating or area personnel for any deficiencies.	.044			✓	✓	✓
2	Check operation of toaster by toasting sample slice of bread.	.034			✓	✓	✓
3	Clean toaster.	.014			✓	✓	✓
4	Check valves and piping for leaks on gas units.	.077			✓	✓	✓
5	Check pilot light and burner flame on gas units; adjust when required.	.017			✓	✓	✓
6	Check insulators, heating elements, connections, and wiring on electric units; tighten connections and repair as necessary.	.134			✓	✓	✓
7	Inspect alignment of baskets and conveyor chains; lubricate chains and align as necessary.	.044			✓	✓	✓
8	Check motor and motor bearings for overheating; lubricate motor bearings.	.086			✓	✓	✓
9	Tighten or replace loose, missing or damaged nuts, bolts, screws.	.005			✓	✓	✓
10	Fill out maintenance checklist and report deficiencies.	.022			✓	✓	✓
	Total labor-hours/period				.477	.477	.477
	Total labor-hours/year				.954	.477	.477

			Cost Each					
				2014 Bare Costs			Total	Total
	Description	Labor-hrs.	Material	Labor	Equip.	Total	In-House	w/O&P
1900	Toaster, rotary, gas / electric, annually	.477	44	22.50		66.50	78	92
1950	Annualized	1.908	70	90.50		160.50	196	235

E10 EQUIPMENT | E1095 930 | Pump, Vacuum

	PM Components	Labor-hrs.	W	M	Q	S	A
PM System E1095 930 1950							
Vacuum							
1	Check with operating or area personnel for deficiencies.	.035				✓	✓
2	Check for leaks on suction and discharge piping, seals, etc.	.077				✓	✓
3	Check pump and motor operation for excessive vibration, noise and overheating.	.022				✓	✓
4	Check alignment and clearances of shaft and coupler.	.260				✓	✓
5	Tighten or replace loose, missing, or damaged nuts, bolts and screws.	.005				✓	✓
6	Lubricate pump and motor as required.	.099				✓	✓
7	Clean pump, motor and surrounding area.	.096				✓	✓
8	Fill out maintenance checklist and report deficiencies.	.022				✓	✓
	Total labor-hours/period					.616	.616
	Total labor-hours/year					.616	.616

			Cost Each					
				2014 Bare Costs			Total	Total
	Description	Labor-hrs.	Material	Labor	Equip.	Total	In-House	w/O&P
1900	Vacuum, annually	.616	8.35	35.50		43.85	53.50	66
1950	Annualized	1.232	16.75	71		87.75	107	132

F10 SPECIAL CONSTR. | F1045 110 | Swimming Pool

PM Components	Labor-hrs.	W	M	Q	S	A
PM System F1045 110 1950						
Swimming pool						
Quantities of chemicals used vary significantly by climate and pool usage, therefore the price of these materials are not provided.						
1 Clean strainer basket.	.050	√	√	√	√	√
2 Backwash pool water filter.	.120	√	√	√	√	√
3 Add soda ash to feeder.	.080	√	√	√	√	√
4 Clean/flush soda ash pump head.	.400		√	√	√	√
5 Check chlorine, change bottles and check for leaks.	.120	√	√	√	√	√
6 Clean chlorine pump, flush with acid, and check for signs of deterioration	.200		√	√	√	√
7 Check acid source, add acid, check for leaks.	.100	√	√	√	√	√
8 Clean acid pump head.	.400		√	√	√	√
9 Check circulating pump for leaks and unusual noises. Lubricate as required.	.050	√	√	√	√	√
10 Check diving board, tighten or replace missing hardware. Inspect for structural defects or deterioration.	.080	√	√	√	√	√
11 Wash exposed stainless steel ladders and scum gutters with tap water to remove oil and dirt.	.300	√	√	√	√	√
12 Fill out maintenance report.	.017	√	√	√	√	√
Total labor-hours/period		.917	1.917	1.917	1.917	1.917
Total labor-hours/year		34.846	15.336	3.834	1.917	1.917

			Cost Each				
			2014 Bare Costs			Total	Total
Description	Labor-hrs.	Material	Labor	Equip.	Total	In-House	w/O&P
1900 Swimming pool outdoor, annually	1.917		90.50		90.50	120	148
1950 Annualized	57.849		2,725		2,725	3,600	4,450

G20 SITE IMPROVEMENTS G2015 610 Traffic Signal Light

PM Components	Labor-hrs.	W	M	Q	S	A
PM System G2015 610 1950						
Traffic signal light						
1 Inspect and clean control cabinet.	.030					√
2 Check wiring for obvious defects and tighten control electrical connections.	.090					√
3 Check operation of cabinet exhaust fan and inspect vents.	.020					√
4 Inspect and tighten power line connections to control cabinet.	.034					√
5 Check operation of cabinet heater.	.037					√
6 Perform the following operational checks of traffic controller. A) Check controller phase timing with stop watch. B) Check controller pedestrian phase timing with stop watch. C) Check controller clocks for correct time and log settings.	.082			√	√	√
7 Walk a circuit of the intersection to ensure all lights are functioning, replace bulbs as required.	.070			√	√	√
8 Check operation of pedestrian push buttons, average two per intersection.	.084			√	√	√
9 Check operation of all informational signs.	.042			√	√	√
10 Check operation of phase detectors (loop amplifiers).	.144			√	√	√
11 Visually inspect road sensors.	.074			√	√	√
12 Check operation of flashers.	.018					√
13 Fill out maintenance checklist and report deficiencies.	.017			√	√	√
Total labor-hours/period				.513	.513	.742
Total labor-hours/year				1.026	.513	.742

			Cost Each				
			2014 Bare Costs			Total	Total
Description		Labor-hrs.	Material	Labor	Equip.	Total	In-House	w/O&P
1900	Traffic signal light, annually	.742		39.50		39.50	49	61.50
1950	Annualized	2.281		122		122	151	188

G20 SITE IMPROVEMENTS G2045 150 Chain Link Fence Gates

PM Components	Labor-hrs.	W	M	Q	S	A
PM System G2045 150 1950						
Manual swing gate						
1 Check gate and hinge alignment.	.325				✓	✓
2 Tighten diagonal brace turnbuckle.	.108				✓	✓
3 Check cane bolt alignment.	.108				✓	✓
4 Grease hinges.	.217				✓	✓
5 Oil latch.	.108				✓	✓
Total labor-hours/period					.866	.866
Total labor-hours/year					.866	.866

			Cost Each				
			2014 Bare Costs			Total	Total
Description	Labor-hrs.	Material	Labor	Equip.	Total	In-House	w/O&P
1900 Manual swing gate, annually	.866	8.35	31.50		39.85	51.50	62.50
1950 Annualized	1.732	8.35	63.50		71.85	94	114

PM Components	Labor-hrs.	W	M	Q	S	A
PM System G2045 150 2950						
Electric swing gate						
1 Check gate and hinge alignment.	.325				✓	✓
2 Tighten diagonal brace turnbuckle.	.108				✓	✓
3 Check cane bolt alignment.	.108				✓	✓
4 Grease hinges.	.217				✓	✓
5 Oil latch.	.108				✓	✓
6 Check electric motor and connections.	.325				✓	✓
7 Grease gearbox.	.217				✓	✓
Total labor-hours/period					1.408	1.408
Total labor-hours/year					1.408	1.408

			Cost Each				
			2014 Bare Costs			Total	Total
Description	Labor-hrs.	Material	Labor	Equip.	Total	In-House	w/O&P
2900 Electric swing gate, annually	1.408	8.35	51.50		59.85	78	95.50
2950 Annualized	2.816	8.35	103		111.35	147	180

PM Components	Labor-hrs.	W	M	Q	S	A
PM System G2045 150 3950						
Manual slide gate						
1 Check alignment of rollers and gate.	.325				✓	✓
2 Grease rollers.	.325				✓	✓
3 Oil latch.	.108				✓	✓
Total labor-hours/period					.758	.758
Total labor-hours/year					.758	.758

			Cost Each				
			2014 Bare Costs			Total	Total
Description	Labor-hrs.	Material	Labor	Equip.	Total	In-House	w/O&P
3900 Manual slide gate, annually	.758	8.35	28		36.35	46	56
3950 Annualized	1.516	8.35	55.50		63.85	83.50	102

G20 SITE IMPROVEMENTS G2045 150 Chain Link Fence Gates

PM Components	Labor-hrs.	W	M	Q	S	A
PM System G2045 150 4950						
Electric slide gate						
1 Check alignment of rollers and gate.	.325				√	√
2 Grease rollers.	.325				√	√
3 Oil latch.	.108				√	√
4 Check electric motor and connections.	.325				√	√
5 Clean and grease chain.	.650			√	√	√
Total labor-hours/period				.650	1.733	1.733
Total labor-hours/year				1.300	1.733	1.733

			Cost Each				
			2014 Bare Costs			Total	Total
Description	Labor-hrs.	Material	Labor	Equip.	Total	In-House	w/O&P
4900 Electric slide gate, annually	1.733	8.35	63.50		71.85	94	114
4950 Annualized	4.766	8.35	175		183.35	242	297

G30 SITE MECH. UTILITIES | G3015 108 | Storage Tank

PM Components	Labor-hrs.	W	M	Q	S	A
PM System G3015 108 1950						
Storage tank, tower						
1 Inspect exterior of tank for leaks or damage, including tank base and base plates.	.151					✓
2 Inspect condition of ladders, sway bracing hardware and structural forms.	.624					✓
3 Tighten and lubricate altitude valve and inspect valve vault.	.039					✓
4 Check operation of storage tank lighting.	.121					✓
5 Check cathodic protection system.	.122					✓
6 Operate elevated tank heat system, if installed, and inspect insulation and coils.	.107					✓
7 Clean up around tank area.	.134					✓
8 Fill out maintenance checklist and report deficiencies.	.022					✓
Total labor-hours/period						1.320
Total labor-hours/year						1.320

			Cost Each				
			2014 Bare Costs			Total	Total
Description	Labor-hrs.	Material	Labor	Equip.	Total	In-House	w/O&P
1900 Storage tank, tower, annually	1.320	16.75	62.50		79.25	101	123
1950 Annualized	1.320	16.75	62.50		79.25	101	123

PM Components	Labor-hrs.	W	M	Q	S	A
PM System G3015 108 2950						
Storage tank, ground level						
1 Inspect exterior of tank for leaks or damage.	.053					✓
2 Check condition of roof and roof hatches including locks; remove trash from roof.	.147					✓
3 Inspect condition of ladders and tighten or replace missing hardware.	.088					✓
4 Inspect valves, fittings, drains and controls for leakage or damage.	.039					✓
5 Check for deterioration of tank foundation.	.077					✓
6 Inspect ground drainage slope, and vent screens; where applicable.	.043					✓
7 Inspect tank lighting system; replace bulbs as needed.	.121					✓
8 Clean up around tank.	.134					✓
9 Fill out maintenance checklist and report deficiencies.	.022					✓
Total labor-hours/period						.724
Total labor-hours/year						.724

			Cost Each				
			2014 Bare Costs			Total	Total
Description	Labor-hrs.	Material	Labor	Equip.	Total	In-House	w/O&P
2900 Storage tank, ground level, annually	.724	16.75	34		50.75	63.50	77
2950 Annualized	.724	16.75	34		50.75	63.50	77

G30 SITE MECH. UTILITIES | G3015 116 | Water Flow Meter

PM Components	Labor-hrs.	W	M	Q	S	A
PM System G3015 116 1950						
Water flow meter, turbine						
1 Inspect meter for leaks, corrosion or broken sight glass; replace broken glass as required.	.035				√	√
2 Check meter operation for unusual noise.	.034				√	√
3 Clean meter to remove mineral buildup.	.039				√	√
4 Clean out trash or sludge from bottom of pit.	.165				√	√
5 Fill out maintenance report and report deficiencies.	.022				√	√
Total labor-hours/period					.295	.295
Total labor-hours/year					.295	.295

		Cost Each					
			2014 Bare Costs			Total	Total
Description	Labor-hrs.	Material	Labor	Equip.	Total	In-House	w/O&P
1900 Water flow meter, turbine, annually	.295	15.85	13.95		29.80	36	42.50
1950 Annualized	.590	15.85	28		43.85	54.50	65.50

G30 SITE MECH. UTILITIES | G3015 118 | Reservoir Controls

PM Components	Labor-hrs.	W	M	Q	S	A
PM System G3015 118 1950						
Fresh water distribution reservoir controls						
1 Check with reservoir personnel for any known deficiencies.	.086			√	√	√
2 Visually inspect and clean interior and exterior of controls cabinet.	.061			√	√	√
3 Inspect electrical system for frayed wires or loose connections; repair as required.	.120			√	√	√
4 Inspect water level receiver at reservoir.	.192			√	√	√
5 Inspect and adjust chronoflo transmitter.	.467			√	√	√
6 Perform operational check of relays, safety switches and electrical contactors.	.542			√	√	√
7 Perform operational check of controls system.	.659			√	√	√
8 Clean up controls area and dispose of debris.	.066			√	√	√
9 Fill out maintenance checklist and report deficiencies.	.022			√	√	√
Total labor-hours/period				2.215	2.215	2.215
Total labor-hours/year				4.430	2.215	2.215

		Cost Each					
			2014 Bare Costs			Total	Total
Description	Labor-hrs.	Material	Labor	Equip.	Total	In-House	w/O&P
1900 Reservoir controls, fresh water distribution, annually	2.215	33.50	105		138.50	175	213
1950 Annualized	8.860	33.50	420		453.50	590	725

G30 SITE MECH. UTILITIES | G3015 126 | Pumps

PM Components	Labor-hrs.	W	M	Q	S	A
PM System G3015 126 0950						
Pump, air lift, well						
1 Check with pump operating personnel for any known deficiencies.	.035				√	√
2 Check operation of compressor and pump.	.033				√	√
3 Check water sample for air or oil contamination.	.009				√	√
4 Check compressor oil level; add oil as necessary and lubricate pump and compressor shaft bearings as applicable.	.121				√	√
5 Inspect electrical system for frayed wires or loose connections; repair as required.	.120				√	√
6 Check and cycle high and low shut off valves.	.130				√	√
7 Inspect and clean air intake filter.	.094				√	√
8 Clean pump body and tighten or replace loose hardware.	.086				√	√
9 Calibrate and adjust pressure gauge.	.191				√	√
10 Clean surrounding area.	.066				√	√
11 Fill out maintenance checklist and report deficiencies.	.022				√	√
Total labor-hours/period					.907	.907
Total labor-hours/year					.907	.907

			Cost Each					
			2014 Bare Costs				Total	Total
Description		Labor-hrs.	Material	Labor	Equip.	Total	In-House	w/O&P
0900	Pump, air lift, well, annually	.907	16.75	43		59.75	75	91
0950	Annualized	1.814	33.50	86		119.50	150	182

PM Components	Labor-hrs.	W	M	Q	S	A
PM System G3015 126 1950						
Pump, centrifugal ejector						
1 Check for proper operation of pump.	.022				√	√
2 Check for leaks on suction and discharge piping, seals, packing glands, etc.; make minor adjustments as required.	.077				√	√
3 Check pump and motor operation for excessive vibration, noise and overheating.	.022				√	√
4 Check alignment of pump and motor; adjust as necessary.	.260				√	√
5 Lubricate pump and motor.	.099				√	√
6 Clean exterior of pump, motor and surrounding area.	.096				√	√
7 Fill out maintenance checklist and report deficiencies.	.022				√	√
Total labor-hours/period					.598	.598
Total labor-hours/year					.598	.598

			Cost Each					
			2014 Bare Costs				Total	Total
Description		Labor-hrs.	Material	Labor	Equip.	Total	In-House	w/O&P
1900	Pump, centrifugal ejector, annually	.598	8.35	28.50		36.85	46.50	56.50
1950	Annualized	1.196	16.75	56.50		73.25	93	113

G30 SITE MECH. UTILITIES | G3015 126 | Pumps

PM Components	Labor-hrs.	W	M	Q	S	A
PM System G3015 126 3950						
Pump, metering (slurry)						
1 Check with pump operating personnel for any obvious deficiencies.	.035				√	√
2 Obtain and put on safety equipment for working with fluoride: goggles, apron, gloves and respirator.	.014				√	√
3 Check oil level in pump with dipstick; add oil as necessary.	.009				√	√
4 Inspect belt condition and tension on pump; adjust as required.	.029				√	√
5 Flush slurry pump with fresh water.	.139				√	√
6 Pressure test pump.	.012				√	√
7 Inspect pump motor.	.039				√	√
8 Operate pump and check for leaks, excessive noise and vibration.	.099				√	√
9 Clean up area.	.066				√	√
10 Fill out maintenance checklist and report deficiencies.	.022				√	√
Total labor-hours/period					.464	.464
Total labor-hours/year					.464	.464

			Cost Each				
			2014 Bare Costs			Total	Total
Description		Labor-hrs.	Material	Labor	Equip.	Total	In-House	w/O&P
3900	Pump, metering (slurry), annually	.464	16.75	22		38.75	47.50	57
3950	Annualized	.928	33.50	44		77.50	95	113

PM Components	Labor-hrs.	W	M	Q	S	A
PM System G3015 126 4950						
Pump, mixed or axial flow velocity						
1 Check with pump operating personnel for any obvious deficiencies.	.035				√	√
2 Clean pump exterior.	.014				√	√
3 Check impeller and shaft for correct alignment and clearances.	.031				√	√
4 Check oil level; add oil as needed.	.009				√	√
5 Inspect packing for leaks and tighten as needed.	.031				√	√
6 Inspect bearings for wear, damage or overheating; lubricate as required.	.053				√	√
7 Inspect float operation and adjust as needed.	.098				√	√
8 Inspect V-belt tension and condition; adjust as required.	.029				√	√
9 Operate pump and check for leaks, excessive noise and vibration.	.099				√	√
10 Tighten loose nuts and bolts and replace missing hardware.	.005				√	√
11 Check electrical system for frayed wires and loose connections.	.120				√	√
12 Clean up area.	.066				√	√
13 Fill out maintenance checklist and report deficiencies.	.022				√	√
Total labor-hours/period					.612	.612
Total labor-hours/year					.612	.612

			Cost Each				
			2014 Bare Costs			Total	Total
Description		Labor-hrs.	Material	Labor	Equip.	Total	In-House	w/O&P
4900	Pump, mixed or axial flow, annually	.612	16.75	29		45.75	56.50	68
4950	Annualized	1.224	33.50	58		91.50	113	136

G30 SITE MECH. UTILITIES | G3015 126 Pumps

PM Components	Labor-hrs.	W	M	Q	S	A
PM System G3015 126 5950						
Pump, reciprocating positive displacement						
1 Check for proper operation of pump.	.022				√	√
2 Check for leaks on suction and discharge piping, seals, packing glands, etc.; make minor adjustments as required.	.077				√	√
3 Check pump and motor operation for excessive vibration, noise and overheating.	.022				√	√
4 Check operation of pressure controls.	.133				√	√
5 Check alignment of pump and motor; adjust as necessary.	.260				√	√
6 Lubricate pump and motor.	.099				√	√
7 Clean exterior of pump, motor and surrounding area.	.094				√	√
8 Fill out maintenance checklist and report deficiencies.	.022				√	√
Total labor-hours/period					.729	.729
Total labor-hours/year					.729	.729

			Cost Each					
			2014 Bare Costs				Total	Total
Description		Labor-hrs.	Material	Labor	Equip.	Total	In-House	w/O&P
5900	Pump, reciprocating displacement, annually	.729	8.35	34.50		42.85	54.50	66.50
5950	Annualized	1.458	16.75	69		85.75	109	133

PM Components	Labor-hrs.	W	M	Q	S	A
PM System G3015 126 6950						
Pump, rotary positive displacement						
1 Check with operating personnel for obvious defects.	.035				√	√
2 Clean surface of pump with solvent.	.014				√	√
3 Operate pump and check for leaks at pipes and connections and for excessive noise and vibration.	.099				√	√
4 Lubricate pump.	.047				√	√
5 Check shaft alignment and clearances and pump rotation.	.031				√	√
6 Inspect packing for leaks and tighten.	.031				√	√
7 Tighten loose bolts and nuts; replace loose and missing hardware as needed.	.009				√	√
8 Inspect electrical system for loose connections and frayed wires; repair as necessary.	.120				√	√
9 Operate pump upon completion of maintenance, read and record pressure gauge readings and check discharge pressure.	.221				√	√
10 Clean up area around pump after repairs.	.066				√	√
11 Fill out maintenance checklist and report deficiencies.	.022				√	√
Total labor-hours/period					.695	.695
Total labor-hours/year					.695	.695

			Cost Each					
			2014 Bare Costs				Total	Total
Description		Labor-hrs.	Material	Labor	Equip.	Total	In-House	w/O&P
6900	Pump, rotary displacement, annually	.695	8.35	33		41.35	52.50	64
6950	Annualized	1.390	16.75	66		82.75	105	128

G30 SITE MECH. UTILITIES G3015 126 Pumps

PM Components	Labor-hrs.	W	M	Q	S	A
PM System G3015 126 7950						
Pump, sump, up to 1 H.P.						
1 Check electrical plug, cord and connections.	.120				✓	✓
2 Activate float switch and check pump for proper operation.	.039				✓	✓
3 Lubricate pump as required.	.047				✓	✓
4 Inspect packing and tighten as required.	.031				✓	✓
5 Check pump for misalignment and bearings for overheating.	.199				✓	✓
6 Clean out trash from sump.	.065				✓	✓
7 Fill out maintenance checklist and report deficiencies.	.022				✓	✓
Total labor-hours/period					.523	.523
Total labor-hours/year					.523	.523

			Cost Each				
			2014 Bare Costs			Total	Total
Description	Labor-hrs.	Material	Labor	Equip.	Total	In-House	w/O&P
7900 Pump, sump, up to 1 H.P., annually	.523	8.35	24.50		32.85	42	51
7950 Annualized	1.046	16.75	49.50		66.25	83.50	101

PM Components	Labor-hrs.	W	M	Q	S	A
PM System G3015 126 8950						
Pump, turbine, well						
1 Check with pump operating personnel for any known deficiencies.	.035				✓	✓
2 Remove turbine and pump from deep well.	.585				✓	✓
3 Inspect pump body, bowls, water passages and impeller for corrosion, wear or pitting; repair as necessary.	.057				✓	✓
4 Check packing for leaks; tighten or replace as necessary.	.325				✓	✓
5 Adjust impeller for optimum operation.	.228				✓	✓
6 Inspect wear ring clearances, thrust and pump bearings for wear; repair as necessary.	.142				✓	✓
7 Lubricate all pump parts as required.	.047				✓	✓
8 Clean suction strainer or replace if damaged.	.338				✓	✓
9 Inspect operation of discharge valve.	.026				✓	✓
10 Inspect electrical system for frayed wires or loose connections; repair as necessary.	.120				✓	✓
11 Check operation of pressure and thermal controls.	.052				✓	✓
12 Paint pump with underwater paint.	.259				✓	✓
13 Run pump and check for noise and vibration.	.022				✓	✓
14 Reinstall pump in deep well and test operation.	.585				✓	✓
15 Fill out maintenance checklist and report deficiencies.	.022				✓	✓
Total labor-hours/period					2.843	2.843
Total labor-hours/year					2.843	2.843

			Cost Each				
			2014 Bare Costs			Total	Total
Description	Labor-hrs.	Material	Labor	Equip.	Total	In-House	w/O&P
8900 Pump, turbine, well, annually	2.843	151	134		285	345	410
8950 Annualized	5.686	167	269		436	540	650

G30 SITE MECH. UTILITIES G3015 126 Pumps

PM Components	Labor-hrs.	W	M	Q	S	A
PM System G3015 126 9950						
Pump, vacuum						
1 Check with pump operating personnel for any known deficiencies.	.035				✓	✓
2 Check for leaks on suction and discharge piping, seals, etc.	.077				✓	✓
3 Check pump and motor operation for vibration, noise, overheating, etc.	.022				✓	✓
4 Check alignment and clearances of shaft and coupler.	.260				✓	✓
5 Tighten or replace loose, missing, or damaged nuts, bolts, and screws.	.005				✓	✓
6 Lubricate pump and motor as required.	.099				✓	✓
7 Clean pump, motor and surrounding area.	.099				✓	✓
8 Fill out maintenance checklist and report deficiencies.	.022				✓	✓
Total labor-hours/period					.619	.619
Total labor-hours/year					.619	.619

			Cost Each				
			2014 Bare Costs			Total	Total
Description		Labor-hrs.	Material	Labor	Equip.	Total	In-House	w/O&P
9900 Pump, vacuum, annually		.619	8.35	29.50		37.85	48	58
9950 Annualized		1.238	16.75	58.50		75.25	95.50	116

G30 SITE MECH. UTILITIES | G3015 310 | Pump, Turbine Well

PM Components	Labor-hrs.	W	M	Q	S	A
PM System G3015 310 1950						
Turbine well pump						
1 Check with operating personnel for any deficiencies.	.035				✓	✓
2 Remove turbine pump from deep well.	.585				✓	✓
3 Inspect pump body, bowls, water passages and impeller for corrosion, wear or pitting; repair as necessary.	.057				✓	✓
4 Check packing for leaks; tighten or repack as necessary.	.330				✓	✓
5 Adjust impeller for optimun operation.	.228				✓	✓
6 Inspect wear ring clearances, thrust and pump bearings for wear; repair as necessary.	.142				✓	✓
7 Lubricate all pump parts as required.	.047				✓	✓
8 Clean suction strainer or replace if damaged.	.338				✓	✓
9 Inspect operation of discharge valve.	.030				✓	✓
10 Inspect electrical system for frayed wires or loose connections; repair as necessary.	.120				✓	✓
11 Check operation of pressure and thermal controls.	.124				✓	✓
12 Paint pump with underwater paint.	.259				✓	✓
13 Run pump and check for noise and vibration.	.022				✓	✓
14 Reinstall pump in deep well and test operation.	.585				✓	✓
15 Fill out maintenance checklist and report deficiencies.	.022				✓	✓
Total labor-hours/period					2.924	2.924
Total labor-hours/year					2.924	2.924

			Cost Each				
			2014 Bare Costs			Total	Total
Description		Labor-hrs.	Material	Labor	Equip.	Total	In-House	w/O&P
1900 Turbine well pump, annually		2.924	83.50	168		251.50	300	365
1950 Annualized		5.848	167	335		502	605	735

G30 SITE MECH. UTILITIES | G3015 320 | Pump, Vertical Lift

PM Components	Labor-hrs.	W	M	Q	S	A
PM System G3015 320 1950						
Vertical lift pump, over 1 H.P.						
1 Check for proper operation of pump.	.130				✓	✓
2 Check for leaks on suction and discharge piping, seals, packing glands, etc.; make minor adjustments as required.	.077				✓	✓
3 Check pump and motor operation for excessive vibration, noise and overheating.	.022				✓	✓
4 Check alignment of pump and motor; adjust as necessary.	.260				✓	✓
5 Lubricate pump and motor.	.099				✓	✓
6 When available, check and record suction or discharge gauge pressure and flow rate.	.022				✓	✓
7 Clean exterior of pump and surrounding area.	.096				✓	✓
8 Fill out maintenance checklist and report deficiencies.	.022				✓	✓
Total labor-hours/period					.728	.728
Total labor-hours/year					.728	.728

			Cost Each				
			2014 Bare Costs			Total	Total
Description		Labor-hrs.	Material	Labor	Equip.	Total	In-House	w/O&P
1900 Vertical lift pump, over 1 H.P., annually		.728	33.50	42		75.50	89	107
1950 Annualized		1.456	67	84		151	178	215

G30 SITE MECH. UTILITIES | G3015 410 | Fire Hydrant

PM Components	Labor-hrs.	W	M	Q	S	A
PM System G3015 410 1950						
Fire hydrant						
1 Remove hydrant caps and check condition of gaskets; replace as required.	.241					✓
2 Flush hydrant, lubricate cap threads and reinstall cap.	.374					✓
3 Fill out maintenance checklist and report deficiencies.	.022					✓
Total labor-hours/period						.637
Total labor-hours/year						.637

			Cost Each				
			2014 Bare Costs			Total	Total
Description		Labor-hrs.	Material	Labor	Equip.	Total	In-House	w/O&P
1900	Fire hydrant, annually	.637	8	35.50		43.50	53	65
1950	Annualized	.637	8	35.50		43.50	53	65

G30 SITE MECH. UTILITIES | G3015 420 | Valve, Post Indicator

PM Components	Labor-hrs.	W	M	Q	S	A
PM System G3015 420 1950						
Valve, post indicator						
1 Remove set screw, cap and wire seal.	.111				✓	✓
2 Apply lubricant to threads, open and close valve.	.293				✓	✓
3 Check operation of and sign and clean glass indicator windows.	.130				✓	✓
4 Install cap and tighten screw, install handle and wire seal with valve open.	.088				✓	✓
5 Clean valve exterior and area around valve.	.066				✓	✓
6 Fill out maintenance checklist and report deficiencies.	.022				✓	✓
Total labor-hours/period					.710	.710
Total labor-hours/year					.710	.710

			Cost Each				
			2014 Bare Costs			Total	Total
Description		Labor-hrs.	Material	Labor	Equip.	Total	In-House	w/O&P
1900	Valve, post indicator, annually	.710	8	39.50		47.50	58	71.50
1950	Annualized	1.420	8	78.50		86.50	107	133

G30 SITE MECH. UTILITIES | G3015 612 | Water Treatment, Filtration Plant

	PM Components	Labor-hrs.	W	M	Q	S	A
PM System G3015 612 1950							
Filter plant							
1	Check daily operating records for unusual occurances.	.035			✓	✓	✓
2	Inspect plant sand or gravel filters.	.694			✓	✓	✓
3	Inspect clear well for cleanliness.	.018			✓	✓	✓
4	Inspect and lubricate flocculator chain and baffle mechanism and waste-line valves.	.163			✓	✓	✓
5	Inspect and adjust rate-of-flow controller.	.036			✓	✓	✓
6	Calibrate and adjust loss-of-head gauge.	.191			✓	✓	✓
7	Inspect clear well vent screen and manhole.	.143			✓	✓	✓
8	Inspect operation of pumps, water mixers, chemical feeders and flow measuring devices.	1.145			✓	✓	✓
9	Check aerator nozzles for obstructions; clean as required.	.065			✓	✓	✓
10	Drain, clean and inspect mixing chambers.	1.455					✓
11	Drain sedimentation basins and clean with fire hose.	2.220					✓
12	Drain, inspect and clean flocculator.	2.401					✓
13	Clean up filter plant area.	.334			✓	✓	✓
14	Fill out maintenance checklist and report deficiencies.	.022			✓	✓	✓
	Total labor-hours/period				2.846	2.846	8.922
	Total labor-hours/year				5.692	2.846	8.922

			Cost Each					
			2014 Bare Costs				Total	Total
	Description	Labor-hrs.	Material	Labor	Equip.	Total	In-House	w/O&P
1900	Filter plant, annually	8.922	8.35	515		523.35	650	810
1950	Annualized	17.460	33.50	1,000		1,033.50	1,300	1,625

G30 SITE MECH. UTILITIES | G3015 630 | Water Treatment, De-Ionization

	PM Components	Labor-hrs.	W	M	Q	S	A
PM System G3015 630 1950							
Water de-ionization station							
1	Check with operating or area personnel for deficiencies.	.035				✓	✓
2	Check bulk storage tanks, pumps, valves and piping for damage, corrosion or leaks.	.130				✓	✓
3	Lubricate pumps.	.047				✓	✓
4	Check venting of bulk storage tanks.	.043				✓	✓
5	Check operation of sodium hydroxide tank and piping heating system.	.124				✓	✓
6	Check operation of resin vent, blower and motor; lubricate as required.	.033				✓	✓
7	Check calibration of meters.	.577				✓	✓
8	Clean and check probes of metering system.	.040				✓	✓
9	Check condition of carbon filter bed; replace annually.	.694				✓	✓
10	Check water softener system for proper operation.	.195				✓	✓
11	Clean equpment and surrounding area.	.066				✓	✓
12	Fill out maintenance checklist and report deficiencies.	.022				✓	✓
	Total labor-hours/period					2.006	2.006
	Total labor-hours/year					2.006	2.006

			Cost Each					
			2014 Bare Costs				Total	Total
	Description	Labor-hrs.	Material	Labor	Equip.	Total	In-House	w/O&P
1900	Water de-ionization system, annually	2.006	325	115		440	505	585
1950	Annualized	4.012	335	231		566	655	780

G30 SITE MECH. UTILITIES | G3015 632 | Reverse Osmosis System

PM Components	Labor-hrs.	W	M	Q	S	A
PM System G3015 632 1950						
Reverse osmosis system, 750 gallons/month						
1 Check with operating personnel for report of treatment effectiveness.	.046			√	√	√
2 Inspect unit, piping and connections for leaks.	.077			√	√	√
3 Check operation of pumps and motors; check shafts for alignment and lubricate as required.	.330			√	√	√
4 Inspect electrical system for frayed wires or loose connections; repair as required.	.120			√	√	√
5 Check filters; replace as required.	.036			√	√	√
6 Check water level controls; adjust as required.	.030			√	√	√
7 Check holding tank temperature control valve operation.	.030			√	√	√
8 Replace charcoal filters as indicated by water test.	.143			√	√	√
9 Check to ensure that chemical supply level is properly maintained in the chemical feed system.	.120			√	√	√
Total labor-hours/period				.932	.932	.932
Total labor-hours/year				1.864	.932	.932

			Cost Each				
			2014 Bare Costs			Total	Total
Description		Labor-hrs.	Material	Labor	Equip.	Total	In-House	w/O&P
1900	Reverse osmosis system, annually	.932	58.50	44		102.50	123	145
1950	Annualized	3.728	184	176		360	435	515

G30 SITE MECH. UTILITIES | G3015 634 | Water Ion Exchange System

PM Components	Labor-hrs.	W	M	Q	S	A
PM System G3015 634 1950						
Water ion exchange system						
1 Check with operating personnel for deficiencies.	.035			√	√	√
2 Clean unit and piping and check for leaks.	.077			√	√	√
3 Clean brine pump and check for proper operation.	.120			√	√	√
4 Clean and check strainers pilot valve, vents and orifices.	.595					√
5 Inspect electrical system including pump contactor and contacts; repair as required.	.120			√	√	√
6 Check regeneration cycle; make adjustments as required.	.325			√	√	√
7 Clean surrounding area.	.066			√	√	√
8 Fill out maintenance report and report deficiencies.	.022			√	√	√
Total labor-hours/period				.765	.765	1.360
Total labor-hours/year				1.530	.765	1.360

			Cost Each				
			2014 Bare Costs			Total	Total
Description		Labor-hrs.	Material	Labor	Equip.	Total	In-House	w/O&P
1900	Water ion exchange system, annually	1.360		64.50		64.50	85	105
1950	Annualized	3.655		173		173	228	282

G30 SITE MECH. UTILITIES | G3015 650 | Water Softener

PM Components	Labor-hrs.	W	M	Q	S	A
PM System G3015 650 1950						
Water softener						
1 Check with building personnel for report of water softener effectiveness.	.086			✓	✓	✓
2 Check pressure gauges for proper operation.	.191			✓	✓	✓
3 Check density of brine solution in salt tank.	.124			✓	✓	✓
4 Check operation of float control in brine.	.040			✓	✓	✓
5 Inspect water softener piping, fittings and valves for leaks.	.077			✓	✓	✓
6 Lubricate valves and motors.	.330			✓	✓	✓
7 Make minor adjustments to water softener controls if required.	.014			✓	✓	✓
8 Inspect softener base and brine tank for corrosion and repair as needed.	.259			✓	✓	✓
9 Check operation of automatic fill valve in brine tank.	.030			✓	✓	✓
10 Check softener electrical wiring and phasing.	.120			✓	✓	✓
11 Clean up area around softener.	.066			✓	✓	✓
12 Fill out maintenance report and report deficiencies.	.022			✓	✓	✓
Total labor-hours/period				1.359	1.359	1.359
Total labor-hours/year				2.718	1.359	1.359

				Cost Each				
			2014 Bare Costs				Total	Total
Description		Labor-hrs.	Material	Labor	Equip.	Total	In-House	w/O&P
1900	Water softener, annually	1.359	15.85	64.50		80.35	102	125
1950	Annualized	5.436	15.85	257		272.85	355	440

G30 SITE MECH. UTILITIES | G3015 660 | Fluoride Saturator

PM Components	Labor-hrs.	W	M	Q	S	A
PM System G3015 660 1950						
Fluoride saturator						
1 Check with operating personnel for any obvious discrepancies.	.035		✓	✓	✓	✓
2 Obtain and put on safety equipment such as goggles, apron, gloves and respirator prior to working on the fluoride saturator.	.014		✓	✓	✓	✓
3 Check oil level in gearbox, agitator and injector pump; add oil as necessary.	.018		✓	✓	✓	✓
4 Check water level in fluoride mixing tank.	.007		✓	✓	✓	✓
5 Check operation of fluoride flow regulator, float switch and solenoid.	.127		✓	✓	✓	✓
6 Check saturator for leaks.	.077		✓	✓	✓	✓
7 Check operation of unit.	.098		✓	✓	✓	✓
8 Clean fluoride tank with brush and detergent.	.085		✓	✓	✓	✓
9 Clean surrounding area.	.066		✓	✓	✓	✓
10 Fill out maintenance checklist and report deficiencies.	.022		✓	✓	✓	✓
Total labor-hours/period			.549	.549	.549	.549
Total labor-hours/year			4.392	1.098	.549	.549

				Cost Each				
			2014 Bare Costs				Total	Total
Description		Labor-hrs.	Material	Labor	Equip.	Total	In-House	w/O&P
1900	Fluoride saturator, annually	.549	8.35	31.50		39.85	48.50	60
1950	Annualized	6.588	33.50	380		413.50	510	630

G30 SITE MECH. UTILITIES   G3015 670 Water Treatment, Chlorine

PM Components	Labor-hrs.	W	M	Q	S	A
PM System G3015 670 1950						
Chlorinator						
1 Check with operating or area personnel for deficiencies.	.035			√	√	√
2 Put on safety equipment prior to checking equipment.	.014			√	√	√
3 Clean chlorinator and check for leaks.	.057			√	√	√
4 Check water trap for proper level and bleed air.	.017			√	√	√
5 Check and clean water strainers.	.675			√	√	√
6 Grease fittings.	.023			√	√	√
7 Make minor adjustments to chlorinator as needed.	.094			√	√	√
8 Fill out maintenance checklist and report deficiencies.	.022			√	√	√
Total labor-hours/period				.937	.937	.937
Total labor-hours/year				1.874	.937	.937

			Cost Each				
			2014 Bare Costs			Total	Total
Description		Labor-hrs.	Material	Labor	Equip.	Total	In-House	w/O&P
1900	Chlorinator, annually	.937	8.35	54		62.35	76.50	94.50
1950	Annualized	3.748	16.75	216		232.75	288	355

PM Components	Labor-hrs.	W	M	Q	S	A
PM System G3015 670 2950						
Chlorine detector						
1 Put on safety equipment prior to checking equipment.	.014			√	√	√
2 Test alarm and inspect fan and sensor wiring.	.252			√	√	√
3 Test detector with chlorine test kit.	.044			√	√	√
4 Flush sensor unit with fresh water and drain and fill reservoir.	.068			√	√	√
5 Check drip rate and reservoir level in sight glass.	.013			√	√	√
6 Check "O" rings; replace as required.	.029			√	√	√
7 Inspect sensor mounting grommet, fan operation, detector mounting bolts and drain tube connection.	.147			√	√	√
8 Check charcoal filter cartridge; replace as needed.	.009			√	√	√
9 Check basket for cracked or worn areas.	.059			√	√	√
10 Check detector for leaks under pressure.	.077			√	√	√
11 Clean unit and surrounding area.	.066			√	√	√
12 Fill out maintenance checklist and report deficiencies.	.022			√	√	√
Total labor-hours/period				.800	.800	.800
Total labor-hours/year				1.600	.800	.800

			Cost Each				
			2014 Bare Costs			Total	Total
Description		Labor-hrs.	Material	Labor	Equip.	Total	In-House	w/O&P
2900	Chlorine detector, annually	.800	33.50	46		79.50	94.50	114
2950	Annualized	3.200	134	184		318	380	455

G30 SITE MECH. UTILITIES   G3025 410   Pump, Sewage Ejector

PM Components	Labor-hrs.	W	M	Q	S	A
PM System G3025 410 1950						
Ejector, pump, sewage						
1 Check for proper operation of pump.	.130				√	√
2 Check for leaks on suction and discharge piping, seals, packing glands, etc.; make minor adjustments as required.	.069				√	√
3 Check ejector and motor operation for excessive vibration, noise and overheating.	.022				√	√
4 Check float or pressure controls for proper operation.	.039				√	√
5 Lubricate ejector and motor.	.099				√	√
6 Clean pump unit and surrounding area.	.096				√	√
7 Fill out maintenance checklist and report deficiencies.	.022				√	√
Total labor-hours/period					.477	.477
Total labor-hours/year					.477	.477

			Cost Each				
			2014 Bare Costs			Total	Total
Description	Labor-hrs.	Material	Labor	Equip.	Total	In-House	w/O&P
1900 Ejector pump, sewage, annually	.477	8.35	27.50		35.85	43.50	53.50
1950 Annualized	.954	16.75	55		71.75	87	106

G30 SITE MECH. UTILITIES | G3025 412 | Ejector Pump

PM Components	Labor-hrs.	W	M	Q	S	A
PM System G3025 412 1950						
Ejector, pump						
1 Check for proper operation of pump.	.130				√	√
2 Check for leaks on suction and discharge piping, seals, packing glands, etc.; make minor adjustments as required.	.077				√	√
3 Check ejector and motor operation for excessive vibration, noise and overheating.	.022				√	√
4 Check float or pressure controls for proper operation.	.039				√	√
5 Lubricate ejector and motor.	.099				√	√
6 Clean pump unit and surrounding area.	.096				√	√
7 Fill out maintenance checklist and report deficiencies.	.022				√	√
Total labor-hours/period					.485	.485
Total labor-hours/year					.485	.485

			Cost Each				
			2014 Bare Costs			Total	Total
Description	Labor-hrs.	Material	Labor	Equip.	Total	In-House	w/O&P
1900 Ejector pump, annually	.485	9.45	23		32.45	40.50	49.50
1950 Annualized	.970	18.95	46		64.95	81.50	98

PM Components	Labor-hrs.	W	M	Q	S	A
PM System G3025 412 2950						
Ejector pump, sump type						
1 Operate float switch, if applicable, and check for proper operation of pump.	.130				√	√
2 Check for leaks on suction and discharge piping, seals, packing glands, etc.; make minor adjustments as required.	.077				√	√
3 Check pump and motor operation for excessive vibration, noise and overheating.	.022				√	√
4 Check alignment of pump and motor; adjust as necessary.	.260				√	√
5 Lubricate motor.	.099				√	√
6 Check electrical wiring and connections and repair as applicable.	.120				√	√
7 Clean exterior of pump and surrounding area.	.096				√	√
8 Fill out maintenance checklist and report deficiencies.	.022				√	√
Total labor-hours/period					.826	.826
Total labor-hours/year					.826	.826

			Cost Each				
			2014 Bare Costs			Total	Total
Description	Labor-hrs.	Material	Labor	Equip.	Total	In-House	w/O&P
2900 Ejector pump, sump type, annually	.826	9.45	39		48.45	62	75.50
2950 Annualized	1.652	18.95	78		96.95	124	151

G30 SITE MECH. UTILITIES G3025 420 Pump, Sewage Lift

PM Components	Labor-hrs.	W	M	Q	S	A
PM System G3025 420 1950						
Sewage lift pump, over 1 H.P.						
1 Check for proper operation of pump.	.130				√	√
2 Check for leaks on suction and discharge piping, seals, packing glands, etc.; make minor adjustments as required.	.077				√	√
3 Check pump and motor operation for excessive vibration, noise and overheating.	.022				√	√
4 Check alignment of pump and motor; adjust as necessary.	.260				√	√
5 Lubricate pump and motor.	.099				√	√
6 When available, check and record suction or discharge gauge pressure and flow rate.	.022				√	√
7 Clean exterior of pump, motor and surrounding area.	.096				√	√
8 Fill out maintenance checklist and report deficiencies.	.022				√	√
Total labor-hours/period					.728	.728
Total labor-hours/year					.728	.728

		Cost Each					
			2014 Bare Costs			Total	Total
Description	Labor-hrs.	Material	Labor	Equip.	Total	In-House	w/O&P
1900 Sewage lift pump, over 1 H.P., annually	.728	8.35	42		50.35	61.50	76
1950 Annualized	1.456	16.75	84		100.75	123	152

G30 SITE MECH. UTILITIES | G3025 530 | Aerator, Floating

PM Components	Labor-hrs.	W	M	Q	S	A
PM System G3025 530 1950						
Aerator, floating						
1 Check with operating or area personnel for deficiencies.	.035			√	√	√
2 Remove aerator from pond.	.390			√	√	√
3 Clean exterior of pump and motor.	.066			√	√	√
4 Check pump and motor operation for excessive vibration, noise and overheating.	.022			√	√	√
5 Check alignment of pump and motor; adjust as necessary.	.260			√	√	√
6 Lubricate pump and motor.	.099			√	√	√
7 Position aerator back in pond.	.390			√	√	√
8 Fill out maintenance checklist and report deficiencies.	.022			√	√	√
Total labor-hours/period				1.284	1.284	1.284
Total labor-hours/year				2.568	1.284	1.284

				Cost Each				
				2014 Bare Costs			Total	Total
	Description	Labor-hrs.	Material	Labor	Equip.	Total	In-House	w/O&P
1900	Aerator, floating, annually	1.284	9.45	60.50		69.95	90.50	111
1950	Annualized	5.136	18.95	243		261.95	340	420

G30 SITE MECH. UTILITIES | G3025 532 | Barminutor

PM Components	Labor-hrs.	W	M	Q	S	A
PM System G3025 532 1950						
Barminutor						
1 Check with operating or area personnel for deficiencies.	.044		√	√	√	√
2 Check condition and clearance of cutting knives and inspect base seal.	.217		√	√	√	√
3 Check oil level in gearbox; add oil as necessary.	.035		√	√	√	√
4 Change oil in gearbox.	.681					√
5 Wire brush and lubricate directional flow valve stem.	.060		√	√	√	√
6 Check for rust and corrosion; scrape, wire brush and spot paint as necessary.	.105		√	√	√	√
7 Fill out maintenance checklist and report deficiencies.	.022		√	√	√	√
Total labor-hours/period			.483	.483	.483	1.164
Total labor-hours/year			3.864	.966	.483	1.164

				Cost Each				
				2014 Bare Costs			Total	Total
	Description	Labor-hrs.	Material	Labor	Equip.	Total	In-House	w/O&P
1900	Barminutor, annually	1.164	33	55		88	109	131
1950	Annualized	6.477	75.50	305		380.50	485	595

G30 SITE MECH. UTILITIES G3025 534 Blower, Aerator

PM Components	Labor-hrs.	W	M	Q	S	A
PM System G3025 534 1950						
Blower, aerator						
1 Check with operating or area personnel for deficiencies.	.035		✓	✓	✓	✓
2 Check blower oil level; add oil as necessary.	.046		✓	✓	✓	✓
3 Change blower oil.	.390				✓	✓
4 Check bolts and tighten as necessary: foundation, cylinder head, belt guard, etc.	.155		✓	✓	✓	✓
5 Check tension, condition, and alignment of V-belts on blower; adjust or replace as necessary.	.030		✓	✓	✓	✓
6 Check pump for excessive heat or vibration.	.042		✓	✓	✓	✓
7 Clean air intake filter on blower; replace as necessary.	.177		✓	✓	✓	✓
8 Perform operation check of air blower and adjust as required.	.221		✓	✓	✓	✓
9 Clean area around equipment.	.066		✓	✓	✓	✓
10 Fill out maintenance checklist and report deficiencies.	.022		✓	✓	✓	✓
Total labor-hours/period			.794	.794	1.184	1.184
Total labor-hours/year			6.352	1.588	1.184	1.184

			Cost Each				
			2014 Bare Costs			Total	Total
Description	Labor-hrs.	Material	Labor	Equip.	Total	In-House	w/O&P
1900 Blower, aerator, annually	1.184	80.50	56		136.50	162	192
1950 Annualized	10.308	151	490		641	810	985

G30 SITE MECH. UTILITIES G3025 536 Clarifier

PM Components	Labor-hrs.	W	M	Q	S	A
PM System G3025 536 1950						
Clarifier						
1 Check with operating or area personnel for deficiencies.	.035		✓	✓	✓	✓
2 Check chain and sprocket for wear, where applicable.	.033		✓	✓	✓	✓
3 Grease drive gear fittings.	.150		✓	✓	✓	✓
4 Check gearbox oil level; add if necessary.	.046		✓	✓	✓	✓
5 Drain and refill gearbox oil reservoir.	.681					✓
6 Check unit for rust and corrosion; scrape, wire brush, and paint as required.	.105					✓
7 Clean clarifier drive and surrounding area.	.098		✓	✓	✓	✓
8 Fill out maintenance checklist and report deficiencies.	.022		✓	✓	✓	✓
Total labor-hours/period			.384	.384	.384	1.170
Total labor-hours/year			3.072	.768	.384	1.170

			Cost Each				
			2014 Bare Costs			Total	Total
Description	Labor-hrs.	Material	Labor	Equip.	Total	In-House	w/O&P
1900 Clarifier, annually	1.170	28.50	55.50		84	104	126
1950 Annualized	5.394	57	255		312	400	485

G30 SITE MECH. UTILITIES G3025 538 Comminutor

PM Components	Labor-hrs.	W	M	Q	S	A
PM System G3025 538 1950						
Comminutor						
1 Check with operating or area personnel for deficiencies.	.035			√	√	√
2 Change gearbox oil and lubricate fittings.	.681					√
3 Check comminutor and motor for excessive noise, vibration and overheating.	.022			√	√	√
4 Lubricate fittings and check oil in gearbox.	.150				√	√
5 Check operation.	.108			√	√	√
6 Fill out maintenance checklist and report deficiencies.	.022			√	√	√
Total labor-hours/period				.337	1.018	1.018
Total labor-hours/year				.674	1.018	1.018

			Cost Each				
			2014 Bare Costs			Total	Total
Description	Labor-hrs.	Material	Labor	Equip.	Total	In-House	w/O&P
1900 Comminutor, annually	1.018	28.50	48		76.50	95	114
1950 Annualized	2.710	75.50	128		203.50	252	305

G30 SITE MECH. UTILITIES G3025 540 Filter, Trickling

PM Components	Labor-hrs.	W	M	Q	S	A
PM System G3025 540 1950						
Filter, trickling						
1 Lubricate fittings and check oil level in oil reservoir.	.035				√	√
2 Check distribution arms, support cables, clevis pins, and turnbuckles for deterioration.	.325				√	√
3 Check for leaks.	.077				√	√
4 Check operation.	.108				√	√
5 Fill out maintenance checklist and report deficiencies.	.022				√	√
Total labor-hours/period					.567	.567
Total labor-hours/year					.567	.567

			Cost Each				
			2014 Bare Costs			Total	Total
Description	Labor-hrs.	Material	Labor	Equip.	Total	In-House	w/O&P
1900 Filter, trickling, annually	.567	18.95	27		45.95	56	67
1950 Annualized	1.134	38	53.50		91.50	113	135

G30 SITE MECH. UTILITIES G3025 542 Grit Drive

PM Components	Labor-hrs.	W	M	Q	S	A
PM System G3025 542 1950						
Grit drive						
1 Check with operating personnel for obvious deficiencies.	.035				✓	✓
2 Lubricate fittings and check oil level in gearbox.	.047				✓	✓
3 Change oil in gearbox.	.681					✓
4 Check chain linkage and operation.	.033				✓	✓
5 Clean area around drive mechanism.	.066				✓	✓
6 Fill out maintenance checklist and report deficiencies.	.022				✓	✓
Total labor-hours/period					.203	.884
Total labor-hours/year					.203	.884

				Cost Each				
				2014 Bare Costs			Total	Total
Description		Labor-hrs.	Material	Labor	Equip.	Total	In-House	w/O&P
1900	Grit drive, annually	.884	28.50	42		70.50	86.50	104
1950	Annualized	1.087	38	51.50		89.50	110	132

G30 SITE MECH. UTILITIES G3025 544 Mixer, Sewage

PM Components	Labor-hrs.	W	M	Q	S	A
PM System G3025 544 1950						
Mixer, sewage						
1 Check with operating or area personnel for deficiencies.	.035			✓	✓	✓
2 Clean drive motor exterior and check for corrosion on motor exterior and base plate.	.053			✓	✓	✓
3 Check motor operation for vibration, noise, overheating; etc.	.022			✓	✓	✓
4 Check alignment and clearances of shaft and coupler.	.031			✓	✓	✓
5 Tighten or replace loose, missing, or damaged nuts, bolts and screws.	.005			✓	✓	✓
6 Lubricate motor and shaft bearings as required.	.047			✓	✓	✓
7 Clean area around motor.	.066			✓	✓	✓
8 Fill out maintenance checklist and report deficiencies.	.022			✓	✓	✓
Total labor-hours/period				.281	.281	.281
Total labor-hours/year				.562	.281	.281

				Cost Each				
				2014 Bare Costs			Total	Total
Description		Labor-hrs.	Material	Labor	Equip.	Total	In-House	w/O&P
1900	Mixer, sewage, annually	.281	9.45	13.30		22.75	28	33.50
1950	Annualized	1.124	18.95	53		71.95	91	110

G30 SITE MECH. UTILITIES G3065 310 | Fuel Oil Tank

PM Components	Labor-hrs.	W	M	Q	S	A
PM System G3065 310 1950						
Fuel oil storage tank, above ground						
1 Inspect tank for corrosion.	.192					√
2 Inspect floating tank roof, lifter or cone and automatic float gauge, if applicable.	.233					√
3 Inspect fuel tank fire protection devices.	.022					√
4 Inspect berms around fuel storage tank for erosion.	.047					√
5 Inspect non-freeze draw valves.	.060					√
6 Inspect storage tank vacuum and pressure vents.	.179					√
7 Inspect vapor recovery line for deterioration.	.082					√
8 Run pump and check operation.	.159					√
9 Calibrate and adjust fuel oil pressure.	.382					√
10 Inspect fuel strainer and clean as needed.	.415					√
11 Inspect tank electrical grounds for breaks or loose fittings.	1.292					√
12 Inspect liquid level gauge for proper operation.	.191					√
13 Fill out maintenance checklist.	.022					√
Total labor-hours/period						3.276
Total labor-hours/year						3.276

			Cost Each					
				2014 Bare Costs			Total	Total
	Description	Labor-hrs.	Material	Labor	Equip.	Total	In-House	w/O&P
1900	Fuel oil storage tank, above ground, annually	3.276	67.50	192		259.50	315	385
1950	Annualized	3.276	67.50	192		259.50	315	385

G40 SITE ELEC. UTILITIES   G4095 110   Cathodic Protection System

PM Components	Labor-hrs.	W	M	Q	S	A
PM System G4095 110 1950						
Cathodic protection system						
1 Inspect exterior of rectifier cabinet, all exposed connecting conduit, and fittings for corrosion, deteriorated paint, mechanical integrity and cleanliness, make minor repairs and clean/touch-up paint as required.	.060				√	√
2 Open and inspect interior of rectifier cabinet, inspect wiring for fraying, signs of overheating, deterioration mechanical damage, and loose connections; make adjustments as required.	.138				√	√
3 Check electrical contacts for overheating and pitting; make adjustments.	.075				√	√
4 Clean interior of rectifier cabinet.	.008				√	√
5 Check panel voltmeter and ampmeter with standard test instruments.	.075				√	√
6 Read and record panel meter voltage and current values.	.007				√	√
7 Measure and record rectifier voltage and current with portable instrument.	.059				√	√
8 Check potential at cathode locations; tighten loose connections.	.286				√	√
9 Fill out maintenance checklist and record deficiencies.	.022				√	√
Total labor-hours/period					.730	.730
Total labor-hours/year					.730	.730

			Cost Each				
			2014 Bare Costs			Total	Total
Description		Labor-hrs.	Material	Labor	Equip.	Total	In-House	w/O&P
1900 Cathodic protection system, annually		.730	6	39		45	55	68
1950 Annualized		1.460	11.95	78		89.95	110	136

General Maintenance

Table of Contents

Landscaping Maintenance
01 93 04.15	Edging	559
01 93 04.20	Flower, Shrub & Tree Care	559
01 93 04.35	Lawn Care	559
01 93 04.40	Lawn Renovation	559
01 93 04.65	Raking	560
01 93 04.85	Walkways and Parking Care	560
01 93 04.90	Yard Waste Disposal	560

Exterior General Maintenance
01 93 05.95	Window Washing	560

Interior General Maintenance
01 93 06.20	Carpet Care	560
01 93 06.25	Ceiling Care	561
01 93 06.40	General Cleaning	561
01 93 06.45	Hard Floor Care	562
01 93 06.80	Vacuuming	565
01 93 06.90	Wall Care	566
01 93 06.95	Window and Accessory Care	566

Facility Maintenance Equipment
01 93 08.50	Equipment	567

How to Use the General Maintenance Pages

The following is a detailed explanation of a sample entry in the General Maintenance Section. Next to each bold number that follows is the described item with the appropriate component of the sample entry following in parentheses.

1 Division Number/Title
(01 93/Facility Maintenance)
Use the General Maintenance Section Table of Contents to locate specific items. This section is classified according to the CSI MasterFormat 2012 system.

2 Line Numbers
(01 93 04.40 0030)
Each General Maintenance unit price line item has been assigned a unique 12-digit code based on the CSI MasterFormat classification.

01 93 Facility Maintenance **1**

01 93 04 – Landscaping Maintenance

01 93 04.15 Edging		Crew	Daily Output	Labor-Hours	Unit	Material	2014 Bare Costs Labor	Equipment	Total	Total In-House	Total Incl O&P
0010	**EDGING**										
0020	Hand edging, at walks	1 Clam	16	.500	C.L.F.		13.75		13.75	17.05	21.50
0030	At planting, mulch or stone beds		7	1.143			31.50		31.50	39.00	48.50
0040	Power edging, at walks		88	.091			2.50		2.50	3.10	3.88
0050	At planting, mulch or stone beds		24	.333			9.15		9.15	11.35	14.20
01 93 04.40 Lawn Renovation		**4**	**5**	**6**					**8**	**9**	
0010	**LAWN RENOVATION**										
0020	Lawn renovations, aerating, 18" walk behind cultivator	1 Clam	95	.084	M.S.F.		2.31		2.31	2.87	3.59
0030	48" tractor drawn cultivator	A-18	750	.011			.29	.46	.75	.82	.96
0040	72" tractor drawn cultivator	"	1100	.007			.20	.31	.51	.56	.65
0050	Fertilizing, dry granular, 4#/M.S.F., drop spreader	1 Clam	24	.333		2.75	9.15		11.90	14.40	17.65
0060	Rotary spreader	"	140	.057		2.75	1.57		4.32	4.98	5.90

3 Description
(Lawn Renovations, etc.)
Each line item is described in detail. Sub-items and additional sizes are indented beneath the appropriate line items. The first line or two after the main item (in boldface) may contain descriptive information that pertains to all line items beneath this boldface listing.

4 Crew (A-18)
The "Crew" column designates the typical trade or crew used for the task. If the task can be accomplished by one trade and requires no power equipment, that trade and the number of workers are listed (for example, "2 Clam"). If the task requires a composite crew, a crew code designation is listed (for example, "A-18"). You'll find full details on all composite crews in the Crew Listings.

- For a complete list of all trades utilized in this book and their abbreviations, see the inside back cover of the book.

Crew A-18	Hr.	Daily	Hr.	Daily	Hr.	Daily	Bare Costs	In House	Incl. O&P
Maintenance Laborer	$27.45	$219.60	$34.10	$272.80	$42.65	$341.20	$27.45	$34.10	$42.65
Farm Tractor w/ Attachment				341.20		341.20	375.32	42.65	42.65	46.91
L.H., Daily Totals		$560.80		$614.00		$716.52	$70.10	$76.75	$89.56

5 Productivity:
Daily Output (750)/ Labor-Hours (.011)
The "Daily Output" represents the typical number of units the designated crew will perform in a normal 8-hour day. To find out the number of days the given crew would require to complete your task, divide your quantity by the daily output. For example:

Quantity	÷	Daily Output	=	Duration
2000 M.S.F.	÷	750 M.S.F./ Crew Day	=	2.7 Crew Days

The "Labor-Hours" figure represents the number of labor-hours required to perform one unit of work. To find out the number of labor-hours required for your particular task, multiply the quantity of the item times the number of labor-hours shown. For example:

Quantity	÷	Productivity Rate	=	Duration
2000 M.S.F.	÷	.011 Labor-Hours/ LF	=	22 Labor-Hours

6 Unit (M.S.F.)
The abbreviated designation indicates the unit of measure upon which the price, production, and crew are based (M.S.F. = Thousand Square Feet). For a complete listing of abbreviations, refer to the Abbreviations Listing in the Reference Section of this book.

7 Bare Costs:
The unit material cost is the "bare" material cost with no overhead and profit included. *Costs shown reflect national average material prices for January of the current year and include delivery to the facility. Small purchases may require an additional delivery charge. No sales taxes are included.*

Labor (.29)
The unit labor cost is derived by multiplying bare labor-hour costs for Crew A-18 by labor-hour units. The bare labor-hour cost is found in the Crew Section under A-18. (If a trade is listed, the hourly labor cost—the wage rate—is found on the inside back cover of the book.)

Labor-Hour CostCrew A-18	x	Labor-Hour Units	=	Labor
$27.45	x	.011	=	$0.29

Equip. (Equipment) (.46)
Equipment costs for each crew are listed in the description of each crew. Tools or equipment whose value justifies purchase or ownership by a contractor are considered overhead as shown on the inside back cover of the book. The unit equipment cost is derived by multiplying the bare equipment hourly cost by the labor-hour units.

Equipment Cost Crew A-18	x	Labor-Hour Units	=	Equip.
$42.65	x	.011	=	$0.46

Total (.75)
The total of the bare costs is the arithmetic total of the three previous columns: mat., labor, and equip

Material	+	Labor	+	Equip.	=	Total
$0.00	+	$0.29	+	$0.46	=	$0.75

8. Total In-House Costs (.82)

"Total In-House Costs" include suggested markups to bare costs for the direct overhead requirements of in-house maintenance staff. The figure in this column is the sum of three components: the bare material cost plus 10% (for purchasing and handling small quantities); the bare labor cost plus workers' compensation and average fixed overhead (per the labor rate table on the inside back cover of the book or, if a crew is listed, from the Crew Listings); and the bare equipment cost.

Material is Bare Material Cost + 10% = $0.00 + $0.00	=	$0.00
Labor for Crew A-18 = Labor-Hour Cost ($34.10) x Labor-Hour Units (0.11)	=	$0.36
Equip. is Bare Equip. Cost	=	$0.46
Total	=	$0.82

9. Total Costs Including O&P (.96)

"Total Costs Including O&P" include suggested markups to bare costs for an outside subcontractor's overhead and profit requirements. The figure in this column is the sum of three components: the bare material cost plus 25% for profit; the bare labor cost plus total overhead and profit (per the labor rate table on the inside back cover of the book or, if a crew is listed, from the Crew Listings); and the bare equipment cost plus 10% for profit.

Material is Bare Material Cost + 25% = $0.00 + $0.00	=	$0.00
Labor for Crew A-18 = Labor-Hour Cost ($42.65) x Labor-Hour Units (0.11)	=	$0.45
Equip. is Bare Equip. Cost + 10% = $0.46 + $0.05	=	$0.51
Total	=	$0.96

01 93 Facility Maintenance

01 93 04 – Landscaping Maintenance

01 93 04.15 Edging

		Crew	Daily Output	Labor-Hours	Unit	Material	2014 Bare Costs Labor	2014 Bare Costs Equipment	Total	Total In-House	Total Incl O&P
0010	**EDGING**										
0020	Hand edging, at walks	1 Clam	16	.500	C.L.F.		13.75		13.75	17.05	21.50
0030	At planting, mulch or stone beds		7	1.143			31.50		31.50	39.00	48.50
0040	Power edging, at walks		88	.091			2.50		2.50	3.10	3.88
0050	At planting, mulch or stone beds	↓	24	.333	↓		9.15		9.15	11.35	14.20

01 93 04.20 Flower, Shrub and Tree Care

		Crew	Daily Output	Labor-Hours	Unit	Material	Labor	Equipment	Total	Total In-House	Total Incl O&P
0010	**FLOWER, SHRUB & TREE CARE**										
0020	Flower or shrub beds, bark mulch, 3" deep hand spreader	1 Clam	100	.080	S.Y.	3.38	2.20		5.58	6.45	7.65
0030	Peat moss, 1" deep hand spreader		900	.009		2.63	.24		2.87	3.19	3.67
0040	Wood chips, 2" deep hand spreader		220	.036	↓	1.45	1		2.45	2.84	3.36
0050	Cleaning		1	8	M.S.F.		220		220	273.00	340
0060	Fertilizing, dry granular, 3 #/C.S.F.		85	.094		6.55	2.58		9.13	10.40	12.20
0070	Weeding, mulched bed		20	.400			11		11	13.65	17.05
0080	Unmulched bed		8	1	↓		27.50		27.50	34.00	42.50
0090	Trees, pruning from ground, 1-1/2" caliper		84	.095	Ea.		2.61		2.61	3.25	4.06
0100	2" caliper		70	.114			3.14		3.14	3.90	4.87
0110	2-1/2" caliper		50	.160			4.39		4.39	5.45	6.80
0120	3" caliper		30	.267			7.30		7.30	9.10	11.35
0130	4" caliper	2 Clam	21	.762			21		21	26.00	32.50
0140	6" caliper		12	1.333			36.50		36.50	45.50	57
0150	9" caliper		7.50	2.133			58.50		58.50	73.00	91
0160	12" caliper	↓	6.50	2.462			67.50		67.50	84.00	105
0170	Fertilize, slow release tablets	1 Clam	100	.080		.57	2.20		2.77	3.36	4.12
0180	Pest control, spray		24	.333		26	9.15		35.15	40.00	46.50
0190	Systemic		48	.167		26.50	4.58		31.08	35.00	40
0200	Watering, under 1-1/2" caliper		34	.235			6.45		6.45	8.00	10.05
0210	1-1/2" to 4" caliper		14.50	.552			15.15		15.15	18.80	23.50
0220	4" caliper and over		10	.800	↓		22		22	27.50	34
0230	Shrubs, prune, entire bed		7	1.143	M.S.F.		31.50		31.50	39.00	48.50
0240	Per shrub, 3' height		190	.042	Ea.		1.16		1.16	1.44	1.80
0250	4' height		90	.089			2.44		2.44	3.03	3.79
0260	6' height		50	.160	↓		4.39		4.39	5.45	6.80
0270	Fertilize, dry granular, 3#/M.S.F.		85	.094	M.S.F.	6.55	2.58		9.13	10.40	12.20
0280	Watering, entire bed		7	1.143	"		31.50		31.50	39.00	48.50
0290	Per shrub	↓	32	.250	Ea.		6.85		6.85	8.55	10.65

01 93 04.35 Lawn Care

		Crew	Daily Output	Labor-Hours	Unit	Material	Labor	Equipment	Total	Total In-House	Total Incl O&P
0010	**LAWN CARE**										
0020	Mowing lawns, power mower, 18"-22"	1 Clam	80	.100	M.S.F.		2.75		2.75	3.41	4.27
0030	22"-30"		120	.067			1.83		1.83	2.27	2.84
0040	30"-32"	↓	140	.057			1.57		1.57	1.95	2.44
0050	Self propelled or riding mower, 36"-44"	A-16	300	.027			.73	.23	.96	1.14	1.39
0060	48"-58"		480	.017			.46	.14	.60	.71	.87
0070	Tractor, 3 gang reel, 7' cut		930	.009			.24	.07	.31	.36	.45
0080	5 gang reel, 12' cut	↓	1200	.007	↓		.18	.06	.24	.29	.34
0090	Edge trimming with weed whacker	1 Clam	5760	.001	L.F.		.04		.04	.05	.06

01 93 04.40 Lawn Renovation

		Crew	Daily Output	Labor-Hours	Unit	Material	Labor	Equipment	Total	Total In-House	Total Incl O&P
0010	**LAWN RENOVATION**										
0020	Lawn renovations, aerating, 18" walk behind cultivator	1 Clam	95	.084	M.S.F.		2.31		2.31	2.87	3.59
0030	48" tractor drawn cultivator	A-18	750	.011			.29	.46	.75	.82	.96
0040	72" tractor drawn cultivator	"	1100	.007			.20	.31	.51	.56	.65
0050	Fertilizing, dry granular, 4#/M.S.F., drop spreader	1 Clam	24	.333		2.75	9.15		11.90	14.40	17.65
0060	Rotary spreader	"	140	.057	↓	2.75	1.57		4.32	4.98	5.90

01 93 Facility Maintenance

01 93 04 – Landscaping Maintenance

01 93 04.40 Lawn Renovation

		Crew	Daily Output	Labor-Hours	Unit	Material	2014 Bare Costs Labor	2014 Bare Costs Equipment	Total	Total In-House	Total Incl O&P
0070	Tractor drawn 8' spreader	A-18	500	.016	M.S.F.	2.75	.44	.68	3.87	4.26	4.87
0080	Tractor drawn 12' spreader	"	800	.010		2.75	.27	.43	3.45	3.80	4.34
0090	Overseeding, utility mix, 7#/M.S.F., drop spreader	1 Clam	10	.800		9.30	22		31.30	37.50	45.50
0100	Tractor drawn spreader	A-18	52	.154		9.30	4.22	6.55	20.07	22.00	25.50
0110	Watering, 1" of water, applied by hand	1 Clam	21	.381			10.45		10.45	13.00	16.25
0120	Soaker hoses	"	82	.098			2.68		2.68	3.33	4.16

01 93 04.65 Raking

		Crew	Daily Output	Labor-Hours	Unit	Material	Labor	Equipment	Total	Total In-House	Total Incl O&P
0010	**RAKING**										
0020	Raking, leaves, by hand	1 Clam	7.50	1.067	M.S.F.		29.50		29.50	36.50	45.50
0030	Power blower		45	.178			4.88		4.88	6.05	7.60
0040	Grass clippings, by hand		7.50	1.067			29.50		29.50	36.50	45.50

01 93 04.85 Walkways and Parking Care

		Crew	Daily Output	Labor-Hours	Unit	Material	Labor	Equipment	Total	Total In-House	Total Incl O&P
0010	**WALKWAYS AND PARKING CARE**										
0020	General cleaning, parking area and walks, sweeping by hand	1 Clam	15	.533	M.S.F.		14.65		14.65	18.20	23
0030	Power vacuum		100	.080			2.20		2.20	2.73	3.41
0040	Power blower		100	.080			2.20		2.20	2.73	3.41
0050	Hose down		30	.267			7.30		7.30	9.10	11.35

01 93 04.90 Yard Waste Disposal

		Crew	Daily Output	Labor-Hours	Unit	Material	Labor	Equipment	Total	Total In-House	Total Incl O&P
0010	**YARD WASTE DISPOSAL**										
0030	On site dump or compost heap	2 Clam	24	.667	C.Y.		18.30		18.30	22.50	28.50
0040	Into 6 C.Y. dump truck, 2 mile haul to compost heap	A-17	125	.064			1.76	2.16	3.92	4.34	5.10
0050	4 mile haul to compost heap, fees not included	"	85	.094			2.58	3.18	5.76	6.40	7.50

01 93 05 – Exterior General Maintenance

01 93 05.95 Window Washing

		Crew	Daily Output	Labor-Hours	Unit	Material	Labor	Equipment	Total	Total In-House	Total Incl O&P
0010	**WINDOW WASHING**										
0030	Minimum productivity	1 Clam	1	8	M.S.F.	4.32	220		224.32	278.00	345
0040	Average productivity		2.50	3.200		4.32	88		92.32	114.00	141
0050	Maximum productivity		4	2		4.32	55		59.32	73.00	91

01 93 06 – Interior General Maintenance

01 93 06.20 Carpet Care

		Crew	Daily Output	Labor-Hours	Unit	Material	Labor	Equipment	Total	Total In-House	Total Incl O&P
0010	**CARPET CARE**										
0100	Carpet cleaning, portable extractor with floor wand	1 Clam	20	.400	M.S.F.	8.30	11		19.30	23.00	27.50
0200	Deep carpet cleaning, self-contained extractor										
0210	24" path @ 100 feet per minute	1 Clam	64	.125	M.S.F.	10.35	3.43		13.78	15.65	18.30
0220	20" path		54	.148		10.35	4.07		14.42	16.45	19.25
0230	16" path		43	.186		10.35	5.10		15.45	17.75	21
0240	12" path		32	.250		10.35	6.85		17.20	19.90	23.50
0300	Carpet pile lifting, self-contained extractor										
0310	24" path at 90 feet per minute	1 Clam	60	.133	M.S.F.	11.40	3.66		15.06	17.10	19.95
0320	20" path		50	.160		11.40	4.39		15.79	18.00	21
0330	16" path		40	.200		11.40	5.50		16.90	19.35	23
0340	12" path		30	.267		11.40	7.30		18.70	21.50	25.50
0400	Carpet restoration cleaning, self-contained extractor										
0410	24" path at 50 feet per minute	1 Clam	32	.250	M.S.F.	12.40	6.85		19.25	22.00	26
0420	20" path		27	.296		12.40	8.15		20.55	23.50	28
0430	16" path		21	.381		12.40	10.45		22.85	26.50	32
0440	12" path		16	.500		12.40	13.75		26.15	30.50	37

01 93 Facility Maintenance

01 93 06 – Interior General Maintenance

01 93 06.25 Ceiling Care

		Crew	Daily Output	Labor-Hours	Unit	Material	2014 Bare Costs Labor	Equipment	Total	Total In-House	Total Incl O&P
0010	**CEILING CARE**										
0020	Washing hard ceiling from ladder, hand	1 Clam	2.05	3.902	M.S.F.	4.58	107		111.58	138.00	172
0030	Machine	"	2.52	3.175	"	4.58	87		91.58	113.00	141
0035	Acoustic tile cleaning, chemical spray, including masking										
0040	Unobstructed floor	4 Clam	5400	.006	S.F.	.06	.16		.22	.27	.33
0045	Obstructed floor	"	4000	.008	"	.06	.22		.28	.34	.42
0048	Chemical spray cleaning and coating, including masking										
0050	Unobstructed floor	4 Clam	4800	.007	S.F.	.13	.18		.31	.37	.44
0055	Obstructed floor	"	3400	.009	"	.13	.26		.39	.46	.56

01 93 06.40 General Cleaning

		Crew	Daily Output	Labor-Hours	Unit	Material	2014 Bare Costs Labor	Equipment	Total	Total In-House	Total Incl O&P
0010	**GENERAL CLEANING**										
0020	Sweep stairs & landings, damp wipe handrails	1 Clam	80	.100	Floor	.09	2.75		2.84	3.51	4.38
0030	Dust mop stairs & landings, damp wipe handrails		62.50	.128		.09	3.51		3.60	4.46	5.55
0040	Damp mop stairs & landings, damp wipe handrails		45	.178		.19	4.88		5.07	6.25	7.85
0050	Vacuum stairs & landings, damp wipe handrails, canister		41	.195		.09	5.35		5.44	6.75	8.40
0060	Vacuum stairs & landings, damp wipe handrails, backpack		58.22	.137		.09	3.77		3.86	4.79	5.95
0070	Vacuum stairs & landings, damp wipe handrails, upright		31.57	.253		.09	6.95		7.04	8.75	10.90
0080	Clean carpeted elevators		48	.167	Ea.	1.04	4.58		5.62	6.80	8.40
0090	Clean uncarpeted elevators		32	.250		.79	6.85		7.64	9.40	11.65
0100	Clean escalator		8	1		.92	27.50		28.42	35.00	43.50
0110	Restrooms, fixture cleaning, toilets		180	.044		.09	1.22		1.31	1.62	2.01
0120	Urinals		192	.042		.08	1.14		1.22	1.51	1.88
0130	Sinks		206	.039		.01	1.07		1.08	1.33	1.67
0140	Utility sink		175	.046		.05	1.25		1.30	1.62	2.01
0150	Bathtub		85	.094		.16	2.58		2.74	3.39	4.21
0160	Shower		85	.094		1.25	2.58		3.83	4.59	5.55
0170	Accessories cleaning, soap dispenser		1280	.006		.01	.17		.18	.22	.28
0180	Sanitary napkin dispenser		640	.013		.01	.34		.35	.44	.54
0190	Paper towel dispenser		1455	.006		.01	.15		.16	.20	.24
0200	Toilet tissue dispenser		1455	.006		.01	.15		.16	.20	.24
0210	Accessories, refilling/restocking, soap dispenser (1 bag)		640	.013		9.70	.34		10.04	11.10	12.65
0220	Sanitary napkin/tampon dispenser (25/50)		384	.021		20.50	.57		21.07	23.50	26.50
0230	Paper towel dispenser, folded (2 pkgs of 250)		545	.015		9.05	.40		9.45	10.45	11.95
0240	Paper towel dispenser, rolled (350' roll)		340	.024		4.58	.65		5.23	5.85	6.75
0250	Toilet tissue dispenser (500 2-ply sheet roll)		1440	.006		1.36	.15		1.51	1.69	1.94
0260	Sanitary seat covers (pkg of 250)		960	.008		5.15	.23		5.38	5.95	6.80
0270	Toilet partition cleaning		96	.083		.34	2.29		2.63	3.21	3.98
0280	Ceramic tile wall cleaning		2400	.003	S.F.	.02	.09		.11	.13	.17
0290	General cleaning, empty to larger portable container, wastebaskets		960	.008	Ea.		.23		.23	.28	.36
0300	Ash trays		2880	.003			.08		.08	.09	.12
0310	Cigarette stands		960	.008			.23		.23	.28	.36
0320	Sand urns		640	.013			.34		.34	.43	.53
0330	Pencil sharpeners		2880	.003			.08		.08	.09	.12
0340	Remove trash to central location		290	.028	Lb.		.76		.76	.94	1.18
0347	Bench		480	.017	Ea.	.05	.46		.51	.63	.77
0350	Chairs		960	.008		.02	.23		.25	.30	.39
0360	Chalkboard erasers		960	.008			.23		.23	.28	.36
0370	Chalkboards		16	.500	M.S.F.	4.58	13.75		18.33	22.00	27
0380	Desk tops		225	.036	Ea.	.07	.98		1.05	1.29	1.61
0390	Glass doors		168	.048	"	.18	1.31		1.49	1.82	2.26
0400	Hard surface audience type seating		40	.200	M.S.F.	5.75	5.50		11.25	13.15	15.75
0409	Locker, metal		320	.025	Ea.	.14	.69		.83	1.00	1.25

01 93 Facility Maintenance

01 93 06 – Interior General Maintenance

01 93 06.40 General Cleaning

		Crew	Daily Output	Labor-Hours	Unit	Material	2014 Bare Costs Labor	2014 Bare Costs Equipment	Total	Total In-House	Total Incl O&P
0410	Microwave ovens	1 Clam	240	.033	Ea.	.03	.91		.94	1.17	1.46
0420	Office partitions, to 48 S.F.		160	.050		.22	1.37		1.59	1.95	2.41
0430	Overhead air vents		175	.046		.02	1.25		1.27	1.58	1.98
0440	Refrigerators		35	.229		.37	6.25		6.62	8.20	10.20
0450	Revolving doors		35	.229		1.08	6.25		7.33	9.00	11.10
0460	Tables		240	.033		.09	.91		1	1.24	1.53
0470	Telephone		288	.028		.01	.76		.77	.96	1.19
0480	Telephone booth		190	.042		.56	1.16		1.72	2.06	2.50
0490	Water fountain		480	.017		.03	.46		.49	.60	.75
0500	Wastebaskets		260	.031		.03	.84		.87	1.08	1.35
0510	Dusting, beds		360	.022			.61		.61	.76	.95
0520	Bedside stand		960	.008			.23		.23	.28	.36
0530	Bookcase		960	.008			.23		.23	.28	.36
0540	Calculator		1920	.004			.11		.11	.14	.18
0550	Chair		1340	.006			.16		.16	.20	.25
0560	Coat rack		400	.020			.55		.55	.68	.85
0570	Couch		320	.025			.69		.69	.85	1.07
0580	Desk		550	.015			.40		.40	.50	.62
0590	Desk lamp		1200	.007			.18		.18	.23	.28
0600	Desk trays		1920	.004			.11		.11	.14	.18
0610	Dictaphone		1920	.004			.11		.11	.14	.18
0620	Dresser		480	.017			.46		.46	.57	.71
0630	File cabinets		960	.008			.23		.23	.28	.36
0640	Lockers		2400	.003			.09		.09	.11	.14
0650	Piano		425	.019			.52		.52	.64	.80
0660	Radio		960	.008			.23		.23	.28	.36
0665	Shelf, storage, 1' x 20'		480	.017			.46		.46	.57	.71
0670	Table		670	.012			.33		.33	.41	.51
0680	Table lamp		960	.008			.23		.23	.28	.36
0690	Telephones		960	.008			.23		.23	.28	.36
0700	Television		960	.008			.23		.23	.28	.36
0710	Typewriter		960	.008			.23		.23	.28	.36
0720	Vacuuming, armless chair		825	.010			.27		.27	.33	.41
0730	Armchair		480	.017			.46		.46	.57	.71
0740	Couch		320	.025			.69		.69	.85	1.07
0750	Entry mats		8	1	M.S.F.		27.50		27.50	34.00	42.50
0760	Overhead air vents		480	.017	Ea.		.46		.46	.57	.71
0770	Upholstered auditorium seating		5.27	1.518	M.S.F.		41.50		41.50	52.00	64.50

01 93 06.45 Hard Floor Care

		Crew	Daily Output	Labor-Hours	Unit	Material	2014 Bare Costs Labor	2014 Bare Costs Equipment	Total	Total In-House	Total Incl O&P
0010	**HARD FLOOR CARE**										
0020	Damp mop w/bucket & wringer, 12 oz. mop head, unobstructed	1 Clam	25	.320	M.S.F.	1.87	8.80		10.67	12.95	16
0030	Obstructed		17	.471		1.87	12.90		14.77	18.10	22.50
0040	20 oz. mop head, unobstructed		29	.276		1.87	7.55		9.42	11.45	14.10
0050	Obstructed		20	.400		1.87	11		12.87	15.70	19.40
0060	24 oz. mop head, unobstructed		35	.229		1.87	6.25		8.12	9.85	12.10
0070	Obstructed		23	.348		1.87	9.55		11.42	13.90	17.20
0080	32 oz. mop head, unobstructed		43	.186		1.87	5.10		6.97	8.40	10.30
0090	Obstructed		29	.276		1.87	7.55		9.42	11.45	14.10
0100	Dust mop, 12" mop, unobstructed		33	.242			6.65		6.65	8.25	10.35
0110	Obstructed		19	.421			11.55		11.55	14.35	17.95
0120	18" mop, unobstructed		49	.163			4.48		4.48	5.55	6.95
0130	Obstructed		28	.286			7.85		7.85	9.75	12.20

01 93 Facility Maintenance

01 93 06 – Interior General Maintenance

01 93 06.45 Hard Floor Care

		Crew	Daily Output	Labor-Hours	Unit	Material	2014 Bare Costs Labor	Equipment	Total	Total In-House	Total Incl O&P
0140	24" mop, unobstructed	1 Clam	61	.131	M.S.F.		3.60		3.60	4.47	5.60
0150	Obstructed		34	.235			6.45		6.45	8.00	10.05
0160	30" mop, unobstructed		74	.108			2.97		2.97	3.69	4.61
0170	Obstructed		42	.190			5.25		5.25	6.50	8.10
0180	36" mop, unobstructed		92	.087			2.39		2.39	2.97	3.71
0190	Obstructed		52	.154			4.22		4.22	5.25	6.55
0200	42" mop, unobstructed		124	.065			1.77		1.77	2.20	2.75
0210	Obstructed		70	.114			3.14		3.14	3.90	4.87
0220	48" mop, unobstructed		185	.043			1.19		1.19	1.47	1.84
0230	Obstructed		104	.077			2.11		2.11	2.62	3.28
0240	60" mop, unobstructed		246	.033			.89		.89	1.11	1.39
0250	Obstructed		138	.058			1.59		1.59	1.98	2.47
0260	72" mop, unobstructed		369	.022			.60		.60	.74	.92
0270	Obstructed		208	.038			1.06		1.06	1.31	1.64
0280	Sweeping, 8" broom, light soil, unobstructed		19	.421			11.55		11.55	14.35	17.95
0290	Obstructed		14	.571			15.70		15.70	19.50	24.50
0300	Heavy soil, unobstructed		11	.727			19.95		19.95	25.00	31
0310	Obstructed		9	.889			24.50		24.50	30.50	38
0320	12" push broom, light soil, unobstructed		32	.250			6.85		6.85	8.55	10.65
0330	Obstructed		24	.333			9.15		9.15	11.35	14.20
0340	Heavy soil, unobstructed		19	.421			11.55		11.55	14.35	17.95
0350	Obstructed		14	.571			15.70		15.70	19.50	24.50
0360	16" push broom, light soil, unobstructed		40	.200			5.50		5.50	6.80	8.55
0370	Obstructed		30	.267			7.30		7.30	9.10	11.35
0380	Heavy soil, unobstructed		24	.333			9.15		9.15	11.35	14.20
0390	Obstructed		18	.444			12.20		12.20	15.15	18.95
0400	18" push broom, light soil, unobstructed		44	.182			4.99		4.99	6.20	7.75
0410	Obstructed		33	.242			6.65		6.65	8.25	10.35
0420	Heavy soil, unobstructed		36	.222			6.10		6.10	7.60	9.50
0430	Obstructed		20	.400			11		11	13.65	17.05
0440	24" push broom, light soil, unobstructed		57	.140			3.85		3.85	4.79	6
0450	Obstructed		43	.186			5.10		5.10	6.35	7.95
0460	Heavy soil, unobstructed		34	.235			6.45		6.45	8.00	10.05
0470	Obstructed		26	.308			8.45		8.45	10.50	13.10
0480	30" push broom, light soil, unobstructed		81	.099			2.71		2.71	3.37	4.21
0490	Obstructed		60	.133			3.66		3.66	4.55	5.70
0500	Heavy soil, unobstructed		48	.167			4.58		4.58	5.70	7.10
0510	Obstructed		36	.222			6.10		6.10	7.60	9.50
0520	36" push broom, light soil, unobstructed		100	.080			2.20		2.20	2.73	3.41
0530	Obstructed		75	.107			2.93		2.93	3.64	4.55
0540	Heavy soil, unobstructed		60	.133			3.66		3.66	4.55	5.70
0550	Obstructed		45	.178			4.88		4.88	6.05	7.60
0560	42" push broom, light soil, unobstructed		133	.060			1.65		1.65	2.05	2.57
0570	Obstructed		100	.080			2.20		2.20	2.73	3.41
0580	Heavy soil, unobstructed		80	.100			2.75		2.75	3.41	4.27
0590	Obstructed		60	.133			3.66		3.66	4.55	5.70
0600	48" push broom, light soil, unobstructed		200	.040			1.10		1.10	1.36	1.71
0610	Obstructed		150	.053			1.46		1.46	1.82	2.27
0620	Heavy soil, unobstructed		120	.067			1.83		1.83	2.27	2.84
0630	Obstructed		90	.089			2.44		2.44	3.03	3.79
0640	Wet mop/rinse w/bucket & wringer, 12 oz. mop, unobstructed		11	.727		4.58	19.95		24.53	30.00	36.50
0650	Obstructed		8	1		4.58	27.50		32.08	39.00	48
0660	20 oz. mop head, unobstructed		14	.571		4.58	15.70		20.28	24.50	30

01 93 Facility Maintenance

01 93 06 – Interior General Maintenance

01 93 06.45 Hard Floor Care

		Crew	Daily Output	Labor-Hours	Unit	Material	2014 Bare Costs Labor	2014 Bare Costs Equipment	Total	Total In-House	Total Incl O&P
0670	Obstructed	1 Clam	10	.800	M.S.F.	4.58	22		26.58	32.50	39.50
0680	24 oz. mop head, unobstructed		20	.400		4.58	11		15.58	18.70	23
0690	Obstructed		15	.533		4.58	14.65		19.23	23.00	28.50
0700	32 oz. mop head, unobstructed		27	.296		4.58	8.15		12.73	15.15	18.40
0710	Obstructed		20	.400		4.58	11		15.58	18.70	23
0720	Dry buffing or polishing, 175 RPM, 12" diameter machine		14.21	.563			15.45		15.45	19.20	24
0730	14" diameter machine		16.51	.485			13.30		13.30	16.50	20.50
0740	17" diameter machine		19.20	.417			11.45		11.45	14.20	17.75
0750	20" diameter machine		22.85	.350			9.60		9.60	11.95	14.95
0760	24" diameter machine		28.99	.276			7.60		7.60	9.40	11.75
0770	350 RPM, 17" diameter machine		30.76	.260			7.15		7.15	8.85	11.10
0780	20" diameter machine		40.60	.197			5.40		5.40	6.70	8.40
0790	24" diameter machine		47.06	.170			4.67		4.67	5.80	7.25
0800	1000 RPM, 17" diameter machine		62.09	.129			3.54		3.54	4.39	5.50
0810	20" diameter machine		67.68	.118			3.24		3.24	4.03	5.05
0820	24" diameter machine		88.79	.090			2.47		2.47	3.07	3.84
0830	27" diameter machine		99.34	.081			2.21		2.21	2.75	3.43
0840	Spray buffing, 175 RPM, 12" diameter		10.66	.750		.83	20.50		21.33	26.50	33
0850	14" diameter machine		11.95	.669		.83	18.40		19.23	23.50	29.50
0860	17" diameter machine		13.80	.580		.83	15.90		16.73	20.50	25.50
0870	20" diameter machine		16	.500		.83	13.75		14.58	17.95	22.50
0880	24" diameter machine		19.05	.420		.83	11.55		12.38	15.25	18.95
0890	350 RPM, 17" diameter machine		17.39	.460		.83	12.65		13.48	16.60	20.50
0900	20" diameter machine		22.13	.362		.83	9.90		10.73	13.25	16.45
0910	24" diameter machine		29.22	.274		.83	7.50		8.33	10.25	12.75
0920	1000 RPM, 17" diameter machine		41.02	.195		.83	5.35		6.18	7.55	9.35
0930	20" diameter machine		44.14	.181		.83	4.98		5.81	7.10	8.80
0940	24" diameter machine		52.21	.153		.83	4.21		5.04	6.15	7.60
0950	27" diameter machine		57.43	.139		.83	3.82		4.65	5.65	7
0960	Burnish, 1000 RPM, 17" diameter machine		60	.133			3.66		3.66	4.55	5.70
0970	2000 RPM, 17" diameter machine		72	.111			3.05		3.05	3.79	4.74
0980	Scrubbing, 12" brush by hand		2	4		1.87	110		111.87	138.00	173
0990	175 RPM, 12" diameter machine		10.05	.796		1.87	22		23.87	29.00	36.50
1000	14" diameter machine		11.62	.688		1.87	18.90		20.77	25.50	32
1010	17" diameter machine		14	.571		1.87	15.70		17.57	21.50	27
1020	20" diameter machine		17.61	.454		1.87	12.45		14.32	17.55	21.50
1030	24" diameter machine		23.58	.339		1.87	9.30		11.17	13.65	16.80
1040	350 RPM, 17" diameter machine		24.12	.332		1.87	9.10		10.97	13.35	16.50
1050	20" diameter machine		28.91	.277		1.87	7.60		9.47	11.50	14.15
1060	24" diameter machine		36.92	.217		1.87	5.95		7.82	9.45	11.60
1070	Automatic, 21" machine		60.75	.132		1.87	3.61		5.48	6.55	7.95
1080	24" machine		83.47	.096		1.87	2.63		4.50	5.35	6.45
1090	32" machine		128.68	.062		1.87	1.71		3.58	4.18	4.99
1100	Stripping, separate wet pick-up required, 175 RPM, 17" diameter		6.86	1.166		2.65	32		34.65	42.50	53
1110	20" diameter machine		7.68	1.042		2.65	28.50		31.15	38.50	48
1120	350 RPM, 17" diameter machine		8.80	.909		2.65	25		27.65	34.00	42.50
1130	20" diameter machine		10.50	.762		2.65	21		23.65	29.00	36
1140	Wet pick-up, 12" opening attachment		18	.444			12.20		12.20	15.15	18.95
1150	20" opening attachment		24.12	.332			9.10		9.10	11.30	14.15
1160	Finish application, mop		14	.571		61.50	15.70		77.20	87.00	101
1170	Gravity fed applicator		21	.381		61.50	10.45		71.95	80.50	93
1180	Lambswool applicator		17	.471		61.50	12.90		74.40	83.50	97
1190	Sealer application, mop		14	.571		24	15.70		39.70	46.00	54.50

01 93 Facility Maintenance

01 93 06 – Interior General Maintenance

01 93 06.45 Hard Floor Care

		Crew	Daily Output	Labor-Hours	Unit	Material	2014 Bare Costs Labor	2014 Bare Costs Equipment	Total	Total In-House	Total Incl O&P
1200	Gravity fed applicator	1 Clam	21	.381	M.S.F.	24	10.45		34.45	39.50	46.50
1210	Lambswool applicator		17	.471		24	12.90		36.90	42.50	50

01 93 06.80 Vacuuming

		Crew	Daily Output	Labor-Hours	Unit	Material	2014 Bare Costs Labor	2014 Bare Costs Equipment	Total	Total In-House	Total Incl O&P
0010	**VACUUMING**										
0020	Vacuum carpet, 12" upright, spot areas	1 Clam	73.68	.109	M.S.F.		2.98		2.98	3.70	4.63
0030	Traffic areas		44.21	.181			4.97		4.97	6.15	7.70
0040	All areas		21.05	.380			10.45		10.45	12.95	16.20
0050	14" upright, spot areas		79.98	.100			2.75		2.75	3.41	4.27
0060	Traffic areas		47.99	.167			4.58		4.58	5.70	7.10
0070	All areas		22.85	.350			9.60		9.60	11.95	14.95
0080	16" upright, spot areas		87.50	.091			2.51		2.51	3.12	3.90
0090	Traffic areas		52.50	.152			4.18		4.18	5.20	6.50
0100	All areas		25	.320			8.80		8.80	10.90	13.65
0110	18" upright, spot areas		96.53	.083			2.28		2.28	2.83	3.53
0120	Traffic areas		57.92	.138			3.79		3.79	4.71	5.90
0130	All areas		27.58	.290			7.95		7.95	9.90	12.35
0140	20" upright, spot areas		107.66	.074			2.04		2.04	2.53	3.17
0150	Traffic areas		64.60	.124			3.40		3.40	4.22	5.30
0160	All areas		30.76	.260			7.15		7.15	8.85	11.10
0170	22" upright, spot areas		121.73	.066			1.80		1.80	2.24	2.80
0180	Traffic areas		73.04	.110			3.01		3.01	3.73	4.67
0190	All areas		34.78	.230			6.30		6.30	7.85	9.80
0200	24" upright, spot areas		140	.057			1.57		1.57	1.95	2.44
0210	Traffic areas		84	.095			2.61		2.61	3.25	4.06
0220	All areas		40	.200			5.50		5.50	6.80	8.55
0230	14" twin motor upright, spot areas		98.81	.081			2.22		2.22	2.76	3.45
0240	Traffic areas		59.28	.135			3.70		3.70	4.60	5.75
0250	All areas		28.23	.283			7.80		7.80	9.65	12.10
0260	18" twin motor upright, spot areas		112	.071			1.96		1.96	2.44	3.05
0270	Traffic areas		67.20	.119			3.27		3.27	4.06	5.10
0280	All areas		32	.250			6.85		6.85	8.55	10.65
0290	Backpack, 12" opening carpet attachment, spot areas		77.77	.103			2.82		2.82	3.51	4.39
0300	Traffic areas		46.66	.171			4.71		4.71	5.85	7.30
0310	All areas		22.22	.360			9.90		9.90	12.30	15.35
0320	14" opening carpet attachment, spot areas		84.84	.094			2.59		2.59	3.22	4.02
0330	Traffic areas		50.90	.157			4.31		4.31	5.35	6.70
0340	All areas		24.24	.330			9.05		9.05	11.25	14.10
0350	16" opening carpet attachment, spot areas		93.35	.086			2.35		2.35	2.92	3.66
0360	Traffic areas		56.01	.143			3.92		3.92	4.87	6.10
0370	All areas		26.67	.300			8.25		8.25	10.25	12.80
0380	18" opening carpet attachment, spot areas		103.67	.077			2.12		2.12	2.63	3.29
0390	Traffic areas		62.20	.129			3.53		3.53	4.39	5.50
0400	All areas		29.62	.270			7.40		7.40	9.20	11.50
0410	20" opening carpet attachment, spot areas		116.66	.069			1.88		1.88	2.34	2.92
0420	Traffic areas		69.99	.114			3.14		3.14	3.90	4.87
0430	All areas		33.33	.240			6.60		6.60	8.20	10.25
0440	22" opening carpet attachment, spot areas		133.32	.060			1.65		1.65	2.05	2.56
0450	Traffic areas		79.99	.100			2.75		2.75	3.41	4.27
0460	All areas		38.09	.210			5.75		5.75	7.15	8.95
0470	24" opening carpet attachment, spot areas		155.54	.051			1.41		1.41	1.75	2.19
0480	Traffic areas		93.32	.086			2.35		2.35	2.92	3.66
0490	All areas		44.44	.180			4.94		4.94	6.15	7.70

01 93 Facility Maintenance

01 93 06 – Interior General Maintenance

01 93 06.80 Vacuuming

		Crew	Daily Output	Labor-Hours	Unit	Material	2014 Bare Costs Labor	2014 Bare Costs Equipment	Total	Total In-House	Total Incl O&P
0500	Tank / cannister, 12" opening carpet attachment, spot areas	1 Clam	70	.114	M.S.F.		3.14		3.14	3.90	4.87
0510	Traffic areas		42	.190			5.25		5.25	6.50	8.10
0520	All areas		20	.400			11		11	13.65	17.05
0530	14" opening carpet attachment, spot areas		75.67	.106			2.90		2.90	3.61	4.51
0540	Traffic areas		45.40	.176			4.84		4.84	6.00	7.50
0550	All areas		21.62	.370			10.15		10.15	12.60	15.80
0560	16" opening carpet attachment, spot areas		82.32	.097			2.67		2.67	3.31	4.14
0570	Traffic areas		49.39	.162			4.45		4.45	5.50	6.90
0580	All areas		23.52	.340			9.35		9.35	11.60	14.50
0590	18" opening carpet attachment, spot areas		90.30	.089			2.43		2.43	3.02	3.78
0600	Traffic areas		54.18	.148			4.05		4.05	5.05	6.30
0610	All areas		25.80	.310			8.50		8.50	10.55	13.20
0620	20" opening carpet attachment, spot areas		100	.080			2.20		2.20	2.73	3.41
0630	Traffic areas		60	.133			3.66		3.66	4.55	5.70
0640	All areas		28.57	.280			7.70		7.70	9.55	11.95
0650	22" opening carpet attachment, spot areas		112	.071			1.96		1.96	2.44	3.05
0660	Traffic areas		67.20	.119			3.27		3.27	4.06	5.10
0670	All areas		32	.250			6.85		6.85	8.55	10.65
0680	24" opening carpet attachment, spot areas		127.26	.063			1.73		1.73	2.14	2.68
0690	Traffic areas		76.36	.105			2.88		2.88	3.57	4.47
0700	All areas		36.36	.220			6.05		6.05	7.50	9.40
0710	Large area machine, no obstructions, 26" opening		44.44	.180			4.94		4.94	6.15	7.70
0720	28" opening		64	.125			3.43		3.43	4.26	5.35
0730	30" opening		80	.100			2.75		2.75	3.41	4.27
0740	32" opening		120	.067			1.83		1.83	2.27	2.84

01 93 06.90 Wall Care

		Crew	Daily Output	Labor-Hours	Unit	Material	Labor	Equipment	Total	Total In-House	Total Incl O&P
0010	**WALL CARE**										
0020	Dusting walls, hand duster	1 Clam	87	.092	M.S.F.		2.52		2.52	3.14	3.92
0030	Treated cloth		40	.200			5.50		5.50	6.80	8.55
0040	Hand held duster vacuum		53	.151			4.14		4.14	5.15	6.45
0050	Cannister vacuum		32	.250			6.85		6.85	8.55	10.65
0060	Backpack vacuum		45	.178			4.88		4.88	6.05	7.60
0070	Washing walls, hand, painted surfaces		2.50	3.200		4.58	88		92.58	114.00	142
0080	Vinyl surfaces		3.25	2.462		4.58	67.50		72.08	89.00	111
0090	Machine, painted surfaces		4.80	1.667		4.58	46		50.58	62.00	76.50
0100	Washing walls from ladder, hand, painted surfaces		1.65	4.848		4.58	133		137.58	170.00	213
0110	Vinyl surfaces		2.15	3.721		4.58	102		106.58	132.00	165
0120	Machine, painted surfaces		3.17	2.524		4.58	69.50		74.08	91.00	114

01 93 06.95 Window and Accessory Care

		Crew	Daily Output	Labor-Hours	Unit	Material	Labor	Equipment	Total	Total In-House	Total Incl O&P
0010	**WINDOW AND ACCESSORY CARE**										
0025	Wash windows, trigger sprayer & wipe cloth, 12 S.F.	1 Clam	4.50	1.778	M.S.F.	4.32	49		53.32	65.50	81.50
0030	Over 12 S.F.		12.75	.627		4.32	17.20		21.52	26.00	32.50
0040	With squeegee & bucket, 12 S.F. and under		5.50	1.455		4.32	40		44.32	54.50	67.50
0050	Over 12 S.F.		15.50	.516		4.32	14.15		18.47	22.50	27.50
0060	Dust window sill		8640	.001	L.F.		.03		.03	.03	.04
0070	Damp wipe, venetian blinds		160	.050	Ea.	.09	1.37		1.46	1.81	2.24
0080	Vacuum, venetian blinds, horizontal		87.27	.092			2.52		2.52	3.13	3.91
0090	Draperies, in place, to 64 S.F.		160	.050			1.37		1.37	1.71	2.13

01 93 Facility Maintenance

01 93 08 – Facility Maintenance Equipment

01 93 08.50 Equipment	Crew	Daily Output	Labor-Hours	Unit	Material	2014 Bare Costs Labor	2014 Bare Costs Equipment	Total	Total In-House	Total Incl O&P
0010 **EQUIPMENT**, Purchase										
0090 Carpet care equipment										
0110 Dual motor vac, 1 H.P., 16" brush				Ea.	1,150			1,150	1,275.00	1,450
0120 Upright vacuum, 12" brush					350			350	385.00	440
0130 14" brush					390			390	430.00	490
0140 Soil extractor, hot water 5' wand, 12" head					3,025			3,025	3,325.00	3,775
0150 Dry foam, 13" brush					1,800			1,800	1,975.00	2,250
0160 24" brush				▼	3,025			3,025	3,325.00	3,775
0240 Floor care equipment										
0260 Polishing, buffing, waxing machine, 175 RPM										
0270 .33 H.P., 20" diam. brush				Ea.	715			715	785.00	895
0280 1 H.P., 16" diam. brush					1,075			1,075	1,175.00	1,350
0290 17" diam. brush					1,350			1,350	1,475.00	1,700
0300 1.5 H.P., 20" diam. brush					1,525			1,525	1,675.00	1,900
0310 20" diam. brush, 1500 RPM				▼	1,100			1,100	1,200.00	1,375
0330 Scrubber, automatic, 2 stage 1 H.P. vacuum motor										
0340 20" diam. brush				Ea.	11,900			11,900	13,100.00	14,900
0350 28" diam. brush				"	17,100			17,100	18,800.00	21,400
1200 Plumbing maintenance equipment										
1220 Kinetic water ram				Ea.	231			231	254.00	289
1240 Cable pipe snake, 104 ft., self feed, electric, .5 H.P.				"	2,575			2,575	2,825.00	3,225
1500 Specialty equipment										
1550 Litterbuggy, 9 H.P. gasoline				Ea.	5,600			5,600	6,150.00	7,000
1560 Propane					3,425			3,425	3,775.00	4,275
1570 Pressure cleaner, hot water, 1000 psi					3,550			3,550	3,900.00	4,450
1580 3500 psi				▼	6,425			6,425	7,075.00	8,025
1590 Vacuum cleaners, steel canister										
1600 Dry only, two stage, 1 H.P.				Ea.	330			330	365.00	415
1610 Wet/dry, two stage, 2 H.P.					500			500	550.00	625
1620 4 H.P.					1,025			1,025	1,125.00	1,275
1630 Wet/dry, three stage, 1.5 H.P.					670			670	735.00	840
1670 Squeegee wet only, 2 stage, 1 H.P.					885			885	975.00	1,100
1680 2 H.P.					1,325			1,325	1,450.00	1,650
1690 Asbestos and hazardous waste dry vacs, 1 H.P.					790			790	870.00	990
1700 2 H.P.					1,550			1,550	1,700.00	1,950
1710 Three stage, 1 H.P.					545			545	600.00	680
1800 Upholstery soil extractor, dry-foam				▼	3,400			3,400	3,750.00	4,250
1900 Snow removal equipment										
1920 Thrower, 3 H.P., 20", single stage, gas				Ea.	635			635	700.00	795
1930 Electric start					750			750	825.00	940
1940 5 H.P., 24", 2 stage, gas					1,250			1,250	1,375.00	1,575
1950 11 H.P., 36"				▼	2,800			2,800	3,075.00	3,500
2000 Tube cleaning equip., for heat exchangers, condensers, evap.										
2010 Tubes/pipes 1/4"-1", 115 volt, .5 H.P. drive				Ea.	1,650			1,650	1,825.00	2,075
2020 Air powered					1,700			1,700	1,875.00	2,125
2040 1" and up, 115 volt, 1 H.P. drive				▼	1,800			1,800	1,975.00	2,250

Facilities Audits

Table of Contents

Auditing the Facility 571
Audit Checklists 575
   Building Data Summary 575
   Facilities Inventory Form 576
   Standard Inspection Form 577
   A10–Foundations Inspection Checklist 578
   A20–Basement Inspection Checklist 579
   B10–Superstructures Inspection Checklist 580
   B20–Exterior Closure Inspection Checklist 582
   B30–Roofing Inspection Checklist 583
   C10–Partitions & Doors Inspection Checklist 585
   C30–Walls & Finishes Inspection Checklist 586
   C30–Floors & Finishes Inspection Checklist 587
   C30–Ceilings & Finishes Inspection Checklist 588
   D10–Conveying Inspection Checklist 589
   D20–Mechanical/Plumbing Inspection Checklist 590
   D30–Mechanical/HVAC Inspection Checklist 591
   D50–Electrical/Service & Distribution Inspection Checklist 593
   G20–Site Work Inspection Checklist 594
   H10–Safety Standards Inspection Checklist 595

Auditing the Facility

Introduction

Capital budget planning begins with compiling a complete inventory of the facility, including equipment, that identifies existing deficiencies. The Facilities Audit is a vehicle for producing such an inventory; it is a system for thoroughly assessing the existing physical condition and functional performance of buildings, grounds, utilities, and equipment. The results of this audit are used to address major and minor, urgent and long-term needs for corrective action, for short- and long-term financial planning. The basic principles presented here can be applied to any scale operation, from a single structure to a facility consisting of multiple building complexes in dispersed locations.

The first step is determining the scope of the audit. Next, the audit team is selected, and a set of forms is designed. A survey team then records field observations on the forms for individual building components. Finally, the information from these forms is summarized, prioritized, presented, and used to:

- compare a building's condition and functional performance to that of other buildings within a facility
- define regular maintenance requirements
- define capital repair and replacement projects in order to eliminate deferred maintenance
- develop cost estimates for capital repair and replacement projects
- restore functionally obsolete facilities to a usable condition
- eliminate conditions that are either potentially damaging to property or present life safety hazards
- identify energy conservation measures

The audit ultimately produces a database on the condition of a facility's buildings, identifying the physical adequacy of construction, material, and equipment, and assessing the functional adequacy of the facility to perform its designated purpose. This comprehensive inventory can be used to guide property management decisions and strategic planning of capital assets, such as determining those expenditures necessary to extend the life of a facility.

The audit procedures and forms included here are for the purpose of identifying the condition of buildings and assessing the cost and priority of their repair. While this is a comprehensive stand-alone process, it is for the limited purposes of assessing costs and establishing priorities for repairs. It represents only a portion of the requirements of a full facilities audit. For further information and guidance on conducting a full audit, contact the RSMeans Engineering staff.

The Rating Forms

The audit approach involves conducting a thorough inspection of a facility, then employing a systematic process of completing individual forms for each component, following the order in which building components are assembled. This is a self-evaluation process designed to fit all types of structures. The audit procedure described is a "generic" approach that can be used as a starting point; the facilities manager should incorporate the special characteristics of an organization and its facilities into an individualized facilities audit.

Audits are best performed by the in-house personnel familiar with the building being audited. However, outside consultants may be required when staff is not available or when special expertise is required to further examine suspected deficiencies.

Four types of forms that can be used to conduct the building audit are described here. A complete set of the blank forms ready for photocopying follows.

1. *Building Data Summary*
   This form can be used to summarize all buildings to be covered by the audit.

2. *Facilities Inventory Form*
   This form is designed to collect the basic data about a specific building's characteristics (name, identification number, location, area), including land data. One form is prepared for each building to be audited as summarized on the *Building Data Summary* form.

3. *Standard Inspection Form*
   The standard form for recording observed deficiencies, it provides a uniform way to comment on, prioritize, and estimate the cost to correct those deficiencies. The *Standard Inspection Form* is designed to make use of cost data from the Maintenance & Repair and Preventive Maintenance sections of this publication. Cost data from publications such as *RSMeans Facilities Construction Cost Data* may be utilized as well.

4. *System Checklists*
   For each major building system from foundations through site work, there is a checklist to guide the audit and record the review process. The checklists are intended as general guides and may not be totally comprehensive. Be sure to make any necessary modifications (additions or deletions) required for the specific building being audited. The forms prompt the user on what to check and provide a place to check off or indicate by "Yes" or "No" the presence of deficiencies requiring further detailed analysis on the *Standard Inspection Form*.

Completing the Inspection Forms

The standard inspection forms are designed to be used in conjunction with the system checklists. A system checklist of typical deficiencies for each building system is completed as part of the inspection process. Then for each building deficiency, a standard inspection report is generated listing the identifying information as required on the form. With the deficiency defined and described, the specific work items required to correct the deficiency can be entered, prioritized, and costed in Part 3 of the form. The priority rating is further described below.

Priority Ratings

Priority rating is the inspector's judgment of the priority of an observed deficiency. The rating provides guidance for facilities managers who must review overall deficiencies and develop maintenance schedules and capital budgets. The following priority rating terms, commonly used in maintenance management, are a suggested guideline for priority values. The scale may be expanded as required to fit specific needs.

1. *Emergency*: This designation indicates work that demands immediate attention to repair an essential operating system or building component, or because life safety or property is endangered. A response should occur within two hours of notification.
2. *Urgent*: Work demanding prompt attention to alleviate temporary emergency repairs or a condition requiring correction within one to five days to prevent a subsequent emergency is classified as urgent.
3. *Routine*: A specific completion date can be requested or required for routine work. This includes work that has no short range effect on life safety, security of personnel, or property. Routine work can be planned in detail and incorporated into a trade backlog for scheduling within twelve months.
4. *Deferred*: Projects that can be deferred into the following year's work planning are classified as deferred.

Cost Estimates for Maintenance

A major characteristic of the audit described here is the application of detailed cost estimates. To prepare cost estimates for work required to correct deficiencies, detailed information about the nature and extent of the facility's condition is required at the beginning of the audit. However, when cost estimates are incorporated, the audit process generates highly useful summaries outlining major maintenance costs and priorities. Including cost estimates allows for comparisons to be made in order to prioritize repair and replacement needs on a project-by-project and building-by-building basis.

Other Considerations

A problem which may have to be settled by the audit leader is what to do with mixed-use buildings, those that have recent additions to the original construction, or buildings that have, under emergency circumstances, been forced to house new activities for which they were not designed. Such problems will have to be dealt with individually by the audit team.

The data gathered in a facilities audit must be kept current, gathered consistently, and regularly updated. An annual cycle of inspections is recommended. Each organization initiating a comprehensive audit and continuing with annual cycles of inspections will benefit from organizing data in a systematic format. Conducting annual audits in a consistent format provides the basis for monitoring the backlog of deferred maintenance and preparing effective long-term plans and budgets for maintenance and repair operations.

The forms are designed for ease of field entry of data and manual preparation of summaries and reports. The forms can also be created using basic word processing and spreadsheet software or database programming. Although the self-evaluation process is readily adaptable to data processing programs, not every organization has the capabilities or resources available. For those without computer capabilities, a manual method of storing and updating must be developed and used.

Summarizing the Audit Data

Information collected on the audit forms is compiled and used to determine the condition of the facility's buildings and what should be done to maintain and/or improve them. Summaries of a building audit can serve several purposes, some of which are listed below.

- *Routine Maintenance Needs:* A current property owner may be interested in planning and estimating a program for the cost of routine operations and maintenance.
- *Major Maintenance:* A current or prospective property owner may want to evaluate major maintenance needs to plan and estimate the cost of a corrective program.
- *Deferred Maintenance:* A backlog of routine and major maintenance can be addressed using a survey of conditions to plan and budget a deferred maintenance program.
- *Renovations and/or Additions:* Prior to developing feasibility studies on alternatives to renovations and/or additions, a survey of existing conditions may be necessary.
- *Capital Budgeting and Planning:* A comprehensive audit of facilities conditions will incorporate major maintenance requirements into overall capital needs.

Conclusion

The information and forms for conducting an audit of the facility are included in *RSMeans Facilities Maintenance & Repair Cost Data* to provide a method for applying the cost data to the actual maintenance and repair needs of a specific facility. The Facility Audit presented here is a subset of a more comprehensive audit program. The more comprehensive approach includes assessment of the building's functional performance, as well as additional forms and procedures to incorporate the cost of maintenance and repair into a comprehensive multi-year facility management plan.

BUILDING DATA SUMMARY

Building Number	Building Name	Building Use*	Location / Address	Ownership** Own/Lea/O-L

*   Principal and Special Uses
** Own = Owned
    Lea = Leased
    O-L = Owned/Leased

FACILITIES INVENTORY FORM

Building Data

1. Organization Name _____
2. Approved by: Name _____ Title _____ Date _____
3. Building Number _____
4. Property Name _____
5. Property Address _____
6. Property Main Use/Special Use _____
7. Gross Area (Sq. Ft.) _____
8. Net Area (Sq. Ft.) _____
9. Type of Ownership: Owned _____ Leased _____
10. Book Value $ _____ Year Appraised _____
11. Replacement Value $ _____
12. Year Built _____ Year Acquired _____
13. Year Major Additions _____

Land Data

14. Type of Ownership: Owned _____ Leased _____ Owned / Leased _____
15. Book Value $ _____ Year Appraised _____
16. Year Acquired _____
17. Size in Acres _____
18. Market Value $ _____

Notes
_____
_____
_____
_____

STANDARD INSPECTION FORM

1. Facility Inspection Data

Facility #: _____   Name: _____
Component #: _____   Name: _____
Inspector: _____   Date: _____

2. Component Description

_____
_____
_____

3. Component Repair Evaluation

Defi-ciency #	Component Description	Line Number	Pri-ority	Unit	Quant.	Unit Cost	Total In-House	Unit Cost	Total W/O&P

A10 FOUNDATIONS INSPECTION CHECKLIST

Deficiencies **Causes**

_____ Settlement, alignment changes or cracks

    _____ Soils - changes in load bearing capacity due to shrinkage, erosion, or compaction. Adjacent construction undermining foundations. Reduced soil cover resulting in frost exposure.

    _____ Design loads - building equipment loads exceeding design loads. Vibration from heavy equipment requiring isolated foundations.

    _____ Structural or occupancy changes - inadequate bearing capacities. Foundation settling.

    _____ Earthquake resistance non-functioning.

_____ Moisture penetration

    _____ Water table changes - inadequate drainage.

    _____ Ineffective drains or sump pump/sump pits.

    _____ Roof drainage - storm sewer connections inadequate or defective. Installation of roof drain restrictors, gutters, and downspouts where required.

    _____ Surface drainage - exterior grades should slope away from building and structures.

    _____ Utilities - broken or improperly functioning utility service lines or drains.

    _____ Leakage - wall cracks, opening of construction joints, inadequate or defective waterproofing.

    _____ Condensation - inadequate ventilation, vapor barrier, and/or dehumidification.

_____ Temperature changes

    _____ Insulation - improperly selected for insulating value, fire ratings, and vermin resistance.

_____ Surface material deterioration

    _____ Concrete, masonry, or stucco - spalling, corrosion of reinforcing, moisture penetration, or chemical reaction between cement and soil.

    _____ Steel or other ferrous metals - corrosion due to moisture or contact with acid-bearing soils.

    _____ Wood - decay due to moisture or insect infestation.

_____ Openings deterioration

    _____ Non-functioning of doors, windows, hatchways, and stairways.

    _____ Utilities penetration due to damage, weather, wear, or other cause.

A20 BASEMENT INSPECTION CHECKLIST

Deficiencies **Causes**

____ Floors, concrete - cracking or arching
- ____ Shrinkage, or settlement of subsoil.
- ____ Inadequate drainage.
- ____ Movement in exterior walls, or frost heave.
- ____ Improper compaction of base. Erosion of base.
- ____ Heaving from hydraulic pressure.

____ Floors, wood - rotting or arching
- ____ Excessive dampness or insect infestation.
- ____ Leak in building exterior.
- ____ Lack of ventilation.

____ Wall deterioration
- ____ Concrete or masonry (see *A10 Foundations*)

____ Crawl space ventilation and maintenance
- ____ Inadequate air circulation due to blockage of openings in foundation walls.
- ____ Moisture barrier ineffective.
- ____ Pest control, housekeeping, and proper drainage.

B10 SUPERSTRUCTURES INSPECTION CHECKLIST

    The primary materials encountered in the superstructure inspection are concrete, steel, and wood. Typical observations of deficiencies will be observed by: failures in the exterior closure systems of exterior walls, openings, and roofs; cracks; movement of materials; moisture penetration; and discoloration. The exterior visual survey will detect failures of surface materials or at openings that will require further inspection to determine whether the cause was the superstructure system.

Deficiencies	Causes
Concrete (columns, walls, and floor and roof slabs)	
_____ Overall alignment	_____ Settlement due to design and construction techniques.
	_____ Under designed for loading conditions (see *A10 Foundations*).
_____ Deflection	_____ Expansion and/or contraction due to changes in design loads.
	_____ Original design deficient.
	_____ Original materials deficient.
_____ Surface Conditions	
_____ Cracks	_____ Inadequate design and/or construction due to changes in design loads.
	_____ Stress concentration.
	_____ Extreme temperature changes; secondary effects of freeze-thaw.
_____ Scaling, spalls, and pop-outs	_____ Extreme temperature changes.
	_____ Reinforcement corrosion.
	_____ Environmental conditions.
	_____ Mechanical damage.
	_____ Poor materials.
_____ Stains	_____ Chemical reaction of reinforcing.
	_____ Reaction of materials in concrete mixture.
	_____ Environmental conditions.
_____ Exposed reinforcing	_____ Corrosion of steel.
	_____ Insufficient cover.
	_____ Mechanical damage.

B10 SUPERSTRUCTURES INSPECTION CHECKLIST

Deficiencies   **Causes**

Steel (structural members, stairs, and connections)

_____ Overall alignment   _____ Settlement due to design and construction techniques; improper fabrication.

_____ Deflection or cracking   _____ Expansion and/or contraction.
    _____ Changes in design loads.
    _____ Fatigue due to vibration or impact.

_____ Corrosion   _____ Electrochemical reaction.
    _____ Failure of protective coating.
    _____ Excessive moisture exposure.

_____ Surface deterioration   _____ Excessive wear.

Wood (structural members and connections)

_____ Overall alignment   _____ Settlement; design and construction techniques.

_____ Deflection or cracking   _____ Expansion and/or contraction.
    _____ Changes in design loads.
    _____ Fatigue due to vibration or impact.
    _____ Failure of compression members.
    _____ Poor construction techniques.
    _____ General material failure.

_____ Rot (Decay)   _____ Direct contact with moisture.
    _____ Condensation.
    _____ Omission or deterioration of vapor barrier.
    _____ Poor construction techniques.
    _____ Damage from rodents or insects.

B20 EXTERIOR CLOSURE INSPECTION CHECKLIST

General Inspection

- Overall appearance
- Displacement
- Paint conditions
- Caulking
- Window & door fit
- Flashing condition
- Material integrity
- Cracks
- Evidence of moisture
- Construction joints
- Hardware condition

Exterior Walls

_____ Wood (Shingles, weatherboard siding, plywood).
    _____ Paint or surface treatment condition   _____ Moisture penetration
    _____ Rot or decay   _____ Loose, cracked, warped, or broken boards and shingles.

_____ Concrete, Masonry, and Tile (Concrete, brick, concrete masonry units, structural tile, glazed tile, stucco, stone).
    _____ Settlement   _____ Structural frame movement causing cracks
    _____ Construction and expansion joints   _____ Condition of caulking and mortar
    _____ Surface deterioration   _____ Efflorescence and staining
    _____ Parapet movement   _____ Tightness of fasteners

_____ Metal (Corrugated iron or steel, aluminum, enamel coated steel, protected metals).
    _____ Settlement   _____ Structural frame movement
    _____ Condition of bracing   _____ Surface damage due to impact
    _____ Tightness of fasteners   _____ Caulking
    _____ Flashings   _____ Corrosion

_____ Finishes (Mineral products, fiberglass, polyester resins, and plastics).
    _____ Settlement   _____ Structural frame movement
    _____ Surface damage due to impact   _____ Cracks
    _____ Stains   _____ Fasteners
    _____ Adhesion to substrate   _____ Caulking
    _____ Flashings   _____ Lead paint
    _____ Asbestos products

_____ Insulation
    _____ Insulation present   _____ Satisfactory condition

Windows and Doors

    _____ Frame fitting   _____ Frame and molding condition
    _____ Paint or surface finish   _____ Putty and weatherstripping
    _____ Hardware and operating parts   _____ Security
    _____ Cleanliness   _____ Material condition (glass, wood, & metal Panels)
    _____ Rot or corrosion   _____ Screens and storm windows

Shading Devices

    _____ Material conditions   _____ Operations
    _____ Cleanliness

B30   ROOFING INSPECTION CHECKLIST

General Appearance

        Good_____   Fair_____   Poor_____

_____ **Water tightness**

      _____ Evidence of leaks on undersurface   _____ Surface weathering
      _____ Faulty material   _____ Faulty design
      _____ Faulty application   _____ Standing water
      _____ Weather damage   _____ Mechanical damage
      _____ Fastening failure   _____ Flashing failure

_____ **Roofing Surface**

    _____ Built Up (Felt or bitumen surfacing)

        _____ Adhesion   _____ Moisture meter readings
        _____ Bare areas   _____ Blisters, wrinkles
        _____ Cracks, holes, tears   _____ Fish mouths
        _____ Alligatoring   _____ Ballast

    _____ Single Ply (Thermosetting, thermoplastic, composites)

        _____ Adhesion   _____ Moisture meter readings
        _____ Bare areas   _____ Blisters, wrinkles
        _____ Cracks   _____ Holes, tears
        _____ Seam conditions   _____ Protective coating
        _____ Ballast

    _____ Metal Roofing (Preformed, formed)

        _____ Corrosion (%)   _____ Protective coating
        _____ Seams   _____ Cracks or breaks
        _____ Holes   _____ Expansion joints

    _____ Shingles & Tiles (Metal, clay, mission, concrete, or others)

        _____ Disintegration   _____ Broken or cracked (%)
        _____ Missing (%)   _____ Fasteners
        _____ Underlayment

    _____ Wood shingles

        _____ Cracked   _____ Curled
        _____ Missing (%)

B30   ROOFING INSPECTION CHECKLIST

B30 ROOFING INSPECTION CHECKLIST

____ **Insulation**

    ____ Full coverage (%)   ____ Full thickness (%)
    ____ Satisfactory condition

____ **Roof Penetrations**

    ____ Weather tightness   ____ Operable

____ **Flashing**

    ____ Deterioration   ____ Open joints
    ____ Holes or damage   ____ Anchoring
    ____ Protective coating

____ **Drainage**

    ____ Alignment   ____ Free flowing
    ____ Clamping rings secure   ____ Screens
    ____ Corrosion

General Inspection

    Overall Appearance

        Good ____   Fair ____   Poor ____

        ____ Evidence of moisture   ____ Settlement
        ____ Cracks   ____ Surface condition
        ____ Cleanliness   ____ Stains
        ____ Discoloration

____ **Paint**

    ____ Repainting necessary   ____ Peeling, cracking, flaking
    ____ Reflectivity   ____ Maintainability
    ____ Lead paint

____ **Coverings & Coatings**

    ____ Replacement necessary   ____ Peeling
    ____ Rips, tears   ____ Holes
    ____ Adhesion   ____ Seams

____ **Interior glazing**

    ____ Cracks   ____ Seals
    ____ Frame condition   ____ Missing panes
    ____ Shading devices

C10 PARTITIONS & DOORS INSPECTION CHECKLIST

General Inspection

- Strength & stability
- Acoustical quality
- Maintainability
- Physical condition
- Adaptability
- Code compliance

Deficiencies

Causes

Wall Material

____	Cracks	____	Settlement
____	Holes	____	Defective material
____	Looseness	____	Operational abuse or vibrations
____	Missing segments	____	Environmental attack
____	Water stains	____	Moisture
____	Joints	____	Structural expansion or contraction
____	Surface appearance	____	Wind pressure

Hardware

____	Overall condition	____	Appearance
____	Keying system	____	Operations
____	Fit	____	Locksets
____	Cylinders	____	Panic devices
____	Maintainability	____	Security operations

C30 WALLS & FINISHES INSPECTION CHECKLIST

General Inspection

    Code compliance _____

    Maintainability _____

    Means of egress _____

    Fire ratings _____

    Audible & visual device condition _____

    Extinguishing systems (see also *08 Mechanical/Plumbing*): _____

        Type _____

    Lighting system (see also *09 Electrical Lighting/Power*): _____

        Type _____

    Handicapped accessibility _____

    Building user comments _____

Exterior Lighting

    Adequacy:

        Good _____   Fair _____   Poor _____

    Condition _____

    Controls (type & location) _____

Fire Alarm Systems

_____	Panel visible	_____	Operational
_____	Pull station condition	_____	Detector conditions

Stairs and Ramps

_____	Exits marked	_____	Hardware operational
_____	Tripping hazards	_____	Surface conditions
_____	Lighting adequate	_____	Handrails

C30 WALLS & FINISHES INSPECTION CHECKLIST

C30 FLOORS & FINISHES INSPECTION CHECKLIST

General Inspection

Overall Appearance

Good____   Fair____   Poor____

- Evidence of moisture
- Irregular surface
- Handicapped hazards
- Visible settlement
- Tripping hazards
- Replacement necessary

Carpet (Tufted, tile)

_____	Age	_____	Excessive wear
_____	Stains	_____	Discoloration
_____	Holes, tears	_____	Seam conditions

Resilient (Asphalt tile, cork tile, linoleum, rubber, vinyl)

_____	Broken tiles	_____	Loose tiles
_____	Shrinkage	_____	Lifting, cupping
_____	Fading	_____	Cuts, holes
_____	Porosity	_____	Asbestos present

Masonry (Stone, brick)

_____	Cracks	_____	Deterioration
_____	Joints	_____	Stains
_____	Porosity	_____	Sealing

Monolithic Topping (Concrete, granolithic, terrazzo, magnesite)

_____	Cracks	_____	Porosity
_____	Joints	_____	Sealing

Wood (Plank, strips, block, parquet)

_____	Shrinkage	_____	Cupping, warpage
_____	Excessive wear	_____	Unevenness
_____	Decay	_____	Sealing

C30 CEILINGS & FINISHES INSPECTION CHECKLIST

General Inspection

Overall appearance

    Good ____   Fair ____   Poor ____

- Settlement or sagging
- Attachment
- Stains, discoloration
- Suitability
- Code compliance
- Alignment
- Evidence of moisture
- Missing units
- Acoustic quality

Exposed Systems (Unpainted, painted, spray-on, decorative)

____ Cracks
____ Surface deterioration
____ Missing elements
____ Adhesion

Applied to Structure & Suspended

General Condition:

    Good ____   Fair ____   Poor ____

____ Fasteners
____ Trim condition

Openings:

____	Panels	____	Inserts
____	Lighting fixtures	____	Air distribution
____	Fire protection	____	Other

D10   CONVEYING INSPECTION CHECKLIST

General Inspection (Passenger Conveying)

Overall appearance (interior)

    Good _____ Fair _____ Poor _____

Overall appearance (exterior)

    Good _____ Fair _____ Poor _____

_____ Maintenance history available

  _____ Regular inspection frequency

  _____ Door operations

  _____ Control systems

  _____ Noise

  _____ Code compliance

  _____ Handicapped access

  _____ Major repairs necessary

  _____ Replacement necessary

D20 MECHANICAL/PLUMBING INSPECTION CHECKLIST

General Inspection

    General appearance

        Good _____   Fair _____   Poor _____

    Leaks, dripping, running faucets and valves _____

    Maintenance history _____

    Supply adequacy _____

    Sanitation hazards _____

    Drain & waste connection _____

    Backflow protection _____

    Cross connections _____

    Fixture quantity _____

    Fixture types & conditions _____

    Handicapped fixtures _____

    Female facilities _____

    Metal pipe & fittings corrosion _____

    Pipe joints & sealing _____

    Pipe insulation _____

    Hanger supports & clamps _____

    Filters _____

    Building user comments _____

_____ **Water System**

    _____ Water pressure adequate   _____ Odors, tastes
    _____ Main cutoff operable   _____ Water heating temperature setting
    _____ Pump condition   _____ Insulation condition

_____ **Sanitary & Storm System**

    _____ Flow adequate   _____ Cleanouts access
    _____ Floor drains   _____ Chemical resistance
    _____ Gradient   _____ On-site disposal system

_____ **Code Requirements**

    _____ EPA/local permits   _____ Other

_____ **Fire protection system**

    _____ Regular inspections   _____ Sprinkler heads operable
    _____ Complies with code   _____ Controls operable
    _____ Hose cabinets functional   _____ Water pressure sufficient

D30 MECHANICAL/HVAC INSPECTION CHECKLIST

General Inspection

    General appearance

        Good _____   Fair _____   Poor _____

    Lubrication: bearings and moving parts _____

    Rust and corrosion _____

    Motors, fans, drive assemblies, and pumps _____

    Wiring and electrical controls _____

    Thermostats and automatic temperature controls _____

    Thermal insulation and protective coatings _____

    Guards, casings, hangers, supports, platforms, and mounting bolts _____

    Piping system identification _____

    Solenoid valves _____

    Burner assemblies _____

    Combustion chambers, smokepipes, and breeching _____

    Electrical heating units _____

    Guards, casings, hangers, supports, platforms, and mounting bolts _____

    Steam and hot water heating equipment _____

    Accessible steam, water, and fuel piping _____

    Traps _____

    Humidifier assemblies and controls _____

    Strainers _____

    Water sprays, weirs, and similar devices _____

    Shell-and tube-type condensers _____

    Self-contained evaporative condensers _____

    Air cooled condensers _____

    Compressors _____

    Liquid receivers _____

    Refrigerant driers, strainers, valves, oil traps, and accessories _____

    Building user comments _____

D30　MECHANICAL/HVAC INSPECTION CHECKLIST

Cleaning, maintenance, repair, and replacement

_____	Registers	_____	Grills
_____	Dampers	_____	Draft diverters
_____	Plenum chambers	_____	Supply and return ducts
_____	Louvers	_____	Fire dampers

Air filters

_____ Correct type _____ Replacement schedule

Heating System Evaluation

_____ Heating capacity _____ Temperature control
_____ Heating:
　　　_____ Seasonal _____ All year
_____ Noise level _____ Energy consumption
_____ Air circulation & ventilation _____ Filtration
_____ Humidity control

Cooling System Evaluation

_____ Cooling capacity _____ Temperature and humidity control
_____ Cooling all season _____ Noise level
_____ Energy consumption _____ Air circulation & ventilation
_____ Filtration _____ Reliability

Ventilation System Evaluation

_____ Air velocity _____ Exhaust air systems
_____ Bag collection _____ Wet collectors
_____ Steam and hot water coils _____ Electrical heating units
_____ Fire hazards _____ Fire protective devices

D50   ELEC./SERVICE & DIST. INSPECTION CHECKLIST

General Inspection

    Safety conditions _____

    Service capacity, % used, and age _____

    Switchgear capacity, % used, and age _____

    Feeder capacity, % used, and age _____

    Panel capacity _____

    Thermo-scanning _____

    Maintenance records available _____

    Convenience outlets _____

    Building user comments _____

Exterior Service

    _____ Line drawing
    _____ Feed source:
        _____ Utility/owned   _____ Above/below ground
    _____ Transformer:
        _____ Transformer tested   _____ Transformer arcing or burning
        _____ Transformer PCB's   _____ Ownership (facility or utility)

Interior Distribution System

    _____ Line drawing   _____ Incoming conduit marked
    _____ Main circuit breaker marked   _____ Panel boards, junction boxes covered
    _____ All wiring in conduit   _____ Conduit properly secured
    _____ Panels marked   _____ Panel schedules
    _____ Missing breakers

Emergency Circuits

    Emergency generator(s):
        _____ Condition and age   _____ Auto start and switchover
        _____ Testing schedule   _____ Test records available
        _____ Service schedule   _____ Service schedule records available
        _____ Circuits appropriate   _____ Cooling & exhaust
        _____ Fuel storage (capacity)

    Emergency lighting/power systems:
        _____ Battery operation   _____ Separate power feed
        _____ Exit signs   _____ Stairways/corridors
        _____ Elevators   _____ Interior areas
        _____ HVAC   _____ Exterior

G20 SITE WORK INSPECTION CHECKLIST

General Inspection

    Overall appearance

        Good _____   Fair _____   Poor _____

    Maintainability   _____

    Repairs/replacements   _____

    Code compliance   _____

Roads, Walks, and Parking Lots

    Surface conditions   _____

    Subsurface conditions   _____

    Settling and uplift   _____

    Cracks, holes   _____

    Drainage and slope   _____

Curbing

    Alignment   _____

    Erosion   _____

    Repairs/replacements   _____

Drainage and Erosion Controls

    Surface drainage   _____

    Manholes, catch basins   _____

    Vegetation   _____

    Channels, dikes   _____

    Retention, detention   _____

    Drains   _____

Parking Lot Controls

    Location   _____

    Operation   _____

    Repairs/replacements   _____

H10 SAFETY INSPECTION CHECKLIST

General Inspection

    Code compliance _____
    Maintainability _____
    Means of egress _____
    Fire ratings _____
    Audible & visual device condition _____
    Extinguishing systems (see also *D20 Mechanical/Plumbing*): _____
        Type _____
    Lighting system (see also *D50 Electrical Lighting/Power*): _____
        Type _____
    Handicapped accessibility _____
    Building user comments _____
    _____

Exterior Lighting

    Adequacy:
        Good _____   Fair _____   Poor _____
    Condition _____
    Controls (type & location) _____

Fire Alarm Systems

    _____ Panel visible   _____ Operational
    _____ Pull station condition   _____ Detector conditions

Stairs and Ramps

    _____ Exits marked   _____ Hardware operational
    _____ Tripping hazards   _____ Surface conditions
    _____ Lighting adequate   _____ Handrails

H10 SAFETY INSPECTION CHECKLIST

H10 SAFETY INSPECTION CHECKLIST

General Inspection

    Code compliance _____

    Maintainability _____

    Means of egress _____

    Fire ratings _____

    Audible & visual device condition _____

    Extinguishing systems (see also *D20 Mechanical/Plumbing*): _____

        Type _____

    Lighting system (see also *D50 Electrical Lighting/Power*): _____

        Type _____

    Handicapped accessibility _____

    Building user comments _____

Exterior Lighting

    Adequacy:

        Good _____   Fair _____   Poor _____

    Condition _____

    Controls (type & location) _____

Fire Alarm Systems

    _____ Panel visible   _____ Operational

    _____ Pull station condition   _____ Detector conditions

Stairs and Ramps

    _____ Exits marked   _____ Hardware operational

    _____ Tripping hazards   _____ Surface conditions

    _____ Lighting adequate   _____ Handrails

Reference Section

All the reference information is in one section, making it easy to find what you need to know ... and easy to use the book on a daily basis. This section is visually identified by a vertical gray bar on the page edges.

In this Reference Section we've included information on Equipment Rental Costs, Crew Listings, Travel Costs, Reference Tables, Life Cycle Costing, and a listing of Abbreviations.

Table of Contents

Equipment Rental Costs	599
Crew Listings	611
Travel Costs	617
Reference Tables	618

H General Conditions
RH1010 Overhead, Design & Engineering, Worker's Compensation	618
RH1020 Overhead & Miscellaneous Data & Overtime	625
RH1030 Sales Tax	625
RH1040 Temporary Supports	626

City Adjustments
RJ1010 Component Indexes	627
RJ1020 Historical Cost Indexes	628
RJ1030 City Cost Indexes	629
RJ1040 Location Factors	642

Reference Aids
RL1010 Live Loads, Dead Loads, Wind Loads & Snow Loads	648
RL1020 Area Requirements & Building Code Requirements	655
RL1030 "R" Values & Thermal Properties	658
RL1040 Construction Classification	660
RL1090 Conversion Factors	662

Life Cycle Costing	663
Abbreviations	668

Equipment Rental Costs

Estimating Tips

- This section contains the average costs to rent and operate hundreds of pieces of construction equipment. This is useful information when estimating the time and material requirements of any particular operation in order to establish a unit or total cost. Equipment costs include not only rental, but also operating costs for equipment under normal use.

Rental Costs

- Equipment rental rates are obtained from industry sources throughout North America–contractors, suppliers, dealers, manufacturers, and distributors.
- Rental rates vary throughout the country, with larger cities generally having lower rates. Lease plans for new equipment are available for periods in excess of six months, with a percentage of payments applying toward purchase.
- Monthly rental rates vary from 2% to 5% of the purchase price of the equipment depending on the anticipated life of the equipment and its wearing parts.
- Weekly rental rates are about 1/3 the monthly rates, and daily rental rates are about 1/3 the weekly rate.
- Rental rates can also be treated as reimbursement costs for contractor-owned equipment. Owned equipment costs include depreciation, loan payments, interest, taxes, insurance, storage, and major repairs.

Operating Costs

- The operating costs include parts and labor for routine servicing, such as repair and replacement of pumps, filters and worn lines. Normal operating expendables, such as fuel, lubricants, tires and electricity (where applicable), are also included.
- Extraordinary operating expendables with highly variable wear patterns, such as diamond bits and blades, are excluded. These costs can be found as material costs in the Unit Price section.
- The hourly operating costs listed do not include the operator's wages.

Equipment Cost/Day

- Any power equipment required by a crew is shown in the Crew Listings with a daily cost.
- The daily cost of equipment needed by a crew is based on dividing the weekly rental rate by 5 (number of working days in the week), and then adding the hourly operating cost times 8 (the number of hours in a day). This "Equipment Cost/Day" is shown in the far right column of the Equipment Rental pages.
- If equipment is needed for only one or two days, it is best to develop your own cost by including components for daily rent and hourly operating cost. This is important when the listed Crew for a task does not contain the equipment needed, such as a crane for lifting mechanical heating/cooling equipment up onto a roof.
- If the quantity of work is less than the crew's Daily Output shown for a Unit Price line item that includes a bare unit equipment cost, it is recommended to estimate one day's rental cost and operating cost for equipment shown in the Crew Listing for that line item.

Mobilization/Demobilization

- The cost to move construction equipment from an equipment yard or rental company to the jobsite and back again is not included in equipment rental costs listed in the Reference section, nor in the bare equipment cost of any Unit Price line item, nor in any equipment costs shown in the Crew listings.
- Mobilization (to the site) and demobilization (from the site) costs can be found in the Unit Price section.
- If a piece of equipment is already at the jobsite, it is not appropriate to utilize mobil./demob. costs again in an estimate.

No part of this publication may be reproduced, stored in a retrieval system, or transmitted in any form or by any means without prior written permission of RSMeans, a division of Reed Construction Data.

01 54 | Construction Aids

01 54 33 | Equipment Rental

			UNIT	HOURLY OPER. COST	RENT PER DAY	RENT PER WEEK	RENT PER MONTH	EQUIPMENT COST/DAY
10	0010	**CONCRETE EQUIPMENT RENTAL** without operators R015433-10						
	0200	Bucket, concrete lightweight, 1/2 C.Y.	Ea.	.80	23	69	207	20.20
	0300	1 C.Y.		.85	27	81	243	23
	0400	1-1/2 C.Y.		1.10	36.50	110	330	30.80
	0500	2 C.Y.		1.20	43.50	130	390	35.60
	0580	8 C.Y.		5.95	252	755	2,275	198.60
	0600	Cart, concrete, self-propelled, operator walking, 10 C.F.		3.40	55	165	495	60.20
	0700	Operator riding, 18 C.F.		5.65	93.50	280	840	101.20
	0800	Conveyer for concrete, portable, gas, 16" wide, 26' long		13.50	122	365	1,100	181
	0900	46' long		13.90	147	440	1,325	199.20
	1000	56' long		14.05	155	465	1,400	205.40
	1100	Core drill, electric, 2-1/2 H.P., 1" to 8" bit diameter		1.50	57	171	515	46.20
	1150	11 H.P., 8" to 18" cores		5.90	115	345	1,025	116.20
	1200	Finisher, concrete floor, gas, riding trowel, 96" wide		14.05	142	425	1,275	197.40
	1300	Gas, walk-behind, 3 blade, 36" trowel		2.35	20	60	180	30.80
	1400	4 blade, 48" trowel		4.70	26.50	80	240	53.60
	1500	Float, hand-operated (Bull float) 48" wide		.08	13.35	40	120	8.65
	1570	Curb builder, 14 H.P., gas, single screw		14.65	250	750	2,250	267.20
	1590	Double screw		15.30	287	860	2,575	294.40
	1600	Floor grinder, concrete and terrazzo, electric, 22" path		2.19	134	402	1,200	97.90
	1700	Edger, concrete, electric, 7" path		1.02	51.50	155	465	39.15
	1750	Vacuum pick-up system for floor grinders, wet/dry		1.47	81.50	245	735	60.75
	1800	Mixer, powered, mortar and concrete, gas, 6 C.F., 18 H.P.		9.20	117	350	1,050	143.60
	1900	10 C.F., 25 H.P.		11.60	140	420	1,250	176.80
	2000	16 C.F.		11.95	162	485	1,450	192.60
	2100	Concrete, stationary, tilt drum, 2 C.Y.		6.80	228	685	2,050	191.40
	2120	Pump, concrete, truck mounted 4" line 80' boom		24.55	855	2,570	7,700	710.40
	2140	5" line, 110' boom		31.75	1,125	3,400	10,200	934
	2160	Mud jack, 50 C.F. per hr.		7.35	127	380	1,150	134.80
	2180	225 C.F. per hr.		9.65	145	435	1,300	164.20
	2190	Shotcrete pump rig, 12 C.Y./hr.		15.10	220	660	1,975	252.80
	2200	35 C.Y./hr.		17.80	237	710	2,125	284.40
	2600	Saw, concrete, manual, gas, 18 H.P.		7.25	43.50	130	390	84
	2650	Self-propelled, gas, 30 H.P.		14.40	100	300	900	175.20
	2675	V-groove crack chaser, manual, gas, 6 H.P.		2.55	17.35	52	156	30.80
	2700	Vibrators, concrete, electric, 60 cycle, 2 H.P.		.43	8.65	26	78	8.65
	2800	3 H.P.		.61	11.65	35	105	11.90
	2900	Gas engine, 5 H.P.		2.20	16	48	144	27.20
	3000	8 H.P.		3.00	15	45	135	33
	3050	Vibrating screed, gas engine, 8 H.P.		2.78	68.50	205	615	63.25
	3120	Concrete transit mixer, 6 x 4, 250 H.P., 8 C.Y., rear discharge		62.50	565	1,690	5,075	838
	3200	Front discharge		73.60	690	2,065	6,200	1,002
	3300	6 x 6, 285 H.P., 12 C.Y., rear discharge		72.65	650	1,950	5,850	971.20
	3400	Front discharge		75.95	695	2,085	6,250	1,025
20	0010	**EARTHWORK EQUIPMENT RENTAL** without operators R015433-10						
	0040	Aggregate spreader, push type 8' to 12' wide	Ea.	3.25	25.50	76	228	41.20
	0045	Tailgate type, 8' wide		3.10	32.50	98	294	44.40
	0055	Earth auger, truck-mounted, for fence & sign posts, utility poles		19.75	445	1,335	4,000	425
	0060	For borings and monitoring wells		47.30	680	2,040	6,125	786.40
	0070	Portable, trailer mounted		3.15	32.50	98	294	44.80
	0075	Truck-mounted, for caissons, water wells		100.90	2,900	8,720	26,200	2,551
	0080	Horizontal boring machine, 12" to 36" diameter, 45 H.P.		24.60	193	580	1,750	312.80
	0090	12" to 48" diameter, 65 H.P.		34.85	335	1,005	3,025	479.80
	0095	Auger, for fence posts, gas engine, hand held		.55	6	18	54	8
	0100	Excavator, diesel hydraulic, crawler mounted, 1/2 C.Y. cap.		25.00	415	1,240	3,725	448
	0120	5/8 C.Y. capacity		33.05	550	1,650	4,950	594.40
	0140	3/4 C.Y. capacity		36.00	625	1,870	5,600	662
	0150	1 C.Y. capacity		50.60	685	2,050	6,150	814.80

01 54 | Construction Aids

01 54 33 | Equipment Rental

		UNIT	HOURLY OPER. COST	RENT PER DAY	RENT PER WEEK	RENT PER MONTH	EQUIPMENT COST/DAY
0200	1-1/2 C.Y. capacity	Ea.	60.80	910	2,725	8,175	1,031
0300	2 C.Y. capacity		68.50	1,050	3,135	9,400	1,175
0320	2-1/2 C.Y. capacity		104.75	1,325	3,995	12,000	1,637
0325	3-1/2 C.Y. capacity		147.35	2,100	6,315	18,900	2,442
0330	4-1/2 C.Y. capacity		176.65	2,725	8,150	24,500	3,043
0335	6 C.Y. capacity		224.40	2,875	8,640	25,900	3,523
0340	7 C.Y. capacity		226.60	2,975	8,915	26,700	3,596
0342	Excavator attachments, bucket thumbs		3.20	245	735	2,200	172.60
0345	Grapples		2.75	193	580	1,750	138
0347	Hydraulic hammer for boom mounting, 5000 ft lb.		14.30	425	1,280	3,850	370.40
0349	11,000 ft lb.		23.20	750	2,250	6,750	635.60
0350	Gradall type, truck mounted, 3 ton @ 15' radius, 5/8 C.Y.		43.45	890	2,670	8,000	881.60
0370	1 C.Y. capacity		47.65	1,050	3,120	9,350	1,005
0400	Backhoe-loader, 40 to 45 H.P., 5/8 C.Y. capacity		14.95	233	700	2,100	259.60
0450	45 H.P. to 60 H.P., 3/4 C.Y. capacity		23.90	290	870	2,600	365.20
0460	80 H.P., 1-1/4 C.Y. capacity		26.20	315	940	2,825	397.60
0470	112 H.P., 1-1/2 C.Y. capacity		41.65	645	1,935	5,800	720.20
0482	Backhoe-loader attachment, compactor, 20,000 lb.		5.75	142	425	1,275	131
0485	Hydraulic hammer, 750 ft lb.		3.25	96.50	290	870	84
0486	Hydraulic hammer, 1200 ft lb.		6.30	217	650	1,950	180.40
0500	Brush chipper, gas engine, 6" cutter head, 35 H.P.		11.95	107	320	960	159.60
0550	Diesel engine, 12" cutter head, 130 H.P.		28.20	285	855	2,575	396.60
0600	15" cutter head, 165 H.P.		33.90	330	985	2,950	468.20
0750	Bucket, clamshell, general purpose, 3/8 C.Y.		1.25	38.50	115	345	33
0800	1/2 C.Y.		1.35	45	135	405	37.80
0850	3/4 C.Y.		1.50	53.50	160	480	44
0900	1 C.Y.		1.55	58.50	175	525	47.40
0950	1-1/2 C.Y.		2.55	80	240	720	68.40
1000	2 C.Y.		2.65	88.50	265	795	74.20
1010	Bucket, dragline, medium duty, 1/2 C.Y.		.75	23.50	70	210	20
1020	3/4 C.Y.		.75	24.50	74	222	20.80
1030	1 C.Y.		.80	26	78	234	22
1040	1-1/2 C.Y.		1.20	40	120	360	33.60
1050	2 C.Y.		1.25	45	135	405	37
1070	3 C.Y.		1.95	61.50	185	555	52.60
1200	Compactor, manually guided 2-drum vibratory smooth roller, 7.5 H.P.		7.35	207	620	1,850	182.80
1250	Rammer/tamper, gas, 8"		2.70	46.50	140	420	49.60
1260	15"		3.00	53.50	160	480	56
1300	Vibratory plate, gas, 18" plate, 3000 lb. blow		2.65	24.50	73	219	35.80
1350	21" plate, 5000 lb. blow		3.30	32.50	98	294	46
1370	Curb builder/extruder, 14 H.P., gas, single screw		14.65	250	750	2,250	267.20
1390	Double screw		15.30	287	860	2,575	294.40
1500	Disc harrow attachment, for tractor		.43	72.50	217	650	46.85
1810	Feller buncher, shearing & accumulating trees, 100 H.P.		42.50	650	1,945	5,825	729
1860	Grader, self-propelled, 25,000 lb.		39.55	595	1,780	5,350	672.40
1910	30,000 lb.		43.25	600	1,805	5,425	707
1920	40,000 lb.		64.65	1,000	3,015	9,050	1,120
1930	55,000 lb.		84.65	1,600	4,800	14,400	1,637
1950	Hammer, pavement breaker, self-propelled, diesel, 1000 to 1250 lb.		30.25	355	1,060	3,175	454
2000	1300 to 1500 lb.		45.38	705	2,120	6,350	787.05
2050	Pile driving hammer, steam or air, 4150 ft lb. @ 225 bpm		10.35	475	1,425	4,275	367.80
2100	8750 ft lb. @ 145 bpm		12.35	660	1,980	5,950	494.80
2150	15,000 ft lb. @ 60 bpm		13.95	795	2,380	7,150	587.60
2200	24,450 ft lb. @ 111 bpm		14.95	880	2,640	7,925	647.60
2250	Leads, 60' high for pile driving hammers up to 20,000 ft lb.		3.30	80.50	242	725	74.80
2300	90' high for hammers over 20,000 ft lb.		4.95	141	424	1,275	124.40
2350	Diesel type hammer, 22,400 ft lb.		26.93	535	1,610	4,825	537.45
2400	41,300 ft lb.		36.30	585	1,760	5,275	642.40

01 54 | Construction Aids

01 54 33 | Equipment Rental

			UNIT	HOURLY OPER. COST	RENT PER DAY	RENT PER WEEK	RENT PER MONTH	EQUIPMENT COST/DAY
20	2450	141,000 ft lb.	Ea.	55.55	1,050	3,160	9,475	1,076
	2500	Vib. elec. hammer/extractor, 200 kW diesel generator, 34 H.P.		56.25	635	1,910	5,725	832
	2550	80 H.P.		103.45	930	2,785	8,350	1,385
	2600	150 H.P.		196.65	1,775	5,335	16,000	2,640
	2800	Log chipper, up to 22" diameter, 600 H.P.		64.95	630	1,895	5,675	898.60
	2850	Logger, for skidding & stacking logs, 150 H.P.		53.45	820	2,460	7,375	919.60
	2860	Mulcher, diesel powered, trailer mounted		24.70	213	640	1,925	325.60
	2900	Rake, spring tooth, with tractor		18.31	335	1,007	3,025	347.90
	3000	Roller, vibratory, tandem, smooth drum, 20 H.P.		8.75	148	445	1,325	159
	3050	35 H.P.		11.25	253	760	2,275	242
	3100	Towed type vibratory compactor, smooth drum, 50 H.P.		26.25	335	1,010	3,025	412
	3150	Sheepsfoot, 50 H.P.		27.60	370	1,115	3,350	443.80
	3170	Landfill compactor, 220 H.P.		91.85	1,500	4,485	13,500	1,632
	3200	Pneumatic tire roller, 80 H.P.		17.00	355	1,065	3,200	349
	3250	120 H.P.		25.45	605	1,815	5,450	566.60
	3300	Sheepsfoot vibratory roller, 240 H.P.		69.65	1,100	3,305	9,925	1,218
	3320	340 H.P.		101.85	1,600	4,830	14,500	1,781
	3350	Smooth drum vibratory roller, 75 H.P.		25.15	595	1,790	5,375	559.20
	3400	125 H.P.		32.85	715	2,150	6,450	692.80
	3410	Rotary mower, brush, 60", with tractor		22.65	305	915	2,750	364.20
	3420	Rototiller, walk-behind, gas, 5 H.P.		2.23	66	198	595	57.45
	3422	8 H.P.		3.30	91.50	275	825	81.40
	3440	Scrapers, towed type, 7 C.Y. capacity		5.70	110	330	990	111.60
	3450	10 C.Y. capacity		6.45	150	450	1,350	141.60
	3500	15 C.Y. capacity		6.85	173	520	1,550	158.80
	3525	Self-propelled, single engine, 14 C.Y. capacity		113.60	1,525	4,600	13,800	1,829
	3550	Dual engine, 21 C.Y. capacity		174.50	2,000	6,015	18,000	2,599
	3600	31 C.Y. capacity		232.10	2,850	8,545	25,600	3,566
	3640	44 C.Y. capacity		287.55	3,650	10,960	32,900	4,492
	3650	Elevating type, single engine, 11 C.Y. capacity		71.20	1,025	3,050	9,150	1,180
	3700	22 C.Y. capacity		138.15	2,225	6,690	20,100	2,443
	3710	Screening plant 110 H.P. w/5' x 10' screen		31.63	435	1,300	3,900	513.05
	3720	5' x 16' screen		34.85	540	1,615	4,850	601.80
	3850	Shovel, crawler-mounted, front-loading, 7 C.Y. capacity		246.75	2,900	8,700	26,100	3,714
	3855	12 C.Y. capacity		354.50	3,950	11,840	35,500	5,204
	3860	Shovel/backhoe bucket, 1/2 C.Y.		2.45	66.50	200	600	59.60
	3870	3/4 C.Y.		2.50	73.50	220	660	64
	3880	1 C.Y.		2.60	83.50	250	750	70.80
	3890	1-1/2 C.Y.		2.75	96.50	290	870	80
	3910	3 C.Y.		3.10	130	390	1,175	102.80
	3950	Stump chipper, 18" deep, 30 H.P.		7.06	176	528	1,575	162.10
	4110	Dozer, crawler, torque converter, diesel 80 H.P.		29.75	405	1,215	3,650	481
	4150	105 H.P.		36.10	505	1,515	4,550	591.80
	4200	140 H.P.		52.10	780	2,335	7,000	883.80
	4260	200 H.P.		77.35	1,175	3,530	10,600	1,325
	4310	300 H.P.		100.55	1,700	5,065	15,200	1,817
	4360	410 H.P.		135.65	2,200	6,620	19,900	2,409
	4370	500 H.P.		174.30	2,875	8,610	25,800	3,116
	4380	700 H.P.		261.35	4,000	11,970	35,900	4,485
	4400	Loader, crawler, torque conv., diesel, 1-1/2 C.Y., 80 H.P.		31.35	455	1,360	4,075	522.80
	4450	1-1/2 to 1-3/4 C.Y., 95 H.P.		34.15	585	1,760	5,275	625.20
	4510	1-3/4 to 2-1/4 C.Y., 130 H.P.		54.60	870	2,615	7,850	959.80
	4530	2-1/2 to 3-1/4 C.Y., 190 H.P.		67.00	1,100	3,275	9,825	1,191
	4560	3-1/2 to 5 C.Y., 275 H.P.		86.10	1,400	4,215	12,600	1,532
	4610	Front end loader, 4WD, articulated frame, diesel 1 to 1-1/4 C.Y., 70 H.P.		19.90	227	680	2,050	295.20
	4620	1-1/2 to 1-3/4 C.Y., 95 H.P.		25.30	285	855	2,575	373.40
	4650	1-3/4 to 2 C.Y., 130 H.P.		29.45	365	1,100	3,300	455.60
	4710	2-1/2 to 3-1/2 C.Y., 145 H.P.		33.35	425	1,280	3,850	522.80

01 54 | Construction Aids

01 54 33 | Equipment Rental

		UNIT	HOURLY OPER. COST	RENT PER DAY	RENT PER WEEK	RENT PER MONTH	EQUIPMENT COST/DAY	
4730	3 to 4-1/2 C.Y., 185 H.P.	Ea.	42.85	550	1,655	4,975	673.80	20
4760	5-1/4 to 5-3/4 C.Y., 270 H.P.		67.80	845	2,535	7,600	1,049	
4810	7 to 9 C.Y., 475 H.P.		116.25	1,775	5,295	15,900	1,989	
4870	9 - 11 C.Y., 620 H.P.		155.75	2,425	7,285	21,900	2,703	
4880	Skid steer loader, wheeled, 10 C.F., 30 H.P. gas		10.90	145	435	1,300	174.20	
4890	1 C.Y., 78 H.P., diesel		20.35	243	730	2,200	308.80	
4892	Skid-steer attachment, auger		.50	83.50	251	755	54.20	
4893	Backhoe		.66	111	332	995	71.70	
4894	Broom		.74	124	372	1,125	80.30	
4895	Forks		.23	38.50	115	345	24.85	
4896	Grapple		.56	92.50	278	835	60.10	
4897	Concrete hammer		1.06	177	531	1,600	114.70	
4898	Tree spade		1.03	172	515	1,550	111.25	
4899	Trencher		.67	112	335	1,000	72.35	
4900	Trencher, chain, boom type, gas, operator walking, 12 H.P.		5.40	46.50	140	420	71.20	
4910	Operator riding, 40 H.P.		20.55	292	875	2,625	339.40	
5000	Wheel type, diesel, 4' deep, 12" wide		87.50	845	2,530	7,600	1,206	
5100	6' deep, 20" wide		93.25	1,925	5,760	17,300	1,898	
5150	Chain type, diesel, 5' deep, 8" wide		38.85	560	1,685	5,050	647.80	
5200	Diesel, 8' deep, 16" wide		156.45	3,550	10,620	31,900	3,376	
5202	Rock trencher, wheel type, 6" wide x 18" deep		21.50	330	985	2,950	369	
5206	Chain type, 18" wide x 7' deep		112.60	2,825	8,445	25,300	2,590	
5210	Tree spade, self-propelled		14.00	267	800	2,400	272	
5250	Truck, dump, 2-axle, 12 ton, 8 C.Y. payload, 220 H.P.		35.15	227	680	2,050	417.20	
5300	Three axle dump, 16 ton, 12 C.Y. payload, 400 H.P.		62.40	320	965	2,900	692.20	
5310	Four axle dump, 25 ton, 18 C.Y. payload, 450 H.P.		73.20	470	1,410	4,225	867.60	
5350	Dump trailer only, rear dump, 16-1/2 C.Y.		5.40	138	415	1,250	126.20	
5400	20 C.Y.		5.85	157	470	1,400	140.80	
5450	Flatbed, single axle, 1-1/2 ton rating		28.65	68.50	205	615	270.20	
5500	3 ton rating		34.25	98.50	295	885	333	
5550	Off highway rear dump, 25 ton capacity		74.55	1,200	3,590	10,800	1,314	
5600	35 ton capacity		83.10	1,350	4,025	12,100	1,470	
5610	50 ton capacity		103.40	1,600	4,800	14,400	1,787	
5620	65 ton capacity		106.30	1,575	4,735	14,200	1,797	
5630	100 ton capacity		154.35	2,775	8,315	24,900	2,898	
6000	Vibratory plow, 25 H.P., walking		9.25	60	180	540	110	
0010	**GENERAL EQUIPMENT RENTAL** without operators	R015433 -10						40
0150	Aerial lift, scissor type, to 15' high, 1000 lb. cap., electric	Ea.	3.10	51.50	155	465	55.80	
0160	To 25' high, 2000 lb. capacity		3.55	66.50	200	600	68.40	
0170	Telescoping boom to 40' high, 500 lb. capacity, diesel		17.00	315	945	2,825	325	
0180	To 45' high, 500 lb. capacity		18.85	355	1,070	3,200	364.80	
0190	To 60' high, 600 lb. capacity		21.25	475	1,430	4,300	456	
0195	Air compressor, portable, 6.5 CFM, electric		.56	12.65	38	114	12.10	
0196	Gasoline		.79	19	57	171	17.70	
0200	Towed type, gas engine, 60 CFM		13.55	48.50	145	435	137.40	
0300	160 CFM		15.80	51.50	155	465	157.40	
0400	Diesel engine, rotary screw, 250 CFM		17.05	110	330	990	202.40	
0500	365 CFM		23.10	133	400	1,200	264.80	
0550	450 CFM		29.35	167	500	1,500	334.80	
0600	600 CFM		51.70	232	695	2,075	552.60	
0700	750 CFM		51.90	240	720	2,150	559.20	
0800	For silenced models, small sizes, add to rent		3%	5%	5%	5%		
0900	Large sizes, add to rent		5%	7%	7%	7%		
0930	Air tools, breaker, pavement, 60 lb.	Ea.	.50	9.65	29	87	9.80	
0940	80 lb.		.50	10.35	31	93	10.20	
0950	Drills, hand (jackhammer) 65 lb.		.60	17.35	52	156	15.20	
0960	Track or wagon, swing boom, 4" drifter		63.45	885	2,660	7,975	1,040	
0970	5" drifter		75.50	1,075	3,255	9,775	1,255	

01 54 | Construction Aids

01 54 33 | Equipment Rental

			UNIT	HOURLY OPER. COST	RENT PER DAY	RENT PER WEEK	RENT PER MONTH	EQUIPMENT COST/DAY	
40	0975	Track mounted quarry drill, 6" diameter drill	Ea.	129.75	1,600	4,830	14,500	2,004	40
	0980	Dust control per drill		.97	24	72	216	22.15	
	0990	Hammer, chipping, 12 lb.		.55	26	78	234	20	
	1000	Hose, air with couplings, 50' long, 3/4" diameter		.03	5	15	45	3.25	
	1100	1" diameter		.04	6.35	19	57	4.10	
	1200	1-1/2" diameter		.05	9	27	81	5.80	
	1300	2" diameter		.07	12	36	108	7.75	
	1400	2-1/2" diameter		.11	19	57	171	12.30	
	1410	3" diameter		.14	23	69	207	14.90	
	1450	Drill, steel, 7/8" x 2'		.05	8.65	26	78	5.60	
	1460	7/8" x 6'		.05	9	27	81	5.80	
	1520	Moil points		.02	3.33	10	30	2.15	
	1525	Pneumatic nailer w/accessories		.47	31	93	279	22.35	
	1530	Sheeting driver for 60 lb. breaker		.03	5.65	17	51	3.65	
	1540	For 90 lb. breaker		.11	7.35	22	66	5.30	
	1550	Spade, 25 lb.		.45	7	21	63	7.80	
	1560	Tamper, single, 35 lb.		.55	36.50	109	325	26.20	
	1570	Triple, 140 lb.		.82	54.50	164	490	39.35	
	1580	Wrenches, impact, air powered, up to 3/4" bolt		.40	12.65	38	114	10.80	
	1590	Up to 1-1/4" bolt		.50	23.50	70	210	18	
	1600	Barricades, barrels, reflectorized, 1 to 99 barrels		.03	4.60	13.80	41.50	3	
	1610	100 to 200 barrels		.02	3.53	10.60	32	2.30	
	1620	Barrels with flashers, 1 to 99 barrels		.03	5.25	15.80	47.50	3.40	
	1630	100 to 200 barrels		.03	4.20	12.60	38	2.75	
	1640	Barrels with steady burn type C lights		.04	7	21	63	4.50	
	1650	Illuminated board, trailer mounted, with generator		3.55	130	390	1,175	106.40	
	1670	Portable barricade, stock, with flashers, 1 to 6 units		.03	5.25	15.80	47.50	3.40	
	1680	25 to 50 units		.03	4.90	14.70	44	3.20	
	1685	Butt fusion machine, wheeled, 1.5 HP electric, 2" - 8" diameter pipe		2.62	167	500	1,500	120.95	
	1690	Tracked, 20 HP diesel, 4"-12" diameter pipe		11.10	490	1,465	4,400	381.80	
	1695	83 HP diesel, 8" - 24" diameter pipe		30.30	975	2,930	8,800	828.40	
	1700	Carts, brick, hand powered, 1000 lb. capacity		.44	72.50	218	655	47.10	
	1800	Gas engine, 1500 lb., 7-1/2' lift		4.10	115	345	1,025	101.80	
	1822	Dehumidifier, medium, 6 lb./hr., 150 CFM		.99	62	186	560	45.10	
	1824	Large, 18 lb./hr., 600 CFM		2.01	126	378	1,125	91.70	
	1830	Distributor, asphalt, trailer mounted, 2000 gal., 38 H.P. diesel		10.15	330	985	2,950	278.20	
	1840	3000 gal., 38 H.P. diesel		11.65	360	1,075	3,225	308.20	
	1850	Drill, rotary hammer, electric		1.00	26.50	80	240	24	
	1860	Carbide bit, 1-1/2" diameter, add to electric rotary hammer		.02	3.61	10.84	32.50	2.35	
	1865	Rotary, crawler, 250 H.P.		150.45	2,100	6,285	18,900	2,461	
	1870	Emulsion sprayer, 65 gal., 5 H.P. gas engine		2.90	98	294	880	82	
	1880	200 gal., 5 H.P. engine		7.85	163	490	1,475	160.80	
	1900	Floor auto-scrubbing machine, walk-behind, 28" path		4.52	293	880	2,650	212.15	
	1930	Floodlight, mercury vapor, or quartz, on tripod, 1000 watt		.41	19.35	58	174	14.90	
	1940	2000 watt		.78	39	117	350	29.65	
	1950	Floodlights, trailer mounted with generator, 1 - 300 watt light		3.65	71.50	215	645	72.20	
	1960	2 - 1000 watt lights		4.90	98.50	295	885	98.20	
	2000	4 - 300 watt lights		4.60	93.50	280	840	92.80	
	2005	Foam spray rig, incl. box trailer, compressor, generator, proportioner		36.53	490	1,465	4,400	585.25	
	2020	Forklift, straight mast, 12' lift, 5000 lb., 2 wheel drive, gas		25.35	203	610	1,825	324.80	
	2040	21' lift, 5000 lb., 4 wheel drive, diesel		21.15	242	725	2,175	314.20	
	2050	For rough terrain, 42' lift, 35' reach, 9000 lb., 110 H.P.		29.95	485	1,460	4,375	531.60	
	2060	For plant, 4 ton capacity, 80 H.P., 2 wheel drive, gas		15.50	93.50	280	840	180	
	2080	10 ton capacity, 120 H.P., 2 wheel drive, diesel		24.30	162	485	1,450	291.40	
	2100	Generator, electric, gas engine, 1.5 kW to 3 kW		3.95	11.35	34	102	38.40	
	2200	5 kW		5.10	15.35	46	138	50	
	2300	10 kW		9.65	36.50	110	330	99.20	
	2400	25 kW		11.00	86.50	260	780	140	

01 54 | Construction Aids

01 54 33 | Equipment Rental

		UNIT	HOURLY OPER. COST	RENT PER DAY	RENT PER WEEK	RENT PER MONTH	EQUIPMENT COST/DAY
2500	Diesel engine, 20 kW	Ea.	12.55	68.50	205	615	141.40
2600	50 kW		24.35	103	310	930	256.80
2700	100 kW		45.15	130	390	1,175	439.20
2800	250 kW		88.85	237	710	2,125	852.80
2850	Hammer, hydraulic, for mounting on boom, to 500 ft lb.		2.55	75	225	675	65.40
2860	1000 ft lb.		4.35	127	380	1,150	110.80
2900	Heaters, space, oil or electric, 50 MBH		1.93	7.65	23	69	20.05
3000	100 MBH		3.61	10.65	32	96	35.30
3100	300 MBH		10.58	38.50	115	345	107.65
3150	500 MBH		17.40	45	135	405	166.20
3200	Hose, water, suction with coupling, 20' long, 2" diameter		.02	3	9	27	1.95
3210	3" diameter		.03	4.33	13	39	2.85
3220	4" diameter		.03	5	15	45	3.25
3230	6" diameter		.11	17.65	53	159	11.50
3240	8" diameter		.20	33.50	100	300	21.60
3250	Discharge hose with coupling, 50' long, 2" diameter		.01	1.33	4	12	.90
3260	3" diameter		.01	2.33	7	21	1.50
3270	4" diameter		.02	3.67	11	33	2.35
3280	6" diameter		.06	9.35	28	84	6.10
3290	8" diameter		.20	33.50	100	300	21.60
3295	Insulation blower		.77	6	18	54	9.75
3300	Ladders, extension type, 16' to 36' long		.14	22.50	68	204	14.70
3400	40' to 60' long		.19	31	93	279	20.10
3405	Lance for cutting concrete		2.52	89	267	800	73.55
3407	Lawn mower, rotary, 22", 5 H.P.		2.10	59	177	530	52.20
3408	48" self propelled		2.92	75	225	675	68.35
3410	Level, electronic, automatic, with tripod and leveling rod		1.49	99	297	890	71.30
3430	Laser type, for pipe and sewer line and grade		.73	48.50	145	435	34.85
3440	Rotating beam for interior control		.97	64	192	575	46.15
3460	Builder's optical transit, with tripod and rod		.10	16.65	50	150	10.80
3500	Light towers, towable, with diesel generator, 2000 watt		4.60	93.50	280	840	92.80
3600	4000 watt		4.90	98.50	295	885	98.20
3700	Mixer, powered, plaster and mortar, 6 C.F., 7 H.P.		2.90	19	57	171	34.60
3800	10 C.F., 9 H.P.		3.05	30.50	91	273	42.60
3850	Nailer, pneumatic		.47	31	93	279	22.35
3900	Paint sprayers complete, 8 CFM		.94	62.50	187	560	44.90
4000	17 CFM		1.57	104	313	940	75.15
4020	Pavers, bituminous, rubber tires, 8' wide, 50 H.P., diesel		31.85	500	1,500	4,500	554.80
4030	10' wide, 150 H.P.		107.95	1,825	5,460	16,400	1,956
4050	Crawler, 8' wide, 100 H.P., diesel		90.45	1,825	5,500	16,500	1,824
4060	10' wide, 150 H.P.		115.10	2,250	6,715	20,100	2,264
4070	Concrete paver, 12' to 24' wide, 250 H.P.		102.05	1,525	4,600	13,800	1,736
4080	Placer-spreader-trimmer, 24' wide, 300 H.P.		147.05	2,475	7,405	22,200	2,657
4100	Pump, centrifugal gas pump, 1-1/2" diam., 65 GPM		3.95	50	150	450	61.60
4200	2" diameter, 130 GPM		5.30	55	165	495	75.40
4300	3" diameter, 250 GPM		5.60	56.50	170	510	78.80
4400	6" diameter, 1500 GPM		29.35	177	530	1,600	340.80
4500	Submersible electric pump, 1-1/4" diameter, 55 GPM		.38	16.65	50	150	13.05
4600	1-1/2" diameter, 83 GPM		.42	19	57	171	14.75
4700	2" diameter, 120 GPM		1.43	23.50	71	213	25.65
4800	3" diameter, 300 GPM		2.55	41.50	125	375	45.40
4900	4" diameter, 560 GPM		10.65	160	480	1,450	181.20
5000	6" diameter, 1590 GPM		15.75	215	645	1,925	255
5100	Diaphragm pump, gas, single, 1-1/2" diameter		1.18	50.50	152	455	39.85
5200	2" diameter		4.25	63.50	190	570	72
5300	3" diameter		4.25	63.50	190	570	72
5400	Double, 4" diameter		6.35	108	325	975	115.80
5450	Pressure washer 5 GPM, 3000 psi		4.80	51.50	155	465	69.40

01 54 | Construction Aids

01 54 33 | Equipment Rental

		Description	UNIT	HOURLY OPER. COST	RENT PER DAY	RENT PER WEEK	RENT PER MONTH	EQUIPMENT COST/DAY	
40	5460	7 GPM, 3000 psi	Ea.	6.25	60	180	540	86	40
	5500	Trash pump, self-priming, gas, 2" diameter		4.50	21.50	64	192	48.80	
	5600	Diesel, 4" diameter		9.30	90	270	810	128.40	
	5650	Diesel, 6" diameter		25.55	150	450	1,350	294.40	
	5655	Grout Pump		26.80	260	780	2,350	370.40	
	5700	Salamanders, L.P. gas fired, 100,000 Btu		3.81	13.65	41	123	38.70	
	5705	50,000 Btu		2.39	10.35	31	93	25.30	
	5720	Sandblaster, portable, open top, 3 C.F. capacity		.55	26.50	80	240	20.40	
	5730	6 C.F. capacity		.95	40	120	360	31.60	
	5740	Accessories for above		.13	21.50	64	192	13.85	
	5750	Sander, floor		.71	17	51	153	15.90	
	5760	Edger		.57	18.35	55	165	15.55	
	5800	Saw, chain, gas engine, 18" long		2.25	22	66	198	31.20	
	5900	Hydraulic powered, 36" long		.75	65	195	585	45	
	5950	60" long		.75	66.50	200	600	46	
	6000	Masonry, table mounted, 14" diameter, 5 H.P.		1.31	56.50	170	510	44.50	
	6050	Portable cut-off, 8 H.P.		2.40	33.50	100	300	39.20	
	6100	Circular, hand held, electric, 7-1/4" diameter		.22	4.67	14	42	4.55	
	6200	12" diameter		.22	8	24	72	6.55	
	6250	Wall saw, w/hydraulic power, 10 H.P.		10.45	60	180	540	119.60	
	6275	Shot blaster, walk-behind, 20" wide		4.80	293	880	2,650	214.40	
	6280	Sidewalk broom, walk-behind		2.32	70	210	630	60.55	
	6300	Steam cleaner, 100 gallons per hour		3.70	76.50	230	690	75.60	
	6310	200 gallons per hour		5.25	95	285	855	99	
	6340	Tar Kettle/Pot, 400 gallons		10.50	76.50	230	690	130	
	6350	Torch, cutting, acetylene-oxygen, 150' hose, excludes gases		.30	15	45	135	11.40	
	6360	Hourly operating cost includes tips and gas		19.00				152	
	6410	Toilet, portable chemical		.12	20.50	62	186	13.35	
	6420	Recycle flush type		.15	24.50	74	222	16	
	6430	Toilet, fresh water flush, garden hose,		.18	29.50	89	267	19.25	
	6440	Hoisted, non-flush, for high rise		.15	24.50	73	219	15.80	
	6465	Tractor, farm with attachment		21.90	277	830	2,500	341.20	
	6500	Trailers, platform, flush deck, 2 axle, 25 ton capacity		5.45	117	350	1,050	113.60	
	6600	40 ton capacity		7.00	160	480	1,450	152	
	6700	3 axle, 50 ton capacity		7.55	177	530	1,600	166.40	
	6800	75 ton capacity		9.35	232	695	2,075	213.80	
	6810	Trailer mounted cable reel for high voltage line work		5.35	255	764	2,300	195.60	
	6820	Trailer mounted cable tensioning rig		10.64	505	1,520	4,550	389.10	
	6830	Cable pulling rig		71.33	2,825	8,510	25,500	2,273	
	6900	Water tank trailer, engine driven discharge, 5000 gallons		6.95	143	430	1,300	141.60	
	6925	10,000 gallons		9.50	198	595	1,775	195	
	6950	Water truck, off highway, 6000 gallons		89.80	775	2,330	7,000	1,184	
	7010	Tram car for high voltage line work, powered, 2 conductor		6.27	138	415	1,250	133.15	
	7020	Transit (builder's level) with tripod		.10	16.65	50	150	10.80	
	7030	Trench box, 3000 lb., 6' x 8'		.56	93.50	280	840	60.50	
	7040	7200 lb., 6' x 20'		.78	130	389	1,175	84.05	
	7050	8000 lb., 8' x 16'		1.08	180	540	1,625	116.65	
	7060	9500 lb., 8' x 20'		1.21	201	603	1,800	130.30	
	7065	11,000 lb., 8' x 24'		1.27	211	633	1,900	136.75	
	7070	12,000 lb., 10' x 20'		1.36	227	680	2,050	146.90	
	7100	Truck, pickup, 3/4 ton, 2 wheel drive		15.25	58.50	175	525	157	
	7200	4 wheel drive		15.55	73.50	220	660	168.40	
	7250	Crew carrier, 9 passenger		21.60	86.50	260	780	224.80	
	7290	Flat bed truck, 20,000 lb. GVW		22.30	125	375	1,125	253.40	
	7300	Tractor, 4 x 2, 220 H.P.		31.05	197	590	1,775	366.40	
	7410	330 H.P.		46.05	272	815	2,450	531.40	
	7500	6 x 4, 380 H.P.		52.85	315	950	2,850	612.80	
	7600	450 H.P.		64.05	385	1,150	3,450	742.40	

01 54 | Construction Aids

01 54 33 | Equipment Rental

			UNIT	HOURLY OPER. COST	RENT PER DAY	RENT PER WEEK	RENT PER MONTH	EQUIPMENT COST/DAY	
40	7610	Tractor, with A frame, boom and winch, 225 H.P.	Ea.	33.95	272	815	2,450	434.60	40
	7620	Vacuum truck, hazardous material, 2500 gallons		12.00	292	875	2,625	271	
	7625	5,000 gallons		14.27	410	1,230	3,700	360.15	
	7650	Vacuum, HEPA, 16 gallon, wet/dry		.83	18	54	162	17.45	
	7655	55 gallon, wet/dry		.77	27	81	243	22.35	
	7660	Water tank, portable		.16	26.50	80	240	17.30	
	7690	Sewer/catch basin vacuum, 14 C.Y., 1500 gallons		18.61	615	1,850	5,550	518.90	
	7700	Welder, electric, 200 amp		3.64	16.35	49	147	38.90	
	7800	300 amp		5.41	20	60	180	55.30	
	7900	Gas engine, 200 amp		13.90	23.50	70	210	125.20	
	8000	300 amp		15.90	24.50	74	222	142	
	8100	Wheelbarrow, any size		.08	13	39	117	8.45	
	8200	Wrecking ball, 4000 lb.		2.30	68.50	205	615	59.40	
50	0010	**HIGHWAY EQUIPMENT RENTAL** without operators							50
	0050	Asphalt batch plant, portable drum mixer, 100 ton/hr. R015433-10	Ea.	79.15	1,450	4,320	13,000	1,497	
	0060	200 ton/hr.		90.32	1,525	4,590	13,800	1,641	
	0070	300 ton/hr.		107.39	1,800	5,390	16,200	1,937	
	0100	Backhoe attachment, long stick, up to 185 H.P., 10.5' long		.35	23	69	207	16.60	
	0140	Up to 250 H.P., 12' long		.37	24.50	74	222	17.75	
	0180	Over 250 H.P., 15' long		.52	34.50	103	310	24.75	
	0200	Special dipper arm, up to 100 H.P., 32' long		1.06	70.50	211	635	50.70	
	0240	Over 100 H.P., 33' long		1.32	87.50	263	790	63.15	
	0280	Catch basin/sewer cleaning truck, 3 ton, 9 C.Y., 1000 gal.		46.25	390	1,170	3,500	604	
	0300	Concrete batch plant, portable, electric, 200 C.Y./hr.		22.08	510	1,525	4,575	481.65	
	0520	Grader/dozer attachment, ripper/scarifier, rear mounted, up to 135 H.P.		3.45	65	195	585	66.60	
	0540	Up to 180 H.P.		4.05	81.50	245	735	81.40	
	0580	Up to 250 H.P.		4.45	93.50	280	840	91.60	
	0700	Pvmt. removal bucket, for hyd. excavator, up to 90 H.P.		1.90	51.50	155	465	46.20	
	0740	Up to 200 H.P.		2.10	73.50	220	660	60.80	
	0780	Over 200 H.P.		2.25	85	255	765	69	
	0900	Aggregate spreader, self-propelled, 187 H.P.		53.75	685	2,055	6,175	841	
	1000	Chemical spreader, 3 C.Y.		3.40	43.50	130	390	53.20	
	1900	Hammermill, traveling, 250 H.P.		77.73	2,050	6,120	18,400	1,846	
	2000	Horizontal borer, 3" diameter, 13 H.P. gas driven		6.25	55	165	495	83	
	2150	Horizontal directional drill, 20,000 lb. thrust, 78 H.P. diesel		31.55	675	2,030	6,100	658.40	
	2160	30,000 lb. thrust, 115 H.P.		39.40	1,025	3,105	9,325	936.20	
	2170	50,000 lb. thrust, 170 H.P.		56.45	1,325	3,970	11,900	1,246	
	2190	Mud trailer for HDD, 1500 gallons, 175 H.P., gas		35.35	153	460	1,375	374.80	
	2200	Hydromulcher, diesel, 3000 gallon, for truck mounting		24.05	247	740	2,225	340.40	
	2300	Gas, 600 gallon		9.20	98.50	295	885	132.60	
	2400	Joint & crack cleaner, walk behind, 25 H.P.		4.35	50	150	450	64.80	
	2500	Filler, trailer mounted, 400 gallons, 20 H.P.		9.60	212	635	1,900	203.80	
	3000	Paint striper, self-propelled, 40 gallon, 22 H.P.		7.60	157	470	1,400	154.80	
	3100	120 gallon, 120 H.P.		24.80	400	1,195	3,575	437.40	
	3200	Post drivers, 6" I-Beam frame, for truck mounting		19.65	380	1,145	3,425	386.20	
	3400	Road sweeper, self-propelled, 8' wide, 90 H.P.		40.30	570	1,715	5,150	665.40	
	3450	Road sweeper, vacuum assisted, 4 C.Y., 220 gallons		77.55	625	1,870	5,600	994.40	
	4000	Road mixer, self-propelled, 130 H.P.		47.80	775	2,320	6,950	846.40	
	4100	310 H.P.		84.40	2,100	6,335	19,000	1,942	
	4220	Cold mix paver, incl. pug mill and bitumen tank, 165 H.P.		98.80	2,325	6,965	20,900	2,183	
	4240	Pavement brush, towed		3.20	93.50	280	840	81.60	
	4250	Paver, asphalt, wheel or crawler, 130 H.P., diesel		98.45	2,300	6,905	20,700	2,169	
	4300	Paver, road widener, gas 1' to 6', 67 H.P.		49.40	905	2,715	8,150	938.20	
	4400	Diesel, 2' to 14', 88 H.P.		63.70	1,050	3,170	9,500	1,144	
	4600	Slipform pavers, curb and gutter, 2 track, 75 H.P.		52.00	805	2,410	7,225	898	
	4700	4 track, 165 H.P.		41.65	730	2,185	6,550	770.20	
	4800	Median barrier, 215 H.P.		54.00	890	2,670	8,000	966	

01 54 | Construction Aids

01 54 33 | Equipment Rental

			UNIT	HOURLY OPER. COST	RENT PER DAY	RENT PER WEEK	RENT PER MONTH	EQUIPMENT COST/DAY	
50	4901	Trailer, low bed, 75 ton capacity	Ea.	10.10	232	695	2,075	219.80	50
	5000	Road planer, walk behind, 10" cutting width, 10 H.P.		4.00	33.50	100	300	52	
	5100	Self-propelled, 12" cutting width, 64 H.P.		11.45	113	340	1,025	159.60	
	5120	Traffic line remover, metal ball blaster, truck mounted, 115 H.P.		48.75	750	2,245	6,725	839	
	5140	Grinder, truck mounted, 115 H.P.		55.45	810	2,435	7,300	930.60	
	5160	Walk-behind, 11 H.P.		4.55	53.50	160	480	68.40	
	5200	Pavement profiler, 4' to 6' wide, 450 H.P.		258.10	3,350	10,045	30,100	4,074	
	5300	8' to 10' wide, 750 H.P.		405.10	4,400	13,180	39,500	5,877	
	5400	Roadway plate, steel, 1" x 8' x 20'		.08	13	39	117	8.45	
	5600	Stabilizer, self-propelled, 150 H.P.		50.10	610	1,835	5,500	767.80	
	5700	310 H.P.		100.25	1,700	5,125	15,400	1,827	
	5800	Striper, truck mounted, 120 gallon paint, 460 H.P.		71.05	485	1,460	4,375	860.40	
	5900	Thermal paint heating kettle, 115 gallons		5.95	25.50	77	231	63	
	6000	Tar kettle, 330 gallon, trailer mounted		8.70	58.50	175	525	104.60	
	7000	Tunnel locomotive, diesel, 8 to 12 ton		32.40	585	1,760	5,275	611.20	
	7005	Electric, 10 ton		26.55	670	2,010	6,025	614.40	
	7010	Muck cars, 1/2 C.Y. capacity		2.05	24.50	74	222	31.20	
	7020	1 C.Y. capacity		2.30	32.50	98	294	38	
	7030	2 C.Y. capacity		2.45	36.50	110	330	41.60	
	7040	Side dump, 2 C.Y. capacity		2.65	45	135	405	48.20	
	7050	3 C.Y. capacity		3.55	51.50	155	465	59.40	
	7060	5 C.Y. capacity		5.10	65	195	585	79.80	
	7100	Ventilating blower for tunnel, 7-1/2 H.P.		2.05	51.50	155	465	47.40	
	7110	10 H.P.		2.22	53.50	160	480	49.75	
	7120	20 H.P.		3.43	69.50	208	625	69.05	
	7140	40 H.P.		5.59	98.50	295	885	103.70	
	7160	60 H.P.		8.54	152	455	1,375	159.30	
	7175	75 H.P.		11.25	207	620	1,850	214	
	7180	200 H.P.		22.56	305	910	2,725	362.50	
	7800	Windrow loader, elevating		55.95	1,325	3,960	11,900	1,240	
60	0010	**LIFTING AND HOISTING EQUIPMENT RENTAL** without operators							60
	0120	Aerial lift truck, 2 person, to 80'	Ea.	26.75	715	2,145	6,425	643	
	0140	Boom work platform, 40' snorkel		16.10	277	830	2,500	294.80	
	0150	Crane, flatbed mounted, 3 ton capacity		16.55	193	580	1,750	248.40	
	0200	Crane, climbing, 106' jib, 6000 lb. capacity, 410 fpm		37.76	1,625	4,860	14,600	1,274	
	0300	101' jib, 10,250 lb. capacity, 270 fpm		44.26	2,050	6,160	18,500	1,586	
	0500	Tower, static, 130' high, 106' jib, 6200 lb. capacity at 400 fpm		41.56	1,875	5,620	16,900	1,456	
	0600	Crawler mounted, lattice boom, 1/2 C.Y., 15 tons at 12' radius		37.52	640	1,920	5,750	684.15	
	0700	3/4 C.Y., 20 tons at 12' radius		50.03	800	2,400	7,200	880.25	
	0800	1 C.Y., 25 tons at 12' radius		66.70	1,075	3,195	9,575	1,173	
	0900	1-1/2 C.Y., 40 tons at 12' radius		66.70	1,075	3,225	9,675	1,179	
	1000	2 C.Y., 50 tons at 12' radius		70.70	1,250	3,765	11,300	1,319	
	1100	3 C.Y., 75 tons at 12' radius		75.65	1,475	4,435	13,300	1,492	
	1200	100 ton capacity, 60' boom		85.65	1,700	5,090	15,300	1,703	
	1300	165 ton capacity, 60' boom		109.15	2,000	5,980	17,900	2,069	
	1400	200 ton capacity, 70' boom		132.10	2,500	7,480	22,400	2,553	
	1500	350 ton capacity, 80' boom		184.60	3,725	11,210	33,600	3,719	
	1600	Truck mounted, lattice boom, 6 x 4, 20 tons at 10' radius		37.31	1,100	3,310	9,925	960.50	
	1700	25 tons at 10' radius		40.35	1,200	3,610	10,800	1,045	
	1800	8 x 4, 30 tons at 10' radius		43.85	1,275	3,840	11,500	1,119	
	1900	40 tons at 12' radius		46.92	1,325	4,010	12,000	1,177	
	2000	60 tons at 15' radius		53.24	1,425	4,240	12,700	1,274	
	2050	82 tons at 15' radius		60.06	1,525	4,540	13,600	1,388	
	2100	90 tons at 15' radius		67.57	1,650	4,940	14,800	1,529	
	2200	115 tons at 15' radius		76.34	1,850	5,520	16,600	1,715	
	2300	150 tons at 18' radius		84.40	1,950	5,815	17,400	1,838	
	2350	165 tons at 18' radius		90.25	2,050	6,160	18,500	1,954	
	2400	Truck mounted, hydraulic, 12 ton capacity		43.00	530	1,595	4,775	663	

Reference codes: R015433-10, R015433-15, R312316-45

01 54 | Construction Aids

01 54 33 | Equipment Rental

		Description	UNIT	HOURLY OPER. COST	RENT PER DAY	RENT PER WEEK	RENT PER MONTH	EQUIPMENT COST/DAY
60	2500	25 ton capacity	Ea.	45.35	640	1,915	5,750	745.80
	2550	33 ton capacity		45.90	655	1,960	5,875	759.20
	2560	40 ton capacity		59.35	765	2,290	6,875	932.80
	2600	55 ton capacity		77.50	860	2,585	7,750	1,137
	2700	80 ton capacity		101.35	1,400	4,205	12,600	1,652
	2720	100 ton capacity		94.95	1,450	4,325	13,000	1,625
	2740	120 ton capacity		109.95	1,550	4,665	14,000	1,813
	2760	150 ton capacity		128.55	2,050	6,135	18,400	2,255
	2800	Self-propelled, 4 x 4, with telescoping boom, 5 ton		17.75	230	690	2,075	280
	2900	12-1/2 ton capacity		32.65	365	1,100	3,300	481.20
	3000	15 ton capacity		33.35	385	1,160	3,475	498.80
	3050	20 ton capacity		36.30	450	1,350	4,050	560.40
	3100	25 ton capacity		37.90	505	1,520	4,550	607.20
	3150	40 ton capacity		46.50	570	1,705	5,125	713
	3200	Derricks, guy, 20 ton capacity, 60' boom, 75' mast		27.18	395	1,190	3,575	455.45
	3300	100' boom, 115' mast		42.63	685	2,050	6,150	751.05
	3400	Stiffleg, 20 ton capacity, 70' boom, 37' mast		29.70	515	1,550	4,650	547.60
	3500	100' boom, 47' mast		45.64	825	2,480	7,450	861.10
	3550	Helicopter, small, lift to 1250 lb. maximum, w/pilot		100.87	3,200	9,610	28,800	2,729
	3600	Hoists, chain type, overhead, manual, 3/4 ton		.10	.33	1	3	1
	3900	10 ton		.75	6	18	54	9.60
	4000	Hoist and tower, 5000 lb. cap., portable electric, 40' high		4.77	228	685	2,050	175.15
	4100	For each added 10' section, add		.11	18	54	162	11.70
	4200	Hoist and single tubular tower, 5000 lb. electric, 100' high		6.47	320	957	2,875	243.15
	4300	For each added 6'-6" section, add		.19	31	93	279	20.10
	4400	Hoist and double tubular tower, 5000 lb., 100' high		6.95	350	1,054	3,150	266.40
	4500	For each added 6'-6" section, add		.21	34.50	103	310	22.30
	4550	Hoist and tower, mast type, 6000 lb., 100' high		7.48	365	1,093	3,275	278.45
	4570	For each added 10' section, add		.13	21.50	64	192	13.85
	4600	Hoist and tower, personnel, electric, 2000 lb., 100' @ 125 fpm		15.90	970	2,910	8,725	709.20
	4700	3000 lb., 100' @ 200 fpm		18.13	1,100	3,290	9,875	803.05
	4800	3000 lb., 150' @ 300 fpm		20.13	1,225	3,690	11,100	899.05
	4900	4000 lb., 100' @ 300 fpm		20.82	1,250	3,760	11,300	918.55
	5000	6000 lb., 100' @ 275 fpm		22.44	1,325	3,950	11,900	969.50
	5100	For added heights up to 500', add	L.F.	.01	1.67	5	15	1.10
	5200	Jacks, hydraulic, 20 ton	Ea.	.05	2	6	18	1.60
	5500	100 ton		.40	11.65	35	105	10.20
	6100	Jacks, hydraulic, climbing w/50' jackrods, control console, 30 ton cap.		1.97	131	394	1,175	94.55
	6150	For each added 10' jackrod section, add		.05	3.33	10	30	2.40
	6300	50 ton capacity		3.17	211	633	1,900	151.95
	6350	For each added 10' jackrod section, add		.06	4	12	36	2.90
	6500	125 ton capacity		8.30	555	1,660	4,975	398.40
	6550	For each added 10' jackrod section, add		.57	37.50	113	340	27.15
	6600	Cable jack, 10 ton capacity with 200' cable		1.65	110	329	985	79
	6650	For each added 50' of cable, add		.20	13	39	117	9.40
70	0010	**WELLPOINT EQUIPMENT RENTAL** without operators						
	0020	Based on 2 months rental						
	0100	Combination jetting & wellpoint pump, 60 H.P. diesel	Ea.	18.14	325	976	2,925	340.30
	0200	High pressure gas jet pump, 200 H.P., 300 psi	"	43.54	278	834	2,500	515.10
	0300	Discharge pipe, 8" diameter	L.F.	.01	.53	1.59	4.77	.40
	0350	12" diameter		.01	.78	2.33	7	.55
	0400	Header pipe, flows up to 150 GPM, 4" diameter		.01	.48	1.44	4.32	.35
	0500	400 GPM, 6" diameter		.01	.56	1.69	5.05	.40
	0600	800 GPM, 8" diameter		.01	.78	2.33	7	.55
	0700	1500 GPM, 10" diameter		.01	.82	2.46	7.40	.55
	0800	2500 GPM, 12" diameter		.02	1.55	4.65	13.95	1.10
	0900	4500 GPM, 16" diameter		.03	1.98	5.95	17.85	1.45

R015433-10

01 54 | Construction Aids

01 54 33 | Equipment Rental

			UNIT	HOURLY OPER. COST	RENT PER DAY	RENT PER WEEK	RENT PER MONTH	EQUIPMENT COST/DAY	
70	0950	For quick coupling aluminum and plastic pipe, add	L.F.	.03	2.05	6.16	18.50	1.45	70
	1100	Wellpoint, 25' long, with fittings & riser pipe, 1-1/2" or 2" diameter		.06	4.10	12.29	37	2.95	
	1200	Wellpoint pump, diesel powered, 4" suction, 20 H.P.		7.71	188	563	1,700	174.30	
	1300	6" suction, 30 H.P.		10.54	233	698	2,100	223.90	
	1400	8" suction, 40 H.P.		14.24	320	957	2,875	305.30	
	1500	10" suction, 75 H.P.		21.97	375	1,119	3,350	399.55	
	1600	12" suction, 100 H.P.		31.24	590	1,770	5,300	603.90	
	1700	12" suction, 175 H.P.		46.78	655	1,970	5,900	768.25	
80	0010	**MARINE EQUIPMENT RENTAL** without operators	R015433 -10						80
	0200	Barge, 400 Ton, 30' wide x 90' long	Ea.	17.50	1,075	3,205	9,625	781	
	0240	800 Ton, 45' wide x 90' long		21.20	1,300	3,900	11,700	949.60	
	2000	Tugboat, diesel, 100 H.P.		38.95	217	650	1,950	441.60	
	2040	250 H.P.		81.55	395	1,185	3,550	889.40	
	2080	380 H.P.		160.00	1,175	3,545	10,600	1,989	
	3000	Small work boat, gas, 16-foot, 50 H.P.		19.15	60	180	540	189.20	
	4000	Large, diesel, 48-foot, 200 H.P.		90.80	1,250	3,745	11,200	1,475	

Crews

Crew No.	Bare Costs		In-House Costs		Incl. Subs O&P		Cost Per Labor-Hour		
Crew A-3H	Hr.	Daily	Hr.	Daily	Hr.	Daily	Bare Costs	In House	Incl. O&P
1 Equip. Oper. (crane)	$50.25	$ 402.00	$64.10	$ 512.80	$79.70	$ 637.60	$ 50.25	$ 64.10	$ 79.70
1 Hyd. Crane, 12 Ton (Daily)		874.00		874.00		961.40	109.25	109.25	120.18
8 L.H., Daily Totals		$1276.00		$1386.80		$1599.00	$159.50	$173.35	$199.88
Crew A-3I	Hr.	Daily	Hr.	Daily	Hr.	Daily	Bare Costs	In House	Incl. O&P
1 Equip. Oper. (crane)	$50.25	$ 402.00	$64.10	$ 512.80	$79.70	$ 637.60	$ 50.25	$ 64.10	$ 79.70
1 Hyd. Crane, 25 Ton (Daily)		1003.00		1003.00		1103.30	125.38	125.38	137.91
8 L.H., Daily Totals		$1405.00		$1515.80		$1740.90	$175.63	$189.47	$217.61
Crew A-3J	Hr.	Daily	Hr.	Daily	Hr.	Daily	Bare Costs	In House	Incl. O&P
1 Equip. Oper. (crane)	$50.25	$ 402.00	$64.10	$ 512.80	$79.70	$ 637.60	$ 50.25	$ 64.10	$ 79.70
1 Hyd. Crane, 40 Ton (Daily)		1240.00		1240.00		1364.00	155.00	155.00	170.50
8 L.H., Daily Totals		$1642.00		$1752.80		$2001.60	$205.25	$219.10	$250.20
Crew A-3K	Hr.	Daily	Hr.	Daily	Hr.	Daily	Bare Costs	In House	Incl. O&P
1 Equip. Oper. (crane)	$50.25	$ 402.00	$64.10	$ 512.80	$79.70	$ 637.60	$ 46.90	$ 59.83	$ 74.38
1 Equip. Oper. (oiler)	43.55	348.40	55.55	444.40	69.05	552.40			
1 Hyd. Crane, 55 Ton (Daily)		1480.00		1480.00		1628.00			
1 P/U Truck, 3/4 Ton (Daily)		182.00		182.00		200.20	103.88	103.88	114.26
16 L.H., Daily Totals		$2412.40		$2619.20		$3018.20	$150.78	$163.70	$188.64
Crew A-3L	Hr.	Daily	Hr.	Daily	Hr.	Daily	Bare Costs	In House	Incl. O&P
1 Equip. Oper. (crane)	$50.25	$ 402.00	$64.10	$ 512.80	$79.70	$ 637.60	$ 46.90	$ 59.83	$ 74.38
1 Equip. Oper. (oiler)	43.55	348.40	55.55	444.40	69.05	552.40			
1 Hyd. Crane, 80 Ton (Daily)		2211.00		2211.00		2432.10			
1 P/U Truck, 3/4 Ton (Daily)		182.00		182.00		200.20	149.56	149.56	164.52
16 L.H., Daily Totals		$3143.40		$3350.20		$3822.30	$196.46	$209.39	$238.89
Crew A-3M	Hr.	Daily	Hr.	Daily	Hr.	Daily	Bare Costs	In House	Incl. O&P
1 Equip. Oper. (crane)	$50.25	$ 402.00	$64.10	$ 512.80	$79.70	$ 637.60	$ 46.90	$ 59.83	$ 74.38
1 Equip. Oper. (oiler)	43.55	348.40	55.55	444.40	69.05	552.40			
1 Hyd. Crane, 100 Ton (Daily)		2200.00		2200.00		2420.00			
1 P/U Truck, 3/4 Ton (Daily)		182.00		182.00		200.20	148.88	148.88	163.76
16 L.H., Daily Totals		$3132.40		$3339.20		$3810.20	$195.78	$208.70	$238.14
Crew A-16	Hr.	Daily	Hr.	Daily	Hr.	Daily	Bare Costs	In House	Incl. O&P
1 Maintenance Laborer	$27.45	$219.60	$34.10	$272.80	$42.65	$341.20	$27.45	$34.10	$42.65
1 Lawn Mower, Riding		68.35		68.35		75.19	8.54	8.54	9.40
8 L.H., Daily Totals		$287.95		$341.15		$416.38	$35.99	$42.64	$52.05
Crew A-17	Hr.	Daily	Hr.	Daily	Hr.	Daily	Bare Costs	In House	Incl. O&P
1 Maintenance Laborer	$27.45	$219.60	$34.10	$272.80	$42.65	$341.20	$27.45	$34.10	$42.65
1 Flatbed Truck, Gas, 1.5 Ton		270.20		270.20		297.22	33.77	33.77	37.15
8 L.H., Daily Totals		$489.80		$543.00		$638.42	$61.23	$67.88	$79.80
Crew A-18	Hr.	Daily	Hr.	Daily	Hr.	Daily	Bare Costs	In House	Incl. O&P
1 Maintenance Laborer	$27.45	$219.60	$34.10	$272.80	$42.65	$341.20	$27.45	$34.10	$42.65
1 Farm Tractor w/ Attachment		341.20		341.20		375.32	42.65	42.65	46.91
8 L.H., Daily Totals		$560.80		$614.00		$716.52	$70.10	$76.75	$89.56
Crew B-1	Hr.	Daily	Hr.	Daily	Hr.	Daily	Bare Costs	In House	Incl. O&P
1 Labor Foreman (outside)	$38.65	$309.20	$51.50	$ 412.00	$63.50	$ 508.00	$37.32	$49.73	$61.30
2 Laborers	36.65	586.40	48.85	781.60	60.20	963.20			
24 L.H., Daily Totals		$895.60		$1193.60		$1471.20	$37.32	$49.73	$61.30

Crews

Crew No.	Bare Costs		In-House Costs		Incl. Subs O&P		Cost Per Labor-Hour		
Crew B-17	Hr.	Daily	Hr.	Daily	Hr.	Daily	Bare Costs	In House	Incl. O&P
2 Laborers	$36.65	$586.40	$48.85	$781.60	$60.20	$963.20	$39.48	$51.79	$64.01
1 Equip. Oper. (light)	47.05	376.40	60.05	480.40	74.60	596.80			
1 Truck Driver (heavy)	37.55	300.40	49.40	395.20	61.05	488.40			
1 Backhoe Loader, 48 H.P.		365.20		365.20		401.72			
1 Dump Truck, 8 C.Y., 220 H.P.		417.20		417.20		458.92	24.45	24.45	26.90
32 L.H., Daily Totals		$2045.60		$2439.60		$2909.04	$63.92	$76.24	$90.91
Crew B-18	Hr.	Daily	Hr.	Daily	Hr.	Daily	Bare Costs	In House	Incl. O&P
1 Labor Foreman (outside)	$38.65	$309.20	$51.50	$412.00	$63.50	$508.00	$37.32	$49.73	$61.30
2 Laborers	36.65	586.40	48.85	781.60	60.20	963.20			
1 Vibrating Plate, Gas, 21"		46.00		46.00		50.60	1.92	1.92	2.11
24 L.H., Daily Totals		$941.60		$1239.60		$1521.80	$39.23	$51.65	$63.41
Crew B-21	Hr.	Daily	Hr.	Daily	Hr.	Daily	Bare Costs	In House	Incl. O&P
1 Labor Foreman (outside)	$38.65	$309.20	$51.50	$412.00	$63.50	$508.00	$42.21	$55.66	$68.74
1 Skilled Worker	47.30	378.40	62.40	499.20	77.05	616.40			
1 Laborer	36.65	293.20	48.85	390.80	60.20	481.60			
.5 Equip. Oper. (crane)	50.25	201.00	64.10	256.40	79.70	318.80			
.5 S.P. Crane, 4x4, 5 Ton		140.00		140.00		154.00	5.00	5.00	5.50
28 L.H., Daily Totals		$1321.80		$1698.40		$2078.80	$47.21	$60.66	$74.24
Crew B-25B	Hr.	Daily	Hr.	Daily	Hr.	Daily	Bare Costs	In House	Incl. O&P
1 Labor Foreman (outside)	$38.65	$309.20	$51.50	$412.00	$63.50	$508.00	$40.90	$53.59	$66.26
7 Laborers	36.65	2052.40	48.85	2735.60	60.20	3371.20			
4 Equip. Oper. (medium)	48.90	1564.80	62.40	1996.80	77.55	2481.60			
1 Asphalt Paver, 130 H.P.		2169.00		2169.00		2385.90			
2 Tandem Rollers, 10 Ton		484.00		484.00		532.40			
1 Roller, Pneum. Whl., 12 Ton		349.00		349.00		383.90	31.27	31.27	34.40
96 L.H., Daily Totals		$6928.40		$8146.40		$9663.00	$72.17	$84.86	$100.66
Crew B-30	Hr.	Daily	Hr.	Daily	Hr.	Daily	Bare Costs	In House	Incl. O&P
1 Equip. Oper. (medium)	$48.90	$391.20	$62.40	$499.20	$77.55	$620.40	$41.33	$53.73	$66.55
2 Truck Drivers (heavy)	37.55	600.80	49.40	790.40	61.05	976.80			
1 Hyd. Excavator, 1.5 C.Y.		1031.00		1031.00		1134.10			
2 Dump Trucks, 12 C.Y., 400 H.P.		1384.40		1384.40		1522.84	100.64	100.64	110.71
24 L.H., Daily Totals		$3407.40		$3705.00		$4254.14	$141.97	$154.38	$177.26
Crew B-34P	Hr.	Daily	Hr.	Daily	Hr.	Daily	Bare Costs	In House	Incl. O&P
1 Pipe Fitter	$58.50	$468.00	$73.15	$585.20	$91.25	$730.00	$47.97	$61.20	$76.05
1 Truck Driver (light)	36.50	292.00	48.05	384.40	59.35	474.80			
1 Equip. Oper. (medium)	48.90	391.20	62.40	499.20	77.55	620.40			
1 Flatbed Truck, Gas, 3 Ton		333.00		333.00		366.30			
1 Backhoe Loader, 48 H.P.		365.20		365.20		401.72	29.09	29.09	32.00
24 L.H., Daily Totals		$1849.40		$2167.00		$2593.22	$77.06	$90.29	$108.05
Crew B-34Q	Hr.	Daily	Hr.	Daily	Hr.	Daily	Bare Costs	In House	Incl. O&P
1 Pipe Fitter	$58.50	$468.00	$73.15	$585.20	$91.25	$730.00	$48.42	$61.77	$76.77
1 Truck Driver (light)	36.50	292.00	48.05	384.40	59.35	474.80			
1 Equip. Oper. (crane)	50.25	402.00	64.10	512.80	79.70	637.60			
1 Flatbed Trailer, 25 Ton		113.60		113.60		124.96			
1 Dump Truck, 8 C.Y., 220 H.P.		417.20		417.20		458.92			
1 Hyd. Crane, 25 Ton		745.80		745.80		820.38	53.19	53.19	58.51
24 L.H., Daily Totals		$2438.60		$2759.00		$3246.66	$101.61	$114.96	$135.28

Crews

Crew No.	Bare Costs		In-House Costs		Incl. Subs O&P		Cost Per Labor-Hour		
Crew B-34R	Hr.	Daily	Hr.	Daily	Hr.	Daily	Bare Costs	In House	Incl. O&P
1 Pipe Fitter	$58.50	$ 468.00	$73.15	$ 585.20	$91.25	$ 730.00	$ 48.42	$ 61.77	$ 76.77
1 Truck Driver (light)	36.50	292.00	48.05	384.40	59.35	474.80			
1 Equip. Oper. (crane)	50.25	402.00	64.10	512.80	79.70	637.60			
1 Flatbed Trailer, 25 Ton		113.60		113.60		124.96			
1 Dump Truck, 8 C.Y., 220 H.P.		417.20		417.20		458.92			
1 Hyd. Crane, 25 Ton		745.80		745.80		820.38			
1 Hyd. Excavator, 1 C.Y.		814.80		814.80		896.28	87.14	87.14	95.86
24 L.H., Daily Totals		$3253.40		$3573.80		$4142.94	$135.56	$148.91	$172.62
Crew B-34S	Hr.	Daily	Hr.	Daily	Hr.	Daily	Bare Costs	In House	Incl. O&P
2 Pipe Fitters	$58.50	$ 936.00	$73.15	$1170.40	$91.25	$1460.00	$ 51.20	$ 64.95	$ 80.81
1 Truck Driver (heavy)	37.55	300.40	49.40	395.20	61.05	488.40			
1 Equip. Oper. (crane)	50.25	402.00	64.10	512.80	79.70	637.60			
1 Flatbed Trailer, 40 Ton		152.00		152.00		167.20			
1 Truck Tractor, 6x4, 380 H.P.		612.80		612.80		674.08			
1 Hyd. Crane, 80 Ton		1652.00		1652.00		1817.20			
1 Hyd. Excavator, 2 C.Y.		1175.00		1175.00		1292.50	112.24	112.24	123.47
32 L.H., Daily Totals		$5230.20		$5670.20		$6536.98	$163.44	$177.19	$204.28
Crew B-34T	Hr.	Daily	Hr.	Daily	Hr.	Daily	Bare Costs	In House	Incl. O&P
2 Pipe Fitters	$58.50	$ 936.00	$73.15	$1170.40	$91.25	$1460.00	$ 51.20	$ 64.95	$ 80.81
1 Truck Driver (heavy)	37.55	300.40	49.40	395.20	61.05	488.40			
1 Equip. Oper. (crane)	50.25	402.00	64.10	512.80	79.70	637.60			
1 Flatbed Trailer, 40 Ton		152.00		152.00		167.20			
1 Truck Tractor, 6x4, 380 H.P.		612.80		612.80		674.08			
1 Hyd. Crane, 80 Ton		1652.00		1652.00		1817.20	75.53	75.53	83.08
32 L.H., Daily Totals		$4055.20		$4495.20		$5244.48	$126.72	$140.47	$163.89
Crew B-35	Hr.	Daily	Hr.	Daily	Hr.	Daily	Bare Costs	In House	Incl. O&P
1 Labor Foreman (outside)	$38.65	$ 309.20	$51.50	$ 412.00	$63.50	$ 508.00	$45.66	$59.06	$73.22
1 Skilled Worker	47.30	378.40	62.40	499.20	77.05	616.40			
1 Welder (plumber)	57.55	460.40	71.95	575.60	89.80	718.40			
1 Laborer	36.65	293.20	48.85	390.80	60.20	481.60			
1 Equip. Oper. (crane)	50.25	402.00	64.10	512.80	79.70	637.60			
1 Equip. Oper. (oiler)	43.55	348.40	55.55	444.40	69.05	552.40			
1 Welder, Electric, 300 amp		55.30		55.30		60.83			
1 Hyd. Excavator, .75 C.Y.		662.00		662.00		728.20	14.94	14.94	16.44
48 L.H., Daily Totals		$2908.90		$3552.10		$4303.43	$60.60	$74.00	$89.65
Crew B-37	Hr.	Daily	Hr.	Daily	Hr.	Daily	Bare Costs	In House	Incl. O&P
1 Labor Foreman (outside)	$38.65	$ 309.20	$51.50	$ 412.00	$63.50	$ 508.00	$38.72	$51.16	$63.15
4 Laborers	36.65	1172.80	48.85	1563.20	60.20	1926.40			
1 Equip. Oper. (light)	47.05	376.40	60.05	480.40	74.60	596.80			
1 Tandem Roller, 5 Ton		159.00		159.00		174.90	3.31	3.31	3.64
48 L.H., Daily Totals		$2017.40		$2614.60		$3206.10	$42.03	$54.47	$66.79
Crew B-47H	Hr.	Daily	Hr.	Daily	Hr.	Daily	Bare Costs	In House	Incl. O&P
1 Skilled Worker Foreman (out)	$49.30	$ 394.40	$65.05	$ 520.40	$80.30	$ 642.40	$47.80	$63.06	$77.86
3 Skilled Workers	47.30	1135.20	62.40	1497.60	77.05	1849.20			
1 Flatbed Truck, Gas, 3 Ton		333.00		333.00		366.30	10.41	10.41	11.45
32 L.H., Daily Totals		$1862.60		$2351.00		$2857.90	$58.21	$73.47	$89.31
Crew B-55	Hr.	Daily	Hr.	Daily	Hr.	Daily	Bare Costs	In House	Incl. O&P
2 Laborers	$36.65	$ 586.40	$48.85	$ 781.60	$60.20	$ 963.20	$36.60	$48.58	$ 59.92
1 Truck Driver (light)	36.50	292.00	48.05	384.40	59.35	474.80			
1 Truck-Mounted Earth Auger		786.40		786.40		865.04			
1 Flatbed Truck, Gas, 3 Ton		333.00		333.00		366.30	46.64	46.64	51.31
24 L.H., Daily Totals		$1997.80		$2285.40		$2669.34	$83.24	$95.22	$111.22

Crews

Crew No.	Bare Costs		In-House Costs		Incl. Subs O&P		Cost Per Labor-Hour		
Crew B-80	Hr.	Daily	Hr.	Daily	Hr.	Daily	Bare Costs	In House	Incl. O&P
1 Labor Foreman (outside)	$38.65	$ 309.20	$51.50	$ 412.00	$63.50	$ 508.00	$39.71	$52.11	$64.41
1 Laborer	36.65	293.20	48.85	390.80	60.20	481.60			
1 Truck Driver (light)	36.50	292.00	48.05	384.40	59.35	474.80			
1 Equip. Oper. (light)	47.05	376.40	60.05	480.40	74.60	596.80			
1 Flatbed Truck, Gas, 3 Ton		333.00		333.00		366.30			
1 Earth Auger, Truck-Mtd.		425.00		425.00		467.50	23.69	23.69	26.06
32 L.H., Daily Totals		$2028.80		$2425.60		$2895.00	$63.40	$75.80	$90.47
Crew B-80C	Hr.	Daily	Hr.	Daily	Hr.	Daily	Bare Costs	In House	Incl. O&P
2 Laborers	$36.65	$ 586.40	$48.85	$ 781.60	$60.20	$ 963.20	$36.60	$48.58	$59.92
1 Truck Driver (light)	36.50	292.00	48.05	384.40	59.35	474.80			
1 Flatbed Truck, Gas, 1.5 Ton		270.20		270.20		297.22			
1 Manual Fence Post Auger, Gas		8.00		8.00		8.80	11.59	11.59	12.75
24 L.H., Daily Totals		$1156.60		$1444.20		$1744.02	$48.19	$60.17	$72.67
Crew D-1	Hr.	Daily	Hr.	Daily	Hr.	Daily	Bare Costs	In House	Incl. O&P
1 Bricklayer	$45.60	$364.80	$59.90	$479.20	$74.05	$ 592.40	$41.30	$54.25	$67.08
1 Bricklayer Helper	37.00	296.00	48.60	388.80	60.10	480.80			
16 L.H., Daily Totals		$660.80		$868.00		$1073.20	$41.30	$54.25	$67.08
Crew D-7	Hr.	Daily	Hr.	Daily	Hr.	Daily	Bare Costs	In House	Incl. O&P
1 Tile Layer	$41.95	$335.60	$53.00	$424.00	$66.05	$528.40	$37.60	$47.52	$59.20
1 Tile Layer Helper	33.25	266.00	42.05	336.40	52.35	418.80			
16 L.H., Daily Totals		$601.60		$760.40		$947.20	$37.60	$47.52	$59.20
Crew D-8	Hr.	Daily	Hr.	Daily	Hr.	Daily	Bare Costs	In House	Incl. O&P
3 Bricklayers	$45.60	$1094.40	$59.90	$1437.60	$74.05	$1777.20	$42.16	$55.38	$68.47
2 Bricklayer Helpers	37.00	592.00	48.60	777.60	60.10	961.60			
40 L.H., Daily Totals		$1686.40		$2215.20		$2738.80	$42.16	$55.38	$68.47
Crew G-1	Hr.	Daily	Hr.	Daily	Hr.	Daily	Bare Costs	In House	Incl. O&P
1 Roofer Foreman (outside)	$41.15	$ 329.20	$61.40	$ 491.20	$74.15	$ 593.20	$36.58	$54.57	$65.92
4 Roofers Composition	39.15	1252.80	58.40	1868.80	70.55	2257.60			
2 Roofer Helpers	29.15	466.40	43.50	696.00	52.55	840.80			
1 Application Equipment		199.20		199.20		219.12			
1 Tar Kettle/Pot		130.00		130.00		143.00			
1 Crew Truck		224.80		224.80		247.28	9.89	9.89	10.88
56 L.H., Daily Totals		$2602.40		$3610.00		$4301.00	$46.47	$64.46	$76.80
Crew G-5	Hr.	Daily	Hr.	Daily	Hr.	Daily	Bare Costs	In House	Incl. O&P
1 Roofer Foreman (outside)	$41.15	$ 329.20	$61.40	$ 491.20	$74.15	$ 593.20	$35.55	$53.04	$64.07
2 Roofers Composition	39.15	626.40	58.40	934.40	70.55	1128.80			
2 Roofer Helpers	29.15	466.40	43.50	696.00	52.55	840.80			
1 Application Equipment		199.20		199.20		219.12	4.98	4.98	5.48
40 L.H., Daily Totals		$1621.20		$2320.80		$2781.92	$40.53	$58.02	$69.55
Crew G-8	Hr.	Daily	Hr.	Daily	Hr.	Daily	Bare Costs	In House	Incl. O&P
1 Roofer Composition	$39.15	$313.20	$58.40	$ 467.20	$70.55	$ 564.40	$ 39.15	$ 58.40	$ 70.55
1 Aerial Lift Truck, 80'		643.00		643.00		707.30	80.38	80.38	88.41
8 L.H., Daily Totals		$956.20		$1110.20		$1271.70	$119.53	$138.78	$158.96
Crew L-2	Hr.	Daily	Hr.	Daily	Hr.	Daily	Bare Costs	In House	Incl. O&P
1 Carpenter	$45.85	$366.80	$61.10	$488.80	$75.35	$ 602.80	$40.25	$53.77	$66.25
1 Carpenter Helper	34.65	277.20	46.45	371.60	57.15	457.20			
16 L.H., Daily Totals		$644.00		$860.40		$1060.00	$40.25	$53.77	$66.25

Crews

Crew No.	Bare Costs		In-House Costs		Incl. Subs O&P		Cost Per Labor-Hour		
	Hr.	Daily	Hr.	Daily	Hr.	Daily	Bare Costs	In House	Incl. O&P
Crew Q-1									
1 Plumber	$57.55	$460.40	$71.95	$575.60	$89.80	$718.40	$51.80	$64.75	$80.83
1 Plumber Apprentice	46.05	368.40	57.55	460.40	71.85	574.80			
16 L.H., Daily Totals		$828.80		$1036.00		$1293.20	$51.80	$64.75	$80.83
Crew Q-2									
2 Plumbers	$57.55	$920.80	$71.95	$1151.20	$89.80	$1436.80	$53.72	$67.15	$83.82
1 Plumber Apprentice	46.05	368.40	57.55	460.40	71.85	574.80	53.72	67.15	83.82
24 L.H., Daily Totals		$1289.20		$1611.60		$2011.60	$53.72	$67.15	$83.82
Crew Q-4									
1 Plumber Foreman (inside)	$58.05	$464.40	$72.55	$580.40	$90.55	$724.40	$54.80	$68.50	$85.50
1 Plumber	57.55	460.40	71.95	575.60	89.80	718.40			
1 Welder (plumber)	57.55	460.40	71.95	575.60	89.80	718.40			
1 Plumber Apprentice	46.05	368.40	57.55	460.40	71.85	574.80			
1 Welder, Electric, 300 amp		55.30		55.30		60.83	1.73	1.73	1.90
32 L.H., Daily Totals		$1808.90		$2247.30		$2796.83	$56.53	$70.23	$87.40
Crew Q-5									
1 Steamfitter	$58.50	$468.00	$73.15	$585.20	$91.25	$730.00	$52.65	$65.83	$82.13
1 Steamfitter Apprentice	46.80	374.40	58.50	468.00	73.00	584.00			
16 L.H., Daily Totals		$842.40		$1053.20		$1314.00	$52.65	$65.83	$82.13
Crew Q-6									
2 Steamfitters	$58.50	$936.00	$73.15	$1170.40	$91.25	$1460.00	$54.60	$68.27	$85.17
1 Steamfitter Apprentice	46.80	374.40	58.50	468.00	73.00	584.00	54.60	68.27	85.17
24 L.H., Daily Totals		$1310.40		$1638.40		$2044.00	$54.60	$68.27	$85.17
Crew Q-7									
1 Steamfitter Foreman (inside)	$59.00	$472.00	$73.75	$590.00	$92.05	$736.40	$55.70	$69.64	$86.89
2 Steamfitters	58.50	936.00	73.15	1170.40	91.25	1460.00			
1 Steamfitter Apprentice	46.80	374.40	58.50	468.00	73.00	584.00			
32 L.H., Daily Totals		$1782.40		$2228.40		$2780.40	$55.70	$69.64	$86.89
Crew Q-8									
1 Steamfitter Foreman (inside)	$59.00	$472.00	$73.75	$590.00	$92.05	$736.40	$55.70	$69.64	$86.89
1 Steamfitter	58.50	468.00	73.15	585.20	91.25	730.00			
1 Welder (steamfitter)	58.50	468.00	73.15	585.20	91.25	730.00			
1 Steamfitter Apprentice	46.80	374.40	58.50	468.00	73.00	584.00			
1 Welder, Electric, 300 amp		55.30		55.30		60.83	1.73	1.73	1.90
32 L.H., Daily Totals		$1837.70		$2283.70		$2841.23	$57.43	$71.37	$88.79
Crew Q-9									
1 Sheet Metal Worker	$54.70	$437.60	$69.30	$554.40	$86.25	$690.00	$49.23	$62.38	$77.63
1 Sheet Metal Apprentice	43.75	350.00	55.45	443.60	69.00	552.00			
16 L.H., Daily Totals		$787.60		$998.00		$1242.00	$49.23	$62.38	$77.63
Crew Q-10									
2 Sheet Metal Workers	$54.70	$875.20	$69.30	$1108.80	$86.25	$1380.00	$51.05	$64.68	$80.50
1 Sheet Metal Apprentice	43.75	350.00	55.45	443.60	69.00	552.00			
24 L.H., Daily Totals		$1225.20		$1552.40		$1932.00	$51.05	$64.68	$80.50
Crew Q-14									
1 Asbestos Worker	$51.15	$409.20	$66.30	$530.40	$82.15	$657.20	$46.02	$59.65	$73.92
1 Asbestos Apprentice	40.90	327.20	53.00	424.00	65.70	525.60			
16 L.H., Daily Totals		$736.40		$954.40		$1182.80	$46.02	$59.65	$73.92

Crews

Crew No.	Bare Costs		In-House Costs		Incl. Subs O&P		Cost Per Labor-Hour		
Crew Q-15	Hr.	Daily	Hr.	Daily	Hr.	Daily	Bare Costs	In House	Incl. O&P
1 Plumber	$57.55	$460.40	$71.95	$575.60	$89.80	$718.40	$51.80	$64.75	$80.83
1 Plumber Apprentice	46.05	368.40	57.55	460.40	71.85	574.80			
1 Welder, Electric, 300 amp		55.30		55.30		60.83	3.46	3.46	3.80
16 L.H., Daily Totals		$884.10		$1091.30		$1354.03	$55.26	$68.21	$84.63
Crew Q-16	Hr.	Daily	Hr.	Daily	Hr.	Daily	Bare Costs	In House	Incl. O&P
2 Plumbers	$57.55	$920.80	$71.95	$1151.20	$89.80	$1436.80	$53.72	$67.15	$83.82
1 Plumber Apprentice	46.05	368.40	57.55	460.40	71.85	574.80			
1 Welder, Electric, 300 amp		55.30		55.30		60.83	2.30	2.30	2.53
24 L.H., Daily Totals		$1344.50		$1666.90		$2072.43	$56.02	$69.45	$86.35
Crew Q-20	Hr.	Daily	Hr.	Daily	Hr.	Daily	Bare Costs	In House	Incl. O&P
1 Sheet Metal Worker	$54.70	$437.60	$69.30	$554.40	$86.25	$690.00	$50.05	$63.10	$78.61
1 Sheet Metal Apprentice	43.75	350.00	55.45	443.60	69.00	552.00			
.5 Electrician	53.35	213.40	66.00	264.00	82.55	330.20			
20 L.H., Daily Totals		$1001.00		$1262.00		$1572.20	$50.05	$63.10	$78.61
Crew R-3	Hr.	Daily	Hr.	Daily	Hr.	Daily	Bare Costs	In House	Incl. O&P
1 Electrician Foreman	$53.85	$430.80	$66.60	$532.80	$83.30	$666.40	$52.93	$65.86	$82.28
1 Electrician	53.35	426.80	66.00	528.00	82.55	660.40			
.5 Equip. Oper. (crane)	50.25	201.00	64.10	256.40	79.70	318.80			
.5 S.P. Crane, 4x4, 5 Ton		140.00		140.00		154.00	7.00	7.00	7.70
20 L.H., Daily Totals		$1198.60		$1457.20		$1799.60	$59.93	$72.86	$89.98
Crew R-26	Hr.	Daily	Hr.	Daily	Hr.	Daily	Bare Costs	In House	Incl. O&P
2 Electricians	$53.35	$853.60	$66.00	$1056.00	$82.55	$1320.80	$53.35	$66.00	$82.55
1 Boom Work Pltfm, 40'		294.80		294.80		324.28	18.43	18.43	20.27
16 L.H., Daily Totals		$1148.40		$1350.80		$1645.08	$71.78	$84.42	$102.82

Travel Costs

The following table is used to estimate the cost of in-house staff or contractor's personnel to travel to and from the job site. The amount incurred must be added to the in-house total labor cost or contractor's billing application to calculate the total cost with travel.

In-House Staff–Travel Times versus Costs

Travel Time Versus Costs	Crew Abbr. Rate with Mark-ups	Skwk $62.40	Clab $48.85	Clam $34.10	Asbe $66.30	Bric $59.90	Carp $61.10	Elec $66.00	Pord $51.20	Plum $71.95	Rofc $58.40	Shee $69.30	Stpi $73.15
Round Trip Travel Time (hours)	Crew Size												
0.50	1	31.20	24.43	17.05	33.15	29.95	30.55	33.00	25.60	35.98	29.20	34.65	36.58
	2	62.40	48.85	34.10	66.30	59.90	61.10	66.00	51.20	71.95	58.40	69.30	73.15
	3	93.60	73.28	51.15	99.45	89.85	91.65	99.00	76.80	107.93	87.60	103.95	109.73
0.75	1	46.80	36.64	25.58	49.73	44.93	45.83	49.50	38.40	53.96	43.80	51.98	54.86
	2	93.60	73.28	51.15	99.45	89.85	91.65	99.00	76.80	107.93	87.60	103.95	109.73
	3	140.40	109.91	76.73	149.18	134.78	137.48	148.50	115.20	161.89	131.40	155.93	164.59
1.00	1	62.40	48.85	34.10	66.30	59.90	61.10	66.00	51.20	71.95	58.40	69.30	73.15
	2	124.80	97.70	68.20	132.60	119.80	122.20	132.00	102.40	143.90	116.80	138.60	146.30
	3	187.20	146.55	102.30	198.90	179.70	183.30	198.00	153.60	215.85	175.20	207.90	219.45
1.50	1	93.60	73.28	51.15	99.45	89.85	91.65	99.00	76.80	107.93	87.60	103.95	109.73
	2	187.20	146.55	102.30	198.90	179.70	183.30	198.00	153.60	215.85	175.20	207.90	219.45
	3	280.80	219.83	153.45	298.35	269.55	274.95	297.00	230.40	323.78	262.80	311.85	329.18
2.00	1	124.80	97.70	68.20	132.60	119.80	122.20	132.00	102.40	143.90	116.80	138.60	146.30
	2	249.60	195.40	136.40	265.20	239.60	244.40	264.00	204.80	287.80	233.60	277.20	292.60
	3	374.40	293.10	204.60	397.80	359.40	366.60	396.00	307.20	431.70	350.40	415.80	438.90

Installing Contractors–Travel Times versus Costs

Travel Time Versus Costs	Crew Abbr. Rate with O & P	Skwk $77.05	Clab $60.20	Clam $42.65	Asbe $82.15	Bric $74.05	Carp $75.35	Elec $82.55	Pord $63.45	Plum $89.80	Rofc $70.55	Shee $86.25	Stpi $91.25
Round Trip Travel Time (hours)	Crew Size												
0.50	1	38.53	30.10	21.33	41.08	37.03	37.68	41.28	31.73	44.90	35.28	43.13	45.63
	2	77.05	60.20	42.65	82.15	74.05	75.35	82.55	63.45	89.80	70.55	86.25	91.25
	3	115.58	90.30	63.98	123.23	111.08	113.03	123.83	95.18	134.70	105.83	129.38	136.88
0.75	1	57.79	45.15	31.99	61.61	55.54	56.51	61.91	47.59	67.35	52.91	64.69	68.44
	2	115.58	90.30	63.98	123.23	111.08	113.03	123.83	95.18	134.70	105.83	129.38	136.88
	3	173.36	135.45	95.96	184.84	166.61	169.54	185.74	142.76	202.05	158.74	194.06	205.31
1.00	1	77.05	60.20	42.65	82.15	74.05	75.35	82.55	63.45	89.80	70.55	86.25	91.25
	2	154.10	120.40	85.30	164.30	148.10	150.70	165.10	126.90	179.60	141.10	172.50	182.50
	3	231.15	180.60	127.95	246.45	222.15	226.05	247.65	190.35	269.40	211.65	258.75	273.75
1.50	1	115.58	90.30	63.98	123.23	111.08	113.03	123.83	95.18	134.70	105.83	129.38	136.88
	2	231.15	180.60	127.95	246.45	222.15	226.05	247.65	190.35	269.40	211.65	258.75	273.75
	3	346.73	270.90	191.93	369.68	333.23	339.08	371.48	285.53	404.10	317.48	388.13	410.63
2.00	1	154.10	120.40	85.30	164.30	148.10	150.70	165.10	126.90	179.60	141.10	172.50	182.50
	2	308.20	240.80	170.60	328.60	296.20	301.40	330.20	253.80	359.20	282.20	345.00	365.00
	3	462.30	361.20	255.90	492.90	444.30	452.10	495.30	380.70	538.80	423.30	517.50	547.50

General Conditions — RH1010-100 Overhead

Table H1010-101 General Contractor's Overhead

There are two distinct types of overhead on a construction project: Project Overhead and Main Office Overhead. Project Overhead includes those costs at a construction site not directly associated with the installation of construction materials. Examples of Project Overhead costs include the following:

1. Superintendent
2. Construction office and storage trailers
3. Temporary sanitary facilities
4. Temporary utilities
5. Security fencing
6. Photographs
7. Clean up
8. Performance and payment bonds

The above Project Overhead items are also referred to as General Requirements and therefore are estimated in Division 1. Division 1 is the first division listed in the CSI MasterFormat but it is usually the last division estimated. The sum of the costs in Divisions 1 through 49 is referred to as the sum of the direct costs.

All construction projects also include indirect costs. The primary components of indirect costs are the contractor's Main Office Overhead and profit. The amount of the Main Office Overhead expense varies depending on the the following:

1. Owner's compensation
2. Project managers and estimator's wages
3. Clerical support wages
4. Office rent and utilities
5. Corporate legal and accounting costs
6. Advertising
7. Automobile expenses
8. Association dues
9. Travel and entertainment expenses

These costs are usually calculated as a percentage of annual sales volume. This percentage can range from 35% for a small contractor doing less than $500,000 to 5% for a large contractor with sales in excess of $100 million.

Table H1010-102 Main Office Expense

A General Contractor's main office expense consists of many items not detailed in the front portion of the book. The percentage of main office expense declines with increased annual volume of the contractor. Typical main office expense ranges from 2% to 20% with the median about 7.2% of total volume. This equals about 7.7% of direct costs. The following are approximate percentages of total overhead for different items usually included in a General Contractor's main office overhead. With different accounting procedures, these percentages may vary.

Item	Typical Range	Average
Managers', clerical and estimators' salaries	40 % to 55 %	48%
Profit sharing, pension and bonus plans	2 to 20	12
Insurance	5 to 8	6
Estimating and project management (not including salaries)	5 to 9	7
Legal, accounting and data processing	0.5 to 5	3
Automobile and light truck expense	2 to 8	5
Depreciation of overhead capital expenditures	2 to 6	4
Maintenance of office equipment	0.1 to 1.5	1
Office rental	3 to 5	4
Utilities including phone and light	1 to 3	2
Miscellaneous	5 to 15	8
Total		100%

General Conditions — RH1010-200 Design & Engineering

Table H1010-201 Architectural Fees

Tabulated below are typical percentage fees by project size, for good professional architectural service. Fees may vary from those listed depending upon degree of design difficulty and economic conditions in any particular area.

Rates can be interpolated horizontally and vertically. Various portions of the same project requiring different rates should be adjusted proportionately. For alterations, add 50% to the fee for the first $500,000 of project cost and add 25% to the fee for project cost over $500,000.

Architectural fees tabulated below include Structural, Mechanical and Electrical Engineering Fees. They do not include the fees for special consultants such as kitchen planning, security, acoustical, interior design, etc.

Building Types	Total Project Size in Thousands of Dollars						
	100	250	500	1,000	5,000	10,000	50,000
Factories, garages, warehouses, repetitive housing	9.0%	8.0%	7.0%	6.2%	5.3%	4.9%	4.5%
Apartments, banks, schools, libraries, offices, municipal buildings	12.2	12.3	9.2	8.0	7.0	6.6	6.2
Churches, hospitals, homes, laboratories, museums, research	15.0	13.6	12.7	11.9	9.5	8.8	8.0
Memorials, monumental work, decorative furnishings	—	16.0	14.5	13.1	10.0	9.0	8.3

Table H1010-202 Engineering Fees

Typical **Structural Engineering Fees** based on type of construction and total project size. These fees are included in Architectural Fees.

Type of Construction	Total Project Size (in thousands of dollars)			
	$500	$500-$1,000	$1,000-$5,000	Over $5000
Industrial buildings, factories & warehouses	Technical payroll times 2.0 to 2.5	1.60%	1.25%	1.00%
Hotels, apartments, offices, dormitories, hospitals, public buildings, food stores		2.00%	1.70%	1.20%
Museums, banks, churches and cathedrals		2.00%	1.75%	1.25%
Thin shells, prestressed concrete, earthquake resistive		2.00%	1.75%	1.50%
Parking ramps, auditoriums, stadiums, convention halls, hangars & boiler houses		2.50%	2.00%	1.75%
Special buildings, major alterations, underpinning & future expansion		Add to above 0.5%	Add to above 0.5%	Add to above 0.5%

For complex reinforced concrete or unusually complicated structures, add 20% to 50%.

Table H1010-203 Mechanical and Electrical Fees

Typical **Mechanical and Electrical Engineering Fees** based on the size of the subcontract. The fee structure for both are shown below. These fees are included in Architectural Fees.

Type of Construction	Subcontract Size							
	$25,000	$50,000	$100,000	$225,000	$350,000	$500,000	$750,000	$1,000,000
Simple structures	6.4%	5.7%	4.8%	4.5%	4.4%	4.3%	4.2%	4.1%
Intermediate structures	8.0	7.3	6.5	5.6	5.1	5.0	4.9	4.8
Complex structures	10.1	9.0	9.0	8.0	7.5	7.5	7.0	7.0

For renovations, add 15% to 25% to applicable fee.

General Conditions — RH1010-300 Individual Insurance Bonds

Table H1010-301 Builder's Risk Insurance

Builder's Risk Insurance is insurance on a building during construction. Premiums are paid by the owner or the contractor. Blasting, collapse and underground insurance would raise total insurance costs above those listed. Floater policy for materials delivered to the job runs $.75 to $1.25 per $100 value. Contractor equipment insurance runs $.50 to $1.50 per $100 value. Insurance for miscellaneous tools to $1,500 value runs from $3.00 to $7.50 per $100 value.

Tabulated below are New England Builder's Risk insurance rates in dollars per $100 value for $1,000 deductible. For $25,000 deductible, rates can be reduced 13% to 34%. On contracts over $1,000,000, rates may be lower than those tabulated. Policies are written annually for the total completed value in place. For "all risk" insurance (excluding flood, earthquake and certain other perils) add $.025 to total rates below.

Coverage	Frame Construction (Class 1)		Brick Construction (Class 4)		Fire Resistive (Class 6)	
	Range	Average	Range	Average	Range	Average
Fire Insurance	$.350 to $.850	$.600	$.158 to $.189	$.174	$.052 to $.080	$.070
Extended Coverage	.115 to .200	.158	.080 to .105	.101	.081 to .105	.100
Vandalism	.012 to .016	.014	.008 to .011	.011	.008 to .011	.010
Total Annual Rate	$.477 to $1.066	$.772	$.246 to $.305	$.286	$.141 to $.196	$.180

Table H1010-302 Performance Bond

This table shows the cost of a Performance Bond for a construction job scheduled to be completed in 12 months. Add 1% of the premium cost per month for jobs requiring more than 12 months to complete. The rates are "standard" rates offered to contractors that the bonding company considers financially sound and capable of doing the work. Preferred rates are offered by some bonding companies based upon financial strength of the contractor. Actual rates vary from contractor to contractor and from bonding company to bonding company. Contractors should prequalify through a bonding agency before submitting a bid on a contract that requires a bond.

Contract Amount	Building Construction Class B Projects	Highways & Bridges Class A New Construction	Class A-1 Highway Resurfacing
First $ 100,000 bid	$25.00 per M	$15.00 per M	$9.40 per M
Next 400,000 bid	$ 2,500 plus $15.00 per M	$ 1,500 plus $10.00 per M	$ 940 plus $7.20 per M
Next 2,000,000 bid	8,500 plus 10.00 per M	5,500 plus 7.00 per M	3,820 plus 5.00 per M
Next 2,500,000 bid	28,500 plus 7.50 per M	19,500 plus 5.50 per M	15,820 plus 4.50 per M
Next 2,500,000 bid	47,250 plus 7.00 per M	33,250 plus 5.00 per M	28,320 plus 4.50 per M
Over 7,500,000 bid	64,750 plus 6.00 per M	45,750 plus 4.50 per M	39,570 plus 4.00 per M

General Conditions — RH1010-400 Workers' Compensation

Table H1010-401 Workers' Compensation Insurance Rates by Trade

The table below tabulates the national averages for workers' compensation insurance rates by trade and type of building. The average "Insurance Rate" is multiplied by the "% of Building Cost" for each trade. This produces the "Workers' Compensation" cost by % of total labor cost, to be added for each trade by building type to determine the weighted average workers' compensation rate for the building types analyzed.

Trade	Insurance Rate (% Labor Cost) Range	Insurance Rate Average	% of Building Cost Office Bldgs.	Schools & Apts.	Mfg.	Workers' Compensation Office Bldgs.	Schools & Apts.	Mfg.
Excavation, Grading, etc.	3.9% to 18.7%	9.7%	4.8%	4.9%	4.5%	0.47%	0.48%	0.44%
Piles & Foundations	5.8 to 37	14.7	7.1	5.2	8.7	1.04	0.76	1.28
Concrete	4.6 to 38.6	13.1	5.0	14.8	3.7	0.66	1.94	0.48
Masonry	5.1 to 35	13.5	6.9	7.5	1.9	0.93	1.01	0.26
Structural Steel	5.8 to 108.2	34.6	10.7	3.9	17.6	3.70	1.35	6.09
Miscellaneous & Ornamental Metals	5.3 to 31.2	12.0	2.8	4.0	3.6	0.34	0.48	0.43
Carpentry & Millwork	5.5 to 34.9	15.4	3.7	4.0	0.5	0.57	0.62	0.08
Metal or Composition Siding	5.8 to 68.3	18.2	2.3	0.3	4.3	0.42	0.05	0.78
Roofing	5.8 to 93.3	31.3	2.3	2.6	3.1	0.72	0.81	0.97
Doors & Hardware	4.2 to 34.8	11.1	0.9	1.4	0.4	0.10	0.16	0.04
Sash & Glazing	5.3 to 36.8	13.1	3.5	4.0	1.0	0.46	0.52	0.13
Lath & Plaster	3.7 to 35	11.6	3.3	6.9	0.8	0.38	0.80	0.09
Tile, Marble & Floors	3.3 to 27.1	8.5	2.6	3.0	0.5	0.22	0.26	0.04
Acoustical Ceilings	3.4 to 36.7	8.1	2.4	0.2	0.3	0.19	0.02	0.02
Painting	5 to 33	11.5	1.5	1.6	1.6	0.17	0.18	0.18
Interior Partitions	5.5 to 34.9	15.4	3.9	4.3	4.4	0.60	0.66	0.68
Miscellaneous Items	2.5 to 150	14.4	5.2	3.7	9.7	0.75	0.53	1.40
Elevators	1.6 to 14.4	5.7	2.1	1.1	2.2	0.12	0.06	0.13
Sprinklers	3.2 to 18.3	7.2	0.5	—	2.0	0.04	—	0.14
Plumbing	2.6 to 17.5	7.1	4.9	7.2	5.2	0.35	0.51	0.37
Heat., Vent., Air Conditioning	4 to 18.8	8.8	13.5	11.0	12.9	1.19	0.97	1.14
Electrical	2.8 to 13.2	5.8	10.1	8.4	11.1	0.59	0.49	0.64
Total	1.6% to 150%	—	100.0%	100.0%	100.0%	14.01%	12.66%	15.81%

Overall Weighted Average 14.16%

Workers' Compensation Insurance Rates by States

The table below lists the weighted average workers' compensation base rate for each state with a factor comparing this with the national average of 14%.

State	Weighted Average	Factor	State	Weighted Average	Factor	State	Weighted Average	Factor
Alabama	20.3%	145	Kentucky	14.9%	106	North Dakota	9.6%	69
Alaska	14.3	102	Louisiana	19.3	138	Ohio	8.6	61
Arizona	13.1	94	Maine	12.7	91	Oklahoma	12.9	92
Arkansas	8.6	61	Maryland	14.6	104	Oregon	12.3	88
California	26.8	191	Massachusetts	12.0	86	Pennsylvania	18.1	129
Colorado	7.5	54	Michigan	16.9	121	Rhode Island	12.6	90
Connecticut	24.6	176	Minnesota	23.5	168	South Carolina	19.6	140
Delaware	12.4	89	Mississippi	13.8	99	South Dakota	15.5	111
District of Columbia	11.5	82	Missouri	15.4	110	Tennessee	12.5	89
Florida	11.1	79	Montana	8.9	64	Texas	9.5	68
Georgia	29.2	209	Nebraska	18.3	131	Utah	9.1	65
Hawaii	8.4	60	Nevada	9.0	64	Vermont	13.6	97
Idaho	10.7	76	New Hampshire	20.8	149	Virginia	9.9	71
Illinois	28.1	201	New Jersey	15.0	107	Washington	10.9	78
Indiana	5.7	41	New Mexico	15.0	107	West Virginia	8.3	59
Iowa	15.1	108	New York	17.9	128	Wisconsin	13.6	97
Kansas	8.7	62	North Carolina	17.4	124	Wyoming	6.3	45

Weighted Average for U.S. is 14.2% of payroll = 100%

Rates in the following table are the base or manual costs per $100 of payroll for workers' compensation in each state. Rates are usually applied to straight time wages only and not to premium time wages and bonuses.

The weighted average skilled worker rate for 35 trades is 14%. For bidding purposes, apply the full value of workers' compensation directly to total labor costs, or if labor is 38%, materials 42% and overhead and profit 20% of total cost, carry 38/80 × 14% =6.6% of cost (before overhead and profit) into overhead. Rates vary not only from state to state but also with the experience rating of the contractor.

Rates are the most current available at the time of publication.

General Conditions — RH1010-400 Workers' Compensation

Table H1010-403 Workers' Compensation Insurance Rates by Trade and State

State	Carpentry — 3 stories or less 5651	Carpentry — interior cab. work 5437	Carpentry — general 5403	Concrete Work — NOC 5213	Concrete Work — flat (flr., sdwk.) 5221	Electrical Wiring — inside 5190	Excavation — earth NOC 6217	Excavation — rock 6217	Glaziers 5462	Insulation Work 5479	Lathing 5443	Masonry 5022	Painting & Decorating 5474	Pile Driving 6003	Plastering 5480	Plumbing 5183
AL	34.03	17.01	24.59	13.79	10.37	8.87	16.42	16.42	19.75	19.84	9.24	23.79	15.91	18.57	15.88	8.73
AK	17.14	9.44	11.58	10.41	9.44	5.63	8.23	8.23	32.03	14.36	11.73	10.86	10.79	21.68	11.31	6.20
AZ	21.40	9.27	20.28	10.31	6.93	6.08	7.07	7.07	10.80	14.99	5.47	13.01	12.78	10.80	8.64	6.84
AR	12.80	4.77	9.18	6.63	5.16	4.03	6.82	6.82	8.14	7.67	3.99	8.30	7.08	8.10	13.00	4.55
CA	34.87	34.83	34.86	16.55	16.55	13.24	18.65	18.65	24.16	16.48	17.83	25.13	23.59	23.18	35.04	17.48
CO	8.95	4.93	6.07	6.68	4.75	2.92	5.49	5.49	6.00	7.27	3.95	8.29	6.56	6.51	5.69	4.16
CT	30.95	18.48	31.07	26.49	14.43	7.89	16.02	16.02	20.24	20.60	9.92	31.53	18.60	20.17	18.75	11.28
DE	12.46	12.46	9.93	10.07	8.80	4.53	7.70	7.70	10.90	9.93	6.60	11.53	13.39	15.03	10.90	6.60
DC	8.92	7.25	6.94	12.65	7.38	4.69	8.02	8.02	9.61	6.63	9.83	9.32	5.82	17.36	8.00	7.80
FL	11.32	8.62	11.92	12.51	6.20	6.17	7.30	7.30	10.56	10.45	5.53	10.92	10.23	17.84	9.68	5.63
GA	68.34	20.32	28.40	21.36	15.09	12.05	17.32	17.32	19.39	18.80	11.94	35.11	33.04	26.44	19.65	12.76
HI	8.51	6.92	12.69	7.05	4.02	4.47	5.40	5.40	8.22	5.20	6.26	7.01	6.19	7.32	8.33	5.31
ID	14.63	6.84	12.43	11.48	5.28	4.43	7.25	7.25	8.81	9.06	4.46	10.55	9.38	7.65	8.43	5.10
IL	33.34	21.75	28.57	38.59	16.83	10.03	14.53	14.53	25.55	21.41	14.32	28.94	18.57	24.53	32.84	12.51
IN	8.36	4.21	6.98	4.56	4.15	2.83	3.92	3.92	5.29	6.15	3.85	5.14	4.97	7.23	3.71	2.56
IA	15.46	11.35	19.21	14.85	11.16	5.20	11.01	11.01	15.76	9.89	6.48	14.30	10.13	11.38	21.73	7.80
KS	14.92	7.39	9.74	8.07	5.70	4.46	5.55	5.55	7.54	8.17	4.09	8.38	7.75	7.56	6.20	5.23
KY	18.75	13.00	20.00	8.50	6.50	6.00	14.00	14.00	22.12	11.15	7.60	7.50	9.00	16.03	12.93	5.00
LA	31.05	16.86	25.02	13.35	12.37	7.85	14.11	14.11	19.19	15.06	7.99	17.31	17.64	18.85	14.84	7.10
ME	12.24	8.87	16.18	16.43	8.70	4.74	12.34	12.34	14.82	12.79	5.09	10.88	12.44	9.00	9.99	5.90
MD	11.40	15.89	15.00	19.22	6.00	5.23	8.11	8.11	13.77	12.42	6.15	10.11	6.40	15.19	9.17	6.65
MA	8.68	5.23	9.61	18.85	6.24	2.84	4.35	4.35	9.58	7.78	5.27	10.55	5.09	12.92	4.68	3.50
MI	18.41	10.45	18.41	18.27	12.58	5.82	13.28	13.28	11.67	10.92	10.45	18.82	13.49	37.00	10.88	7.18
MN	20.27	20.82	34.13	11.58	13.15	5.83	11.00	11.00	36.85	18.37	11.55	15.59	17.55	27.58	11.58	8.09
MS	22.76	9.99	13.34	9.60	6.67	6.18	9.49	9.49	14.94	10.15	5.94	13.12	11.01	13.14	10.43	8.17
MO	19.66	11.10	12.81	12.38	10.34	6.15	9.96	9.96	9.47	11.55	7.44	15.13	11.42	13.38	12.15	9.00
MT	9.67	6.59	10.80	6.22	5.67	3.71	7.41	7.41	7.50	8.67	4.15	7.62	8.22	7.95	6.80	4.79
NE	21.80	13.15	21.48	18.23	11.20	8.03	17.38	17.38	12.05	21.53	8.35	18.73	13.00	13.93	13.68	8.73
NV	10.41	6.82	12.39	10.83	5.26	4.33	8.75	8.75	6.55	7.55	3.41	6.63	7.10	8.43	6.21	6.76
NH	24.24	11.06	19.63	28.40	16.47	5.63	12.61	12.61	14.12	12.63	7.30	16.19	20.93	18.53	11.20	9.49
NJ	18.69	13.18	18.69	15.51	11.69	5.13	10.83	10.83	11.21	12.63	15.13	18.71	11.28	15.79	15.13	6.96
NM	25.01	9.56	18.92	11.73	9.43	6.05	7.22	7.22	10.95	10.56	6.11	15.74	12.91	12.66	14.20	6.70
NY	12.54	10.74	19.18	23.71	17.15	7.85	12.16	12.16	18.72	10.88	12.86	21.24	14.99	20.13	10.02	10.83
NC	30.12	9.54	15.76	16.11	9.31	10.56	15.41	15.41	14.28	13.03	8.11	13.05	14.08	16.61	14.12	10.73
ND	9.54	4.31	9.54	5.43	5.43	3.75	4.13	4.13	10.50	10.50	9.93	6.52	6.00	10.59	9.93	6.11
OH	7.73	4.31	5.52	4.67	4.92	3.43	4.79	4.79	8.83	11.38	36.65	6.93	8.15	6.99	7.80	3.52
OK	18.82	10.32	12.49	13.50	7.66	6.82	9.48	9.48	11.79	11.31	5.13	14.63	10.66	11.24	14.01	6.82
OR	22.30	8.52	11.92	10.51	10.42	5.35	9.08	9.08	12.64	10.88	6.09	14.37	11.81	11.44	10.81	6.35
PA	19.22	19.22	15.60	19.54	13.58	7.39	11.11	11.11	14.34	15.60	14.34	16.04	17.13	19.44	14.34	9.19
RI	9.84	11.55	13.75	12.56	9.68	4.58	8.16	8.16	10.61	12.48	5.17	10.07	8.82	19.30	8.69	6.75
SC	29.32	13.80	17.10	12.84	8.21	9.82	12.01	12.01	12.92	14.13	9.57	15.39	16.10	15.67	14.47	9.40
SD	20.86	17.96	20.35	16.10	10.49	5.08	9.75	9.75	10.84	14.71	6.95	14.65	11.22	12.63	10.35	10.02
TN	22.71	10.14	11.54	10.75	7.08	5.49	10.05	10.05	11.07	10.34	6.60	11.69	8.27	12.08	11.04	5.17
TX	10.12	7.33	10.12	8.08	5.99	5.98	7.60	7.60	8.25	9.27	4.15	8.69	7.28	10.26	7.28	5.27
UT	14.72	6.59	7.50	7.34	6.57	3.56	6.79	6.79	7.45	7.15	6.41	8.96	7.73	6.72	5.56	4.17
VT	15.69	10.21	13.77	13.60	11.57	4.37	11.83	11.83	12.40	15.34	5.79	11.92	8.78	10.43	9.35	7.38
VA	12.38	8.14	8.89	10.51	5.76	4.50	7.40	7.40	8.02	7.27	5.45	7.98	8.93	8.31	6.69	5.07
WA	9.18	9.18	9.18	8.55	8.55	3.18	6.68	6.68	13.99	6.79	9.18	13.26	12.26	21.68	10.43	4.33
WV	14.85	5.54	8.34	7.68	4.51	3.53	6.46	6.46	6.74	6.29	4.33	8.39	6.03	8.09	6.54	3.52
WI	12.52	13.14	17.06	11.98	7.59	5.08	7.84	7.84	13.88	11.22	4.00	13.69	12.43	28.24	10.16	5.95
WY	5.82	5.82	5.82	5.82	5.82	5.82	5.82	5.82	5.82	5.82	5.82	5.82	5.82	5.82	5.82	5.82
AVG.	18.19	11.07	15.38	13.15	8.92	5.83	9.69	9.69	13.15	11.67	8.12	13.49	11.54	14.65	11.55	7.08

General Conditions — RH1010-400 Workers' Compensation

R013113-60 Workers' Compensation Insurance Rates by Trade and State (cont.)

State	Roofing 5551	Sheet Metal Work (HVAC) 5538	Steel Erection — door & sash 5102	Steel Erection — inter., ornam. 5102	Steel Erection — structure 5040	Steel Erection — NOC 5057	Tile Work — (interior ceramic) 5348	Waterproofing 9014	Wrecking 5701	Boiler-making 3726	Mill-wrights 3724	Structural Painters 5037	Elevator Constructor 5160	Truck Drivers 7219	Sprinkler Installer 5188
AL	54.71	10.34	12.52	12.52	40.75	21.56	12.35	9.27	40.75	11.93	12.83	55.64	10.09	23.31	8.40
AK	25.49	8.05	7.82	7.82	29.28	26.40	5.33	5.77	29.28	8.56	8.24	31.37	4.06	11.75	5.13
AZ	29.22	10.31	13.61	13.61	26.91	12.89	4.50	6.21	26.91	9.07	7.41	59.76	5.09	8.01	5.48
AR	16.03	4.74	8.34	8.34	15.40	11.28	6.69	2.58	15.40	4.94	5.76	25.69	2.73	10.96	5.72
CA	71.13	18.78	18.41	18.41	35.28	19.69	14.59	23.59	19.69	8.98	13.58	35.28	5.07	27.13	18.32
CO	15.60	4.16	5.36	5.36	23.64	10.59	4.10	3.23	18.40	4.37	3.87	22.88	2.78	7.22	3.85
CT	58.74	12.93	21.24	21.24	65.13	29.83	15.63	6.43	65.13	19.44	14.12	66.07	11.07	22.67	10.49
DE	26.71	5.55	13.16	13.16	23.27	13.16	8.00	11.53	23.27	6.17	6.19	23.27	6.19	11.13	6.60
DC	17.05	6.45	23.33	23.33	26.02	7.91	6.08	4.38	26.02	15.31	7.62	35.16	9.65	14.34	4.00
FL	18.17	7.69	9.91	9.91	23.04	12.59	5.63	5.26	23.04	9.07	6.05	39.44	3.54	9.24	6.41
GA	93.33	16.61	22.17	22.17	60.89	25.88	13.50	11.64	60.89	19.54	14.79	72.60	7.53	19.39	15.17
HI	17.24	4.93	7.60	7.60	21.77	6.88	6.52	5.40	21.77	4.82	7.71	24.83	2.28	10.54	3.49
ID	24.73	7.46	12.31	12.31	25.42	10.44	6.15	4.84	25.42	5.30	5.61	29.98	3.26	9.46	4.65
IL	47.55	15.21	31.16	31.16	84.68	19.36	27.09	7.88	84.68	23.23	15.10	101.23	14.42	19.73	16.65
IN	11.57	4.32	5.32	5.32	11.59	4.80	3.26	2.77	11.59	3.85	3.70	13.20	1.58	5.75	3.20
IA	28.01	8.53	9.15	9.15	52.79	14.30	9.38	6.15	25.61	14.19	8.59	87.69	6.14	13.60	7.75
KS	16.26	5.64	6.71	6.71	18.97	11.93	5.23	4.63	11.96	6.01	7.60	26.96	3.22	7.88	5.30
KY	34.00	12.00	11.72	11.72	36.00	11.02	16.73	3.50	36.00	12.15	8.35	150.00	9.56	10.00	10.32
LA	43.53	14.84	20.14	20.14	44.26	12.79	9.99	7.91	44.20	15.48	10.94	55.57	6.11	23.25	11.22
ME	22.72	6.77	8.92	8.92	37.18	12.21	5.76	4.64	37.18	8.08	6.58	54.29	4.39	8.15	6.70
MD	29.05	11.90	11.92	11.92	59.02	17.92	6.49	5.84	32.91	14.58	9.95	43.60	4.46	16.02	11.88
MA	30.99	5.72	6.89	6.89	54.08	33.00	5.81	2.48	23.75	15.42	5.98	23.51	4.21	8.28	4.13
MI	41.18	9.83	12.28	12.28	37.00	11.27	11.73	6.46	37.00	9.41	8.67	11.27	5.13	9.09	8.21
MN	67.71	17.30	9.82	9.82	108.16	7.70	14.69	7.95	7.70	10.92	12.24	36.60	6.41	11.39	7.62
MS	31.61	7.46	15.08	15.08	38.45	7.53	8.45	5.85	38.45	7.77	9.97	38.74	4.78	52.03	7.53
MO	36.73	8.64	12.33	12.33	47.55	24.29	10.41	5.92	47.55	14.45	9.66	60.85	6.20	12.44	8.43
MT	29.42	4.80	6.98	6.98	18.72	5.89	5.47	4.64	5.89	5.03	5.91	28.63	3.01	6.06	3.67
NE	35.08	12.88	14.80	14.80	59.03	14.13	8.68	6.73	67.83	16.33	11.35	66.18	7.13	15.98	10.30
NV	16.59	7.34	7.06	7.06	19.32	12.20	4.42	4.23	12.20	4.79	6.15	28.67	4.30	10.30	5.47
NH	36.82	8.93	14.28	14.28	95.03	28.57	12.35	7.17	95.03	8.91	9.04	54.18	6.29	36.82	10.68
NJ	40.10	7.22	12.87	12.87	17.25	13.36	9.55	6.63	16.86	4.18	9.20	23.37	7.00	15.75	5.89
NM	35.07	9.63	11.79	11.79	45.52	16.76	7.73	7.53	45.52	8.00	8.00	42.38	8.41	14.60	7.75
NY	35.96	13.56	19.25	19.25	35.26	19.08	10.35	7.49	18.10	21.41	10.22	42.40	12.20	14.37	6.45
NC	35.39	11.87	14.43	14.43	45.05	22.54	7.62	6.10	45.05	18.41	9.71	78.32	8.90	17.84	10.60
ND	17.63	6.11	10.59	10.59	10.90	10.59	10.50	17.63	8.18	3.33	10.50	10.59	3.33	8.79	6.11
OH	14.89	4.00	6.58	6.58	11.33	5.84	4.85	4.58	5.84	2.76	4.52	24.79	2.13	10.02	3.68
OK	22.44	9.62	10.98	10.98	29.21	15.15	6.69	6.41	32.94	9.25	6.21	52.38	3.88	11.26	6.40
OR	27.22	6.85	7.08	7.08	22.04	13.44	9.10	6.73	22.04	9.10	9.28	11.81	3.70	14.41	6.02
PA	38.35	8.92	19.10	19.10	30.01	19.10	11.69	16.04	30.01	7.95	8.85	30.01	8.85	16.69	9.19
RI	23.59	8.48	7.37	7.37	35.11	24.46	5.33	4.98	22.15	11.33	6.32	39.60	4.85	10.13	5.85
SC	46.45	10.88	15.24	15.24	31.33	24.73	7.98	5.42	31.33	11.24	10.59	66.07	8.64	15.07	10.12
SD	34.39	7.29	14.74	14.74	33.52	23.05	6.19	5.30	33.52	22.67	11.38	66.74	6.69	10.62	8.14
TN	29.78	8.13	13.42	13.42	16.07	18.88	7.46	4.09	16.07	8.10	7.46	40.34	4.92	10.23	6.10
TX	16.22	11.69	7.60	7.60	22.99	9.48	4.68	5.57	7.13	4.30	5.95	7.34	3.34	12.75	5.27
UT	23.93	5.78	5.71	5.71	24.61	11.81	5.75	4.31	20.14	5.91	4.77	21.36	3.71	7.05	4.16
VT	27.01	7.30	12.01	12.01	34.39	18.99	7.01	7.90	34.39	8.67	7.28	48.29	6.07	13.40	6.27
VA	25.97	5.47	7.46	7.46	29.39	10.80	5.76	3.30	29.39	4.92	7.38	27.90	5.39	9.03	5.30
WA	21.14	4.47	8.09	8.09	8.09	8.09	7.86	21.14	8.09	2.77	6.15	12.26	2.46	8.52	4.70
WV	19.62	5.62	7.13	7.13	19.39	9.55	4.86	3.11	20.70	5.62	6.64	35.86	4.23	8.93	3.76
WI	28.04	9.43	12.08	12.08	20.72	16.12	13.78	5.07	20.72	9.68	8.34	24.08	4.11	10.15	4.64
WY	5.82	5.82	5.82	5.82	5.82	5.82	5.82	5.82	5.82	5.82	5.82	5.82	5.82	5.82	5.82
AVG.	31.29	8.79	11.95	11.95	34.64	15.21	8.54	6.86	29.21	9.87	8.39	41.47	5.67	13.67	7.24

General Conditions — RH1010-400 Workers' Compensation

Table H1010-404 Workers' Compensation (Canada in Canadian dollars)

Province		Alberta	British Columbia	Manitoba	Ontario	New Brunswick	Newfndld. & Labrador	Northwest Territories	Nova Scotia	Prince Edward Island	Quebec	Saskatchewan	Yukon
Carpentry—3 stories or less	Rate	5.71	3.27	3.98	4.55	3.17	7.81	4.99	6.85	6.42	12.03	3.43	6.63
	Code	42143	721028	40102	723	238130	4226	4-41	4226	403	80110	B1226	202
Carpentry—interior cab. work	Rate	1.89	2.80	3.98	4.55	3.17	4.07	4.99	5.87	3.94	12.03	3.43	6.63
	Code	42133	721021	40102	723	238350	4279	4-41	4274	402	80110	B1127	202
CARPENTRY—general	Rate	5.71	3.27	3.98	4.55	3.17	4.07	4.99	6.85	6.42	12.03	3.43	6.63
	Code	42143	721028	40102	723	238130	4299	4-41	4226	403	80110	B1202	202
CONCRETE WORK—NOC	Rate	3.88	3.57	6.50	18.31	3.17	7.81	4.99	5.75	3.94	1.92	2.99	6.63
	Code	42104	721010	40110	748	238110	4224	4-41	4224	402	80100	B1314	203
CONCRETE WORK—flat (flr. sidewalk)	Rate	3.88	3.57	6.50	18.31	3.17	7.81	4.99	5.75	3.94	11.92	2.99	6.63
	Code	42104	721010	40110	748	238110	4224	4-41	4224	402	80100	B1314	203
ELECTRICAL Wiring—inside	Rate	1.56	1.18	1.65	3.69	1.28	2.29	2.44	1.97	3.94	5.04	1.83	3.89
	Code	42124	721019	40203	704	238210	4261	4-46	4261	402	80170	B1105	206
EXCAVATION—earth NOC	Rate	1.86	3.79	4.00	5.29	3.17	2.93	3.88	3.89	3.94	6.40	2.19	3.89
	Code	40604	721031	40706	711	238190	4214	4-43	4214	402	80030	R1106	207
EXCAVATION—rock	Rate	1.86	3.79	4.00	5.29	3.17	2.93	3.88	3.89	3.94	6.40	2.19	3.89
	Code	40604	721031	40706	711	238190	4214	4-43	4214	402	80030	R1106	207
GLAZIERS	Rate	2.23	3.79	3.98	10.25	3.17	5.10	4.99	5.22	3.94	15.02	2.99	3.89
	Code	42121	715020	40109	751	238150	4233	4-41	4233	402	80150	B1304	212
INSULATION WORK	Rate	1.62	9.87	3.98	10.25	3.17	5.10	4.99	5.22	3.94	12.03	3.43	6.63
	Code	42184	721029	40102	751	238310	4234	4-41	4234	402	80110	B1207	202
LATHING	Rate	4.20	7.84	3.98	4.55	3.17	4.07	4.99	5.87	3.94	12.03	2.99	6.63
	Code	42135	721042	40102	723	238390	4279	4-41	4271	402	80110	B1316	202
MASONRY	Rate	3.88	3.30	3.98	12.70	3.17	5.10	4.99	5.22	3.94	11.92	2.99	6.63
	Code	42102	721037	40102	741	238140	4231	4-41	4231	402	80100	B1318	202
PAINTING & DECORATING	Rate	3.16	3.62	3.11	7.51	3.17	4.07	4.99	5.87	3.94	12.03	3.43	6.63
	Code	42111	721041	40105	719	238320	4275	4-41	4275	402	80110	B1201	202
PILE DRIVING	Rate	3.88	3.42	4.00	7.03	3.17	2.93	3.88	5.75	6.42	6.40	2.99	6.63
	Code	42159	722004	40706	732	238190	4129	4-43	4221	403	80030	B1310	202
PLASTERING	Rate	4.20	7.84	4.45	7.51	3.17	4.07	4.99	5.87	3.94	12.03	3.43	6.63
	Code	42135	721042	40108	719	238390	4271	4-41	4271	402	80110	B1221	202
PLUMBING	Rate	1.56	2.55	2.78	4.16	1.28	2.47	2.44	2.81	3.94	5.89	1.83	3.89
	Code	42122	721043	40204	707	238220	4241	4-46	4241	401	80160	B1101	214
ROOFING	Rate	4.71	5.42	7.22	14.80	3.17	7.81	4.99	9.43	3.94	18.52	2.99	6.63
	Code	42118	721036	40403	728	238160	4236	4-41	4236	402	80130	B1320	202
SHEET METAL WORK (HVAC)	Rate	1.56	2.55	7.22	4.16	1.28	2.47	2.44	2.81	3.94	5.89	1.83	3.89
	Code	42117	721043	40402	707	238220	4244	4-46	4244	401	80160	B1107	208
STEEL ERECTION—door & sash	Rate	1.62	10.62	6.82	18.31	3.17	7.81	4.99	5.22	3.94	16.70	2.99	6.63
	Code	42106	722005	40502	748	238120	4227	-37	4227	402	80080	B1322	202
STEEL ERECTION—inter., ornam.	Rate	1.62	10.62	6.82	18.31	3.17	7.81	4.99	5.22	3.94	16.70	2.99	6.63
	Code	42106	722005	40502	748	238120	4227	4-41	4227	402	80080	B1322	202
STEEL ERECTION—structure	Rate	1.62	10.62	6.82	18.31	3.17	7.81	4.99	5.22	3.94	16.70	2.99	6.63
	Code	42106	722005	40502	748	238120	4227	4-41	4227	402	80080	B1322	202
STEEL ERECTION—NOC	Rate	1.62	10.62	6.82	18.31	3.17	7.81	4.99	5.22	3.94	16.70	2.99	6.63
	Code	42106	722005	40502	748	238120	4227	4-41	4227	402	80080	B1322	202
TILE WORK—inter. (ceramic)	Rate	3.02	3.65	1.86	7.51	3.17	4.07	4.99	5.87	3.94	12.03	2.99	6.63
	Code	42113	721054	40103	719	238340	4276	4-41	4276	402	80110	B1301	202
WATERPROOFING	Rate	3.16	3.62	3.98	4.55	3.17	4.07	4.99	5.22	3.94	18.52	3.13	6.63
	Code	42139	721016	40102	723	238190	4299	4-41	4239	402	80130	B1217	202
WRECKING	Rate	1.86	5.64	6.90	18.31	3.17	2.93	3.88	3.89	3.94	12.03	2.19	6.63
	Code	40604	721005	40106	748	238190	4211	4-43	4211	402	80110	R1125	202

General Conditions — RH1020-300 Overhead & Miscellaneous Data

Unemployment Taxes and Social Security Taxes

State unemployment tax rates vary not only from state to state, but also with the experience rating of the contractor. The federal unemployment tax rate is 6.0% of the first $7,000 of wages. This is reduced by a credit of up to 5.4% for timely payment to the state. The minimum federal unemployment tax is 0.6% after all credits.

Social security (FICA) for 2014 is estimated at time of publication to be 7.65% of wages up to $115,800.

General Conditions — RH1020-400 Overtime

Overtime

One way to improve the completion date of a project or eliminate negative float from a schedule is to compress activity duration times. This can be achieved by increasing the crew size or working overtime with the proposed crew.

To determine the costs of working overtime to compress activity duration times, consider the following examples. Below is an overtime efficiency and cost chart based on a five, six, or seven day week with an eight through twelve hour day. Payroll percentage increases for time and one half and double time are shown for the various working days.

Days per Week	Hours per Day	Production Efficiency					Payroll Cost Factors	
		1 Week	2 Weeks	3 Weeks	4 Weeks	Average 4 Weeks	@ 1-1/2 Times	@ 2 Times
5	8	100%	100%	100%	100%	100 %	100 %	100 %
	9	100	100	95	90	96.25	105.6	111.1
	10	100	95	90	85	91.25	110.0	120.0
	11	95	90	75	65	81.25	113.6	127.3
	12	90	85	70	60	76.25	116.7	133.3
6	8	100	100	95	90	96.25	108.3	116.7
	9	100	95	90	85	92.50	113.0	125.9
	10	95	90	85	80	87.50	116.7	133.3
	11	95	85	70	65	78.75	119.7	139.4
	12	90	80	65	60	73.75	122.2	144.4
7	8	100	95	85	75	88.75	114.3	128.6
	9	95	90	80	70	83.75	118.3	136.5
	10	90	85	75	65	78.75	121.4	142.9
	11	85	80	65	60	72.50	124.0	148.1
	12	85	75	60	55	68.75	126.2	152.4

General Conditions — RH1030-100 General

Sales Tax by State

State sales tax on materials is tabulated below (5 states have no sales tax). Many states allow local jurisdictions, such as a county or city, to levy additional sales tax.

Some projects may be sales tax exempt, particularly those constructed with public funds.

State	Tax (%)	State	Tax (%)	State	Tax (%)	State	Tax (%)
Alabama	4	Illinois	6.25	Montana	0	Rhode Island	7
Alaska	0	Indiana	7	Nebraska	5.5	South Carolina	6
Arizona	5.6	Iowa	6	Nevada	6.85	South Dakota	4
Arkansas	6	Kansas	6.3	New Hampshire	0	Tennessee	7
California	7.25	Kentucky	6	New Jersey	7	Texas	6.25
Colorado	2.9	Louisiana	4	New Mexico	5.13	Utah	5.95
Connecticut	6.35	Maine	5	New York	4	Vermont	6
Delaware	0	Maryland	6	North Carolina	4.75	Virginia	5
District of Columbia	6	Massachusetts	6.25	North Dakota	5	Washington	6.5
Florida	6	Michigan	6	Ohio	5.5	West Virginia	6
Georgia	4	Minnesota	6.88	Oklahoma	4.5	Wisconsin	5
Hawaii	4	Mississippi	7	Oregon	0	Wyoming	4
Idaho	6	Missouri	4.23	Pennsylvania	6	Average	5.06 %

General Conditions — RH1040-100 Scaffolding

Table H1040-101 Steel Tubular Scaffolding

On new construction, tubular scaffolding is efficient up to 60' high or five stories. Above this it is usually better to use a hung scaffolding if construction permits. Swing scaffolding operations may interfere with tenants. In this case, the tubular is more practical at all heights.

In repairing or cleaning the front of an existing building, the cost of tubular scaffolding per S.F. of building front increases as the height increases above the first tier. The first tier cost is relatively high due to leveling and alignment.

The minimum efficient crew for erection is three workers. For heights over 50', a crew of four is more efficient. Use two or more on top and two at the bottom for handing up or hoisting. Four workers can erect and dismantle about nine frames per hour up to five stories. From five to eight stories, they will average six frames per hour. With 7' horizontal spacing, this will run about 400 S.F. and 265 S.F. of wall surface, respectively. Time for placing planks must be added to the above. On heights above 50', five planks can be placed per labor-hour.

The cost per 1,000 S.F. of building front in the table below was developed by pricing the materials required for a typical tubular scaffolding system eleven frames long and two frames high. Planks were figured five wide for standing plus two wide for materials.

Frames are 5' wide and usually spaced 7' O.C. horizontally. Sidewalk frames are 6' wide. Rental rates will be lower for jobs over three months' duration.

For jobs under twenty-five frames, add 50% to rental cost. These figures do not include accessories which are listed separately below. Large quantities for long periods can reduce rental rates by 20%.

Item	Unit	Monthly Rent	Per 1,000 S.F. of Building Front	
			No. of Pieces	Rental per Month
5' Wide Standard Frame, 6'-4" High	Ea.	$ 5.15	24	$123.60
Leveling Jack & Plate		2.00	24	48.00
Cross Brace		1.00	44	44.00
Side Arm Bracket, 21"		2.00	12	24.00
Guardrail Post		1.00	12	12.00
Guardrail, 7' section		1.00	22	22.00
Stairway Section		35.50	2	71.00
Walk-Thru Frame Guardrail		2.50	2	5.00
			Total	$349.60
			Per C.S.F., 1 Use/Mo.	$ 34.96

Scaffolding is often used as falsework over 15' high during construction of cast-in-place concrete beams and slabs. Two-foot wide scaffolding is generally used for heavy beam construction. The span between frames depends upon the load to be carried, with a maximum span of 5'.

Heavy duty shoring frames with a capacity of 10,000#/leg can be spaced up to 10' O. C., depending upon form support design and loading.

Scaffolding used as horizontal shoring requires less than half the material required with conventional shoring.

On new construction, erection is done by carpenters.

Rolling towers supporting horizontal shores can reduce labor and speed the job. For maintenance work, catwalks with spans up to 70' can be supported by the rolling towers.

City Adjustments   RJ1010-010

General: The following information on current city cost indexes is calculated for over 930 zip code locations in the United States and Canada. Index figures for both material and installation are based upon the 30 major city average of 100 and represent the cost relationship on July 1, 2013.

In addition to index adjustment, the user should consider:
1. productivity
2. management efficiency
3. competitive conditions
4. automation
5. restrictive union practices
6. unique local requirements
7. regional variations due to specific building codes

The weighted-average index is calculated from about 66 materials and equipment types and 21 building trades. The component contribution of these in a model building is tabulated in Table J1010-011 below.

If the systems component distribution of a building is unknown, the weighted-average index can be used to adjust the cost for any city.

Table J1010-011 Labor, Material, and Equipment Cost Distribution for Weighted Average Listed by System Division

Division No.	Building System	Percentage	Division No.	Building System	Percentage
A	Substructure	6.2%	D10	Services: Conveying	4.3%
B10	Shell: Superstructure	18.8	D20-40	Mechanical	22.7
B20	Exterior Closure	11.6	D50	Electrical	11.5
B30	Roofing	2.9	E	Equipment & Furnishings	2.3
C	Interior Construction	15.5	G	Site Work	4.2
				Total weighted average (Div. A-G)	100.0%

How to Use the Component Indexes: Table J1010-012 below shows how the average costs obtained from this book for each division should be adjusted for your particular building in your city. The example is a building adjusted for New York, NY. Indexes for other cities are tabulated in Division RJ1030. These indexes should also be used when compiling data on the form in Division K1010.

Table J1010-012 Adjustment of "Book" Costs to a Particular City

| Systems Division Number | System Description | New York, NY | | | | | | | |
		City Cost Index	Cost Factor	Book Cost	Adjusted Cost	City Cost Index	Cost Factor	Book Cost	Adjusted Cost
A	Substructure	143.5	1.435	$ 100,800	$ 144,600				
B10	Shell: Superstructure	132.6	1.326	278,400	369,200				
B20	Exterior Closure	137.8	1.378	188,700	260,000				
B30	Roofing	129.3	1.293	41,600	53,800				
C	Interior Construction	131.8	1.318	268,600	354,000				
D10	Services: Conveying	108.5	1.085	53,700	58,300				
D20-40	Mechanical	126.4	1.264	333,300	421,300				
D50	Electrical	141.1	1.411	186,500	263,200				
E	Equipment	104.1	1.041	28,200	29,400				
G	Site Work	122.4	1.224	79,300	97,100				
Total Cost, (Div. A-G)				$1,559,100	$2,050,900				

Alternate Method: Use the weighted-average index for your city rather than the component indexes. **For example:**

(Total weighted-average index A-G)/100 x (total cost) = Adjusted city cost
New York, NY 1.311 x $1,559,100 = $2,044,000

City Adjustments   RJ1020-010

City to City: To convert known or estimated costs from one city to another, cost indexes can be used as follows:

$$\text{Unknown City Cost} = \text{Known Cost} \times \frac{\text{Unknown City Index}}{\text{Known City Index}}$$

For example: If the building cost in Boston, MA, is $2,000,000, how much would a duplicated building cost in Los Angeles, CA?

$$\text{L.A. Cost} = \$2,000,000 \times \frac{(\text{Los Angeles})\ 107.3}{(\text{Boston})\ 117.6} = \$1,825,000$$

The table below lists both the Means City Cost Index based on January 1, 1993 = 100 as well as the computed value of an index based on January 1, 2014 costs. Since the January 1, 2014 figure is estimated, space is left to write in the actual index figures as they become available through either the quarterly *Means Construction Cost Indexes* or as printed in the *Engineering News-Record*. To compute the actual index based on January 1, 2014 = 100, divide the Quarterly City Cost Index for a particular year by the actual January 1, 2014 Quarterly City Cost Index. Space has been left to advance the index figures as the year progresses.

Table J1020-011 Historical Cost Indexes

Year	Historical Cost Index Jan. 1, 1993 = 100		Current Index Based on Jan. 1, 2014 = 100		Year	Historical Cost Index Jan. 1, 1993 = 100	Current Index Based on Jan. 1, 2014 = 100		Year	Historical Cost Index Jan. 1, 1993 = 100	Current Index Based on Jan. 1, 2014 = 100	
	Est.	Actual	Est.	Actual		Actual	Est.	Actual		Actual	Est.	Actual
Oct 2014*					July 1999	117.6	58.0		July 1981	70.0	34.5	
July 2014*					1998	115.1	56.8		1980	62.9	31.0	
April 2014*					1997	112.8	55.6		1979	57.8	28.5	
Jan 2014*	202.7		100.0	100.0	1996	110.2	54.4		1978	53.5	26.4	
July 2013		201.2	99.3		1995	107.6	53.1		1977	49.5	24.4	
2012		194.6	96.0		1994	104.4	51.5		1976	46.9	23.1	
2011		191.2	94.3		1993	101.7	50.2		1975	44.8	22.1	
2010		183.5	90.5		1992	99.4	49.1		1974	41.4	20.4	
2009		180.1	88.9		1991	96.8	47.8		1973	37.7	18.6	
2008		180.4	89.0		1990	94.3	46.5		1972	34.8	17.2	
2007		169.4	83.6		1989	92.1	45.5		1971	32.1	15.8	
2006		162.0	79.9		1988	89.9	44.3		1970	28.7	14.2	
2005		151.6	74.8		1987	87.7	43.3		1969	26.9	13.3	
2004		143.7	70.9		1986	84.2	41.6		1968	24.9	12.3	
2003		132.0	65.1		1985	82.6	40.8		1967	23.5	11.6	
2002		128.7	63.5		1984	82.0	40.4		1966	22.7	11.2	
2001		125.1	61.7		1983	80.2	39.5		1965	21.7	10.7	
2000		120.9	59.6		▼ 1982	76.1	37.6		1964	21.2	10.5	

To find the **current cost** from a project built previously in either the same city or a different city, the following formula is used:

$$\text{Present Cost (City X)} = \frac{\text{Current HCI} \times \text{CCI (City X)}}{\text{Previous HCI} \times \text{CCI (City Y)}} \times \text{Former Cost (City Y)}$$

For example: Find the construction cost of a building to be built in San Francisco, CA, as of January 1, 2014 when the identical building cost $500,000 in Boston, MA, on July 1, 1968.

$$\text{Jan. 1, 2014 (San Francisco)} = \frac{(\text{San Francisco})\ 202.7 \times 122.5}{(\text{Boston})\ 24.9 \times 117.6} \times \$500,000 = \$4,240,000$$

Note: The City Cost Indexes for Canada can be used to convert U.S. national averages to local costs in Canadian dollars.

To Project Future Construction Costs: Using the results of the last five years average percentage increase as a basis, an average increase of 4.4% could be used.

The historical index figures above are compiled from the Means Construction Index Service.

*Historical Cost Index updates and other resources are provided on the following website.
http://www.reedconstructiondata.com/rsmeans/chgnotice/456321

City Cost Indexes

How to Use the City Cost Indexes

What you should know before you begin

RSMeans City Cost Indexes (CCI) are an extremely useful tool to use when you want to compare costs from city to city and region to region.

This publication contains average construction cost indexes for 317 U.S. and Canadian cities covering over 930 three-digit zip code locations, as listed directly under each city.

Keep in mind that a City Cost Index number is a percentage ratio of a specific city's cost to the national average cost of the same item at a stated time period.

In other words, these index figures represent relative construction factors (or, if you prefer, multipliers) for Material and Installation costs, as well as the weighted average for Total In Place costs for each CSI MasterFormat division. Installation costs include both labor and equipment rental costs. When estimating equipment rental rates only, for a specific location, use 01 54 33 EQUIPMENT RENTAL COSTS in the Reference Section.

The 30 City Average Index is the average of 30 major U.S. cities and serves as a National Average.

Index figures for both material and installation are based on the 30 major city average of 100 and represent the cost relationship as of July 1, 2013. The index for each division is computed from representative material and labor quantities for that division. The weighted average for each city is a weighted total of the components listed above it, but does not include relative productivity between trades or cities.

As changes occur in local material prices, labor rates, and equipment rental rates, (including fuel costs) the impact of these changes should be accurately measured by the change in the City Cost Index for each particular city (as compared to the 30 City Average).

Therefore, if you know (or have estimated) building costs in one city today, you can easily convert those costs to expected building costs in another city.

In addition, by using the Historical Cost Index, you can easily convert National Average building costs at a particular time to the approximate building costs for some other time. The City Cost Indexes can then be applied to calculate the costs for a particular city.

Quick Calculations

Location Adjustment Using the City Cost Indexes:

$$\frac{\text{Index for City A}}{\text{Index for City B}} \times \text{Cost in City B} = \text{Cost in City A}$$

Time Adjustment for the National Average Using the Historical Cost Index:

$$\frac{\text{Index for Year A}}{\text{Index for Year B}} \times \text{Cost in Year B} = \text{Cost in Year A}$$

Adjustment from the National Average:

$$\frac{\text{Index for City A}}{100} \times \text{National Average Cost} = \text{Cost in City A}$$

Since each of the other RSMeans publications contains many different items, any *one* item multiplied by the particular city index may give incorrect results. However, the larger the number of items compiled, the closer the results should be to actual costs for that particular city.

The City Cost Indexes for Canadian cities are calculated using Canadian material and equipment prices and labor rates, in Canadian dollars. Therefore, indexes for Canadian cities can be used to convert U.S. National Average prices to local costs in Canadian dollars.

How to use this section

1. Compare costs from city to city.

In using the RSMeans Indexes, remember that an index number is not a fixed number but a ratio: It's a percentage ratio of a building component's cost at any stated time to the National Average cost of that same component at the same time period. Put in the form of an equation:

$$\frac{\text{Specific City Cost}}{\text{National Average Cost}} \times 100 = \text{City Index Number}$$

Therefore, when making cost comparisons between cities, do not subtract one city's index number from the index number of another city and read the result as a percentage difference. Instead, divide one city's index number by that of the other city. The resulting number may then be used as a multiplier to calculate cost differences from city to city.

The formula used to find cost differences between cities for the purpose of comparison is as follows:

$$\frac{\text{City A Index}}{\text{City B Index}} \times \text{City B Cost (Known)} = \text{City A Cost (Unknown)}$$

In addition, you can use RSMeans CCI to calculate and compare costs division by division between cities using the same basic formula. (Just be sure that you're comparing similar divisions.)

2. Compare a specific city's construction costs with the National Average.

When you're studying construction location feasibility, it's advisable to compare a prospective project's cost index with an index of the National Average cost.

For example, divide the weighted average index of construction costs of a specific city by that of the 30 City Average, which = 100.

$$\frac{\text{City Index}}{100} = \% \text{ of National Average}$$

As a result, you get a ratio that indicates the relative cost of construction in that city in comparison with the National Average.

3. Convert U.S. National Average to actual costs in Canadian City.

$$\frac{\text{Index for Canadian City}}{100} \times \text{National Average Cost} = \text{Cost in Canadian City in \$ CAN}$$

4. **Adjust construction cost data based on a National Average.**
When you use a source of construction cost data which is based on a National Average (such as RSMeans cost data publications), it is necessary to adjust those costs to a specific location.

$$\frac{\text{City Index}}{100} \times \text{"Book" Cost Based on National Average Costs} = \text{City Cost (Unknown)}$$

5. **When applying the City Cost Indexes to demolition projects, use the appropriate division installation index.** For example, for removal of existing doors and windows, use Division 8 (Openings) index.

What you might like to know about how we developed the Indexes

The information presented in the CCI is organized according to the Construction Specifications Institute (CSI) MasterFormat 2010.

To create a reliable index, RSMeans researched the building type most often constructed in the United States and Canada. Because it was concluded that no one type of building completely represented the building construction industry, nine different types of buildings were combined to create a composite model.

The exact material, labor, and equipment quantities are based on detailed analyses of these nine building types, then each quantity is weighted in proportion to expected usage. These various material items, labor hours, and equipment rental rates are thus combined to form a composite building representing as closely as possible the actual usage of materials, labor and equipment used in the North American Building Construction Industry.

The following structures were chosen to make up that composite model:

1. Factory, 1 story
2. Office, 2-4 story
3. Store, Retail
4. Town Hall, 2-3 story
5. High School, 2-3 story
6. Hospital, 4-8 story
7. Garage, Parking
8. Apartment, 1-3 story
9. Hotel/Motel, 2-3 story

For the purposes of ensuring the timeliness of the data, the components of the index for the composite model have been streamlined. They currently consist of:

- specific quantities of 66 commonly used construction materials;
- specific labor-hours for 21 building construction trades; and
- specific days of equipment rental for 6 types of construction equipment (normally used to install the 66 material items by the 21 trades.) Fuel costs and routine maintenance costs are included in the equipment cost.

A sophisticated computer program handles the updating of all costs for each city on a quarterly basis. Material and equipment price quotations are gathered quarterly from 731 cities in the United States and Canada. These prices and the latest negotiated labor wage rates for 21 different building trades are used to compile the quarterly update of the City Cost Index.

The 30 major U.S. cities used to calculate the National Average are:

Atlanta, GA
Baltimore, MD
Boston, MA
Buffalo, NY
Chicago, IL
Cincinnati, OH
Cleveland, OH
Columbus, OH
Dallas, TX
Denver, CO
Detroit, MI
Houston, TX
Indianapolis, IN
Kansas City, MO
Los Angeles, CA

Memphis, TN
Milwaukee, WI
Minneapolis, MN
Nashville, TN
New Orleans, LA
New York, NY
Philadelphia, PA
Phoenix, AZ
Pittsburgh, PA
St. Louis, MO
San Antonio, TX
San Diego, CA
San Francisco, CA
Seattle, WA
Washington, DC

What the CCI does not indicate

The weighted average for each city is a total of the divisional components weighted to reflect typical usage, but it does not include the productivity variations between trades or cities.

In addition, the CCI does not take into consideration factors such as the following:

- managerial efficiency
- competitive conditions
- automation
- restrictive union practices
- unique local requirements
- regional variations due to specific building codes

City Cost Indexes — RJ1030-010 Building Systems

ALABAMA

DIV. NO.	BUILDING SYSTEMS	BIRMINGHAM MAT.	INST.	TOTAL	HUNTSVILLE MAT.	INST.	TOTAL	MOBILE MAT.	INST.	TOTAL	MONTGOMERY MAT.	INST.	TOTAL	TUSCALOOSA MAT.	INST.	TOTAL
A	Substructure	96.4	79.7	86.2	89.1	77.4	82.0	99.6	70.0	81.5	101.3	64.7	79.0	91.4	70.2	78.5
B10	Shell: Superstructure	97.1	85.3	92.4	97.4	82.4	91.5	98.4	77.7	90.2	97.9	75.5	89.1	97.3	78.6	89.9
B20	Exterior Closure	92.0	77.1	84.9	92.2	68.3	80.8	93.7	57.7	76.5	92.3	53.8	73.9	91.7	67.9	80.3
B30	Roofing	95.9	82.1	90.7	92.9	78.7	87.6	99.8	73.0	89.7	100.2	68.9	88.5	93.0	74.0	85.9
C	Interior Construction	97.1	69.7	85.4	97.3	71.6	86.4	97.8	55.5	79.8	98.4	48.3	77.1	97.3	52.5	78.2
D10	Services: Conveying	100.0	89.9	97.3	100.0	86.8	96.4	100.0	87.8	96.7	100.0	84.1	95.7	100.0	85.9	96.2
D20-40	Mechanical	100.0	62.8	85.1	100.0	52.9	81.1	100.0	61.5	84.5	100.1	34.8	73.9	100.0	34.6	73.8
D50	Electrical	99.7	62.8	80.4	95.7	66.5	80.4	93.4	58.5	75.2	93.9	62.3	77.4	95.1	62.8	78.2
E	Equipment & Furnishings	100.0	67.9	98.4	100.0	73.2	98.6	100.0	53.7	97.6	100.0	43.7	97.1	100.0	45.9	97.2
G	Site Work	89.2	94.8	93.0	82.3	93.7	90.0	94.1	89.0	90.6	94.0	88.9	90.6	82.8	93.9	90.3
A-G	WEIGHTED AVERAGE	97.5	74.1	87.3	96.7	70.8	85.4	97.8	65.9	83.9	97.9	58.1	80.5	96.7	62.0	81.6

ALASKA / ARIZONA

DIV. NO.	BUILDING SYSTEMS	ANCHORAGE MAT.	INST.	TOTAL	FAIRBANKS MAT.	INST.	TOTAL	JUNEAU MAT.	INST.	TOTAL	FLAGSTAFF MAT.	INST.	TOTAL	MESA/TEMPE MAT.	INST.	TOTAL
A	Substructure	132.5	119.5	124.6	129.7	119.5	123.5	130.5	119.4	123.8	93.4	77.9	84.0	97.9	79.4	86.7
B10	Shell: Superstructure	118.3	109.2	114.7	121.6	109.1	116.7	120.4	109.1	115.9	95.6	73.6	87.0	93.9	74.8	86.4
B20	Exterior Closure	161.8	120.3	141.9	146.2	120.4	133.8	153.8	120.3	137.8	117.8	64.5	92.3	107.8	64.7	87.2
B30	Roofing	170.6	115.1	149.7	169.9	116.4	149.8	164.0	115.1	145.6	96.6	69.2	86.3	98.5	67.3	86.8
C	Interior Construction	130.2	119.3	125.6	126.5	119.8	123.7	126.7	119.3	123.6	103.8	61.5	85.8	99.6	65.6	85.1
D10	Services: Conveying	100.0	111.8	103.2	100.0	111.8	103.2	100.0	111.8	103.2	100.0	85.7	96.1	100.0	85.9	96.2
D20-40	Mechanical	100.2	104.3	101.9	100.2	106.7	102.8	100.1	100.6	100.3	100.2	80.0	92.1	100.1	79.0	91.6
D50	Electrical	120.5	115.6	117.9	134.7	115.6	124.7	125.5	115.6	120.3	101.1	65.8	82.7	95.3	65.8	79.9
E	Equipment & Furnishings	100.0	118.2	100.9	100.0	117.7	100.9	100.0	118.2	100.9	100.0	65.6	98.3	100.0	69.4	98.4
G	Site Work	127.8	129.3	128.8	118.7	129.3	125.9	135.2	129.3	131.2	84.0	96.8	92.7	91.7	98.6	96.4
A-G	WEIGHTED AVERAGE	121.3	114.3	118.3	120.7	115.0	118.2	120.6	113.5	117.5	101.0	72.9	88.8	98.8	73.8	87.9

ARIZONA / ARKANSAS

DIV. NO.	BUILDING SYSTEMS	PHOENIX MAT.	INST.	TOTAL	PRESCOTT MAT.	INST.	TOTAL	TUCSON MAT.	INST.	TOTAL	FORT SMITH MAT.	INST.	TOTAL	JONESBORO MAT.	INST.	TOTAL
A	Substructure	97.8	79.6	86.7	91.5	73.2	80.3	95.2	79.4	85.6	89.5	70.0	77.6	87.2	65.4	73.9
B10	Shell: Superstructure	95.1	75.3	87.3	95.4	69.5	85.2	94.1	74.8	86.5	98.9	67.9	86.7	92.9	67.4	82.9
B20	Exterior Closure	101.8	66.3	84.8	103.4	63.9	84.5	100.4	64.7	83.3	96.2	51.2	74.7	94.9	46.5	71.7
B30	Roofing	98.1	69.3	87.3	95.4	64.6	83.8	99.5	66.5	87.1	101.5	54.4	83.8	105.4	51.4	85.1
C	Interior Construction	101.9	66.5	86.8	102.6	50.9	80.6	96.6	64.7	83.0	93.4	59.1	78.8	91.6	50.1	73.9
D10	Services: Conveying	100.0	85.9	96.2	100.0	85.7	96.1	100.0	85.9	96.2	100.0	79.0	94.3	100.0	46.0	85.3
D20-40	Mechanical	100.0	80.0	92.0	100.2	78.9	91.7	100.1	79.0	91.6	100.2	50.1	80.1	100.4	52.9	81.4
D50	Electrical	102.2	68.3	84.5	100.7	65.8	82.5	97.6	65.6	81.0	92.3	66.1	78.6	99.0	64.0	80.7
E	Equipment & Furnishings	100.0	69.6	98.5	100.0	49.6	97.4	100.0	69.4	98.4	100.0	60.1	98.0	100.0	49.4	97.4
G	Site Work	92.0	98.8	96.6	73.4	95.5	89.1	87.6	98.6	95.0	80.3	86.2	84.3	98.8	100.0	99.6
A-G	WEIGHTED AVERAGE	99.4	74.8	88.7	98.9	69.7	86.2	97.6	73.6	87.1	96.8	61.7	81.5	96.3	59.4	80.2

ARKANSAS / CALIFORNIA

DIV. NO.	BUILDING SYSTEMS	LITTLE ROCK MAT.	INST.	TOTAL	PINE BLUFF MAT.	INST.	TOTAL	TEXARKANA MAT.	INST.	TOTAL	ANAHEIM MAT.	INST.	TOTAL	BAKERSFIELD MAT.	INST.	TOTAL
A	Substructure	90.9	70.2	78.3	82.0	70.1	74.8	88.9	56.1	68.9	92.8	117.8	108.1	97.3	116.8	109.2
B10	Shell: Superstructure	98.8	67.2	86.4	94.8	67.2	83.9	90.4	57.4	77.4	101.4	109.7	104.6	103.9	109.1	106.0
B20	Exterior Closure	97.9	51.5	75.7	111.6	51.5	82.9	96.4	37.1	68.0	98.3	116.1	106.8	99.4	114.2	106.5
B30	Roofing	97.7	54.8	81.6	96.4	54.8	80.8	97.1	46.3	78.0	97.9	124.3	107.8	104.0	113.5	107.6
C	Interior Construction	95.2	61.1	80.7	92.2	61.1	78.9	95.7	42.4	73.0	100.7	119.0	108.5	99.4	118.3	107.5
D10	Services: Conveying	100.0	79.0	94.3	100.0	79.0	94.3	100.0	32.4	81.6	100.0	110.0	102.7	100.0	110.0	102.7
D20-40	Mechanical	100.0	56.4	82.5	100.2	52.4	81.1	100.2	54.6	81.9	100.0	118.0	107.2	100.1	108.3	103.4
D50	Electrical	99.3	69.5	83.7	91.6	69.5	80.1	93.4	36.8	63.9	91.4	105.9	99.0	102.0	102.3	102.2
E	Equipment & Furnishings	100.0	63.5	98.1	100.0	63.5	98.1	100.0	41.9	97.0	100.0	121.7	101.1	100.0	121.3	101.1
G	Site Work	89.4	86.2	87.3	81.1	86.2	84.6	91.6	86.9	88.4	97.7	107.8	104.6	97.3	105.5	102.9
A-G	WEIGHTED AVERAGE	98.0	63.8	83.1	96.9	62.9	82.1	95.7	49.9	75.7	98.9	114.1	105.6	100.8	110.6	105.1

CALIFORNIA

DIV. NO.	BUILDING SYSTEMS	FRESNO MAT.	INST.	TOTAL	LOS ANGELES MAT.	INST.	TOTAL	OAKLAND MAT.	INST.	TOTAL	OXNARD MAT.	INST.	TOTAL	REDDING MAT.	INST.	TOTAL
A	Substructure	95.5	119.1	109.9	99.5	117.2	110.3	114.3	126.6	121.8	99.5	116.7	110.0	108.5	118.3	114.5
B10	Shell: Superstructure	104.0	111.8	107.1	100.4	110.6	104.4	103.2	116.7	108.5	100.6	109.5	104.1	101.9	111.6	105.7
B20	Exterior Closure	103.1	119.7	111.0	96.3	119.3	107.3	111.1	131.1	120.7	100.6	116.3	108.1	114.4	117.3	115.8
B30	Roofing	93.9	112.5	100.9	100.2	123.1	108.8	101.3	136.2	114.4	103.3	122.0	110.3	108.8	119.4	112.8
C	Interior Construction	100.0	131.1	113.2	100.5	119.2	108.4	100.1	144.5	119.0	97.8	118.6	106.7	106.3	126.7	115.0
D10	Services: Conveying	100.0	119.6	105.3	100.0	109.9	102.7	100.0	121.4	105.8	100.0	110.2	102.8	100.0	119.6	105.3
D20-40	Mechanical	100.2	109.6	104.0	100.0	118.0	107.2	100.1	144.7	118.0	100.1	118.0	107.3	100.1	109.8	104.0
D50	Electrical	91.7	102.9	97.6	103.6	120.6	112.5	108.3	142.6	126.2	96.2	109.9	103.3	97.8	109.9	104.1
E	Equipment & Furnishings	100.0	134.4	101.8	100.0	122.2	101.1	100.0	153.4	102.7	100.0	121.7	101.1	100.0	131.9	101.6
G	Site Work	99.6	105.0	103.2	100.5	105.9	104.1	127.4	106.1	112.9	99.8	103.6	102.3	110.3	104.6	106.4
A-G	WEIGHTED AVERAGE	100.0	114.5	106.3	100.1	116.5	107.3	103.9	133.1	116.6	99.6	114.2	105.9	103.6	114.6	108.4

CALIFORNIA

DIV. NO.	BUILDING SYSTEMS	RIVERSIDE MAT.	INST.	TOTAL	SACRAMENTO MAT.	INST.	TOTAL	SAN DIEGO MAT.	INST.	TOTAL	SAN FRANCISCO MAT.	INST.	TOTAL	SAN JOSE MAT.	INST.	TOTAL
A	Substructure	96.9	117.4	109.4	98.2	121.1	112.2	100.4	109.4	105.9	119.6	128.9	125.2	112.2	125.1	120.1
B10	Shell: Superstructure	102.5	109.7	105.3	97.0	109.6	102.0	99.8	105.6	102.1	109.1	122.4	114.4	100.0	120.3	108.0
B20	Exterior Closure	94.6	115.8	104.7	101.6	120.4	110.6	101.9	113.6	107.5	112.2	135.4	123.3	116.8	131.7	123.9
B30	Roofing	98.1	124.3	107.9	108.7	120.9	113.3	99.7	107.1	102.5	103.1	138.6	116.4	97.4	139.0	113.0
C	Interior Construction	100.8	119.0	108.6	106.7	129.9	116.6	99.0	110.6	103.9	103.6	145.6	121.5	95.1	144.4	116.1
D10	Services: Conveying	100.0	110.0	102.7	100.0	120.1	105.5	100.0	109.1	102.5	100.0	121.9	106.0	100.0	120.9	105.7
D20-40	Mechanical	100.0	118.0	107.2	100.0	121.9	108.7	100.0	116.3	106.5	100.0	171.6	128.7	100.0	143.3	117.3
D50	Electrical	91.6	106.0	99.1	103.4	111.3	107.5	102.2	98.9	100.5	108.3	161.4	136.1	101.0	153.8	128.6
E	Equipment & Furnishings	100.0	121.7	101.1	100.0	134.8	101.8	100.0	109.1	100.5	100.0	154.4	102.8	100.0	154.0	102.7
G	Site Work	96.4	105.8	102.8	103.3	111.8	109.1	99.7	103.3	102.2	128.8	111.8	117.3	130.6	99.5	109.6
A-G	WEIGHTED AVERAGE	98.9	114.0	105.5	101.2	118.6	108.8	100.2	109.1	104.1	106.1	143.8	122.5	102.2	134.6	116.3

City Cost Indexes — RJ1030-010 Building Systems

DIV. NO.	BUILDING SYSTEMS	CALIFORNIA SANTA BARBARA			CALIFORNIA STOCKTON			CALIFORNIA VALLEJO			COLORADO COLORADO SPRINGS			COLORADO DENVER		
		MAT.	INST.	TOTAL	MAT.	INST.	TOTAL	MAT.	INST.	TOTAL	MAT.	INST.	TOTAL	MAT.	INST.	TOTAL
A	Substructure	99.3	117.1	110.1	99.4	119.2	111.4	104.5	126.6	117.9	101.8	83.3	90.5	98.8	85.2	90.5
B10	Shell: Superstructure	100.9	109.5	104.3	99.8	112.4	104.8	101.0	114.6	106.4	100.7	83.6	94.0	102.5	83.7	95.1
B20	Exterior Closure	99.6	117.1	108.0	104.6	120.8	112.4	88.7	131.5	109.2	104.4	74.9	90.3	101.8	75.1	89.0
B30	Roofing	100.2	121.3	108.1	101.1	118.0	107.4	103.2	133.2	114.4	105.3	80.7	96.1	103.7	74.0	92.6
C	Interior Construction	98.4	118.0	106.8	101.5	128.9	113.2	107.4	143.9	122.9	102.8	73.9	90.5	104.9	79.5	94.1
D10	Services: Conveying	100.0	110.2	102.8	100.0	119.6	105.3	100.0	119.9	105.4	100.0	91.2	97.6	100.0	90.8	97.5
D20-40	Mechanical	100.1	118.0	107.3	100.0	108.8	103.5	100.0	128.8	111.6	100.2	72.7	89.2	100.0	79.9	91.9
D50	Electrical	89.4	111.8	101.1	97.9	110.9	104.7	100.6	125.1	113.4	100.5	83.0	91.4	102.1	85.2	93.3
E	Equipment & Furnishings	100.0	121.7	101.1	100.0	134.8	101.8	100.0	152.4	102.7	100.0	78.8	98.9	100.0	79.0	98.9
G	Site Work	99.7	105.5	103.7	99.7	104.6	103.0	103.9	111.7	109.2	95.2	93.2	93.8	94.5	101.3	99.1
A-G	WEIGHTED AVERAGE	98.9	114.6	105.7	100.5	115.5	107.0	100.6	127.2	112.2	101.3	79.3	91.7	101.6	82.6	93.3

DIV. NO.	BUILDING SYSTEMS	COLORADO FORT COLLINS			COLORADO GRAND JUNCTION			COLORADO GREELEY			COLORADO PUEBLO			CONNECTICUT BRIDGEPORT		
		MAT.	INST.	TOTAL	MAT.	INST.	TOTAL	MAT.	INST.	TOTAL	MAT.	INST.	TOTAL	MAT.	INST.	TOTAL
A	Substructure	109.2	81.6	92.4	110.8	81.5	92.9	96.5	77.1	84.7	102.3	83.2	90.6	106.4	120.7	115.1
B10	Shell: Superstructure	101.0	79.7	92.6	100.8	79.1	92.3	98.5	76.4	89.8	100.2	84.3	93.9	102.4	125.3	111.4
B20	Exterior Closure	112.6	77.8	96.0	122.0	74.3	99.2	103.5	54.9	80.3	100.8	75.1	88.5	101.7	133.5	116.9
B30	Roofing	104.9	70.3	91.9	104.1	67.3	90.3	104.3	68.8	91.0	103.6	80.2	94.8	100.3	127.5	110.5
C	Interior Construction	101.3	71.1	88.5	107.3	73.6	92.9	100.7	73.3	89.1	102.7	76.3	91.5	104.7	126.7	114.0
D10	Services: Conveying	100.0	90.8	97.5	100.0	91.0	97.6	100.0	90.8	97.5	100.0	91.8	97.8	100.0	108.9	102.4
D20-40	Mechanical	100.1	77.5	91.0	100.0	84.3	93.7	100.1	77.5	91.0	100.0	72.8	89.1	100.0	116.0	106.4
D50	Electrical	96.9	85.1	90.7	95.9	55.1	74.6	96.9	85.1	90.7	96.9	74.3	85.1	99.2	108.7	104.2
E	Equipment & Furnishings	100.0	78.6	98.9	100.0	77.1	98.8	100.0	84.0	99.2	100.0	78.3	98.9	100.0	121.9	101.1
G	Site Work	106.1	94.7	98.4	127.3	95.6	105.8	93.2	94.1	93.8	118.9	91.7	100.5	109.8	104.2	106.0
A-G	WEIGHTED AVERAGE	102.1	79.6	92.3	104.4	76.7	92.4	99.7	76.1	89.4	100.9	78.6	91.2	101.8	120.2	109.8

DIV. NO.	BUILDING SYSTEMS	CONNECTICUT BRISTOL			CONNECTICUT HARTFORD			CONNECTICUT NEW BRITAIN			CONNECTICUT NEW HAVEN			CONNECTICUT NORWALK		
		MAT.	INST.	TOTAL	MAT.	INST.	TOTAL	MAT.	INST.	TOTAL	MAT.	INST.	TOTAL	MAT.	INST.	TOTAL
A	Substructure	102.4	120.7	113.6	102.1	120.7	113.4	103.4	120.7	113.9	104.4	120.8	114.4	105.4	121.3	115.1
B10	Shell: Superstructure	101.5	125.3	110.9	106.2	125.3	113.7	99.1	125.3	109.4	99.4	125.3	109.6	102.2	125.3	111.3
B20	Exterior Closure	98.4	133.5	115.2	99.4	133.5	115.7	99.0	133.5	115.5	112.3	133.5	122.4	98.3	133.5	115.2
B30	Roofing	100.4	125.0	109.6	104.8	125.0	112.4	100.4	125.0	109.6	100.5	124.7	109.6	100.5	127.8	110.7
C	Interior Construction	104.7	126.7	114.1	103.9	126.7	113.6	104.7	126.7	114.1	104.7	126.7	114.1	104.7	126.7	114.1
D10	Services: Conveying	100.0	108.9	102.4	100.0	108.9	102.4	100.0	108.9	102.4	100.0	108.9	102.4	100.0	108.9	102.4
D20-40	Mechanical	100.0	116.0	106.4	100.0	116.0	106.4	100.0	116.0	106.4	100.0	116.0	106.4	100.0	116.1	106.4
D50	Electrical	99.2	111.3	105.5	98.6	112.3	105.7	99.3	111.3	105.6	99.2	111.3	105.5	99.2	156.4	129.1
E	Equipment & Furnishings	100.0	122.0	101.1	100.0	122.1	101.1	100.0	122.0	101.1	100.0	122.0	101.1	100.0	124.2	101.2
G	Site Work	108.9	104.2	105.7	105.1	104.2	104.5	109.1	104.2	105.8	108.9	104.9	106.2	109.6	104.2	105.9
A-G	WEIGHTED AVERAGE	101.1	120.4	109.5	102.0	120.6	110.1	100.8	120.4	109.3	102.2	120.5	110.2	101.4	126.9	112.5

DIV. NO.	BUILDING SYSTEMS	CONNECTICUT STAMFORD			CONNECTICUT WATERBURY			D.C. WASHINGTON			DELAWARE WILMINGTON			FLORIDA DAYTONA BEACH		
		MAT.	INST.	TOTAL	MAT.	INST.	TOTAL	MAT.	INST.	TOTAL	MAT.	INST.	TOTAL	MAT.	INST.	TOTAL
A	Substructure	106.5	121.3	115.5	106.3	120.7	115.1	115.3	87.0	98.0	91.5	104.4	99.4	91.8	76.6	82.6
B10	Shell: Superstructure	102.4	125.8	111.6	102.4	125.3	111.4	103.2	96.5	100.6	100.8	110.1	104.4	95.2	83.3	90.5
B20	Exterior Closure	98.6	133.5	115.3	98.6	133.5	115.3	96.2	83.3	90.0	102.3	101.6	102.0	93.3	68.3	81.3
B30	Roofing	100.4	127.8	110.7	100.4	124.7	109.5	102.4	86.8	96.6	102.2	112.3	106.0	99.6	75.2	90.4
C	Interior Construction	104.8	126.7	114.1	104.7	126.7	114.0	101.0	82.2	93.0	98.6	102.8	100.4	98.9	71.1	87.1
D10	Services: Conveying	100.0	108.9	102.4	100.0	108.9	102.4	100.0	101.1	100.3	100.0	88.2	96.8	100.0	85.2	96.0
D20-40	Mechanical	100.0	116.1	106.4	100.0	116.0	106.4	100.1	92.9	97.2	100.2	115.8	106.4	100.0	77.8	91.1
D50	Electrical	99.2	163.8	132.9	98.7	108.7	103.9	102.7	102.6	102.6	96.6	109.0	103.1	94.4	56.2	74.5
E	Equipment & Furnishings	100.0	124.5	101.2	100.0	121.9	101.1	100.0	77.9	98.9	100.0	101.0	100.0	100.0	69.7	98.5
G	Site Work	110.2	104.2	106.1	109.6	104.2	105.9	107.1	93.7	98.0	92.6	112.6	106.2	105.1	90.2	95.0
A-G	WEIGHTED AVERAGE	101.5	127.9	113.0	101.4	120.1	109.6	101.6	91.6	97.2	99.4	108.1	103.2	97.4	74.4	87.3

DIV. NO.	BUILDING SYSTEMS	FLORIDA FORT LAUDERDALE			FLORIDA JACKSONVILLE			FLORIDA MELBOURNE			FLORIDA MIAMI			FLORIDA ORLANDO		
		MAT.	INST.	TOTAL	MAT.	INST.	TOTAL	MAT.	INST.	TOTAL	MAT.	INST.	TOTAL	MAT.	INST.	TOTAL
A	Substructure	91.5	75.2	81.5	92.3	69.3	78.3	103.0	77.7	87.6	94.0	75.6	82.8	105.2	77.5	88.3
B10	Shell: Superstructure	94.7	83.1	90.1	94.2	75.3	86.7	103.5	84.1	95.9	95.6	82.7	90.5	97.0	83.4	91.7
B20	Exterior Closure	92.0	70.2	81.6	93.3	63.7	79.1	95.1	72.8	84.4	92.3	73.1	83.1	95.3	68.1	82.3
B30	Roofing	96.9	84.3	92.2	99.8	63.0	85.9	99.9	78.0	91.7	98.2	77.2	90.3	98.8	75.9	90.2
C	Interior Construction	97.8	69.5	85.8	98.9	57.3	81.2	98.0	73.5	87.6	100.4	70.5	87.7	101.0	74.4	89.6
D10	Services: Conveying	100.0	88.9	97.0	100.0	85.1	95.9	100.0	86.8	96.4	100.0	89.5	97.1	100.0	85.2	96.0
D20-40	Mechanical	100.0	68.0	87.2	100.0	65.5	86.2	100.0	79.9	92.0	100.0	67.9	87.1	100.0	57.8	83.1
D50	Electrical	92.6	72.7	82.2	94.1	63.7	78.2	95.7	69.5	82.0	96.6	75.5	85.6	98.9	61.4	79.3
E	Equipment & Furnishings	100.0	68.6	98.4	100.0	53.2	97.6	100.0	70.3	98.5	100.0	68.8	98.4	100.0	75.5	98.8
G	Site Work	93.4	77.9	82.9	105.2	89.5	94.6	112.4	89.5	96.9	96.3	77.7	83.7	107.6	89.4	95.2
A-G	WEIGHTED AVERAGE	96.4	73.9	86.6	97.2	67.7	84.3	99.9	77.9	90.3	97.7	74.4	87.6	99.3	71.4	87.1

DIV. NO.	BUILDING SYSTEMS	FLORIDA PANAMA CITY			FLORIDA PENSACOLA			FLORIDA ST. PETERSBURG			FLORIDA TALLAHASSEE			FLORIDA TAMPA		
		MAT.	INST.	TOTAL	MAT.	INST.	TOTAL	MAT.	INST.	TOTAL	MAT.	INST.	TOTAL	MAT.	INST.	TOTAL
A	Substructure	97.3	62.9	76.3	110.3	68.6	84.8	98.5	69.7	80.9	97.9	61.9	76.0	97.0	81.2	87.4
B10	Shell: Superstructure	95.8	72.2	86.5	99.3	76.4	90.3	98.0	79.6	90.8	93.9	70.5	84.7	97.1	90.1	94.3
B20	Exterior Closure	100.5	49.8	76.2	107.6	58.1	83.9	115.4	54.4	86.2	95.1	56.3	76.5	94.0	78.2	86.5
B30	Roofing	100.0	58.6	84.5	99.9	63.5	86.3	96.8	59.7	82.8	105.6	74.4	93.9	97.0	86.3	93.0
C	Interior Construction	98.1	46.7	76.2	97.4	56.1	79.8	98.1	54.3	79.5	102.2	49.1	79.6	99.5	73.8	88.6
D10	Services: Conveying	100.0	48.0	85.8	100.0	46.9	85.5	100.0	58.1	88.6	100.0	70.9	92.1	100.0	85.7	96.1
D20-40	Mechanical	100.0	53.4	81.3	100.0	53.6	81.4	100.0	59.7	83.9	100.0	39.8	75.9	100.0	62.6	93.0
D50	Electrical	93.1	60.7	76.2	97.0	56.8	76.0	92.8	63.0	77.3	99.2	60.8	79.1	92.6	63.1	77.2
E	Equipment & Furnishings	100.0	43.0	97.1	100.0	52.3	97.5	100.0	51.2	97.5	100.0	42.6	97.1	100.0	77.9	98.9
G	Site Work	117.0	88.3	97.6	117.3	88.9	98.0	106.3	89.1	94.6	104.1	88.6	93.6	106.5	89.5	95.0
A-G	WEIGHTED AVERAGE	98.5	59.1	81.3	100.8	62.4	84.1	100.2	64.7	84.7	98.7	58.0	81.0	97.9	79.8	90.0

City Cost Indexes — RJ1030-010 Building Systems

DIV. NO.	BUILDING SYSTEMS	GEORGIA ALBANY			GEORGIA ATLANTA			GEORGIA AUGUSTA			GEORGIA COLUMBUS			GEORGIA MACON		
		MAT.	INST.	TOTAL	MAT.	INST.	TOTAL	MAT.	INST.	TOTAL	MAT.	INST.	TOTAL	MAT.	INST.	TOTAL
A	Substructure	92.3	61.0	73.2	102.9	79.5	88.6	99.0	69.5	81.0	92.1	64.9	75.5	92.0	70.6	78.9
B10	Shell: Superstructure	95.6	71.0	85.9	94.0	76.7	87.2	92.2	68.7	83.0	95.1	74.5	87.0	91.8	75.4	85.3
B20	Exterior Closure	94.1	51.1	73.5	93.3	68.3	81.4	90.7	47.0	69.8	94.1	59.4	77.5	97.9	48.3	74.2
B30	Roofing	96.6	63.7	84.2	93.4	73.4	85.9	93.2	60.1	80.8	96.5	65.8	85.0	94.9	65.9	84.0
C	Interior Construction	99.8	44.6	76.3	94.9	74.7	86.3	91.7	63.2	79.6	99.7	55.1	80.7	94.9	51.9	76.6
D10	Services: Conveying	100.0	85.7	96.1	100.0	86.8	96.4	100.0	81.8	95.0	100.0	85.7	96.1	100.0	84.0	95.6
D20-40	Mechanical	100.0	69.5	87.7	100.0	71.9	88.7	100.0	62.6	85.0	100.0	64.7	85.8	100.0	68.2	87.2
D50	Electrical	96.1	60.1	77.3	98.8	72.7	85.2	100.1	63.2	80.9	96.2	71.3	83.2	95.2	62.7	78.2
E	Equipment & Furnishings	100.0	39.6	96.9	100.0	76.7	98.8	100.0	71.1	98.5	100.0	53.7	97.6	100.0	52.3	97.6
G	Site Work	95.5	79.4	84.6	90.4	94.8	93.4	87.5	93.0	91.2	95.4	79.5	84.6	95.6	94.6	94.9
A-G	WEIGHTED AVERAGE	97.5	62.5	82.3	96.8	75.3	87.5	95.6	64.8	82.2	97.4	66.6	84.0	96.2	65.9	83.0

DIV. NO.	BUILDING SYSTEMS	GEORGIA SAVANNAH			GEORGIA VALDOSTA			HAWAII HONOLULU			HAWAII STATES & POSS., GUAM			IDAHO BOISE		
		MAT.	INST.	TOTAL	MAT.	INST.	TOTAL	MAT.	INST.	TOTAL	MAT.	INST.	TOTAL	MAT.	INST.	TOTAL
A	Substructure	98.9	62.8	76.9	91.7	61.2	73.1	140.5	124.0	130.5	165.3	81.8	114.4	92.5	86.5	88.9
B10	Shell: Superstructure	94.2	70.8	85.0	94.1	70.5	84.8	124.8	118.3	122.3	143.0	74.9	116.2	99.7	81.7	92.6
B20	Exterior Closure	93.1	52.8	73.9	101.4	51.6	77.6	127.1	127.9	127.5	143.6	45.9	96.9	113.1	80.0	97.2
B30	Roofing	95.6	59.3	82.0	96.7	66.2	85.2	123.5	124.5	123.9	124.9	71.4	104.8	94.8	81.9	90.0
C	Interior Construction	102.8	50.0	80.3	96.5	44.7	74.5	119.2	137.0	126.8	147.0	59.7	109.8	98.1	76.7	89.0
D10	Services: Conveying	100.0	84.4	95.7	100.0	84.4	95.7	100.0	113.6	103.7	100.0	78.6	94.2	100.0	91.4	97.7
D20-40	Mechanical	100.0	62.9	85.2	100.0	71.4	88.5	100.2	109.4	103.8	102.1	38.1	76.4	100.0	72.5	89.0
D50	Electrical	101.0	58.7	78.9	94.2	57.6	75.1	106.5	124.0	115.7	151.3	42.2	94.3	96.6	74.6	85.1
E	Equipment & Furnishings	100.0	47.2	97.3	100.0	39.3	96.9	100.0	137.6	101.9	100.0	59.0	97.9	100.0	77.5	98.9
G	Site Work	97.2	80.9	86.1	103.8	79.5	87.3	137.9	106.6	116.7	161.7	102.5	121.6	83.9	97.1	92.9
A-G	WEIGHTED AVERAGE	98.4	62.0	82.6	97.5	62.6	82.3	114.9	121.0	117.6	131.2	58.9	99.7	99.8	79.4	90.9

DIV. NO.	BUILDING SYSTEMS	IDAHO LEWISTON			IDAHO POCATELLO			IDAHO TWIN FALLS			ILLINOIS CHICAGO			ILLINOIS DECATUR		
		MAT.	INST.	TOTAL	MAT.	INST.	TOTAL	MAT.	INST.	TOTAL	MAT.	INST.	TOTAL	MAT.	INST.	TOTAL
A	Substructure	102.3	88.4	93.8	94.7	86.4	89.6	97.3	71.6	81.6	101.6	142.2	126.4	96.3	109.8	104.5
B10	Shell: Superstructure	96.5	87.2	92.8	106.3	81.5	96.5	106.7	71.1	92.7	99.7	145.5	115.5	97.8	111.4	103.2
B20	Exterior Closure	122.7	82.4	103.4	110.0	78.3	94.8	118.2	76.7	98.3	99.7	157.0	127.1	88.5	114.3	100.9
B30	Roofing	148.4	81.4	123.2	94.9	73.6	86.9	95.6	75.0	87.9	99.2	147.5	117.4	104.1	110.0	106.3
C	Interior Construction	140.2	83.3	116.0	99.0	76.8	89.5	100.5	51.8	79.7	100.1	159.3	125.3	101.3	116.7	107.8
D10	Services: Conveying	100.0	90.8	97.5	100.0	91.4	97.7	100.0	45.0	85.0	100.0	120.4	105.6	100.0	101.3	100.3
D20-40	Mechanical	100.7	87.2	95.3	100.0	72.4	88.9	100.0	70.8	88.3	99.9	136.1	114.4	100.0	98.9	99.3
D50	Electrical	86.7	80.8	83.6	94.1	70.9	82.0	90.2	61.6	75.2	96.1	137.1	117.5	98.4	91.3	94.7
E	Equipment & Furnishings	100.0	81.6	99.1	100.0	77.3	98.8	100.0	49.6	97.4	100.0	159.4	103.0	100.0	115.8	100.8
G	Site Work	88.9	93.0	91.7	83.8	97.1	92.8	90.1	98.0	95.5	104.8	95.5	98.5	93.0	98.4	96.7
A-G	WEIGHTED AVERAGE	108.3	85.5	98.4	100.8	78.4	91.1	101.9	68.5	87.4	98.9	141.9	117.6	98.2	105.9	101.6

DIV. NO.	BUILDING SYSTEMS	ILLINOIS EAST ST. LOUIS			ILLINOIS JOLIET			ILLINOIS PEORIA			ILLINOIS ROCKFORD			ILLINOIS SPRINGFIELD		
		MAT.	INST.	TOTAL	MAT.	INST.	TOTAL	MAT.	INST.	TOTAL	MAT.	INST.	TOTAL	MAT.	INST.	TOTAL
A	Substructure	93.7	112.3	105.0	101.7	142.2	126.4	97.0	112.5	106.5	97.6	122.8	113.0	95.7	109.1	103.8
B10	Shell: Superstructure	94.8	118.5	104.1	94.4	143.1	113.6	99.3	115.2	105.6	99.5	128.7	111.0	95.4	111.1	101.6
B20	Exterior Closure	82.7	118.3	99.7	101.0	150.8	124.9	105.5	119.0	111.9	94.4	133.7	113.2	92.8	117.2	104.5
B30	Roofing	98.0	112.7	103.5	99.1	145.9	116.7	101.2	114.3	106.1	103.6	128.7	113.0	105.3	113.0	108.2
C	Interior Construction	94.8	114.3	103.1	98.5	160.6	124.9	100.4	120.6	109.0	100.4	130.2	113.1	104.0	115.1	108.8
D10	Services: Conveying	100.0	105.7	101.6	100.0	120.3	105.5	100.0	102.5	100.7	100.0	109.4	102.6	100.0	101.7	100.5
D20-40	Mechanical	99.9	98.4	99.3	100.0	132.7	113.1	99.9	105.8	102.3	100.0	114.9	106.0	99.9	103.4	101.3
D50	Electrical	95.0	104.3	99.9	95.4	141.2	119.3	95.5	97.4	96.5	95.8	127.0	112.1	101.0	92.8	96.8
E	Equipment & Furnishings	100.0	112.5	100.6	100.0	164.5	103.3	100.0	117.8	100.9	100.0	128.5	101.4	100.0	115.9	100.8
G	Site Work	102.6	98.9	100.1	104.6	94.8	98.0	99.3	97.7	98.2	98.7	98.7	98.7	98.6	98.5	98.5
A-G	WEIGHTED AVERAGE	95.5	109.3	101.5	98.4	140.7	116.8	99.9	110.4	104.5	98.9	123.5	109.6	99.0	107.2	102.5

DIV. NO.	BUILDING SYSTEMS	INDIANA ANDERSON			INDIANA BLOOMINGTON			INDIANA EVANSVILLE			INDIANA FORT WAYNE			INDIANA GARY		
		MAT.	INST.	TOTAL	MAT.	INST.	TOTAL	MAT.	INST.	TOTAL	MAT.	INST.	TOTAL	MAT.	INST.	TOTAL
A	Substructure	100.7	84.1	90.6	95.9	83.7	88.4	94.8	92.4	93.3	104.3	82.8	91.2	103.6	110.0	107.5
B10	Shell: Superstructure	94.9	85.6	91.2	96.5	79.5	89.8	90.6	83.7	87.9	95.7	82.6	90.5	95.4	110.5	101.3
B20	Exterior Closure	88.5	80.6	84.7	102.8	77.0	90.0	99.7	82.9	91.6	90.0	77.4	84.0	89.1	114.4	101.2
B30	Roofing	105.6	76.8	94.8	95.2	80.7	89.8	99.3	85.3	94.1	105.4	79.1	95.5	104.2	109.0	106.0
C	Interior Construction	94.0	81.1	88.5	100.6	80.8	92.2	96.5	81.9	90.3	93.9	76.0	86.3	93.6	117.5	103.7
D10	Services: Conveying	100.0	92.8	98.0	100.0	90.1	97.3	100.0	95.9	98.9	100.0	91.4	97.7	100.0	104.0	101.1
D20-40	Mechanical	100.0	75.8	90.3	99.8	79.9	91.4	100.0	80.2	92.0	100.0	72.4	88.9	100.0	105.9	102.3
D50	Electrical	85.9	88.4	87.2	98.2	83.5	90.5	94.5	86.3	90.2	86.6	78.4	82.3	97.7	110.2	104.2
E	Equipment & Furnishings	100.0	81.3	99.0	100.0	81.7	99.1	100.0	80.4	99.0	100.0	75.1	98.7	100.0	115.4	100.8
G	Site Work	91.8	96.0	94.6	83.5	94.9	91.2	88.6	123.8	112.4	93.4	95.9	95.1	92.4	99.2	97.0
A-G	WEIGHTED AVERAGE	95.5	83.1	90.1	98.7	81.5	91.2	96.5	86.6	92.1	96.0	79.2	88.7	96.8	110.1	102.6

DIV. NO.	BUILDING SYSTEMS	INDIANA INDIANAPOLIS			INDIANA MUNCIE			INDIANA SOUTH BEND			INDIANA TERRE HAUTE			IOWA CEDAR RAPIDS		
		MAT.	INST.	TOTAL	MAT.	INST.	TOTAL	MAT.	INST.	TOTAL	MAT.	INST.	TOTAL	MAT.	INST.	TOTAL
A	Substructure	100.4	88.3	93.0	99.9	83.7	90.0	97.4	84.0	89.2	93.1	92.0	92.4	105.5	85.4	93.2
B10	Shell: Superstructure	95.2	83.2	90.5	98.4	85.4	93.3	94.1	90.8	92.8	90.8	85.1	88.6	92.8	88.4	91.1
B20	Exterior Closure	94.9	81.4	88.4	98.1	80.6	89.8	88.7	80.4	84.7	107.4	80.9	94.7	98.7	78.2	88.9
B30	Roofing	101.0	83.2	94.3	97.9	77.8	90.4	97.6	82.5	91.9	99.4	82.1	92.9	104.9	80.8	95.8
C	Interior Construction	99.4	85.4	93.4	95.4	80.9	89.2	91.5	81.8	87.4	96.8	79.6	89.5	104.8	80.8	94.6
D10	Services: Conveying	100.0	94.0	98.4	100.0	92.2	97.9	100.0	92.2	97.9	100.0	95.4	98.8	100.0	91.5	97.7
D20-40	Mechanical	99.9	80.1	92.0	99.8	75.8	90.2	99.9	77.4	90.9	100.0	80.4	92.1	100.0	80.7	92.4
D50	Electrical	100.7	88.3	94.2	90.2	80.3	85.0	97.6	86.6	91.9	92.6	83.6	87.9	98.2	81.3	89.3
E	Equipment & Furnishings	100.0	85.7	99.3	100.0	81.0	99.0	100.0	79.5	99.0	100.0	76.2	98.8	100.0	80.2	99.0
G	Site Work	91.1	99.4	96.7	83.7	94.8	91.2	93.2	95.6	94.8	90.7	124.5	113.6	97.9	96.0	96.6
A-G	WEIGHTED AVERAGE	98.3	85.1	92.6	97.3	81.8	90.6	95.7	84.2	90.7	97.2	85.8	92.2	99.4	83.4	92.4

City Cost Indexes — RJ1030-010 Building Systems

DIV. NO.	BUILDING SYSTEMS	IOWA COUNCIL BLUFFS MAT.	INST.	TOTAL	DAVENPORT MAT.	INST.	TOTAL	DES MOINES MAT.	INST.	TOTAL	DUBUQUE MAT.	INST.	TOTAL	SIOUX CITY MAT.	INST.	TOTAL
A	Substructure	108.3	78.2	90.0	102.7	94.4	97.7	94.3	86.7	89.7	101.8	87.5	93.1	106.4	72.8	85.9
B10	Shell: Superstructure	96.3	82.5	90.8	92.3	98.7	94.8	95.1	87.9	92.3	90.3	88.9	89.8	92.7	78.2	87.2
B20	Exterior Closure	100.3	77.1	89.2	97.4	86.2	92.0	95.1	81.9	88.8	98.0	75.6	87.3	95.9	57.2	77.4
B30	Roofing	104.3	66.7	90.2	104.4	88.5	98.4	100.7	79.6	92.8	104.6	71.4	92.1	104.4	67.3	90.5
C	Interior Construction	99.9	72.4	88.2	102.8	93.6	98.9	103.6	80.4	93.7	102.0	73.9	90.0	103.6	66.3	87.7
D10	Services: Conveying	100.0	91.4	97.7	100.0	93.5	98.2	100.0	92.0	97.8	100.0	90.9	97.5	100.0	89.4	97.1
D20-40	Mechanical	100.2	75.8	90.4	100.2	93.2	97.4	100.0	79.8	91.9	100.2	75.4	90.2	100.2	76.3	90.6
D50	Electrical	103.7	81.6	92.1	95.5	91.2	93.2	106.9	81.3	93.6	102.2	78.7	89.8	98.2	74.7	85.9
E	Equipment & Furnishings	100.0	69.9	98.5	100.0	90.1	99.5	100.0	77.4	98.8	100.0	71.6	98.6	100.0	67.6	98.4
G	Site Work	101.3	93.1	95.2	96.3	99.3	98.3	95.4	99.7	98.3	96.3	93.2	94.2	105.9	95.7	99.0
A-G	WEIGHTED AVERAGE	100.2	78.8	90.9	98.4	93.4	96.2	99.4	83.9	92.7	98.5	80.4	90.6	99.1	73.6	88.0

DIV. NO.	BUILDING SYSTEMS	IOWA WATERLOO MAT.	INST.	TOTAL	KANSAS DODGE CITY MAT.	INST.	TOTAL	KANSAS CITY MAT.	INST.	TOTAL	SALINA MAT.	INST.	TOTAL	TOPEKA MAT.	INST.	TOTAL
A	Substructure	101.0	66.5	80.0	110.4	66.7	83.7	92.2	96.6	94.9	99.5	65.7	78.9	97.0	62.3	75.9
B10	Shell: Superstructure	93.7	75.3	86.5	101.4	68.9	88.6	102.6	99.2	101.3	99.1	70.0	87.7	100.5	75.7	90.8
B20	Exterior Closure	95.0	71.5	83.8	111.0	59.2	86.3	103.3	97.6	100.6	111.6	58.8	86.3	99.2	59.7	80.3
B30	Roofing	105.6	71.8	92.9	101.8	63.1	87.3	97.4	99.9	98.3	101.4	63.6	87.2	100.6	68.9	88.7
C	Interior Construction	101.1	54.2	81.2	99.7	54.5	80.5	96.8	93.5	95.4	99.1	54.8	80.2	100.8	48.1	78.4
D10	Services: Conveying	100.0	91.4	97.7	100.0	40.4	83.7	100.0	59.2	88.9	100.0	54.3	87.5	100.0	49.7	86.3
D20-40	Mechanical	100.2	80.0	92.1	100.0	68.7	87.4	99.9	96.8	98.6	100.0	68.7	87.5	100.0	69.0	87.5
D50	Electrical	97.0	57.0	76.1	94.7	73.4	83.6	95.5	97.3	96.4	94.5	73.3	83.4	94.2	73.1	83.2
E	Equipment & Furnishings	100.0	38.8	96.9	100.0	59.2	97.9	100.0	96.7	99.8	100.0	53.7	97.6	100.0	38.9	96.9
G	Site Work	98.5	96.0	96.8	112.0	94.3	100.0	94.9	92.8	93.5	102.6	94.6	97.2	97.5	90.9	93.0
A-G	WEIGHTED AVERAGE	98.3	70.9	86.4	101.7	66.5	86.4	99.4	95.7	97.7	100.4	67.0	85.9	99.4	66.5	85.1

DIV. NO.	BUILDING SYSTEMS	KANSAS WICHITA MAT.	INST.	TOTAL	KENTUCKY BOWLING GREEN MAT.	INST.	TOTAL	LEXINGTON MAT.	INST.	TOTAL	LOUISVILLE MAT.	INST.	TOTAL	OWENSBORO MAT.	INST.	TOTAL
A	Substructure	96.3	65.2	77.3	86.2	91.5	89.4	93.5	88.5	90.4	94.5	85.7	89.1	89.1	97.2	94.0
B10	Shell: Superstructure	100.4	71.0	88.8	92.3	87.8	90.5	96.0	86.5	92.2	97.6	86.0	93.0	86.6	89.1	87.6
B20	Exterior Closure	98.3	53.6	76.9	100.6	80.4	90.9	94.3	74.0	84.6	93.8	79.4	86.9	106.1	91.7	99.2
B30	Roofing	100.0	57.6	84.1	86.5	82.7	85.1	99.1	88.8	95.2	98.0	79.7	91.1	99.4	93.9	97.3
C	Interior Construction	102.3	54.3	81.9	95.3	84.1	90.6	98.2	74.1	87.9	98.6	83.9	92.4	94.3	84.7	90.2
D10	Services: Conveying	100.0	50.9	86.6	100.0	56.7	88.2	100.0	90.8	97.5	100.0	87.5	96.6	100.0	103.5	101.0
D20-40	Mechanical	99.8	64.3	85.6	100.0	84.3	93.7	100.0	80.6	92.2	100.0	82.6	93.0	100.0	78.4	91.3
D50	Electrical	99.0	73.3	85.6	93.3	82.1	87.4	93.4	82.1	87.5	100.0	82.1	90.6	92.7	82.2	87.3
E	Equipment & Furnishings	100.0	52.5	97.6	100.0	83.8	99.2	100.0	70.4	98.5	100.0	83.7	99.2	100.0	80.4	99.0
G	Site Work	98.2	93.1	94.7	75.9	97.9	90.8	86.5	101.0	96.3	75.9	98.0	90.9	88.9	124.4	112.9
A-G	WEIGHTED AVERAGE	99.9	65.1	84.8	95.5	84.8	90.8	97.0	82.4	90.7	97.8	84.2	91.9	95.6	88.9	92.7

DIV. NO.	BUILDING SYSTEMS	LOUISIANA ALEXANDRIA MAT.	INST.	TOTAL	BATON ROUGE MAT.	INST.	TOTAL	LAKE CHARLES MAT.	INST.	TOTAL	MONROE MAT.	INST.	TOTAL	NEW ORLEANS MAT.	INST.	TOTAL
A	Substructure	92.2	57.5	71.1	93.3	66.3	76.8	97.7	67.2	79.1	92.0	58.9	71.8	99.9	71.8	82.8
B10	Shell: Superstructure	89.7	60.4	78.2	103.3	66.0	88.6	97.1	66.2	84.9	89.6	59.9	77.9	107.4	70.3	92.8
B20	Exterior Closure	108.6	54.1	82.5	100.1	55.4	78.7	98.2	59.0	79.4	106.4	50.0	79.4	99.7	62.1	81.7
B30	Roofing	97.5	52.9	80.7	95.7	62.4	83.2	96.4	58.0	82.0	97.5	54.8	81.4	95.9	70.0	86.2
C	Interior Construction	95.9	49.5	76.2	99.3	62.2	83.5	98.0	60.0	81.8	95.8	46.8	74.9	98.4	66.6	84.9
D10	Services: Conveying	100.0	52.5	87.1	100.0	83.0	95.4	100.0	83.3	95.4	100.0	47.5	85.7	100.0	85.2	96.0
D20-40	Mechanical	100.2	56.5	82.7	100.0	59.5	83.8	100.1	61.7	84.7	100.2	54.1	81.7	100.0	65.8	86.3
D50	Electrical	90.5	58.2	73.6	103.0	60.8	81.0	98.9	66.6	82.0	92.2	59.6	75.2	103.2	72.9	87.4
E	Equipment & Furnishings	100.0	42.7	97.1	100.0	64.1	98.2	100.0	58.5	97.9	100.0	43.3	97.1	100.0	67.4	98.3
G	Site Work	97.0	86.6	90.0	95.5	86.0	89.1	100.6	85.8	90.6	97.0	85.9	89.5	104.4	88.4	93.6
A-G	WEIGHTED AVERAGE	96.8	57.8	79.8	100.4	63.6	84.4	98.6	65.0	84.0	96.7	56.4	79.2	101.5	69.8	87.7

DIV. NO.	BUILDING SYSTEMS	LOUISIANA SHREVEPORT MAT.	INST.	TOTAL	MAINE AUGUSTA MAT.	INST.	TOTAL	BANGOR MAT.	INST.	TOTAL	LEWISTON MAT.	INST.	TOTAL	PORTLAND MAT.	INST.	TOTAL
A	Substructure	98.1	58.4	73.9	97.0	89.4	92.4	84.9	104.8	97.0	90.8	104.8	99.3	97.4	104.8	101.9
B10	Shell: Superstructure	94.3	60.0	80.8	97.8	88.7	94.2	87.8	97.0	91.4	91.6	97.0	93.7	99.1	97.0	98.3
B20	Exterior Closure	103.0	51.3	78.3	105.1	66.7	86.7	108.8	98.7	104.0	99.2	98.7	99.0	103.0	98.7	100.9
B30	Roofing	97.2	51.3	79.9	104.3	66.6	90.1	100.2	83.4	93.9	100.1	83.4	93.8	101.5	83.4	94.7
C	Interior Construction	97.7	47.5	76.4	103.3	90.3	97.8	100.9	96.0	98.8	104.0	96.0	100.6	104.7	96.0	101.0
D10	Services: Conveying	100.0	81.7	95.0	100.0	103.0	100.8	100.0	111.7	103.2	100.0	111.7	103.2	100.0	111.7	103.2
D20-40	Mechanical	100.0	58.9	83.5	99.9	65.6	86.2	100.1	75.9	90.4	100.1	75.9	90.4	100.0	75.9	90.3
D50	Electrical	97.7	67.2	81.8	99.3	80.0	89.2	98.5	77.2	87.4	100.6	80.1	89.9	103.3	80.1	91.2
E	Equipment & Furnishings	100.0	44.1	97.2	100.0	104.5	100.2	100.0	96.7	99.8	100.0	96.9	99.8	100.0	96.9	99.8
G	Site Work	100.7	85.9	90.7	88.9	100.9	97.0	90.3	101.1	97.6	86.8	101.1	96.5	86.5	101.1	96.4
A-G	WEIGHTED AVERAGE	98.4	59.6	81.5	100.3	80.8	91.8	97.6	90.8	94.7	98.3	91.2	95.2	100.8	91.2	96.6

DIV. NO.	BUILDING SYSTEMS	MARYLAND BALTIMORE MAT.	INST.	TOTAL	HAGERSTOWN MAT.	INST.	TOTAL	MASSACHUSETTS BOSTON MAT.	INST.	TOTAL	BROCKTON MAT.	INST.	TOTAL	FALL RIVER MAT.	INST.	TOTAL
A	Substructure	107.1	80.9	91.1	88.5	82.6	84.9	105.0	138.8	125.6	100.7	134.2	121.1	98.7	135.5	121.1
B10	Shell: Superstructure	102.9	85.9	96.2	97.0	85.5	92.5	101.6	138.9	116.3	99.8	135.0	113.6	99.4	137.6	114.4
B20	Exterior Closure	96.7	76.4	87.0	92.7	84.2	88.6	109.5	158.1	132.7	106.0	155.6	129.7	106.4	154.4	129.4
B30	Roofing	99.1	79.8	91.9	99.3	77.8	91.2	104.0	148.5	120.7	101.7	144.3	117.7	101.6	141.5	116.6
C	Interior Construction	98.6	74.1	88.2	95.5	77.7	87.9	100.0	150.1	121.3	98.2	140.6	116.3	98.2	141.2	116.5
D10	Services: Conveying	100.0	87.9	96.7	100.0	91.5	97.7	100.0	115.6	104.2	100.0	114.8	104.0	100.0	115.5	104.2
D20-40	Mechanical	100.0	81.7	92.7	100.0	81.4	92.5	100.1	127.0	110.9	100.1	111.2	104.5	100.1	111.3	104.6
D50	Electrical	101.4	91.7	96.3	96.4	81.6	88.7	100.9	138.6	120.6	100.7	100.7	100.7	100.6	100.7	100.6
E	Equipment & Furnishings	100.0	74.4	98.7	100.0	74.5	98.7	100.0	141.6	102.1	100.0	127.9	101.4	100.0	127.8	101.4
G	Site Work	101.0	94.4	96.5	90.8	91.2	91.0	100.3	108.3	105.7	95.2	104.9	101.8	94.1	105.1	101.5
A-G	WEIGHTED AVERAGE	100.5	82.8	92.8	96.9	82.8	90.7	101.8	138.1	117.6	100.4	126.3	111.7	100.2	126.8	111.8

City Cost Indexes — RJ1030-010 Building Systems

MASSACHUSETTS (Hyannis, Lawrence, Lowell, New Bedford, Pittsfield)

DIV. NO.	BUILDING SYSTEMS	HYANNIS MAT.	INST.	TOTAL	LAWRENCE MAT.	INST.	TOTAL	LOWELL MAT.	INST.	TOTAL	NEW BEDFORD MAT.	INST.	TOTAL	PITTSFIELD MAT.	INST.	TOTAL
A	Substructure	89.3	135.3	117.3	100.5	132.2	119.8	95.2	132.0	117.6	91.8	135.5	118.4	96.3	116.0	108.3
B10	Shell: Superstructure	94.5	137.2	111.3	98.8	131.8	111.8	97.6	130.2	110.4	98.0	137.6	113.5	97.7	113.4	103.9
B20	Exterior Closure	102.9	154.4	127.6	106.6	155.1	129.8	97.2	151.2	123.1	105.5	154.4	128.9	97.5	128.4	112.3
B30	Roofing	101.2	142.4	116.7	101.4	143.8	117.3	101.2	142.5	116.7	101.6	141.5	116.6	101.2	117.2	107.2
C	Interior Construction	94.5	141.0	114.3	98.6	140.0	116.2	103.4	140.0	119.0	98.1	141.2	116.5	103.4	117.5	109.4
D10	Services: Conveying	100.0	114.8	104.0	100.0	114.8	104.0	100.0	114.8	104.0	100.0	115.5	104.2	100.0	103.9	101.1
D20 - 40	Mechanical	100.1	111.2	104.5	100.0	120.4	108.2	100.0	118.1	107.3	100.1	111.3	104.6	100.0	100.0	100.0
D50	Electrical	98.1	100.7	99.4	99.4	129.2	115.0	99.9	129.2	115.2	101.5	100.7	101.1	99.9	92.8	96.2
E	Equipment & Furnishings	100.0	127.9	101.4	100.0	129.2	101.5	100.0	129.2	101.5	100.0	127.8	101.4	100.0	108.7	100.4
G	Site Work	92.7	104.9	101.0	94.3	105.0	101.5	93.1	104.9	101.1	92.6	105.1	101.1	94.2	103.3	100.4
A-G	WEIGHTED AVERAGE	97.6	126.7	110.3	100.1	131.0	113.7	99.4	130.1	112.8	99.6	126.8	111.4	99.6	109.6	103.9

MASSACHUSETTS (Springfield, Worcester) / MICHIGAN (Ann Arbor, Dearborn, Detroit)

DIV. NO.	BUILDING SYSTEMS	SPRINGFIELD MAT.	INST.	TOTAL	WORCESTER MAT.	INST.	TOTAL	ANN ARBOR MAT.	INST.	TOTAL	DEARBORN MAT.	INST.	TOTAL	DETROIT MAT.	INST.	TOTAL
A	Substructure	97.4	116.1	108.8	97.1	130.8	117.7	88.1	109.4	101.1	86.9	112.0	102.2	93.7	112.0	104.8
B10	Shell: Superstructure	100.0	113.5	105.3	100.0	130.8	112.1	92.9	115.7	101.8	92.7	117.5	102.4	95.8	108.0	100.6
B20	Exterior Closure	97.3	128.8	112.4	97.1	151.4	123.1	98.8	108.5	103.5	98.8	113.1	105.6	98.2	111.8	104.7
B30	Roofing	101.2	117.4	107.3	101.2	136.0	114.2	100.1	108.5	103.3	98.6	113.9	104.4	97.4	113.9	103.6
C	Interior Construction	103.4	117.6	109.5	103.4	136.4	117.4	95.6	111.8	102.5	95.6	114.9	103.8	98.2	114.9	105.3
D10	Services: Conveying	100.0	104.1	101.1	100.0	106.8	101.9	100.0	105.3	101.5	100.0	106.7	101.8	100.0	106.6	101.8
D20 - 40	Mechanical	100.0	100.2	100.1	100.0	113.1	105.3	100.0	99.9	100.0	100.0	108.8	103.5	100.0	109.0	103.6
D50	Electrical	99.9	92.8	96.2	99.9	104.2	102.1	96.0	110.0	103.4	96.0	105.2	100.8	97.2	105.1	101.3
E	Equipment & Furnishings	100.0	108.7	100.4	100.0	121.8	101.1	100.0	111.5	100.6	100.0	115.7	100.8	100.0	115.7	100.8
G	Site Work	93.6	103.3	100.2	93.5	104.9	101.2	79.4	96.8	91.2	79.2	97.0	91.3	91.4	99.1	96.7
A-G	WEIGHTED AVERAGE	100.0	109.8	104.3	100.0	124.6	110.7	96.4	107.9	101.4	96.2	110.9	102.6	97.9	109.3	102.8

MICHIGAN (Flint, Grand Rapids, Kalamazoo, Lansing, Muskegon)

DIV. NO.	BUILDING SYSTEMS	FLINT MAT.	INST.	TOTAL	GRAND RAPIDS MAT.	INST.	TOTAL	KALAMAZOO MAT.	INST.	TOTAL	LANSING MAT.	INST.	TOTAL	MUSKEGON MAT.	INST.	TOTAL
A	Substructure	87.0	98.0	93.7	98.4	90.4	93.5	95.2	91.5	92.9	99.0	96.7	97.6	94.0	90.3	91.7
B10	Shell: Superstructure	93.1	107.2	98.7	94.2	89.0	92.2	95.8	89.4	93.3	93.2	106.0	98.2	93.8	89.3	92.0
B20	Exterior Closure	98.9	97.7	98.3	97.6	86.5	92.3	99.9	85.0	92.8	98.0	94.7	96.4	94.8	89.0	92.0
B30	Roofing	97.9	97.7	97.8	99.4	75.9	90.6	96.6	84.9	92.2	98.7	90.0	95.4	95.7	78.2	89.1
C	Interior Construction	95.3	91.6	93.7	101.6	83.7	94.0	96.2	84.4	91.2	98.2	88.3	94.0	94.0	84.2	89.8
D10	Services: Conveying	100.0	94.5	98.5	100.0	100.8	100.2	100.0	100.3	100.1	100.0	94.1	98.4	100.0	100.2	100.1
D20 - 40	Mechanical	100.0	90.6	96.2	100.0	83.8	93.5	100.0	81.6	92.6	99.9	88.9	95.5	99.9	83.1	93.1
D50	Electrical	96.0	98.3	97.2	95.2	72.3	83.3	90.2	83.7	86.8	99.2	91.4	95.1	90.7	77.2	83.6
E	Equipment & Furnishings	100.0	89.6	99.5	100.0	79.0	98.9	100.0	82.5	99.1	100.0	87.4	99.4	100.0	81.1	99.0
G	Site Work	70.2	96.1	87.8	89.7	88.1	88.6	90.0	88.1	88.7	90.6	96.2	94.4	87.4	88.0	87.8
A-G	WEIGHTED AVERAGE	96.1	96.8	96.4	98.1	84.5	92.2	97.1	85.9	92.2	97.7	94.1	96.1	95.6	85.5	91.2

MICHIGAN (Saginaw) / MINNESOTA (Duluth, Minneapolis, Rochester, Saint Paul)

DIV. NO.	BUILDING SYSTEMS	SAGINAW MAT.	INST.	TOTAL	DULUTH MAT.	INST.	TOTAL	MINNEAPOLIS MAT.	INST.	TOTAL	ROCHESTER MAT.	INST.	TOTAL	SAINT PAUL MAT.	INST.	TOTAL
A	Substructure	86.2	96.3	92.4	104.8	107.6	106.5	105.5	118.3	113.3	104.1	105.1	104.7	111.8	116.3	114.5
B10	Shell: Superstructure	92.8	105.7	97.9	95.8	117.3	104.2	100.2	126.7	110.6	97.3	118.4	105.6	97.8	125.8	108.8
B20	Exterior Closure	99.0	89.8	94.6	99.5	116.5	107.6	104.6	125.5	114.6	99.5	114.6	106.7	107.7	125.4	116.2
B30	Roofing	98.7	95.0	97.4	100.2	116.8	106.4	102.1	130.7	112.9	103.6	100.5	102.4	101.8	130.2	112.4
C	Interior Construction	94.3	88.1	91.6	99.7	116.1	106.7	97.8	129.0	111.1	96.1	111.8	102.8	98.4	126.9	110.5
D10	Services: Conveying	100.0	93.2	98.2	100.0	98.5	99.6	100.0	106.9	101.9	100.0	102.3	100.6	100.0	106.6	101.8
D20 - 40	Mechanical	100.0	84.2	93.7	99.9	97.2	98.8	99.9	116.9	106.7	100.0	96.0	98.4	99.9	114.9	105.9
D50	Electrical	93.5	91.1	92.2	98.4	104.9	101.8	101.8	109.0	105.6	98.2	89.2	93.5	95.2	111.4	103.7
E	Equipment & Furnishings	100.0	87.9	99.4	100.0	112.2	100.6	100.0	126.1	101.3	100.0	105.1	100.3	100.0	123.2	101.2
G	Site Work	72.8	96.0	88.5	96.1	100.3	98.9	98.2	105.6	103.2	97.4	99.9	99.1	99.5	101.0	100.5
A-G	WEIGHTED AVERAGE	95.7	92.4	94.2	99.0	108.6	103.2	100.6	119.9	109.0	98.8	104.9	101.5	100.2	118.8	108.3

MINNESOTA (St. Cloud) / MISSISSIPPI (Biloxi, Greenville, Jackson, Meridian)

DIV. NO.	BUILDING SYSTEMS	ST. CLOUD MAT.	INST.	TOTAL	BILOXI MAT.	INST.	TOTAL	GREENVILLE MAT.	INST.	TOTAL	JACKSON MAT.	INST.	TOTAL	MERIDIAN MAT.	INST.	TOTAL
A	Substructure	96.2	115.3	107.8	103.8	63.5	79.2	102.3	68.4	81.6	99.0	65.9	78.8	99.8	68.7	80.8
B10	Shell: Superstructure	92.6	123.6	104.8	94.4	70.2	84.9	93.1	74.5	85.8	98.5	71.6	87.9	91.6	73.5	84.4
B20	Exterior Closure	95.1	122.4	108.2	92.6	49.3	71.9	116.2	52.8	85.9	95.4	52.3	74.8	90.6	57.7	74.9
B30	Roofing	104.9	121.5	111.1	97.1	56.0	81.6	96.2	62.4	83.5	97.2	61.6	83.8	96.8	64.3	84.6
C	Interior Construction	90.1	123.5	104.3	98.5	54.5	79.8	96.6	65.1	83.2	100.7	59.8	83.3	95.8	65.1	82.8
D10	Services: Conveying	100.0	105.7	101.6	100.0	57.0	88.3	100.0	58.9	88.8	100.0	58.9	88.8	100.0	60.5	89.2
D20 - 40	Mechanical	99.7	116.3	106.3	100.0	50.1	80.0	100.0	48.6	79.4	100.0	53.8	83.3	100.0	60.4	84.1
D50	Electrical	101.9	111.4	106.9	98.5	57.6	77.1	95.6	64.2	79.2	100.1	64.2	81.3	97.3	58.2	76.9
E	Equipment & Furnishings	100.0	118.5	100.9	100.0	56.6	97.8	100.0	68.6	98.4	100.0	60.5	98.0	100.0	65.7	98.3
G	Site Work	91.2	105.1	100.6	98.9	89.2	92.4	101.4	89.3	93.2	95.4	89.3	91.3	95.0	89.6	91.3
A-G	WEIGHTED AVERAGE	96.3	117.8	105.7	97.7	59.1	80.9	99.4	63.1	83.6	99.1	63.5	83.6	96.2	65.3	82.7

MISSOURI (Cape Girardeau, Columbia, Joplin, Kansas City, Springfield)

DIV. NO.	BUILDING SYSTEMS	CAPE GIRARDEAU MAT.	INST.	TOTAL	COLUMBIA MAT.	INST.	TOTAL	JOPLIN MAT.	INST.	TOTAL	KANSAS CITY MAT.	INST.	TOTAL	SPRINGFIELD MAT.	INST.	TOTAL
A	Substructure	93.5	85.3	88.5	91.7	91.0	91.3	104.3	85.2	92.7	97.5	106.8	103.2	96.7	86.1	90.2
B10	Shell: Superstructure	96.1	95.0	95.7	95.5	104.8	99.2	100.9	91.8	97.3	105.8	112.3	108.4	100.3	95.8	98.5
B20	Exterior Closure	107.1	80.1	94.2	107.8	92.3	100.4	96.4	86.3	91.6	99.9	109.4	104.4	95.6	86.8	91.4
B30	Roofing	103.1	84.3	96.0	95.4	87.6	92.5	97.0	77.6	89.7	97.1	108.4	101.4	99.9	79.2	92.1
C	Interior Construction	98.2	75.6	88.6	95.8	82.7	90.2	101.8	75.6	90.7	102.9	108.9	105.5	101.4	77.0	91.1
D10	Services: Conveying	100.0	94.9	98.6	100.0	98.1	99.5	100.0	92.0	97.8	100.0	100.5	100.1	100.0	95.8	98.8
D20 - 40	Mechanical	99.9	96.6	98.6	99.9	97.8	99.0	100.1	72.7	89.1	100.0	106.7	102.7	99.9	73.0	89.1
D50	Electrical	98.0	103.9	101.1	98.4	86.6	92.2	91.5	69.7	80.1	102.0	100.1	101.0	102.9	70.3	85.9
E	Equipment & Furnishings	100.0	72.9	98.6	100.0	76.6	98.8	100.0	71.5	98.5	100.0	107.1	100.4	100.0	72.3	98.6
G	Site Work	92.4	94.6	93.9	100.8	96.2	97.7	105.8	99.2	101.3	99.3	97.9	98.3	97.5	94.2	95.3
A-G	WEIGHTED AVERAGE	99.1	90.6	95.4	98.6	93.5	96.4	99.5	81.0	91.5	101.6	106.7	103.8	99.9	82.1	92.1

City Cost Indexes — RJ1030-010 Building Systems

DIV. NO.	BUILDING SYSTEMS	MISSOURI — ST. JOSEPH			MISSOURI — ST. LOUIS			MONTANA — BILLINGS			MONTANA — BUTTE			MONTANA — GREAT FALLS		
		MAT.	INST.	TOTAL	MAT.	INST.	TOTAL	MAT.	INST.	TOTAL	MAT.	INST.	TOTAL	MAT.	INST.	TOTAL
A	Substructure	97.3	97.7	97.6	93.1	104.4	100.0	110.5	78.5	91.0	119.5	79.4	95.0	124.0	77.3	95.5
B10	Shell: Superstructure	102.6	104.7	103.4	100.3	113.5	105.5	108.3	82.5	98.1	103.9	83.2	95.8	107.9	82.0	97.7
B20	Exterior Closure	99.6	98.3	99.0	98.2	114.0	105.8	105.4	73.8	90.3	100.4	76.7	89.1	105.2	75.0	90.8
B30	Roofing	97.6	96.9	97.4	103.2	107.0	104.6	108.3	70.7	94.2	108.2	71.4	94.4	108.7	68.6	93.6
C	Interior Construction	102.3	93.2	98.5	100.6	102.9	101.6	97.9	73.4	87.4	97.4	72.3	86.7	100.6	73.9	89.3
D10	Services: Conveying	100.0	97.9	99.4	100.0	102.4	100.7	100.0	99.5	99.9	100.0	99.5	99.9	100.0	99.5	99.9
D20-40	Mechanical	100.1	89.5	95.8	100.0	107.6	103.0	100.1	74.9	90.0	100.2	70.7	88.4	100.2	72.2	89.0
D50	Electrical	100.5	80.5	90.1	100.5	107.1	103.9	96.8	73.9	84.8	104.4	72.2	87.6	96.9	72.2	84.0
E	Equipment & Furnishings	100.0	88.8	99.4	100.0	101.4	100.1	100.0	69.9	98.5	100.0	72.2	98.6	100.0	69.8	98.5
G	Site Work	100.9	93.6	96.0	92.1	98.1	96.2	96.3	97.0	96.8	102.9	96.8	98.8	106.4	96.8	99.9
A-G	WEIGHTED AVERAGE	100.7	93.9	97.8	99.6	107.6	103.1	102.2	77.9	91.6	102.0	77.2	91.2	103.4	77.1	92.0

DIV. NO.	BUILDING SYSTEMS	MONTANA — HELENA			MONTANA — MISSOULA			NEBRASKA — GRAND ISLAND			NEBRASKA — LINCOLN			NEBRASKA — NORTH PLATTE		
		MAT.	INST.	TOTAL	MAT.	INST.	TOTAL	MAT.	INST.	TOTAL	MAT.	INST.	TOTAL	MAT.	INST.	TOTAL
A	Substructure	106.4	75.3	87.4	97.1	77.6	85.2	112.7	77.5	91.2	98.7	76.3	85.0	110.5	76.3	89.7
B10	Shell: Superstructure	103.5	80.0	94.2	97.1	82.3	91.3	97.4	80.5	90.8	96.3	79.0	89.5	98.2	79.6	90.9
B20	Exterior Closure	104.0	75.6	90.4	103.6	75.9	90.4	103.1	77.2	90.7	97.3	68.1	83.4	96.6	87.2	92.1
B30	Roofing	105.6	71.4	92.7	107.5	80.7	97.5	104.6	73.2	92.8	100.4	70.7	89.3	100.8	81.2	93.5
C	Interior Construction	100.2	64.5	85.0	97.1	71.8	86.3	96.5	77.6	88.5	102.7	71.7	89.5	96.2	76.0	87.6
D10	Services: Conveying	100.0	99.7	99.9	100.0	99.5	99.9	100.0	91.5	97.7	100.0	91.5	97.7	100.0	68.0	91.3
D20-40	Mechanical	100.1	70.9	88.4	100.2	69.9	88.0	100.0	78.5	91.4	99.9	78.5	91.3	99.9	77.1	90.8
D50	Electrical	103.8	72.7	87.6	102.2	69.3	85.0	90.7	65.9	77.8	103.2	65.9	83.7	92.0	78.9	85.2
E	Equipment & Furnishings	100.0	64.1	98.2	100.0	69.6	98.5	100.0	73.0	98.6	100.0	66.7	98.3	100.0	73.5	98.7
G	Site Work	99.4	96.5	97.5	85.8	96.5	93.1	105.0	92.6	96.6	93.9	92.6	93.0	104.6	91.7	95.9
A-G	WEIGHTED AVERAGE	102.0	75.1	90.2	99.4	76.4	89.4	99.1	77.8	89.9	99.5	75.3	89.0	98.5	79.6	90.2

DIV. NO.	BUILDING SYSTEMS	NEBRASKA — OMAHA			NEVADA — CARSON CITY			NEVADA — LAS VEGAS			NEVADA — RENO			NEW HAMPSHIRE — MANCHESTER		
		MAT.	INST.	TOTAL	MAT.	INST.	TOTAL	MAT.	INST.	TOTAL	MAT.	INST.	TOTAL	MAT.	INST.	TOTAL
A	Substructure	95.0	79.1	85.3	102.5	96.6	98.9	98.1	109.2	104.9	103.5	97.0	99.5	107.8	102.0	104.3
B10	Shell: Superstructure	95.4	77.3	88.3	92.9	100.3	95.8	100.0	110.3	104.0	96.6	101.2	98.4	104.7	97.4	101.8
B20	Exterior Closure	97.7	79.4	88.9	107.5	93.8	101.0	104.3	105.4	104.8	106.8	91.9	99.6	100.6	99.1	99.8
B30	Roofing	99.2	79.1	91.6	102.7	90.8	98.2	112.6	102.9	108.9	99.0	90.8	95.9	107.0	93.3	101.9
C	Interior Construction	110.0	73.4	94.4	100.0	97.5	99.0	98.7	112.6	104.6	98.0	97.7	97.9	103.2	97.0	100.6
D10	Services: Conveying	100.0	90.1	97.3	100.0	91.5	97.7	100.0	105.6	101.5	100.0	91.5	97.7	100.0	106.5	101.8
D20-40	Mechanical	99.8	78.0	91.1	100.0	81.4	92.6	100.2	105.1	102.1	100.0	81.5	92.6	99.9	86.2	94.4
D50	Electrical	97.9	82.4	89.8	100.4	97.4	98.8	102.7	120.6	112.0	98.8	97.4	98.0	101.6	80.4	90.6
E	Equipment & Furnishings	100.0	72.5	98.6	100.0	96.5	99.8	100.0	111.4	100.6	100.0	96.5	99.8	100.0	91.8	99.6
G	Site Work	85.5	91.8	89.8	78.5	99.6	92.8	70.7	101.7	91.7	68.2	99.6	89.5	93.8	101.6	99.1
A-G	WEIGHTED AVERAGE	99.6	79.3	90.7	99.1	93.8	96.8	100.2	109.4	104.2	99.0	93.8	96.7	102.1	93.6	98.4

DIV. NO.	BUILDING SYSTEMS	NEW HAMPSHIRE — NASHUA			NEW HAMPSHIRE — PORTSMOUTH			NEW JERSEY — CAMDEN			NEW JERSEY — ELIZABETH			NEW JERSEY — JERSEY CITY		
		MAT.	INST.	TOTAL	MAT.	INST.	TOTAL	MAT.	INST.	TOTAL	MAT.	INST.	TOTAL	MAT.	INST.	TOTAL
A	Substructure	96.9	102.0	100.0	87.5	102.2	96.4	90.6	122.7	110.2	91.1	122.8	110.4	89.3	122.7	109.7
B10	Shell: Superstructure	100.0	97.4	99.0	95.1	97.8	96.1	98.5	116.7	105.7	96.1	120.1	105.6	98.3	119.1	106.5
B20	Exterior Closure	99.1	99.1	99.1	95.4	98.0	96.6	98.1	124.3	110.6	103.7	125.2	114.0	94.1	125.0	108.9
B30	Roofing	103.7	93.3	99.8	103.3	117.2	108.5	104.6	121.7	111.0	101.8	130.8	112.7	101.2	130.8	112.3
C	Interior Construction	103.7	97.0	100.8	101.6	97.0	99.6	102.7	131.2	114.8	105.1	134.2	117.5	102.4	134.2	116.0
D10	Services: Conveying	100.0	106.5	101.8	100.0	106.5	101.8	100.0	107.9	102.2	100.0	111.4	103.1	100.0	111.4	103.1
D20-40	Mechanical	100.1	86.2	94.5	100.0	86.2	94.5	100.0	119.7	107.9	100.0	122.8	109.1	100.0	123.4	109.3
D50	Electrical	102.2	80.4	90.8	100.1	80.4	89.8	97.9	141.4	120.6	91.5	137.6	115.6	95.5	139.7	118.6
E	Equipment & Furnishings	100.0	91.8	99.6	100.0	91.8	99.6	100.0	130.2	101.5	100.0	130.2	101.5	100.0	130.3	101.5
G	Site Work	95.7	101.6	99.7	90.2	101.7	98.0	97.4	104.9	102.5	111.7	105.6	107.6	96.4	105.6	102.7
A-G	WEIGHTED AVERAGE	100.6	93.6	97.6	98.2	94.2	96.4	99.4	123.6	109.9	99.5	125.3	110.7	98.5	125.5	110.2

DIV. NO.	BUILDING SYSTEMS	NEW JERSEY — NEWARK			NEW JERSEY — PATERSON			NEW JERSEY — TRENTON			NEW MEXICO — ALBUQUERQUE			NEW MEXICO — FARMINGTON		
		MAT.	INST.	TOTAL	MAT.	INST.	TOTAL	MAT.	INST.	TOTAL	MAT.	INST.	TOTAL	MAT.	INST.	TOTAL
A	Substructure	108.3	122.8	117.1	103.4	122.7	115.2	100.4	122.1	113.6	96.2	76.9	84.4	98.7	76.9	85.4
B10	Shell: Superstructure	103.5	120.4	110.2	97.0	120.3	106.2	100.3	115.8	106.4	102.6	78.8	93.3	101.3	78.8	92.4
B20	Exterior Closure	99.5	125.2	111.8	97.1	125.2	110.6	99.1	124.0	111.0	103.6	62.7	84.0	109.2	62.7	86.9
B30	Roofing	102.0	130.8	112.8	101.7	122.4	109.5	104.3	122.1	111.0	99.3	73.2	89.5	99.5	73.2	89.6
C	Interior Construction	103.9	134.2	116.8	107.2	134.2	118.7	104.2	134.0	116.9	99.5	66.4	85.4	100.1	66.4	85.8
D10	Services: Conveying	100.0	111.4	103.1	100.0	111.4	103.1	100.0	107.8	102.1	100.0	77.7	93.9	100.0	77.7	93.9
D20-40	Mechanical	100.0	122.8	109.1	100.0	123.4	109.4	100.0	122.5	109.0	100.2	70.7	88.4	100.1	70.7	88.3
D50	Electrical	99.2	139.7	120.3	95.5	137.6	117.5	101.4	138.7	120.8	88.5	72.9	80.4	86.8	72.9	79.6
E	Equipment & Furnishings	100.0	130.3	101.5	100.0	130.2	101.5	100.0	130.2	101.5	100.0	65.3	98.2	100.0	65.3	98.2
G	Site Work	113.2	105.6	108.1	109.6	105.6	106.9	96.0	105.5	102.5	83.0	103.9	97.2	90.3	103.9	99.5
A-G	WEIGHTED AVERAGE	101.9	125.6	112.2	100.0	125.2	111.1	100.8	124.0	110.9	99.2	73.8	88.0	99.7	73.6	88.3

DIV. NO.	BUILDING SYSTEMS	NEW MEXICO — LAS CRUCES			NEW MEXICO — ROSWELL			NEW MEXICO — SANTA FE			NEW YORK — ALBANY			NEW YORK — BINGHAMTON		
		MAT.	INST.	TOTAL	MAT.	INST.	TOTAL	MAT.	INST.	TOTAL	MAT.	INST.	TOTAL	MAT.	INST.	TOTAL
A	Substructure	93.8	69.5	79.0	97.8	76.9	85.1	101.8	76.9	86.6	91.7	101.5	97.7	98.8	100.6	99.9
B10	Shell: Superstructure	96.3	73.9	87.5	99.2	78.8	91.2	98.8	78.8	90.9	99.7	104.6	101.7	96.9	109.1	101.7
B20	Exterior Closure	87.6	62.0	75.4	114.3	62.7	89.6	103.3	62.7	83.8	100.0	99.7	99.9	96.9	122.3	109.0
B30	Roofing	86.2	67.9	79.3	99.6	73.2	89.7	101.6	73.2	91.0	99.4	96.8	98.4	107.2	88.0	100.0
C	Interior Construction	97.7	65.6	84.0	97.9	66.4	84.5	102.1	66.4	86.9	93.9	96.0	98.3	93.9	83.9	89.6
D10	Services: Conveying	100.0	74.7	93.1	100.0	77.7	93.9	100.0	77.7	93.9	100.0	96.3	99.0	100.0	95.3	98.7
D20-40	Mechanical	100.4	70.5	88.4	100.0	70.7	88.3	100.1	70.7	88.3	100.0	98.8	99.5	100.4	87.3	95.1
D50	Electrical	89.9	72.9	81.0	89.6	72.9	80.9	100.8	72.9	86.3	96.0	99.3	97.7	98.9	107.9	103.6
E	Equipment & Furnishings	100.0	64.7	98.2	100.0	65.3	98.2	100.0	65.3	98.2	100.0	93.8	99.7	100.0	80.3	99.0
G	Site Work	92.3	83.7	86.5	95.0	103.9	101.0	94.3	103.9	100.8	74.7	107.0	96.6	92.6	93.5	93.2
A-G	WEIGHTED AVERAGE	95.8	70.4	84.7	99.8	73.6	88.4	100.5	73.6	88.8	98.6	100.2	99.3	98.1	99.4	98.6

City Cost Indexes — RJ1030-010 Building Systems

DIV. NO.	BUILDING SYSTEMS	NEW YORK BUFFALO			HICKSVILLE			NEW YORK			RIVERHEAD			ROCHESTER		
		MAT.	INST.	TOTAL	MAT.	INST.	TOTAL	MAT.	INST.	TOTAL	MAT.	INST.	TOTAL	MAT.	INST.	TOTAL
A	Substructure	103.2	110.8	107.8	95.0	151.1	129.2	107.4	166.6	143.5	96.6	151.1	129.8	97.0	98.1	97.7
B10	Shell: Superstructure	100.3	104.3	101.8	103.2	146.4	120.2	112.5	163.5	132.6	104.1	146.4	120.8	101.1	99.7	100.6
B20	Exterior Closure	103.9	115.4	109.4	108.1	163.6	134.6	103.0	175.7	137.8	109.8	163.6	135.5	107.1	97.8	102.6
B30	Roofing	101.5	109.6	104.5	108.2	151.1	124.3	108.6	163.7	129.3	109.0	151.1	124.8	100.4	98.1	99.5
C	Interior Construction	95.9	115.2	104.1	94.2	144.2	115.5	96.7	179.2	131.8	94.5	144.2	115.6	100.0	93.6	97.3
D10	Services: Conveying	100.0	104.9	101.3	100.0	122.8	106.2	100.0	131.3	108.5	100.0	122.8	106.2	100.0	97.9	99.4
D20 - 40	Mechanical	100.0	97.1	98.8	99.8	151.8	120.6	100.1	165.7	126.4	100.0	151.8	120.7	100.0	87.9	95.1
D50	Electrical	97.5	101.9	99.8	98.0	135.8	117.7	104.2	175.0	141.1	99.7	135.8	118.5	99.2	94.3	96.7
E	Equipment & Furnishings	100.0	114.6	100.7	100.0	142.3	102.2	100.0	181.2	104.1	100.0	142.3	102.2	100.0	90.7	99.5
G	Site Work	99.2	98.3	98.6	111.7	127.9	122.7	118.8	124.2	122.4	113.4	127.9	123.2	78.1	109.6	99.4
A-G	WEIGHTED AVERAGE	99.7	105.8	102.4	100.7	146.6	120.7	103.8	166.4	131.1	101.4	146.6	121.1	100.3	95.6	98.3

DIV. NO.	BUILDING SYSTEMS	NEW YORK SCHENECTADY			SYRACUSE			UTICA			WATERTOWN			WHITE PLAINS		
		MAT.	INST.	TOTAL	MAT.	INST.	TOTAL	MAT.	INST.	TOTAL	MAT.	INST.	TOTAL	MAT.	INST.	TOTAL
A	Substructure	96.3	101.5	99.4	94.5	99.3	97.4	88.0	93.3	91.2	96.0	94.1	94.9	96.1	133.4	118.8
B10	Shell: Superstructure	98.9	104.6	101.2	98.7	100.7	99.5	96.2	96.7	96.4	97.1	98.0	97.4	99.7	135.8	113.9
B20	Exterior Closure	98.3	99.7	99.0	101.2	102.3	101.8	98.3	96.3	97.3	105.7	98.6	102.3	96.2	130.4	112.5
B30	Roofing	94.2	96.8	95.2	102.5	96.7	100.3	90.9	92.9	91.7	91.2	93.0	91.9	109.3	134.6	118.8
C	Interior Construction	96.8	96.0	96.5	94.0	90.8	92.7	95.6	84.2	90.8	94.5	88.0	91.7	90.3	139.5	111.3
D10	Services: Conveying	100.0	96.3	99.0	100.0	97.4	99.3	100.0	93.6	98.2	100.0	94.8	98.6	100.0	115.9	104.3
D20 - 40	Mechanical	100.2	98.8	99.6	100.2	91.2	96.6	100.2	88.8	95.6	100.2	81.5	92.7	100.4	118.6	107.7
D50	Electrical	95.5	99.3	97.5	99.1	97.7	98.4	97.3	97.7	97.5	99.1	90.5	94.6	94.4	158.9	128.1
E	Equipment & Furnishings	100.0	93.8	99.7	100.0	88.1	99.4	100.0	79.7	99.0	100.0	86.1	99.3	100.0	132.6	101.7
G	Site Work	74.7	107.0	96.6	91.5	106.5	101.6	70.1	104.4	93.3	77.3	106.0	96.7	112.1	119.1	116.9
A-G	WEIGHTED AVERAGE	97.8	100.2	98.8	98.5	97.0	97.9	96.7	93.2	95.1	97.8	91.9	95.4	98.0	133.3	113.4

DIV. NO.	BUILDING SYSTEMS	NEW YORK YONKERS			NORTH CAROLINA ASHEVILLE			CHARLOTTE			DURHAM			FAYETTEVILLE		
		MAT.	INST.	TOTAL	MAT.	INST.	TOTAL	MAT.	INST.	TOTAL	MAT.	INST.	TOTAL	MAT.	INST.	TOTAL
A	Substructure	105.8	133.4	122.6	104.8	56.2	75.2	109.7	55.7	76.8	101.5	57.3	74.6	108.6	63.1	80.9
B10	Shell: Superstructure	108.8	135.8	119.4	96.3	67.0	84.8	98.0	66.1	85.5	107.2	65.5	90.8	112.2	70.0	95.6
B20	Exterior Closure	102.8	130.5	116.0	89.4	46.3	68.8	93.1	53.0	73.9	91.1	41.8	67.5	89.7	43.4	67.6
B30	Roofing	109.6	134.6	119.0	103.8	43.9	81.3	98.4	45.6	78.6	106.6	47.7	84.5	103.6	47.3	82.5
C	Interior Construction	94.2	139.5	113.5	95.6	42.6	73.0	98.3	44.4	75.4	98.6	44.8	75.7	96.1	57.2	79.5
D10	Services: Conveying	100.0	118.9	105.1	100.0	83.1	95.4	100.0	83.4	95.5	100.0	77.7	93.9	100.0	77.5	93.9
D20 - 40	Mechanical	100.4	118.6	107.7	100.4	53.5	81.6	100.0	55.3	82.1	100.3	53.1	81.4	100.3	52.8	81.3
D50	Electrical	102.4	158.9	131.9	96.8	57.3	76.0	95.7	59.9	77.1	94.4	45.6	68.9	95.6	51.9	72.8
E	Equipment & Furnishings	100.0	132.6	101.7	100.0	40.9	97.0	100.0	42.6	97.1	100.0	44.6	97.2	100.0	64.4	98.2
G	Site Work	118.3	119.1	118.9	103.0	76.0	84.7	106.2	76.0	85.8	102.8	84.8	90.6	102.4	84.8	90.5
A-G	WEIGHTED AVERAGE	102.5	133.4	116.0	97.6	55.9	79.4	98.7	57.6	80.8	100.2	54.4	80.2	100.9	58.5	82.5

DIV. NO.	BUILDING SYSTEMS	NORTH CAROLINA GREENSBORO			RALEIGH			WILMINGTON			WINSTON-SALEM			NORTH DAKOTA BISMARCK		
		MAT.	INST.	TOTAL	MAT.	INST.	TOTAL	MAT.	INST.	TOTAL	MAT.	INST.	TOTAL	MAT.	INST.	TOTAL
A	Substructure	100.8	57.6	74.5	105.8	59.5	77.5	104.8	57.8	76.1	102.5	59.2	76.3	107.7	62.2	80.0
B10	Shell: Superstructure	101.7	65.8	87.6	97.1	66.9	85.2	96.1	67.0	84.7	100.0	67.0	87.0	99.7	68.8	87.5
B20	Exterior Closure	89.9	44.8	68.4	91.8	45.3	69.6	85.1	45.6	66.2	90.0	42.6	67.3	103.4	58.4	81.8
B30	Roofing	106.8	43.5	82.8	100.6	48.3	80.9	103.8	49.1	83.3	106.3	45.1	83.3	109.6	48.9	86.8
C	Interior Construction	98.7	44.0	75.4	98.1	46.7	76.2	96.0	48.7	75.9	98.7	47.8	77.0	105.3	44.2	79.3
D10	Services: Conveying	100.0	82.1	95.1	100.0	78.1	94.0	100.0	79.1	94.3	100.0	83.2	95.4	100.0	91.1	97.6
D20 - 40	Mechanical	100.2	53.3	81.4	100.0	53.6	81.4	100.4	55.1	82.2	100.2	53.5	81.5	100.2	61.3	84.6
D50	Electrical	93.4	57.0	74.4	96.5	41.3	67.7	97.0	51.9	73.4	93.4	57.0	74.4	102.2	73.3	87.1
E	Equipment & Furnishings	100.0	45.3	97.2	100.0	47.1	97.3	100.0	49.2	97.4	100.0	49.6	97.4	100.0	37.3	96.8
G	Site Work	102.7	84.8	90.6	103.9	84.9	91.0	104.4	76.2	85.3	103.0	84.9	90.7	100.0	97.5	98.3
A-G	WEIGHTED AVERAGE	98.8	56.3	80.3	98.2	55.1	79.4	97.2	56.6	79.5	98.5	57.1	80.5	102.0	64.0	85.5

DIV. NO.	BUILDING SYSTEMS	NORTH DAKOTA FARGO			GRAND FORKS			MINOT			OHIO AKRON			CANTON		
		MAT.	INST.	TOTAL	MAT.	INST.	TOTAL	MAT.	INST.	TOTAL	MAT.	INST.	TOTAL	MAT.	INST.	TOTAL
A	Substructure	100.0	63.1	77.5	107.8	58.1	77.5	107.9	70.8	85.3	93.7	96.0	95.1	94.3	89.5	91.4
B10	Shell: Superstructure	99.1	69.5	87.5	97.2	63.6	84.0	97.4	75.9	88.9	94.6	87.5	91.8	94.8	79.6	88.8
B20	Exterior Closure	104.1	58.4	82.2	104.9	61.8	84.2	102.9	65.1	84.8	96.0	93.9	95.0	95.5	83.1	89.6
B30	Roofing	109.0	50.1	86.9	107.1	44.4	83.5	106.9	53.2	86.7	111.3	96.0	105.5	112.4	96.0	105.0
C	Interior Construction	104.6	48.4	80.7	104.7	35.6	75.3	104.2	66.3	88.1	108.5	92.5	101.7	104.8	80.9	94.6
D10	Services: Conveying	100.0	91.1	97.6	100.0	33.5	81.9	100.0	91.1	97.6	100.0	95.9	98.9	100.0	94.7	98.6
D20 - 40	Mechanical	100.1	77.6	91.1	100.4	33.9	73.8	100.4	61.4	84.7	100.0	94.3	97.7	100.0	82.5	93.0
D50	Electrical	104.8	67.7	85.5	103.3	54.8	78.0	106.9	77.5	91.5	99.3	92.8	95.9	98.5	87.8	92.9
E	Equipment & Furnishings	100.0	37.5	96.8	100.0	33.0	96.6	100.0	68.4	98.4	100.0	91.0	99.5	100.0	80.6	99.0
G	Site Work	97.4	97.5	97.5	106.2	93.6	97.6	103.8	97.5	99.5	97.0	102.7	100.8	97.1	102.2	100.6
A-G	WEIGHTED AVERAGE	101.7	67.6	86.9	101.8	51.6	79.9	101.8	71.0	88.4	99.8	93.4	97.0	99.2	85.0	93.0

DIV. NO.	BUILDING SYSTEMS	OHIO CINCINNATI			CLEVELAND			COLUMBUS			DAYTON			LORAIN		
		MAT.	INST.	TOTAL	MAT.	INST.	TOTAL	MAT.	INST.	TOTAL	MAT.	INST.	TOTAL	MAT.	INST.	TOTAL
A	Substructure	93.7	89.6	91.2	92.8	100.4	97.4	94.0	91.2	92.3	89.5	86.7	87.8	91.0	95.0	93.4
B10	Shell: Superstructure	95.6	84.5	91.2	95.7	91.5	94.0	95.6	83.9	91.0	94.2	79.3	88.3	94.6	87.0	91.6
B20	Exterior Closure	94.9	85.2	90.2	96.5	103.1	99.7	100.0	92.2	96.3	94.7	81.4	88.3	93.7	97.3	95.5
B30	Roofing	99.5	90.6	96.1	109.9	109.0	109.5	101.8	95.7	99.5	104.6	87.3	98.1	112.3	100.9	108.0
C	Interior Construction	100.3	81.2	92.2	102.2	98.1	100.5	100.1	85.5	93.9	100.4	78.8	91.2	104.7	88.9	98.0
D10	Services: Conveying	100.0	93.1	98.1	100.0	103.0	100.8	100.0	95.1	98.7	100.0	92.7	98.0	100.0	100.5	100.1
D20 - 40	Mechanical	100.0	81.3	92.5	100.0	99.8	99.9	100.0	92.3	96.9	100.8	81.7	93.1	100.0	85.2	94.0
D50	Electrical	96.1	79.7	87.6	98.8	106.2	102.6	97.1	88.1	92.4	94.7	82.7	88.4	98.7	86.7	92.4
E	Equipment & Furnishings	100.0	78.7	98.9	100.0	95.4	99.8	100.0	84.1	99.2	100.0	77.6	98.9	100.0	81.7	99.1
G	Site Work	93.7	103.6	100.4	96.9	103.2	101.2	95.3	99.4	98.1	92.6	103.8	100.2	96.4	101.6	99.9
A-G	WEIGHTED AVERAGE	97.8	84.8	92.1	98.9	100.0	99.4	98.6	89.7	94.7	97.5	83.2	91.3	98.8	90.4	95.2

City Cost Indexes — RJ1030-010 Building Systems

DIV. NO.	BUILDING SYSTEMS	OHIO SPRINGFIELD			OHIO TOLEDO			OHIO YOUNGSTOWN			OKLAHOMA ENID			OKLAHOMA LAWTON		
		MAT.	INST.	TOTAL	MAT.	INST.	TOTAL	MAT.	INST.	TOTAL	MAT.	INST.	TOTAL	MAT.	INST.	TOTAL
A	Substructure	90.9	87.6	88.9	93.9	95.6	95.0	93.2	92.9	93.0	93.7	57.0	71.3	92.1	61.1	73.2
B10	Shell: Superstructure	94.4	79.9	88.7	95.4	90.1	93.3	94.6	83.7	90.3	94.8	58.7	80.6	97.5	61.6	83.4
B20	Exterior Closure	94.3	80.8	87.8	103.7	95.7	99.9	95.3	90.0	92.8	102.1	57.2	80.6	97.6	57.6	78.5
B30	Roofing	104.5	88.0	98.3	103.4	101.1	102.5	112.5	92.9	105.1	106.4	58.3	88.3	106.2	60.0	88.9
C	Interior Construction	99.5	82.5	92.3	98.4	93.5	96.3	104.7	86.3	96.9	93.4	37.1	69.5	95.2	45.9	74.2
D10	Services: Conveying	100.0	92.7	98.0	100.0	96.9	99.2	100.0	95.3	98.7	100.0	83.1	95.4	100.0	83.1	95.4
D20-40	Mechanical	100.8	81.7	93.1	100.1	99.2	99.7	100.0	88.4	95.3	100.3	66.3	86.6	100.3	66.3	86.7
D50	Electrical	94.7	85.6	90.0	97.2	101.6	99.5	98.7	82.9	90.5	90.7	70.4	80.1	92.3	70.4	80.9
E	Equipment & Furnishings	100.0	83.2	99.1	100.0	92.0	99.6	100.0	83.8	99.2	100.0	31.4	96.5	100.0	44.6	97.2
G	Site Work	93.0	102.4	99.4	94.6	99.9	98.2	96.8	103.2	101.2	98.9	90.4	93.1	95.1	91.8	92.9
A-G	WEIGHTED AVERAGE	97.5	84.2	91.7	98.7	96.4	97.7	99.1	88.3	94.4	97.2	60.9	81.4	97.5	63.3	82.6

DIV. NO.	BUILDING SYSTEMS	OKLAHOMA MUSKOGEE			OKLAHOMA OKLAHOMA CITY			OKLAHOMA TULSA			OREGON EUGENE			OREGON MEDFORD		
		MAT.	INST.	TOTAL	MAT.	INST.	TOTAL	MAT.	INST.	TOTAL	MAT.	INST.	TOTAL	MAT.	INST.	TOTAL
A	Substructure	87.7	56.1	68.4	94.2	64.9	76.3	93.2	58.9	72.3	95.0	102.0	99.3	96.8	102.0	100.0
B10	Shell: Superstructure	92.3	64.3	81.3	92.6	64.6	81.6	95.7	66.6	84.3	92.3	98.8	94.9	91.9	98.7	94.6
B20	Exterior Closure	104.0	47.2	76.8	100.4	57.5	79.9	98.4	60.5	80.3	97.3	104.3	100.7	101.7	104.3	103.0
B30	Roofing	100.0	48.2	80.6	98.8	61.3	84.7	100.0	61.1	85.4	103.9	91.5	99.2	104.5	93.1	100.2
C	Interior Construction	94.4	35.9	69.5	96.1	54.4	78.4	95.1	41.6	72.3	97.4	98.8	98.0	99.8	97.7	98.9
D10	Services: Conveying	100.0	82.3	95.2	100.0	82.9	95.3	100.0	84.7	95.8	100.0	102.1	100.6	100.0	102.1	100.6
D20-40	Mechanical	100.3	60.5	84.3	100.0	66.0	86.4	100.3	63.5	85.5	100.0	102.6	101.0	100.0	102.6	101.0
D50	Electrical	92.7	53.7	72.3	97.3	70.4	83.3	94.7	66.1	79.8	94.5	96.3	95.4	97.9	81.3	89.2
E	Equipment & Furnishings	100.0	31.7	96.5	100.0	57.7	97.9	100.0	38.4	96.9	100.0	100.2	100.0	100.0	99.5	100.0
G	Site Work	89.4	86.9	87.7	94.0	92.2	92.8	95.1	87.4	89.9	96.9	102.9	101.0	104.1	102.9	103.3
A-G	WEIGHTED AVERAGE	96.6	56.3	79.0	97.3	65.4	83.4	97.4	62.2	82.1	97.1	100.4	98.5	98.4	98.2	98.3

DIV. NO.	BUILDING SYSTEMS	OREGON PORTLAND			OREGON SALEM			PENNSYLVANIA ALLENTOWN			PENNSYLVANIA ALTOONA			PENNSYLVANIA ERIE		
		MAT.	INST.	TOTAL	MAT.	INST.	TOTAL	MAT.	INST.	TOTAL	MAT.	INST.	TOTAL	MAT.	INST.	TOTAL
A	Substructure	96.8	102.1	100.0	93.8	102.1	98.8	89.1	108.4	100.8	94.1	91.8	92.7	94.5	93.4	93.8
B10	Shell: Superstructure	93.5	99.0	95.6	97.2	98.9	97.9	97.9	117.4	105.6	93.1	104.0	97.4	94.1	104.3	98.1
B20	Exterior Closure	97.0	104.4	100.5	100.7	104.4	102.4	98.4	104.3	101.2	91.0	70.9	81.4	84.9	93.5	89.0
B30	Roofing	103.9	96.9	101.3	101.7	94.7	99.1	102.5	123.4	110.3	101.8	94.7	99.1	102.3	93.3	98.9
C	Interior Construction	96.3	98.3	97.2	98.0	98.8	98.3	92.9	109.5	100.0	88.4	87.3	88.0	89.5	90.6	90.0
D10	Services: Conveying	100.0	102.1	100.6	100.0	102.1	100.6	100.0	103.0	100.8	100.0	97.0	99.2	100.0	102.8	100.8
D20-40	Mechanical	100.0	102.6	101.0	100.0	102.6	101.1	100.2	110.6	104.4	99.8	84.3	93.6	99.8	93.1	97.1
D50	Electrical	94.7	103.2	99.1	102.1	96.3	99.1	98.2	98.9	98.6	88.3	111.1	100.2	90.1	87.4	88.7
E	Equipment & Furnishings	100.0	100.5	100.0	100.0	100.2	100.0	100.0	115.7	100.8	100.0	83.1	99.1	100.0	86.6	99.3
G	Site Work	99.8	102.9	101.9	93.6	102.9	99.9	90.3	105.2	100.4	92.5	105.2	101.1	89.5	105.7	100.5
A-G	WEIGHTED AVERAGE	97.2	101.4	99.1	99.0	100.5	99.7	97.6	108.8	102.4	94.3	92.7	93.6	94.1	95.0	94.5

DIV. NO.	BUILDING SYSTEMS	PENNSYLVANIA HARRISBURG			PENNSYLVANIA PHILADELPHIA			PENNSYLVANIA PITTSBURGH			PENNSYLVANIA READING			PENNSYLVANIA SCRANTON		
		MAT.	INST.	TOTAL	MAT.	INST.	TOTAL	MAT.	INST.	TOTAL	MAT.	INST.	TOTAL	MAT.	INST.	TOTAL
A	Substructure	96.0	97.4	96.9	97.1	128.2	116.1	101.7	102.3	102.0	81.5	99.3	92.4	91.4	97.2	94.9
B10	Shell: Superstructure	104.4	108.4	106.0	99.9	134.3	113.4	99.5	112.7	104.7	96.0	108.0	100.7	100.0	108.6	103.4
B20	Exterior Closure	97.7	92.5	95.2	98.3	132.0	114.4	101.8	105.3	103.5	98.4	96.4	97.5	98.6	96.7	97.7
B30	Roofing	106.2	114.6	109.4	102.6	137.0	115.5	104.1	101.0	102.9	104.2	116.1	108.7	102.4	95.8	99.9
C	Interior Construction	97.3	91.3	94.7	95.6	142.7	115.6	97.7	102.9	99.9	92.9	92.4	92.7	94.0	95.3	94.5
D10	Services: Conveying	100.0	97.5	99.3	100.0	117.7	104.8	100.0	103.0	100.8	100.0	100.4	100.1	100.0	100.5	100.1
D20-40	Mechanical	100.0	90.7	96.3	100.0	130.5	112.2	99.9	101.2	100.4	100.2	107.3	103.0	100.2	98.8	99.6
D50	Electrical	96.8	88.4	92.4	98.6	148.7	124.7	96.7	111.1	104.2	98.7	91.1	94.7	98.2	97.2	97.7
E	Equipment & Furnishings	100.0	88.1	99.4	100.0	141.9	102.1	100.0	98.4	99.9	100.0	88.1	99.4	100.0	88.3	99.4
G	Site Work	86.0	104.3	98.4	99.4	99.3	99.3	103.2	106.7	105.6	96.0	111.5	106.5	90.8	105.7	100.9
A-G	WEIGHTED AVERAGE	99.6	95.9	98.0	98.9	133.4	113.9	99.7	105.9	102.3	97.1	101.1	98.8	98.3	99.7	98.9

DIV. NO.	BUILDING SYSTEMS	PENNSYLVANIA YORK			PUERTO RICO SAN JUAN			RHODE ISLAND PROVIDENCE			SOUTH CAROLINA CHARLESTON			SOUTH CAROLINA COLUMBIA		
		MAT.	INST.	TOTAL	MAT.	INST.	TOTAL	MAT.	INST.	TOTAL	MAT.	INST.	TOTAL	MAT.	INST.	TOTAL
A	Substructure	86.3	97.7	93.3	123.8	37.9	71.4	98.7	122.0	112.9	96.3	65.5	77.5	98.3	59.1	74.4
B10	Shell: Superstructure	98.1	108.6	102.2	117.6	28.8	82.6	105.1	126.6	113.6	94.7	72.1	85.8	93.2	66.7	82.7
B20	Exterior Closure	100.4	92.6	96.7	97.4	17.8	59.3	104.9	127.8	115.9	91.9	46.8	70.3	90.6	43.2	67.9
B30	Roofing	102.5	114.7	107.1	129.1	22.2	89.0	102.4	118.5	108.4	104.1	49.1	83.5	99.9	45.3	79.4
C	Interior Construction	91.2	91.4	91.2	181.2	17.1	111.4	99.4	125.9	110.7	98.6	65.0	84.3	98.7	47.7	77.0
D10	Services: Conveying	100.0	97.6	99.3	100.0	16.6	77.3	100.0	104.1	101.1	100.0	70.9	92.1	100.0	70.9	92.1
D20-40	Mechanical	100.2	90.8	96.4	102.8	14.2	67.3	99.9	108.4	103.3	100.4	54.9	82.1	100.0	53.2	81.2
D50	Electrical	90.9	88.4	89.6	126.9	13.5	67.7	102.5	100.4	101.4	99.0	90.9	94.8	99.0	61.1	79.3
E	Equipment & Furnishings	100.0	88.1	99.4	100.0	16.3	95.7	100.0	118.5	100.9	100.0	69.6	98.5	100.0	45.6	97.2
G	Site Work	81.8	104.3	97.0	132.4	90.9	104.3	91.6	104.4	100.3	95.7	85.5	88.8	96.4	85.5	89.0
A-G	WEIGHTED AVERAGE	96.5	96.0	96.3	122.1	24.6	79.7	101.5	116.6	108.1	97.7	66.4	84.1	97.2	57.3	79.8

DIV. NO.	BUILDING SYSTEMS	SOUTH CAROLINA FLORENCE			SOUTH CAROLINA GREENVILLE			SOUTH CAROLINA SPARTANBURG			SOUTH DAKOTA ABERDEEN			SOUTH DAKOTA PIERRE		
		MAT.	INST.	TOTAL	MAT.	INST.	TOTAL	MAT.	INST.	TOTAL	MAT.	INST.	TOTAL	MAT.	INST.	TOTAL
A	Substructure	88.2	59.0	70.4	88.5	56.8	69.2	88.8	58.6	70.4	105.5	53.3	73.6	98.5	57.8	73.7
B10	Shell: Superstructure	91.3	66.8	81.7	92.1	62.8	80.6	92.3	66.2	82.0	100.0	53.7	81.8	99.5	64.2	85.6
B20	Exterior Closure	91.4	46.1	69.7	88.5	45.5	67.9	89.4	46.0	68.6	99.6	54.8	78.2	99.3	55.2	78.2
B30	Roofing	104.5	46.2	82.6	104.4	46.2	82.5	104.4	46.2	82.6	104.5	45.5	82.3	105.2	43.9	82.2
C	Interior Construction	95.4	48.1	75.3	96.2	49.2	76.2	96.6	49.8	76.7	99.6	39.2	73.9	103.7	40.0	76.6
D10	Services: Conveying	100.0	70.9	92.1	100.0	70.9	92.1	100.0	70.9	92.1	100.0	42.4	84.3	100.0	83.2	95.4
D20-40	Mechanical	100.4	53.3	81.5	100.4	53.1	81.4	100.4	53.3	81.5	100.2	40.6	76.3	100.0	64.6	85.8
D50	Electrical	96.8	61.1	78.2	99.0	58.0	77.6	99.0	58.0	77.6	99.2	52.4	74.7	104.6	52.4	77.4
E	Equipment & Furnishings	100.0	45.6	97.2	100.0	45.4	97.2	100.0	45.4	97.2	100.0	36.9	96.8	100.0	37.6	96.8
G	Site Work	104.7	85.5	91.7	100.0	85.2	89.9	99.7	85.2	89.9	94.9	94.3	94.5	92.7	94.4	93.9
A-G	WEIGHTED AVERAGE	96.2	57.8	79.4	96.2	56.5	78.9	96.4	57.4	79.5	100.1	50.6	78.6	100.8	59.2	82.7

City Cost Indexes — RJ1030-010 Building Systems

DIV. NO.	BUILDING SYSTEMS	SOUTH DAKOTA						TENNESSEE								
		RAPID CITY			SIOUX FALLS			CHATTANOOGA			JACKSON			JOHNSON CITY		
		MAT.	INST.	TOTAL	MAT.	INST.	TOTAL	MAT.	INST.	TOTAL	MAT.	INST.	TOTAL	MAT.	INST.	TOTAL
A	Substructure	102.5	57.4	75.0	102.0	59.2	75.9	98.5	68.8	80.4	96.8	61.9	75.5	86.4	60.2	70.4
B10	Shell: Superstructure	100.9	64.0	86.4	101.5	64.9	87.2	97.7	74.6	88.6	94.2	70.5	84.9	92.2	69.5	83.3
B20	Exterior Closure	98.8	57.6	79.1	95.9	55.4	76.5	100.2	51.9	77.1	105.6	51.0	79.5	122.6	48.1	86.9
B30	Roofing	104.8	45.7	82.6	109.6	46.6	85.9	100.1	58.4	84.4	96.0	54.8	80.5	95.2	53.6	79.6
C	Interior Construction	101.5	45.0	77.4	101.1	48.0	78.5	98.3	57.7	81.0	95.6	46.4	74.7	98.2	40.4	73.6
D10	Services: Conveying	100.0	83.5	95.5	100.0	83.2	95.4	100.0	40.4	83.7	100.0	43.9	84.7	100.0	45.2	85.1
D20-40	Mechanical	100.2	65.1	86.1	100.1	39.4	75.8	100.2	61.9	84.9	100.1	69.5	87.8	100.0	60.4	84.1
D50	Electrical	95.7	52.5	73.1	97.4	76.9	86.7	101.9	69.8	85.1	99.5	59.9	78.8	91.9	47.4	68.7
E	Equipment & Furnishings	100.0	34.4	96.7	100.0	41.2	97.0	100.0	58.5	97.9	100.0	46.2	97.3	100.0	37.5	96.8
G	Site Work	93.0	94.5	94.0	91.9	96.0	94.7	102.5	99.4	100.4	95.5	97.7	97.0	109.8	88.2	95.1
A-G	WEIGHTED AVERAGE	100.0	60.3	82.7	100.0	58.8	82.1	99.5	65.5	84.7	98.3	62.5	82.7	99.2	56.7	80.7

DIV. NO.	BUILDING SYSTEMS	TENNESSEE									TEXAS					
		KNOXVILLE			MEMPHIS			NASHVILLE			ABILENE			AMARILLO		
		MAT.	INST.	TOTAL	MAT.	INST.	TOTAL	MAT.	INST.	TOTAL	MAT.	INST.	TOTAL	MAT.	INST.	TOTAL
A	Substructure	92.6	58.0	71.5	93.8	71.1	80.0	92.6	73.8	81.1	96.7	54.3	70.8	93.2	61.1	73.6
B10	Shell: Superstructure	97.3	68.5	86.0	97.9	77.9	90.0	98.2	78.2	90.3	100.2	57.0	83.2	95.8	61.8	82.4
B20	Exterior Closure	88.3	54.0	71.9	96.5	64.4	81.1	92.0	61.1	77.2	98.7	56.3	78.4	102.0	58.1	81.0
B30	Roofing	93.3	59.1	80.4	95.5	63.3	83.4	97.2	61.9	83.9	101.2	46.1	80.5	101.9	50.6	82.6
C	Interior Construction	93.5	45.0	72.9	98.9	60.3	82.5	98.7	66.7	85.1	93.9	45.3	73.3	96.0	56.2	79.1
D10	Services: Conveying	100.0	48.0	85.8	100.0	84.3	95.7	100.0	84.8	95.9	100.0	81.2	94.9	100.0	69.4	91.7
D20-40	Mechanical	100.0	65.7	86.2	100.0	74.9	90.0	100.0	83.2	93.3	100.3	46.0	78.5	100.1	50.8	80.3
D50	Electrical	97.8	57.3	76.6	100.9	64.9	82.1	97.6	63.1	79.6	94.9	48.7	70.7	98.7	64.6	80.9
E	Equipment & Furnishings	100.0	38.4	96.9	100.0	65.7	98.3	100.0	65.8	98.3	100.0	37.1	96.8	100.0	54.4	97.7
G	Site Work	90.0	88.6	89.1	92.2	94.3	93.6	98.0	100.3	99.5	95.0	86.6	89.4	93.1	87.7	89.5
A-G	WEIGHTED AVERAGE	96.2	60.5	80.6	98.5	71.3	86.7	97.9	74.1	87.5	98.3	53.6	78.8	98.2	60.0	81.6

DIV. NO.	BUILDING SYSTEMS	TEXAS														
		AUSTIN			BEAUMONT			CORPUS CHRISTI			DALLAS			EL PASO		
		MAT.	INST.	TOTAL	MAT.	INST.	TOTAL	MAT.	INST.	TOTAL	MAT.	INST.	TOTAL	MAT.	INST.	TOTAL
A	Substructure	93.0	63.5	75.0	88.4	57.7	69.7	107.7	53.3	74.5	98.3	65.9	78.6	96.4	53.4	70.2
B10	Shell: Superstructure	96.5	61.0	82.6	94.5	57.3	79.9	94.8	60.4	81.3	100.0	70.8	88.5	95.7	54.0	79.3
B20	Exterior Closure	97.9	48.3	74.2	104.1	56.6	81.4	94.8	51.6	74.2	103.2	57.4	81.3	99.3	47.6	74.6
B30	Roofing	101.1	48.9	81.5	103.6	53.5	84.8	102.3	46.6	81.4	92.0	62.8	81.0	99.0	53.8	82.0
C	Interior Construction	100.6	51.2	79.6	91.6	51.5	74.5	105.0	44.6	79.3	102.4	60.9	84.7	94.6	46.3	74.0
D10	Services: Conveying	100.0	81.1	94.8	100.0	83.4	95.5	100.0	83.5	95.5	100.0	84.0	95.6	100.0	81.4	94.9
D20-40	Mechanical	100.1	56.1	82.5	100.2	59.7	83.9	100.2	40.0	76.0	100.0	62.1	84.8	100.0	34.2	73.6
D50	Electrical	97.0	62.1	78.8	92.1	70.8	81.0	91.8	61.4	76.0	95.7	65.2	79.7	92.7	54.9	73.0
E	Equipment & Furnishings	100.0	57.6	97.8	100.0	48.9	97.4	100.0	39.1	96.9	100.0	66.5	98.3	100.0	45.3	97.2
G	Site Work	96.7	86.7	89.9	90.5	87.5	88.5	135.1	82.3	99.3	107.4	87.9	94.2	95.6	86.1	89.2
A-G	WEIGHTED AVERAGE	98.6	59.1	81.4	96.7	61.2	81.2	99.7	53.7	79.6	100.1	65.8	85.2	97.2	50.6	76.9

DIV. NO.	BUILDING SYSTEMS	TEXAS														
		FORT WORTH			HOUSTON			LAREDO			LUBBOCK			ODESSA		
		MAT.	INST.	TOTAL	MAT.	INST.	TOTAL	MAT.	INST.	TOTAL	MAT.	INST.	TOTAL	MAT.	INST.	TOTAL
A	Substructure	97.7	64.1	77.2	92.7	71.3	79.7	86.3	57.5	68.7	100.8	58.8	75.2	96.9	58.5	73.5
B10	Shell: Superstructure	98.7	64.0	85.1	97.5	78.9	90.2	93.3	56.2	78.7	103.1	66.9	88.9	99.6	59.6	83.9
B20	Exterior Closure	102.7	56.6	80.7	101.1	67.7	85.1	100.7	56.9	79.7	99.0	56.2	78.6	98.7	43.6	72.4
B30	Roofing	98.5	51.4	80.8	94.0	67.7	84.2	98.1	48.8	79.6	90.6	49.7	75.2	101.2	46.1	80.5
C	Interior Construction	101.8	59.6	83.9	100.2	64.8	85.1	96.9	44.2	74.5	100.6	49.6	78.9	93.9	51.1	75.7
D10	Services: Conveying	100.0	83.7	95.6	100.0	86.6	96.4	100.0	81.2	94.9	100.0	81.0	94.8	100.0	81.5	95.0
D20-40	Mechanical	100.0	57.7	83.1	100.0	66.1	86.4	100.1	38.8	75.5	99.8	45.2	77.9	100.3	45.3	78.2
D50	Electrical	94.5	62.4	77.8	93.7	67.8	80.2	94.2	59.9	76.3	93.6	49.6	70.6	95.0	40.3	66.4
E	Equipment & Furnishings	100.0	66.3	98.3	100.0	66.3	98.3	100.0	38.9	96.9	100.0	53.6	97.6	100.0	53.0	97.6
G	Site Work	98.0	87.1	90.6	105.5	87.5	93.3	99.9	86.7	91.0	114.1	84.6	94.1	95.0	87.0	89.6
A-G	WEIGHTED AVERAGE	99.6	62.5	83.5	98.7	70.9	86.6	97.1	53.7	78.2	100.0	56.3	81.0	98.2	52.5	78.3

DIV. NO.	BUILDING SYSTEMS	TEXAS									UTAH					
		SAN ANTONIO			WACO			WICHITA FALLS			LOGAN			OGDEN		
		MAT.	INST.	TOTAL	MAT.	INST.	TOTAL	MAT.	INST.	TOTAL	MAT.	INST.	TOTAL	MAT.	INST.	TOTAL
A	Substructure	88.3	65.6	74.5	89.8	59.3	71.2	93.5	52.9	68.7	92.4	74.2	81.3	91.6	74.2	81.0
B10	Shell: Superstructure	94.4	63.4	82.2	97.8	58.5	82.4	98.6	56.1	81.9	99.2	72.4	88.7	99.7	72.4	89.0
B20	Exterior Closure	101.2	65.2	84.0	101.1	54.1	78.6	101.3	56.3	79.8	114.7	59.6	88.3	100.7	59.6	81.0
B30	Roofing	95.8	63.1	83.5	100.5	45.5	79.9	100.5	47.2	80.5	97.8	62.4	84.5	96.8	62.4	83.9
C	Interior Construction	97.4	54.6	79.2	89.1	39.9	68.2	89.3	43.9	70.0	95.4	56.9	79.0	94.5	56.9	78.5
D10	Services: Conveying	100.0	84.3	95.7	100.0	83.7	95.6	100.0	69.1	91.6	100.0	53.8	87.4	100.0	53.8	87.4
D20-40	Mechanical	100.0	64.5	85.8	100.2	54.2	81.8	100.2	48.4	79.5	100.0	69.0	87.6	100.0	69.0	87.6
D50	Electrical	94.3	60.0	76.4	97.4	65.1	80.5	99.1	61.0	79.2	95.4	74.3	84.4	95.7	74.3	84.5
E	Equipment & Furnishings	100.0	53.1	97.6	100.0	36.2	96.8	100.0	35.5	96.7	100.0	56.9	97.8	100.0	56.9	97.8
G	Site Work	102.2	91.4	94.9	95.3	87.1	89.7	96.1	86.6	89.7	93.3	96.2	95.2	82.8	96.2	91.9
A-G	WEIGHTED AVERAGE	97.5	64.5	83.1	97.2	57.3	79.8	97.8	55.0	79.2	99.7	68.8	86.2	97.9	68.8	85.2

DIV. NO.	BUILDING SYSTEMS	UTAH						VERMONT						VIRGINIA		
		PROVO			SALT LAKE CITY			BURLINGTON			RUTLAND			ALEXANDRIA		
		MAT.	INST.	TOTAL	MAT.	INST.	TOTAL	MAT.	INST.	TOTAL	MAT.	INST.	TOTAL	MAT.	INST.	TOTAL
A	Substructure	93.7	73.7	81.5	97.2	74.1	83.1	100.9	97.8	99.0	92.1	97.7	95.5	101.1	81.7	89.3
B10	Shell: Superstructure	98.2	72.3	88.1	104.6	72.3	91.9	103.6	91.9	99.0	99.5	91.9	96.5	102.1	88.3	96.7
B20	Exterior Closure	116.4	59.6	89.2	122.4	59.6	92.4	105.0	93.8	99.6	96.7	93.8	95.3	94.0	74.9	84.8
B30	Roofing	99.5	62.4	85.6	103.7	62.4	88.2	104.3	79.6	95.0	98.2	79.6	91.2	102.8	83.9	95.7
C	Interior Construction	97.7	55.7	79.8	96.2	55.7	79.0	104.0	91.7	98.8	102.9	91.7	98.1	99.2	75.3	89.1
D10	Services: Conveying	100.0	53.8	87.4	100.0	53.8	87.4	100.0	101.8	100.5	100.0	101.8	100.5	100.0	90.6	97.4
D20-40	Mechanical	100.0	69.0	87.6	100.1	69.0	87.6	99.9	73.3	89.3	100.1	73.3	89.3	100.3	87.7	95.2
D50	Electrical	96.2	61.2	77.9	98.6	61.2	79.1	98.0	62.6	79.5	98.6	62.6	79.8	95.8	99.2	97.6
E	Equipment & Furnishings	100.0	56.3	97.8	100.0	56.3	97.8	100.0	87.7	99.4	100.0	87.6	99.4	100.0	72.5	98.6
G	Site Work	90.2	94.5	93.1	82.5	96.1	91.7	86.8	100.2	95.9	85.8	99.9	95.4	115.4	88.9	97.5
A-G	WEIGHTED AVERAGE	100.1	66.7	85.6	102.2	66.8	86.8	101.5	85.1	94.4	99.2	85.1	93.0	99.8	85.4	93.5

City Cost Indexes

RJ1030-010 Building Systems

VIRGINIA

DIV. NO.	BUILDING SYSTEMS	ARLINGTON MAT.	ARLINGTON INST.	ARLINGTON TOTAL	NEWPORT NEWS MAT.	NEWPORT NEWS INST.	NEWPORT NEWS TOTAL	NORFOLK MAT.	NORFOLK INST.	NORFOLK TOTAL	PORTSMOUTH MAT.	PORTSMOUTH INST.	PORTSMOUTH TOTAL	RICHMOND MAT.	RICHMOND INST.	RICHMOND TOTAL
A	Substructure	103.1	80.9	89.5	101.9	72.5	84.0	107.7	72.8	86.4	100.2	68.4	80.8	99.8	66.6	79.6
B10	Shell: Superstructure	101.0	88.3	96.0	102.0	79.9	93.3	103.1	80.0	94.0	100.2	76.5	91.0	104.9	76.8	93.8
B20	Exterior Closure	103.6	73.6	89.3	93.1	57.1	75.9	96.7	57.1	77.8	95.6	56.6	76.9	94.1	56.3	76.0
B30	Roofing	104.8	82.5	96.5	105.3	71.5	92.6	101.9	71.6	90.5	105.3	68.4	91.5	103.6	70.6	91.2
C	Interior Construction	98.2	74.7	88.2	98.3	64.3	83.8	96.7	67.4	84.2	96.9	57.7	80.2	101.5	62.4	84.9
D10	Services: Conveying	100.0	87.8	96.7	100.0	76.6	93.6	100.0	76.6	93.6	100.0	76.6	93.6	100.0	80.8	94.8
D20-40	Mechanical	100.3	86.8	94.9	100.2	64.1	85.7	100.0	64.7	85.8	100.2	64.7	86.0	100.0	66.9	86.7
D50	Electrical	93.4	99.2	96.5	93.1	60.3	76.0	96.9	59.7	77.4	91.4	59.7	74.9	96.2	70.8	83.0
E	Equipment & Furnishings	100.0	72.5	98.6	100.0	65.5	98.2	100.0	65.5	98.2	100.0	51.3	97.5	100.0	58.5	97.9
G	Site Work	124.7	86.8	99.1	105.9	88.1	93.8	105.6	89.1	94.4	104.2	87.9	93.1	103.5	88.2	93.1
A-G	WEIGHTED AVERAGE	100.6	84.7	93.6	99.2	68.2	85.7	100.0	68.8	86.4	98.6	66.1	84.5	100.4	68.9	86.7

VIRGINIA / WASHINGTON

DIV. NO.	BUILDING SYSTEMS	ROANOKE MAT.	ROANOKE INST.	ROANOKE TOTAL	EVERETT MAT.	EVERETT INST.	EVERETT TOTAL	RICHLAND MAT.	RICHLAND INST.	RICHLAND TOTAL	SEATTLE MAT.	SEATTLE INST.	SEATTLE TOTAL	SPOKANE MAT.	SPOKANE INST.	SPOKANE TOTAL
A	Substructure	110.3	67.7	84.3	102.3	103.1	102.8	97.8	84.6	89.7	105.0	103.6	104.1	100.7	84.4	90.7
B10	Shell: Superstructure	103.8	75.4	92.7	104.5	95.9	101.1	90.6	82.9	87.6	104.8	97.2	101.8	93.4	82.6	89.2
B20	Exterior Closure	93.8	55.9	75.7	101.6	99.3	100.5	113.0	80.6	97.5	102.4	100.1	101.3	113.4	81.2	98.0
B30	Roofing	104.3	68.7	90.9	100.4	100.8	100.6	144.6	79.3	120.0	97.5	101.3	98.9	141.0	78.4	117.5
C	Interior Construction	98.7	62.6	83.3	105.5	97.0	101.9	112.6	76.6	97.3	107.0	98.4	103.3	111.7	78.1	97.4
D10	Services: Conveying	100.0	74.0	92.9	100.0	99.7	99.9	100.0	88.7	96.9	100.0	100.2	100.1	100.0	88.7	96.9
D20-40	Mechanical	100.3	65.8	86.5	100.1	100.3	100.2	100.3	109.7	104.1	100.0	117.1	106.8	100.2	85.9	94.5
D50	Electrical	95.5	55.7	74.7	103.0	98.5	100.7	92.1	94.4	93.3	102.4	107.0	104.8	90.5	79.1	84.5
E	Equipment & Furnishings	100.0	62.4	98.1	100.0	99.2	100.0	100.0	76.4	98.8	100.0	99.7	100.0	100.0	75.6	98.8
G	Site Work	105.7	85.5	92.0	89.5	110.7	103.8	103.5	90.7	94.8	94.1	108.9	104.1	102.4	90.7	94.5
A-G	WEIGHTED AVERAGE	100.2	66.0	85.3	102.1	99.6	101.0	102.2	89.6	96.7	102.5	104.8	103.5	102.5	82.7	93.8

WASHINGTON / WEST VIRGINIA

DIV. NO.	BUILDING SYSTEMS	TACOMA MAT.	TACOMA INST.	TACOMA TOTAL	VANCOUVER MAT.	VANCOUVER INST.	VANCOUVER TOTAL	YAKIMA MAT.	YAKIMA INST.	YAKIMA TOTAL	CHARLESTON MAT.	CHARLESTON INST.	CHARLESTON TOTAL	HUNTINGTON MAT.	HUNTINGTON INST.	HUNTINGTON TOTAL
A	Substructure	103.4	103.7	103.6	112.2	96.6	102.7	107.8	94.6	99.8	100.2	94.9	97.0	105.2	96.3	99.8
B10	Shell: Superstructure	105.7	96.1	101.9	105.1	93.6	100.6	105.1	88.4	98.5	100.1	97.4	99.0	103.9	98.5	101.8
B20	Exterior Closure	100.5	100.0	100.3	105.7	97.0	101.5	98.9	79.1	89.4	93.0	93.7	93.4	92.9	97.7	95.2
B30	Roofing	100.2	98.7	99.6	100.6	91.6	97.3	100.3	81.5	93.2	103.6	89.5	98.3	107.3	90.0	100.8
C	Interior Construction	105.5	97.2	102.0	100.7	91.8	96.9	104.8	87.9	97.6	99.7	95.7	98.0	97.8	98.7	98.2
D10	Services: Conveying	100.0	100.2	100.1	100.0	58.0	88.6	100.0	97.1	99.2	100.0	98.4	99.6	100.0	99.0	99.7
D20-40	Mechanical	100.1	100.8	100.4	100.2	88.6	95.5	100.1	109.1	103.7	100.1	92.9	97.2	100.4	89.3	95.9
D50	Electrical	102.8	99.7	101.2	107.5	101.0	104.1	105.9	94.4	99.9	96.3	91.9	94.0	94.9	96.1	95.5
E	Equipment & Furnishings	100.0	99.3	100.0	100.0	92.8	99.6	100.0	98.5	99.9	100.0	92.5	99.6	100.0	95.4	99.8
G	Site Work	92.1	110.7	104.7	101.3	98.0	99.1	94.8	110.1	105.2	99.3	89.6	92.7	103.0	90.1	94.3
A-G	WEIGHTED AVERAGE	102.3	100.0	101.3	103.1	93.2	98.8	102.5	94.4	99.0	99.0	94.1	96.9	99.8	95.2	97.8

WEST VIRGINIA / WISCONSIN

DIV. NO.	BUILDING SYSTEMS	PARKERSBURG MAT.	PARKERSBURG INST.	PARKERSBURG TOTAL	WHEELING MAT.	WHEELING INST.	WHEELING TOTAL	EAU CLAIRE MAT.	EAU CLAIRE INST.	EAU CLAIRE TOTAL	GREEN BAY MAT.	GREEN BAY INST.	GREEN BAY TOTAL	KENOSHA MAT.	KENOSHA INST.	KENOSHA TOTAL
A	Substructure	101.1	91.5	95.3	101.2	93.4	96.4	99.2	101.3	100.5	102.6	99.5	100.7	107.8	103.5	105.2
B10	Shell: Superstructure	100.4	94.3	98.0	100.6	95.6	98.6	94.5	101.8	97.3	97.4	99.4	98.2	98.0	101.9	99.6
B20	Exterior Closure	91.5	89.6	90.6	101.5	91.0	96.5	93.1	101.2	97.0	108.8	100.9	105.0	97.2	113.9	105.2
B30	Roofing	104.3	89.7	98.8	104.7	89.0	98.8	103.4	85.9	96.8	105.2	84.7	97.5	103.5	107.8	105.1
C	Interior Construction	97.7	92.9	95.7	98.2	95.1	96.9	101.2	103.1	102.0	99.8	102.2	100.8	100.0	104.4	101.8
D10	Services: Conveying	100.0	98.2	99.5	100.0	97.7	99.4	100.0	98.2	99.5	100.0	97.7	99.4	100.0	101.9	100.5
D20-40	Mechanical	100.2	88.1	95.4	100.3	92.8	97.3	100.1	87.5	95.1	100.4	84.3	93.9	100.2	96.9	98.9
D50	Electrical	96.0	92.8	94.3	93.1	96.8	95.0	103.9	86.9	95.0	98.2	82.9	90.2	103.4	99.6	101.4
E	Equipment & Furnishings	100.0	88.8	99.4	100.0	92.7	99.6	100.0	101.0	100.1	100.0	100.8	100.0	100.0	99.4	100.0
G	Site Work	108.6	89.6	95.7	109.4	89.5	96.0	94.9	103.5	100.7	97.5	99.6	98.9	101.6	105.6	104.3
A-G	WEIGHTED AVERAGE	98.9	91.4	95.6	99.8	93.8	97.2	98.7	96.4	97.7	100.5	94.1	97.7	100.2	102.9	101.4

WISCONSIN / WYOMING

DIV. NO.	BUILDING SYSTEMS	LA CROSSE MAT.	LA CROSSE INST.	LA CROSSE TOTAL	MADISON MAT.	MADISON INST.	MADISON TOTAL	MILWAUKEE MAT.	MILWAUKEE INST.	MILWAUKEE TOTAL	RACINE MAT.	RACINE INST.	RACINE TOTAL	CASPER MAT.	CASPER INST.	CASPER TOTAL
A	Substructure	91.2	99.4	96.2	96.5	102.1	99.9	101.4	109.4	106.3	99.8	104.4	102.6	101.2	67.4	80.6
B10	Shell: Superstructure	92.3	98.8	94.8	94.2	99.1	96.1	98.8	103.7	100.7	96.7	101.9	98.7	98.7	64.1	85.1
B20	Exterior Closure	89.8	97.7	93.6	99.1	104.5	101.7	100.3	118.2	108.9	98.4	113.9	105.8	102.5	38.4	71.8
B30	Roofing	102.9	84.9	96.1	106.7	92.1	101.2	100	112.1	104.5	103.5	107.3	104.9	105.3	52.5	85.4
C	Interior Construction	99.7	101.5	100.5	103.6	103.2	103.4	105.1	120.0	111.5	102.0	104.4	103.0	104.0	41.4	77.4
D10	Services: Conveying	100.0	98.2	99.5	100.0	97.8	99.4	100.0	102.7	100.7	100.0	101.9	100.5	100.0	95.2	98.7
D20-40	Mechanical	100.1	87.4	95.0	100.0	96.3	98.5	100.0	102.5	101.0	100.0	96.9	98.8	100.0	65.4	86.1
D50	Electrical	104.2	87.1	95.3	103.6	94.3	98.8	102.9	101.9	102.4	102.6	95.2	98.7	96.0	65.6	80.1
E	Equipment & Furnishings	100.0	101.0	100.1	100.0	101.4	100.1	100.0	120.7	101.1	100.0	99.2	100.0	100.0	41.6	97.0
G	Site Work	89.0	103.5	98.8	91.2	108.6	103.0	98.2	98.9	98.7	95.8	109.4	105.0	96.8	95.1	95.7
A-G	WEIGHTED AVERAGE	97.2	95.0	96.3	99.5	99.8	99.6	100.9	107.9	104.0	99.7	102.6	101.0	100.4	60.2	83.1

WYOMING / CANADA

DIV. NO.	BUILDING SYSTEMS	CHEYENNE MAT.	CHEYENNE INST.	CHEYENNE TOTAL	ROCK SPRINGS MAT.	ROCK SPRINGS INST.	ROCK SPRINGS TOTAL	CALGARY, ALBERTA MAT.	CALGARY, ALBERTA INST.	CALGARY, ALBERTA TOTAL	EDMONTON, ALBERTA MAT.	EDMONTON, ALBERTA INST.	EDMONTON, ALBERTA TOTAL	HALIFAX, NOVA SCOTIA MAT.	HALIFAX, NOVA SCOTIA INST.	HALIFAX, NOVA SCOTIA TOTAL
A	Substructure	94.4	71.5	80.4	94.8	68.7	78.9	171.1	98.5	126.8	176.1	98.5	128.8	148.1	81.9	107.7
B10	Shell: Superstructure	100.3	68.0	87.6	97.4	67.3	85.6	139.1	92.4	120.7	139.3	92.4	120.9	134.0	79.4	112.5
B20	Exterior Closure	102.9	49.9	77.6	126.5	50.3	90.1	158.3	88.9	125.1	153.8	88.9	122.8	146.9	81.8	115.8
B30	Roofing	99.6	54.9	82.8	101.3	59.6	85.6	121.8	95.7	112.0	127.2	95.7	115.4	118.5	77.2	103.0
C	Interior Construction	102.2	53.8	81.6	103.2	61.6	85.5	110.9	97.1	105.0	113.1	97.1	106.3	101.2	79.8	92.1
D10	Services: Conveying	100.0	95.2	98.7	100.0	84.7	95.8	131.1	98.7	122.3	131.1	98.7	122.3	131.1	68.6	114.1
D20-40	Mechanical	100.0	65.5	86.2	100.0	64.2	85.6	101.3	92.2	97.6	101.2	92.2	97.6	100.7	77.5	91.5
D50	Electrical	99.1	70.2	84.0	95.0	64.4	79.0	114.2	100.8	107.2	110.3	100.8	105.4	120.1	80.5	99.4
E	Equipment & Furnishings	100.0	55.5	97.7	100.0	67.1	98.3	131.1	96.9	129.4	131.1	96.9	129.4	131.1	77.4	128.4
G	Site Work	90.0	95.1	93.5	87.9	94.6	92.5	127.7	105.7	112.8	146.8	105.7	118.9	109.8	98.7	102.3
A-G	WEIGHTED AVERAGE	100.1	65.7	85.2	101.8	65.4	85.9	124.7	95.4	111.9	125.0	95.4	112.1	119.9	80.6	102.8

City Cost Indexes — RJ1030-010 Building Systems

CANADA

DIV. NO.	BUILDING SYSTEMS	HAMILTON, ONTARIO			KITCHENER, ONTARIO			LAVAL, QUEBEC			LONDON, ONTARIO			MONTREAL, QUEBEC		
		MAT.	INST.	TOTAL	MAT.	INST.	TOTAL	MAT.	INST.	TOTAL	MAT.	INST.	TOTAL	MAT.	INST.	TOTAL
A	Substructure	144.6	100.0	117.4	125.2	92.8	105.4	130.8	93.9	108.3	145.5	95.5	115.0	153.6	97.3	119.2
B10	Shell: Superstructure	126.7	94.1	113.9	117.0	90.6	106.6	111.1	90.0	102.8	128.1	91.8	113.8	131.7	94.2	116.9
B20	Exterior Closure	146.3	98.8	123.6	120.5	94.5	108.1	133.1	85.7	110.4	148.6	96.0	123.5	136.9	90.8	114.8
B30	Roofing	124.2	94.0	112.8	112.0	90.2	103.8	106.5	92.4	101.2	125.2	91.6	112.6	118.3	96.7	110.2
C	Interior Construction	109.6	95.9	103.7	100.2	89.1	95.5	104.5	91.6	99.0	108.4	90.7	100.9	107.3	95.7	102.3
D10	Services: Conveying	131.1	101.3	123.0	131.1	100.0	122.6	131.1	87.0	119.1	131.1	100.6	122.8	131.1	88.1	119.4
D20-40	Mechanical	101.4	89.4	95.1	100.5	87.6	95.3	100.5	93.0	97.5	101.5	82.8	94.0	101.3	93.8	98.3
D50	Electrical	111.6	96.0	103.5	114.4	93.6	103.6	114.4	74.3	93.5	109.7	93.6	101.3	113.8	86.5	99.6
E	Equipment & Furnishings	131.1	93.9	129.2	131.1	85.8	128.8	131.1	89.2	129.0	131.1	86.9	128.9	131.1	92.8	129.2
G	Site Work	120.3	112.9	115.3	104.6	105.1	105.0	100.4	100.5	100.4	120.1	105.3	110.1	112.4	99.9	103.9
A-G	WEIGHTED AVERAGE	119.3	95.3	108.8	111.6	92.0	103.1	112.4	89.2	102.3	119.5	91.9	107.5	119.2	93.4	107.9

DIV. NO.	BUILDING SYSTEMS	OSHAWA, ONTARIO			OTTAWA, ONTARIO			QUEBEC, QUEBEC			REGINA, SASKATCHEWAN			SASKATOON, SASKATCHEWAN		
		MAT.	INST.	TOTAL	MAT.	INST.	TOTAL	MAT.	INST.	TOTAL	MAT.	INST.	TOTAL	MAT.	INST.	TOTAL
A	Substructure	152.5	93.2	116.3	147.7	96.7	116.6	134.5	97.5	111.9	152.3	92.7	116.0	128.0	92.7	106.5
B10	Shell: Superstructure	116.0	91.3	106.3	129.2	92.7	114.9	127.7	94.4	114.6	131.2	88.0	114.2	107.4	88.0	99.8
B20	Exterior Closure	129.7	95.6	113.4	138.7	96.1	118.3	137.1	93.4	116.2	146.3	90.8	119.8	128.0	90.8	110.2
B30	Roofing	112.9	87.9	103.5	127.5	91.5	114.0	118.7	96.9	110.6	117.9	84.2	105.2	106.6	84.2	98.2
C	Interior Construction	104.1	94.3	99.9	111.3	92.1	103.1	109.5	95.7	103.6	107.3	86.9	98.6	103.0	86.9	96.1
D10	Services: Conveying	131.1	100.9	122.9	131.1	98.0	122.1	131.1	88.2	119.4	131.1	71.1	114.8	131.1	71.1	114.8
D20-40	Mechanical	100.5	102.9	101.5	101.5	83.2	94.2	101.2	93.8	98.3	100.7	84.2	94.1	100.6	84.2	94.0
D50	Electrical	115.6	92.6	103.6	109.0	94.4	101.4	110.9	86.5	98.2	115.4	91.4	102.9	117.1	91.4	103.7
E	Equipment & Furnishings	131.1	90.2	129.1	131.1	90.7	129.1	131.1	92.8	129.2	131.1	91.4	129.1	131.1	91.4	129.1
G	Site Work	116.8	104.6	108.5	114.7	105.0	108.1	107.6	99.9	102.4	128.1	98.4	107.9	111.9	98.4	102.7
A-G	WEIGHTED AVERAGE	114.6	96.1	106.5	119.2	92.5	107.5	117.5	93.8	107.2	120.3	88.4	106.4	111.3	88.4	101.3

DIV. NO.	BUILDING SYSTEMS	ST CATHARINES, ONTARIO			ST JOHNS, NEWFOUNDLAND			THUNDER BAY, ONTARIO			TORONTO, ONTARIO			VANCOUVER, BRITISH COLUMBIA		
		MAT.	INST.	TOTAL	MAT.	INST.	TOTAL	MAT.	INST.	TOTAL	MAT.	INST.	TOTAL	MAT.	INST.	TOTAL
A	Substructure	121.2	97.3	106.6	153.2	86.9	112.8	128.3	95.6	108.4	146.9	102.5	119.8	146.4	96.4	115.9
B10	Shell: Superstructure	108.7	93.8	102.9	132.1	83.0	112.8	110.3	92.1	103.2	129.4	96.7	116.5	142.3	91.9	122.5
B20	Exterior Closure	120.2	98.4	109.7	156.8	86.7	123.3	123.6	97.4	111.1	146.1	103.2	125.6	149.3	92.8	122.3
B30	Roofing	112.0	95.0	105.6	118.6	84.9	105.9	112.2	92.0	104.6	124.2	98.9	114.7	130.4	88.0	114.3
C	Interior Construction	98.5	95.3	97.1	109.6	77.8	96.1	100.5	92.9	97.3	109.3	101.0	105.7	111.2	92.2	103.1
D10	Services: Conveying	131.1	74.0	115.6	131.1	72.3	115.1	131.1	74.2	115.6	131.1	102.5	123.3	131.1	97.7	122.0
D20-40	Mechanical	100.5	84.4	94.1	100.8	73.9	90.0	100.5	84.3	94.0	101.2	94.6	98.6	101.3	82.3	93.7
D50	Electrical	116.0	94.5	104.7	115.9	72.4	93.2	114.4	93.0	103.2	111.5	96.6	103.7	111.3	81.3	95.6
E	Equipment & Furnishings	131.1	93.8	129.2	131.1	80.5	128.6	131.1	89.9	129.0	131.1	98.8	129.5	131.1	90.9	129.1
G	Site Work	105.8	101.8	103.1	121.8	99.8	106.9	110.7	101.7	104.6	138.2	105.8	116.3	122.7	108.1	112.8
A-G	WEIGHTED AVERAGE	109.7	93.0	102.4	122.0	80.4	103.9	110.9	91.7	102.6	120.2	99.0	111.0	123.3	90.0	108.8

DIV. NO.	BUILDING SYSTEMS	WINDSOR, ONTARIO			WINNIPEG, MANITOBA											
		MAT.	INST.	TOTAL	MAT.	INST.	TOTAL	MAT.	INST.	TOTAL	MAT.	INST.	TOTAL	MAT.	INST.	TOTAL
A	Substructure	123.4	96.0	106.7	160.5	76.2	109.1									
B10	Shell: Superstructure	109.4	92.5	102.8	147.3	71.8	117.6									
B20	Exterior Closure	120.0	96.6	108.8	141.0	68.4	106.3									
B30	Roofing	112.0	90.9	104.1	118.6	70.5	100.5									
C	Interior Construction	99.2	92.3	96.3	109.8	67.5	91.8									
D10	Services: Conveying	131.1	73.6	115.5	131.1	67.9	113.9									
D20-40	Mechanical	100.5	83.9	93.9	101.2	65.7	87.0									
D50	Electrical	119.3	94.5	106.4	112.9	66.9	88.9									
E	Equipment & Furnishings	131.1	89.5	129.0	131.1	68.2	127.9									
G	Site Work	100.5	101.6	101.3	123.4	101.2	108.3									
A-G	WEIGHTED AVERAGE	110.2	91.7	102.2	123.6	70.8	100.6									

641

Location Factors — RJ1040-010 Building Systems

Costs shown in RSMeans cost data publications are based on national averages for materials and installation. To adjust these costs to a specific location, simply multiply the base cost by the factor and divide by 100 for that city. The data is arranged alphabetically by state and postal zip code numbers. For a city not listed, use the factor for a nearby city with similar economic characteristics.

STATE/ZIP	CITY	MAT.	INST.	TOTAL
ALABAMA				
350-352	Birmingham	97.5	74.1	87.3
354	Tuscaloosa	96.7	62.0	81.6
355	Jasper	97.0	56.7	79.4
356	Decatur	96.6	59.1	80.3
357-358	Huntsville	96.7	70.8	85.4
359	Gadsden	96.7	60.8	81.1
360-361	Montgomery	97.9	58.1	80.5
362	Anniston	96.5	63.0	81.9
363	Dothan	96.9	50.5	76.7
364	Evergreen	96.5	51.9	77.1
365-366	Mobile	97.8	65.9	83.9
367	Selma	96.7	50.8	76.7
368	Phenix City	97.3	55.7	79.2
369	Butler	96.8	50.5	76.6
ALASKA				
995-996	Anchorage	121.3	114.3	118.3
997	Fairbanks	120.7	115.0	118.2
998	Juneau	120.6	113.5	117.5
999	Ketchikan	132.0	113.5	124.0
ARIZONA				
850,853	Phoenix	99.4	74.8	88.7
851,852	Mesa/Tempe	98.8	73.8	87.9
855	Globe	99.0	71.2	86.9
856-857	Tucson	97.6	73.6	87.1
859	Show Low	99.1	73.5	88.0
860	Flagstaff	101.0	72.9	88.8
863	Prescott	98.9	69.7	86.2
864	Kingman	97.3	73.2	86.8
865	Chambers	97.3	71.9	86.3
ARKANSAS				
716	Pine Bluff	96.9	62.9	82.1
717	Camden	94.5	50.4	75.3
718	Texarkana	95.7	49.9	75.7
719	Hot Springs	93.8	52.3	75.8
720-722	Little Rock	98.0	63.8	83.1
723	West Memphis	95.6	62.7	81.2
724	Jonesboro	96.3	59.4	80.2
725	Batesville	94.2	55.1	77.2
726	Harrison	95.5	51.1	76.1
727	Fayetteville	93.0	52.9	75.6
728	Russellville	94.1	54.4	76.8
729	Fort Smith	96.8	61.7	81.5
CALIFORNIA				
900-902	Los Angeles	100.1	116.5	107.3
903-905	Inglewood	95.7	113.4	103.4
906-908	Long Beach	97.1	113.4	104.2
910-912	Pasadena	96.3	113.9	104.0
913-916	Van Nuys	99.1	113.9	105.5
917-918	Alhambra	98.2	113.9	105.0
919-921	San Diego	100.2	109.1	104.1
922	Palm Springs	96.7	111.6	103.2
923-924	San Bernardino	94.6	111.2	101.9
925	Riverside	98.9	114.0	105.5
926-927	Santa Ana	96.5	111.6	103.1
928	Anaheim	98.9	114.1	105.6
930	Oxnard	99.6	114.2	105.9
931	Santa Barbara	98.9	114.6	105.7
932-933	Bakersfield	100.8	110.6	105.1
934	San Luis Obispo	99.3	111.6	104.7
935	Mojave	96.8	108.6	101.9
936-938	Fresno	100.0	114.5	106.3
939	Salinas	100.2	120.3	108.9
940-941	San Francisco	106.1	143.8	122.5
942,956-958	Sacramento	101.2	118.6	108.8
943	Palo Alto	99.5	133.1	114.2
944	San Mateo	101.7	132.7	115.2
945	Vallejo	100.6	127.2	112.2
946	Oakland	103.9	133.1	116.6
947	Berkeley	103.4	133.1	116.4
948	Richmond	102.5	131.2	115.0
949	San Rafael	104.2	133.6	117.0
950	Santa Cruz	104.1	120.4	111.2

STATE/ZIP	CITY	MAT.	INST.	TOTAL
CALIFORNIA (CONT'D)				
951	San Jose	102.2	134.6	116.3
952	Stockton	100.5	115.5	107.0
953	Modesto	100.4	114.9	106.7
954	Santa Rosa	100.3	133.4	114.7
955	Eureka	101.6	111.6	106.0
959	Marysville	100.8	115.1	107.0
960	Redding	103.6	114.6	108.4
961	Susanville	102.6	114.6	107.8
COLORADO				
800-802	Denver	101.6	82.6	93.3
803	Boulder	98.6	82.0	91.3
804	Golden	100.6	79.6	91.4
805	Fort Collins	102.1	79.6	92.3
806	Greeley	99.7	76.1	89.4
807	Fort Morgan	99.1	79.6	90.6
808-809	Colorado Springs	101.3	79.3	91.7
810	Pueblo	100.9	78.6	91.2
811	Alamosa	102.0	74.6	90.1
812	Salida	101.7	76.2	90.6
813	Durango	102.4	76.8	91.2
814	Montrose	101.2	76.5	90.5
815	Grand Junction	104.4	76.7	92.4
816	Glenwood Springs	102.2	79.3	92.2
CONNECTICUT				
060	New Britain	100.8	120.4	109.3
061	Hartford	102.0	120.6	110.1
062	Willimantic	101.4	120.6	109.7
063	New London	97.7	120.5	107.6
064	Meriden	99.6	120.5	108.7
065	New Haven	102.2	120.5	110.2
066	Bridgeport	101.8	120.2	109.8
067	Waterbury	101.4	120.1	109.6
068	Norwalk	101.4	126.9	112.5
069	Stamford	101.5	127.9	113.0
D.C.				
200-205	Washington	101.6	91.6	97.2
DELAWARE				
197	Newark	98.8	108.1	102.8
198	Wilmington	99.4	108.1	103.2
199	Dover	99.7	108.1	103.3
FLORIDA				
320,322	Jacksonville	97.2	67.7	84.3
321	Daytona Beach	97.4	74.4	87.3
323	Tallahassee	98.7	58.0	81.0
324	Panama City	98.5	59.1	81.3
325	Pensacola	100.8	62.4	84.1
326,344	Gainesville	98.7	68.0	85.3
327-328,347	Orlando	99.3	71.4	87.1
329	Melbourne	99.9	77.9	90.3
330-332,340	Miami	97.7	74.6	87.6
333	Fort Lauderdale	96.4	73.9	86.6
334,349	West Palm Beach	95.3	72.3	85.3
335-336,346	Tampa	97.9	79.8	90.0
337	St. Petersburg	100.2	64.7	84.7
338	Lakeland	97.1	79.3	89.4
339,341	Fort Myers	96.4	74.1	86.7
342	Sarasota	99.3	75.3	88.8
GEORGIA				
300-303,399	Atlanta	96.8	75.3	87.5
304	Statesboro	96.5	56.9	79.2
305	Gainesville	95.2	62.8	81.1
306	Athens	94.7	65.1	81.8
307	Dalton	96.4	60.1	80.6
308-309	Augusta	95.6	64.8	82.2
310-312	Macon	96.2	65.9	83.0
313-314	Savannah	98.4	62.0	82.6
315	Waycross	97.3	60.7	81.3
316	Valdosta	97.5	62.6	82.3
317,398	Albany	97.5	62.5	82.3
318-319	Columbus	97.4	66.6	84.0

Location Factors — RJ1040-010 Building Systems

STATE/ZIP	CITY	MAT.	INST.	TOTAL
HAWAII				
967	Hilo	110.4	121.0	115.0
968	Honolulu	114.9	121.0	117.6
STATES & POSS.				
969	Guam	131.2	58.9	99.7
IDAHO				
832	Pocatello	100.8	78.4	91.1
833	Twin Falls	101.9	68.5	87.4
834	Idaho Falls	99.6	75.5	89.1
835	Lewiston	108.3	85.5	98.4
836-837	Boise	99.8	79.4	90.9
838	Coeur d'Alene	107.6	83.3	97.0
ILLINOIS				
600-603	North Suburban	98.5	135.9	114.8
604	Joliet	98.4	140.7	116.8
605	South Suburban	98.5	135.9	114.8
606-608	Chicago	98.9	141.9	117.6
609	Kankakee	94.8	130.8	110.5
610-611	Rockford	98.9	123.5	109.6
612	Rock Island	96.6	99.8	98.0
613	La Salle	97.7	125.1	109.6
614	Galesburg	97.6	107.6	102.0
615-616	Peoria	99.9	110.4	104.5
617	Bloomington	96.9	109.9	102.6
618-619	Champaign	100.2	108.0	103.6
620-622	East St. Louis	95.5	109.3	101.5
623	Quincy	96.9	101.6	99.0
624	Effingham	96.3	106.3	100.6
625	Decatur	98.2	105.9	101.6
626-627	Springfield	99.0	107.2	102.5
628	Centralia	94.1	108.4	100.4
629	Carbondale	93.9	106.2	99.2
INDIANA				
460	Anderson	95.5	83.1	90.1
461-462	Indianapolis	98.3	85.1	92.6
463-464	Gary	96.8	110.1	102.6
465-466	South Bend	95.7	84.2	90.7
467-468	Fort Wayne	96.0	79.2	88.7
469	Kokomo	93.3	82.0	88.4
470	Lawrenceburg	93.3	79.3	87.2
471	New Albany	94.4	77.0	86.8
472	Columbus	96.6	80.9	89.7
473	Muncie	97.3	81.8	90.6
474	Bloomington	98.7	81.5	91.2
475	Washington	95.0	83.3	89.9
476-477	Evansville	96.5	86.6	92.1
478	Terre Haute	97.2	85.8	92.2
479	Lafayette	96.3	82.9	90.5
IOWA				
500-503,509	Des Moines	99.4	83.9	92.7
504	Mason City	96.5	66.6	83.5
505	Fort Dodge	96.6	58.8	80.2
506-507	Waterloo	98.3	70.9	86.4
508	Creston	96.7	77.9	88.5
510-511	Sioux City	99.1	73.6	88.0
512	Sibley	97.7	57.5	80.2
513	Spencer	99.2	57.5	81.0
514	Carroll	96.4	72.2	85.9
515	Council Bluffs	100.2	78.8	90.9
516	Shenandoah	97.1	72.1	86.2
520	Dubuque	98.5	80.4	90.6
521	Decorah	97.4	65.6	83.5
522-524	Cedar Rapids	99.4	83.4	92.4
525	Ottumwa	97.3	73.0	86.7
526	Burlington	96.6	76.9	88.0
527-528	Davenport	98.4	93.4	96.2
KANSAS				
660-662	Kansas City	99.4	95.7	97.7
664-666	Topeka	99.4	66.5	85.1
667	Fort Scott	97.7	74.3	87.5
668	Emporia	97.8	70.0	85.7
669	Belleville	99.4	65.6	84.7
670-672	Wichita	99.9	65.1	84.8
673	Independence	99.9	70.3	87.0
674	Salina	100.4	67.0	85.9
675	Hutchinson	95.8	65.4	82.5
676	Hays	99.6	65.6	84.8
677	Colby	100.2	65.6	85.1

STATE/ZIP	CITY	MAT.	INST.	TOTAL
KANSAS (CONT'D)				
678	Dodge City	101.7	66.5	86.4
679	Liberal	99.5	66.4	85.1
KENTUCKY				
400-402	Louisville	97.8	84.2	91.9
403-405	Lexington	97.0	82.4	90.7
406	Frankfort	97.9	82.4	91.1
407-409	Corbin	94.4	73.3	85.2
410	Covington	95.2	93.2	94.3
411-412	Ashland	94.0	100.1	96.6
413-414	Campton	95.1	80.2	88.6
415-416	Pikeville	96.2	90.0	93.5
417-418	Hazard	94.6	79.9	88.2
420	Paducah	93.2	86.2	90.2
421-422	Bowling Green	95.5	84.8	90.8
423	Owensboro	95.6	88.9	92.7
424	Henderson	93.0	87.9	90.8
425-426	Somerset	92.7	81.5	87.8
427	Elizabethtown	92.2	83.7	88.5
LOUISIANA				
700-701	New Orleans	101.5	69.8	87.7
703	Thibodaux	98.7	64.9	84.0
704	Hammond	96.3	54.9	78.3
705	Lafayette	98.5	62.7	82.9
706	Lake Charles	98.6	65.0	84.0
707-708	Baton Rouge	100.4	63.6	84.4
710-711	Shreveport	98.4	59.6	81.5
712	Monroe	96.7	56.4	79.2
713-714	Alexandria	96.8	57.8	79.8
MAINE				
039	Kittery	95.1	85.7	91.0
040-041	Portland	100.8	91.2	96.6
042	Lewiston	98.3	91.2	95.2
043	Augusta	100.3	80.8	91.8
044	Bangor	97.6	90.8	94.7
045	Bath	96.1	83.0	90.4
046	Machias	95.7	86.5	91.7
047	Houlton	95.9	86.6	91.8
048	Rockland	95.0	86.5	91.3
049	Waterville	96.2	80.8	89.5
MARYLAND				
206	Waldorf	97.5	82.2	90.9
207-208	College Park	97.6	83.9	91.7
209	Silver Spring	96.9	83.8	91.2
210-212	Baltimore	100.5	82.8	92.8
214	Annapolis	100.3	81.8	92.3
215	Cumberland	96.0	81.9	89.9
216	Easton	97.4	69.3	85.2
217	Hagerstown	96.9	82.8	90.7
218	Salisbury	97.8	63.3	82.8
219	Elkton	95.1	81.9	89.4
MASSACHUSETTS				
010-011	Springfield	100.0	109.8	104.3
012	Pittsfield	99.6	109.6	103.9
013	Greenfield	97.5	110.2	103.0
014	Fitchburg	96.3	124.6	108.6
015-016	Worcester	100.0	124.6	110.7
017	Framingham	95.5	131.0	111.0
018	Lowell	99.4	130.1	112.8
019	Lawrence	100.1	131.3	113.7
020-022, 024	Boston	101.8	138.1	117.6
023	Brockton	100.4	126.3	111.7
025	Buzzards Bay	94.9	126.7	108.8
026	Hyannis	97.6	126.7	110.3
027	New Bedford	99.6	126.8	111.4
MICHIGAN				
480,483	Royal Oak	94.2	107.5	100.0
481	Ann Arbor	96.4	107.9	101.4
482	Detroit	97.9	109.3	102.8
484-485	Flint	96.1	96.8	96.4
486	Saginaw	95.7	92.4	94.2
487	Bay City	95.8	92.5	94.4
488-489	Lansing	97.7	94.1	96.1
490	Battle Creek	96.8	87.4	92.7
491	Kalamazoo	97.1	85.9	92.2
492	Jackson	95.0	95.9	95.4
493,495	Grand Rapids	98.1	84.5	92.2
494	Muskegon	95.6	85.5	91.2

643

Location Factors — RJ1040-010 Building Systems

STATE/ZIP	CITY	MAT.	INST.	TOTAL
MICHIGAN (CONT'D)				
496	Traverse City	94.6	81.0	88.6
497	Gaylord	95.5	84.4	90.7
498-499	Iron mountain	97.5	87.4	93.1
MINNESOTA				
550-551	Saint Paul	100.2	118.8	108.3
553-555	Minneapolis	100.6	119.9	109.0
556-558	Duluth	99.0	108.6	103.2
559	Rochester	98.8	104.9	101.5
560	Mankato	96.6	102.4	99.1
561	Windom	95.3	93.3	94.4
562	Willmar	94.9	103.0	98.4
563	St. Cloud	96.3	117.8	105.7
564	Brainerd	96.3	102.3	98.9
565	Detroit Lakes	98.3	96.7	97.6
566	Bemidji	97.7	99.5	98.5
567	Thief River Falls	97.3	93.8	95.7
MISSISSIPPI				
386	Clarksdale	95.8	53.8	77.5
387	Greenville	99.4	63.1	83.6
388	Tupelo	97.3	56.1	79.4
389	Greenwood	97.0	53.3	78.0
390-392	Jackson	99.1	63.5	83.6
393	Meridian	96.2	65.3	82.7
394	Laurel	97.2	57.1	79.7
395	Biloxi	97.7	59.1	80.9
396	Mccomb	95.7	52.9	77.1
397	Columbus	97.1	55.1	78.8
MISSOURI				
630-631	St. Louis	99.6	107.6	103.1
633	Bowling Green	98.1	94.7	96.6
634	Hannibal	97.1	92.6	95.1
635	Kirksville	100.3	90.0	95.8
636	Flat River	99.0	95.8	97.6
637	Cape Girardeau	99.1	90.6	95.4
638	Sikeston	97.5	90.6	94.5
639	Poplar Bluff	97.0	90.3	94.1
640-641	Kansas City	101.6	106.7	103.8
644-645	St. Joseph	100.7	93.9	97.8
646	Chillicothe	97.7	96.2	97.1
647	Harrisonville	97.3	102.2	99.5
648	Joplin	99.5	81.0	91.5
650-651	Jefferson City	98.4	93.4	96.2
652	Columbia	98.6	93.5	96.4
653	Sedalia	98.1	92.6	95.7
654-655	Rolla	96.3	95.6	96.0
656-658	Springfield	99.9	82.1	92.1
MONTANA				
590-591	Billings	102.2	77.9	91.6
592	Wolf Point	101.7	74.9	90.0
593	Miles City	99.5	74.7	88.7
594	Great Falls	103.4	77.1	92.0
595	Havre	100.6	74.2	89.1
596	Helena	102.0	75.1	90.2
597	Butte	102.0	77.2	91.2
598	Missoula	99.4	76.4	89.4
599	Kalispell	98.4	76.3	88.8
NEBRASKA				
680-681	Omaha	99.6	79.3	90.7
683-685	Lincoln	99.5	75.3	89.0
686	Columbus	97.5	77.9	88.9
687	Norfolk	99.2	77.5	89.8
688	Grand Island	99.1	77.8	89.9
689	Hastings	98.5	81.2	90.9
690	Mccook	98.1	72.1	86.8
691	North Platte	98.5	79.6	90.2
692	Valentine	100.6	73.6	88.9
693	Alliance	100.4	71.6	87.9
NEVADA				
889-891	Las Vegas	100.2	109.4	104.2
893	Ely	98.7	103.4	100.8
894-895	Reno	99.0	93.8	96.7
897	Carson City	99.1	93.8	96.8
898	Elko	97.6	86.1	92.6
NEW HAMPSHIRE				
030	Nashua	100.6	93.6	97.6
031	Manchester	102.1	93.6	98.4

STATE/ZIP	CITY	MAT.	INST.	TOTAL
NEW HAMPSHIRE (CONT'D)				
032-033	Concord	100.8	92.3	97.1
034	Keene	97.4	74.0	87.2
035	Littleton	97.1	77.2	88.4
036	Charleston	96.9	70.1	85.2
037	Claremont	96.0	70.1	84.7
038	Portsmouth	98.2	94.2	96.4
NEW JERSEY				
070-071	Newark	101.9	125.6	112.2
072	Elizabeth	99.5	125.3	110.7
073	Jersey City	98.5	125.5	110.2
074-075	Paterson	100.2	125.2	111.1
076	Hackensack	98.4	125.5	110.2
077	Long Branch	97.8	124.1	109.2
078	Dover	98.7	125.4	110.3
079	Summit	98.7	125.3	110.3
080,083	Vineland	97.5	123.4	108.8
081	Camden	99.4	123.6	109.9
082,084	Atlantic City	98.1	123.5	109.2
085-086	Trenton	100.8	124.0	110.9
087	Point Pleasant	99.2	121.4	108.8
088-089	New Brunswick	99.7	125.2	110.8
NEW MEXICO				
870-872	Albuquerque	99.2	73.6	88.0
873	Gallup	99.3	73.6	88.1
874	Farmington	99.7	73.6	88.3
875	Santa Fe	100.5	73.6	88.8
877	Las Vegas	97.7	73.6	87.2
878	Socorro	97.3	73.6	87.0
879	Truth/Consequences	97.0	70.4	85.4
880	Las Cruces	95.8	70.4	84.7
881	Clovis	98.1	73.5	87.4
882	Roswell	99.8	73.6	88.4
883	Carrizozo	100.1	73.6	88.5
884	Tucumcari	98.8	73.5	87.8
NEW YORK				
100-102	New York	103.8	166.4	131.1
103	Staten Island	99.7	163.9	127.6
104	Bronx	98.0	163.9	126.7
105	Mount Vernon	97.9	133.3	113.3
106	White Plains	98.0	133.3	113.4
107	Yonkers	102.5	133.4	116.0
108	New Rochelle	98.2	133.3	113.5
109	Suffern	98.0	126.2	110.3
110	Queens	100.8	162.9	127.8
111	Long Island City	102.2	162.9	128.6
112	Brooklyn	102.6	162.9	128.9
113	Flushing	102.3	162.9	128.7
114	Jamaica	100.6	162.9	127.7
115,117,118	Hicksville	100.7	146.6	120.7
116	Far Rockaway	102.4	162.9	128.7
119	Riverhead	101.4	146.6	121.1
120-122	Albany	98.6	100.2	99.3
123	Schenectady	97.8	100.2	98.8
124	Kingston	100.7	118.3	108.4
125-126	Poughkeepsie	100.0	127.5	112.0
127	Monticello	99.4	116.5	106.8
128	Glens Falls	92.6	95.0	93.6
129	Plattsburgh	97.3	88.4	93.4
130-132	Syracuse	98.5	97.0	97.9
133-135	Utica	96.7	93.2	95.1
136	Watertown	98.1	91.9	95.4
137-139	Binghamton	98.1	99.4	98.6
140-142	Buffalo	99.7	105.8	102.4
143	Niagara Falls	96.9	103.9	100.0
144-146	Rochester	100.3	95.6	98.3
147	Jamestown	96.0	87.8	92.4
148-149	Elmira	95.9	93.1	94.7
NORTH CAROLINA				
270,272-274	Greensboro	98.8	56.3	80.3
271	Winston-Salem	98.5	57.1	80.5
275-276	Raleigh	98.2	55.1	79.4
277	Durham	100.2	54.4	80.2
278	Rocky Mount	96.0	53.2	77.4
279	Elizabeth City	96.9	52.2	77.4
280	Gastonia	98.4	56.1	80.0
281-282	Charlotte	98.7	57.6	80.8
283	Fayetteville	100.9	58.5	82.5
284	Wilmington	97.2	56.6	79.5
285	Kinston	95.3	53.7	77.2

Location Factors — RJ1040-010 Building Systems

STATE/ZIP	CITY	MAT.	INST.	TOTAL
NORTH CAROLINA (CONT'D)				
286	Hickory	95.6	54.9	77.9
287-288	Asheville	97.6	55.9	79.4
289	Murphy	96.4	49.2	75.8
NORTH DAKOTA				
580-581	Fargo	101.7	67.6	86.9
582	Grand Forks	101.8	51.6	79.9
583	Devils Lake	101.0	58.6	82.5
584	Jamestown	101.0	45.1	76.7
585	Bismarck	102.0	64.0	85.5
586	Dickinson	101.9	57.9	82.7
587	Minot	101.8	71.0	88.4
588	Williston	100.3	58.0	81.9
OHIO				
430-432	Columbus	98.6	89.7	94.7
433	Marion	95.0	83.7	90.1
434-436	Toledo	98.7	96.4	97.7
437-438	Zanesville	95.5	85.6	91.2
439	Steubenville	96.8	94.5	95.8
440	Lorain	98.8	90.4	95.2
441	Cleveland	98.9	100.0	99.4
442-443	Akron	99.8	93.4	97.0
444-445	Youngstown	99.1	88.3	94.4
446-447	Canton	99.2	85.0	93.0
448-449	Mansfield	96.6	85.6	91.8
450	Hamilton	97.4	84.5	91.8
451-452	Cincinnati	97.8	84.8	92.1
453-454	Dayton	97.5	83.2	91.3
455	Springfield	97.5	84.2	91.7
456	Chillicothe	96.4	92.2	94.5
457	Athens	99.1	80.6	91.1
458	Lima	99.6	84.4	93.0
OKLAHOMA				
730-731	Oklahoma City	97.3	65.4	83.4
734	Ardmore	95.2	62.3	80.9
735	Lawton	97.5	63.3	82.6
736	Clinton	96.5	63.6	82.2
737	Enid	97.2	60.9	81.4
738	Woodward	95.3	63.8	81.6
739	Guymon	96.3	60.5	80.7
740-741	Tulsa	97.4	62.2	82.1
743	Miami	94.0	68.3	82.8
744	Muskogee	96.6	56.3	79.0
745	Mcalester	93.7	61.1	79.5
746	Ponca City	94.3	62.5	80.5
747	Durant	94.3	64.0	81.1
748	Shawnee	95.7	62.2	81.1
749	Poteau	93.5	62.1	79.8
OREGON				
970-972	Portland	97.2	101.4	99.1
973	Salem	99.0	100.5	99.7
974	Eugene	97.1	100.4	98.5
975	Medford	98.4	98.2	98.3
976	Klamath Falls	98.2	98.2	98.2
977	Bend	97.3	100.5	98.7
978	Pendleton	93.1	102.5	97.2
979	Vale	91.3	89.8	90.7
PENNSYLVANIA				
150-152	Pittsburgh	99.7	105.8	102.3
153	Washington	96.7	103.2	99.5
154	Uniontown	97.1	102.2	99.4
155	Bedford	98.0	92.6	95.7
156	Greensburg	98.0	101.6	99.5
157	Indiana	96.9	100.1	98.3
158	Dubois	98.4	97.1	97.8
159	Johnstown	97.9	97.3	97.7
160	Butler	92.1	102.2	96.5
161	New Castle	92.1	97.2	94.4
162	Kittanning	92.5	103.4	97.3
163	Oil City	92.1	97.0	94.3
164-165	Erie	94.1	95.0	94.5
166	Altoona	94.3	92.7	93.6
167	Bradford	95.3	96.8	96.0
168	State College	94.9	93.0	94.1
169	Wellsboro	95.9	92.3	94.3
170-171	Harrisburg	99.6	95.9	98.0
172	Chambersburg	96.0	89.7	93.2
173-174	York	96.5	96.0	96.3
175-176	Lancaster	94.8	89.3	92.4

STATE/ZIP	CITY	MAT.	INST.	TOTAL
PENNSYLVANIA (CONT'D)				
177	Williamsport	93.6	79.3	87.4
178	Sunbury	95.5	96.3	95.8
179	Pottsville	94.7	98.0	96.1
180	Lehigh Valley	95.4	112.5	102.9
181	Allentown	97.6	108.8	102.4
182	Hazleton	95.0	98.8	96.7
183	Stroudsburg	94.9	102.0	98.0
184-185	Scranton	98.3	99.7	98.9
186-187	Wilkes-Barre	94.7	99.2	96.7
188	Montrose	94.4	97.1	95.6
189	Doylestown	94.6	120.8	106.0
190-191	Philadelphia	98.9	133.4	113.9
193	Westchester	95.8	121.7	107.1
194	Norristown	94.9	129.5	110.0
195-196	Reading	97.1	101.1	98.8
PUERTO RICO				
009	San Juan	122.1	24.6	79.7
RHODE ISLAND				
028	Newport	99.0	116.6	106.6
029	Providence	101.5	116.6	108.1
SOUTH CAROLINA				
290-292	Columbia	97.2	57.3	79.8
293	Spartanburg	96.4	57.4	79.5
294	Charleston	97.7	66.4	84.1
295	Florence	96.2	57.8	79.4
296	Greenville	96.2	56.5	78.9
297	Rock Hill	95.6	51.2	76.3
298	Aiken	96.4	70.4	85.1
299	Beaufort	97.1	45.7	74.7
SOUTH DAKOTA				
570-571	Sioux Falls	100.0	58.8	82.1
572	Watertown	98.5	50.2	77.5
573	Mitchell	97.5	49.5	76.6
574	Aberdeen	100.1	50.6	78.6
575	Pierre	100.8	59.2	82.7
576	Mobridge	98.1	49.8	77.1
577	Rapid City	100.0	60.3	82.7
TENNESSEE				
370-372	Nashville	97.9	74.1	87.5
373-374	Chattanooga	99.5	65.5	84.7
375,380-381	Memphis	98.5	71.3	86.7
376	Johnson City	99.2	56.7	80.7
377-379	Knoxville	96.2	60.5	80.6
382	Mckenzie	96.6	60.4	80.9
383	Jackson	98.3	62.5	82.7
384	Columbia	95.3	67.5	83.2
385	Cookeville	96.4	60.4	80.7
TEXAS				
750	Mckinney	99.3	48.9	77.3
751	Waxahackie	99.3	56.4	80.6
752-753	Dallas	100.1	65.8	85.2
754	Greenville	99.4	42.2	74.5
755	Texarkana	98.4	49.7	77.2
756	Longview	99.0	40.5	73.5
757	Tyler	98.9	54.6	79.6
758	Palestine	95.4	55.8	78.2
759	Lufkin	95.9	59.1	79.9
760-761	Fort Worth	99.6	62.5	83.5
762	Denton	98.2	46.6	75.7
763	Wichita Falls	97.8	55.0	79.2
764	Eastland	96.5	49.1	75.9
765	Temple	95.0	52.7	76.6
766-767	Waco	97.2	57.3	79.8
768	Brownwood	98.5	48.4	76.7
769	San Angelo	98.2	49.5	77.0
770-772	Houston	98.7	70.9	86.6
773	Huntsville	97.2	56.9	79.6
774	Wharton	98.1	53.8	78.8
775	Galveston	96.3	69.7	84.7
776-777	Beaumont	96.7	61.2	81.2
778	Bryan	94.1	63.0	80.5
779	Victoria	98.2	42.7	74.1
780	Laredo	97.1	53.7	78.2
781-782	San Antonio	97.5	64.5	83.1
783-784	Corpus Christi	99.7	53.7	79.6
785	Mc Allen	99.7	46.6	76.6
786-787	Austin	98.6	59.1	81.4

Location Factors — RJ1040-010 Building Systems

STATE/ZIP	CITY	MAT.	INST.	TOTAL
TEXAS (CONT'D)				
788	Del Rio	99.4	46.2	76.3
789	Giddings	96.4	55.6	78.6
790-791	Amarillo	98.2	60.0	81.6
792	Childress	97.9	56.9	80.0
793-794	Lubbock	100.0	56.3	81.0
795-796	Abilene	98.3	53.6	78.8
797	Midland	99.9	53.6	79.7
798-799,885	El Paso	97.2	50.6	76.9
UTAH				
840-841	Salt Lake City	102.2	66.8	86.8
842,844	Ogden	97.9	68.8	85.2
843	Logan	99.7	68.8	86.2
845	Price	100.1	65.3	84.9
846-847	Provo	100.1	66.7	85.6
VERMONT				
050	White River Jct.	97.9	79.7	90.0
051	Bellows Falls	96.6	94.7	95.8
052	Bennington	96.9	95.3	96.2
053	Brattleboro	97.2	94.5	96.0
054	Burlington	101.5	85.1	94.4
056	Montpelier	99.2	84.3	92.7
057	Rutland	99.2	85.1	93.0
058	St. Johnsbury	97.9	79.2	89.8
059	Guildhall	96.8	79.2	89.1
VIRGINIA				
220-221	Fairfax	99.3	84.4	92.8
222	Arlington	100.6	84.7	93.6
223	Alexandria	99.8	85.4	93.5
224-225	Fredericksburg	98.2	83.0	91.6
226	Winchester	98.7	82.1	91.5
227	Culpeper	98.6	80.8	90.9
228	Harrisonburg	98.8	68.1	85.5
229	Charlottesville	99.2	67.3	85.3
230-232	Richmond	100.4	68.9	86.7
233-235	Norfolk	100.0	68.8	86.4
236	Newport News	99.2	68.2	85.7
237	Portsmouth	98.6	66.1	84.5
238	Petersburg	98.6	68.9	85.6
239	Farmville	97.8	53.3	78.5
240-241	Roanoke	100.2	66.0	85.3
242	Bristol	98.1	55.6	79.6
243	Pulaski	97.7	58.3	80.6
244	Staunton	98.4	62.7	82.9
245	Lynchburg	98.6	66.1	84.5
246	Grundy	98.0	53.4	78.6
WASHINGTON				
980-981,987	Seattle	102.5	104.8	103.5
982	Everett	102.1	99.6	101.0
983-984	Tacoma	102.3	100.0	101.3
985	Olympia	101.0	100.0	100.6
986	Vancouver	103.1	93.2	98.8
988	Wenatchee	102.0	87.1	95.5
989	Yakima	102.5	94.4	99.0
990-992	Spokane	102.5	82.7	93.8
993	Richland	102.2	89.6	96.7
994	Clarkston	100.6	80.7	92.0
WEST VIRGINIA				
247-248	Bluefield	97.1	91.7	94.7
249	Lewisburg	98.4	91.4	95.4
250-253	Charleston	99.0	94.1	96.9
254	Martinsburg	98.3	86.1	93.0
255-257	Huntington	99.8	95.2	97.8
258-259	Beckley	96.9	92.0	94.8
260	Wheeling	99.8	93.8	97.2
261	Parkersburg	98.9	91.4	95.6
262	Buckhannon	98.2	93.6	96.2
263-264	Clarksburg	98.8	93.5	96.5
265	Morgantown	98.8	92.6	96.1
266	Gassaway	98.2	92.8	95.8
267	Romney	98.2	90.4	94.8
268	Petersburg	98.0	90.1	94.6
WISCONSIN				
530,532	Milwaukee	100.9	107.9	104.0
531	Kenosha	100.2	102.9	101.4
534	Racine	99.7	102.6	101.0
535	Beloit	99.6	101.9	100.6
537	Madison	99.5	99.8	99.6

STATE/ZIP	CITY	MAT.	INST.	TOTAL
WISCONSIN (CONT'D)				
538	Lancaster	97.4	93.5	95.7
539	Portage	96.1	97.9	96.9
540	New Richmond	96.2	95.9	96.1
541-543	Green Bay	100.5	94.1	97.7
544	Wausau	95.7	94.4	95.2
545	Rhinelander	98.7	92.4	95.9
546	La Crosse	97.2	95.0	96.3
547	Eau Claire	98.7	96.4	97.7
548	Superior	96.1	97.6	96.7
549	Oshkosh	96.3	92.4	94.6
WYOMING				
820	Cheyenne	100.1	65.7	85.2
821	Yellowstone Nat'l Park	97.7	65.2	83.5
822	Wheatland	98.7	64.6	83.9
823	Rawlins	100.2	67.8	86.1
824	Worland	98.3	64.1	83.4
825	Riverton	99.3	64.9	84.3
826	Casper	100.4	60.7	83.1
827	Newcastle	98.2	67.8	84.9
828	Sheridan	100.9	65.2	85.4
829-831	Rock Springs	101.8	65.4	85.9
CANADIAN FACTORS (reflect Canadian currency)				
ALBERTA				
	Calgary	124.7	95.4	111.9
	Edmonton	125.0	95.4	112.1
	Fort McMurray	124.1	94.2	111.1
	Lethbridge	118.7	93.7	107.8
	Lloydminster	113.9	90.3	103.6
	Medicine Hat	114.0	89.6	103.4
	Red Deer	114.5	89.6	103.6
BRITISH COLUMBIA				
	Kamloops	114.9	92.4	105.1
	Prince George	115.9	91.7	105.3
	Vancouver	123.3	90.0	108.8
	Victoria	115.9	88.1	103.8
MANITOBA				
	Brandon	118.7	77.3	100.7
	Portage la Prairie	113.9	75.6	97.2
	Winnipeg	123.6	70.8	100.6
NEW BRUNSWICK				
	Bathurst	112.1	68.6	93.1
	Dalhousie	112.1	68.8	93.2
	Fredericton	118.1	74.0	98.9
	Moncton	112.3	71.4	94.5
	Newcastle	112.1	69.3	93.5
	St. John	114.6	78.3	98.8
NEWFOUNDLAND				
	Corner Brook	122.2	69.0	99.0
	St Johns	122.0	80.4	103.9
NORTHWEST TERRITORIES				
	Yellowknife	128.8	85.8	110.1
NOVA SCOTIA				
	Bridgewater	113.4	76.1	97.2
	Dartmouth	119.7	76.1	100.7
	Halifax	119.9	80.6	102.8
	New Glasgow	117.7	76.1	99.6
	Sydney	115.6	76.1	98.4
	Truro	113.0	76.1	96.9
	Yarmouth	117.6	76.1	99.6
ONTARIO				
	Barrie	117.3	94.9	107.5
	Brantford	115.5	98.8	108.2
	Cornwall	115.5	95.1	106.7
	Hamilton	119.3	95.3	108.8
	Kingston	116.6	95.2	107.3
	Kitchener	111.6	92.0	103.1
	London	119.5	91.9	107.5
	North Bay	120.5	92.9	108.5
	Oshawa	114.6	96.1	106.5
	Ottawa	119.2	92.5	107.5
	Owen Sound	117.4	93.0	106.8
	Peterborough	115.5	94.9	106.5
	Sarnia	115.3	99.5	108.4

Location Factors — RJ1040-010 Building Systems

STATE/ZIP	CITY	MAT.	INST.	TOTAL
ONTARIO (CONT'D)				
	Sault Ste Marie	110.7	93.1	103.0
	St. Catharines	109.7	93.0	102.4
	Sudbury	109.9	91.7	101.9
	Thunder Bay	110.9	91.7	102.6
	Timmins	115.6	92.9	105.7
	Toronto	120.2	99.0	111.0
	Windsor	110.2	91.7	102.2
PRINCE EDWARD ISLAND				
	Charlottetown	120.5	64.0	95.9
	Summerside	119.5	64.1	95.4
QUEBEC				
	Cap-de-la-Madeleine	112.8	89.4	102.6
	Charlesbourg	112.8	89.4	102.6
	Chicoutimi	111.9	93.1	103.7
	Gatineau	112.5	89.2	102.3
	Granby	112.8	89.1	102.4
	Hull	112.6	89.2	102.4
	Joliette	113.0	89.4	102.7
	Laval	112.4	89.2	102.3
	Montreal	119.2	93.4	107.9
	Quebec	117.5	93.8	107.2
	Rimouski	112.3	93.1	103.9
	Rouyn-Noranda	112.5	89.2	102.3
	Saint Hyacinthe	111.9	89.2	102.0
	Sherbrooke	112.8	89.2	102.5
	Sorel	113.0	89.4	102.7
	St Jerome	112.5	89.2	102.3
	Trois Rivieres	117.9	89.4	105.5
SASKATCHEWAN				
	Moose Jaw	111.2	70.8	93.6
	Prince Albert	110.3	69.0	92.3
	Regina	120.3	88.4	106.4
	Saskatoon	111.3	88.4	101.3
YUKON				
	Whitehorse	126.3	69.8	101.7

Reference Aids — RL1010-100 Live Loads

L1010-101 Minimum Design Live Loads in Pounds per S.F. for Various Building Codes

Occupancy	Description	Minimum Live Loads, Pounds per S.F. IBC
Access Floors	Office use	50
	Computer use	100
Armories		150
Assembly	Fixed seats	60
	Movable seats	100
	Platforms or stage floors	125
Commercial & Industrial	Light manufacturing	125
	Heavy manufacturing	250
	Light storage	125
	Heavy storage	250
	Stores, retail, first floor	100
	Stores, retail, upper floors	75
	Stores, wholesale	125
Dance halls	Ballrooms	100
Dining rooms	Restaurants	100
Fire escapes	Other than below	100
	Single family residential	40
Garages	Passenger cars only	40
Gymnasiums	Main floors and balconies	100
Hospitals	Operating rooms, laboratories	60
	Private room	40
	Corridors, above first floor	80
Libraries	Reading rooms	60
	Stack rooms	150
	Corridors, above first floor	80
Marquees		75
Office Buildings	Offices	50
	Lobbies	100
	Corridors, above first floor	80
Residential	Multi family private apartments	40
	Multi family, public rooms	100
	Multi family, corridors	100
	Dwellings, first floor	40
	Dwellings, second floor & habitable attics	30
	Dwellings, uninhabitable attics	20
	Hotels, guest rooms	40
	Hotels, public rooms	100
	Hotels, corridors serving public rooms	100
	Hotels, corridors	100
Schools	Classrooms	40
	Corridors	100
Sidewalks	Driveways, etc. subject to trucking	250
Stairs	Exits	100
Theaters	Aisles, corridors and lobbies	See Assembly
	Orchestra floors	↓
	Balconies	
	Stage floors	125
Yards	Terrace, pedestrian	100

UBC = Uniform Building Code, International Conference of Building Officials, 2006 Edition

Reference Aids — RL1010-200 Dead Load, Material Weights

Table L1010-201 Design Weight Per S.F. for Walls and Partitions

Type	Wall Thickness	Description	Weight Per S.F.	Type	Wall Thickness	Description	Weight Per S.F.
Brick	4"	Clay brick, high absorption	34 lb.	Clay tile	2"	Split terra cotta furring	10 lb.
		Clay brick, medium absorption	39			Non load bearing clay tile	11
		Clay brick, low absorption	46		3"	Split terra cotta furring	12
		Sand-lime brick	38			Non load bearing clay tile	18
		Concrete brick, heavy aggregate	46		4"	Non load bearing clay tile	20
		Concrete brick, light aggregate	33			Load bearing clay tile	24
	8"	Clay brick, high absorption	69		6"	Non load bearing clay tile	30
		Clay brick, medium absorption	79			Load bearing clay tile	36
		Clay brick, low absorption	89		8"	Non load bearing clay tile	36
		Sand-lime brick	74			Load bearing clay tile	42
		Concrete brick, heavy aggregate	89		12"	Non load bearing clay tile	46
		Concrete brick, light aggregate	68			Load bearing clay tile	58
	12"	Common brick	120	Gypsum block	2"	Hollow gypsum block	9.5
		Pressed brick	130			Solid gypsum block	12
		Sand-lime brick	105		3"	Hollow gypsum block	10
		Concrete brick, heavy aggregate	130			Solid gypsum block	18
		Concrete brick, light aggregate	98		4"	Hollow gypsum block	15
	16"	Clay brick, high absorption	134			Solid gypsum block	24
		Clay brick, medium absorption	155		5"	Hollow gypsum block	18
		Clay brick, low absorption	173		6"	Hollow gypsum block	24
		Sand-lime brick	138	Structural facing tile	2"	Facing tile	15
		Concrete brick, heavy aggregate	174		4"	Facing tile	25
		Concrete brick, light aggregate	130		6"	Facing tile	38
Concrete block	4"	Solid conc. block, stone aggregate	45	Glass	4"	Glass block	18
		Solid conc. block, lightweight	34		1"	Structural glass	15
		Hollow conc. block, stone aggregate	30	Plaster	1"	Gypsum plaster (1 side)	5
		Hollow conc. block, lightweight	20			Cement plaster (1 side)	10
	6"	Solid conc. block, stone aggregate	50			Gypsum plaster on lath	8
		Solid conc. block, lightweight	37			Cement plaster on lath	13
		Hollow conc. block, stone aggregate	42	Plaster partition (2 finished faces)	2"	Solid gypsum on metal lath	18
		Hollow conc. block, lightweight	30			Solid cement on metal lath	25
	8"	Solid conc. block, stone aggregate	67			Solid gypsum on gypsum lath	18
		Solid conc. block, lightweight	48			Gypsum on lath & metal studs	18
		Hollow conc. block, stone aggregate	55		3"	Gypsum on lath & metal studs	19
		Hollow conc. block, lightweight	38		4"	Gypsum on lath & metal studs	20
	10"	Solid conc. block, stone aggregate	84		6"	Gypsum on lath & wood studs	18
		Solid conc. block, lightweight	62	Concrete	6"	Reinf concrete, stone aggregate	75
		Hollow conc. block, stone aggregate	55			Reinf. concrete, lightweight	36-60
		Hollow conc. block, lightweight	38		8"	Reinf. concrete, stone aggregate	100
	12"	Solid conc. block, stone aggregate	108			Reinf. concrete, lightweight	48-80
		Solid conc. block, lightweight	72		10"	Reinf. concrete, stone aggregate	125
		Hollow conc. block, stone aggregate	85			Reinf. concrete, lightweight	60-100
		Hollow conc. block, lightweight	55		12"	Reinf concrete stone aggregate	150
Drywall	6"	Drywall on wood studs	10			Reinf. concrete, lightweight	72-120

Table L1010-202 Design Weight per S.F. for Roof Coverings

Type		Description	Weight lb. Per S.F.	Type	Wall Thickness	Description	Weight lb. Per S.F.
Sheathing	Gypsum	1" thick	4	Metal	Aluminum	Corr. & ribbed, .024" to .040"	.4-.8
	Wood	¾" thick	3		Copper	or tin	1.5-2.5
Insulation	per 1"	Loose	.5		Steel	Corrugated, 29 ga. to 12 ga.	.6-5.0
		Poured in place	2	Shingles	Asphalt	Strip shingles	1.7-2.8
		Rigid	1.5		Clay	Tile	8-16
Built-up	Tar & gravel	3 ply felt	5.5		Slate	¼" thick	9.5
		5 ply felt	6.5		Wood		2-3

Reference Aids — RL1010-200 Dead Load, Material Weights

Table L1010-203 Design Weight per Square Foot for Floor Fills and Finishes

Type		Description	Weight lb. per S.F.	Type		Description	Weight lb. per S.F.
Floor fill	per 1"	Cinder fill	5	Wood	Single 7/8"	On sleepers, light concrete fill	16
		Cinder concrete	9			On sleepers, stone concrete fill	25
		Lightweight concrete	3-9		Double 7/8"	On sleepers, light concrete fill	19
		Stone concrete	12			On sleepers, stone concrete fill	28
		Sand	8		3"	Wood block on mastic, no fill	15
		Gypsum	6			Wood block on ½" mortar	16
Terrazzo Marble	1" and mortar	Terrazzo, 2" stone concrete on stone concrete fill	25 33		per 1"	Hardwood flooring (25/32") Underlayment (Plywood per 1")	4 3
Resilient	1/16"-1/4"	Linoleum, asphalt, vinyl tile	2	Asphalt	1-1/2"	Mastic flooring	18
Tile	3/4"	Ceramic or quarry	10		2"	Block on ½" mortar	30

Table L1010-204 Design Weight Per Cubic Foot for Miscellaneous Materials

Type		Description	Weight lb. per C.F.	Type		Description	Weight lb. per C.F.
Bituminous	Coal, piled	Anthracite	47-58	Masonry	Ashlar	Granite	165
		Bituminous	40-54			Limestone, crystalline	165
		Peat, turf, dry	47			Limestone, oolitic	135
		Coke	75			Marble	173
	Petroleum	Unrefined	54			Sandstone	144
		Refined	50		Rubble, in mortar	Granite	155
		Gasoline	42			Limestone, crystalline	147
	Pitch		69			Limestone, oolitic	138
	Tar	Bituminous	75			Marble	156
Concrete	Plain	Stone aggregate	144			Sandstone & Bluestone	130
		Slag aggregate	132		Brick	Pressed	140
		Expanded slag aggregate	100			Common	120
		Haydite (burned clay agg.)	90			Soft	100
		Vermiculite & perlite, load bearing	70-105		Cement	Portland, loose	90
		Vermiculite & perlite, non load bear	35-50			Portland set	183
	Reinforced	Stone aggregate	150		Lime	Gypsum, loose	53-64
		Slag aggregate	138		Mortar	Set	103
		Lightweight aggregates	30-120	Metals	Aluminum	Cast, hammered	165
Earth	Clay	Dry	63		Brass	Cast, rolled	534
		Damp, plastic	110		Bronze	7.9 to 14% Sn	509
		and gravel, dry	100		Copper	Cast, rolled	556
	Dry	Loose	76		Iron	Cast, pig	450
		Packed	95			Wrought	480
	Moist	Loose	78		Lead		710
		Packed	96		Monel		556
	Mud	Flowing	108		Steel	Rolled	490
		Packed	115		Tin	Cast, hammered	459
	Riprap	Limestone	80-85		Zinc	Cast rolled	440
		Sandstone	90	Timber	Cedar	White or red	24.2
		Shale	105		Fir	Douglas	23.7
	Sand & gravel	Dry, loose	90-105			Eastern	25
		Dry, packed	100-120		Maple	Hard	44.5
		Wet	118-120			White	33
Gases	Air	0C., 760 mm.	.0807		Oak	Red or Black	47.3
	Gas	Natural	.0385			White	47.3
Liquids	Alcohol	100%	49		Pine	White	26
	Water	4°C., maximum density	62.5			Yellow, long leaf	44
		Ice	56			Yellow, short leaf	38
		Snow, fresh fallen	8		Redwood	California	26
		Sea water	64		Spruce	White or black	27

Reference Aids — RL1010-220 Dead Load, System Weights

Table L1010-225 Design Weight Per S.F. for Structural Floor and Roof Systems

Type Slab		Description	Weight in Pounds per S.F. — Slab Depth in Inches												
Concrete Slab	Reinforced	Stone aggregate	1"	12.5	2"	25	3"	37.5	4"	50	5"	62.5	6"	75
		Lightweight sand aggregate		9.5		19		28.5		38		47.5		57
		All lightweight aggregate		9.0		18		27.0		36		45.0		54
	Plain, nonreinforced	Stone aggregate		12.0		24		36.0		48		60.0		72
		Lightweight sand aggregate		9.0		18		27.0		36		45.0		54
		All lightweight aggregate		8.5		17		25.5		34		42.5		51
Concrete Waffle	19" x 19"	5" wide ribs @ 24" O.C.	6+3	77	8+3	92	10+3	100	12+3	118				
			6+4½	96	8+4½	110	10+4½	119	12+4½	136				
	30" x 30"	6" wide ribs @ 36" O.C.	8+3	83	10+3	95	12+3	109	14+3	118	16+3	130	20+3	155
			8+4½	101	10+4½	113	12+4½	126	14+4½	137	16+4½	149	20+4½	173
Concrete Joist	20" wide form	5" wide rib	8+3	60	10+3	67	12+3	74	14+3	81				
		6" wide rib		63		70		78		86	16+3	94	20+3	111
		7" wide rib										99		118
		5" wide rib	8+4½	79	10+4½	85	12+4½	92	14+4½	99				
		6" wide rib		82		89		97		104	16+4½	113	20+4½	130
		7" wide rib										118		136
	30" wide form	5" wide rib	8+3	54	10+3	58	12+3	63	14+3	68				
		6" wide rib		56		61		67		72	16+3	78	20+3	91
		7" wide rib										83		96
		5" wide rib	8+4½	72	10+4½	77	12+4½	82	14+4½	87				
		6" wide rib		75		80		85		91	16+4½	97	20+4½	109
		7" wide rib										101		115
Wood Joists	Incl. Subfloor	12" O.C.	2x6	6	2x8	6	2x10	7	2x12	8	3x8	8	3x12	11
		16" O.C.		5		6		6		7		7		9

Table L1010-226 Superimposed Dead Load Ranges

Component	Load Range (PSF)
Ceiling	5-10
Partitions	20-30
Mechanical	4-8

Reference Aids — RL1010-300 Wind Loads

Table L1010-301 Design Loads for Structures for Wind Load

Wind Loads: Structures are designed to resist the wind force from any direction. Usually 2/3 is assumed to act on the windward side, 1/3 on the leeward side.

For more than 1/3 openings, add 10 psf for internal wind pressure or 5 psf for suction, whichever is critical.

For buildings and structures, use psf values from Table L1010-301.

For glass over 4 S.F. use values in Table L1010-302 after determining 30' wind velocity from Table L1010-303.

Type Structure	Height Above Grade	Horizontal Load in Lb. per S.F.
Buildings	Up to 50 ft.	15
	50 to 100 ft.	20
	Over 100 ft.	20 + .025 per ft.
Ground signs & towers	Up to 50 ft.	15
	Over 50 ft.	20
Roof structures		30
Glass	See Table below	

Table L1010-302 Design Wind Load in PSF for Glass at Various Elevations (BOCA Code)

Height From Grade	Vel.	PSF	Vel.	PSF	Vel.	PSF	Vel.	PSF	Vel.	PSF	Vel.	PSF	Vel.	PSF	Vel.	PSF	Vel.	PSF	Vel.	PSF	Vel.	PSF
To 10 ft.	42	6	46	7	49	8	52	9	55	10	59	11	62	12	66	14	69	15	76	19	83	22
10-20	52	9	58	11	61	11	65	14	70	16	74	18	79	20	83	22	87	24	96	30	105	35
20-30*	60	12	67	14	70	16	75	18	80	20	85	23	90	26	95	29	100	32	110	39	120	46
30-60	66	14	74	18	77	19	83	22	88	25	94	28	99	31	104	35	110	39	121	47	132	56
60-120	73	17	82	21	85	12	92	27	98	31	104	35	110	39	116	43	122	48	134	57	146	68
120-140	81	21	91	26	95	29	101	33	108	37	115	42	122	48	128	52	135	48	149	71	162	84
240-480	90	26	100	32	104	35	112	40	119	45	127	51	134	57	142	65	149	71	164	86	179	102
480-960	98	31	110	39	115	42	123	49	131	55	139	62	148	70	156	78	164	86	180	104	197	124
Over 960	98	31	110	39	115	42	123	49	131	55	139	62	148	70	156	78	164	86	180	104	197	124

*Determine appropriate wind at 30' elevation Fig. L1010-303 below.

Table L1010-303 Design Wind Velocity at 30 Ft. Above Ground

SPEEDS ARE FOR NORMAL EXPOSURE WHERE SURFACE FRICTION IS RELATIVELY UNIFORM FOR A FETCH OF ABOUT 25 MILES. IF THE EXPOSURE IS ELEVATED, SUBJECT TO CHANNELING, OR OTHER SPECIAL CONDITIONS AFFECTING THE EXTREME WIND SPEEDS, ADJUSTMENTS MUST BE MADE TO THE MAP VALUES.

Reference Aids — RL1010-400 Snow Loads

Table L1010-401 Snow Load in Pounds Per Square Foot on the Ground

Based on 50 year storm for INDUSTRIAL BUSINESS, MERCANTILE, & RESIDENTIAL

Table L1010-402 Snow Load in Pounds Per Square Foot on the Ground

Based on 100 year storm for ASSEMBLY INSTITUTIONAL, HIGH HAZARD MOTELS

Reference Aids — RL1010-400 Snow Loads

Table L1010-403 Snow Loads

To convert the ground snow loads on the previous page to roof snow loads, the ground snow loads should be multiplied by the following factors depending upon the roof characteristics.

Note, in all cases $\dfrac{\alpha - 30}{50}$ is valid only for > 30 degrees.

Description	Sketch	Formula	Angle	Conversion Factor = C_f Sheltered	Exposed
Simple flat and shed roofs	(sketch; Load Diagram)	For $\alpha > 30°$ $C_f = 0.8 - \dfrac{\alpha - 30}{50}$	0° to 30°	0.8	0.6
			40°	0.6	0.45
			50°	0.4	0.3
			60°	0.2	0.15
			70° to 90°	0	0

Description	Sketch	Formula	Angle	Case I	Case II
Simple gable and hip roofs	(sketch; Case I Load Diagram; Case II Load Diagram)	$C_f = 0.8 - \dfrac{\alpha - 30}{50}$ $C_f = 1.25\left(0.8 - \dfrac{\alpha - 30}{50}\right)$	10°	0.8	—
			20°	0.8	—
			30°	0.8	1.0
			40°	0.6	0.75
			50°	0.4	0.5
			60°	0.2	0.25

Description	Sketch	Formula	Condition
Valley areas of Two span roofs	(sketch; Case I, Case II)	$\beta = \dfrac{\alpha_1 + \alpha_2}{2}$ $C_f = 0.8 - \dfrac{\alpha - 30}{50}$	$\beta \le 10°$ use Case I only $\beta > 10°$ $\beta < 20°$ use Case I & II $\beta \ge 20°$ use Case I, II & III
Lower level of multi level roofs (or on an adjacent building not more then 15 ft. away)	(sketch; $W = 2h$)	$C_f = 15 \dfrac{h}{g}$ h = difference in roof height in feet g = ground snow load in psf w = width of drift For $h < 5$, $w = 10$ $h > 15$, $w = 30$	When $15 \dfrac{h}{g} < .8$, use 0.8 When $15 \dfrac{h}{g} > 3.0$, use 3.0

Summary of above:

1. For flat roofs or roofs up to 30°, use 0.8 x ground snow load for the roof snow load.
2. For roof pitches in excess of 30°, conversion factor becomes lower than 0.8.
3. For exposed roofs there is a further 25% reduction of conversion factor.
4. For steep roofs a more highly loaded half span must be considered.
5. For shallow roof valleys conversion factor is 0.8.
6. For moderate roof valleys, conversion factor is 1.0 for half the span.
7. For steep roof valleys, conversion factor is 1.5 for one quarter of the span.
8. For roofs adjoining vertical surfaces the conversion factor is up to 3.0 for part of the span.
9. If snow load is less than 30 psf, use water load on roof for clogged drain condition.

Reference Aids   RL1020-100 Area Requirements

Table L1020-101 Floor Area Ratios

Table below lists commonly used gross to net area and net to gross area ratios expressed in % for various building types.

Building Type	Gross to Net Ratio	Net to Gross Ratio	Building Type	Gross to Net Ratio	Net to Gross Ratio
Apartment	156	64	School Buildings (campus type)		
Bank	140	72	Administrative	150	67
Church	142	70	Auditorium	142	70
Courthouse	162	61	Biology	161	62
Department Store	123	81	Chemistry	170	59
Garage	118	85	Classroom	152	66
Hospital	183	55	Dining Hall	138	72
Hotel	158	63	Dormitory	154	65
Laboratory	171	58	Engineering	164	61
Library	132	76	Fraternity	160	63
Office	135	75	Gymnasium	142	70
Restaurant	141	70	Science	167	60
Warehouse	108	93	Service	120	83
			Student Union	172	59

The gross area of a building is the total floor area based on outside dimensions.

The net area of a building is the usable floor area for the function intended and excludes such items as stairways, corridors, and mechanical rooms.

In the case of a commercial building, it might be considered as the "leasable area."

Reference Aids

RL1020-200 Partition/Door Guidelines

Table L1020-201 Partition/Door Density

Building Type		Stories	Partition/Density	Doors	Description of Partition
Apartments		1 story	9 SF/LF	90 SF/door	Plaster, wood doors & trim
		2 story	8 SF/LF	80 SF/door	Drywall, wood studs, wood doors & trim
		3 story	9 SF/LF	90 SF/door	Plaster, wood studs, wood doors & trim
		5 story	9 SF/LF	90 SF/door	Plaster, metal studs, wood doors & trim
		6-15 story	8 SF/LF	80 SF/door	Drywall, metal studs, wood doors & trim
Bakery		1 story	50 SF/LF	500 SF/door	Conc. block, paint, door & drywall, wood studs
		2 story	50 SF/LF	500 SF/door	Conc. block, paint, door & drywall, wood studs
Bank		1 story	20 SF/LF	200 SF/door	Plaster, wood studs, wood doors & trim
		2-4 story	15 SF/LF	150 SF/door	Plaster, metal studs, wood doors & trim
Bottling Plant		1 story	50 SF/LF	500 SF/door	Conc. block, drywall, metal studs, wood trim
Bowling Alley		1 story	50 SF/LF	500 SF/door	Conc. block, wood & metal doors, wood trim
Bus Terminal		1 story	15 SF/LF	150 SF/door	Conc. block, ceramic tile, wood trim
Cannery		1 story	100 SF/LF	1000 SF/door	Drywall on metal studs
Car Wash		1 story	18 SF/LF	180 SF/door	Concrete block, painted & hollow metal door
Dairy Plant		1 story	30 SF/LF	300 SF/door	Concrete block, glazed tile, insulated cooler doors
Department Store		1 story	60 SF/LF	600 SF/door	Drywall, metal studs, wood doors & trim
		2-5 story	60 SF/LF	600 SF/door	30% concrete block, 70% drywall, wood studs
Dormitory		2 story	9 SF/LF	90 SF/door	Plaster, concrete block, wood doors & trim
		3-5 story	9 SF/LF	90 SF/door	Plaster, concrete block, wood doors & trim
		6-15 story	9 SF/LF	90 SF/door	Plaster, concrete block, wood doors & trim
Funeral Home		1 story	15 SF/LF	150 SF/door	Plaster on concrete block & wood studs, paneling
		2 story	14 SF/LF	140 SF/door	Plaster, wood studs, paneling & wood doors
Garage Sales & Service		1 story	30 SF/LF	300 SF/door	50% conc. block, 50% drywall, wood studs
Hotel		3-8 story	9 SF/LF	90 SF/door	Plaster, conc. block, wood doors & trim
		9-15 story	9 SF/LF	90 SF/door	Plaster, conc. block, wood doors & trim
Laundromat		1 story	25 SF/LF	250 SF/door	Drywall, wood studs, wood doors & trim
Medical Clinic		1 story	6 SF/LF	60 SF/door	Drywall, wood studs, wood doors & trim
		2-4 story	6 SF/LF	60 SF/door	Drywall, metal studs, wood doors & trim
Motel		1 story	7 SF/LF	70 SF/door	Drywall, wood studs, wood doors & trim
		2-3 story	7 SF/LF	70 SF/door	Concrete block, drywall on metal studs, wood paneling
Movie Theater	200-600 seats	1 story	18 SF/LF	180 SF/door	Concrete block, wood, metal, vinyl trim
	601-1400 seats		20 SF/LF	200 SF/door	Concrete block, wood, metal, vinyl trim
	1401-22000 seats		25 SF/LF	250 SF/door	Concrete block, wood, metal, vinyl trim
Nursing Home		1 story	8 SF/LF	80 SF/door	Drywall, metal studs, wood doors & trim
		2-4 story	8 SF/LF	80 SF/door	Drywall, metal studs, wood doors & trim
Office		1 story	20 SF/LF	200-500 SF/door	30% concrete block, 70% drywall on wood studs
		2 story	20 SF/LF	200-500 SF/door	30% concrete block, 70% drywall on metal studs
		3-5 story	20 SF/LF	200-500 SF/door	30% concrete block, 70% movable partitions
		6-10 story	20 SF/LF	200-500 SF/door	30% concrete block, 70% movable partitions
		11-20 story	20 SF/LF	200-500 SF/door	30% concrete block, 70% movable partitions
Parking Ramp (Open)		2-8 story	60 SF/LF	600 SF/door	Stair and elevator enclosures only
Parking garage		2-8 story	60 SF/LF	600 SF/door	Stair and elevator enclosures only
Pre-Engineered	Steel	1 story	0		
	Store	1 story	60 SF/LF	600 SF/door	Drywall on metal studs, wood doors & trim
	Office	1 story	15 SF/LF	150 SF/door	Concrete block, movable wood partitions
	Shop	1 story	15 SF/LF	150 SF/door	Movable wood partitions
	Warehouse	1 story	0		
Radio & TV Broadcasting		1 story	25 SF/LF	250 SF/door	Concrete block, metal and wood doors
& TV Transmitter		1 story	40 SF/LF	400 SF/door	Concrete block, metal and wood doors
Self Service Restaurant		1 story	15 SF/LF	150 SF/door	Concrete block, wood and aluminum trim
Cafe & Drive-in Restaurant		1 story	18 SF/LF	180 SF/door	Drywall, metal studs, ceramic & plastic trim
Restaurant with seating		1 story	25 SF/LF	250 SF/door	Concrete block, paneling, wood studs & trim
Supper Club		1 story	25 SF/LF	250 SF/door	Concrete block, paneling, wood studs & trim
Bar or Lounge		1 story	24 SF/LF	240 SF/door	Plaster or gypsum lath, wooded studs
Retail Store or Shop		1 story	60 SF/LF	600 SF/door	Drywall metal studs, wood doors & trim
Service Station	Masonry	1 story	15 SF/LF	150 SF/door	Concrete block, paint, door & drywall, wood studs
	Metal panel	1 story	15 SF/LF	150 SF/door	Concrete block, paint, door & drywall, wood studs
	Frame	1 story	15 SF/LF	150 SF/door	Drywall, wood studs, wood doors & trim
Shopping Center	(strip)	1 story	30 SF/LF	300 SF/door	Drywall, metal studs, wood doors & trim
	(group)	1 story	40 SF/LF	400 SF/door	50% concrete block, 50% drywall, wood studs
		2 story	40 SF/LF	400 SF/door	50% concrete block, 50% drywall, wood studs
Small Food Store		1 story	30 SF/LF	300 SF/door	Concrete block drywall, wood studs, wood trim
Store/Apt. above	Masonry	2 story	10 SF/LF	100 SF/door	Plaster, metal studs, wood doors & trim
	Frame	2 story	10 SF/LF	100 SF/door	Plaster, metal studs, wood doors & trim
	Frame	3 story	10 SF/LF	100 SF/door	Plaster, metal studs, wood doors & trim
Supermarkets		1 story	40 SF/LF	400 SF/door	Concrete block, paint, drywall & porcelain panel
Truck Terminal		1 story	0		
Warehouse		1 story	0		

RL1020-300 Building Code Requirements

Table L1020-301 Occupancy Determinations

Description		S.F. Required per Person
		IBC
Assembly Areas	Fixed seats**	# fixed seats
	Without fixed seats	
	Concentrated (chairs only)	7 net
	Standing Space	5 net
	Unconcentrated (tables and chairs)	15 net
	Gaming floors (Keno, slots, etc.)	11 gross
Educational	Classroom area	20 net
	Shops and other vocational room areas	50 net
Institutional	Inpatient treatment areas	240 gross
	Outpatient areas	100 gross
	Sleeping Areas	120 gross
Mercantile	Areas on other floors	60 gross
	Basement and grade floor areas	30 gross
	Storage, stock, shipping areas	300 gross
Business Areas		100 gross

IBC = *International Building Code*, 2006 Edition

** The occupancy load for assembly area with fixed seats shall be determined by the number of fixed seats installed.

Table L1020-302 Length of Exitway Access Travel (ft.)

Use Group	Without Fire Suppression System	With Fire Suppression System
Assembly	200	250
Business	200	300
Factory and industrial	300	400
High hazard	—	75
Institutional	200	250
Mercantile	200	250
Residential	200	250
Storage, low hazard	300	400
Storage, moderate hazard	200	250

Note: The maximum length of exitway access travel in unlimited area buildings shall be 400'.

Table L1020-303 Capacity Per Unit Egress Width*

	Without Fire Suppression System (Inches per Person)*		With Fire Suppression System (Inches per Person)*	
Use Group	Stairways	Doors, Ramps and Corridors	Stairways	Doors, Ramps and Corridors
Assembly, Business, Educational, Factory Industrial, Mercantile, Residential, Storage	0.3	0.2	0.2	0.15
Institutional—1	0.3	0.2	0.2	0.15
Institutional—2	—	0.7	0.3	0.2
Institutional—3	0.3	0.2	0.2	0.15
High Hazard	0.7	0.4	0.3	0.2

* 1" = 25.4 mm

Reference Aids — RL1030-400 Thermal Properties of Materials

Table L1030-401 Resistances ("R") of Building and Insulating Materials

Material	Wt./Lbs. per C.F.	R per Inch	R Listed Size
Air Spaces and Surfaces			
Enclosed non-reflective spaces, E=0.82			
50° F mean temp., 30°/10° F diff.			
.5"			.90/.91
.75"			.94/1.01
1.50"			.90/1.02
3.50"			.91/1.01
Inside vert. surface (still air)			0.68
Outside vert. surface (15 mph wind)			0.17
Building Boards			
Asbestos cement, 0.25" thick	120		0.06
Gypsum or plaster, 0.5" thick	50		0.45
Hardboard regular	50	1.37	
Tempered	63	1.00	
Laminated paper	30	2.00	
Particle board	37	1.85	
	50	1.06	
	63	0.85	
Plywood (Douglas Fir), 0.5" thick	34		0.62
Shingle backer, .375" thick	18		0.94
Sound deadening board, 0.5" thick	15		1.35
Tile and lay-in panels, plain or			
acoustical, 0.5" thick	18		1.25
Vegetable fiber, 0.5" thick	18		1.32
	25		1.14
Wood, hardwoods	48	0.91	
Softwoods	32	1.25	
Flooring Carpet with fibrous pad			2.08
With rubber pad			1.23
Cork tile, 1/8" thick			0.28
Terrazzo			0.08
Tile, resilient			0.05
Wood, hardwood, 0.75" thick			0.68
Subfloor, 0.75" thick			0.94
Glass			
Insulation, 0.50" air space			2.04
Single glass			0.91
Insulation Blanket or Batt, mineral, glass			
or rock fiber, approximate thickness			
3.0" to 3.5" thick			11
3.5" to 4.0" thick			13
6.0" to 6.5" thick			19
6.5" to 7.0" thick			22
8.5" to 9.0" thick			30
Boards			
Cellular glass	8.5	2.63	
Fiberboard, wet felted			
Acoustical tile	21	2.70	
Roof insulation	17	2.94	
Fiberboard, wet molded			
Acoustical tile	23	2.38	
Mineral fiber with resin binder	15	3.45	
Polystyrene, extruded,			
cut cell surface	1.8	4.00	
smooth skin surface	2.2	5.00	
	3.5	5.26	
Bead boards	1.0	3.57	
Polyurethane	1.5	6.25	
Wood or cane fiberboard, 0.5" thick			1.25

Material	Wt./Lbs. per C.F.	R per Inch	R Listed Size
Insulation Loose Fill			
Cellulose	2.3	3.13	
	3.2	3.70	
Mineral fiber, 3.75" to 5" thick	2-5		11
6.5" to 8.75" thick			19
7.5" to 10" thick			22
10.25" to 13.75" thick			30
Perlite	5-8	2.70	
Vermiculite	4-6	2.27	
Wood fiber	2-3.5	3.33	
Masonry Brick, Common	120	0.20	
Face	130	0.11	
Cement mortar	116	0.20	
Clay tile, hollow			
1 cell wide, 3" width			0.80
4" width			1.11
2 cells wide, 6" width			1.52
8" width			1.85
10" width			2.22
3 cells wide, 12" width			2.50
Concrete, gypsum fiber	51	0.60	
Lightweight	120	0.19	
	80	0.40	
	40	0.86	
Perlite	40	1.08	
Sand and gravel or stone	140	0.08	
Concrete block, lightweight			
3 cell units, 4"-15 lbs. ea.			1.68
6"-23 lbs. ea.			1.83
8"-28 lbs. ea.			2.12
12"-40 lbs. ea.			2.62
Sand and gravel aggregates,			
4"-20 lbs. ea.			1.17
6"-33 lbs. ea.			1.29
8"-38 lbs. ea.			1.46
12"-56 lbs. ea.			1.81
Plastering Cement Plaster,			
Sand aggregate	116	0.20	
Gypsum plaster, Perlite aggregate	45	0.67	
Sand aggregate	105	0.18	
Vermiculite aggregate	45	0.59	
Roofing			
Asphalt, felt, 15 lb.			0.06
Rolled roofing	70		0.15
Shingles	70		0.44
Built-up roofing .375" thick	70		0.33
Cement shingles	120		0.21
Vapor-permeable felt			0.06
Vapor seal, 2 layers of			
mopped 15 lb. felt			0.12
Wood, shingles 16"-7.5" exposure			0.87
Siding			
Aluminum or steel (hollow backed)			
oversheathing			0.61
With .375" insulating backer board			1.82
Foil backed			2.96
Wood siding, beveled, ½" x 8"			0.81

Reference Aids   RL1030-500 Weather Information

Table L1030-502 Maximum Depth of Frost Penetration in Inches

Reference Aids — RL1040-100 Construction Classification

Table L1040-101 Fire-Resisting Ratings of Structural Elements (in hours)

Description of the Structural Element	No. 1 Fireproof		No. 2 Non Combustible			No. 3 Exterior Masonry Wall			No. 4 Frame	
			Protected		Unprotected	Heavy Timber	Ordinary		Protected	Unprotected
							Protected	Unprotected		
	1A	1B	2A	2B	2C	3A	3B	3C	4A	4B
Exterior, Bearing Walls	4	3	2	1½	1	2	2	2	1	1
Nonbearing Walls	2	2	1½	1	1	2	2	2	1	1
Interior Bearing Walls and Partitions	4	3	2	1	0	2	1	0	1	0
Fire Walls and Party Walls	4	3	2	2	2	2	2	2	2	2
Fire Enclosure of Exitways, Exit Hallways and Stairways	2	2	2	2	2	2	2	2	1	1
Shafts other than Exitways, Hallways and Stairways	2	2	2	2	2	2	2	2	1	1
Exitway access corridors and Vertical separation of tenant space	1	1	1	1	0	1	1	0	1	0
Columns, girders, trusses (other than roof trusses) and framing: Supporting more than one floor	4	3	2	1	0	—	1	0	1	0
Supporting one floor only	3	2	1½	1	0	—	1	0	1	0
Structural members supporting wall	3	2	1½	1	0	1	1	0	1	0
Floor construction including beams	3	2	1½	1	0	—	1	0	1	0
Roof construction including beams, trusses and framing arches and roof deck 15′ or less in height to lowest member	2	1½	1	1	0	—	1	0	1	0

Note:
a. Codes include special requirements and exceptions that are not included in the table above.
b. Each type of construction has been divided into sub-types which vary according to the degree of fire resistance required. Sub-types (A) requirements are more severe than those for sub-types (B).
c. Protected construction means all structural members are chemically treated, covered or protected so that the unit has the required fire resistance.

Type No. 1, Fireproof Construction — Buildings and structures of fireproof construction are those in which the walls, partitions, structural elements, floors, ceilings, and roofs, and the exitways are protected with approved noncombustible materials to afford the fire-resistance rating specified in Table L1040-101; except as otherwise specifically regulated. Fire-resistant treated wood may be used as specified.

Type No. 2, Noncombustible Construction — Buildings and structures of noncombustible construction are those in which the walls, partitions, structural elements, floors, ceilings, roofs and the exitways are approved noncombustible materials meeting the fire-resistance rating requirements specified in Table L1040-101; except as modified by the fire limit restrictions. Fire-retardant treated wood may be used as specified.

Type No. 3, Exterior Masonry Wall Construction — Buildings and structures of exterior masonry wall construction are those in which the exterior, fire and party walls are masonry or other approved noncombustible materials of the required fire-resistance rating and structural properties. The floors, roofs, and interior framing are wholly or partly wood or metal or other approved construction. The fire and party walls are ground-supported; except that girders and their supports, carrying walls of masonry shall be protected to afford the same degree of fire-resistance rating of the supported walls. All structural elements have the required fire-resistance rating specified in Table L1040-101.

Type No. 4, Frame Construction — Buildings and structures of frame construction are those in which the exterior walls, bearing walls, partitions, floor and roof construction are wholly or partly of wood stud and joist assemblies with a minimum nominal dimension of two inches or of other approved combustible materials. Fire stops are required at all vertical and horizontal draft openings in which the structural elements have required fire-resistance ratings specified in Table L1040-101.

Reference Aids — RL1040-200 Fire Hazard Classification

Table L1040-201 Fire Grading of Building in Hours for Fire Walls and Fire Separation Walls

Fire Hazard for Fire Walls

The degree of fire hazard of buildings relating to their intended use is defined by "Fire Grading" the buildings. Such a grading system is listed in the Table L1040-201 below. This type of grading determines the requirements for fire walls and fire separation walls (exterior fire exposure). For mixed use occupancy, use the higher Fire Grading requirement of the components.

Building Type	Hours	Building Type	Hours
Businesses	2	Recreation Centers	2
Churches	1½	Residential Hotels	2
Factories	3	Residential, Multi-family Dwellings	1½
High Hazard*	4	Residential, One & Two Family Dwellings	¾
Industrial	3	Restaurants	2
Institutional, Incapacitated Occupants	2	Schools	1½
Institutional, Restrained Occupants	3	Storage, Low Hazard*	2
Lecture Halls	2	Storage, Moderate Hazard*	3
Mercantile	3	Terminals	2
Night Club	3	Theatres	3

Note: *The difference in "Fire Hazards" is determined by their occupancy and use.

High Hazard: Industrial and storage buildings in which the combustible contents might cause fires to be unusually intense or where explosives, combustible gases or flammable liquids are manufactured or stored.

Moderate Hazard: Mercantile buildings, industrial and storage buildings in which combustible contents might cause fires of moderate intensity.

Low Hazard: Business buildings that ordinarily do not burn rapidly.

Table L1040-202 Interior Finish Classification

Flame Spread for Interior Finishes

The flame spreadability of a material is the burning characteristic of the material relative to the fuel contributed by its combustion and the density of smoke developed. The flame spread classification of a material is based on a ten minute test on a scale of 0 to 100. Cement asbestos board is assigned a rating of 0 and select red oak flooring a rating of 100. The four classes are listed in Table L1040-202.

The flame spread ratings for interior finish walls and ceilings shall not be greater than the Class listed in the Table L1040-203 below.

Class of Material	Surface Burning Characteristics
I	0 to 25
II	26 to 75
III	76 to 200
IV	201 to 5000

Table L1040-203 Interior Finish Requirements by Class

Building Type	Vertical Exitways & Passageways	Corridors Providing Exitways	Rooms or Enclosed Spaces	Building Type	Vertical Exitways & Passageways	Corridors Providing Exitways	Rooms or Enclosed Space
Assembly Halls	I	I	II	Mercantile Walls	I	II	III
Businesses	I	II	III	Night Clubs	I	I	II
Churches	I	I	II	Residential Hotels	I	II	III
Factory	I	II	III	Residential, Multi-family	I	II	III
High Hazard	I	II	II	Residential, 1 & 2 Family	IV	IV	IV
Industrial	I	II	III	Restaurants	I	I	II
Institutional, Incapacitated	I	II	I	Storage, Low Hazard	I	II	III
Institutional, Restrained	I	I	I	Storage, Moderate Hazard	I	II	III
Mercantile Ceilings	I	II	II	Terminals	I	I	II

Reference Aids — RL1090-100 Conversion Factors

Description: This table is primarily for converting customary U.S. units in the left hand column to SI metric units in the right hand column. In addition, conversion factors for some commonly encountered Canadian and non-SI metric units are included.

Table L1090-101 Metric Conversion Factors

	If You Know		Multiply By		To Find
Length	Inches	x	25.4[a]	=	Millimeters
	Feet	x	0.3048[a]	=	Meters
	Yards	x	0.9144[a]	=	Meters
	Miles (statute)	x	1.609	=	Kilometers
Area	Square inches	x	645.2	=	Square millimeters
	Square feet	x	0.0929	=	Square meters
	Square yards	x	0.8361	=	Square meters
Volume (Capacity)	Cubic inches	x	16,387	=	Cubic millimeters
	Cubic feet	x	0.02832	=	Cubic meters
	Cubic yards	x	0.7646	=	Cubic meters
	Gallons (U.S. liquid)[b]	x	0.003785	=	Cubic meters[c]
	Gallons (Canadian liquid)[b]	x	0.004546	=	Cubic meters[c]
	Ounces (U.S. liquid)[b]	x	29.57	=	Milliliters[c,d]
	Quarts (U.S. liquid)[b]	x	0.9464	=	Liters[c,d]
	Gallons (U.S. liquid)[b]	x	3.785	=	Liters[c,d]
Force	Kilograms force[d]	x	9.807	=	Newtons
	Pounds force	x	4.448	=	Newtons
	Pounds force	x	0.4536	=	Kilograms force[d]
	Kips	x	4448	=	Newtons
	Kips	x	453.6	=	Kilograms force[d]
Pressure, Stress, Strength (Force per unit area)	Kilograms force per square centimeter[d]	x	0.09807	=	Megapascals
	Pounds force per square inch (psi)	x	0.006895	=	Megapascals
	Kips per square inch	x	6.895	=	Megapascals
	Pounds force per square inch (psi)	x	0.07031	=	Kilograms force per square centimeter[d]
	Pounds force per square foot	x	47.88	=	Pascals
	Pounds force per square foot	x	4.882	=	Kilograms force per square meter[d]
Bending Moment Or Torque	Inch-pounds force	x	0.01152	=	Meter-kilograms force[d]
	Inch-pounds force	x	0.1130	=	Newton-meters
	Foot-pounds force	x	0.1383	=	Meter-kilograms force[d]
	Foot-pounds force	x	1.356	=	Newton-meters
	Meter-kilograms force[d]	x	9.807	=	Newton-meters
Mass	Ounces (avoirdupois)	x	28.35	=	Grams
	Pounds (avoirdupois)	x	0.4536	=	Kilograms
	Tons (metric)	x	1000	=	Kilograms
	Tons, short (2000 pounds)	x	907.2	=	Kilograms
	Tons, short (2000 pounds)	x	0.9072	=	Megagrams[e]
Mass per Unit Volume	Pounds mass per cubic foot	x	16.02	=	Kilograms per cubic meter
	Pounds mass per cubic yard	x	0.5933	=	Kilograms per cubic meter
	Pounds mass per gallon (U.S. liquid)[b]	x	119.8	=	Kilograms per cubic meter
	Pounds mass per gallon (Canadian liquid)[b]	x	99.78	=	Kilograms per cubic meter
Temperature	Degrees Fahrenheit	(F−32)/1.8		=	Degrees Celsius
	Degrees Fahrenheit	(F+459.67)/1.8		=	Degrees Kelvin
	Degrees Celsius	C+273.15		=	Degrees Kelvin

[a] The factor given is exact
[b] One U.S. gallon = 0.8327 Canadian gallon
[c] 1 liter = 1000 milliliters = 1000 cubic centimeters
   1 cubic decimeter = 0.001 cubic meter
[d] Metric but not SI unit
[e] Called "tonne" in England and "metric ton" in other metric countries

Life Cycle Costing (LCC) for Evaluating Economic Performance of Building Investments

Introduction

A complete analysis of maintenance costs requires consideration of life cycle costs, or LCC, as it is often called. Under current conditions, facilities managers have to wisely spend their limited operating and maintenance budgets. When deciding how to allocate limited budgets, maintenance professionals are constantly faced with the economic decision of when to stop maintaining and start replacing. Once the decision to replace has been made, LCC analysis can enhance the choice of alternatives as well.

The American Society of Testing and Materials (ASTM) Subcommittee E06.81 on Building Economics has developed a series of standards for evaluating the economic performance of building investments. These standards, which have been endorsed by the American Association of Cost Engineers (AACE), are intended to provide uniformity in both the terminology and calculation methods utilized for Life Cycle Costing. Without these standards as a basis for calculations, life cycle cost comparisons could be misleading or incomplete. This chapter is based on the current ASTM standards which are listed in the references at the end of this section.

Definition of Life Cycle Costing
(ASTM Standard Terminology of Building Economics E833-89)

A technique of economic evaluation that sums over a given study period the cost of initial investment less resale value, replacements, operations, energy use, and maintenance and repair of an investment decision. The costs are expressed in either lump sum present value terms or in equivalent uniform annual values.

This technique for life cycle costing is important to the facilities manager, as it is a valuable tool for evaluating investment alternatives and supports his or her decision-making.

To understand Life Cycle Costing, one should be familiar with the notions of compounding, discounting, present value, and equivalent uniform annual value.

Compounding

Definition of Compounding *(Webster's New Collegiate Dictionary)*
The process of computing the value of an original principal sum based on interest calculated on the sum of the original principal and accrued interest.

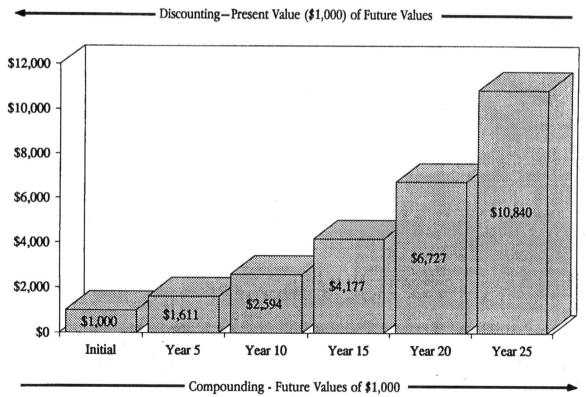

Figure 1: Compounding and Discounting at 10% Per Annum

Figure 1 illustrates compounding. As shown, a principal sum of $1,000 at a compound interest rate of 10% per annum increases to $1,611 in 5 years, $2,594 in 10 years, $4,177 in 15 years, $6,727 in 20 years, and $10,840 in 25 years; this process (compounding) is usually well understood in that it represents the amount accumulated in a bank account for a specific amount at a given interest rate and calculated on the sum of the original principal and accrued interest.

Discounting

ASTM Definition of Discounting

A technique for converting cash flows that occur over time to equivalent amounts at a common time (base time).

It is also evident from Figure 1 that if one had to pay out or meet an obligation of either $1,611 in 5 years, $2,594 in 10 years, $4,177 in 15 years, $6,727 in 20 years, or $10,840 in 25 years, the amount to be invested initially at 10% interest would be $1,000. This amount is referred to as the present value (PV) of a future amount and it is calculated by "discounting" a future value (F) in a specific year at a given rate of interest.

Present Value

ASTM Definition of Present Value

The value of a benefit or cost found by discounting future cash flows to the base time. (Syn. present worth)

The basic mathematics for calculating present values are relatively simple:

Given:

- P = Present sum of money
- F = Future sum of money
- i = Interest or discount rate (expressed as a decimal and not a percentage)
- n = Number of years

The future value (F) of a present sum of money (P) may be computed as shown in Table 1, a relatively simple process.

Year	Future Value (F)
1	$P(1+i)$
2	$P(1+i)(1+i) = P(1+i)^2$
3	$P(1+i)^2(1+i) = P(1+i)^3$
n	$P(1+i)^n$

Table 1: The Future Value of a Present Sum of Money

Since $\quad F = P(1+i)^n$

Therefore $\quad P = \dfrac{F}{(1+i)^n} = F \times \dfrac{1}{(1+i)^n}$

Example

What is the present value (P) of an anticipated maintenance expense of $300 in year 3 (F) if the interest rate is 10%?

$$P = F \times \frac{1}{(1+i)^n} = 300 \times \frac{1}{(1+0.10)^n} = 300 \times \frac{1}{1.33}$$

$$= 300 \times 0.75$$

$$= \$225$$

From an interest standpoint, this tells us that $225 is the amount that would have to be deposited today into an account paying 10% interest per annum in order to provide $300 at the end of year 3 to meet the anticipated expense. From the economist's point of view, discounting using the investor's minimum acceptable rate of return (10% in this case) makes the economic outlook the same, whether it is the future amount of $300 or a present value of $225.

Note that $(1+i)^n$ is commonly referred to as the *Single Compound Amount (SCA)* factor, and $1/(1+i)^n$ as the *Single Present Value Factor (SPV)*. Both factors may be readily found in standard discount factor tables for a wide range of interest rates and time periods, thus simplifying calculations. The use of these factors in calculating life cycle costs by the formula method is discussed later in this section.

A life cycle cost analysis can be performed in *constant dollars* (dollars of constant purchasing power, net of general inflation) or *current dollars* (actual prices inclusive of general inflation). If you use constant dollars, you must use a "real" discount rate (i.e., a rate which reflects the time value of money, net of general inflation) and differential escalation rates (also net of general inflation) for future cost items. Constant dollars are generally tied to a specific year. If you use current dollars, the discount rates and escalation rate must include general inflation (i.e., they must be "nominal" or "market" rates). In this chapter, all costs are expressed in current dollars and rates in nominal terms. When taxes are included in the analysis, the current dollar approach is generally preferred.

The Cash Flow Method for Calculating Life Cycle Costs

There are basically two approaches to calculating life cycle costs—the *cash flow* method and the *formula* method which, when applicable, is somewhat more simple.

To illustrate the cash flow method, a relatively simple example follows.

Example

A facility manager is considering the purchase of maintenance equipment required for a four-year period for which initial costs, energy costs and maintenance costs of alternative proposals vary.

The financial criteria on which the economic evaluation will be based are the following:

Interest/Discount Rate	– 10%
Energy Escalation Rate	– 8% per annum
Labor and Material Escalation Rate	– 4% per annum
Period of Study	– 4 years

The following data is provided in one of the alternative proposals to be analyzed:

- Initial Capital Cost – $1,000
- Maintenance Costs – A fixed annual cost of $100 per year quoted by the supplier (no escalation to be considered)
- Annual Energy Costs – Initially $100 per year and subject to 8% annual escalation – i.e., $108 for year one.
- Salvage Value – At the end of the four-year period, the equipment has no further useful life and the supplier agrees to purchase it as scrap for $50.

What is the present value of this alternative proposal?

The steps to resolve this problem are the following:

    Step 1 – Prepare a Cash Flow Diagram.
    Step 2 – Establish a Time Schedule of Costs.
    Step 3 – Calculate Annual Net Cash Flows.
    Step 4 – Calculate Present Value Factors.
    Step 5 – Calculate Present Values of Annual Cash Flows and Life Cycle Costs.

Step 1: Prepare a Cash Flow Diagram

To better understand the problem, cash flow diagrams are often prepared. Revenues are noted as vertical lines above a horizontal time axis, and disbursements are noted as vertical lines below it.

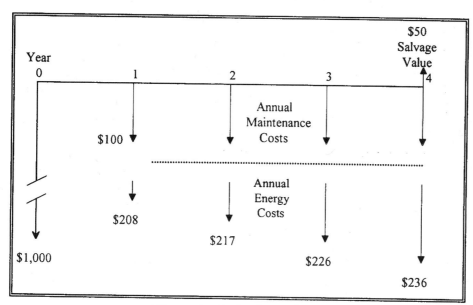

Figure 2: Cash Flow Diagram

Step 2: Establish a Time Schedule of Costs

Table 2 below presents revenues and disbursements for each year of the study:

A Year	B Capital Costs	C Maintenance Costs	D Energy Cost (8% escalation)
0	1,000	–	–
1	–	100	108
2	–	100	117
3	–	100	126
4	(50)	100	136

Table 2

Note that the disbursements in this example are considered positive, and revenues for the salvage value are negative.

Step 3: Calculate Annual Net Cash Flows

Table 3 below provides the net cash flow (NCF) for each year, which could be a disbursement or a revenue.

A Year	B Capital Costs	C Maintenance Costs	D Energy Cost (8% escalation)	E NCF (B + C + D)
0	1,000	–	–	1,000
1	–	100	108	208
2	–	100	117	217
3	–	100	126	226
4	(50)	100	136	186

Table 3

Step 4: Calculate Present Value Factors

As in a previous example, present value factors based on 10% interest will be calculated for each year to convert the annual net cash flows to present values.

Year n	PV Factor	$\dfrac{1}{(1+i)^n}$
0		1
1		$\dfrac{1}{(1+0.1)^1} = 0.91$
2		$\dfrac{1}{(1+0.1)^2} = 0.83$
3		$\dfrac{1}{(1+0.1)^3} = 0.75$
4		$\dfrac{1}{(1+0.1)^4} = 0.68$

Table 4

Step 5: Calculate Present Values of Annual Cash Flows and Life Cycle Costs

Annual net cash flows are converted to present values of the base year. Their sum (Total Present Value) will represent the life cycle costs of the alternative.

Year n	NCF	PV Factor	PV $
0	1,000	1	1,000
1	208	0.91	189
2	217	0.83	180
3	226	0.75	170
4	186	0.68	126
PV Life Cycle Cost	–	–	1,665

Table 5

The total life cycle cost of this alternative in terms of today's dollar (if this year is the base year) is $1,665. This means that if a present sum of $1,665 were deposited today at an interest rate of 10%, all expenses could be paid over a four-year period, at which time the bank balance would be zero.

Once the costs of the other proposals have been calculated in present value dollars, the facilities manager will be in a position to quantify the differences in life cycle costs and decide which proposal represents the lowest cost of ownership in the long run.

The approach outlined is a practical one when considering the purchase of equipment such as elevators, controls, etc.; quotations should not only include capital costs, but also the costs of long-term maintenance contracts which could make a higher initial cost proposal lower over the life cycle of the equipment.

The Formula Method for Calculating Life Cycle Costs

It is also possible to solve the previous problem utilizing the formula method if the problem is adaptable to this technique. Business calculators or discount factor tables such as those published by ASTM can be employed to generate the present value factors, and results may be obtained more quickly.

In some cases, the complexity of the calculations (such as when taxes on net revenues are considered) will necessitate the "cash flow" method, with calculations done manually or using a spreadsheet program.

With the formula method approach, the calculations for the life cycle cost of the maintenance equipment alternative analyzed would be as follows:

1.	PV of Initial Investment	$1,000
2.	PV of Maintenance Costs Annual Cost (A) x UPV factor 100 x 3.17	$ 317
3.	PV of Energy Costs Base Year Cost (A) x UPV* 100 x 3.82	$ 382
4.	Salvage (Scrap) Value Future Value (F) x SPV (50) x 0.68	(34)
5.	Life Cycle Cost / Total Present Value	$1,665

Note: The above factors can be found in Discount Factor Tables.

UPV–Uniform Present Value
SPV–Single Present Value
UPV*–Uniform Present Value Modified (*)

The calculated life cycle cost of $1,665 is identical to that obtained with the cash flow method, but is arrived at with less effort.

Equivalent Uniform Annual Value

ASTM Definition of Equivalent Uniform Annual Value

A uniform annual amount equivalent to the project costs or benefits, taking into account the time value of money throughout the study period. (Syn. annual value)

As noted previously in the definition of Life Cycle Costs, these may be expressed as a lump sum present value with a base time reference, or in equivalent uniform annual values (EUAV) over the study period; decision-makers may prefer one or the other, or require both for the analysis of investment alternatives.

To express life cycle costs in annual values, it is necessary to first determine the lump sum present value of an alternative and multiply it by the Uniform Capital Recovery (UCR) factor given the interest rate and time period. This process is identical to calculating annual payments for a mortgage.

Example

The present value life cycle cost for one of the maintenance equipment alternatives analyzed amounted to $1,665; what is the equivalent uniform annual value "A"?

Given:	P	(a principal sum or present value life cycle cost)	= $1,665
	i		= 10%
	n		= 4 years
	UCR Factor		= 0.32 (from standard factor tables for i)
			= 10%, n = 4 years
	A		= equivalent uniform annual value – to be determined
Solution:	A		= P x UCR
			= $1,665 x 0.32
			= $530

The life cycle cost of the alternative may be expressed as a lump sum, $1,665, or an annual value of $530 based on a four-year time period.

Other Methods of Evaluating Economic Performance

While calculating life cycle costs is a helpful tool for the facilities manager evaluating alternatives, there are other approaches for evaluating economic performance of building investments. Among these other methods are *payback, net benefits,* and *internal rate of return;* these are described in detail in an ASTM compilation of *Building Economics Standards* which are endorsed by the American Association of Cost Engineers (AACE). These standards, as well as adjuncts available such as discount factor tables and a life cycle cost analysis computer program, are included in the References at the end of this section.

ASTM has also developed a Standard Classification of Building Elements and Related Site Work entitled UNIFORMAT II (Designation E-1557-93). One of the many applications for this classification for facilities managers is its use to structure operating and maintenance cost data by building elements or systems, thus facilitating life cycle costing studies. This classification can also be the basis of life cycle cost plans for new facilities that should be developed during the design period with the assistance of the facility manager.

References (Annual Book of ASTM Standards Vol. 04.07)

ASTM Standards

E 833 Terminology of Building Economics

E 917 Practice for Measuring Life Cycle Costs of Buildings and Building Systems

E 964 Practice for Measuring Benefit-to-Cost and Savings-to-Investment Ratios for Buildings and Building Systems

E 1057 Practice for Measuring Internal Rates of Return and Adjusted Internal Rates of Return for Investments in Buildings and Building Systems

E 1074 Practice for Measuring Net Benefits for Investments in Buildings and Building Systems

E 1121 Practice for Measuring Payback for Investments in Buildings and Building Systems

E 1185 Guide for Selecting Economic Methods for Evaluating Investments in Buildings and Building Systems

E 1369 Guide for Selecting Techniques for Treating Uncertainty and Risk in the Economic Evaluation of Buildings and Building Systems

E 1557 Standard Classification for Building Elements and Related Site Work–UNIFORMAT II

ASTM Adjuncts

Discount Factor Tables, Adjunct to Practices E 917, E 964, E 1057, and E 1074.

Computer Program and User's Guide to Building Maintenance, Repair, and Replacement Database for Life Cycle Cost Analysis, Adjunct to Practices E 917, E 964, E 1057, and E 1121.

Note: The Standards noted above, other than E 1557, are available in one publication entitled *ASTM Standards on Building Economics, Second Edition.*

Abbreviations

A	Area Square Feet; Ampere	Brk., brk	Brick	Csc	Cosecant
AAFES	Army and Air Force Exchange Service	brkt	Bracket	C.S.F.	Hundred Square Feet
ABS	Acrylonitrile Butadiene Stryrene; Asbestos Bonded Steel	Brng.	Bearing	CSI	Construction Specifications Institute
A.C., AC	Alternating Current; Air-Conditioning; Asbestos Cement; Plywood Grade A & C	Brs.	Brass	CT	Current Transformer
		Brz.	Bronze	CTS	Copper Tube Size
		Bsn.	Basin	Cu	Copper, Cubic
		Btr.	Better	Cu. Ft.	Cubic Foot
ACI	American Concrete Institute	Btu	British Thermal Unit	cw	Continuous Wave
ACR	Air Conditioning Refrigeration	BTUH	BTU per Hour	C.W.	Cool White; Cold Water
ADA	Americans with Disabilities Act	Bu.	bushels	Cwt.	100 Pounds
AD	Plywood, Grade A & D	BUR	Built-up Roofing	C.W.X.	Cool White Deluxe
Addit.	Additional	BX	Interlocked Armored Cable	C.Y.	Cubic Yard (27 cubic feet)
Adj.	Adjustable	°C	degree centegrade	C.Y./Hr.	Cubic Yard per Hour
af	Audio-frequency	c	Conductivity, Copper Sweat	Cyl.	Cylinder
AFUE	Annual Fuel Utilization Efficiency	C	Hundred; Centigrade	d	Penny (nail size)
AGA	American Gas Association	C/C	Center to Center, Cedar on Cedar	D	Deep; Depth; Discharge
Agg.	Aggregate	C-C	Center to Center	Dis., Disch.	Discharge
A.H., Ah	Ampere Hours	Cab	Cabinet	Db	Decibel
A hr.	Ampere-hour	Cair.	Air Tool Laborer	Dbl.	Double
A.H.U., AHU	Air Handling Unit	Cal.	caliper	DC	Direct Current
A.I.A.	American Institute of Architects	Calc	Calculated	DDC	Direct Digital Control
AIC	Ampere Interrupting Capacity	Cap.	Capacity	Demob.	Demobilization
Allow.	Allowance	Carp.	Carpenter	d.f.t.	Dry Film Thickness
alt., alt	Alternate	C.B.	Circuit Breaker	d.f.u.	Drainage Fixture Units
Alum.	Aluminum	C.C.A.	Chromate Copper Arsenate	D.H.	Double Hung
a.m.	Ante Meridiem	C.C.F.	Hundred Cubic Feet	DHW	Domestic Hot Water
Amp.	Ampere	cd	Candela	DI	Ductile Iron
Anod.	Anodized	cd/sf	Candela per Square Foot	Diag.	Diagonal
ANSI	American National Standards Institute	CD	Grade of Plywood Face & Back	Diam., Dia	Diameter
		CDX	Plywood, Grade C & D, exterior glue	Distrib.	Distribution
APA	American Plywood Association	Cefi.	Cement Finisher	Div.	Division
Approx.	Approximate	Cem.	Cement	Dk.	Deck
Apt.	Apartment	CF	Hundred Feet	D.L.	Dead Load; Diesel
Asb.	Asbestos	C.F.	Cubic Feet	DLH	Deep Long Span Bar Joist
A.S.B.C.	American Standard Building Code	CFM	Cubic Feet per Minute	dlx	Deluxe
Asbe.	Asbestos Worker	CFRP	Carbon Fiber Reinforced Plastic	Do.	Ditto
ASCE.	American Society of Civil Engineers	c.g.	Center of Gravity	DOP	Dioctyl Phthalate Penetration Test (Air Filters)
A.S.H.R.A.E.	American Society of Heating, Refrig. & AC Engineers	CHW	Chilled Water; Commercial Hot Water	Dp., dp	Depth
		C.I., CI	Cast Iron	D.P.S.T.	Double Pole, Single Throw
ASME	American Society of Mechanical Engineers	C.I.P., CIP	Cast in Place	Dr.	Drive
		Circ.	Circuit	DR	Dimension Ratio
ASTM	American Society for Testing and Materials	C.L.	Carload Lot	Drink.	Drinking
		CL	Chain Link	D.S.	Double Strength
Attchmt.	Attachment	Clab.	Common Laborer	D.S.A.	Double Strength A Grade
Avg., Ave.	Average	Clam	Common maintenance laborer	D.S.B.	Double Strength B Grade
AWG	American Wire Gauge	C.L.F.	Hundred Linear Feet	Dty.	Duty
AWWA	American Water Works Assoc.	CLF	Current Limiting Fuse	DWV	Drain Waste Vent
Bbl.	Barrel	CLP	Cross Linked Polyethylene	DX	Deluxe White, Direct Expansion
B&B, BB	Grade B and Better; Balled & Burlapped	cm	Centimeter	dyn	Dyne
		CMP	Corr. Metal Pipe	e	Eccentricity
B&S	Bell and Spigot	CMU	Concrete Masonry Unit	E	Equipment Only; East; emissivity
B.&W.	Black and White	CN	Change Notice	Ea.	Each
b.c.c.	Body-centered Cubic	Col.	Column	EB	Encased Burial
B.C.Y.	Bank Cubic Yards	CO_2	Carbon Dioxide	Econ.	Economy
BE	Bevel End	Comb.	Combination	E.C.Y	Embankment Cubic Yards
B.F.	Board Feet	comm.	Commercial, Communication	EDP	Electronic Data Processing
Bg. cem.	Bag of Cement	Compr.	Compressor	EIFS	Exterior Insulation Finish System
BHP	Boiler Horsepower; Brake Horsepower	Conc.	Concrete	E.D.R.	Equiv. Direct Radiation
		Cont., cont	Continuous; Continued, Container	Eq.	Equation
B.I.	Black Iron	Corr.	Corrugated	EL	elevation
bidir.	bidirectional	Cos	Cosine	Elec.	Electrician; Electrical
Bit., Bitum.	Bituminous	Cot	Cotangent	Elev.	Elevator; Elevating
Bit., Conc.	Bituminous Concrete	Cov.	Cover	EMT	Electrical Metallic Conduit; Thin Wall Conduit
Bk.	Backed	C/P	Cedar on Paneling		
Bkrs.	Breakers	CPA	Control Point Adjustment	Eng.	Engine, Engineered
Bldg., bldg	Building	Cplg.	Coupling	EPDM	Ethylene Propylene Diene Monomer
Blk.	Block	CPM	Critical Path Method		
Bm.	Beam	CPVC	Chlorinated Polyvinyl Chloride	EPS	Expanded Polystyrene
Boil.	Boilermaker	C.Pr.	Hundred Pair	Eqhv.	Equip. Oper., Heavy
bpm	Blows per Minute	CRC	Cold Rolled Channel	Eqlt.	Equip. Oper., Light
BR	Bedroom	Creos.	Creosote	Eqmd.	Equip. Oper., Medium
Brg.	Bearing	Crpt.	Carpet & Linoleum Layer	Eqmm.	Equip. Oper., Master Mechanic
Brhe.	Bricklayer Helper	CRT	Cathode-ray Tube	Eqol.	Equip. Oper., Oilers
Bric.	Bricklayer	CS	Carbon Steel, Constant Shear Bar Joist	Equip.	Equipment
				ERW	Electric Resistance Welded

Abbreviations

E.S.	Energy Saver	H	High Henry	Lath.	Lather
Est.	Estimated	HC	High Capacity	Lav.	Lavatory
esu	Electrostatic Units	H.D., HD	Heavy Duty; High Density	lb.; #	Pound
E.W.	Each Way	H.D.O.	High Density Overlaid	L.B., LB	Load Bearing; L Conduit Body
EWT	Entering Water Temperature	HDPE	High density polyethelene plastic	L. & E.	Labor & Equipment
Excav.	Excavation	Hdr.	Header	lb./hr.	Pounds per Hour
excl	Excluding	Hdwe.	Hardware	lb./L.F.	Pounds per Linear Foot
Exp., exp	Expansion, Exposure	H.I.D., HID	High Intensity Discharge	lbf/sq.in.	Pound-force per Square Inch
Ext., ext	Exterior; Extension	Help.	Helper Average	L.C.L.	Less than Carload Lot
Extru.	Extrusion	HEPA	High Efficiency Particulate Air Filter	L.C.Y.	Loose Cubic Yard
f.	Fiber stress			Ld.	Load
F	Fahrenheit; Female; Fill	Hg	Mercury	LE	Lead Equivalent
Fab., fab	Fabricated; fabric	HIC	High Interrupting Capacity	LED	Light Emitting Diode
FBGS	Fiberglass	HM	Hollow Metal	L.F.	Linear Foot
F.C.	Footcandles	HMWPE	high molecular weight polyethylene	L.F. Nose	Linear Foot of Stair Nosing
f.c.c.	Face-centered Cubic			L.F. Rsr	Linear Foot of Stair Riser
f'c.	Compressive Stress in Concrete; Extreme Compressive Stress	HO	High Output	Lg.	Long; Length; Large
		Horiz.	Horizontal	L & H	Light and Heat
F.E.	Front End	H.P., HP	Horsepower; High Pressure	LH	Long Span Bar Joist
FEP	Fluorinated Ethylene Propylene (Teflon)	H.P.F.	High Power Factor	L.H.	Labor Hours
		Hr.	Hour	L.L., LL	Live Load
F.G.	Flat Grain	Hrs./Day	Hours per Day	L.L.D.	Lamp Lumen Depreciation
F.H.A.	Federal Housing Administration	HSC	High Short Circuit	lm	Lumen
Fig.	Figure	Ht.	Height	lm/sf	Lumen per Square Foot
Fin.	Finished	Htg.	Heating	lm/W	Lumen per Watt
FIPS	Female Iron Pipe Size	Htrs.	Heaters	LOA	Length Over All
Fixt.	Fixture	HVAC	Heating, Ventilation & Air-Conditioning	log	Logarithm
FJP	Finger jointed and primed			L-O-L	Lateralolet
Fl. Oz.	Fluid Ounces	Hvy.	Heavy	long.	longitude
Flr.	Floor	HW	Hot Water	L.P., LP	Liquefied Petroleum; Low Pressure
FM	Frequency Modulation; Factory Mutual	Hyd.; Hydr.	Hydraulic	L.P.F.	Low Power Factor
		Hz	Hertz (cycles)	LR	Long Radius
Fmg.	Framing	I.	Moment of Inertia	L.S.	Lump Sum
FM/UL	Factory Mutual/Underwriters Labs	IBC	International Building Code	Lt.	Light
Fdn.	Foundation	I.C.	Interrupting Capacity	Lt. Ga.	Light Gauge
FNPT	Female National Pipe Thread	ID	Inside Diameter	L.T.L.	Less than Truckload Lot
Fori.	Foreman, Inside	I.D.	Inside Dimension; Identification	Lt. Wt.	Lightweight
Foro.	Foreman, Outside	I.F.	Inside Frosted	L.V.	Low Voltage
Fount.	Fountain	I.M.C.	Intermediate Metal Conduit	M	Thousand; Material; Male; Light Wall Copper Tubing
fpm	Feet per Minute	In.	Inch		
FPT	Female Pipe Thread	Incan.	Incandescent	M²CA	Meters Squared Contact Area
Fr	Frame	Incl.	Included; Including	m/hr.; M.H.	Man-hour
F.R.	Fire Rating	Int.	Interior	mA	Milliampere
FRK	Foil Reinforced Kraft	Inst.	Installation	Mach.	Machine
FSK	Foil/scrim/kraft	Insul., insul	Insulation/Insulated	Mag. Str.	Magnetic Starter
FRP	Fiberglass Reinforced Plastic	I.P.	Iron Pipe	Maint.	Maintenance
FS	Forged Steel	I.P.S., IPS	Iron Pipe Size	Marb.	Marble Setter
FSC	Cast Body; Cast Switch Box	IPT	Iron Pipe Threaded	Mat; Mat'l.	Material
Ft., ft	Foot; Feet	I.W.	Indirect Waste	Max.	Maximum
Ftng.	Fitting	J	Joule	MBF	Thousand Board Feet
Ftg.	Footing	J.I.C.	Joint Industrial Council	MBH	Thousand BTU's per hr.
Ft lb.	Foot Pound	K	Thousand; Thousand Pounds; Heavy Wall Copper Tubing, Kelvin	MC	Metal Clad Cable
Furn.	Furniture			MCC	Motor Control Center
FVNR	Full Voltage Non-Reversing	K.A.H.	Thousand Amp. Hours	M.C.F.	Thousand Cubic Feet
FVR	Full Voltage Reversing	kcmil	Thousand Circular Mils	MCFM	Thousand Cubic Feet per Minute
FXM	Female by Male	KD	Knock Down	M.C.M.	Thousand Circular Mils
Fy.	Minimum Yield Stress of Steel	K.D.A.T.	Kiln Dried After Treatment	MCP	Motor Circuit Protector
g	Gram	kg	Kilogram	MD	Medium Duty
G	Gauss	kG	Kilogauss	MDF	Medium-density fibreboard
Ga.	Gauge	kgf	Kilogram Force	M.D.O.	Medium Density Overlaid
Gal., gal.	Gallon	kHz	Kilohertz	Med.	Medium
gpm, GPM	Gallon per Minute	Kip	1000 Pounds	MF	Thousand Feet
Galv., galv	Galvanized	KJ	Kiljoule	M.F.B.M.	Thousand Feet Board Measure
GC/MS	Gas Chromatograph/Mass Spectrometer	K.L.	Effective Length Factor	Mfg.	Manufacturing
		K.L.F.	Kips per Linear Foot	Mfrs.	Manufacturers
Gen.	General	Km	Kilometer	mg	Milligram
GFI	Ground Fault Interrupter	KO	Knock Out	MGD	Million Gallons per Day
GFRC	Glass Fiber Reinforced Concrete	K.S.F.	Kips per Square Foot	MGPH	Thousand Gallons per Hour
Glaz.	Glazier	K.S.I.	Kips per Square Inch	MH, M.H.	Manhole; Metal Halide; Man-Hour
GPD	Gallons per Day	kV	Kilovolt	MHz	Megahertz
gpf	Gallon per flush	kVA	Kilovolt Ampere	Mi.	Mile
GPH	Gallons per Hour	kVAR	Kilovar (Reactance)	MI	Malleable Iron; Mineral Insulated
GPM	Gallons per Minute	KW	Kilowatt	MIPS	Male Iron Pipe Size
GR	Grade	KWh	Kilowatt-hour	mj	Mechanical Joint
Gran.	Granular	L	Labor Only; Length; Long; Medium Wall Copper Tubing	m	Meter
Grnd.	Ground			mm	Millimeter
GVW	Gross Vehicle Weight	Lab.	Labor	Mill.	Millwright
GWB	Gypsum wall board	lat	Latitude	Min., min.	Minimum, minute

Abbreviations

Misc.	Miscellaneous	PDCA	Painting and Decorating Contractors of America	SC	Screw Cover
ml	Milliliter, Mainline			SCFM	Standard Cubic Feet per Minute
M.L.F.	Thousand Linear Feet	P.E., PE	Professional Engineer; Porcelain Enamel; Polyethylene; Plain End	Scaf.	Scaffold
Mo.	Month			Sch., Sched.	Schedule
Mobil.	Mobilization			S.C.R.	Modular Brick
Mog.	Mogul Base	P.E.C.I.	Porcelain Enamel on Cast Iron	S.D.	Sound Deadening
MPH	Miles per Hour	Perf.	Perforated	SDR	Standard Dimension Ratio
MPT	Male Pipe Thread	PEX	Cross linked polyethylene	S.E.	Surfaced Edge
MRGWB	Moisture Resistant Gypsum Wallboard	Ph.	Phase	Sel.	Select
		P.I.	Pressure Injected	SER, SEU	Service Entrance Cable
MRT	Mile Round Trip	Pile.	Pile Driver	S.F.	Square Foot
ms	Millisecond	Pkg.	Package	S.F.C.A.	Square Foot Contact Area
M.S.F.	Thousand Square Feet	Pl.	Plate	S.F. Flr.	Square Foot of Floor
Mstz.	Mosaic & Terrazzo Worker	Plah.	Plasterer Helper	S.F.G.	Square Foot of Ground
M.S.Y.	Thousand Square Yards	Plas.	Plasterer	S.F. Hor.	Square Foot Horizontal
Mtd., mtd., mtd	Mounted	plf	Pounds Per Linear Foot	SFR	Square Feet of Radiation
Mthe.	Mosaic & Terrazzo Helper	Pluh.	Plumbers Helper	S.F. Shlf.	Square Foot of Shelf
Mtng.	Mounting	Plum.	Plumber	S4S	Surface 4 Sides
Mult.	Multi; Multiply	Ply.	Plywood	Shee.	Sheet Metal Worker
M.V.A.	Million Volt Amperes	p.m.	Post Meridiem	Sin.	Sine
M.V.A.R.	Million Volt Amperes Reactance	Pntd.	Painted	Skwk.	Skilled Worker
MV	Megavolt	Pord.	Painter, Ordinary	SL	Saran Lined
MW	Megawatt	pp	Pages	S.L.	Slimline
MXM	Male by Male	PP, PPL	Polypropylene	Sldr.	Solder
MYD	Thousand Yards	P.P.M.	Parts per Million	SLH	Super Long Span Bar Joist
N	Natural; North	Pr.	Pair	S.N.	Solid Neutral
nA	Nanoampere	P.E.S.B.	Pre-engineered Steel Building	SO	Stranded with oil resistant inside insulation
NA	Not Available; Not Applicable	Prefab.	Prefabricated		
N.B.C.	National Building Code	Prefin.	Prefinished	S-O-L	Socketolet
NC	Normally Closed	Prop.	Propelled	sp	Standpipe
NEMA	National Electrical Manufacturers Assoc.	PSF, psf	Pounds per Square Foot	S.P.	Static Pressure; Single Pole; Self-Propelled
		PSI, psi	Pounds per Square Inch		
NEHB	Bolted Circuit Breaker to 600V.	PSIG	Pounds per Square Inch Gauge	Spri.	Sprinkler Installer
NFPA	National Fire Protection Association	PSP	Plastic Sewer Pipe	spwg	Static Pressure Water Gauge
NLB	Non-Load-Bearing	Pspr.	Painter, Spray	S.P.D.T.	Single Pole, Double Throw
NM	Non-Metallic Cable	Psst.	Painter, Structural Steel	SPF	Spruce Pine Fir; Sprayed Polyurethane Foam
nm	Nanometer	P.T.	Potential Transformer		
No.	Number	P. & T.	Pressure & Temperature	S.P.S.T.	Single Pole, Single Throw
NO	Normally Open	Ptd.	Painted	SPT	Standard Pipe Thread
N.O.C.	Not Otherwise Classified	Ptns.	Partitions	Sq.	Square; 100 Square Feet
Nose.	Nosing	Pu	Ultimate Load	Sq. Hd.	Square Head
NPT	National Pipe Thread	PVC	Polyvinyl Chloride	Sq. In.	Square Inch
NQOD	Combination Plug-on/Bolt on Circuit Breaker to 240V.	Pvmt.	Pavement	S.S.	Single Strength; Stainless Steel
		PRV	Pressure Relief Valve	S.S.B.	Single Strength B Grade
N.R.C., NRC	Noise Reduction Coefficient/ Nuclear Regulator Commission	Pwr.	Power	sst, ss	Stainless Steel
		Q	Quantity Heat Flow	Sswk.	Structural Steel Worker
N.R.S.	Non Rising Stem	Qt.	Quart	Sswl.	Structural Steel Welder
ns	Nanosecond	Quan., Qty.	Quantity	St.; Stl.	Steel
nW	Nanowatt	Q.C.	Quick Coupling	STC	Sound Transmission Coefficient
OB	Opposing Blade	r	Radius of Gyration	Std.	Standard
OC	On Center	R	Resistance	Stg.	Staging
OD	Outside Diameter	R.C.P.	Reinforced Concrete Pipe	STK	Select Tight Knot
O.D.	Outside Dimension	Rect.	Rectangle	STP	Standard Temperature & Pressure
ODS	Overhead Distribution System	recpt.	receptacle	Stpi.	Steamfitter, Pipefitter
O.G.	Ogee	Reg.	Regular	Str.	Strength; Starter; Straight
O.H.	Overhead	Reinf.	Reinforced	Strd.	Stranded
O&P	Overhead and Profit	Req'd.	Required	Struct.	Structural
Oper.	Operator	Res.	Resistant	Sty.	Story
Opng.	Opening	Resi.	Residential	Subj.	Subject
Orna.	Ornamental	RF	Radio Frequency	Subs.	Subcontractors
OSB	Oriented Strand Board	RFID	Radio-frequency identification	Surf.	Surface
OS&Y	Outside Screw and Yoke	Rgh.	Rough	Sw.	Switch
OSHA	Occupational Safety and Health Act	RGS	Rigid Galvanized Steel	Swbd.	Switchboard
		RHW	Rubber, Heat & Water Resistant; Residential Hot Water	S.Y.	Square Yard
Ovhd.	Overhead			Syn.	Synthetic
OWG	Oil, Water or Gas	rms	Root Mean Square	S.Y.P.	Southern Yellow Pine
Oz.	Ounce	Rnd.	Round	Sys.	System
P.	Pole; Applied Load; Projection	Rodm.	Rodman	t.	Thickness
p.	Page	Rofc.	Roofer, Composition	T	Temperature; Ton
Pape.	Paperhanger	Rofp.	Roofer, Precast	Tan	Tangent
P.A.P.R.	Powered Air Purifying Respirator	Rohe.	Roofer Helpers (Composition)	T.C.	Terra Cotta
PAR	Parabolic Reflector	Rots.	Roofer, Tile & Slate	T & C	Threaded and Coupled
P.B., PB	Push Button	R.O.W.	Right of Way	T.D.	Temperature Difference
Pc., Pcs.	Piece, Pieces	RPM	Revolutions per Minute	Tdd	Telecommunications Device for the Deaf
P.C.	Portland Cement; Power Connector	R.S.	Rapid Start		
P.C.F.	Pounds per Cubic Foot	Rsr	Riser	T.E.M.	Transmission Electron Microscopy
PCM	Phase Contrast Microscopy	RT	Round Trip	temp	Temperature, Tempered, Temporary
		S.	Suction; Single Entrance; South	TFFN	Nylon Jacketed Wire
		SBS	Styrene Butadiere Styrene		

Abbreviations

TFE	Tetrafluoroethylene (Teflon)	U.L., UL	Underwriters Laboratory	w/	With
T. & G.	Tongue & Groove; Tar & Gravel	Uld.	unloading	W.C., WC	Water Column; Water Closet
Th., Thk.	Thick	Unfin.	Unfinished	W.F.	Wide Flange
Thn.	Thin	UPS	Uninterruptible Power Supply	W.G.	Water Gauge
Thrded	Threaded	URD	Underground Residential Distribution	Wldg.	Welding
Tilf.	Tile Layer, Floor	US	United States	W. Mile	Wire Mile
Tilh.	Tile Layer, Helper	USGBC	U.S. Green Building Council	W-O-L	Weldolet
THHN	Nylon Jacketed Wire	USP	United States Primed	W.R.	Water Resistant
THW.	Insulated Strand Wire	UTMCD	Uniform Traffic Manual For Control Devices	Wrck.	Wrecker
THWN	Nylon Jacketed Wire			W.S.P.	Water, Steam, Petroleum
T.L., TL	Truckload	UTP	Unshielded Twisted Pair	WT., Wt.	Weight
T.M.	Track Mounted	V	Volt	WWF	Welded Wire Fabric
Tot.	Total	VA	Volt Amperes	XFER	Transfer
T-O-L	Threadolet	V.C.T.	Vinyl Composition Tile	XFMR	Transformer
tmpd	Tempered	VAV	Variable Air Volume	XHD	Extra Heavy Duty
TPO	Thermoplastic Polyolefin	VC	Veneer Core	XHHW, XLPE	Cross-Linked Polyethylene Wire Insulation
T.S.	Trigger Start	VDC	Volts Direct Current	XLP	Cross-linked Polyethylene
Tr.	Trade	Vent.	Ventilation	Xport	Transport
Transf.	Transformer	Vert.	Vertical	Y	Wye
Trhv.	Truck Driver, Heavy	V.F.	Vinyl Faced	yd	Yard
Trlr	Trailer	V.G.	Vertical Grain	yr	Year
Trlt.	Truck Driver, Light	VHF	Very High Frequency	Δ	Delta
TTY	Teletypewriter	VHO	Very High Output	%	Percent
TV	Television	Vib.	Vibrating	~	Approximately
T.W.	Thermoplastic Water Resistant Wire	VLF	Vertical Linear Foot	Ø	Phase; diameter
		VOC	Volitile Organic Compound	@	At
UCI	Uniform Construction Index	Vol.	Volume	#	Pound; Number
UF	Underground Feeder	VRP	Vinyl Reinforced Polyester	<	Less Than
UGND	Underground Feeder	W	Wire; Watt; Wide; West	>	Greater Than
UHF	Ultra High Frequency			Z	zone
U.I.	United Inch				

Notes

Notes

Notes

Notes

Notes

Notes

Notes

Reed Construction Data/RSMeans—
a tradition of excellence in construction cost information and services since 1942

Table of Contents
Annual Cost Guides
Seminars
New Titles
Reference Books
Order Form

For more information visit the RSMeans website at **www.rsmeans.com**

Unit prices according to the latest MasterFormat!

Book Selection Guide — The following table provides definitive information on the content of each cost data publication. The number of lines of data provided in each unit price or assemblies division, as well as the number of reference tables and crews, is listed for each book. The presence of other elements such as equipment rental costs, historical cost indexes, city cost indexes, square foot models, or cross-referenced indexes is also indicated. You can use the table to help select the RSMeans book that has the quantity and type of information you most need in your work.

Unit Cost Divisions	Building Construction	Mechanical	Electrical	Commercial Renovation	Square Foot	Site Work Landsc.	Green Building	Interior	Concrete Masonry	Open Shop	Heavy Construction	Light Commercial	Facilities Construction	Plumbing	Residential
1	582	395	411	508		521	196	319	463	581	511	244	1054	411	180
2	789	279	85	741		1005	180	399	222	788	746	481	1233	289	257
3	1689	347	231	1087		1477	995	329	2040	1688	1685	479	1787	317	385
4	963	21	0	932		729	179	620	1159	931	621	521	1176	0	435
5	1873	158	155	1068		811	1779	1073	716	1873	1034	954	1874	204	723
6	2404	18	18	2062		110	560	1523	278	2400	126	2085	2076	22	2612
7	1531	215	128	1569		584	747	516	509	1528	26	1282	1633	227	1018
8	2107	81	45	2530		268	1140	1768	112	2093	0	2109	2842	0	1511
9	1929	72	26	1726		313	449	2014	387	1870	15	1644	2179	54	1430
10	1006	17	10	623		215	27	818	158	1006	29	511	1098	235	224
11	1026	208	165	489		126	54	878	28	1010	0	230	1046	169	110
12	557	0	2	327		248	151	1669	12	545	19	364	1718	23	317
13	729	117	114	246		347	128	253	66	728	253	82	756	71	80
14	277	36	0	224		0	0	261	0	277	0	12	297	16	6
21	90	0	16	37		0	0	250	0	87	0	68	427	432	220
22	1159	7329	156	1170		1578	1099	849	20	1123	1687	861	7171	9082	712
23	1182	6901	586	886		160	898	741	38	1106	111	759	5013	1830	434
26	1379	499	10134	1039		816	661	1142	55	1351	562	1305	10083	444	621
27	72	0	298	34		13	0	60	0	72	39	52	280	0	4
28	97	58	137	72		0	21	79	0	100	0	41	152	44	26
31	1498	735	610	803		3206	289	7	1208	1443	3285	600	1560	660	611
32	782	54	8	859		4396	352	362	286	753	1824	407	1660	166	455
33	509	1020	542	209		2036	41	0	231	508	2024	120	1585	1211	118
34	109	0	47	4		192	0	0	31	64	195	0	130	0	0
35	18	0	0	0		327	0	0	0	18	442	0	83	0	0
41	60	0	0	32		0	0	22	0	60	30	0	67	14	0
44	75	83	0	0		0	0	0	0	0	0	0	75	75	0
46	23	16	0	0		274	151	0	0	23	264	0	33	33	0
48	12	0	27	0		0	27	0	0	12	19	12	12	0	12
Totals	**24527**	**18659**	**13951**	**19277**		**19752**	**10124**	**15952**	**8019**	**24038**	**15547**	**15223**	**49100**	**16029**	**12501**

Assem Div	Building Construction	Mechanical	Electrical	Commercial Renovation	Square Foot	Site Work Landscape	Assemblies	Green Building	Interior	Concrete Masonry	Heavy Construction	Light Commercial	Facilities Construction	Plumbing	Asm Div	Residential
A		15	0	188	150	577	598	0	0	536	571	154	24	0	1	368
B		0	0	848	2498	0	5658	56	329	1975	368	2089	174	0	2	211
C		0	0	647	872	0	1251	0	1576	146	0	762	249	0	3	588
D		1067	943	712	1856	72	2532	330	822	0	0	1342	1106	1081	4	851
E		0	0	86	260	0	300	0	5	0	0	257	5	0	5	392
F		0	0	0	114	0	114	0	0	0	0	114	3	0	6	357
G		522	442	318	262	3740	663	0	0	534	1558	199	288	758	7	307
															8	760
															9	80
															10	0
															11	0
															12	0
Totals		**1604**	**1385**	**2799**	**6012**	**4389**	**11116**	**386**	**2732**	**3191**	**2497**	**4917**	**1849**	**1839**		**3914**

Reference Section	Building Construction Costs	Mechanical	Electrical	Commercial Renovation	Square Foot	Site Work Landscape	Assem.	Green Building	Interior	Concrete Masonry	Open Shop	Heavy Construction	Light Commercial	Facilities Construction	Plumbing	Resi.
Reference Tables	yes	yes	yes	yes	no	yes	yes	yes	yes	yes	yes	yes	yes	yes	yes	yes
Models					111			25					50			28
Crews	561	561	561	540		561		558	561	561	538	561	538	540	561	538
Equipment Rental Costs	yes	yes	yes	yes		yes		yes	yes	yes	yes	yes	yes	yes	yes	yes
Historical Cost Indexes	yes	yes	yes	yes		yes		yes	yes	yes	yes	yes	yes	yes	yes	no
City Cost Indexes	yes	yes	yes	yes		yes		yes	yes	yes	yes	yes	yes	yes	yes	yes

679

Annual Cost Guides

Cost guides also available in CD an at rsmeansonline.co

For more information visit the RSMeans website at **www.rsmeans.com**

Unit prices according to the latest MasterForma

RSMeans Building Construction Cost Data 2014

Offers you unchallenged unit price reliability in an easy-to-use format. Whether used for verifying complete, finished estimates or for periodic checks, it supplies more cost facts better and faster than any comparable source. More than 24,000 unit prices have been updated for 2014. The City Cost Indexes and Location Factors cover more than 930 areas, for indexing to any project location in North America. Order and get *RSMeans Quarterly Update Service* FREE.

$194.95 | Available Sept. 2013 | Catalog no. 60014

RSMeans Green Building Cost Data 2014

Estimate, plan, and budget the costs of green building for both new commercia construction and renovation work with this fourth edition of *RSMeans Green Building Cost Data*. More than 10,000 unit costs for a wide array of green building products plus assemblies costs Easily identified cross references to LEED and Green Globes building rating systems criteria.

$160.95 | Available Nov. 2013 | Catalog no. 60554

RSMeans Mechanical Cost Data 2014

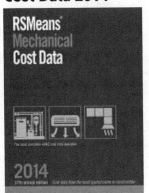

Total unit and systems price guidance for mechanical construction . . . materials, parts, fittings, and complete labor cost information. Includes prices for piping, heating, air conditioning, ventilation, and all related construction.

Plus new 2014 unit costs for:

- Thousands of installed HVAC/controls assemblies components
- "On-site" Location Factors for more than 930 cities and towns in the U.S. and Canada
- Crews, labor, and equipment

$191.95 | Available Oct. 2013 | Catalog no. 60024

RSMeans Facilities Construction Cost Data 2014

For the maintenance and construction commercial, industrial, municipal, and institutional properties. Costs are show for new and remodeling construction a are broken down into materials, labor, equipment, and overhead and profit. Special emphasis is given to sections on mechanical, electrical, furnishings, site work, building maintenance, finish work, and demolition.

More than 49,000 unit costs, plus assemblies costs and a comprehensive Reference Section are included.

$496.95 | Available Nov. 2013 | Catalog no. 60204

RSMeans Square Foot Costs 2014
Accurate and Easy-to-Use

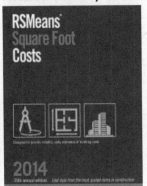

- **Updated price information** based on nationwide figures from suppliers, estimators, labor experts, and contractors
- New green building models
- Realistic graphics, offering true-to-life illustrations of building projects
- Extensive information on using square foot cost data, including sample estimates and alternate pricing methods

$205.95 | Available Oct. 2013 | Catalog no. 60054

RSMeans Commercial Renovation Cost Data 2014
Commercial/Multifamily Residential

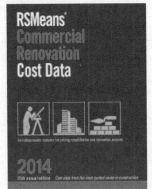

Use this valuable tool to estimate commercial and multifamily residentia renovation and remodeling.

Includes: New costs for hundreds of unique methods, materials, and conditions that only come up in repair and remodeling, PLUS:

- Unit costs for more than 19,000 construction components
- Installed costs for more than 2,000 assemblies
- More than 930 "on-site" localizatior factors for the U.S. and Canada

$156.95 | Available Oct. 2013 | Catalog no. 60044

Annual Cost Guides

Cost guides also available in CD and at rsmeansonline.com

For more information visit the RSMeans website at www.rsmeans.com

Unit prices according to the latest MasterFormat!

RSMeans Electrical Cost Data 2014

Pricing information for every part of electrical cost planning. More than 13,000 unit and systems costs with design tables; clear specifications and drawings; engineering guides; illustrated estimating procedures; complete labor-hour and materials costs for better scheduling and procurement; and the latest electrical products and construction methods.

- A variety of special electrical systems, including cathodic protection
- Costs for maintenance, demolition, HVAC/mechanical, specialties, equipment, and more

$196.95 | Available Oct. 2013 | Catalog no. 60034

RSMeans Electrical Change Order Cost Data 2014

RSMeans Electrical Change Order Cost Data provides you with electrical unit prices exclusively for pricing change orders—based on the recent, direct experience of contractors and suppliers. Analyze and check your own change order estimates against the experience others have had doing the same work. It also covers productivity analysis and change order cost justifications. With useful information for calculating the effects of change orders and dealing with their administration.

$187.95 | Available Dec. 2013 | Catalog no. 60234

RSMeans Assemblies Cost Data 2014

RSMeans Assemblies Cost Data takes the guesswork out of preliminary or conceptual estimates. Now you don't have to try to calculate the assembled cost by working up individual component costs. We've done all the work for you. Presents detailed illustrations, descriptions, specifications, and costs for every conceivable building assembly—over 350 types in all—arranged in the easy-to-use UNIFORMAT II system. Each illustrated "assembled" cost includes a complete grouping of materials and associated installation costs, including the installing contractor's overhead and profit.

$315.95 | Available Sept. 2013 | Catalog no. 60064

RSMeans Open Shop Building Construction Cost Data 2014

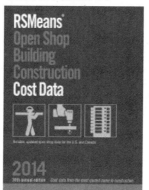

The latest costs for accurate budgeting and estimating of new commercial and residential construction . . . renovation work . . . change orders . . . cost engineering.

RSMeans Open Shop "BCCD" will assist you to:

- Develop benchmark prices for change orders
- Plug gaps in preliminary estimates and budgets
- Estimate complex projects
- Substantiate invoices on contracts
- Price ADA-related renovations

$165.95 | Available Dec. 2013 | Catalog no. 60154

RSMeans Residential Cost Data 2014

Contains square foot costs for 28 basic home models with the look of today, plus hundreds of custom additions and modifications you can quote right off the page. Includes costs for more than 700 residential systems. Complete with blank estimating forms, sample estimates, and step-by-step instructions.

Contains line items for cultured stone and brick, PVC trim lumber, and TPO roofing.

$139.95 | Available Oct. 2013 | Catalog no. 60174

RSMeans Site Work & Landscape Cost Data 2014

Includes unit and assemblies costs for earthwork, sewerage, piped utilities, site improvements, drainage, paving, trees and shrubs, street openings/repairs, underground tanks, and more. Contains 78 types of assemblies costs for accurate conceptual estimates.

Includes:

- Estimating for infrastructure improvements
- Environmentally-oriented construction
- ADA-mandated handicapped access
- Hazardous waste line items

$187.95 | Available Dec. 2013 | Catalog no. 60284

Annual Cost Guides

Cost guides also available in CD and at rsmeansonline.com

For more information visit the RSMeans website at **www.rsmeans.com**

Unit prices according to the latest MasterFormat

RSMeans Facilities Maintenance & Repair Cost Data 2014

RSMeans Facilities Maintenance & Repair Cost Data gives you a complete system to manage and plan your facility repair and maintenance costs and budget efficiently. Guidelines for auditing a facility and developing an annual maintenance plan. Budgeting is included, along with reference tables on cost and management, and information on frequency and productivity of maintenance operations.

The only nationally recognized source of maintenance and repair costs. Developed in cooperation with the Civil Engineering Research Laboratory (CERL) of the Army Corps of Engineers.

$434.95 | Available Nov. 2013 | Catalog no. 60304

RSMeans Light Commercial Cost Data 2014

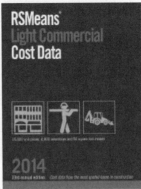

Specifically addresses the light commercial market, which is a specialized niche in the construction industry. Aids you, the owner/designer/contractor, in preparing all types of estimates—from budgets to detailed bid. Includes new advances in methods and materials.

Assemblies Section allows you to evaluate alternatives in the early stages of design/planning.

More than 15,000 unit costs ensure that you have the prices you need . . . when you need them.

$144.95 | Available Nov. 2013 | Catalog no. 60184

RSMeans Concrete & Masonry Cost Data 2014

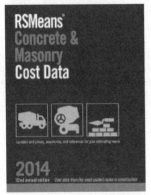

Provides you with cost facts for virtually all concrete/masonry estimating needs, from complicated formwork to various sizes and face finishes of brick and block—all in great detail. The comprehensive Unit Price Section contains more than 8,000 selected entries. Also contains an Assemblies [Cost] Section, and a detailed Reference Section that supplements the cost data.

$177.95 | Available Dec. 2013 | Catalog no. 60114

RSMeans Labor Rates for the Construction Industry 2014

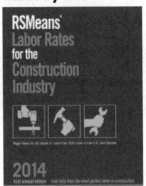

Complete information for estimating labor costs, making comparisons, and negotiating wage rates by trade for more than 300 U.S. and Canadian cities. With 46 construction trades listed by local union number in each city, and historic wage rates included for comparison. Each city chart lists the county and is alphabetically arranged with handy visual flip tabs for quick reference.

$424.95 | Available Dec. 2013 | Catalog no. 60124

RSMeans Construction Cost Indexes 2014

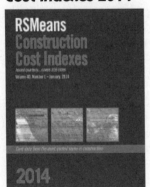

What materials and labor costs will change unexpectedly this year? By how much?

- Breakdowns for 318 major cities
- National averages for 30 key cities
- Expanded five major city indexes
- Historical construction cost indexes

$362.00 per year (subscription) | Catalog no. 50144
$93.50 individual quarters | Catalog no. 60144 A,B,C,D

RSMeans Interior Cost Data 2014

Provides you with prices and guidance needed to make accurate interior work estimates. Contains costs on materials, equipment, hardware, custom installations, furnishings, and labor costs . . . for new and remodel commercial and industrial interior construction, including updated information on office furnishings, and reference information.

$196.95 | Available Nov. 2013 | Catalog no. 60094

Annual Cost Guides

Cost guides also available in CD and at rsmeansonline.com

For more information visit the RSMeans website at www.rsmeans.com

Unit prices according to the latest MasterFormat!

RSMeans Heavy Construction Cost Data 2014

A comprehensive guide to heavy construction costs. Includes costs for highly specialized projects such as tunnels, dams, highways, airports, and waterways. Information on labor rates, equipment, and materials costs is included. Features unit price costs, systems costs, and numerous reference tables for costs and design.

$192.95 | Available Dec. 2013 | Catalog no. 60164

RSMeans Plumbing Cost Data 2014

Comprehensive unit prices and assemblies for plumbing, irrigation systems, commercial and residential fire protection, point-of-use water heaters, and the latest approved materials. This publication and its companion, *RSMeans Mechanical Cost Data*, provide full-range cost estimating coverage for all the mechanical trades.

Now contains updated costs for potable water and radiant heat systems and high efficiency fixtures.

$194.95 | Available Oct. 2013 | Catalog no. 60214

2014 RSMeans Seminar Schedule

Note: call for exact dates and details.

Location	Dates	Location	Dates
Seattle, WA	January and August	Jacksonville FL	September
Dallas/Ft. Worth, TX	January	Dallas, TX	September
Austin, TX	February	Charleston, SC	October
Anchorage, AK	March and September	Houston, TX	October
Las Vegas, NV	March and October	Atlanta, GA	November
Washington, DC	April and September	Baltimore, MD	November
Phoenix, AZ	April	Orlando, FL	November
Kansas City, MO	April	San Diego, CA	December
Toronto	May	San Antonio, TX	December
Denver, CO	May	Raleigh, NC	December
San Francisco, CA	June		
Bethesda, MD	June		
Columbus, GA	June		
El Segundo, CA	August		

1-800-334-3509, Press 1

Professional Development

For more information visit the RSMeans website at www.rsmeans.com

eLearning Training Sessions

Learn how to use *RSMeans Online*™ or *RSMeans CostWorks*® CD from the convenience of your home or office. Our eLearning training sessions let you join a training conference call and share the instructors' desktops, so you can view the presentation and step-by-step instructions on your own computer screen. The live webinars are held from 9 a.m. to 4 p.m. eastern standard time, with a one-hour break for lunch.

For these sessions, you must have a computer with high speed Internet access and a compatible Web browser. Learn more at www.rsmeansonline.com or call for a schedule: 1-800-334-3509 and press 1. Webinars are generally held on selected Wednesdays each month.

RSMeans Online™
$299 per person

RSMeans CostWorks® CD
$299 per person

Professional Development

For more information visit the RSMeans website at **www.rsmeans.com**

RSMeans Online™ Training

Construction estimating is vital to the decision-making process at each state of every project. RSMeansOnline works the way you do. It's systematic, flexible and intuitive. In this one day class you will see how you can estimate any phase of any project faster and better.

Some of what you'll learn:
- Customizing RSMeansOnline
- Making the most of RSMeans "Circle Reference" numbers
- How to integrate your cost data
- Generate reports, exporting estimates to MS Excel, sharing, collaborating and more

Also available: RSMeans Online™ training webinar

Facilities Construction Estimating

In this *two-day* course, professionals working in facilities management can get help with their daily challenges to establish budgets for all phases of a project.

Some of what you'll learn:
- Determining the full scope of a project
- Identifying the scope of risks and opportunities
- Creative solutions to estimating issues
- Organizing estimates for presentation and discussion
- Special techniques for repair/remodel and maintenance projects
- Negotiating project change orders

Who should attend: facility managers, engineers, contractors, facility tradespeople, planners, and project managers.

Scheduling with Microsoft Project

Two days of hands-on training gives you a look at the basics to putting it all together with MS Project to create a complete project schedule. Learn to work better and faster, manage changes and updates, enhance tracking, generate actionable reports, and boost control of your time and budget.

Some of what you'll learn:
- Defining task/activity relationships: link tasks, establish predecessor/successor relationships and create logs
- Baseline scheduling: the essential what/when
- Using MS Project to determine the critical path
- Integrate RSMeans data with MS Project
- The dollar-loaded schedule: enhance the reliability of this "must-know" technique for scheduling public projects.

Who should attend: contractors' project-management teams, architects and engineers, project owners and their representatives, and those interested in improving their project planning and scheduling and management skills.

Maintenance & Repair Estimating for Facilities

This *two-day* course teaches attendees how to plan, budget, and estimate the cost of ongoing and preventive maintenance and repair for existing buildings and grounds.

Some of what you'll learn:
- The most financially favorable maintenance, repair, and replacement scheduling and estimating
- Auditing and value engineering facilities
- Preventive planning and facilities upgrading
- Determining both in-house and contract-out service costs
- Annual, asset-protecting M&R plan

Who should attend: facility managers, maintenance supervisors, buildings and grounds superintendents, plant managers, planners, estimators, and others involved in facilities planning and budgeting.

Practical Project Management for Construction Professionals

In this *two-day* course, acquire the essential knowledge and develop the skills to effectively and efficiently execute the day-to-day responsibilities of the construction project manager.

Covers:
- General conditions of the construction contract
- Contract modifications: change orders and construction change directives
- Negotiations with subcontractors and vendors
- Effective writing: notification and communications
- Dispute resolution: claims and liens

Who should attend: architects, engineers, owner's representatives, project managers.

Mechanical & Electrical Estimating

This *two-day* course teaches attendees how to prepare more accurate and complete mechanical/electrical estimates, avoiding the pitfalls of omission and double-counting, while understanding the composition and rationale within the RSMeans mechanical/electrical database.

Some of what you'll learn:
- The unique way mechanical and electrical systems are interrelated
- M&E estimates—conceptual, planning, budgeting, and bidding stages
- Order of magnitude, square foot, assemblies, and unit price estimating
- Comparative cost analysis of equipment and design alternatives

Who should attend: architects, engineers, facilities managers, mechanical and electrical contractors, and others who need a highly reliable method for developing, understanding, and evaluating mechanical and electrical contracts.

Professional Development

For more information visit the RSMeans website at **www.rsmeans.com**

Unit Price Estimating

This interactive *two-day* seminar teaches attendees how to interpret project information and process it into final, detailed estimates with the greatest accuracy level.

The most important credential an estimator can take to the job is the ability to visualize construction and estimate accurately.

Some of what you'll learn:
- Interpreting the design in terms of cost
- The most detailed, time-tested methodology for accurate pricing
- Key cost drivers—material, labor, equipment, staging, and subcontracts
- Understanding direct and indirect costs for accurate job cost accounting and change order management

Who should attend: corporate and government estimators and purchasers, architects, engineers, and others who need to produce accurate project estimates.

RSMeans CostWorks® CD Training

This one-day course helps users become more familiar with the functionality of *RSMeans CostWorks* program. Each menu, icon, screen, and function found in the program is explained in depth. Time is devoted to hands-on estimating exercises.

Some of what you'll learn:
- Searching the database using all navigation methods
- Exporting RSMeans data to your preferred spreadsheet format
- Viewing crews, assembly components, and much more
- Automatically regionalizing the database

This training session requires you to bring a laptop computer to class. When you register for this course you will receive an outline for your laptop requirements.

Also offering web training for CostWorks CD!

Facilities Est. Using RSMeans CostWorks® CD

Combines hands-on skill building with best estimating practices and real-life problems. Brings you up-to-date with key concepts, and provides tips, pointers, and guidelines to save time and avoid cost oversights and errors.

Some of what you'll learn:
- Estimating process concepts
- Customizing and adapting RSMeans cost data
- Establishing scope of work to account for all known variables
- Budget estimating: when, why, and how
- Site visits: what to look for—what you can't afford to overlook
- How to estimate repair and remodeling variables

This training session requires you to bring a laptop computer to class.

Who should attend: facility managers, architects, engineers, contractors, facility tradespeople, planners, project managers and anyone involved with JOC, SABRE, or IDIQ.

Conceptual Estimating Using RSMeans CostWorks® CD

This *two day* class uses the leading industry data and a powerful software package to develop highly accurate conceptual estimates for your construction projects. All attendees must bring a laptop computer loaded with the current year *Square Foot Model Costs* and the *Assemblies Cost Data* CostWorks titles.

Some of what you'll learn:
- Introduction to conceptual estimating
- Types of conceptual estimates
- Helpful hints
- Order of magnitude estimating
- Square foot estimating
- Assemblies estimating

Who should attend: architects, engineers, contractors, construction estimators, owner's representatives, and anyone looking for an electronic method for performing square foot estimating.

Assessing Scope of Work for Facility Construction Estimating

This *two-day* practical training program addresses the vital importance of understanding the SCOPE of projects in order to produce accurate cost estimates for facility repair and remodeling.

Some of what you'll learn:
- Discussions of site visits, plans/specs, record drawings of facilities, and site-specific lists
- Review of CSI divisions, including means, methods, materials, and the challenges of scoping each topic
- Exercises in SCOPE identification and SCOPE writing for accurate estimating of projects
- Hands-on exercises that require SCOPE, take-off, and pricing

Who should attend: corporate and government estimators, planners, facility managers, and others who need to produce accurate project estimates.

Unit Price Est. Using RSMeans CostWorks® CD

Step-by-step instruction and practice problems to identify and track key cost drivers—material, labor, equipment, staging, and subcontractors—for each specific task. Learn the most detailed, time-tested methodology for accurately "pricing" these variables, their impact on each other and on total cost.

Some of what you'll learn:
- Unit price cost estimating
- Order of magnitude, square foot, and assemblies estimating
- Quantity takeoff
- Direct and indirect construction costs
- Development of contractor's bill rates
- How to use *RSMeans Building Construction Cost Data*

This training session requires you to bring a laptop computer to class.

Who should attend: architects, engineers, corporate and government estimators, facility managers, and government procurement staff.

Professional Development

For more information visit the RSMeans website at **www.rsmeans.com**

Registration Information

Register early ... Save up to $100! Register 30 days before the start date of a seminar and save $100 off your total fee. *Note: This discount can be applied only once per order. It cannot be applied to team discount registrations or any other special offer.*

How to register Register by phone today! The RSMeans toll-free number for making reservations is **1-800-334-3509, Press 1.**

Two day seminar registration fee - $935. One-day *RSMeans CostWorks*® training registration fee - $375. To register by mail, complete the registration form and return, with your full fee, to: RSMeans Seminars, 700 Longwater Drive, Norwell, MA 02061.

Government pricing All federal government employees save off the regular seminar price. Other promotional discounts cannot be combined with the government discount.

Team discount program for two to four seminar registrations. Call for pricing: 1-800-334-3509, Press 1

Multiple course discounts When signing up for two or more courses, call for pricing.

Refund policy Cancellations will be accepted up to ten business days prior to the seminar start. There are no refunds for cancellations received later than ten working days prior to the first day of the seminar. A $150 processing fee will be applied for all cancellations. Written notice of cancellation is required. Substitutions can be made at any time before the session starts. **No-shows are subject to the full seminar fee.**

AACE approved courses Many seminars described and offered here have been approved for 14 hours (1.4 recertification credits) of credit by the AACE International Certification Board toward meeting the continuing education requirements for recertification as a Certified Cost Engineer/Certified Cost Consultant.

AIA Continuing Education We are registered with the AIA Continuing Education System (AIA/CES) and are committed to developing quality learning activities in accordance with the CES criteria. Many seminars meet the AIA/CES criteria for Quality Level 2. AIA members may receive (14) learning units (LUs) for each two-day RSMeans course.

Daily course schedule The first day of each seminar session begins at 8:30 a.m. and ends at 4:30 p.m. The second day begins at 8:00 a.m. and ends at 4:00 p.m. Participants are urged to bring a hand-held calculator, since many actual problems will be worked out in each session.

Continental breakfast Your registration includes the cost of a continental breakfast, and a morning and afternoon refreshment break. These informal segments allow you to discuss topics of mutual interest with other seminar attendees. (You are free to make your own lunch and dinner arrangements.)

Hotel/transportation arrangements RSMeans arranges to hold a block of rooms at most host hotels. To take advantage of special group rates when making your reservation, be sure to mention that you are attending the RSMeans seminar. You are, of course, free to stay at the lodging place of your choice. (**Hotel reservations and transportation arrangements should be made directly by seminar attendees.**)

Important Class sizes are limited, so please register as soon as possible.

Note: Pricing subject to change.

Registration Form

ADDS-100

Call 1-800-334-3509, Press 1 to register or FAX this form 1-800-632-6732. Visit our website: www.rsmeans.com

Please register the following people for the RSMeans construction seminars as shown here. We understand that we must make our own hotel reservations if overnight stays are necessary.

☐ Full payment of $_____ enclosed.

☐ Bill me.

Please print name of registrant(s).
(To appear on certificate of completion)

_____
_____
_____

P.O. #: _____
GOVERNMENT AGENCIES MUST SUPPLY PURCHASE ORDER NUMBER OR TRAINING FORM.

Firm name_____

Address_____

City/State/Zip_____

Telephone no._____ Fax no._____

E-mail address_____

Charge registration(s) to: ☐ MasterCard ☐ VISA ☐ American Express

Account no._____ Exp. date_____

Cardholder's signature_____

Seminar name_____

Seminar City_____

Please mail check to: RSMeans Seminars, 700 Longwater Drive, Norwell, MA 02061 USA

New Titles - Reference Books

For more information visit the RSMeans website at **www.rsmeans.com**

Solar Energy: Technologies & Project Delivery for Buildings

by Andy Walker

An authoritative reference on the design of solar energy systems in building projects, with applications, operating principles, and simple tools for the construction, engineering, and design professionals, the book simplifies the solar design and engineering process and provides sample documentation for the complete design of a solar energy system for buildings.

$85.00 | 320 pages, hardcover | Catalog no. 67365

Project Control: Integrating Cost & Schedule in Construction

by Wayne J. Del Pico

Written by a seasoned professional in the field, *Project Control: Integrating Cost and Schedule in Construction* fills a void in the area of project control as applied in the construction industry today. It demonstrates how productivity models for an individual project are created, monitored, and controlled, and how corrective actions are implemented as deviations from the baseline occur.

$95.00 | 240 pages, softcover | Catalog no. 67366

Project Scheduling & Management for Construction, 4th Edition

by David R. Pierce

First published in 1988 by RSMeans, the new edition of *Project Scheduling and Management for Construction* has been substantially revised for both professionals and students enrolled in construction management and civil engineering programs. While retaining its emphasis on developing practical, professional-level scheduling skills, the new edition is a relatable, real world case study.

$95.00 | 272 pages, softcover | Catalog no. 67367

How to Estimate with Means Data & CostWorks, 4th Edition

by RSMeans and Saleh A. Mubarak, Ph.D.

This step-by-step guide takes you through all the major construction items with extensive coverage of site work, concrete and masonry, wood and metal framing, doors and windows, and other divisions. The only construction cost estimating handbook that uses the most popular source of construction data, RSMeans, this indispensible guide features access to the instructional version of CostWorks in electronic form, enabling you to practice techniques to solve real-world estimating problems.

$75.00 | 292 pages, softcover | Includes CostWorks CD
Catalog no. 67324C

Estimating Building Costs, 2nd Edition
for the Residential & Light Commercial Construction Professional

by Wayne J. Del Pico

This book guides readers through the entire estimating process, explaining in detail how to put together a reliable estimate that can be used not only for budgeting, but also for developing a schedule, managing a project, dealing with contingencies, and ultimately making a profit.

Completely revised and updated to reflect the CSI MasterFormat 2010 system, this practical guide describes estimating techniques for each building system and how to apply them according to the latest industry standards.

$75.00 | 398 pages, softcover | Catalog no. 67343A

RSMeans Cost Data, Student Edition

This book provides a thorough introduction to cost estimating in a self-contained print and online package. With clear explanations and a hands-on, example-driven approach, it is the ideal reference for students and new professionals.

Features include:

- Commercial and residential construction cost data in print and online formats
- Complete how-to guidance on the essentials of cost estimating
- A supplemental website with plans, problem sets, and a full sample estimate

$99.00 | 512 pages, softcover | Catalog no. 67363

Reference Books

For more information visit the RSMeans website at **www.rsmeans.com**

Value Engineering: Practical Applications
For Design, Construction, Maintenance & Operations

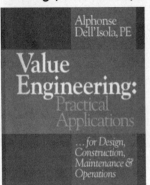

by Alphonse Dell'Isola, PE

A tool for immediate application—for engineers, architects, facility managers, owners, and contractors. Includes making the case for VE—the management briefing; integrating VE into planning, budgeting, and design; conducting life cycle costing; using VE methodology in design review and consultant selection; case studies; VE workbook; and a life cycle costing program on disk.

$85.00 | Over 450 pages, illustrated, softcover | Catalog no. 67319A

Risk Management for Design & Construction

Introduces risk as a central pillar of project management and shows how a project manager can prepare to deal with uncertainty. Experts in the field use clear, straightforward terminology to demystify the concepts of project uncertainty and risk. Includes:

- Integrated cost and schedule risk analysis
- An introduction to a ready-to-use system of analyzing a project's risks and tools to proactively manage risks
- A methodology that was developed and used by the Washington State Department of Transportation
- Case studies and examples on the proper application of principles
- Information about combining value analysis with risk analysis

$125.00 | Over 250 pages, softcover | Catalog no. 67359

How Your House Works, 2nd Edition

by Charlie Wing

Knowledge of your home's systems helps you control repair and construction costs and makes sure the correct elements are being installed or replaced. This book uncovers the mysteries behind just about every major appliance and building element in your house. See-through, cross-section drawings in full color show you exactly how these things should be put together and how they function, including what to check if they don't work. It just might save you having to call in a professional.

$22.95 | 192 pages, softcover | Catalog no. 67351A

Complete Book of Framing, 2nd Edition

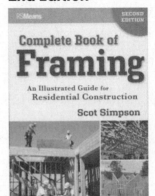

by Scot Simpson

This updated, easy-to-learn guide to rough carpentry and framing is written by an expert with more than thirty years of framing experience. Starting with the basics, this book begins with types of lumber, nails, and what tools are needed, followed by detailed, fully illustrated steps for framing each building element. Framer-Friendly Tips throughout the book show how to get a task done right—and more easily.

$29.95 | 352 pages, softcover | Catalog no. 67353A

Green Home Improvement

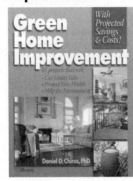

by Daniel D. Chiras, PhD

With energy costs rising and environmental awareness increasing, people are looking to make their homes greener. This book, with 65 projects and actual costs and projected savings help homeowners prioritize their green improvements.

Projects range from simple water savers that cost only a few dollars, to bigger-ticket items such as HVAC systems. With color photos and cost estimates, each project compares options and describes the work involved, the benefits, and the savings.

$34.95 | 320 pages, illustrated, softcover | Catalog no. 67355

Life Cycle Costing for Facilities

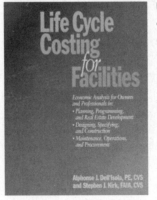

by Alphonse Dell'Isola and Dr. Steven Kirk

Guidance for achieving higher quality design and construction projects at lower costs! Cost-cutting efforts often sacrifice quality to yield the cheapest product. Life cycle costing enables building designers and owners to achieve both. The authors of this book show how LCC can work for a variety of projects — from roads to HVAC upgrades to different types of buildings.

$99.95 | 396 pages, hardcover | Catalog no. 67341

Reference Books

For more information visit the RSMeans website at **www.rsmeans.com**

Builder's Essentials: Plan Reading & Material Takeoff
For Residential and Light Commercial Construction

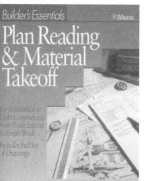

by Wayne J. DelPico

A valuable tool for understanding plans and specs, and accurately calculating material quantities. Step-by-step instructions and takeoff procedures based on a full set of working drawings.

$45.00 | Over 420 pages, softcover | Catalog no. 67307

Means Unit Price Estimating Methods, 4th Edition

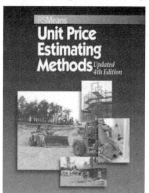

Includes cost data and estimating examples, updated to reflect changes to the CSI numbering system and new features of RSMeans cost data. It describes the most productive, universally accepted ways to estimate, and uses checklists and forms to illustrate shortcuts and timesavers. A model estimate demonstrates procedures. A new chapter explores computer estimating alternatives.

$75.00 | Over 350 pages, illustrated, softcover | Catalog no. 67303B

Concrete Repair & Maintenance Illustrated

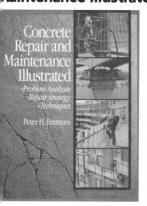

by Peter Emmons

Hundreds of illustrations show users how to analyze, repair, clean, and maintain concrete structures for optimal performance and cost effectiveness. From parking garages to roads and bridges to structural concrete, this comprehensive book describes the causes, effects, and remedies for concrete wear and failure. Invaluable for planning jobs, selecting materials, and training employees, this book is a must-have for concrete specialists, general contractors, facility managers, civil and structural engineers, and architects.

$75.00 | 300 pages, illustrated, softcover | Catalog no. 67146

Square Foot & UNIFORMAT Assemblies Estimating Methods, 3rd Edition

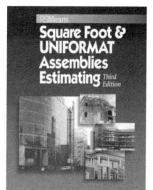

Develop realistic square foot and assemblies costs for budgeting and construction funding. The new edition features updated guidance on square foot and assemblies estimating using UNIFORMAT II. An essential reference for anyone who performs conceptual estimates.

$75.00 | Over 300 pages, illustrated, softcover | Catalog no. 67145B

Electrical Estimating Methods, 3rd Edition

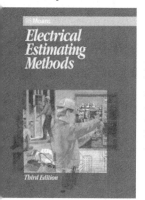

Expanded edition includes sample estimates and cost information in keeping with the latest version of the CSI MasterFormat and UNIFORMAT II. Complete coverage of fiber optic and uninterruptible power supply electrical systems, broken down by components, and explained in detail. Includes a new chapter on computerized estimating methods. A practical companion to *RSMeans Electrical Cost Data*.

$75.00 | Over 325 pages, hardcover | Catalog no. 67230B

Residential & Light Commercial Construction Standards, 3rd Edition

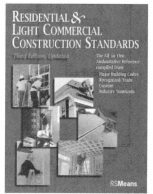

Updated by RSMeans and contributing authors

This book provides authoritative requirements and recommendations compiled from leading professional associations, industry publications, and building code organizations. This all-in-one reference helps establish a standard for workmanship, quickly resolve disputes, and avoid defect claims. Updated third edition includes new coverage of green building, seismic, hurricane, and mold-resistant construction.

$65.00 | Over 550 pages, illustrated, softcover | Catalog no. 67322B

Reference Books

For more information visit the RSMeans website at www.rsmeans.com

Construction Business Management

by Nick Ganaway

Only 43% of construction firms stay in business after four years. Make sure your company thrives with valuable guidance from a pro with 25 years of success as a commercial contractor. Find out what it takes to build all aspects of a business that is profitable, enjoyable, and enduring. With a bonus chapter on retail construction.

$55.00 | 200 pages, softcover | Catalog no. 67352

Project Scheduling & Management for Construction, 3rd Edition

by David R. Pierce, Jr.

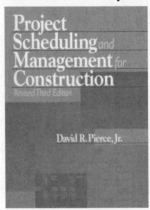

A comprehensive yet easy-to-follow guide to construction project scheduling and control—from vital project management principles through the latest scheduling, tracking, and controlling techniques. The author is a leading authority on scheduling, with years of field and teaching experience at leading academic institutions. Spend a few hours with this book and come away with a solid understanding of this essential management topic.

$64.95 | Over 300 pages, illustrated, hardcover | Catalog no. 67247B

The Practice of Cost Segregation Analysis

by Bruce A. Desrosiers and Wayne J. DelPico

This expert guide walks you through the practice of cost segregation analysis, which enables property owners to defer taxes and benefit from "accelerated cost recovery" through depreciation deductions on assets that are properly identified and classified.

With a glossary of terms, sample cost segregation estimates for various building types, key information resources, and updates via a dedicated website, this book is a critical resource for anyone involved in cost segregation analysis.

$105.00 | Over 225 pages | Catalog no. 67345

Preventive Maintenance for Multi-Family Housing

by John C. Maciha

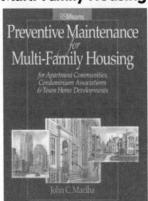

Prepared by one of the nation's leading experts on multi-family housing.

This complete PM system for apartment and condominium communities features expert guidance, checklists for buildings and grounds maintenance tasks and their frequencies, a reusable wall chart to track maintenance, and a dedicated website featuring customizable electronic forms. A must-have for anyone involved with multi-family housing maintenance and upkeep.

$95.00 | 225 pages | Catalog no. 67346

Green Building: Project Planning & Cost Estimating, 3rd Edition

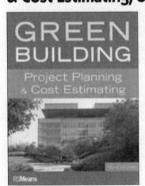

Since the widely read first edition of this book, green building has gone from a growing trend to a major force in design and construction.

This new edition has been updated with the latest in green building technologies, design concepts, standards, and costs. Full-color with all new case studies—plus a new chapter on commercial real estate.

$110.00 | Over 450 pages, softcover | Catalog no. 67338B

Job Order Contracting: Expediting Construction Project Delivery

by Allen L. Henderson

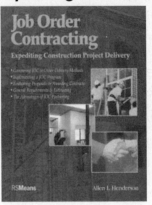

Expert guidance to help you implement JOC—fast becoming the preferred project delivery method for repair and renovation, minor new construction, and maintenance projects in the public sector and in many states and municipalities. The author, a leading JOC expert and practitioner, shows how to:

- Establish a JOC program
- Evaluate proposals & award contracts
- Handle general requirements and estimating
- Partner for maximum benefits

$89.95 | 192 pages, illustrated, hardcover | Catalog no. 67348

Reference Books

For more information visit the RSMeans website at **www.rsmeans.com**

Construction Supervision

This brand new title inspires supervisory excellence with proven tactics and techniques applied by thousands of construction supervisors over the past decade. Recognizing the unique and critical role the supervisor plays in project success, the book's leadership guidelines carve out a practical blueprint for motivating work performance and increasing productivity through effective communication.

Features:

- A unique focus on field supervision and crew management

Coverage of supervision from the foreman to the superintendent level

An overview of technical skills whose mastery will build confidence and success for the supervisor

A detailed view of "soft" management and communication skills

95.00 | Over 450 pages, softcover | Catalog no. 67358

The Homeowner's Guide to Mold

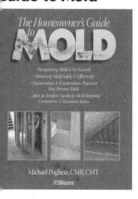

By Michael Pugliese

Expert guidance to protect your health and your home.

Mold, whether caused by leaks, humidity or flooding, is a real health and financial issue—for homeowners and contractors.

This full-color book explains:

- Construction and maintenance practices to prevent mold
- How to inspect for and remove mold
- Mold remediation procedures and costs
- What to do after a flood
- How to deal with insurance companies

21.95 | 144 pages, softcover | Catalog no. 67344

Landscape Estimating Methods, 5th Edition

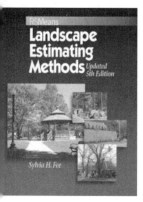

Answers questions about preparing competitive landscape construction estimates, with up-to-date cost estimates and the new MasterFormat classification system. Expanded and revised to address the latest materials and methods, including new coverage on approaches to green building. Includes:

- Step-by-step explanation of the estimating process
- Sample forms and worksheets that save time and prevent errors

5.00 | Over 350 pages, softcover | Catalog no. 67295C

Building & Renovating Schools

This all-inclusive guide covers every step of the school construction process—from initial planning, needs assessment, and design, right through moving into the new facility. A must-have resource for anyone concerned with new school construction or renovation. With square foot cost models for elementary, middle, and high school facilities, and real-life case studies of recently completed school projects.

The contributors to this book—architects, construction project managers, contractors, and estimators who specialize in school construction—provide start-to-finish, expert guidance on the process.

$99.95 | Over 425 pages, hardcover | Catalog no. 67342

Universal Design Ideas for Style, Comfort & Safety

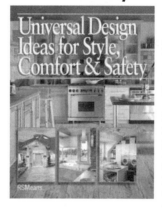

by RSMeans and Lexicon Consulting, Inc.

Incorporating universal design when building or remodeling helps people of any age and physical ability more fully and safely enjoy their living spaces. This book shows how universal design can be artfully blended into the most attractive homes. It discusses specialized products like adjustable countertops and chair lifts, as well as simple ways to enhance a home's safety and comfort. With color photos and expert guidance, every area of the home is covered. Includes budget estimates that give an idea how much projects will cost.

$21.95 | 160 pages, illustrated, softcover | Catalog no. 67354

Plumbing Estimating Methods, 3rd Edition

by Joseph J. Galeno and Sheldon T. Greene

Updated and revised! This practical guide walks you through a plumbing estimate, from basic materials and installation methods through change order analysis. *Plumbing Estimating Methods* covers residential, commercial, industrial, and medical systems, and features sample takeoff and estimate forms and detailed illustrations of systems and components.

$65.00 | 330+ pages, softcover | Catalog no. 67283B

Reference Books

For more information visit the RSMeans website at **www.rsmeans.com**

Understanding & Negotiating Construction Contracts

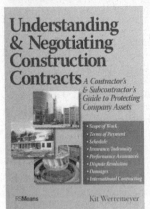

by Kit Werremeyer

Take advantage of the author's 30 years' experience in small-to-large (including international) construction projects. Learn how to identify, understand, and evaluate high risk terms and conditions typically found in all construction contracts—then negotiate to lower or eliminate the risk, improve terms of payment, and reduce exposure to claims and disputes. The author avoids "legalese" and gives real-life examples from actual projects.

$80.00 | 300 pages, softcover | Catalog no. 67350

The Gypsum Construction Handbook

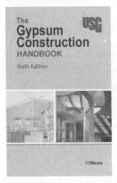

by USG

An invaluable reference of construction procedures for gypsum drywall, cement board, veneer plaster, and conventional plaster. This new edition includes the newest product developments, installation methods, fire- and sound-rated construction information, illustrated framing-to-finish application instructions, estimating and planning information, and more. Great for architects and engineers; contractors, builders, and dealers; apprentices and training programs; building inspectors and code officials; and anyone interested in gypsum construction. Features information on tools and safety practices, a glossary of construction terms, and a list of important agencies and associations. Also available in Spanish.

$40.00 | Over 575 pages, illustrated, softcover | Catalog no. 67357A

Mechanical Estimating Methods, 4th Edition

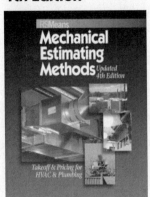

Completely updated, this guide assists you in making a review of plans, specs, and bid packages, with suggestions for takeoff procedures, listings, substitutions, and pre-bid scheduling for all components of HVAC. Includes suggestions for budgeting labor and equipment usage. Compares materials and construction methods to allow you to select the best option.

$70.00 per copy | Over 350 pages, illustrated, softcover | Catalog no. 67294B

Means Illustrated Construction Dictionary

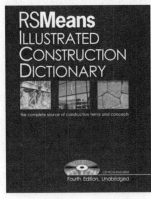

Unabridged 4th Edition, with CD-ROM

Long regarded as the industry's finest, *Means Illustrated Construction Dictionary* is now even better. With nearly 20,000 terms and more than 1,400 illustrations and photos, it is the clear choice for the most comprehensive and current information. The companion CD-ROM that comes with this new edition adds extra features such as: larger graphics and expanded definitions.

$99.95 | Over 790 pages, illust., hardcover | Catalog no. 67292B

RSMeans Estimating Handbook, 3rd Edition

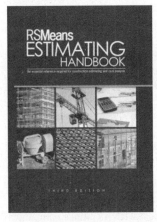

Widely used in the industry for tasks ranging from routine estimates to special cost analysis projects, this handbook has been completely updated and reorganized with new and expanded technical information.

RSMeans Estimating Handbook will help construction professionals:

- Evaluate architectural plans and specifications
- Prepare accurate quantity takeoffs
- Compare design alternatives and cost
- Perform value engineering
- Double-check estimates and quotes
- Estimate change orders

$110.00 | Over 900 pages, hardcover | Catalog No. 67276B

2014 Order Form

ORDER TOLL FREE 1-800-334-3509
OR FAX 1-800-632-6732

For more information visit the RSMeans website at www.rsmeans.com

Qty.	Book no.	COST ESTIMATING BOOKS	Unit Price	Total
	60064	Assemblies Cost Data 2014	$315.95	
	60014	Building Construction Cost Data 2014	194.95	
	60044	Commercial Renovaion Cost Data 2014	156.95	
	60114	Concrete & Masonry Cost Data 2014	177.95	
	50144	Construction Cost Indexes 2014 (subscription)	362.00	
	60144A	Construction Cost Index—January 2014	93.50	
	60144B	Construction Cost Index—April 2014	93.50	
	60144C	Construction Cost Index—July 2014	93.50	
	60144D	Construction Cost Index—October 2014	93.50	
	60344	Contr. Pricing Guide: Resid. R & R Costs 2014	39.95	
	60234	Electrical Change Order Cost Data 2014	187.95	
	60034	Electrical Cost Data 2014	196.95	
	60204	Facilities Construction Cost Data 2014	496.95	
	60304	Facilities Maintenance & Repair Cost Data 2014	434.95	
	60554	Green Building Cost Data 2014	160.95	
	60164	Heavy Construction Cost Data 2014	192.95	
	60094	Interior Cost Data 2014	196.95	
	60124	Labor Rates for the Const. Industry 2014	424.95	
	60184	Light Commercial Cost Data 2014	144.95	
	60024	Mechanical Cost Data 2014	191.95	
	60154	Open Shop Building Const. Cost Data 2014	165.95	
	60214	Plumbing Cost Data 2014	194.95	
	60174	Residential Cost Data 2014	139.95	
	60284	Site Work & Landscape Cost Data 2014	187.95	
	60054	Square Foot Costs 2014	205.95	
	62013	Yardsticks for Costing (2013)	185.95	
	62014	Yardsticks for Costing (2014)	189.95	
		REFERENCE BOOKS		
	67329	Bldrs Essentials: Best Bus. Practices for Bldrs	40.00	
	67307	Bldrs Essentials: Plan Reading & Takeoff	35.95	
	67342	Building & Renovating Schools	99.95	
	67261A	The Building Prof. Guide to Contract Documents	75.00	
	67339	Building Security: Strategies & Costs	95.00	
	67353A	Complete Book of Framing, 2nd Ed.	29.95	
	67146	Concrete Repair & Maintenance Illustrated	75.00	
	67352	Construction Business Management	55.00	
	67364	Construction Law	110.00	
	67358	Construction Supervision	95.00	
	67363	Cost Data Student Edition	99.00	
	67314	Cost Planning & Est. for Facil. Maint.	95.00	
	67230B	Electrical Estimating Methods, 3rd Ed.	75.00	
	67343A	Est. Bldg. Costs for Resi. & Lt. Comm., 2nd Ed.	75.00	
	67276B	Estimating Handbook, 3rd Ed.	110.00	
	67318	Facilities Operations & Engineering Reference	115.00	
	67301	Facilities Planning & Relocation	89.95	
	67338B	Green Building: Proj. Planning & Cost Est., 3rd Ed.	110.00	
	67355	Green Home Improvement	34.95	
	67357A	The Gypsum Construction Handbook	40.00	
	67148	Heavy Construction Handbook	110.00	
	67308E	Home Improvement Costs–Int. Projects, 9th Ed.	24.95	
	67309E	Home Improvement Costs–Ext. Projects, 9th Ed.	24.95	

Qty.	Book no.	REFERENCE BOOKS (Cont.)	Unit Price	Total
	67344	Homeowner's Guide to Mold	21.95	
	67324C	How to Est.w/Means Data & CostWorks, 4th Ed.	75.00	
	67351A	How Your House Works, 2nd Ed.	22.95	
	67282A	Illustrated Const. Dictionary, Condensed, 2nd Ed.	65.00	
	67292B	Illustrated Const. Dictionary, w/CD-ROM, 4th Ed.	99.95	
	67362	Illustrated Const. Dictionary, Student Ed.	55.00	
	67348	Job Order Contracting	89.95	
	67295C	Landscape Estimating Methods, 5th Ed.	75.00	
	67341	Life Cycle Costing for Facilities	99.95	
	67294B	Mechanical Estimating Methods, 4th Ed.	70.00	
	67283B	Plumbing Estimating Methods, 3rd Ed.	65.00	
	67345	Practice of Cost Segregation Analysis	105.00	
	67366	Project Control: Integrating Cost & Sched.	95.00	
	67346	Preventive Maint. for Multi-Family Housing	95.00	
	67367	Project Scheduling & Manag. for Constr., 4th Ed.	95.00	
	67322B	Resi. & Light Commercial Const. Stds., 3rd Ed.	65.00	
	67359	Risk Management for Design & Construction	125.00	
	67365	Solar Energy: Technology & Project Delivery	85.00	
	67145B	Sq. Ft. & Assem. Estimating Methods, 3rd Ed.	75.00	
	67321	Total Productive Facilities Management	85.00	
	67350	Understanding and Negotiating Const. Contracts	80.00	
	67303B	Unit Price Estimating Methods, 4th Ed.	75.00	
	67354	Universal Design	21.95	
	67319A	Value Engineering: Practical Applications	85.00	

MA residents add 6.25% state sales tax _____

Shipping & Handling** _____

Total (U.S. Funds)* _____

Prices are subject to change and are for U.S. delivery only. *Canadian customers may call for current prices. **Shipping & handling charges: Add 7% of total order for check and credit card payments. Add 9% of total order for invoiced orders.

Send order to: ADDV-1000

Name (please print) _____

Company _____

☐ Company
☐ Home Address _____
_____

City/State/Zip _____

Phone # _____ P.O. # _____

(Must accompany all orders being billed)

Mail to: RSMeans, 700 Longwater Drive, Norwell, MA 02061

RSMeans Project Cost Report

By filling out this report, your project data will contribute to the database that supports the RSMeans® Project Cost Square Foot Data. When you fill out this form, RSMeans will provide a $50 discount off one of the RSMeans products advertised in the preceding pages. Please complete the form including all items where you have cost data, and all the items marked (☑).

$50.00 Discount per product for each report you submit

Project Description (NEW construction only, please)

☑ Building Use (Office School . . .) _____

☑ Address (City, State) _____

_____

☑ Total Building Area (SF) _____

☑ Ground Floor (SF) _____

☑ Frame (Wood, Steel . . .) _____

☑ Exterior Wall (Brick, Tilt-up . . .) _____

☑ Basement: (check one)   ☐ Full   ☐ Partial   ☐ None

☑ Number of Stories _____

☑ Floor-to-Floor Height _____

☑ Volume (C.F.) _____

% Air Conditioned _____ Tons _____

Total Project Cost $ _____

Owner _____

Architect _____

General Contractor _____

☑ Bid Date _____

☑ Typical Bay Size _____

☑ Occupant Capacity _____

☑ Labor Force: ____ % Union   ____ % Non-Union

☑ Project Description (Circle one number in each line.)

   1. Economy   2. Average   3. Custom   4. Luxury

   1. Square   2. Rectangular   3. Irregular   4. Very Irregular

Comments _____

_____

A	☑	General Conditions	$
B	☑	Site Work	$
C	☑	Concrete	$
D	☑	Masonry	$
E	☑	Metals	$
F	☑	Wood & Plastics	$
G	☑	Thermal & Moisture Protection	$
GR		Roofing & Flashing	$
H	☑	Doors and Windows	$
J	☑	Finishes	$
JP		Painting & Wall Covering	$

K	☑	Specialties	$
L	☑	Equipment	$
M	☑	Furnishings	$
N	☑	Special Construction	$
P	☑	Conveying Systems	$
Q	☑	Mechanical	$
QP		Plumbing	$
QB		HVAC	$
R	☑	Electrical	$
S	☑	Mech./Elec. Combined	$

Please specify the RSMeans product you wish to receive. Complete the address information.

Product Name _____

Product Number _____

Name _____

Street Address _____

888-492-6770.

Method of Payment:

Credit Card # _____

Expiration Date _____

Check _____

Purchase Order _____

Phone Number _____
(if you would prefer a sales call)

RSMeans
a division of Reed Construction Data
Square Foot Costs Department
700 Longwater Drive
Norwell, MA 02061
www.RSMeans.com

2014 Order Form

ORDER TOLL FREE 1-800-334-3509
OR FAX 1-800-632-6732

For more information visit the RSMeans website at **www.rsmeans.com**

QTY.	Book no.	COST ESTIMATING BOOKS	Unit Price	Total
	60064	Assemblies Cost Data 2014	$315.95	
	60014	Building Construction Cost Data 2014	194.95	
	60044	Commercial Renovaion Cost Data 2014	156.95	
	60114	Concrete & Masonry Cost Data 2014	177.95	
	50144	Construction Cost Indexes 2014 (subscription)	362.00	
	60144A	Construction Cost Index–January 2014	93.50	
	60144B	Construction Cost Index–April 2014	93.50	
	60144C	Construction Cost Index–July 2014	93.50	
	60144D	Construction Cost Index–October 2014	93.50	
	60344	Contr. Pricing Guide: Resid. R & R Costs 2014	39.95	
	60234	Electrical Change Order Cost Data 2014	187.95	
	60034	Electrical Cost Data 2014	196.95	
	60204	Facilities Construction Cost Data 2014	496.95	
	60304	Facilities Maintenance & Repair Cost Data 2014	434.95	
	60554	Green Building Cost Data 2014	160.95	
	60164	Heavy Construction Cost Data 2014	192.95	
	60094	Interior Cost Data 2014	196.95	
	60124	Labor Rates for the Const. Industry 2014	424.95	
	60184	Light Commercial Cost Data 2014	144.95	
	60024	Mechanical Cost Data 2014	191.95	
	60154	Open Shop Building Const. Cost Data 2014	165.95	
	60214	Plumbing Cost Data 2014	194.95	
	60174	Residential Cost Data 2014	139.95	
	60284	Site Work & Landscape Cost Data 2014	187.95	
	60054	Square Foot Costs 2014	205.95	
	62013	Yardsticks for Costing (2013)	185.95	
	62014	Yardsticks for Costing (2014)	189.95	
		REFERENCE BOOKS		
	67329	Bldrs Essentials: Best Bus. Practices for Bldrs	40.00	
	67307	Bldrs Essentials: Plan Reading & Takeoff	35.95	
	67342	Building & Renovating Schools	99.95	
	67261A	The Building Prof. Guide to Contract Documents	75.00	
	67339	Building Security: Strategies & Costs	95.00	
	67353A	Complete Book of Framing, 2nd Ed.	29.95	
	67146	Concrete Repair & Maintenance Illustrated	75.00	
	67352	Construction Business Management	55.00	
	67364	Construction Law	110.00	
	67358	Construction Supervision	95.00	
	67363	Cost Data Student Edition	99.00	
	67314	Cost Planning & Est. for Facil. Maint.	95.00	
	67230B	Electrical Estimating Methods, 3rd Ed.	75.00	
	67343A	Est. Bldg. Costs for Resi. & Lt. Comm., 2nd Ed.	75.00	
	67276B	Estimating Handbook, 3rd Ed.	110.00	
	67318	Facilities Operations & Engineering Reference	115.00	
	67301	Facilities Planning & Relocation	89.95	
	67338B	Green Building: Proj. Planning & Cost Est., 3rd Ed.	110.00	
	67355	Green Home Improvement	34.95	
	67357A	The Gypsum Construction Handbook	40.00	
	67148	Heavy Construction Handbook	110.00	
	67308E	Home Improvement Costs–Int. Projects, 9th Ed.	24.95	
	67309E	Home Improvement Costs–Ext. Projects, 9th Ed.	24.95	

QTY.	Book no.	REFERENCE BOOKS (Cont.)	Unit Price	Total
	67344	Homeowner's Guide to Mold	21.95	
	67324C	How to Est.w/Means Data & CostWorks, 4th Ed.	75.00	
	67351A	How Your House Works, 2nd Ed.	22.95	
	67282A	Illustrated Const. Dictionary, Condensed, 2nd Ed.	65.00	
	67292B	Illustrated Const. Dictionary, w/CD-ROM, 4th Ed.	99.95	
	67362	Illustrated Const. Dictionary, Student Ed.	55.00	
	67348	Job Order Contracting	89.95	
	67295C	Landscape Estimating Methods, 5th Ed.	75.00	
	67341	Life Cycle Costing for Facilities	99.95	
	67294B	Mechanical Estimating Methods, 4th Ed.	70.00	
	67283B	Plumbing Estimating Methods, 3rd Ed.	65.00	
	67345	Practice of Cost Segregation Analysis	105.00	
	67366	Project Control: Integrating Cost & Sched.	95.00	
	67346	Preventive Maint. for Multi-Family Housing	95.00	
	67367	Project Scheduling & Manag. for Constr., 4th Ed.	95.00	
	67322B	Resi. & Light Commercial Const. Stds., 3rd Ed.	65.00	
	67359	Risk Management for Design & Construction	125.00	
	67365	Solar Energy: Technology & Project Delivery	85.00	
	67145B	Sq. Ft. & Assem. Estimating Methods, 3rd Ed.	75.00	
	67321	Total Productive Facilities Management	85.00	
	67350	Understanding and Negotiating Const. Contracts	80.00	
	67303B	Unit Price Estimating Methods, 4th Ed.	75.00	
	67354	Universal Design	21.95	
	67319A	Value Engineering: Practical Applications	85.00	

MA residents add 6.25% state sales tax

Shipping & Handling**

Total (U.S. Funds)*

Prices are subject to change and are for U.S. delivery only. *Canadian customers may call for current prices. **Shipping & handling charges: Add 7% of total order for check and credit card payments. Add 9% of total order for invoiced orders.

Send order to: ADDV-1000

Name (please print) _____

Company _____

☐ Company
☐ Home Address _____

_____

City/State/Zip _____

Phone # _____ P.O. # _____

(Must accompany all orders being billed)

Mail to: RSMeans, 700 Longwater Drive, Norwell, MA 02061

RSMeans Project Cost Report

By filling out this report, your project data will contribute to the database that supports the RSMeans® Project Cost Square Foot Data. When you fill out this form, RSMeans will provide a $50 discount off one of the RSMeans products advertised in the preceding pages. Please complete the form including all items where you have cost data, and all the items marked (☑).

$50.00 Discount per product for each report you submit

Project Description (NEW construction only, please)

☑ Building Use (Office School . . .) _____

☑ Address (City, State) _____

_____

☑ Total Building Area (SF) _____

☑ Ground Floor (SF) _____

☑ Frame (Wood, Steel . . .) _____

☑ Exterior Wall (Brick, Tilt-up . . .) _____

☑ Basement: (check one)   ☐ Full   ☐ Partial   ☐ None

☑ Number of Stories _____

☑ Floor-to-Floor Height _____

☑ Volume (C.F.) _____

% Air Conditioned _____ Tons _____

Total Project Cost $ _____

Owner _____

Architect _____

General Contractor _____

☑ Bid Date _____

☑ Typical Bay Size _____

☑ Occupant Capacity _____

☑ Labor Force: _____ % Union _____ % Non-Union

☑ Project Description (Circle one number in each line.)

    1. Economy   2. Average   3. Custom   4. Luxury

    1. Square   2. Rectangular   3. Irregular   4. Very Irregular

Comments _____

_____

A	☑	General Conditions	$
B	☑	Site Work	$
C	☑	Concrete	$
D	☑	Masonry	$
E	☑	Metals	$
F	☑	Wood & Plastics	$
G	☑	Thermal & Moisture Protection	$
GR		Roofing & Flashing	$
H	☑	Doors and Windows	$
J	☑	Finishes	$
JP		Painting & Wall Covering	$

K	☑	Specialties	$
L	☑	Equipment	$
M	☑	Furnishings	$
N	☑	SpecialConstruction	$
P	☑	ConveyingSystems	$
Q	☑	Mechanical	$
QP		Plumbing	$
QB		HVAC	$
R	☑	Electrical	$
S	☑	Mech./Elec. Combined	$

Please specify the RSMeans product you wish to receive.
Complete the address information.

Product Name _____

Product Number _____

Your Name _____

Title _____

Company _____

    ☐ Company
    ☐ Home   Street Address _____

City, State, Zip _____

Email Address _____

Return by mail or fax 888-492-6770.

Method of Payment:

Credit Card # _____

Expiration Date _____

Check _____

Purchase Order _____

Phone Number _____
(if you would prefer a sales call)

RSMeans
a division of Reed Construction Data
Square Foot Costs Department
700 Longwater Drive
Norwell, MA 02061
www.RSMeans.com